D1542421

AIChE Symposium Series No. 312
Volume 92, 1996

First International Conference on

INTELLIGENT SYSTEMS in PROCESS ENGINEERING

Proceedings of the Conference held at
Snowmass, Colorado, July 9-14, 1995

Editors

James F. Davis
Ohio State University

George Stephanopoulos
MIT

Venkat Venkatasubramanian
Purdue University

Production Editor, CACHE Publications
Brice Carnahan
University of Michigan

CACHE
American Institute of Chemical Engineers

1996

Neither AIChE nor CACHE shall be responsible for statements or opinions advanced in their papers or printed in their publications.

Library of Congress Cataloging-in-Publication Data

International Conference on Intelligent Systems in Process Engineering (1995 : Snowmass, Colo.)
International Conference on Intelligent Systems in Process Engineering / James F. Davis, George Stephanopoulos and Venkat Venkatasubramanian, volume editors. Brice Carnahan, production editor.
p. cm. — (AIChE symposium series ; v. 92, no. 312)
Includes bibliographical references and index.
ISBN 0-8169-0707-2
1. Chemical process control — Congresses. 2. Intelligent control systems — Congresses. I. Davis, James F. II. Stephanopoulos, George. III. Venkatasubramanian, Venkat. IV. Title. V. Series: AIChE symposium series ; no. 312.
TP 155.75.1554 1995 96-17881
660'.2815—dc20 CIP

PREFACE

This volume contains the proceedings of the First International Conference on *Intelligent Systems in Process Engineering*, ISPE-95, held in Snowmass, Colorado (USA) from July 9th to 14th 1995. It was sponsored jointly by the Computing and Systems Technology (CAST) Division of the American Institute of Chemical Engineers, the non-profit CACHE Corporation, and the American Association of Artificial Intelligence (AAAI).

Since the early 80s, we have witnessed a broadly-based and intense effort to bring ideas and methodologies from artificial intelligence into the scope of process systems engineering problems. Academic research and industrial practice have generated an impressive amount of work, which has spanned, (i) a multitude of theoretical developments, (ii) a large variety of paradigms, and (iii) virtually every aspect of process engineering work, e.g. product and process development and design, process operations monitoring-diagnosis, process control, operations planning and scheduling, operator training, process hazards analysis and risk assessment, etc.

In the early 90s it became evident that it was very important to pose for a week and critically examine what happened during the previous years, evaluate the work carried out in several areas, identify successes and failures, and debate directions for future work. Many researchers and industrial practitioners believed that the field was at an important junction, especially in view of the fact that work in the area of intelligent systems had matured significantly and was becoming well-integrated into main-stream activities of process engineering.

In response to these challenges, the CAST Division of AIChE took the initiative to foster the organization of ISPE'95, which was also encouraged by AAAI through its official sponsorship and financial support. CACHE Corporation took responsibility for all organizational aspects of ISPE'95. The objectives of the conference were articulated by the programming committee as follows:

1. Bring together a diverse group of people with interests in modeling and simulation, design, operations/control, process management, computer science and technology, operations research, statistics, and systems and control theory, all of which are components in the solution of process engineering problems.
2. Present state-of-the-art developments in intelligent systems for various areas of process engineering.
3. Discuss the experience gained from numerous applications of intelligent systems in the process industries.
4. Provide a forum for in-depth discussions between researchers and practitioners on the theoretical and practical challenges in developing and deploying intelligent systems in industry.
5. Define the directions of future research and industrial applications in this very important field of engineering.

The Conference program was structured in such a way as to offer a multitude of diverse experiences and to allow meaningful and substantive interaction among the participants. It included the following elements:

- invited, state-of-the-art lectures from academia and industry,
- presentation of state-of-the-art industrial applications,
- contributed papers on theory and implementation,
- extensive discussion periods with general-audience participation, and
- demonstrations of commercial products.

ISPE'95 was very successful in attracting a large and diverse audience from around the world, representing academic and industrial experience over a wide range of process systems engineering activities. With about 60% of the participants from the US and the remaining from over 20 countries, ISPE'95 blended about 40% academics with 60% industrial researchers and practitioners, thus providing a healthy representation of many current directions in research, development and application of intelligent systems for process development and design, process operations monitoring/analysis/diagnosis, process operations planning and scheduling, process control, and software engineering for computer-aided engineering environments. The industrial participants came from a broad spectrum of processing industries, e.g. refining and petrochemicals, specialty chemicals and pharmaceuticals, pulp and paper, food, mining and minerals, and a variety of software and engineering services companies, e.g. control vendors, simulation companies, data analysis systems vendors, real-time expert system software and service providers, and others.

The keynote address, by Patrick Winston, highlighted the growing up of artificial intelligence from wild dreams to high-impact applications and the identification of new scientific frontiers. The subsequent technical sessions focused the attention to a series of distinct, although occasionally interrelated and interacting areas, covering the following themes:

1. *Monitoring, Analysis and Synthesis of Process Operations*
2. *Intelligent Control*
3. *Intelligence in Integrated Manufacturing*
4. *Knowledge-Based Product and Process Design*
5. *Intelligence in Modeling and Numerical Computing*
6. *Human-Machine Interaction*
7. *Knowledge and CAD Environments in Engineering Design.*

The session on *Paradigms of Intelligent Systems in Industrial Process Operations* provided the forum for the in-depth presentation and discussion of significant industrial applications, while a poster session hosted 40 papers with recent developments in the theory and practice of intelligent systems for:

1. *Product and Process Design*
2. *Computer-Aided Engineering and Data Management*
3. *Modeling, Learning and Prediction*
4. *Monitoring, Diagnosis and Control*
5. *Modeling, Design and Verification of Process Control Systems*
6. *Planning, Scheduling and Decision Support of Process Operations*

By general conference consensus, ISPE'95 indeed met key objectives of providing a forum that delineated the current state of Intelligent Systems (IS) technology and implementation and set forth future directions of focus that would continue moving the technology into the mainstream of beneficial application in the process industry. As Drs. Reklaitis and Koppel asserted, the ultimate benefits and therefore measures of the effectiveness of IS technologies are achievement oriented: how much capital costs are reduced in both equipment and inventory, the degree of responsiveness in meeting more stringent quality and order-to-delivery times, and the amount of flexibility inherent in processes. ISPE'95 re-confirmed that the process industry exhibits considerable opportunity.

The conference reached a number of key conclusions:

1. The process industry is clearly undergoing significant business challenges and change which require a sophisticated balance between the squeeze on profit margins, budget reduction, manpower reduction, the need to optimize operation with increased reliability and reduced maintenance costs while maintaining safety and regulatory compliance. The pace and intensity of change has provided an enormous impetus for the utilization of new approaches and technology. In particular, the need to simultaneously improve manufacturing efficiency, product quality, and customer responsiveness has motivated an integrated view of manufacturing that encompasses suppliers through customers and a spectrum of time scales from data acquisition through long range corporate planning. IS technologies are key to industry successfully addressing these changes toward broad integrated approaches.

2. Probably the strongest single message from ISPE'95 centered around integration. The need to develop and deliver tools that support an integrated view of process and manufacturing has resulted in the use of a wide variety of IS technologies as part of broad array of multi-disciplinary computer-based tools. As an integrating technology, IS must address issues such as adaptation and learning, planning under large uncertainty, and coping with large amounts of data. IS is no longer a stand-alone technology. The technology has evolved into an indispensable part of the existing tools for supporting manufacturing.

3. In process and manufacturing where there is rapid change and where substantial changes in technology are needed, past experience will be less focused, and there will be an elevated need for new and more fundamental design methods. We are indeed in a period of development that one would call design engineering, namely, the engineering of the design process itself.

4. Cross-fertilization between design and operations disciplines is important. Decision support systems for design provide the necessary crosslinks. What better framework to confront difficult operating problems, even for operations people who have to deal with problems in plants that have already been built, than a thoroughly engineered and understood "map" of all the entities (static and dynamic) and relations that arise in the description of the plant? Resources invested in engineering the design process may be repaid in many different ways through reuse for control, optimization, monitoring, diagnosis.

5. Modeling is recognized as a frontier for IS. Only a small amount of effort has gone into the human component associated with modeling within the design

task. Future modeling software must be able to support the actual modeling process in a way that goes beyond the present descriptive languages and graphical tools.

6. Practice has proven that human involvement in IS design and operation systems is a marked advantage, rather than a disadvantage. Consequently the man-machine interface plays a key role. Indeed, the acceptance and operational success of an intelligent system relies on whether or not it has a truly user-friendly interface. In future research and development, there is a strong need to move away from interfaces that force users to think like computers and move toward interfaces that are responsive to the reasoning processes of the user.

7. IS technologies are solving practical problems, growing in application and becoming mainstream approaches. Nevertheless, the practical problems encountered in the implementation of an intelligent system are many, and they appear even when starting to discuss a potential application. Is this application going to be profitable? How is the profit going to be predicted and measured? How is the application going to obtain live data? Who is going to provide the knowledge, maintain and support the application? How is the operator training going to be implemented? These issues are often limiting steps in the development of an application.

8. The future promises an even larger role for IS technologies. However the technology will become less and less explicit as it is integrated into a mainstream set of technologies. A key role for the technology will be that of integrator. Consider that fully achieving integrated manufacturing requires a tight coordination among low-level activities such as data acquisition, monitoring, and diagnosis; intermediate-level activities such as process control and scheduling; and high-level activities such as long range planning and supply chain coordination. While the integration of low-level manufacturing activities has benefited from the type of automated support provided by IS technologies, the integration of intermediate- and high-level manufacturing activities has barely begun. Future tools for supporting manufacturing activities will have to use representations that readily provide performance metrics and still quickly respond to changing manufacturing conditions and large volumes of data.

9. Perhaps the greatest contribution of research on IS technologies has been and will be the introduction of new and very different approaches to solving problems.

ISPE'95 was blessed from the beginning with an extraordinary group of individuals from around the world, who offered, through their participation in the International Programming Committee, an invaluable breadth and depth of scope in shaping the Conference program. These individuals are:

Antsaklis, P.	Notre Dame University
Arzén, K.-E.	Lund Institute of Tech.
Beach, D.W.	BP Oil Co.
Bhalodia, M.	EXXON Research & Eng.
Chandrasekaran, B.	Ohio State University
Cordingley, R.	Monsanto Chemical Co.
Depeyre, D.	Ecole Centrale, Paris
Fan, L.T.	Kansas State University
Greenberg, H.	Univ. of Colorado, Denver
Hashimoto, I.	Kyoto University
Himmelblau, D.M.	University of Texas
Kramer, M.A.	Gensym Corporation
Kuipers, B.	University of Texas
Lien, K.	University of Trondheim
Liu, Y.A.	Virginia Polytechnic Inst.
Macchietto, S.	Imperial College
Marquardt, W.	Aachen Tech. University
Matsuyama, H.	Kyushu University
Mavrovouniotis, M.	Northwestern University
McAvoy, T.	University of Maryland
Miller, E.	Air Products and Chemicals
Nishitani, H.	Nara Inst. Science & Tech.
Patsidou, L.	Shell Develop. Co.
Piovoso, M.	DuPont Co.
Ponton, J.	University of Edinburgh
Romagnoli, J.	University of Sydney
Samdani, G.	Chemical Engin. Magazine
Schwenzer, G.	Mobil R&D Corp.
Simmrock, K.	University of Dortmund
Suzuki, G.	Toyo Engin. Corp.
Tomita, S.	Miyazaki University
Ungar, L.	University of Pennsylvania
Westerberg, A.W.	Carnegie Mellon University
Yoon, E.-S.	Seoul National University

Many people and organizations contributed a great deal in making IPSE'95 a notable success. We would like to take this opportunity and extend to them the sincere thanks of the participants and undersigned co-chairs:

1. The CAST Division of AIChE for its continuing boldness in undertaking new initiatives and expanding the scope of intellectual and practical challenges and services to our profession.
2. The AAAI for its financial support and for reaching out to the Process Systems Engineering community and building bridges from which we can all benefit.
3. The CACHE Corporation for its continued leadership in organizing world-class

conferences. We would like to single out Dr. Jeffrey J. Siirola, the Chair of the CACHE Conferences Committee, for his unwavering support of the idea from very early on, and his tireless efforts throughout the end of the conference. Particular thanks are also due to David Himmelblau and the staff at the CACHE office, Janet Sandy and Margaret Beam, as well as to Robin Craven for planning the conference and ensuring that everything ran smoothly on site.

4. The National Science Foundation for its financial support of ISPE'95, which made possible the participation of new academic researchers, brave enough to undertake new ventures in pushing the envelope of process systems engineering methodologies.

One can never overestimate the importance of the financial support offered so generously by a number of industrial companies. We hope that the success of ISPE'95 has justified their trust. These companies are:

◊ *Gensym Corporation*, for sponsoring the ISPE'95 Opening Reception.

◊ *Shell Oil Company*, for sponsoring the ISPE'95 Banquet.

◊ *Eastman Chemical Company*, for sponsoring the "Knowledge-Based Product and Process Design" session.

◊ *Aspen Technology, Inc.*, for sponsoring the "Knowledge and CAD Environments in Engineering Design" session.

◊ *EXXON Research and Engineering Company*, for sponsoring the "Intelligence in Integrated Manufacturing" session.

◊ *Mitsubishi Chemical Company*, for sponsoring the "Monitoring, Analysis and Synthesis of Process Operations" session.

◊ *Oil Systems Company*, for sponsoring the "Intelligence in Modeling and Numerical Computing" session.

◊ *UOP*, for sponsoring the "Intelligent Control" session.

◊ *Weyerhaeuser Corporation*, for sponsoring the "Man-Machine Interaction" session.

Considerable recognition must be given to Jack L. Marchio, a Ph.D. graduate student at Ohio State University who took nearly full responsibility for the administration, organization and editing of the invited and contributed papers and ultimately the proceedings. Jack, through his meticulous attention to detail and his broad background in IS technologies, was instrumental in ensuring the content and quality of the papers that appear in this proceedings. He organized and implemented all processes associated with producing the proceedings including communications, reviews, editing and follow-up; he read each paper, taking responsibility for the finalized details of the electronic copies of the papers; and he provided an important layer of review for technical content. Quite simply this proceedings was made possible by Jack's efforts.

Jack worked closely with Brice Carnahan, the Chair of the Publications Committee for CACHE and the person responsible for the production of the proceedings. Brice's key contribution was seen from the beginning in the guidelines for publishing the conference papers and the templates for electronic copies. It is through Brice's efforts and those of Matthew Smart, a University of Michigan undergraduate who assumed much of the responsibility for final assembly and indexing of the volume, that this proceedings was able to be published completely as an electronic document. We are indebted to Jack, Brice, and Matt for their monumental efforts in producing this important record of ISPE'95.

As we look into the crystal ball for future developments, one aspect remains indisputable, although a little fuzzy in its outlines; the pervasiveness of the computers in process systems engineering work will expand and deepen. It is reasonable, then, to expect that the directions of research and industrial applications, as defined from the ISPE'95 proceedings, will be strengthened in technical content and expanded in scope, thus setting the stage and providing the necessity and justification for future ISPE Conferences.

The ISPE'95 Co-Chairmen

George Stephanopoulos	MIT
James F. Davis	Ohio State University
Venkat Venkatasubramanian	Purdue University

TABLE OF CONTENTS

Monitoring, Analysis and Support of Process Operations

Intelligent Control

Intelligence in Integrated Manufacturing

Intelligence in Modeling and Numerical Computing

Knowledge-Based Product and Process Design

Knowledge and CAD Environments in Engineering Design

Human-Machine Interaction

Paradigms of Intelligent Systems in Industrial Process Operations

Poster Session

Product and Process Design

Computer-Aided Engineering and Data Management

Modeling, Learning and Prediction

Monitoring, Diagnosis and Control

Modeling, Design and Verification of Process Control Systems

Planning, Scheduling and Decision Support of Process Operations

PROCESS MONITORING, DATA ANALYSIS AND DATA INTERPRETATION

James F. Davis and Bhavik R. Bakshi
Ohio State University
Columbus, OH 43210

Karlene A. Kosanovich
University of South Carolina
Columbia, SC 29208

Michael J. Piovoso
DuPont Company
Wilmington, DE 19880

Abstract

Advances in automation and distributed control make possible the collection of large quantities of data that in principle contain a full description of the operating status of the plant at any time. Every modern industrial site believes this data to be a mine of information if only the important and relevant information could be extracted and interpreted, especially in real-time. Automated data interpretation faces many challenges including process uncertainty and noise, limited data, abundance of data, changing process conditions, unknown but feasible conditions, access to interpretation knowledge, inherent uncertainties, undetected sensor failures, uncalibrated and misplaced sensors, and lack of integrity of the data historian. This vast set of considerations has led to a broad base of approaches that perform data interpretation without the benefit of explicit quantitative behavioral models. These approaches encompass a wide range of technologies that include signal processing, statistics, neural networks and knowledge-based systems. While much work remains in the development of a comprehensive and adaptive interpretation capability, many systems addressing aspects of the overall problem are being used successfully in industry. By taking a pattern recognition viewpoint where data analysis and interpretation are considered as feature extraction and label assignment steps, respectively, this paper provides a unifying perspective that reduces the large number of apparently disparate methods into a small set of categories defined by key performance characteristics. This perspective is the basis for reviewing the current state of the art and providing a roadmap for future development of process monitoring systems.

Keywords

Process monitoring, Data analysis, Data interpretation, Pattern recognition, Feature extraction, Knowledge-based systems, Multivariate statistical methods, Artificial neural networks, Clustering.

Introduction

Any action taken on a process, whether it is to improve product quality, take corrective action in response to an abnormal situation, improve efficiency, or react to unsafe conditions, relies on an explicit or implicit description of the state of the operation or events that are occurring. Although there may be hundreds of measurements in a typical chemical process, there are relatively few events occurring. The data from these measurements must therefore be mapped into meaningful descriptions of the event(s) occurring — a difficult task without adequate tools.

Data analysis, data interpretation and monitoring refer collectively to the recognition and assignment of specific state descriptions or labels to discrete data points primarily from process measurements. Data analysis is a term used to describe how data are manipulated and processed to produce the features of interest. Data interpretation describes the mechanism by which labels are assigned. Monitoring refers to the combination of the two steps to produce a machine-based system capable of mapping process data to labels of practical use. In practice, monitoring refers to the detection of abnormal situations and the isolation of the faults. In this context, there are many labels of interest including, *state descriptions* (e.g. normal, high), *trends* (e.g. increasing, pulsing), *landmarks* (e.g. process change), *shape descriptions* (e.g. skewed, tail), and *fault descriptions* (e.g. flooding, contamination). The problem is complex when viewed as a single feature extraction step and a single interpretation step because practical considerations often demand an integrated approach. The overall problem can be conveniently decomposed into three distinct components that distinguish the strategy for solving the monitoring problem.

Numeric-symbolic interpretation, where numeric data from the process are mapped into useful labels, is the essential component of a comprehensive interpretation system. Feature extraction is intended to produce the features that are resolved into the labels. The problem boundary is defined backwards from the label of interest so that feature extraction is associated only with the input requirement for a given approach to produce the immediate features needed to generate the label. The distinctions of "immediate features" and "required input data" are made because a separate data preprocessing component may be necessary to condition the input to the numeric-symbolic interpreter so that it performs more effectively. Data preprocessing is a form of numeric-numeric feature extraction which removes irrelevant information and conditions the input data to allow easier subsequent feature extraction and increased resolution. On the output side of a numeric-symbolic interpreter, there are situations where insufficient information is available to develop feature extraction and assignment systems for some specific labels. When the knowledge is available, symbolic-symbolic interpreters such as knowledge-based systems can be used to extend intermediate interpretations from the numeric-symbolic interpreter to labels of interest. This paper develops a perspective on the roles of these three components while focusing on the core numeric-symbolic intepretations.

Although all numeric-symbolic techniques broadly encompass feature extraction and intepretation, important differences manifest themselves based on whether or not the approach emphasizes the development of an explicit predictive behavioral model. When an explicit model is emphasized, the focus of attention becomes the development of the model. The features for label assignment are typically some form of residuals generated from either comparison of predicted and observed process variables or changes in the interpreted model's parameters that drive the residuals to zero. The key to these approaches is the development of models that can predict measured process variables with sufficient accuracy that residuals can be appropriately generated and interpreted. The body of literature and application has grown from the development of model matching techniques (Kramer and Mah, 1994), and from the development of first principles and empirical modeling approaches including a variety of statistical and neural network approaches (Bulsari, 1995; Piovoso and Kosanovich, 1995).

When explicit behavioral models are not emphasized, the focus shifts to identifying those features in the process data that are direct manifestations of various operating states or situations of interest. Rather than focusing on comparisons between observed data and predicted plant behavior, these approaches focus on distinguishing patterns such as shapes or characteristic signatures as well as movement or changes in patterns relative to a reference set. The distinctions between using distinguishing pattern features and model generated residuals become important and manifest themselves in a number of important practical considerations, generally external to the algorithm: (1) Inductive learning to relate labels to patterns and the requirement of a variety of known (labeled) operating situations are emphasized in approaches that focus on distinguishing features. (2) Even though the same algorithm might be used, the output results are interpreted differently. (3) Selection of methods focuses on the representation of the extracted features and the ability of these representations to form readily distinguishable data classes. To illustrate, partial least squares regression is a multivariate technique that can be used to produce an input-output behavioral model suitable for control or inferred sensor applications. PLS also produces statistical features related to the variance of the data that can be used directly for label assignment. Similarly, backpropagation neural networks can be used for either modeling or direct data interpretation. While the algorithm is exactly the same, both the data used to train the networks and the interpretations of the output are different.

In this paper, we focus on those approaches that produce and represent distinguishing features directly from patterns of data associated with known (labeled) operating situations. The objective is to review state-of-the-art techniques for data interpretation and process monitoring and to provide perspective on the large number of techniques. A key reference related to this paper is, "Model-Based Monitoring," by Kramer and Mah (1994). Emphasizing a statistical view and organized according to the three primary activities in monitoring (data rectification, fault detection and fault diagnosis) the paper offers excellent descriptions of many methods that we must also discuss in the context of data interpretation. This paper provides a complementary perspective by focusing on the theory, development and performance of approaches that combine data analysis and data interpretation into an automated mechanism for process monitoring via feature extraction and label assignment.

Figure 1 offers a unifying perspective for data interpretation and simultaneously serves as the organization for the main sections of the paper. As shown, assignment of labels to time series of data can involve three distinct components — *numeric-numeric* interpretation (data preprocessing), *numeric-symbolic* interpretation and *symbolic-symbolic* interpretation. The paper begins by developing a pattern recognition viewpoint and the motivation for the three possible components. Particular emphasis is placed on the essential numeric-symbolic component which establishes many of the critical performance expectations.

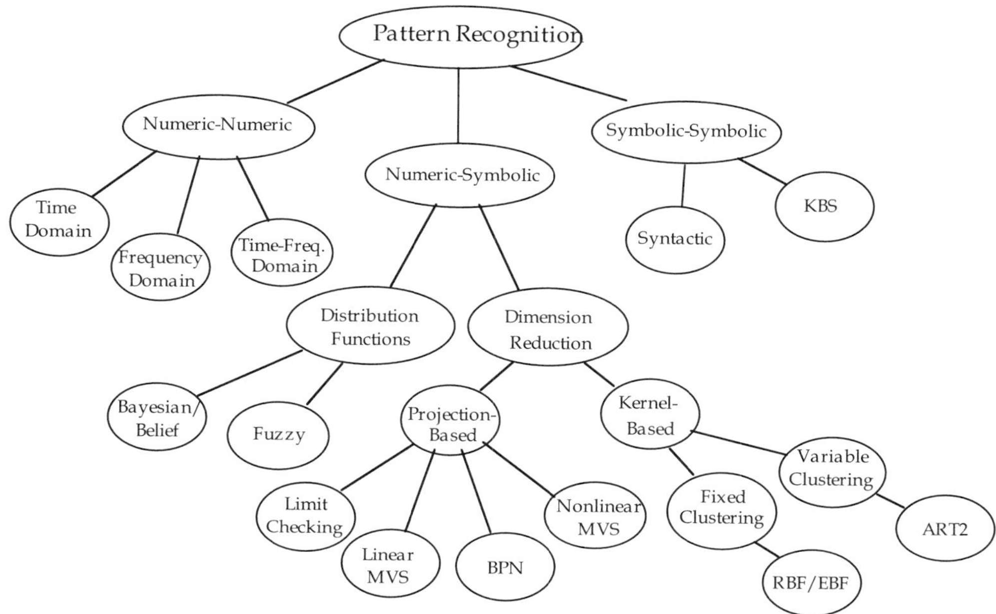

Figure 1. Unifying view of data analysis and interpretation methods for process monitoring.

Process Monitoring as Pattern Recognition

At the most fundamental level, the assignment of state descriptive labels is a pattern recognition problem that involves a two-step procedure: feature extraction and label assignment (interpretation). Extension of the classical pattern recognition view beyond that in the computer science literature offers a unifying basis for a methods-based analysis of a diverse array of process monitoring approaches. The principle objective is the determination of class membership. Performance is ultimately driven by the appropriateness of the discriminant definition in resolving the state labels of interest. A pattern **X** is defined by a number of features or attributes, x_i:

$$\mathbf{X} = (x_1, x_2, \ldots, x_i, \ldots, x_d) \tag{1}$$

The representation space Ξ consists of the set of all possible patterns, $\mathbf{X} \in \Xi$. The set of all possible class labels ω_j forms the interpretation space Ω:

$$\Omega = (\omega_1, \omega_2, \ldots, \omega_j, \ldots \omega_p) \tag{2}$$

Pattern recognition then corresponds to learning the mapping ζ from the representation space to the interpretation space:

$$\zeta : \Xi \rightarrow \Omega \text{ or } \zeta\text{: } (x_1, x_2, \ldots x_d) \rightarrow \omega_j \tag{3}$$

Typically, determining ζ is easier after extracting relevant features, **Y**, from **X** and make the following mapping possible:

$$\zeta\text{:}\{\zeta' : (x_1, x_2, \ldots x_d) \rightarrow (y_1, y_2, \ldots y_p)\} \rightarrow \omega_j \tag{4}$$

Using this notation, **X** corresponds to any time series of data with x_i being a sampled value, and **Y** represents the processed forms of the data, i.e. a pattern. The y_i are the pattern features, ω_j is the appropriate label or interpretation and ζ is the mapping that must be developed. The development of a mapping scheme is inherently an inductive process requiring a good knowledge source. Determination of ζ, therefore, requires a reference that consists of one or more known patterns with correct class labels, or *pattern exemplars*. A training set comprises a collection of pattern exemplars.

In feature extraction, distinguishing features are identified and expressed either explicitly or implicitly. In process monitoring, this generally means transforming time series data into a static collection of features that represent a view of the operation at some time. In the interpretation phase, labels are assigned based on some discriminant relative to the extracted features.

Labels that require no reference information are *context free* while those that do are *context dependent* and therefore require simultaneous consideration of time records from more than one process variable. As a result, generating context dependent state, shape, and fault descriptions are more complicated than generating context free trend and landmark descriptions. The knowledge source requirements for context dependent interpretations contrasts strongly with context free or single variable identifications. In the latter case, methods exist for many types of common yet useful interpretations. However, for context dependent interpretations, many practical issues grow from the requirements for process knowledge. The availability, the coverage, and the distribution of labeled data from known process situations become key performance issues.

Other practical issues include the dynamic nature of chemical processes and the curse of dimensionality. The dynamic nature of processes demands an adaptive interpretation ability. Process information and its interpretations are affected by production rates, quality targets, feed compositions, equipment conditions and many additional factors, all of which change frequently. Context dependency leads to large-scale input, and requires reduction of the problem scale or dimension. The complexity of the induction process and of the learned discriminant functions, increases with the number of measurements, resulting in a curse of dimensionality. Complexity may be significantly reduced if redundancy in the data is eliminated, and only the most relevant features are used for the mapping. Reducing the dimensionality may also improve system performance, since the number of training examples, needed to achieve a desired error rate increases with the number of measured variables or features (Raudys and Jain, 1991).

A pattern recognition view allows us to generalize on several characteristics of chemical processes that strongly affect the performance of approaches:

1. Chemical processes are highly controlled and usually operate normally. Thus, the number and type of labeled sensor patterns available for learning purposes will be limited for certain classes of labels and will be abundant for others. There will be an abundance of normal operation exemplars, a very limited supply of abnormal operation exemplars and an even smaller number of fault exemplars.
2. Since processes operate over a continuum and the measured data are corrupted by noise and errors, feature extraction generally produces distinguishing features that exist over a continuum. Label assignment, therefore, requires construction of decision surfaces through regions of overlapping features. This necessarily implies a region of uncertainty between data classes. As a result, it is impossible to define completely distinguishing criteria for the patterns, and uncertainty must be addressed.
3. Performance depends heavily on the known patterns for inductive learning, on their distribution, particularly in the vicinity of pattern class boundaries, on the capacity for the approach to represent decision boundaries in the context of some set of features, and on the mechanisms available to deal with uncertainty in the boundary regions.

The motivation for numeric-symbolic and symbolic-symbolic categories is illustrated in Figs. 2a and b. Figure 2a shows a system that maps directly from a conditioned input data pattern, X', to the desired set of labels, Ω. Consisting only of a feature extractor and feature interpreter, the system reflects the availability of a sufficient number of labeled training patterns, X', from the operation such that the labels can be assigned with certainty. The implication is that if there is sufficient operating history to provide the labeled data, then it is desirable to rely on this information for data interpretation since it reflects the actual operation. With sufficient operating history, the burden of label assignment is on feature extraction as indicated by the large box. Figure 2b, however, shows a system where there is not sufficient operating history to map with certainty. Under these circumstances, a successful system requires a knowledge-based system (KBS) for symbolic-symbolic mapping in lieu of available operating data. The KBS, however, is not sufficient in itself. It must also define intermediate interpretations that can be generated with certainty from a numeric-symbolic mapper. The burden of interpretation becomes distributed between the numeric-symbolic and symbolic-symbolic interpreters.

General considerations of data availability lead immediately to the recognition that detection systems are likely to be designed as comprehensive numeric-symbolic systems as shown in Fig. 2a. State descriptions may be configured either way. Fault classification systems are most likely to require symbolic-symbolic mapping to compensate for limited data.

Numeric-Numeric Feature Extraction

Data preprocessing is essential before the application of most feature extraction or input-output modeling methods. In addition to the actual variation of the measured variable, measured data consists of contributions from several extraneous or irrelevant events such as noise, outliers, and missing sensors or samples. Extraction of the actual variation of the measured variable can significantly simplify the data interpretation task.

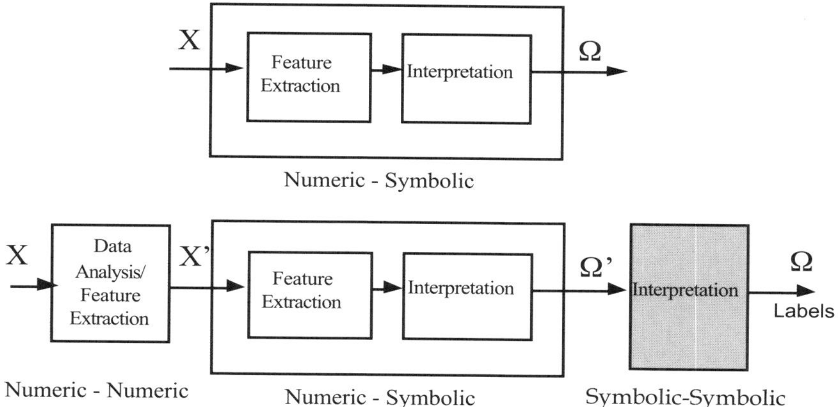

Figure 2. Motivation for N-N, N-S, S-S categorization.

Data preprocessing methods may be classified into two categories. Model-based methods utilize a fundamental model for detecting gross errors and then reconciling the data by constraining the variables to satisfy the model. Approaches in this category have been reviewed recently by Kramer and Mah (1994). In this paper, our emphasis is on non-model-based methods. These methods rectify data based on assumptions about the underlying nature of the measured signal. Most approaches are based on filtering the data to remove contributions in the time domain, frequency domain, or time-frequency domain. The ideal filtering technique should remove irrelevant features with minimal distortion of the relevant signal features.

Symbolic-Symbolic Mapping

Knowledge-based systems (KBS) provide the necessary symbolic-symbolic mapping techniques for context-dependent interpretations when operating histories do not provide sufficient exemplars to develop an adequate map directly from the input. They have been shown to be particularly useful for fault and malfunction descriptions where the data exemplars relating the malfunctions directly to the measured data are very limited and in many cases non-existent. By using high certainty intermediate data labels, the KBS can map to more robust and accurate conclusions. This implies a shift to interpretation where either there is insufficient data for inductive learning or the characteristics of the intermediate data lower the uncertainty in the boundary regions.

In lieu of data exemplars, KBS approaches use expert descriptions of behavioral relations to accomplish the mapping. Generally speaking, KBS approaches are tasked to take symbolic input data values and resolve appropriate labels through explicit fragments of knowledge that provide relation-based pathways between combinations of input values and possible output labels. A KBS approach constitutes a model description of how the presence or absence of various features (with degrees of certainty) will map to a set of desired labels. The mapping can be direct, i.e. features-to-labels to form a table look-up approach or it can be in the form of a semantic network that can provide a high degree of generalization. KBS approaches not only provide the means of capturing this mapping knowledge, but also offer a medium for exploiting efficient strategies used by experts.

KBS approaches are knowledge intensive to construct, and not currently adaptable. Use of a KBS approach needs to be justified on the value of the output classifications given there is insufficient training data to develop an alternative approach. It is unlikely that a KBS approach can be justified if sufficient training data does exist. This explains the wide interest in use of KBSs for fault classification and corrective action systems.

Numeric-Symbolic Mapping

Numeric-symbolic pattern recognition constitutes the core process monitoring techniques, and employs deterministic and statistical methods when patterns can be represented as vectors or distribution functions. These methods are also called *geometric* approaches since they identify regions in a mathematical or statistical representation space that are used to distinguish classes of data. These approaches often can be characterized by closely linked feature extraction and mapping steps.

Numeric-symbolic approaches are particularly important in process applications since the time series of data is by far the predominant form of input data. They are the method of choice if the data exist to develop the system. With a heavy dependence on the number of training exemplars, numeric-symbolic interpreters can be used to assign labels directly. As the number and coverage of available training exemplars diminishes for the given label of interest, there generally is a need to integrate geometrical and knowledge-based approaches where the geometrical approaches are used to produce intermediate labels for the knowledge-based approaches.

There are a large number of numeric-symbolic approaches, far too many to address individually here. In this paper we offer a decomposition by category that brings out the practical advantages and limitations and provides a useful map for selecting approaches for particular problems.

Referring to Fig. 1, *Numeric-symbolic* mapping first breaks down in terms of direct use of *distribution functions* (DF) to calculate class membership probabilities or possibilities. This branch considers those approaches that attempt to determine the probability of a particular label given a particular pattern. Recognizing extreme problems in developing DFs for interpretation problems of high dimensionality with few exemplars, the *dimension reduction* branch relates those methods concerned with fighting the curse of dimensionality. It is these methods that have grown most in practical importance and application.

As modeling techniques, dimension reduction methods may be generally represented as a weighted sum of basis functions,

$$y_p = \sum_{i=1}^{K} \beta_i \theta_i(x_1, x_2, \ldots, x_n; \alpha_j) \tag{5}$$

where, the x_i are the inputs, y_p, the output(s), θ_i are the basis functions, and α_j, β_i are adjustable parameters. The nature of the extracted features depends on the way in which the input space is represented by the basis functions for reducing the dimension. The types of features and the strategies for dimensionality reduction may be separated as projection-based and kernel-based methods. Projection-based methods fight dimensionality by globally projecting the inputs on a non-local hypersurface to exploit correlation among variables, whereas kernel-based methods consider local regions of the input space as defined by a localized kernel, as illustrated in Fig. 3. Projection-based methods include simple univariate approaches that capture the tradition of SPC and limit checking; linear multivariate statistical methods such as, principal component analysis and partial least squares regression; artificial neural networks such as backpropagation networks; and nonlinear multivariate statistical methods such as, nonlinear versions of PCA, PCR, PLS, and projection pursuit regression (PPR). Kernel-based methods are distinguished as fixed and variable cluster methods. These include, nearest neighbor clustering, radial basis function networks (RBFN), adaptive resonance theory (ART).

The model determined by projection-based or non-local approaches is infinite along at least one direction, which may be desirable for extrapolation, but can also lead to dangerous misclassifications without warning about the unreliable nature of the predicted outputs. The potential for extrapolation error must be managed by estimating the reliability of the predicted values. This is particularly important for process applications where data associated with specific operational classes can be limited or where process units can transition into novel operating states. Local methods do not permit arbitrary extrapolation due to the localized nature of their activation functions but can readily recognize situations outside a given structure of the data.

As in any pattern recognition problem, the model represented by Eqn. (5) may be considered to consist of feature extraction and mapping. The extracted features consist of the projected inputs for projection-based methods, or the localized regions for kernel-based methods. The input-output mapping is determined by the nature of the projections, the basis functions, θ_i, and the corresponding weights, β_i.

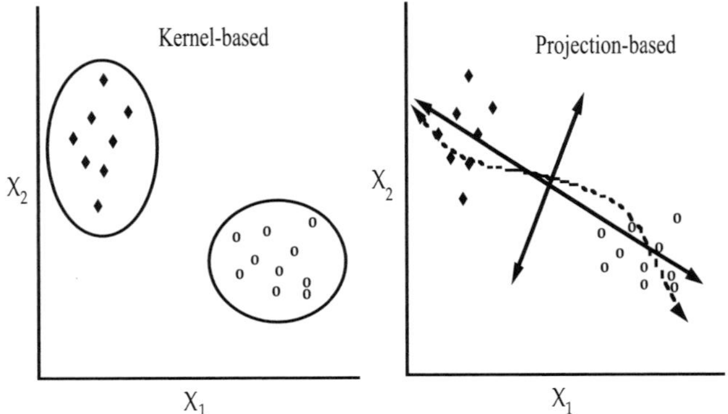

Figure 3. Projection-based vs. kernel-based methods.

Probability Density Function Methods

Bayesian Decision Theory. The ideal situation would be to use Bayesian Decision Theory since it represents an optimal classifier. From a theoretical perspective, Bayesian Decision Theory offers a general definition of the pattern recognition problem and, with appropriate assumptions, it can be shown to be the basis of many of the so called non-probability density function approaches (Pao, 1989). In practice, however, it is typically treated as a separate method since it has strong data availability requirements for direct use compared to other approaches.

The Bayesian approach uses PDFs of pattern classes to establish class membership. Feature extraction corresponds to calculation of the *a posteriori* conditional probability or joint probability using Bayes formula which expresses the probability that a particular pattern label can be associated with a particular pattern:

$$P(w_i|x_k) = [P(x_k|w_i)\, P(w_i)] / P(x_k) \tag{6}$$

Class membership is assigned using a decision rule, typically some inequality test performed on $P(w_i|x_k)$.

The knowledge required to implement Bayes formula is daunting in that *a priori* as well as class conditional probabilities must be known. Some reduction in requirements can be accomplished by using joint probability distributions in place of the *a priori* and class conditional probabilities. Even with this simplification, few interpretation problems are so well posed that the required information is available. As a general rule considerable work is required for this approach to be viable and it is presently unrealistic to expect to be able to adequately recreate the underlying densities.

Bayesian Belief Networks. We include Bayesian Belief Networks (Kramer, 1993) in this category because it combines the Bayesian classification approach with a knowledge-based approach in the form of a semantic network. Ultimately, the approach attempts to calculate the probability of events and assign labels to those events with the highest probabilities. Events are linked based on their causal relationship. Data interpretation is performed by calculating and propagating probabilities exhaustively through the network. While an exact computation may be intractable for realistic problems, several approximate inferencing methods have been proposed.

Even though a number of desirable properties of the belief network have been demonstrated and discussed, the major consideration in practice is the availability of probability information. Considerable probability information is required that simply is not existent. Prototype demonstration results have indicated that rough and subjective estimates of probabilities may suffice and therefore data requirements may be somewhat reduced (Kramer and Mah, 1994).

Fuzzy-Sets. Fuzzy set theory provides a mathematical basis to relate numerical measurements with qualitative descriptions of the measured feature. At the heart of the fuzzy set approach is the membership function, $\mu(x)$, which relates the compatibility of feature x with the qualitative description of the feature. The value of $\mu(x)$ is referred to as the grade of membership and is typically normalized to the range [0,1].

Performance is ultimately attributed to the membership functions for each of the possible interpretations. It is tempting to interpret the membership functions as subjective probability distributions but a better description is the possibility distribution. Establishing a function is a subjective process that requires selecting a function and parameter values that reflect the "expert's" feel for the qualitative interpretation. It is therefore difficult to specify. No formal method exists and one is forced to use heuristic approaches. The unique distribution and limited availability of process data patterns aggravates the situation. In general, the fuzzy set approach suffers from the same problems as the Bayesian approach.

Projection-Based, or Non-Local Methods

Projection-based methods may be further divided based on the linearity of the activation function, θ_i, and the linearity of the projection. Due to their popularity and wide use, this paper focuses primarily on methods based on linear projection, with linear or nonlinear activation functions, which includes, limit checking; linear PCA, PCR, PLS; nonlinear PCR, PLS with linear projection but nonlinear activation functions; PPR; and BPN with a single hidden layer. The features extracted by these linear projection methods are linear combinations of the inputs,

$$f = \sum_{j=1}^{n} \alpha_j x_j \tag{7}$$

These features may be determined independently of the outputs, or through relations to the outputs as,

$$y_p = \sum_{i=1}^{K} \beta_i \theta_i \left(\sum_{j=1}^{n} \alpha_j x_j \right) \tag{8}$$

A unified view describing a general methodology and selection criteria for methods based on linear projection have been described by Utojo and Bakshi (1995), and forms the basis of the description in the rest of this sub-section.

Univariate limit checking is among the simplest and most commonly used methods and comprises a broad family of statistical approaches. The most commonly implemented method of limit checking is the "absolute value check" described by Iserman (1984):

$$\text{if } \left(X_{i,min} < x_i(t) < X_{i,max} \right) \text{ then class } \omega_j \tag{9}$$

where, x(t) is the value of the measured variable at any time t, and X_{min} and X_{max} are the lower and upper mapping limits, respectively. Limit checking is the simplest specialization of Eqn. (7) as $f = x_i$, i.e., each input is directly considered as a feature, resulting in a discriminant surface perpendicular to an input axis. This form of limit checking is easy to implement, but, because of its univariate nature, other variables can take on arbitrary values that can lead to misclassification when the data are outside the range used to establish the limits.

In the extended family of limit checking approaches there are a variety of definitions for the features, f, including, individual point, statistically averaged data points, integral of the absolute error (IAE), moving average filtering, cumulative sum (CUSUM), Shewart Charts and exponentially weighted moving average (EWMA) models (MacGregor and Kourti, 1995). All definitions retain the univariate character (based on a single measured variable), and are designed to render a time series of data points into a feature that can be appropriately labeled with static limits.

While usable in many situations, limit checking methods have limited applicability. Due to their univariate nature, these approaches work very well during periods of pseudo steady-state behavior where various averaging methods represent the true behavior exhibited by the data. However, without the ability to make use of information during transient periods, it becomes difficult to interpret information during critical periods like start-up, shutdown, and process transitions. Furthermore, these methods do not work for data interpretation requiring changing limits, such as when process conditions change.

Linear Multivariate Statistical (MVS) Methods. These methods may be divided into two categories depending on whether the features are extracted in an unsupervised manner, without considering input-output relationships, or in a supervised manner, trying to get the best input-output relationship.

Unsupervised feature extraction methods simply determine the features as a linear sum of the inputs as given by Eqn. (7). In *principal component analysis*, the weights, α_j, are referred to as principal component loadings and maximize the captured input covariance while maintaining orthogonal directions, as depicted in Fig. 4. The collection of these loadings forms an orthogonal set of pseudo-measurements containing the significant variations in the input data. The weighted sum in Eqn. (7), called the latent variable, principal component, or *score*, may be used directly as features for process monitoring.

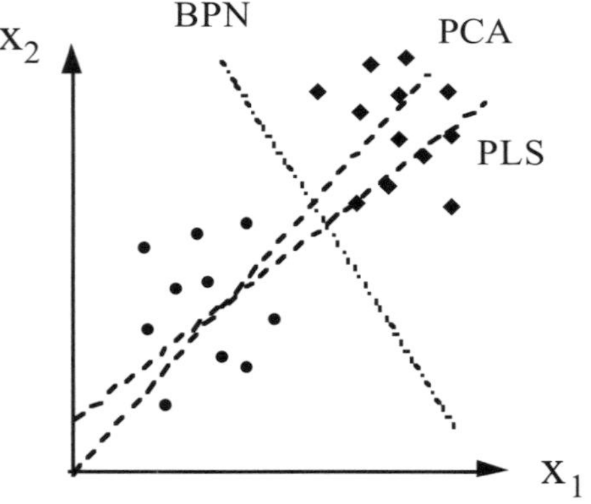

Figure 4. Projection-based methods.

Supervised feature extraction techniques extract the features by finding the best input-output relationship. Thus, they may be used as feature extractors or as empirical models. This category includes linear PCR and PLS. These methods may be represented by specializing Eqn. (8) to the form,

$$y_p = \sum_{i=1}^{K} \beta_i \sum_{j=1}^{n} \alpha_j x_j \tag{10}$$

Different methods in this category are characterized by different objective functions for estimating the parameters, α and β.

Principal component regression is a simple extension of PCA, where the α_j are computed via PCA. The β_i that minimize the output prediction error are then determined. For process monitoring, PCR provides features that are identical to those determined via PCA.

The *partial least squares regression* model is also linear, but the model parameters, α and β are determined to simultaneously maximize the input covariance captured by each linear combination of inputs, while minimizing the output approximation error. The approach was developed by Wold (1982) to overcome the parameter identification and convergence problems faced by standard techniques when dealing with a modest number of observations, highly collinear variables and data with noise in the inputs as well as the outputs. The principal component loadings are rotated to maximize the correlation with the outputs resulting in better prediction, as illustrated in Fig. 4.

For most process data, depending on the correlation between the measurements, the number of principal component loadings necessary for capturing the data variance is significantly smaller than the number of measured variables. This dimensionality reduction strategy has been very successful for multivariate statistical process monitoring (Kresta, et al., 1991; Piovoso and Kosanovich, 1994), and for overcoming some of the disadvantages of traditional SPC based on univariate limit checking.

Backpropagation Networks. The input-output model learned by BPN with a single hidden layer may be represented by Eqn. (8) with activation functions restricted to be of fixed shape, usually sigmoid. The model parameters are determined simultaneously by minimizing the overall least squares error of approximation of the outputs only. The features extracted by BPN are aimed at optimizing the discriminant surface, as depicted in Fig. 4. These networks are known to be universal approximators, and can approximate arbitrary nonlinearities, given enough nodes. The hidden layers perform feature extraction via linear projection if restricted to a single hidden layer, or via nonlinear projection, if multiple hidden layers are allowed. The output layer implements a form of linear discriminant classification. It can be readily shown that the numerical output of the ANN corresponds to the distance a data vector is from each of the linear decision surfaces (Whiteley, 1995). In order to be usable in limited data situations, methods for managing extrapolation errors are necessary.

Nonlinear MVS Methods. Superior feature extraction, greater dimensionality reduction, and better approximation ability may be obtained by allowing for nonlinear input-output relationships. Nonlinear models may be derived by applying nonlinear activation functions to linear input projections. This category includes certain nonlinear PCR and PLS and projection pursuit regression (PPR). The nonlinear activation function is derived by fitting a univariate function to the projected data using ANN (Qin and McAvoy, 1992), polynomials (Wold, et al., 1989), or variable span smoothers (Friedman and Stuetzle, 1991; Frank, 1990). The resulting input-output model is of the form given by Eqn. (8). Nonlinear PCR, PLS and PPR differ in the objective functions to be satisfied for determining α, and β, and the nature of θ (Utojo and Bakshi, 1995). Nonlinear MVS based on projection by nonlinear combination of inputs have also been developed (Kramer, 1992; Hastie and Stuetzle, 1989; Dong and McAvoy, 1994).

Kernel-Based, or Local Methods

Kernel-based methods may be related to Eqn. (5) by combining the inputs to cover a localized region in the input space, as shown in Fig. 3. The underlying assumption is that patterns in a common pattern class exhibit similar features and that this similarity can be measured using an appropriate proximity index. The training objective, therefore, is to define an appropriate prototype vector description for a given set of sufficiently similar input data patterns and to establish a topological mapping of the prototypes such that each has a set of neighbors. Feature extraction is associated with forming the prototype descriptions and then placing them in an abstract feature space. Mapping is based on the similarity or nearness of a given pattern with labeled prototype descriptions as shown in Fig. 5. Clustering lacks the strict Bayesian requirement that input patterns be identifiable with a known prototype for which underlying joint distributions or other statistical information is known. Rather, the clustering approach leverages similarities between pattern features so that patterns which have never been seen before and for which no statistical data are available can still be processed (Whiteley, 1991).

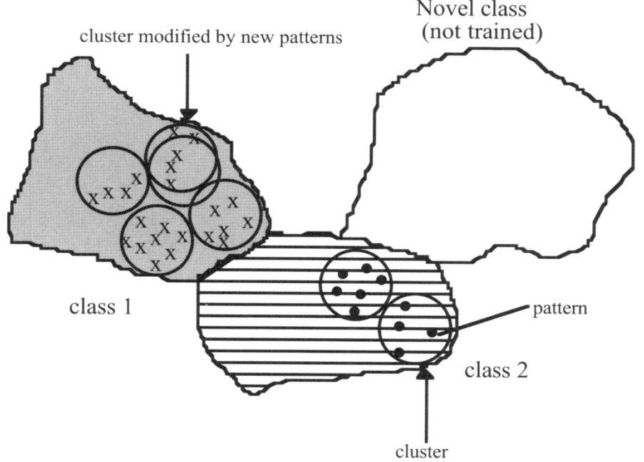

Figure 5. Kernel-based methods.

For process systems one of the key practical advantages of clustering is the ability to work with limited and poorly distributed pattern data, typical with many labels of interest. Rather than attempting to partition unknown regions of the representation space as with projection-based methods, clustering approaches simply identify the structure in the existing representation space based on available data. Kernel-based methods are, therefore, more robust to unreliable extrapolation, but they also give up potential benefits. The notion of proximity or distance can be used to make additional decisions about the data patterns such as novel pattern identification or to establish gradations of uncertainty (Davis and Wang, 1995).

It is possible to group approaches along several possible performance dimensions including distance versus directional similarity measures, variations of how matching clusters are determined. Fixed versus variable clusters appears to have the greatest practical significance. We therefore consider these as the two primary categories.

Fixed Cluster Approaches. This category represents those approaches that require a fixed number of clusters to be specified *a priori*. From an adaptation standpoint, fixed cluster approaches can adapt the prototype definition, but their number cannot be changed. Probably the most commonly discussed approach is *Radial Basis Function Networks* (RBF) (Moody and Darken, 1989, Leonard and Kramer, 1991). A topological structure is formed by specifying the number of clusters that will be used and using training data to converge on an optimal set of cluster centers via K-means clustering. The selection of cluster number, which is generally not know beforehand, represents the primary performance criteria. Optimization of performance therefore requires trial-and-error adjustment of the number of clusters.

Ellipsoidal Basis Functions Networks (EBFN) represent an approach to addressing some of the shortcomings of RBFNs (Kavuri and Venkatasubramanian, 1993). An EBFN is structured via a version of K-means clustering. Rather than establishing a cluster center based only on the winning cluster, EBFN uses a fuzzy membership cluster neighborhood idea that is an extension of the Kohonen algorithm.

Variable Cluster Approaches. Variable cluster approaches have the important capacity for adding new clusters or removing unused or unimportant clusters. To be effective, the kernel-based approach must be adaptive to the situations that have not previously occurred while maintaining information already residing in the system. In practice, this means that the systems must be designed to evolve as more information about a particular process becomes available. This requires that the system be plastic so that it can incorporate new knowledge while maintaining a level of stability relative to the existing knowledge. This implies that the number of clusters be variable and that exiting clusters be adjustable (Davis and Wang, 1995, Whiteley and Davis, 1994).

Adaptive Resonance Theory (ART) and in particular ART2, developed for both binary and analog input patterns, is a kernel-based architecture that addresses variable clustering capacity and supports the information management components that are required to decide whether a new cluster is needed or whether an existing cluster should be modified (Whiteley and Davis, 1994). ART2 forms clusters from training patterns by first computing a measure of similarity (directional rather than distance) of each pattern vector to a cluster prototype vector and then comparing this measure to an arbitrarily specified proximity criterion, the vigilance. It is a modified winner-take-all strategy in that a winning cluster is identified, but unlike RBFNs and EBFNs, the algorithm asks and answers the question about the pattern being close enough.

Conclusions

Since the early days of statistical process control several new and powerful techniques have been developed for process monitoring, data analysis and data interpretation. The original univariate limit checking methods are giving way to new and powerful analysis and interpretation techniques. In response to much more stringent demands on production, quality and product flexibility, these techniques are beginning to prove their economic worth in industrial application. The complexity of the problem has led to comprehensive approaches that involve integrated techniques that can be classified as numeric-numeric feature extraction, numeric-symbolic interpretation, and symbolic-symbolic interpretation. Numeric-symbolic interpretation is pivotal, and drives the overall performance of any data interpretation system.

With a focus on numeric-symbolic interpretation, the various approaches can be largely viewed in terms of a progression of more sophisticated feature extraction and interpretation techniques. The quantum jump in terms of the existing technology is from univariate to multivariate techniques. Within these multivariate techniques, the progression proceeds from feature extraction via linear projections of inputs, nonlinear projections of inputs, and localized regions of the input space. This progressive view provides a unifying perspective for an apparently disparate array of approaches.

In practice, decisions must be made on what approach is most appropriate for a particular application. We recognize that the critical performance issues are availability of data, reliability requirements for classification and adaptability. For these purposes the progression is conveniently divided into projection-based methods and kernel-based methods. Projection-based methods depend heavily on good coverage of the input space, and should be used very carefully in limited data situations. The non-local projection brings a potentially valuable extrapolation capability, but only if techniques are available to manage the inherent uncertainty of interpretation. In limited data situations, the kernel-based approaches offer significant advantages because the techniques are based on the local structure of the data no matter how little there is. Classification performance certainly deteriorates as the number of exemplars becomes smaller, however, kernel-based approaches are associated with clear definitions of the knowledge limitations. Patterns that cannot be classified are recognized as novel patterns. Uncertainty can be handled through the values of the proximity measures.

For a given set of data, no group of methods emerges as the best, and selection of the appropriate method is still subjective, requiring trial-and-error. With increasing insight, such as the view provided in this paper, it may be possible to select the best method depending on the data and problem characteristics, and may be the next leap in techniques for monitoring, data analysis and interpretation.

Acknowledgments

We gratefully acknowledge a technical report written by Dr. J. R. Whiteley, now at Oklahoma State University, and input by S. McVey, D. C. Miller, C. Wang and U. Utojo, graduate students at Ohio State University.

References

Bulsari, A. B. (Ed.) (1995). *Neural Networks for Chemical Engineers*. Elsevier Science, Amsterdam.

Davis, J. F. and C. M. Wang (1995). Pattern-based interpretation of on-line process data. In A. B. Bulsari (Ed.), *Neural Networks for Chemical Engineers*. Elsevier, Amsterdam. pp. 443-470.

DeVeaux, R. D., D. C. Psichogios and L. H. Ungar (1993). A comparison of two nonparametric estimation schemes: MARS and neural networks. *Comp. Chem. Eng.*, **17**, 819-837.

Dong D. and T. J. McAvoy (1994). Nonlinear principal component analysis-based on principal curves and neural networks. *Proc. Amer. Cont. Conf.*

Frank, I. E. (1990). A nonlinear PLS model. *Chemom. Intel. Lab. Sys.*, **8**, 109-119.

Friedman, J. H. (1991). Multivariate adaptive regression splines. *Annals Stat.*, **19**(1), 1-141.

Friedman, J. H. and W. Stuetzle (1981). Projection pursuit regression. *J. Amer. Stat. Assoc.*, **76**(376), 817-823.

Hastie T. and W. Stuetzle (1989). Principal curves. *J. Amer. Stat. Assoc.*, **84**(406), 1284-1288.

Kavuri, S. N. and V. Venkatasubramanian (1993). Using fuzzy clustering with ellipsoidal units in neural networks for robust fault classification. *Comp. Chem. Eng.*, **17**, 765-784.

Kramer, M. A. (1992). Autoassociative neural networks. *Comp. Chem. Eng.*, **16**, 313-328.

Kramer, M. A. and R. S. H. Mah (1994). Model-based monitoring. In D. Rippin, J. Hale and J. Davis (Eds.), *Proc. Second Int. Conf. on Foundations of Computer Aided Process Operations*. CACHE.

Leonard, J. A. and M. A. Kramer (1991). Radial basis function networks for classifying process faults. *IEEE Control Systems*, April, 31-38.

MacGregor, J. F. and T. Kourti, (1995). Statistical process control of multivariate processes. *Cont. Eng. Prac.* **3**, 404-414.

Moody, J. and C. J. Darken (1989). Fast learning in networks of locally-tuned processing units. *Neural Comp.*, **1**, 281-294.

Oja, E. (1995). Unsupervised neural learning. In A. B. Bulsari (Ed.), *Neural Networks for Chemical Engineers*. Elsevier, Amsterdam. pp. 21-32.

Pao, Y.-H. (1989). *Adaptive Pattern Recognition and Neural Networks*. Addison-Wesley, New York.

Piovoso, M. J. and K. A. Kosanovich (1994). Applications of multivariate statistical methods to process monitoring and controller design, *Int. J. Cont.*, 59, **3**, pp. 743-765.

Piovoso, M. J. and K. A. Kosanovich, (1995). The use of multivariate statistics in process control. In W. S. Levine and R. C. Dort (Eds.), *The Control Handbook*. CRC Press, Boca Raton, FL.

Qin, S. J. and T. J. McAvoy (1992). Nonlinear PLS modeling using neural networks. *Comp. Chem. Eng.*, **16**, 379-391.

Utojo, U. and B. R. Bakshi (1995). A unified view of artificial neural networks and multivariate statistical methods. *Proc. Intel. Sys. Proc. Engg*. Snowmass, CO.

Whiteley, J. R. (1991). *Knowledge-based Intepretation of Process Sensor Patterns*. PhD Dissertation, Dept. Chem. Eng., Ohio State University.

Whiteley, J. R. (1995). Interpretation of output from fault detection and other feedforward classification networks. *Technical Report*, Oklahoma State Univ.

Whiteley, J. R. and J. F. Davis (1994). A similarity-based approach to interpretation of sensor data using adaptive resonance theory. *Comp. Chem. Eng.*, **18**, 637-661.

Wold, S. (1982). Soft modeling. The basic design and some extensions. In K. Joreskog and H. Wold (Eds.), *Systems Under Indirect Observation*. Elsevier, Amsterdam

FAULT DIAGNOSIS AND COMPUTER-AIDED DIAGNOSTIC ADVISORS

Mark A. Kramer
Gensym Corporation
Cambridge, MA 02140

Roar Fjellheim
Computas Expert Systems A.S.
Sandvika, Norway

Abstract

Virtually all modern process plants include computerized information systems that centralize, present, and alarm sensor data. Yet there remains tremendous scope for improving operator support by adopting computer-based fault detection, identification and supervision systems to satisfy incentives for human safety, environmental safeguards, equipment protection and product quality. This paper reviews requirements, recent progress and remaining challenges in detecting, identifying and correcting faults in process plants, and samples a number of architectures, tools, and industrial applications.

Keywords

Fault diagnosis, Fault detection, Fault identification, Pattern recognition, Neural networks, Inductive learning, Expert systems, Operator support systems.

Introduction

Every industrial process has the potential of deviating outside its normal and intended range of behavior. Unless contained, process deviations may have a serious impact on process economy, safety, product quality and pollution level. Proper mechanisms for preventing, detecting, diagnosing and correcting abnormal process behavior should therefore be an important part of the supervisory control system of any plant. While this is widely recognized, in practice the diagnosis and response tasks are too often characterized by manual, ill-documented or *ad hoc* operator procedures. There is tremendous scope for improvement by adopting computer-based fault diagnosis and advisory systems. The direct benefits to be gained include:

- Increased safety and reduced costs by vigilant monitoring of multiple safety and economic parameters;
- Decreases in emissions, material and energy waste associated with excursions from normal operation;
- Increased product quality by rapid detection and correction of incipient disturbances;
- Reductions in human error due to mis-assessment of the process condition or failure to follow standard procedures;
- Increased plant lifetime by reduction of the duration and severity of out-of-control episodes.

These incentives have stimulated a large number of academic and industrial activities over the last decade. Perusal of the literature reveals a broad, almost bewildering array of proposed diagnostic approaches, spawned from artificial intelligence (expert systems, neural networks, qualitative simulation, case-based reasoning), probability and statistics (statistical process control, chemometrics, Bayesian networks), systems theory (estimators, observers, analytical redundancy), and safety and reliability (fault trees, causal reasoning). A growing number of industrial systems are also represented, each displaying a unique architecture, knowledge representation, solution algorithm, and human-machine interaction. Meanwhile, the process industries have not widely adopted any diagnostic technique (with the possible exception of SPC for fault detection), and there are few standard vendor-supported tools available to support off-the-shelf solutions. The prospects for this dynamic field are indeed difficult to discern.

In this paper, we examine the challenges underlying the design of diagnostic systems, beginning with a perspective on the role and functions of diagnosis systems, moving to design considerations, followed by a review of diagnostic methodologies, and concluding with a sampling of architectures, tools and environments, industrial applications, and future challenges.

Role of Monitoring and Diagnostic Systems

Diagnostic systems fit into the hierarchy of plant management at the execution supervision layer, above regulatory control layer, and below the process planning layer (Fig. 1). The general goal at this level is to assure the success of the planned operations by monitoring the performance of the system and its regulatory controls. At minimum this implies keeping operators, managers, and maintenance personnel better informed about what is going on in the process. The design and implementation of computerized fault diagnosis advisors must be driven by the needs of these user groups.

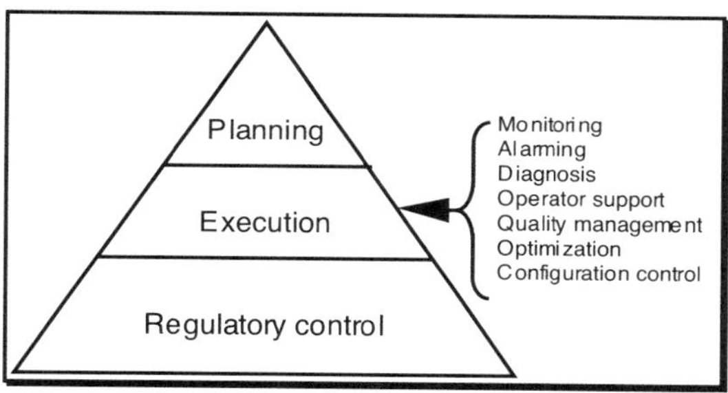

Figure 1. Plant management hierarchy.

The functions and information flow in an on-line diagnostic system are shown in Fig. 2. The diagram assumes a process monitored and controlled by a SCADA system, and supervised by an operator or operational team. Because many factors can influence the precise requirements, diagnostic applications may involve only selected elements of these functionalities, and additional non-diagnostic functions may play an important role. Nor do these functions necessarily occur sequentially in the order shown; for example, isolation and identification are not necessarily required for prognosis and compensation. The following functions may be included:

- Feature Extraction. The diagnostic process starts from process data and alarms supplied by a SCADA system, manual inputs from the operator, and information from the plant database (such as laboratory results). Normally it is desirable to process this data to extract features relevant to the fault detection and diagnosis tasks, rather than relying on raw data alone.
- *Prediction*: Ideally, one would like to prevent faults before they occur. This implies a predictive ability that might be realized through monitoring cumulative wear and shocks, composition analysis of lubrication fluids, extrapolation of real-time trends, etc.
- *Detection:* Detecting that a fault has actually occurred is not always obvious, due to slow induction or compensation by the control system. Early detection may provide invaluable warning on emerging problems, thus enabling the operator to issue actions that avoid serious process upsets.
- *Isolation:* In order to handle the fault, it is necessary to narrow the location of origin. If the physical unit or functional subsystem that is the origin of the fault can be isolated, this may be sufficient for error handling.
- *Identification:* This step involves determining the identity of the fault, usually from a pre-enumerated set of possibilities. Identification may imply one or more of the following:
  (a) *Classification*: Determining the type of fault (e.g. a leak), without necessarily providing other details (extent, time of occurrence, precise location, root cause, etc.).
  (b) *Estimation*: Determining the extent of the fault and other parameter values quantifying the fault.
  (c) *Diagnosis*: Determining the underlying cause for the fault (e.g. corrosion as the cause of a leak).
- *Prognosis:* When an abnormal event is in progress, we would like to know the potential outcome of the event and the time window available for affecting that outcome.
- *Compensation*: In many cases, the first-level response to a fault is to mitigate its negative consequence, for example, by starting a back-up unit.
- *Correction*: Finally, the system may be corrected by repairing or replacing the faulted component or ingredient. The fault handling functions may issue advice to the operator, or take direct actions via the SCADA system.

This diversity of functions implies that there cannot be an all-encompassing approach to diagnosis. As is clear from the literature on the subject, the contributions of AI, statistics and systems theory all play important roles in solving different aspects of the problem. While individual elements of the problem may be fairly well understood, integrating these elements into an overall operating system presents significant engineering challenges.

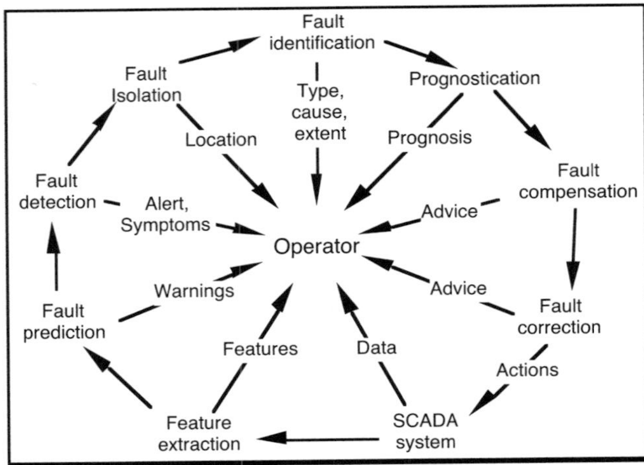

Figure 2. Functions and information flow in diagnostic systems.

Considerations in Diagnostic System Design

In this section, we discuss requirements and raise issues to be considered in the design of a diagnostic system. Each diagnostic function in Fig. 2 is characterized by information flows to the operator. Early in the design and planning of a diagnostic system, there are several strategic decisions that must be made regarding these information flows. These include:

- Which diagnostic functions should be performed autonomously by the system, which on an on-demand basis, and which should be left to the users?
- Should the system or the user initiate and guide the diagnostic reasoning process? Under what circumstances should manual input be depended upon?
- What information should the system display on a continuous basis, and what information should be accessible on demand?
- What strategy should be used to prioritize and control the flow of information from the system to the user?
- How should status information be presented so the user always knows the status of the system at a glance? How should the system focus user attention?
- How should the system alert the operators when a deviation is detected, for example, via text, color coding, sound, animation, or a combination thereof?
- What should the end-user screens look like with respect to color, layout, information density, etc.? How should the user navigate between the screens? What measures should be taken to make sure important information is not missed if the user is "on the wrong page?"

Design decisions concerning the technical scope and approach to the diagnostic system include (see Stephanopoulos and Han, 1994):

- The sources of faults to be considered, including sensors, actuators, controllers, process equipment, process parameters, raw material properties, and mis-operation;
- The failure modes for each fault source, including type of fault, extent, and temporal characteristics (e.g. step versus drift);
- The decomposition boundaries for applying monitoring models and algorithms;
- The representation of normal process behavior used for detecting faults (steady state, dynamic, linear, nonlinear, deterministic, stochastic, etc.);
- The sources of data to consider, including on-line measurements and possibly off-line manual measurements and laboratory analyses;
- Features to be extracted and examined in the data (signal properties, equation residuals, parameter estimates, dimensional projections, symbolic or Boolean discretizations, etc.);
- The type of fault detection tests to be applied (statistical, logical, univariate or multivariate);
- The description of fault behavior used to determine fault identity (algebraic or differential equations, order-of-magnitude relationships, causal graphs, probabilistic models, nonlinear mappings, rules, etc.);
- The algorithm for applying fault description to determine the fault identity.

Of course, the technical approach cannot be considered independent of the type of knowledge that is available to support it, or that its creators are willing to invest the time and effort to create. Possible knowledge sources for developing the diagnostic system include:

- Process and instrumentation diagrams;
- Equipment and material specifications;
- Operational specifications, including standard operating plans, alarm limits, and the like;
- Design and operability analyses such as fault trees, HAZOP studies, etc.;
- Analytical models, for either normal and/or faulty operations;
- Historical data, possibly including both normal and abnormal data;
- Knowledge of operating personnel.

The quality and extent of these knowledge resources will guide the technical approach and may critically impede or facilitate the development and deployment of the diagnostic system.

From an systems integration and software engineering viewpoints, additional requirements may include:

- Interfaces with the plant information systems for accessing real-time, historical and off-line laboratory data, and existing reporting and documentation systems;
- Utilization of a common database of information on plant layout, standard operating procedures, operating ranges, hazard and operability study results, etc.;
- Recording and archiving of data, conclusions, and user inputs;
- Knowledge representation in a form that is transparent, verifiable, and easy to maintain;
- Use of algorithms that are scaleable in terms of computer power, memory, modeling effort, etc., and capable of operating on the same time scale of the disturbance.

Fault Detection, Isolation and Identification

Of the monitoring and diagnosis functions defined above, the core problems of fault detection and identification (FDI) have received the most research attention. Associated problems such as prioritization of symptomatic information, fault prediction, explanation, and response are generally regarded as highly process-specific and resistant to generalized formalization and solution, although there seems to be no fundamental justification for this perception. In this section, diagnosis is described as a three-stage process involving fault detection, isolation and identification. We first discuss these problems in general terms, and then focus on five specific classes of techniques. Additional reviews of FDI are provided by Isermann (1993), Frank (1992), Kim (1994), Kramer and Mah (1993) and Stephanopoulos and Han (1994).

Fault Detection

Fault detection is a model-based task that involves comparison of the observed behavior of the process to a reference model representing fault-free behavior, and detecting significant discrepancies. Fault detection methods are defined by the model used to represent normal behavior, and the nature of the test used to detect deviation from the normal model. No information concerning failure modes and effects is required in the fault detection step.

A general representation of the fault detection problem is shown in Fig. 3. The axes represent features used to detect faults, z_i, which are generally operating conditions and product quality measurements, but may also involve derived quantities such as estimated values of parameters and states. Additionally, features can include operating history through the use of delayed measurements, trends, and explicit elapsed times (e.g. the time from the beginning of a batch). In this space, two types of conditions define normal operation. First, the process must obey certain constraints $g(z) = 0$, shown as a surface in Fig. 3, which represent the governing equations of the process, such as mass balances or process dynamics. Second, the process can be characterized in terms of a range of normal variability within the subspace defined by the model equations, shown as a contour in Fig. 3.

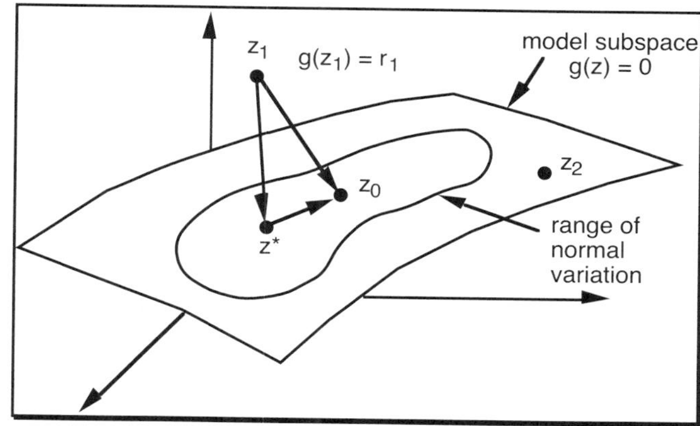

Figure 3. Fault detection.

Fault detection requires both the detection of departures from the model subspace, indicative of changes in the equations governing the process (e.g. the point z_1), and detection of excessive variability within the model subspace (e.g. point z_2), indicating an excursion from normal operating conditions. Different fault detection methods can be interpreted in terms of this general framework:

- *Alarm limits.* Alarms set at the safe operating bounds of measured variables respond only to the very large excursions from normal operation, and represent a particularly insensitive method of fault detection.
- *Univariate statistical tests.* Limit checks on individual variables, such as 3σ limits, collectively represent a hyperrectangle enclosing the normal region which may grossly overestimate the extent of the normal region, causing low sensitivity to faults.
- *Multivariate statistical tests.* Multivariate statistical models can form relatively accurate representations of the normal region. The Hotellings T^2 test, for example, defines a hyperellipse whose axes are small in dimensions orthogonal to the model subspace, while approximating variations within the model subspace.
- *Model residual tests.* Techniques based on evaluating model residuals $r = g(z)$, such as gross error detection methods and tests based on the innovations of (extended) Kalman filters, are capable of detecting departures from the model subspace. These tests, however, neglect variation within the model subspace.

- *Compound tests*. PCA and PLS, discussed below, are examples of methods that support compound tests involving both model violation and variability, though restricted to linear constraints on the features determined by regression of historical data. Similar techniques for general nonlinear analytical models are currently lacking.

The decision metrics associated with fault detection methods are frequently based on two auxiliary quantities: the rectified value z* (see Fig. 3), defined as the most likely point in the model subspace given z (which, under typical assumptions, is the mapping of z to the closest point in the model subspace), and the mean value in the model subspace z_0. These metrics may include:

a. The distance from z to z_0, or the magnitude of the components $z_i - z_{i0}$;
b. The distance from z to z*, or the magnitude of the components $z_i - z^*_i$;
c. The distance from z* to z_0, often measured in a coordinate system defined in the model subspace.
d. The magnitude of the model residual r, or of its components, r_i. Because $r = A\ (z - z^*)$, where A is the local linearization of g(z) around z*, this test is closely related to (b).

For example, the global and nodal tests involve metric (d), while a third gross error detection test, the measurement test, involves metric (b). PCA and PLS-based fault detection use both (b) and (c), while multivariate SPC tests use metric (a). For all fault detection tests, determining appropriate detection thresholds requires balancing sensitivity against false alarm rate.

Fault Isolation

As a rule, it is desirable to apply fault isolation to the maximum possible extent before attempting fault identification, due to the fact that *fault isolation can be performed without fault models, by performing fault detection over different plant subsystems*. As a result, fault isolation presents much more modest knowledge requirements than fault identification. By isolating the fault, it may be possible to formulate a response without further fault identification.

The three approaches to fault isolation are: disaggregation, hierarchical decomposition, and topological search. In the disaggregated approach, the process is decomposed into several non-interacting subsystems. Each subsystem requires a separate fault detection test, and the fault is located by continuously monitoring each subsystem for faults. This approach is suitable for processes where there are decoupled processing centers.

In the hierarchical approach, the plant is broken into successively smaller subsystems. Each level of aggregation requires a fault detection test, although not all subsystems need to be monitored continuously. When a fault is detected at the top level of the hierarchy, diagnostic focus can be shifted to inferior subsystems to refine the fault origin by applying fault detection to those subsystems. Examples of this approach include McDowell and Davis (1991), Miller, et al. (1994) and Ramesh, et al. (1992). It should be noted that this approach lends itself well to an object-oriented representation and an interface featuring "zooming in" on the fault origin through successive layers of detail.

The topological approach involves modeling subsystems and their causal precedence. Instead of assuming non-interacting subsystems, if faults appear in several subsystems, the subsystem farthest upstream is assumed to be the origin of the fault, since it is possible that the effects of the fault can propagate to subsequent subsystems. This approach can work even in the case of strongly connected subsystems (involving recycles or feedback loops), so long as there is sufficient time delay in the loop to make the fault origin clear before the effects of feedback obscure the origin. However, if the fault detection sensitivity in different subsystems are not equal, the point of earliest detection may not be the subsystem containing the fault.

Fault Identification

Fault identification requires information on how possible faults relate to observable symptoms (features). In diagnostic systems, we are interested in deducing faults from symptoms, which is opposite to the natural causal directionality of predictive models. Diagnosis is therefore an inverse problem, and as such, there are issues of solution uniqueness, solution conditioning (particularly with respect to model uncertainty), and the algorithmic efficiency of the inversion process. Modeling effort and accuracy is also of particular concern, since there are potentially a large number of fault modes. The literature contains a wealth of qualitative, logical and semi-quantitative model representations developed to help address these issues.

To better understand the motivations for non-numerical models in diagnosis, it is instructive to consider some potential limitations of direct parameter estimation as a means of "inverting" a numerical model to determine faults given observed variable trajectories. First, the parameter estimation problem may have more than one local optimum, particularly when the set of possible faults is large, includes sensor failures, or contains faults with similar effects. Parameter estimation will converge arbitrarily to one of the local optima, missing other solutions that may be more likely, or of interest as alternative hypotheses. Many qualitative models used in diagnosis are designed specifically to yield a ranked list of possible faults without the expense of global optimization. Second, modeling errors may have unpredictable effects on estimated parameter values. On the other hand, abstract or approximate models can be "right" in the sense of matching the actual fault behavior more often, by making less precise or even incomplete predictions, thus increasing robustness. Third, numerical estimation may be computationally intractable for on-line application. Fourth, the modeling effort involved with numerical modeling of all fault modes might be prohibitive. These factors account for the interest in causal, semi-quantitative, and similar model representations in fault diagnosis.

In general, there are three approaches to utilizing a predictive model for on-line diagnosis:

1. Invert the model using an algorithm specific to the model form, such as parameter estimation in the case of numerical models, graphical search in the case of causal networks, or belief updating in the case of Bayesian network.
2. Apply a general technique such as hypothesis-test or comparative simulation to effectively invert the model. In hypothesis-test, a conjecture is made about the identity of the fault, and the model is used to yield a prediction of the expected features, which is then compared to the observed symptoms. The process is repeated until the best match is found. Comparative simulation is similar, except that multiple models are run in parallel with the process.
3. Off-line, create a database of predicted symptoms for each possible fault, and use inductive learning techniques on this database to yield a pattern classifier, such as a neural network or decision tree. This approach is often referred to as "model compilation."

Of the three general approaches, the second is the least efficient. The appeal of the third approach is superior on-line efficiency. However, compiled forms of knowledge such as pattern classifiers provide a weak basis for explanation and are not useful for supporting ancillary tasks such as prognosis and correction. Therefore, in practice is may be desirable to use pattern classifiers as a supplement to, rather than a replacement for, predictive models.

Specific FDI Approaches

In this section, we review recent progress in five classes of approaches to FDI: statistical process control, parameter estimation, analytical redundancy, causal model analysis, and pattern recognition. These methods are summarized in Table 1 in terms of their approaches to feature generation, fault detection, and fault identification.

Statistical Process Control Approaches

Statistical process control techniques such as Shewhart charts, Cusum charts and Hotelling's T^2 are natural candidates for detecting faults. These methodologies utilize normal data to build a statistical characterization of the normal operating region of the process that can be used to detect abnormal events.

Projection techniques such as principal component analysis (PCA) and partial least squares (PLS) have recently attracted attention. In these approaches, fault detection is accomplished by establishing control limits on: (1) variation in projection (score) space, and (2) mapping distance between the measured point and the point projected onto the plane of constraints. Extensions to batch process have been investigated by Nomikos and MacGregor (1995) and Kourti, et al. (1995). Applications are discussed in Piovoso and Kosanovich (1993).

Table 1. Categories of FDI Approaches and Their Approaches to Feature Extraction, Fault Detection and Identification.

Approach	Features generated	Detection method	Identification method
SPC (including PCA, PLS)	*sample mean, range, scores, contributions*	*3σ limits, T^2, etc.*	*pattern recognition on scores, contributions*
Estimation (including parallel models)	*estimated parameters and states*	*parameter bounds, innovation statistics*	*likelihood ratio (parallel models), classification of parameters*
Analytical redundancy (incl. parity space, input observers)	*model equation residuals*	*statistical checks of residuals*	*pattern recognition or causal analysis of residuals*
Causal analysis (SDG, fault tree, belief net, etc.)	*discretized or fuzzified measure-ments and residuals*	*simple range checks*	*graphical, logical, and abductive approaches*
Pattern recognition (neural net, decision tree, rules, etc.)	*any features*	*membership in normal class (if represented)*	*classifier output*

Fault isolation using PCA and PLS can be carried out by decomposing global PLS models into separate blocks representing different process units (MacGregor, et al., 1994). These authors also introduce contribution plots as a way to analyze the measurement sources of abnormal behavior within each block. Although contribution plots do not strictly qualify as fault identification techniques since they do not identify the fault, they may help focus operator attention.

To carry out true diagnosis with pre-defined fault modes, fault data is used to build PCA or PLS models characterizing fault behavior, either pooling data from all faults or by building a separate PLS/PCA model for each fault. Fault identification is carried out via pattern recognition on the scores (for combined models), or by comparative model analysis (for separate fault models). Vinson, et al. (1994) have provided a very interesting critique of the combined approach using the Tennessee Eastman problem; their main difficulty was unique classification of the patterns in the score space.

The appeal of statistical process control approaches lies in their simplicity, rather than in any unique modeling or statistical properties they might possess. PCA and PLS are in essence regression techniques producing linear algebraic models that characterize normal operation. Thus PCA and PLS can be considered a simple case of analytical redundancy with linear algebraic models. Fault detection with PCA and PLS is also closely related to gross error detection techniques for linear systems, which have been extensively studied. Discussion of analogies between gross error detection and fault detection in PCA is given in Kramer (1992) and Kramer and Mah (1993).

Parameter and State Estimation Approaches

Faults associated with continuous parametric changes can be effectively diagnosed using parameter estimation techniques if the system is observable and appropriate mathematical models can be formulated. Beginning in the 1970's, many authors have applied extended Kalman filters and related approaches to this problem, recent examples including Li and Olson (1991), Fathi, et al. (1993), Isermann and Freyermuth (1991), Isermann (1993), and Ku, et al. (1992). To accomplish fault identification, many of these approaches apply pattern classification or causal analysis to the estimated states and parameters, which can be considered the features extracted in this approach.

Aside from the rather stringent modeling requirements, there are two main limitations in the parameter estimation approach. First, as the number of faults represented by undetermined parameters in the model grows large, observability may be violated. Second, structural (integer) parameters cannot typically be included in the estimation models. To overcome both limitations, it is necessary to use a bank of parallel estimators to reduce the number of adjustable parameters per model and/or replace structural parameterizations with explicitly enumerated structural alternatives. Not only does this multiply the modeling and computation work, but this strategy also introduces an additional model discrimination step (most often approached using the generalized likelihood ratio criterion) to determine which model best matches the process.

Approaches Based on Analytical Redundancy

Another large class of model-based methods is based on analytical redundancy. The basic approach is to compare actual behavior with that predicted by a model, and use the resulting differences (residuals) as feature inputs to fault identification via logical, causal, or pattern recognition techniques.

In the simplest form, measurements are directly substituted into model equations, and the resulting residual pattern is analyzed. The residual patterns are typically discretized or given a quantitative degree of violation before being causally related to faults. Recent examples of this type of approach include Chang, et al. (1994), Howell (1994), Lee (1994), Ning and Chou (1992), Petti and Dhurjati (1991), Petti, et al. (1990), and Tsai and Chou (1993). These methods differ mainly in terms of how process faults are associated with residual patterns, rather than the method of residual generation.

Some recent researchers, including Gertler and Singer (1990), Frank (1990), Frank and Ding (1994) and Patton and Chen (1993), have developed much more powerful methods, known as *structured residuals* and *unknown input observers*, that generate residuals with important properties such as optimal robustness to model uncertainty, decoupling from input disturbances, and incidence structures tailored to reveal the identity of specific faults. The theoretical basis of these methods, derived from traditional control and identification, is very sound. This line of work is advancing rapidly to include extensions to nonlinear processes, multiple faults, and structured model errors.

Approaches Based on Causal Analysis

Many contributions have been based on the concept of causal modeling of fault-symptom relationships. The relationships in these causal models have taken many different forms, including qualitative and semi-quantitative relationships, logical and probabilistic relations. Causal models have primarily been used for fault identification.

Graphical cause-and-effect models, exemplified by the signed directed graph (SDG), continue to appear frequently in the literature. In a SDG, nodes represent the system state variables and malfunctions, and arcs represent causal relationships. Recent work involves representing gains and delays, the use of fuzzy logic, diagnosis of multiple faults, increasing robustness and efficiency, and learning of fuzzy membership functions (see Chang and Yu, 1990; Finch, et al., 1990; Han, et al., 1994; Hsu and Yu, 1992; Mohindra and Clark, 1993; Wilcox and Himmelblau, 1994; Park and Seong, 1994; Qian, 1990; Yu and Lee, 1991).

Nuclear engineering has primarily adopted fault trees and similar representations for modeling causal knowledge. Because the relationship between faults and symptoms generally forms a graph, not a tree, the fault trees developed for each potential deviation or process alarm are not independent, and must be integrated through a pre-processing or run-time algorithm. Recent developments involve real-time use of fault trees (Gmytrasiewicz, et al., 1990; Zhang, 1994, Zhang, et al., 1994), goal trees and success trees (Chen and Modarres, 1992; Kim, et al., 1990; Nordvik, et al., 1994), and cause-consequence information generated from HAZOP or similar design-stage studies (Heino, et al., 1994; Martinez, et al., 1992).

Although many industrial diagnostic systems have incorporated elements of causal reasoning, from a theoretical viewpoint this area suffers from a multiplicity of modeling techniques and adoption of *ad hoc* criteria for identifying possible fault origins. Another persistent problem is the treatment of temporally-varying measurements. Because causal modeling deals with relating states of symbolic variables, the logical approach to unifying this area is through probability theory. Probability theory is sufficiently powerful to represent many types of causal influences, and also supports graphical analysis in the form of Bayesian belief networks (Rojas-Guzman and Kramer, 1993, 1994; Chu, 1993).

Pattern Recognition Methods

Pattern recognition uses associations between data patterns and fault classes without explicit modeling of internal process states or structure. Although model-based techniques are more flexible, there are several potential reasons for adopting a pattern recognition approach to fault identification:

- To capture human fault recognition rules and diagnostic associations that are not readily translated into mathematical or causal models;
- To capture the diagnostic information contained in fault data;
- As compilations of model-based descriptions for faster on-line response.

The first case suggests the application of rule-based systems. Although research interest in this approach has declined, diagnostic rules can nonetheless provide a compact and effective representation of simple diagnostic heuristics. The free-form character of this approach is both a potential benefit and a liability. Whether rule-based systems can progress beyond their current limited niche is an open question.

In contrast, the last five years has seen considerable research on training of pattern recognition systems from examples of fault behavior, addressing the second motivation given above. Since there is rarely any prior knowledge about the form of the probability distribution of the symptoms conditioned on the faults (e.g. Gaussian), non-parametric classifiers such as linear discriminants, nearest-neighbor methods, decision trees, or neural networks are usually applied. Although results obtained from these approaches are often similar in terms of accuracy, there is an advantage to methods that output the classification as a probability to permit inclusion of prior fault probabilities, control of false alarm rates, etc. Another important property to be maintained is the ability to detect novel situations for which the classifier is not trained.

Recent work in the application of neural networks to fault isolation include Becraft and Lee (1993), Fan, et al. (1993), Farell and Roat (1994), Hoskins, et al. (1991), Kavuri and Venkatasubramanian (1993, 1994), Kramer and Leonard (1990), Leonard and Kramer (1991), Marseguerra and Zio (1994), Sorsa and Kiovo (1991, 1993), Srinivasan and Batur (1994), Venkatasubramanian, et al. (1990), Watanabe, et al. (1994), and Xing and Okrent (1994). In spite of the high level of activity, the applicability of these methods is *seriously limited in practice by the availability of ample, representative, well-documented fault data*.

Decision trees are an alternative way to induce a classifier from training cases. Significant theory has been developed on the statistical interpretation and optimization of inductive decision trees in the presence of noisy features (Quinlan, 1990). Although originally derived for discrete feature vectors, decision trees can also be applied to a mixture of continuous and discrete inputs (Saraiva and Stephanopoulos, 1992). Decision trees possess certain advantages relative to neural networks, including the automatic selection of inputs and the transparency of the resulting classifier structure. However, the decision regions are limited to hyper-rectangular shapes.

For dynamic systems, pattern recognition in time is an important issue. The methods discussed so far identify faults given features developed from a moving time window of fixed length. Cheung and Stephanopoulos (1990), Konstantinov and Yoshida (1992), and Whiteley and Davis (1992) give methods for deriving qualitative features from dynamic trajectory. Wavelet transformations have also been suggested (Bakshi and Stephanopoulos, 1992). To introduce memory into the classification, an architecture like recurrent neural networks can be used, where the current classification is an input for future classifications. Alternately, Leonard and Kramer (1993) and Smyth (1994) present techniques which combine over time the instantaneous estimates of the classifier using knowledge of the statistical properties of the failure modes of the system.

Finally, pattern recognition can also be used to enhance run-time efficiency of model-based diagnosis by "compiling" the model (learning with data simulated using the model) since most classifiers run extremely fast once they have been trained. However, the compiled form will be much less maintainable than the model, and will not support functions like prognosis and explanation. An example of a system that compiles diagnostic knowledge is Far and Nakamichi (1993).

Architectures, Environments, and Applications

Thus far, we have discussed the theory and specific methods for fault diagnosis. In this section, we look at the integration of these building blocks into functional diagnostic systems, through a discussion of architectures, tools and environments, and applications.

Diagnostic System Architectures

Three R&D projects that have been carried out under the auspices of the European Commission exemplify the architectures of diagnostic systems: ARTIST, REAKT, and CommonKADS. Although by no means spanning the range of possibilities, these systems are representative of the state of the art, in which AI-derived methods are featured prominently.

The purpose of the ARTIST project (Leitch, et al., 1992) was to build a generic architecture for model-based diagnosis, and to verify the architecture on different applications. The architecture separates different types of knowledge: process knowledge, hypothesis generation knowledge, and diagnostic strategy knowledge. The resulting ARTIST architecture is shown in Fig. 4, where the role of the different modules are:

- *Predictor*: Produces behavioral predictions based on explicit models of the physical system from observations and detects discrepancies between observed and predicted behavior and/or different predictions.
- *Candidate proposer*: Generates diagnostic candidates based on discrepancies, ranks candidates according to some criterion, refines candidates with respect to structure and behavior, and discriminates between candidates.
- *Diagnostic strategist*: Controls the diagnostic process by evaluating the performance of the diagnostic process with respect to goals and resources, determining the foci of attention and suspicion, and determining the next diagnostic action.

ARTIST has been used in different applications, including diagnostic systems for the steam condenser and the boiler of thermal power plants (Angeli, et al., 1994).

In the REAKT project (Fjellheim, et al., 1994), an advanced tool for real time AI applications was developed. Its main features include a blackboard architecture, multiple cooperating agents, and predictable execution times for critical tasks. For the purpose of this paper, the main interest of REAKT lies in its support for a diagnostic/alarm handling application at an oil refinery. The general philosophy behind the application (called MORSAF) is to manage alarm situations as far as possible by *anticipation*, i.e. to compare actual with expected behavior. The expectations are alarms predicted by occurrence of previous alarms or "pre-alarm" situations. Of course, not all alarms can be anticipated, and unexpected alarms must be handled as well. A second principle of MORSAF is to base diagnosis of a fault (alarm) situation on *causal knowledge*, expressed in terms of causal networks. These networks are also used for prediction.

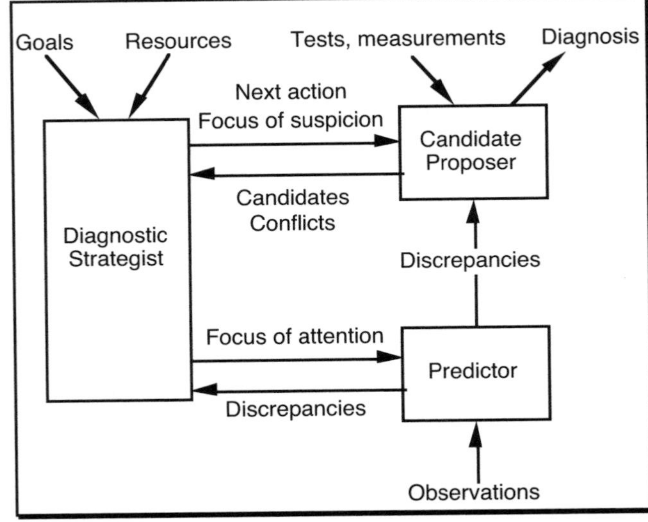

Figure 4. ARTIST architecture.

The overall architecture is illustrated by the flow diagram in Fig. 5. Incoming process data is analyzed by MORSAF, and used to detect potentially alarming situations, as well as updating the causal networks that represent state information to be used by causal diagnosis. Alarms are processed and filtered by matching earlier predicted alarms. Genuinely new (not expected) alarms are used to predict later alarms, while the expected ones (confirmed alarms) drive action suggestion. The latter also requires explanations provided by diagnosis, which is triggered by new alarms. The advice so generated is presented to the operator.

We include a brief description of CommonKADS (Schreiber, et al., 1994) here, because of its dominant position as a *de facto* standard methodology for development of knowledge based systems in Europe, and because the diagnostic task has been described in a systematic manner in this approach. A major theme in CommonKADS is *knowledge engineering as modeling*. In contrast to the more traditional view, where knowledge acquisition was seen as somehow "extracting" the knowledge from the head of an expert, CommonKADS stresses the active cooperation between the expert and the knowledge engineer in *modeling* the domain of expertise. The methods, notations and tools for supporting modeling are important ingredients of CommonKADS. Among several models, the *expertise model* is prominent. It contains domain knowledge, inference knowledge, and control knowledge.

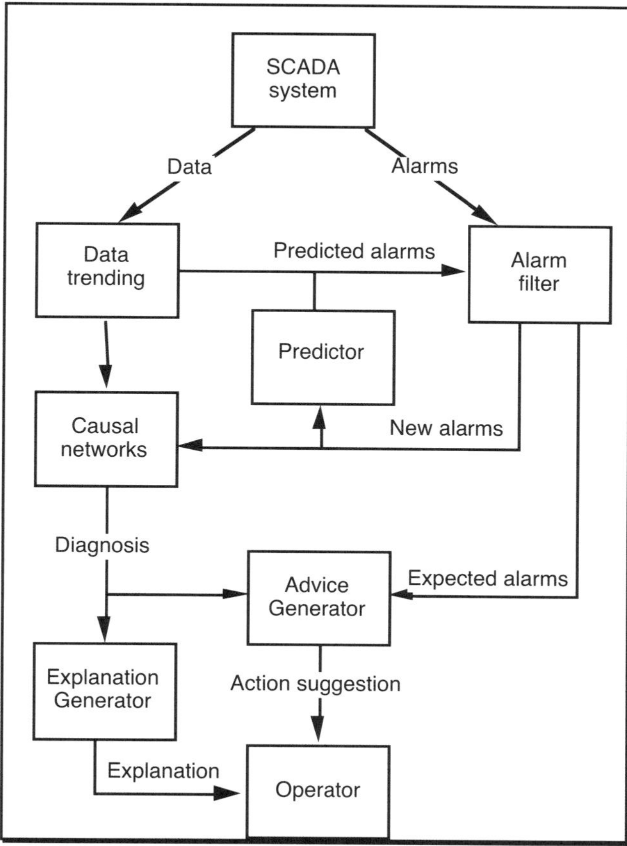

Figure 5. MORSAF architecture.

An extremely valuable part of CommonKADS is a library of generic and reusable Expertise Models (Breuker and Van de Velde, 1994). The library contains models for different types of diagnosis, including model-based diagnosis and "shallow" diagnosis (empirical, associational). With the help of the library, a new application need not start from scratch, but may rely on the structure provided by an appropriately selected library model for diagnosis. REAKT, mentioned above, embodies this philosophy in the domain of real-time diagnosis applications.

Tools and Environments

Implementation of diagnostic systems of the type outlined here requires high-level programming environments and powerful knowledge representation tools. The foundation needed to support these systems includes objects, methods, if-then rules, procedures, inheritance, hierarchies, relations, event triggering, and multi-tasking (for dealing with any number of problems simultaneously). Specific solution technologies, such as simulators, classifiers, identification packages, belief networks, statistical tools, fuzzy logic, etc. constitute a useful tool layer above the basic representations. Advanced intelligent automation systems environments, such as Gensym's G2, include representational and reasoning primitives (rules and objects), a procedural language, a graphical interface, a multi-threaded evaluation engine, and links to external code, databases, simulators, distributed control systems, and data historians.

Graphics also play a key role in how knowledge can be represented and used. Visual presentation can be enormously valuable; once gathered, reasoning about relationships in otherwise scattered data can have tremendous economic potential. In addition, many high-level solutions are most clearly represented by graphical languages. Connecting and configuring graphical objects expresses the desired behavior, and with the right software support, the graphic itself becomes the program. These are often combined with schematic representations of the system to be monitored. Gensym's G2 Diagnostic Assistant (Fraleigh, et al., 1992) and MFM (Larsson, 1994) typify graphical languages designed for diagnostic purposes.

Applications

Literally hundreds of industrial applications could be mentioned here, but only a few can be singled out for lack of space. Some excellent examples found in the literature include:

- Diagnosis of PreussenElectra's Staudinger power plant (Neupert and Schlee, 1994);
- Monitoring system for a cogeneration plant (Padalkar, et al., 1991);
- Operator support system for hydrogen peroxide production plant (Turunen, et al., 1992);
- Discharge reduction at a fertilizer plant (Saelid, et al., 1992).

The latter system exemplifies the potential benefits from diagnostic systems. Operating on one of Norsk Hydro's fertilizer plants since the beginning of 1993, the performance has been tracked and analyzed with very encouraging results. The introduction of the system led to a decrease of nitrogen discharge to the sewer from the ammonia stripper by a factor of ten, from an average of 9.6 kg/h to less than 1 kg/h.

Conclusions

As this paper has shown, the theoretical basis for computerized fault diagnosis has advanced considerably over the past few years. Still, this technology is not yet an established engineering discipline with a common terminology and a framework for systematically relating one's own problems with those reported by others. As a consequence, a practicing engineer will have no firm ground if given the assignment to design and implement a diagnosis system for a specific plant. He will be faced with an ill-defined problem, an array of available methods, and no reliable procedure for going from the problem to the solution. What is missing is a sound *engineering theory for diagnostic systems*.

On a more specific level, a number of open challenges remain within various approaches to diagnosis:

- Fault isolation is fundamentally important because it yields important diagnostic information without requiring models of faulty behavior. Topological, functional and compositional decomposition strategies for fault isolation need to be better characterized and their limits better understood.
- We would like to see better theoretical grounding of causal techniques through the application of probability theory, and integration of these methods with those derived from parameter estimation and analytical redundancy, thus profiting from the strong features of each.
- Techniques for determining the class of fault (as opposed to specific fault identity) have not received adequate study. Together with fault isolation techniques, fault classification could provide an attractive alternative to detailed identification based on pre-enumerated faults.
- Knowledge-based techniques are under-utilized due to the lack of understanding on how to integrate heuristic rules into maintainable, verifiable FDI systems.
- Because design-stage safety analyses such as HAZOP, cause-consequence analysis, etc. overlap many of the issues faced by monitoring and diagnostic systems, it seems reasonable to expect some re-use of information. Specifying a common knowledge format adequate for both off-line and on-line tasks should be a high priority.
- More attention should be paid to the ancillary problems of fault prediction, prognostication, and response. The theory in these areas lags considerably behind practical needs.

In terms of implementation, the requirements for success of diagnostic technology in industrial plants are not very different from those that apply to other operational systems. Some of the important issues are:

- The diagnostic system should not be separated from other tools used by the operator, by running on a separate computer, using a different user interface, etc. Instead, the system should be an integral part of the operator's normal working environment.
- The user interface must be up to the standards that one now takes for granted, i.e. intuitive graphics, zooming in/out on details.
- Models, in one way or another, will be part of diagnostic systems. High-level tools exist for modular construction, modification and reuse must be developed.
- Diagnostic systems will be critical to plant operations. User organizations will demand that they are delivered by vendors with a good track record, that are responsive to user needs, and provide support and maintenance services.

As we have shown earlier in this paper, a number of operational diagnosis systems have been successfully deployed and accepted by industrial users. However, many efforts have terminated with the shelving of unfinished prototypes. In general, computerized fault diagnosis systems are not as widely used today as one might expect, given the potential benefits. The reasons for this lack of acceptance, in our opinion, do not go much further than the issues listed in this section. If the research community is able to start developing engineering principles for building diagnostic systems, and vendors deliver these principles in robust, user-friendly software, user acceptance will follow. In an ever-more competitive business environment, industry has no choice other than adopting new technology that will give competitive advantage.

References

Angeli, F., L. Capetta, M. Gallant and L. Mazzocchi (1994). On-line performance and diagnosis of a steam condenser in a thermal power plant. *POWER-GEN Europe*. Cologne, Germany.

Bakshi, B.R. and G. Stephanopoulos (1992). Temporal representation of process trends for diagnosis and control. *IFAC Symp. on On-Line Fault Detection and Supervision in the Chemical Process Industries*. Newark, DE. pp. 69-74.

Becraft, W.R. and P.L. Lee (1993). An integrated neural network/expert system approach for fault diagnosis. *Comput. Chem. Engng.*, **17**, 1001-1014.

Breuker, J. and W. Van de Velde (Eds.) (1994). *CommonKADS Library for Expertise Modeling*. IOS Press, Amsterdam.

Chang, C. and C. Yu (1990). On-line fault diagnosis using the signed directed graph. *Ind. Eng. Chem. Res.*, **29**, 1290-1299.

Chang, I-C., C. Yu and C. Liou (1994). Model-based approach for fault diagnosis. 1. principles of deep model algorithm. *Ind. Eng. Chem. Res.*, **33**, 1542-1555.

Chen, L.W. and M. Modarres (1992). Hierarchical decision process for fault administration. *Comput. Chem. Engng.*, **16**, 425-448.

Cheung, J.T.-Y. and G. Stephanopoulos (1990). Representation of process trends-I and II. *Comput. Chem. Engng.*, **14**, 495-539.

Chu, B. (1993). Fault diagnosis with continuous system models. *IEEE Trans. Sys. Man Cyber.*, **23**, 55-64.

Fan, J.Y., M. Niolaou and R.E. White (1993). An approach to fault diagnosis of chemical processes via neural networks. *AIChE J.*, **39**, 82-88.

Far, B.H. and M. Nakamichi (1993). Qualitative fault diagnosis in systems with nonintermittent concurrent faults: a subjective approach. *IEEE Trans. Sys. Man Cyber.*, **23**, 14-30.

Farell, A.E. and S.D. Roat (1994). Framework for enhancing fault diagnosis capabilities of artificial neural networks. *Comput. Chem. Engng.*, **18**, 613-635.

Fathi, Z., W.F. Ramirez and J. Korbicz (1993). Analytical and knowledge-based redundancy for fault diagnosis in process plants. *AIChE J.*, **39**, 42-56.

Finch, F.E., O.O. Oyeleye and M.A. Kramer (1990). A robust event-oriented methodology for diagnosis of dynamic process systems. *Comput. Chem. Engng.*, **14**, 1379-1396.

Fjellheim, R., T.B. Pettersen, B. Christoffersen and A. Nicholls (1994). Application methodology for REAKT systems. *2nd IFAC Symp. on Artif. Intell. in Real Time Control*. Valencia, Spain.

Fraleigh, S.P., F.E. Finch and G.M. Stanley (1992). Integrating dataflow and sequential control in a graphical diagnostic language. *IFAC Symp. on On-Line Fault Detection and Supervision in the Chemical Process Industries*. Newark, DE. pp. 49-56.

Frank, P.M. (1990). Fault diagnosis in dynamic systems using analytical and knowledge-based redundancy — a survey and some new results. *Automatica*, **26**, 459-474.

Frank, P.M. (1992). Robust model-based fault detection in dynamic systems. *IFAC Symp. on On-Line Fault Detection and Supervision in the Chemical Process Industries*. Newark, DE. pp. 1-13.

Frank, P.M. and X. Ding (1994). Frequency domain approach to optimally robust residual generation and evaluation for model-based fault diagnosis. *Automatica*, **30**, 789-804.

Gertler, J. and D. Singer (1990). A new structural framework for parity equation based failure detection and isolation. *Automatica*, **26**, 381-388.

Gmytrasiewicz, P., J.A. Hassberger and J.C. Lee (1990). Fault tree based diagnostics using fuzzy logic. *Trans. Patt. Anal. Mach. Intell.*, **12**, 1115-1119.

Han, C., R. Shih and L. Lee (1994). Quantifying signed directed graphs with the fuzzy set for fault diagnosis resolution improvement. *Ind. Eng. Chem. Res.*, **33**, 1943-1954.

Heino, P., I. Karvonen, T. Pettersen, R. Wennersten and T. Andersen (1994). Monitoring and analysis of hazards using HAZOP-based plant safety model. *Reliab. Eng. Sys. Safety*, **44**, 335-343.

Hoskins, J.C., K.M. Kaliyur and D.M. Himmelblau (1991). Fault diagnosis in complex chemical plants using artificial neural networks. *AIChE J.*, **37**, 137-141.

Howell, J. (1994). Model-based fault detection in information poor plants. *Automatica*, **30**, 929-943.

Hsu, Y. and C. Yu (1992). A self-learning fault diagnosis system based on reinforcement learning. *Ind. Eng. Chem. Res.*, **31**, 1937-1946.

Isermann, R. and B. Freyermuth (1991). Process fault diagnosis based on process model knowledge — part 1: principles for fault diagnosis with parameter estimation. *Trans. ASME*, **113**, 620-626.

Isermann, R. (1993). Fault diagnosis of machines via parameter estimation and knowledge processing — tutorial paper. *Automatica*, **29**, 815-835.

Kavuri, S.N. and V. Venkatasubramanian (1993). Representing bounded fault classes using neural networks with ellipsoidal activation functions. *Comput. Chem. Engng.*, **17**, 139-164.

Kavuri, S.N. and V. Venkatasubramanian (1994). Neural network decomposition strategies for large-scale fault diagnosis. *Int. J. Control*, **59**, 767-792.

Kim, I.S., M. Modarres and R.N.M. Hunt (1990). A model-based approach to on-line process disturbance management: the models. *Reliab. Eng. Sys. Safety*, **28**, 265-305.

Kim, I.S. (1994). Computerized systems for on-line management of failures: a state-of-the-art discussion of alarm systems and diagnostic systems applied in the nuclear industry. *Reliab. Eng. Sys. Safety*, **44**, 279-295.

Konstantinov, K.B. and T. Yoshida (1992). A method for on-line reasoning about the time-profiles of process variables. *IFAC Symp. on On-Line Fault Detection and Supervision in the Chemical Process Industries*. Newark, DE. pp. 93-98.

Kourti, T., P. Nomikos and J.F. MacGregor (1995). Analysis, monitoring and fault diagnosis of batch processes using multiblock and multiway PLS. *J. Process Control (to appear)*.

Kramer, M.A. (1992). Autoassociative neural networks. *Comput. Chem. Engng.*, **16**, 313-328.

Kramer, M.A. and J.L. Leonard (1990). Diagnosis using backpropagation neural networks: analysis and criticism. *Comput. Chem. Engng.*, **14**, 1323-1338.

Kramer, M.A. and R.S.H. Mah (1993). Model-based monitoring. In *Proc. Second Conf. on Foundations of Computer-Aided Process Operations*. CACHE, Austin, TX. pp. 45-68.

Ku, W., R.H. Storer and C. Georgakis (1992). Disturbance detection and identification in statistical process control. *AIChE Natl. Meeting*. Miami Beach, FL.

Larsson, J.E. (1994). Diagnostic reasoning strategies for means-end models. *Automatica*, **30**, 775-787.

Lee, S.C. (1994). Sensor value validation based on systematic exploration of the sensor redundancy for fault diagnosis KBS. *IEEE Trans. Sys. Man Cyber.*, **24**, 594-605.

Leitch, R., H. Freitag, Q. Shen, P. Struss and G. Tornielli (1992). ARTIST: a methdological approach to specifying model based diagnostic systems. In Guida and Stefanini (Eds.), *Industrial Applications of Knowledge-Based Diagnosis*.

Leonard, J.A. and M.A. Kramer (1991) Radial basis function networks for classifying process faults. *IEEE Control Systems*, **11**, 31-38.

Leonard, J.A. and M.A. Kramer (1993). Diagnosing dynamic faults using modular neural nets. *IEEE Expert*, **8**(2), 44-53.

Li, R. and J.H. Olson (1991). Fault detection and diagnosis in a closed-loop nonlinear distillation process: application of extended Kalman filters. *Ind. Eng. Chem. Res.*, **30**, 898-908.

MacGregor, J.F., C. Jaeckle, C. Keparissides and M. Koutoudi (1994). Process monitoring and diagnosis by multiblock PLS methods. *AIChE J.*, **40**, 826-838.

Marseguerra, M. and E. Zio (1994). Fault diagnosis via neural networks: the Boltzmann machine. *Nucl. Sci. Engng.*, **117**, 194-200.

Martinez, E., L. Beltramini, H. Leone, C.A. Ruiz and E. Huete (1992). Knowledge elicitation and structuring for a real-time expert system monitoring a butadiene extraction system. *Comput. Chem. Engng.*, **16**, S345-S352.

McDowell, J.K. and J.F. Davis (1991). Managing qualitative simulation in knowledge-based chemical diagnosis. *AIChE J.*, **37**, 569-580.

Miller, D.W., J.W. Hines, B.K. Hajek, L. Khartabill, C.R. Hardy, M.A. Haas and L. Robbins (1994). Experience with the hierarchical method for diagnosis of faults in nuclear power plant systems. *Reliab. Eng. Sys. Safety*, **44**, 297-311.

Mohindra, S. and P.A. Clark (1993). A distributed fault diagnosis method based on digraph models: steady-state analysis. *Comput. Chem. Engng.*, **17**, 193-210.

Neupert, D. and M. Schlee (1994). Staudinger power plant uses expert control system to increase efficiency. *Power Engng. Intl.*, Nov/Dec.

Ning, J.N. and H.P. Chou (1992). Construction and evaluation of fault detection network for signal validation. *IEEE Trans. Nucl. Sci.*, **39**, 943-947.

Nomikos, P. and J.F. MacGregor (1995). Multivariate SPC charts for monitoring batch processes. *Technometrics*, **37**, 41-59.

Nordvik, J.P., N. Mitchison and M. Wilkens (1994). The role of the goal tree-success tree model in the real-time supervision of hazardous plants. *Reliab. Eng. Sys. Safety*, **44**, 345-360.

Padalkar, S., G. Karsai, C. Biegl, J. Sztipanovits, K. Okuda and N. Miyasaka (1991). Real-time fault diagnosis. *IEEE Expert*, **6**(3), 75-85.

Park, J.H. and P.H. Seong (1994). Nuclear power plant pressurizer fault diagnosis using fuzzy signed-digraph and spurious faults eliminations methods. *Ann. Nucl. Energy*, **21**, 357-369.

Patton, R.J. and J. Chen (1993). Optimal unknown input disturbance matrix selection in robust fault diagnosis. *Automatica*, **29**, 837-841.

Petti, T.F. and P.S. Dhurjati (1991). Object-based automated fault diagnosis. *Chem. Eng. Comm.*, **102**, 107-126.

Petti, T.F., J. Klein and P.S. Dhurjati (1990). Diagnostic model processor: using deep knowledge for process fault diagnosis. *AIChE J.*, **36**, 565-575.

Piovoso, M.J. and K.A. Kosanovich (1993). Applications of multivariate statistical methods to process monitoring and controller design. *Int. J. Control*, **59**, 743-765.

Qian, D.Q. (1990). An improved method for fault location of chemical plants. *Comput. Chem. Engng.*, **14**, 41-48.

Quinlan, J.R. (1990). Decision trees and decisionmaking. *IEEE Trans. on Sys. Man Cybernetics*, **20**, 339-346.

Ramesh, T.S., J.F. Davis and G.M. Schwenzer (1992). Knowledge-based diagnostic systems for continuous process operations based upon the task framework. *Comput. Chem. Engng.*, **16**, 109-127.

Rojas-Guzmán, C. and M.A. Kramer (1993). Comparison of belief networks and rule-based expert systems for fault diagnosis of chemical processes. *Engng. Applic. Artif. Intell.*, **6**, 191-202.

Rojas-Guzmán, C. and M.A. Kramer (1994). Multi-stage Bayesian networks subsume digraph and residual-pattern approaches to fault diagnosis. *Proc. 5th Intl. Symp. on Process Systems Engng.* Kyongju, Korea. pp. 947-952.

Saelid, S., A. Mjaavatten and K. Fjalestad (1992). An object oriented operator support system based on process models and an expert system shell. *Comput. Chem. Engng.*, **16**, S97-S108.

Saraiva, P. and G. Stephanopoulos (1992). Continuous process improvement through inductive and analogical learning. *AIChE J.*, **38**, 161-183.

Schreiber, G., B. Wielinga, R. de Hoog, H. Akkermans and W. Van de Velde (1994). CommonKADS: a comprehensive methodology for KBS development. *IEEE Expert*, **9**(6), 28-37.

Smyth, P. (1994). Hidden Markov models for fault detection in dynamic systems. *Pattern Recog.*, **27**, 149-164.

Sorsa, T. and H.N. Koivo (1991). Neural networks in process fault diagnosis. *IEEE Trans. Sys. Man Cyber.*, **21**, 815-825.

Sorsa, T. and H.N. Koivo (1993). Application of artificial neural networks in process fault diagnosis. *Automatica*, **29**, 843-849.

Srinivasan, A. and C. Batur (1994). Hopfield/ART-1 neural network-based fault detection and isolation. *IEEE Trans. Neural Networks*, **5**, 890-899.

Stephanopoulos, G. and C. Han (1994). Intelligent systems in process engineering: a review. *Proc. 5th Intl. Symp. on Process Systems Engineering*. Kyongju, Korea. pp. 1339-1366.

Tsai, T.M. and H.P. Chou (1993). Sensor fault detection with single sensor parity relation. *Nucl. Sci. Engng.*, **114**, 141-148.

Turunen, I., M. Piironen and K. Westerstrahle (1992). Expero — an advanced support system for hydrogen peroxide process control. *Comput. Chem. Engng.*, **16**, S531-S538.

Venkatasubramanian, V., R. Vaidyanathan and Y. Yamamoto (1990). Process fault detection and diagnosis using neural networks-I. steady-state processes. *Comput. Chem. Engng.*, **14**, 699-712.

Vinson, J.M., L.H. Ungar and R.D. DeVeaux (1994). Using PLS for fault analysis. *AIChE Natl. Meeting*. San Francisco, CA.

Watanabe, K., S. Hirota, L. Hou and D.M. Himmelblau (1994). Diagnosis of multiple simultaneous fault via hierarchical artificial neural networks. *AIChE J.*, **40**, 839-848.

Whiteley, J.R. and J.F. Davis (1992). Qualitative interpretation of sensor patterns using a similarity-based method. *IFAC Symp. on On-Line Fault Detection and Supervision in the Chemical Process Industries*. Newark, DE. pp. 75-80.

Wilcox, N.A. and D.M. Himmelblau (1994). The possible cause and effect graphs model for fault diagnosis. *Comput. Chem. Engng.*, **18**, 103-128.

Xing, L. and D. Okrent (1994). The use of neural network and a prototype expert system in BWR ATWS accidents diagnosis. *Reliab. Eng. Sys. Safety*, **44**, 361-372.

Yu, C.C. and C. Lee. (1991). Fault diagnosis based on qualitative/quantitative knowledge. *AIChE J.*, **37**, 617-628.

Zhang, Q. (1994). A frequency and knowledge tree/causality diagram based expert system approach for fault diagnosis. *Reliab. Eng. Sys. Safety*, **43**, 17-28.

Zhang, Q., X. An, J. Gu, B. Zhao, D. Xu and S. Xi (1994). Application of FBOLES — a prototype expert system for fault diagnosis in nuclear power plants. *Reliab. Eng. Sys. Safety*, **44**, 225-235.

SYNTHESIS AND VERIFICATION OF HIGH INTEGRITY OPERATING PROCEDURES

Scott T. Probst, Vital Aelion and Gary J. Powers
Department of Chemical Engineering
Carnegie Mellon University
5000 Forbes Avenue, Pittsburgh, PA 15213

Abstract

A formal verification method has been applied to the evaluation of chemical process maintenance and operating procedures. The evaluation is based on temporal logic models (including failure modes) of the procedure, equipment and human behaviors. The temporal model is formally checked, using Symbolic Model Checking, against sets of specifications that describe the safety, reliability and operability of the procedure and equipment. This approach allows for the synthesis of improved high integrity procedures and systems. An example procedure for checking valve tightness in a combustion system gas train is used to illustrate the approach.

Keywords

Symbolic verification, Binary decision diagrams, Operating procedure synthesis, Safety evaluation.

Introduction

The synthesis of high integrity operating procedures has been an important feature of the design of chemical processes for many decades. The recent emphasis of the AIChE Center for Chemical Process Safety, OSHA and various pending federal bills on Risk Assessment has been on the establishment of engineering guidelines for the synthesis and evaluation of such procedures. It has been repeatedly acknowledged that the operating and maintenance procedures should be conceived early in the design of the process, evolved in concert with the equipment and control system design, and continuously improved by testing with the operations and maintenance staffs.

The synthesis of operating procedures for chemical process systems has been studied by several groups using a wide range of technologies. Rivas and Rudd (1974) and O'Shima (1978) used logic models of chemical processes to synthesize feasible sequences of valve operations. Foulkes, et al. (1988) used similar models and an algorithm implemented in PROLOG to synthesize more complex procedures. Ivanov, et al. (1980a, b) modeled process phenomena and operations as state transition graphs with weighted arcs in order to synthesize optimal startup procedures. Kinoshita, et al. (1982) decomposed chemical processes and used state transition models to synthesize procedures on subsytems. The final synthesis step was integration of the subsystem procedures. Aelion and Powers (1991, 1993), Fusillo and Powers (1987, 1988), and Lakshmanan and Stephanopoulos (1988a, b, 1990) all used artificial intelligence planning techniques to synthesize operating procedures. Finally, Crooks and Macchietto (1992), Crooks, et al. (1993), and Rotstein, et al. (1992a, b) synthesized operating procedures using mixed-integer optimization methods.

In nearly all of these approaches, the detailed evaluation of the procedures was not attempted. The importance of the safety, reliability, maintainability, readability and flexibility of the procedures were addressed but not quantified. The main goal of many of the projects was to automate the synthesis of feasible plans and the evaluation of alternate plans was to be done by the writer of the procedures. Fusillo and Aelion stressed the importance of having stationary states in the procedures to allow time for the system to be checked for safety, reliability and quality. Aelion developed the concept of using fault tree analysis to quantitatively evaluate alternate procedures and to minimize the risk of the procedures with respect to the equipment utilized. We have tested a new method, Symbolic Model Checking, for the verification of procedures that offers a means for automatically evaluating the logical correctness of a set of procedures. We indicate how this approach can be combined with heuristic or algorithmic methods to develop improved procedures. An example procedure for testing valve tightness in a combustion system gas train is used to illustrate the approach.

Symbolic Model Checking

Model checking is a formal verification technique which may be used to certify that an operating procedure is safe with respect to its environment. In this case, the model is a discrete event representation of the steps in the procedure, process hardware, physical phenomena, and nondeterministic failure modes. More formally, a model is a structure $M = (S, R, L)$ in which S is a set of states, R is a transition relation which defines legal state transitions in the structure, and L is a function which labels each state with the atomic propositions which hold true (Clarke, et al., 1986). Algorithms have been developed which can check whether or not the model satisfies temporal logic properties specified by the user. If violated specifications are found, counterexamples may be generated which demonstrate how the specifications may be violated (Clarke, et al., 1993).

Symbolic model checking is a variation of model checking in which sets of states and the transition relation are represented implicitly using Boolean functions rather than explicitly using a graph structure (Burch, et al., 1994). A set of states may be represented by Boolean formulas which hold on those states. Transitions may be represented by a relation $R(\mathbf{v}, \mathbf{v}')$ in which $\mathbf{v}$ is the vector of current state assignments and $\mathbf{v}'$ is the vector of next state assignments. An edge exists in the graph If R is true for two state vectors $\mathbf{v}$ and $\mathbf{v}'$. If the set of initial states and the transition relation are given for a model, the entire reachable state space may be found.

Computation Tree Logic (CTL)

Properties to be verified are expressed in a propositional, branching-time temporal logic named Computation Tree Logic (CTL) (Clarke, et al., 1986). Any propositional logic formula is a CTL formula. CTL formulas may also contain *path quantifiers* followed by *temporal operators*. The path quantifier **E** specifies some path from the current state while the path quantifier **A** specifies all paths from the current state. The temporal operators are **X**, the *next time* operator, **U** the *until* operator, and **G** the *always* operator. $\mathbf{X}\psi$ specifies that ψ holds on the next state along the path. $\varphi\mathbf{U}\psi$ specifies that φ holds on every state along the path until ψ holds. $\mathbf{G}\psi$ specifies that ψ holds on every state along the path.

Given the above information, CTL can be defined by the following statements:

- Every atomic proposition is a CTL formula. An atomic proposition is the constant *true* or a state variable assignment.
- If φ and ψ are formulas, then $\neg\varphi$ and $\varphi\wedge\psi$ are formulas.
- If φ and ψ are formulas, then $\mathbf{EX}\varphi$, $\mathbf{E}(\varphi\mathbf{U}\psi)$, and $\mathbf{EG}\psi$ are formulas.

The formulas *false*, $\varphi\vee\psi$, $\varphi\rightarrow\psi$, $\varphi\oplus\psi$, $\varphi\leftrightarrow\psi$, $\mathbf{EF}\varphi$, $\mathbf{AX}\varphi$, $\mathbf{AG}\varphi$, $\mathbf{AF}\varphi$, $\mathbf{A}(\varphi\mathbf{U}\psi)$ can be derived from the fundamental formulas given above. $\mathbf{F}\varphi$ means that φ will hold at some future state along the path. The following are some sample CTL formulas with their English explanations.

- $\neg\mathbf{EF}(\varphi\wedge\psi)$: There does not exist a future state on which φ and ψ are simultaneously true.
- $\mathbf{EF}\ \mathbf{EG}\ (\neg\varphi)$: Some future state will be on a path in which φ is never true.
- $\mathbf{AG}\varphi$: φ holds on every reachable state.

CTL Fixpoint Characterizations

CTL formulas involving **EU** and **EG** can be characterized as fixpoints of monotonic functionals (Tarski, 1955; McMillan, 1992). A fixpoint represents all states in the model which satisfy a CTL formula. Monotonic functionals will transform a predicate in one of the following two ways:

- The set of states represented by the new predicate will contain the set of states represented by the old predicate.
- The set of states represented by the new predicate will be a subset of the set of states represented by the old predicate.

If the monotonic functional does not change the predicate at all, then a fixpoint has been reached.

Equation (1) gives the fixpoint formula for $\mathbf{E}(\varphi\mathbf{U}\psi)$.

$$\mathbf{E}(\varphi\mathbf{U}\psi) = \text{lfp}\ Z[\psi \vee (\varphi \wedge \mathbf{EX}\ Z)] \tag{1}$$

In Eqn. (1), "lfp" stands for the *least fixpoint* and Z is the fixpoint predicate. The least fixpoint is computed by initially setting Z = *false* and successively applying the functional (in this case $\psi \vee (\varphi \wedge \mathbf{EX}\ Z)$) until the fixpoint is reached. At each step states are added to the fixpoint approximation until all states which satisfy $\mathbf{E}(\varphi\mathbf{U}\psi)$ are found.

Equation (2) gives the fixpoint formula for $\mathbf{EG}\psi$.

$$\mathbf{EG}\psi = \text{gfp}\ Z[\psi \wedge \mathbf{EX}\ Z] \tag{2}$$

In Eqn. (2), "gfp" stands for the *greatest fixpoint* and Z is the fixpoint predicate. The greatest fixpoint is computed by initially setting Z = *true* and successively applying the functional (in this case $\psi \wedge \mathbf{EX}\ Z$) until the fixpoint is reached. At each step states are removed from the fixpoint approximation until just the states which satisfy $\mathbf{EG}\psi$ are left.

If the fixpoint for a CTL formula contains the set of designated initial states in the model, then the CTL formula is true. Otherwise, the formula is false and a counterexample demonstrating why the formula is false may be generated[1]

Computing the Relational Product

EX appears in both of the fixpoint formulas given above. It is computed in the relational product given in Eqn. (3).

$$\mathbf{EX}\varphi(\mathbf{v}) = \exists\mathbf{v}'[\varphi(\mathbf{v}') \wedge R(\mathbf{v},\mathbf{v}')] \tag{3}$$

In other words, $\mathbf{EX}\varphi$ is true if φ holds on some state $\mathbf{v}'$ and $\mathbf{v}'$ is an immediate successor of $\mathbf{v}$. Symbolic representations of $\varphi(\mathbf{v}') \wedge R(\mathbf{v},\mathbf{v}')$ can be large, but algorithms exist which compute the relational product without having to compute $\varphi(\mathbf{v}') \wedge R(\mathbf{v},\mathbf{v}')$ directly (Burch, et al., 1991). Efficiency in computing the relational product is crucial since this function is called repeatedly when computing fixpoints.

Symbolic Model Checking Implementation

The CTL model checking method has been implemented in a package named SMV (for Symbolic Model Verifier) (McMillan, 1992). SMV has its own language for defining finite-state concurrent systems (i.e. the transition relation and the set of initial states). Once the transition relation is built, SMV executes the symbolic model checking procedures for the user-provided CTL properties. The SMV input language has provisions for representing systems which are hierarchical, modular, and nondeterministic. SMV also provides counterexamples whenever necessary or possible.

SMV uses Ordered Binary Decision Diagrams (OBDDs) for efficient representation and manipulation of Boolean formulas. OBDDs are directed, acyclic graphs which are more compact than other currently used representations (Bryant, 1986). They are canonical with respect to a fixed variable ordering so identical functions are isomorphic (useful for recognizing fixpoints). The number of nodes required to represent a function is sensitive to the variable ordering. Any Boolean operation can be applied to OBDDs, and they support the implementation of existential quantification routines.

SMV has been used to verify electrical and computer hardware including synchronous and asynchronous circuits and cache coherency protocols for distributed multiprocessors (Burch, et al., 1990; McMillan and Schwalbe, 1991). It also has been used to verify programmable logic controller software in chemical process environments (Moon, et al., 1992; Moon, 1992, 1994; Probst and Powers, 1995). In this paper, we will show how SMV has been used to verify a maintenance procedure for a combustion system.

A Furnace Leak Testing Procedure

Figure 1 is a diagram of a furnace which has a pilot burner and a main burner. Natural gas flows through hand valve V9 and lockable ball valve LBV8, and then the stream is split into a main line and a pilot line. On the pilot line, there are two hand valves, BV10 and BV15, and two automatic blocking valves, LS12 and BS11. The automatic blocking valves open when the furnace is being started and remain open while the furnace is operating, but they close whenever the flame detector in the firebox indicates that there is no flame. This is a safety measure to prevent the formation of an explosive mixture of natural gas and air in the firebox. The main line has two automatic blocking valves, L14 and B13, which operate in the same manner. There is also a hand valve BV7 at the end of the main burner line.

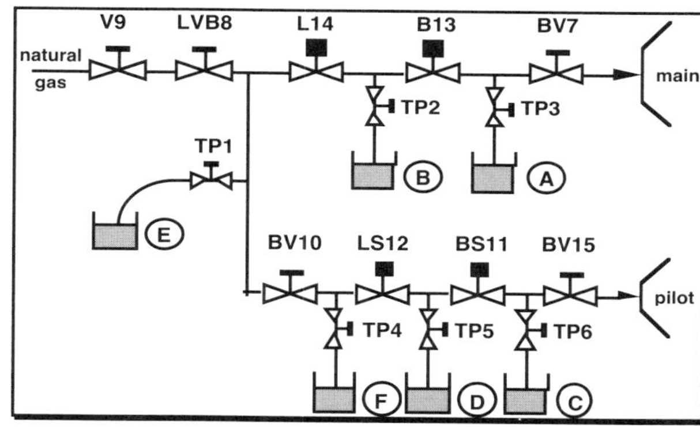

Figure 1. A Furnace with a main line, pilot line, and leak test valves.

The furnace system has a number of smaller valves, TP1 through TP6, which are connected to pipe segments between valves in the pilot line and the main line. These valves are used to check the tightness (i.e. leak across status) of the valves in the furnace, and will be referred to as *leak test valves*.

The leak test procedure is based on the following concepts. If gas is trapped between two valves, it will remain trapped if the valves are tight. If a valve is not tight and the pressure on the other side of the valve is lower than the trapped pressure, gas will escape across the valve. If the gas remains trapped and the leak test valve (TP1 through TP6) opened, then the gas will flow through the leak test valve. Each leak test valve has a length of tubing attached

[1] Counterexamples cannot be generated for all CTL formulas. For example, to demonstrate that $\mathbf{EF}\varphi$ is false, one would have to show that φ never holds on any path. If there are millions of paths in the computation tree, this would be impossible to demonstrate.

to it, and gas flowing through these valves can be detected by placing the end of the tubing in a cup of water. If bubbles are seen in the water, it means that either gas remained trapped in the pipe or gas is leaking into that pipe segment. If no bubbles are seen, natural gas is assumed to have leaked out of the pipe segment.

Leak Test Procedure Description

An operating procedure for checking the tightness of the valves (i.e. checking for leaks *across* valves) in the furnace system has been devised based on the concepts described above. The following is a description of the leak test procedure.

1. If a leaking valve is discovered during the procedure, it must be repaired or replaced immediately. Before repairing or replacing the valve, the surrounding pipe fitting joints and the interconnecting piping must be inspected to insure that they are not causing the leak. Once the repair is finished, the procedure must be re-started from the beginning.
2. Before starting the test, the main burner should be firing. The pilot is interruptible, so it may or may not be firing.
3. Close valves BV7 and BV15 . An alarm will sound and blocking valves L14, B13, LS12, and BS11 will close immediately after the flames are extinguished.
4. Close valve BV10.
5. Wait at least 5 minutes. (This amount of time defines acceptable level of tightness.)
6. Open TP3. If no bubbling occurs at point **A** (see Fig. 1), then valve BV7 is leaking (repair/replace).
7. Open TP2. If no bubbling occurs at point **B**, B13 is faulty (repair/replace). If the bubbling does not stop at point **B**, L14 is leaking (repair/replace).
8. Open BV10 and close TP2 and TP3. Valves BV7 and BV15 remain closed.
9. Attempt to re-light the furnace in pilot mode. Blocking valves LS12 and BS11 will open, but then they will close immediately since no flame is possible with BV15 closed. This action re-pressurizes pipe segments in the pilot line.
10. Open TP6. If no bubbling is observed at point **C**, BV15 is defective (repair/replace).
11. Open TP5. If no bubbling occurs at point **D**, BS11 is defective (repair/replace). If the bubbling continues indefinitely at point **D**, LS12 is defective (repair/replace).
12. Close BV10.
13. Open TP4. If bubbling continues indefinitely at point **F**, BV10 is leaking (repair/replace).
14. Close LBV8.
15. Open TP1. If bubbling continues indefinitely at point **E**, LBV8 is defective (repair / replace).

Creating the System Model

Models of this system were constructed to describe both the steps in the procedure and the behavior of the natural gas in the furnace piping. The valve and pipe segment models were developed in a modular fashion to facilitate their repeated use in the system.

Model Variables

The models which have been constructed make use of both Boolean and multivalued variables. Boolean variables represent whether or not natural gas is leaking across a valve. For example, the variable BV7_LEAK is 0 if the valve is not leaking and 1 if the valve is leaking. Such a variable exists for all valves which are tested in the procedure (i.e. all valves except the leak test valves). In the base case model, all leak test valves are assumed tight, but in other models the possibility that these valves may leak is captured.

Boolean variables also represent gas pressure in the pipe sections, and the furnace system is divided into 8 sections. Variables P1 through P8 represent whether or not positive pressure exists in a section. For example, P3 is 1 if there is a positive pressure in section 3 and it is 0 if not.

Multivalued variables represent the STATUS of the furnace and the current STEP of the operating procedure. The values STATUS may assume are *pilotstart*, *mainstart*, *burn*, *test*, *repair*, and *repairdone*. STEP can assume integer values between 0 and 17. These integers correspond roughly to the steps in the procedure given above.

Valve positions are not variables in the model. They may be open or closed depending on the values of STATUS and STEP.

Pressure Definitions

Boolean pressure variables exist in the model which define whether or not a positive pressure exists in each of the eight pipeline sections. Therefore, pressures may be either one or zero, and differences between positive pressures (i.e. pressure gradients) are not represented.

Sections which are connected by a leak or an open valve are assumed to be in pressure equilibrium, and when two or more sections are in equilibrium, they are treated as a single group. Natural gas flows in the system when a group with positive pressure is connected, via a leak or an open valve, to another group with zero pressure or to the atmosphere. If any one of a group of sections in equilibrium are leaking or vented to the atmosphere, the entire group will have zero pressure by the next step of the procedure unless one of the sections is the natural gas source. In that case, the group will maintain a positive pressure. If gas flows into a zero pressure group which is isolated from the atmosphere, that group will be appended to the group which provided the natural gas. The pressure in any one of the original eight sections is assumed to be sufficient to provide a measurable pressure (i.e. a positive indication of pressure) in any number of previously unpressurized adjacent sections. If a group of sections is completely isolated from the atmosphere and adjacent groups, the entire group will maintain its current pressure value.

Other Components of the Transition Relation

The transition relation for a model is built from a collection of rules expressed in the SMV language. The rules which govern pressure variable transitions were described in the previous section. The following is a description of the rules which define the transition relation for the other state variables.

Variables which represent whether or not a valve is leaking are nondeterministically assigned an initial value so each valve can either be leaking or not leaking before the procedure begins. The transition relation is defined so that these variables retain their initial value (1 or 0) throughout the procedure unless a valve is repaired or replaced. If a valve is repaired or replaced, it is assumed that the new or repaired valve does not leak (always takes value 0). It is also assumed that once the procedure is started, a leaking valve will not become tight and a tight valve will not begin to leak during the period of the test. In other words, intermittent faults are not modeled.

The transition rules for STEP depend upon the variable STATUS and upon itself. If STATUS ∈ {*pilotstart*, *mainstart*, *burn*, *repairdone*} in the current state, then STEP = 0 in the next state. When STATUS = *repair*, STEP retains its value in the next state. This modeling convention ensures that the valve being repaired is known since at most one valve is tested in each step. When STATUS = *test*, STEP will be incremented by 1 if the current step is successfully completed. If the end of the procedure is reached (STEP = 17), STEP is reset to 0 in the next state.

The transition rules for STATUS also depend upon STEP. If STEP = 0 and there is no pilot flame in the current step, STATUS = *pilotstart* in the next state. Also, STATUS = *pilotstart* in the next state if the test is completed (STEP = 17) or when the pilot line is re-pressurized. STATUS = *test* in the next state indicating that the test resumes after the pilot line is re-pressurized. STATUS = *mainstart* after the pilot burner is ignited, STATUS = *burn* once the main burner has been lit, and STATUS = *test* in the next state after STATUS = *burn*. Whenever a leak is discovered during the procedure, STATUS = *repair*, and in the following state, STATUS = *repairdone*.

Properties Specified

The operating procedure should meet the following specifications:

- All leaks, except intermittent ones, should be detected by the procedure.
- A valve that is not faulty should not be replaced unnecessarily.

Formula 4 shows how the first specification can be expressed in CTL. It states that there should not exist a future state in which STEP = 17 is true and one or more of the leak variables is true.

$$\neg\mathbf{EF}((\text{STEP} = 17) \wedge (\text{LBV8\_LEAK} \vee \text{B13\_LEAK} \vee \text{BV7\_LEAK} \vee \text{BV10\_LEAK} \vee \text{LS12\_LEAK} \vee \text{BS11\_LEAK} \vee \text{BV15\_LEAK})) \quad (4)$$

The second specification requires that a CTL formula be written for each valve in the pipeline. Each formula states that a valve should not be replaced unnecessarily. Formula 5 is the specification written for BV15. It states that there should not exist a future state in which STEP = 8, STATUS = *repair*, and ¬BV15_LEAK are simultaneously true.

$$\neg\mathbf{EF}((\text{STEP} = 8) \wedge (\text{STATUS} = \text{repair}) \wedge \neg\text{BV15\_LEAK}) \quad (5)$$

Models Verified

The above specifications were checked against seven different models of the furnace system. In each model, the procedure was the same and the equipment model was changed by introducing failure modes. The failure modes that were modeled represented the possibility that the leak test valves themselves could leak. For each model checked, Table 2 lists the number of Boolean variables in the model[2], the size of the reachable state space, the number of OBDD nodes in the transition relation, and the computation time.

No failure modes were included in the model *base*. In other models, the leak test valves themselves could leak. In model *TP1*, only valve TP1 could possibly leak while in model *TP1-6*, all six of the leak test valves could possibly leak. Every time a failure mode was added, an additional state variable was included in the model. If state variables are unrestricted in the model, the number of reachable states will be 2^n where n is the number of state variables. The variables representing whether or not a leak test valve is leaking are unrestricted in the initial state, therefore the reachable state space roughly doubles every time a failure

---

[2] Multivalued variables which may take n values are counted as $\lceil \log_2 n \rceil$ Boolean variables.

mode is added. The addition of these extra leaks also increases the size of the transition relation because the pressure models must now include the possibility that a leaking test valve can cause a pipe section to lose pressure.

Table 1. Computational results from leak test models containing increasing nondeterminism. Computations were performed on an Hewlett-Packard 712/80 workstation.

Model Name	Boolean Vars.	State Space	Trans. Rel. (nodes)	CPU Time (s)
base	24	5,944	11,211	4.81
TP1	25	11,859	14,804	2.03
TP1-2	26	22,263	18,734	3.90
TP1-3	27	42,154	22,961	5.15
TP1-4	28	84,313	28,215	27.88
TP1-5	29	169,049	33,701	39.26
TP1-6	30	340,193	39,087	51.66

Counterexamples

All specifications proved true in models *base* and *TP1*. This means that all pathways in those models satisfied the specifications. When verifying the other models, counterexamples were produced which showed that it is possible to for valves BV7 and B13 to be replaced unnecessarily in the presence of failure modes. The specification that no leak (of pipeline valves) should go undetected was true in every model.

BV7 can be replaced unnecessarily in the following situation. Initially, valves B13 and TP2 are leaking. When the furnace is shut down to begin the test, the pressures P4 and P5 (the two segments at the end of the main line) are in equilibrium since they are connected by the leak in B13. Therefore, gas from both segments escapes through the leak in TP2 during the 5 minute pause in the procedure. When TP3 is opened to test valve BV7, no bubbles appear at point **A** because the pressure has been depleted. However, the operator will infer that there are no bubbles at **A** because gas leaked through BV7 into the firebox. Because of this inference, BV7 will be repaired/replaced even though it is not leaking.

If TP2 is leaking, valve B13 can also be replaced unnecessarily. If TP2 is the only leaking valve, the gas trapped between L14 and B13 (P4) will escape before the operator opens TP2 to check for bubbles. When no bubbles are seen at point **B**, the operator will assume that B13 is leaking, even though it is not.

BV7 can also be replaced unnecessarily when TP3 is leaking. If TP3 is leaking, the pressure P5 will be depleted during the 5 minute pause. Therefore, no bubbles will be seen at **A** and the operator will assume that BV7 is leaking.

Similar cases in which BV15 and BS11 can be repaired/replaced unnecessarily were discovered when the leak test valves in the pilot line (TP5 and TP6) were allowed to leak.

Recommendations

Our analysis shows that the following should be added to the leak test procedure.

- Valve V9 should be tested in the procedure. To conduct this test, V9 should be closed, LBV8 should be opened, and the operator should check if bubbling eventually stops at point **E**.
- The leak test valves should also be checked for tightness. This can be done by checking if bubbles can be seen at any test point when the test valve is closed and both the pilot burner and the main burner are firing.
- The leak test valves should be checked regularly for obstructions. If a leak test valve is plugged, no bubbles will appear at a test point even though positive pressure exists in the line. This is another failure mode that will cause the operator to infer that a pipeline valve is leaking even though it is not.

These alterations to the leak test procedure indicate how formal verification methods can be used with heuristic synthesis techniques to guide the generation of higher integrity operating procedures.

Discussion

We have introduced symbolic model checking as a verification tool which may be applied to the synthesis and evaluation of operating procedures. A procedure was modeled by assigning actions to procedural steps, and this model was tested in the context of an independent model which represented physical phenomena (i.e. pressure changes in the combustion piping). The integrated model describing the interaction between the procedure and the hardware was verified with respect to safety and operability specifications expressed in CTL. We specified that no leak should go undetected (safety) and that valves should not be repaired/replaced unnecessarily (operability). By introducing failure modes into the model (leaking test valves) we found cases in which valves could be replaced unnecessarily.

These cases were described by counterexamples generated by SMV.

The only failure mode that was explored in this paper was the possibility of leaking test valves. Other failure modes such as blockages in the leak test valves or human error such as omitting steps, repeating steps, or ignoring directions were not included. Such failure modes could have been included through nondeterminism.

The models discussed in this paper were created by hand, but portions of models like these could be built using automated tools. The rules which define how pressure changes in the system have a very regular pattern which corresponds to the layout of the pipeline, so there is potential for automation. Automating the procedure modeling may be more difficult. Integers can be assigned to steps in the procedure in a straightforward way, but the actions that are performed at each step are unique to each procedure. Model construction for large systems can be tedious so automation is desirable.

Since Boolean models provide no information about pressure gradients, it was not possible to reason locally about natural gas flow. In other words, the pressure in a pipeline section (i.e. the length of piping between two valves) in the next state cannot be accurately defined by the current pressure of the section and the pressure of adjacent sections. This is because a model based on local rules will not contain some important information about adjacent sections.

In a model using local reasoning, two adjacent sections which are connected by a leak or an open valve can serve as sources of natural gas for each other. A typical rule in a local model might state that if a section has positive pressure and its neighbor does not, then both sections will have a positive pressure in the next state. If one of the sections is vented to the atmosphere, the other section will serve as a source of pressurized gas for the venting section. In a local model, the fact that the pocket of trapped gas has a finite mass is not considered. Therefore, the model predicts that gas will be vented indefinitely.

To avoid this problem, our models used more global reasoning. The rules in our model found which sections were connected, assumed that they were in equilibrium, and then assigned pressure values to the next state. As a result, we were able to create a Boolean model which represented physical behavior suitable for the scope of our verification exercise.

Acknowledgment

This research has been supported by the National Science Foundation (grant CCR-8722633).

References

Aelion V. and G. Powers (1991). A unified strategy for the retrofit synthesis of flowsheet structure for attaining or improving operating procedures. *Computers & Chemical Engineering*, **15**, 349-360.

Aelion, V. and G.J. Powers (1993). Risk reduction of operating procedures and process flowsheets. *Industrial & Engineering Chemistry Research*, **32**, 82-90.

Bryant, R.E. (1986). Graph-based algorithms for boolean function manipulation. *IEEE Transactions on Computers*, **C-35**, 677-691.

Burch, J.R., E.M. Clarke and D.E. Long (1991). Representing circuits more efficiently in symbolic model checking. *28th ACM/IEEE Design Automation Conference*. San Francisco, CA.

Burch, J.R., E.M. Clarke, D.E. Long, K.L. McMillan and D.L. Dill (1994). Symbolic model checking for sequential circuit verification. *IEEE Transactions on Computer-Aided Design of Integrated Circuits and Systems*, **13**, 401-424.

Burch, J.R., E.M. Clarke, K.L. McMillan, D.L. Dill and J. Hwang (1990). Symbolic model checking: 10^{20} states and beyond. *Proceedings of the Fifth Annual IEEE Symposium on Logic in Computer Science*.

Clarke, E.M., A. Emerson and A.P. Sistla (1986). Automatic verification of finite-state concurrent systems using temporal logic specifications. *ACM Trans. on Programming Lang. and Sys.*, **8**, 244-263.

Clarke, E.M., O. Grumberg and D.E. Long (1993). Verification tools for finite-state concurrent systems. *Lecture Notes in Computer Science*, **803**, 124-175, Springer-Verlag.

Crooks, C.A. and S. Macchietto (1992). A combined MILP and logic-based approach to the synthesis of operating procedures for batch plants. *Chemical Engineering Communications*, **114**, 117-144.

Crooks, C.A., S.F. Evans and S. Macchietto (1993). An application of automated operating procedure synthesis in the nuclear industry. *ESCAPE-3*. Graz, Austria, July 5-7.

Foulkes, N.R., M.J. Walton, P.K. Andow and M. Gulls (1988). Computer-aided synthesis of complex pump and valve operations. *Computers & Chemical Engineering*, **12**, 1035-1044.

Fusillo, R.H. and G.J. Powers (1987). A synthesis method for chemical plant operating procedures. *Computers & Chemical Engineering*, **11**, 369-382.

Fusillo, R.H. and G.J. Powers (1988) Operating procedure synthesis using local models and distributed goals. *Computers & Chemical Engineering*, **12**, 1023-1034.

Ivanov, V.A., V.V. Kafarov, V.L. Kafarov and A.A. Reznichenko (1980a). Design principles for chemical production startup algorithms. *Automation and Remote Control*, **41**, 1023-1032.

Ivanov, V.A., V.V. Kafarov, V.L. Kafarov and A.A. Reznichenko (1980b). On algorithmization of the startup of chemical productions. *Engineering Cybernetics*, **18**, 104-110.

Kinoshita, A., T. Umeda and E. O'Shima (1982). An approach for determination of operational procedures for chemical plants. *PSE*, **1**, 114.

Lakshmanan, R. and G. Stephanopoulos (1988a). Synthesis of operating procedures for complete chemical plants. part I: hierarchical, structured modeling for nonlinear planning. *Computers & Chemical Engineering*, **12**, 985-1002.

Lakshmanan, R. and G. Stephanopoulos (1988b). Synthesis of operating procedures for complete chemical plants. part II: a nonlinear planning methodology. *Computers & Chemical Engineering*, **12**, 1003-1021.

Lakshmanan, R. and G. Stephanopoulos (1990). Synthesis of operating procedures for complete chemical plants. part III: planning in the presence of qualitative mixing constraints. *Computers & Chemical Engineering*, **14**, 301-317.

McMillan, K.L. (1992). *Symbolic Model Checking — An Approach to the State Explosion Problem*. Ph.D. Thesis, School of Computer Science, Carnegie Mellon University, Pittsburgh, PA.

McMillan, K.L. and J. Schwalbe (1991). Formal verification of the encore gigamax cache consistency protocol. *Proceedings of the International Symposium on Shared Memory Multiprocessors*.

Moon, I. (1994). Modeling programmable logic controllers for logic verification. *IEEE Control Systems Magazine*, **14**(2), 53-59.

Moon, I. (1992). *Automatic Verification of Discrete-Event Chemical Process Control Systems*. Ph.D. Thesis, Department of Chemical Engineering, Carnegie Mellon University, Pittsburgh, PA.

Moon, I., G.J. Powers, J.R. Burch and E.M. Clarke (1992). Automatic verification of sequential control systems using temporal logic. *AIChE Journal*, **38**, 67-75.

O'Shima, E. (1978). Safety supervision of valve operations. *Journal of Chem. Eng. of Japan*, **11**, 390-395.

Probst, S.T., G.J. Powers, D.E. Long and I. Moon (1995). verification of a logically controlled, solids transport system using symbolic model checking. *Computers & Chemical Engineering*, to appear.

Rivas, J.R. and D.F. Rudd (1974). Synthesis of failure-safe operations. *AIChE Journal*, **20**, 320-325.

Rotstein, G.E., R. Lavie and D.R. Lewin (1992). A qualitative process-oriented approach for chemical plant operations — the generation of feasible operating procedures. *ESCAPE-1*. Elsimore, Denmark, May 24-28.

Rotstein, G.E., R. Lavie and D.R. Lewin (1992). Automatic synthesis of chemical batch plant procedures — a qualitative process-centered approach for chemical plant operations. *IFAC Symposium On-Line Fault Detection and Supervision in the Chemical Process Industries*. Newark, Delaware, April 22-24.

Tarski, A. (1955). A lattice-theoretical fixpoint theorem and its applications. *Pacific J. Math*, 5, 285-309.

SESSION SUMMARY: MONITORING, ANALYSIS AND SUPPORT OF PROCESS OPERATIONS

G.M. Schwenzer
Mobil Research and Development Corporation
Princeton, NJ 08540

Benjamin Kuipers
University of Texas at Austin
Austin, TX 78712

Schwenzer on Technology Acceptance

The process industry is undergoing significant business challenges and change which requires a sophisticated balancing between the squeeze on profit margins, budget reduction, manpower reduction, the need to optimize operation with increased reliability and reduced maintenance cost while maintaining safety and regulatory compliance.

There are three approaches to meeting these challenges. Each approach offers strengths and weakness, which must be balanced successfully. The first method is through an experienced workforce which is available to diligently monitor the process and respond to problems and take proactive action when required. However, at the same time this staff is required we see pressures to reduce staff, cut back on training budgets, etc. One pitfall of this approach is making staff available around the clock for a global business. Computer technology is one enabling technology to facilitate global access to people. The second pitfall is that the process industry requires one to monitor and interpret an excessive amount of data. Normally, staff can react to problems, however, being proactive requires much more data analysis time than available.

The second approach is to meet plant needs by process hardware improvements to increase reliability or to facilitate plant modifications for improved performance. This approach requires capital expenditure which is being limited only to the most pressing situations.

The third method is the use of computer software and hardware to perform the diligent data collection and monitoring of plant information. This alternative provides around-the-clock supervision and the ability to absorb large quantities of data on a routine basis. The approach suffers from the rapid obsolescence of computer technology, the need for computer knowledgeable staff, and lack of vendor available tools for turnkey solutions. This approach has also suffered from failed expectations from past product deliveries. Often this resulted from the problem being harder to solve than expected. However, computer technology is vital to providing the underlying enabling networks for a distributed workforce and the collection of raw data and information for problem solving. The ability to computerize the sifting of data for the attention of experts to focus on the most critical problems will also be vital for rapid problem solving with a limited workforce.

The hierarchy of plant applications is shown in Fig. 1. At the foundation is the raw data gathering mechanisms which is then turned into data historians and presentation tools. A single raw value is usually not of much value, it must be interpreted over time and in conjunction with other variables. The sensor and data validation layer is responsible for interpreting raw data. This foundation provides the infrastructure for applications in performance improvement, training and procedures, performance monitoring and operational planning. Figure 1 highlights those areas where intelligent systems in process engineering apply and those which are addressed in this session.

While computer systems may be a viable alternative to meeting business needs, there are forces that resist this change and those that continue to necessitate the ultimate use of such systems. The change initiators to move forward in the use of intelligent systems in process operations are the cultural factor of experience erosion; budget requirements to focus maintenance expense while increasing reliability, manpower reduction and the need to optimize profit. Regulatory requirements which continue to increase limits of environmental requirements and the routine need for safety improvements also will necessitate application of this technology. The infrastructure is now allowing a larger volume of data to be collected which at the same time is critical to effective problem solving.

However, it requires more processing and interpretation by resources that are not available. Software solutions are often low cost alternatives when considered against other approaches to meeting these conflicting business demands.

The resistance to change is high and it comes from the same areas of culture, budget, infrastructure and technology. The cultural effects are the reluctance to share expertise, and the user acceptance barrier to systems which continues to rise in expectations. The target audience which must be addressed by systems are the first line operational staff. There is a conflict in understanding by those who use traditional mathematical approaches and those who use some of the non-traditional approaches of the operational research and artificial intelligence community for problem solution. Budget constraints result in a lack of resources in personal and computer expense. The infrastructure itself becomes a resistance to change. There is a lack of low cost instrumentation. The vendor industry is slow to move to open systems and this requires an infrastructure change by the operating companies. The software vendors have been slow to provide open, integratable tools offering the capability to have shareable knowledge bases. There are also a host of disjoint methods which have not been turned into commercial tools thus requiring knowledgeable staff to implement. Overall there is an awareness that the problems are harder to solve than expected.

The papers in this session address process monitoring, diagnosis and design. They address the current state of the technology and highlight some areas for future work to bring this technology to practical application. They highlight the gap between academic and industrial strength use.

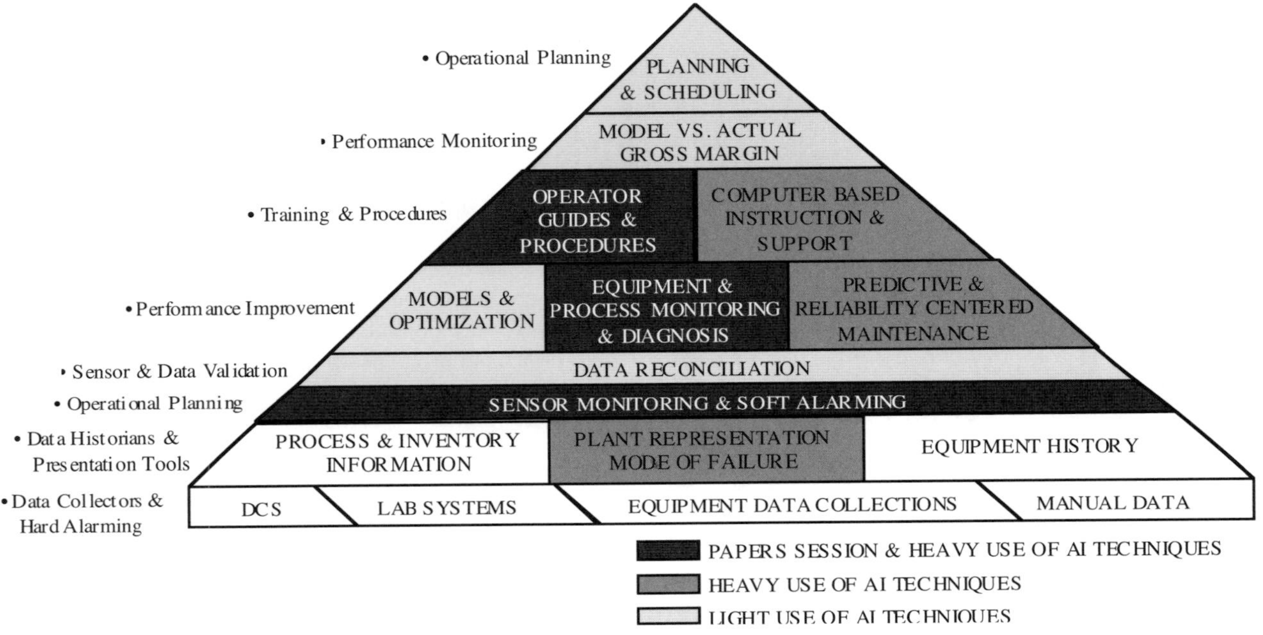

Figure 1. Plant application hierarchy.

Kuipers on Temporal Logic

Probst, Aelion and Powers derive and analyze all possible behaviors of a discrete control system, using temporal logic to prove properties of the system. They use Binary Decision Diagrams (Bryant, 1986) to provide a compact representation for the discrete control logic and its possible behaviors.

Their approach is novel and unconventional, and I believe holds much promise for the validation of discrete control systems. In our laboratory, we are doing the same thing with continuous control systems by combining qualitative simulation with temporal logic model checking.

Qualitative simulation using QSIM predicts a tree of qualitative states representing all possible behaviors of a dynamic system (Kuipers, 1986, 1994). Like Probst, Aelion and Powers, we use the temporal logics CTL and CTL* (Emerson, 1990) to express the desired properties of our control system. The behavior tree predicted by QSIM can then be interpreted as a temporal structure for the purpose of model-checking statements in CTL and CTL* (Kuipers and Shults, 1994).

A branching-time temporal logic like CTL includes modal connectives (necessarily, possibly), temporal connectives (until, next, eventually, always), as well as the standard logical connectives (not, and, or, implies), and primitive relations provided by QSIM to describe qualitative states (qval, status). The proof of a heterogeneous control law (Kuipers and Åström, 1994) includes checking the validity of a statement such as the following, which asserts that every behavior terminates at a stable quiescent state in which the value of the variable X is within the interval (B,C),

```
(necessarily (eventually (and  (qval X ((B,C) std))
                               (status quiescent)
                               (status stable)))).
```

It may be possible to handle hybrid control systems by combining these two methods for using temporal logic to reason about discrete and continuous control systems.

References

Bryant, R. E. (1986). Graph-based algorithms for Boolean function manipulation. *IEEE Trans. on Computers*, **C-35**(8).

Emerson, E. A. (1990). Temporal and modal logic. In J. van Leeuwen (Ed.), *Handbook of Theoretical Computer Science*. Elsevier Science Publ. B. V./MIT Press. pp. 995-1072.

Kuipers, B. J. (1986). Qualitative simulation. *Artificial Intelligence*, **29**, 298-338.

Kuipers, B. J. (1994). *Qualitative Reasoning: Modeling and Simulation with Incomplete Knowledge*. MIT Press, Cambridge, MA.

Kuipers, B. J. and K. Åström (1994). The composition and validation of heterogeneous control laws. *Automatica*, **30**, 233-249.

Kuipers, B. J. and Shults, B. (1994). Reasoning in logic about continuous systems. In J. Doyle, E. Sandwell and P. Torasso (Eds.), *Principles of Knowledge Representation and Reasoning: Proceedings of the Fourth International Conference (KR-94)*. Morgan Kaufmann, San Mateo, CA.

INTELLIGENT CONTROL FOR HIGH AUTONOMY PROCESS CONTROL SYSTEMS

Panos J. Antsaklis[1] and Jeffrey C. Kantor[2]
Departments of Electrical and Chemical Engineering
University of Notre Dame
Notre Dame, IN 46556

Abstract

Highly demanding control requirements in chemical processes coupled with the complexity and uncertainty of the models require the use of sophisticated control methods. To meet demanding control specifications in complex system, a number of methods have been developed that are collectively known as intelligent control methodologies. They enhance and extend traditional control methods. Hybrid control systems that combine continuous and discrete dynamics are important in the analysis and design of intelligent control, and in their own right. The work of the authors in this area is discussed briefly in this paper.

Keywords

Intelligent control, Hybrid systems.

Introduction

Intelligent controllers are envisioned emulating human mental faculties such as adaptation and learning, planning under large uncertainty, coping with large amounts of data, etc. in order to effectively control complex processes; and this is the justification for the use of the term intelligent, since these mental faculties are considered to be important attributes of human intelligence; see for example Antsaklis and Passino (1993a, 1993b), Antsaklis (1994), and Antsaklis, Lemmon and Stiver (1995) and the references therein. An alternative term used is intelligent autonomous control. It emphasizes the fact that an intelligent controller typically aims to attain higher degrees of autonomy in accomplishing and even setting control goals, rather than stressing the (intelligent) methodology that achieves those goals.

In the following, an introduction to the area of Intelligent Control is given first. A research area important to intelligent control is the area of hybrid control systems and approaches to modeling and control of such systems are discussed. This is a brief version of the paper which was presented and distributed at the conference (Antsaklis and Kantor, 1995).

[1, 2] *antsaklis.1@nd.edu* and *kantor.1@nd.edu*.

On Intelligent Autonomous Control Systems

It is perhaps appropriate to first explain further what is meant by the term Intelligent Autonomous Control: In the design of controllers for complex dynamic systems there are needs today that cannot be successfully addressed with the existing conventional control theory. Heuristic methods may be needed to tune the parameters of an adaptive control law. New control laws to perform novel control functions to meet new objectives should be designed while the system is in operation. Learning from past experience and planning control actions may be necessary. Failure detection and identification is needed. Such functions have been performed in the past by human operators. To increase the speed of response, to relieve operators from mundane tasks, to protect them from hazards, a high degree of autonomy is desired. To achieve this, high level decision making techniques for reasoning under uncertainty and taking actions must be utilized. These techniques, if used by humans, may be attributed to intelligent behavior. Hence, one way to achieve a high degree of autonomy is to utilize high level decision making techniques, intelligent methods, in the autonomous controller. In our view, *higher autonomy is the objective, and intelligent controllers are one way to achieve it*. The need for quantitative methods to model and analyze the dynamic behavior of such autonomous systems presents significant challenges well beyond current capabilities. It is clear that the development of autonomous controllers requires significant interdisciplinary research effort as it integrates concepts and methods from areas such as Control, Identification, Estimation, and Communication Theory, Computer Science, Artificial Intelligence, and Operations Research.

Control systems have a long history. Mathematical modeling has played a central role in its development in the last century and today conventional control theory is based on firm theoretical foundations. Designing control systems with higher degrees of autonomy has been a strong driving force in the evolution of control systems for a long time. What is new today is that with the advances in computing machines we are closer to realizing highly autonomous control systems than ever before.

When the uncertainties in the plant and environment are large, the fixed feedback controllers may not be adequate, and adaptive controllers are used. Note that adaptive control in conventional control theory has a specific and rather narrow meaning. In particular, it typically refers to adapting to variations in the constant coefficients in the equations describing the linear plant; these new coefficient values are identified and then used, directly or indirectly, to reassign the values of the constant coefficients in the equations describing the linear controller. Adaptive controllers provide for wider operating ranges than fixed controllers and so conventional adaptive control systems can be considered to have higher degrees of autonomy than control systems employing fixed feedback controllers.

There are cases where we need to significantly increase the operating range. We must be able to deal effectively with significant uncertainties in models of increasingly complex dynamic systems in addition to increasing the validity range of our control methods. We need to cope with significant unmodeled and unanticipated changes in the plant, in the environment and in the control objectives. This will involve the use of intelligent decision making processes to generate control actions so that a certain performance level is maintained even though there are drastic changes in the operating conditions.

In view of the above, it is quite clear that in the control of systems there are requirements today that cannot be successfully addressed with the existing conventional control theory. They mainly pertain to the area of uncertainty, present because of poor models due to lack of knowledge, or due to high level models used to avoid excessive computational complexity.

The need to use intelligent autonomous control stems from the need for an increased level of autonomous decision making abilities in achieving complex control tasks. Note that intelligent methods are not necessary for increasing control system autonomy. It is possible to attain higher degrees of autonomy by using methods that are not considered intelligent. It appears, however, that to achieve the highest degrees of autonomy, intelligent methods are necessary.

We recommend at this point that the interested reader refer to a recent report on "Defining Intelligent Control" (Antsaklis, 1994).

An Approach to Hybrid Control Systems Modeling and Design

Hybrid control systems contain two distinct types of systems, systems with continuous dynamics and systems with discrete dynamics, that interact with each other. Their study is essential in designing sequential supervisory controllers for continuous systems, and it is central in designing intelligent control systems with a high degree of autonomy (Antsaklis, 1994; Antsaklis, et al., 1993). Our group has made a number of contributions in this area (Stiver, et al., 1994, 1995a, 1995b; Antsaklis, et al., 1993, 1994; Lemmon, et al., 1993; Lemmon and Antsaklis, 1995; Yang, et al., 1994; Stiver and Antsaklis, 1992, 1993).

Hybrid control systems typically arise when continuous processes interact with, or are supervised by, sequential machines. Examples of hybrid control systems are common in practice and are found in such applications as flexible manufacturing, chemical process control, electric power distribution, and computer communication networks. A simple example of a hybrid control system is the heating and cooling system of a typical home. The furnace and air conditioner, along with the heat flow characteristics of the home, form a continuous-time system which is to be controlled. The thermostat is a simple discrete event system which basically handles the symbols *too hot*, *too cold* and *normal*. The temperature of the room is translated into these representations in the thermostat and the thermostat's response is translated back to electrical currents which control the furnace, air conditioner, blower, etc.

The hybrid control systems of interest here consist of a continuous-state system to be controlled, also called the plant, and a discrete-state controller connected to the plant via an interface. It is generally assumed that the dynamic behavior of the plant is governed by a set of known nonlinear ordinary differential equations, although our development (Stiver, et al., 1995b) is based on the state trajectories of the plant rather than the particular mechanism in force that generated those trajectories; that is, our results apply, under certain assumptions, to systems where a differential or difference equation description is not known or may not even exist. The plant contains all continuous-state subsystems of the hybrid control system, such as any conventional continuous-state controller that may have been developed, a clock if time and synchronous operations are to be modeled, etc. The controller is an event driven, asynchronous discrete event system (DES), described here by a finite state automaton. The hybrid control system also contains an interface that provides the means for communication between the continuous-state plant and the DES controller. The interface receives information from the plant in the form of a measurement of a continuous variable, such as the continuous state, and issues a sequence of symbols to the DES controller. It also receives a sequence of control symbols from the controller and issues (piecewise) continuous input commands to the plant.

The interface plays a key role in determining the dynamic behavior of the hybrid control system. Understanding how the interface affects the properties of the hybrid system is one of the fundamental issues in the theory of hybrid control systems. In Stiver, et al. (1994, 1995a, 1995b), Antsaklis, et al. (1993, 1994), Lemmon, et al. (1993), Lemmon and Antsaklis (1995), Yang, et al. (1994) and Stiver and Antsaklis (1992, 1993), the interface has been chosen to be simply a partitioning of the state space. If memory is necessary to derive an effective control, it is included in the DES controller and not in the interface. Also, the piecewise continuous command signal issued by the interface is simply a staircase signal, not unlike the output of a zero-order hold in a digital control system. Including an appropriate continuous system at (the input of) the plant, signals such as ramps, sinusoids, etc. can be generated if desired. So, the simple interface used is without loss of generality and it allows analysis of the hybrid control system, with development of properties such as controllability (Stiver and Antsaklis, 1993), stability (Lemmon and Antsaklis, 1995) and determinism (Stiver, et al., 1995b; Antsaklis, et al., 1993), in addition to control design methodologies (Stiver, et al., 1995a, 1995b). The simplicity of the interface, with the resulting benefits in identifying central issues and concepts in hybrid control systems, is perhaps the main characteristic of the approach in Stiver, et al. (1994, 1995a, 1995b), Antsaklis, et al. (1993, 1994), Lemmon, et al. (1993), Lemmon and Antsaklis (1995), Yang, et al. (1994), and Stiver and Antsaklis (1992, 1993). It is also what has been distinguishing this approach from other approaches (with more complex interfaces or with restrictions on the class of systems studied) since early versions of the model first appeared in 1991.

In general, the design of the interface depends not only on the plant to be controlled, but also on the control policies available, as well as, on the control goals. Depending on the control goals, one may need, for example, detailed state information or not, corresponding to small or large regions in the partition of the measured signal space (or greater or lower granularity). That is, of course, not surprising as it is rather well known that to stabilize a system, for example, requires less detailed information about the system's dynamic behavior than to do tracking. The fewer the distinct regions in the partitioned signal space, the simpler (fewer states) the resulting DES plant model and the simpler the DES controller design. Since the systems to be controlled via hybrid controllers are typically complex, it is important to make every effort to use only the necessary information to derive the control goals; as this leads to simpler interfaces that issue only the necessary number of distinct symbols, and to simpler DES plant models and controllers. The question of systematically determining the minimum amount of information needed from the plant in order to achieve particular control goals via a number of specialized control policies is an important and still open question; recent results in Stiver, et al. (1995a) partially resolves this question.

In Stiver, et al. (1995b), first a model is developed for hybrid control system analysis and design (see Fig. 1). This general model is characterized by a simple interface. This allows the development of a DES plant model, including the continuous-state plant together with the interface, the properties of which are discussed. In particular, DES plant properties such as determinism, quasi-determinism and observability are introduced and discussed, as well as, the notion of controllability, which is an extension of the corresponding controllability notion in logical DES's. Given control goals, such as forbidden states, a design methodology is presented that leads to a DES controller. This design method is an extension of the corresponding method in logical DES's. An alternative approach based on the invariant surfaces of the plant model with different control inputs has also been developed (Stiver, et al., 1995a).

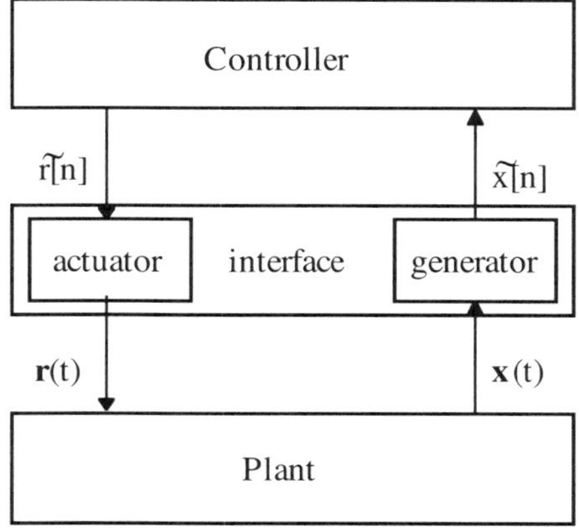

Figure 1. Hybrid control system.

It is important to note that the core issue in hybrid control systems is the way in which the interface relates the continuous plant to the DES plant. This issue is rather fundamental and quite difficult to resolve. Its solution will depend on the answer to the question of what is the minimum amount of information about the plant that will allow, with the controls available, the accomplishment of the control objectives.

Note that the approach taken in Stiver, et al. (1995a, 1995b), where a DES plant model is derived to describe the dynamic behavior of the continuous plant together with the interface, is similar to one of the main approaches in digital control (sampled data) systems. There the continuous-time plant is combined with the A/D and D/A converters, typically a sampler and a zero-order hold, to obtain a discrete-time plant model. Then the discrete-time controller is designed using discrete-time control design techniques. Of course attention should be paid to the intersample behavior of the continuous system and the selection of the sampling period. There is an alternative approach to digital control design that may lead to an alternative approach to hybrid controller design. In this approach, the discrete-time controller is an approximation of an existing continuous-time controller; that is, the control design is carried out in the continuous domain and then appropriate approximations (similar to the ones used in digital filter design) are employed to derive the discrete-time controller. This is a convenient design approach, however, it does not take full advantage of the discrete nature of the controller, since the discrete controller can be, at best, as good as the continuous controller and behavior such as deadbeat is not attained; furthermore this approach typically requires higher sampling rates than the previous ones.

Finally, in Stiver, et al. (1995b) and below, finite automata models were used to represent the DES plant model. This was done because the purpose of this work was to use the tools from the logical theory of DES, with appropriate modifications, to design controllers for hybrid systems. Note that the controller used is also a finite automaton. It is of course possible to use other models to represent the DES plant, such as Petri nets. Details of the approach described in Stiver, et al. (1995a, 1995b) can also be found in the full version of this paper (Antsaklis and Kantor, 1995).

Discrete event system theory is important in intelligent control, as it can be used, for example, to study planning the different control tasks. DES's have been studied in connection to hybrid systems; see Passino and Antsaklis (1989, 1992, 1994) and Passino, et al. (1994) for additional contributions. Recently, a very promising approach to design feedback Petri net controllers for DES's described by Petri nets has been developed (Yamalidou, et al., 1994a). Petri nets are very powerful and flexible graphical and mathematical modeling tools. As a graphical tool, Petri nets can be used as a visual communication aid similar to flow charts, block diagrams and networks and for simulation of DES's. As a mathematical tool, it is possible to set up state equations that describe the behavior of the system. In the past, their use in control has been somewhat limited, the main reason being the lack of appropriate methodologies to control systems described via Petri nets. Recently, an approach to feedback control of systems described via Petri nets was developed that uses the concept of place invariants of the net and it is simple and transparent. It appears that for the first time one will be able to systematically derive feedback controllers for real practical discrete event systems (Yamalidou, et al., 1994a).

An Alternative Approach to Hybrid Control and Supervision

The term "hybrid systems" is not well defined, but is generally understood as a class of dynamic systems in which the state space includes discrete and continuous components. In the prior discussion of this paper, hybrid systems were the consequence of quantizing the state and input spaces of an otherwise continuous process model. The quantization may be a result of physical considerations, such as the on/off quantization of measurements and control in the conventional home furnace controlled by a simple thermostat. In other cases, the quantization may be imposed by the analyst for reasons of abstracting the process behavior.

In this section, we further discuss notions of supervision for DES and hybrid systems, and survey existing approaches to these problems. To illustrate some of the issues in the supervision of hybrid systems, we consider a class of linear *discrete-time* models where some variables are discrete. A simple, one-event-ahead (OEA) supervisor is developed and applied to a fuel/air combustor example. Representative types of decomposition results are listed.

The Notion of a Hybrid Supervisor

Supervision is a weak form of feedback control that has undergone extensive development in the context of DES systems (Ramadge and Wonham, 1989; Holloway and Krogh, 1990; Balemi, et al., 1993; Scheuring and Wehlan, 1993; Chung, et al., 1992; Hadj-Alouane, et al., 1994; Li and Wonham, 1994). A supervisor directly or indirectly monitors the system state, then uses that information to enable or disable possible control actions. The final choice of the control action is determined by some other agent, the purpose of the supervisor is essentially to prohibit "bad" choices from being made. Supervisors, therefore, are not usually designed to achieve particularly closed-loop setpoint or disturbance rejection properties, but rather to permit maximum permissiveness in process operation without violating certain constraints. The jargon for this field includes suggestive terminology like "maximally permissive supervisors."

For DES systems modeled as finite state automata, Petri nets, or vector addition systems, the computation of a DES supervisor is of polynomial complexity in the number of system states. Note, however, that the number of states is typically exponential in the number of interacting elements that make up a processing system. Consequently, the direct synthesis of DES supervisors is computationally difficult for larger process systems, though promising special computational procedures have been developed and applied to realistic process models (Holloway and Krogh, 1990; Scheuring and Wehlan, 1993; Balemi, et al., 1993; del Sagrado Corazón Sanchez Carmona, 1994).

Look-ahead supervision has been proposed for systems with infinite or large state spaces, or where specifications are not completely known a priori (Chung, et al., 1992; Hadj-Alouane, et al., 1994). The look-ahead notion is similar to the ideas of "receding horizon" or "predictive" control that are part of the standard process control lexicon. Look-ahead supervision is basically an N event ahead projection of possible system behavior. The search is conducted under "conservative" and "optimistic" attitudes regarding unexpanded states at the computational horizon. The worst case complexity is quadratic in the number of expanded states. A number of desirable closed-loop properties are known. Overall, look-ahead supervision shows promise for intelligent process control.

A third approach to supervisor synthesis for Petri net models has been independently proposed by Guia, et al. (1992, 1993) and Yamalidou, et al. (1994b). The special value of the method is that it does not expand the state space of the Petri net, and involves only simple linear algebra computations. However, its application to DES systems has only been developed for Petri nets subject to convex constraints (termed generalized mutual exclusion constraints). If some of the transitions are not controllable, then the class of systems is further restricted to *safe* and *conservative* Petri nets. Unfortunately, the more general *forbidden marking* problem is outside the scope of this computationally attractive technique.

By contrast, the supervision, control, and verification of hybrid systems is in a much earlier stage of theoretical and computational development. Oversimplifying, the available approaches can be broken into three categories that are mainly distinguished by type of models used to represent the hybrid system in question:

- **Simulation/Optimization**
  The familiar commercial modeling packages, such as Simulink and MatrixX, include elements that can be used to simulate the behavior of hybrid systems. Newer tools include gPROMS (Pantelides, 1995), dstools (Back, et al., 1993), and other packages. Supporting analysis is typically based on existing theories for variable structure systems (e.g. Guckenheimer, 1995) and quantized systems (e.g. Delchamps, 1990).
- **Logic/Formal Systems**
  Several different approaches towards a logic for hybrid systems have been proposed, such a TLA+ (Lamport, 1993), Duration calculus (Chaochen, et al., 1993), and Hybrid Automata (Alur, et al., 1993; Tittus and Egardt, 1994). SIGNAL is an ambitious attempt to couple a workable simulation environment for hybrid systems with a strong logical foundation (Benveniste and Guernic, 1990).
- **DES encoding**
  A number of results have been presented wherein hybrid systems are essentially "encoded" as DES systems with DES methods applied. For examples of this approach see Antsaklis, et al. (1993), Peleties and DeCarlo (1994), and Raisch and O'Young (1994).

There would appear to be significant but unexploited overlap among the different approaches cited above. For example, it would seem the special class of Linear Hybrid Automata studied effectively by Tittus, et al. (1994) could also be a very efficient basis for the simulation/optimization approach based on gPROMS (Pantelides, 1995).

A broad range of potential applications for hybrid systems exists in the process industries. These include batch processing systems, safety interlock systems, pipe and valve network operations, and intelligent control. However, it is not yet clear which of these methods will lead to practical tools for control design and analysis.

A Class of Linear Hybrid Models with Exclusion Constraints

In this section, we consider in more depth a simple class of linear hybrid models in discrete time. This class generalizes standard condition/event Petri nets to include continuous and discrete state variables. As a subset, it includes the continuous Petri net formulation suggested by David and Alla (1994). This model can be exploited very effectively to develop simple hybrid supervisors, and yield strong analytical results.

For modeling purposes, the system state prior to the kth event is represented by an n element vector $x(k)$. The individual elements are defined as either real, integer or 0-1 quantities. The state is an element of a state space $\mathcal{S}$ constructed as a direct sum

$$\mathcal{S} \equiv \mathcal{R}^{n_r} \oplus \mathcal{Z}^{n_i} \oplus \mathcal{B}^{n-n_r-n_i} \tag{1}$$

where $\mathcal{R}$ is the set of reals, $\mathcal{Z}$ is the set of integers, and $\mathcal{B}$ is the boolean set $\{0,1\}$.

The state of a hybrid system model changes in response to events. An m element vector $q(k)$ represents the kth event. The elements of $q(k)$ represent components of an event, which are individually represented as either real or integer, or 0-1 variables. The vector $q(k)$ is an element of an event space $\mathcal{E}$ constructed as a direct sum

$$\mathcal{E} \equiv \mathcal{R}^{m_r} \oplus \mathcal{Z}^{m_i} \oplus \mathcal{B}^{m-m_r-m_i} \tag{2}$$

The evolution of the state of a hybrid system can be modeled in many different ways, depending on the application and the purpose for building the model. In the present case, we restrict our attention to models in which the next state is a linear function of the current state and event, subject to a system of linear constraints. We will call this the Linear Hybrid Model and refer to it with an acronym LHM.

Definition 1 (Linear Hybrid Model) *A linear hybrid model (LHM) is given in the form*

$$x(k+1) = Ax(k) + Bq(k) \tag{3}$$
$$Px(k) + Qq(k) \le R \tag{4}$$

where $x(k) \in \mathcal{S}$ *is the state vector, and* $q(k) \in \mathcal{E}$ *is the event vector. A, B, P, Q, and R are conformable matrix or vector valued parameters.*

Not all choices of matrix parameters A, B are compatible with a given state space $\mathcal{S}$ and event space $\mathcal{E}$. This situation motivates the following definition.

Definition 2 (Well formed LHM) *A linear hybrid model is well formed if* $Ax(k) + Bq(k) \in \mathcal{S}$ *for all* $x(k) \in \mathcal{S}$ *and* $q(k) \in \mathcal{E}$.

In the sequel, we will assume that we are discussing well-formed LHM's. It is a straightforward exercise to construct LHM's exhibiting chaos and other forms of complex nonlinear dynamics, consult the full version of this paper for examples (Antsaklis and Kantor, 1995). Here we present some examples relevant to the literature on DES and hybrid systems.

Examples of Linear Hybrid Models

Condition-Event Petri Nets

The LHM specializes to a condition-event Petri net. Consider the case where the elements of $x(k)$ and $q(k)$ are 0-1 variables, and the model parameters are integer arrays of the form

$$A = I_n \tag{5}$$
$$B = D^+ - D^- \tag{6}$$
$$P = \begin{bmatrix} -I_n \\ 0_{1\times n} \end{bmatrix} \tag{7}$$
$$Q = \begin{bmatrix} D^- \\ 1_{1\times n} \end{bmatrix} \tag{8}$$
$$R = \begin{bmatrix} 0_{n\times 1} \\ 1 \end{bmatrix} \tag{9}$$

where $D^{\pm} \ge 0$, and the notations $0_{r\times s}$ and $1_{r\times s}$ denote r x s matrices of zeros and ones, respectively. These definitions yield a model

$$x(k+1) = x(k) + (D^+ - D^-)q(k) \tag{10}$$
$$x(k) \ge D^- q(k) \tag{11}$$
$$1 \ge \sum_{i=1}^{m} q_i(k) \tag{12}$$

A non-negative initial condition $x(0) \ge 0$ will result in $x(k) \ge 0$ for all $k > 0$. The condition $\sum_{i=1}^{m} q_i(k) \le 1$ is introduced here to yield a model where one event transition occurs per increment of the counter k. This is a standard representation of a condition-event Petri net where the matrices D^- and D^+ describe the net topology (Cassandras and Ramadge, 1990).

Vector Discrete-Event Systems

Li and Wonham (1993, 1994) introduced Vector Discrete-Event Systems (VDES) as an analytically tractable device for modeling discrete-event systems. The state space of a VDES is a set of n dimensional integer vectors, i.e. $\mathcal{S} \equiv \mathcal{Z}^n$. A set of possible events is denoted by Σ which is further partitioned into controllable and uncontrollable event sets. A state transition is modeled as $\delta : \Sigma \times \mathcal{S} \to \mathcal{S}$ denoted by

$$\delta(\alpha, x) = x + E_\alpha \tag{13}$$

where $E_\alpha \in \mathcal{S}$ is the displacement vector for event $\alpha \in \Sigma$. In the formulation of Li and Wonham, there is a state transition function written for each event. A state transition is defined, denoted by $\delta(\alpha, x)!$, if $x \ge F_\alpha$, where F_α is called the occurrence-condition vector for event α.

A VDES model can be directly translated to an instance of an LHM. First, we consider α as index into the m elements i of the event set Σ. Then $q_\alpha(k) = 1$ if event $\alpha \in \Sigma$ is the kth event to occur, otherwise $q_\alpha(k) = 0$. In the obvious way, a VDES translated into an LHM is given by

$$x(k+1) = x(k) + \begin{bmatrix} E_1 & \cdots & E_m \end{bmatrix} \begin{bmatrix} q_1(k) \\ \vdots \\ q_m(k) \end{bmatrix} \tag{14}$$

$$-x(k) + \begin{bmatrix} F_1 & \cdots & F_m \end{bmatrix} \begin{bmatrix} q_1(k) \\ \vdots \\ q_m(k) \end{bmatrix} \le 0 \tag{15}$$

An additional constraint $\sum_{i=1}^{m} q_i(k) \le 1$ assures that only one event takes place per unit "time." The result is very similar to the case of a condition-event Petri net, with the main difference being that the state vector is defined over the set of integers (including non-positive integers) rather than 0-1 variables.

Linear Discrete Time Systems

A specialization of LHM to conventional linear time-invariant discrete time systems is constructed very simply. Let $P = Q = R = []$, and let $x(k) \in \mathcal{R}$, then

$$x(k+1) = Ax(k) + Bq(k) \tag{16}$$

where $q(k)$ may be interpreted as exogenous inputs.

The inequalities can be used to model systems with feedback. For example, choosing $P = \begin{bmatrix} K^T & K^T \end{bmatrix}^T$, $Q = \begin{bmatrix} I_m & I_m \end{bmatrix}^T$, and $R = 0_{sm \times n}$ is equivalent to the conventional LTI system above with the feedback constraint $q(k) = -Kx(k)$.

A One-Event Ahead Hybrid Supervisor

Here we demonstrate the construction of a simple supervisor for an LHM subject to mutual exclusion constraints. The technique follows along the lines developed by Guia, et al. (1992, 1993) and Yamalidou, et al. (1994b) for Petri nets, but incorporating certain elements of look-ahead supervision due to Lafortune and Yoo (1990).

For application to LHM's, it is first necessary to resolve the difference between time and events. In the present discussion, the counter k refers to time measured as ticks on a master clock. The control horizon, N, is chosen to look ahead a fixed amount of time horizon. Over this time horizon, 0 or more events might occur as indicated by the discrete elements of the input vector $q(k)$ on the interval from k to $k+N-1$. The one-event ahead supervisor is developed assuming that only one controllable event occurs on this interval at time k. For the limiting case where the LHM represents a Petri net, we show this is equivalent to the situation presented by Guia and Yamalidou. In the other limit of continuous variable linear discrete-time systems, this is roughly analogous to one-step ahead model predictive control on a horizon N.

Consider an LHM for which the exclusion constraints depend on state alone, that is $Q=0$

$$x(k+1) = Ax(k) + Bq(k) \tag{17}$$
$$Px(k) \leq R \tag{18}$$

with initial conditions $x(0) = x_0$. Following Yamalidou, et al. (1994b), a conformable vector of slack variables $s(k)$ is introduced so that the constraints become

$$px(k) + s(k) = R \tag{19}$$

The constraints $s(k) \geq 0$ are to hold at all future time steps. Considering just the next time step yields the recursion

$$\begin{bmatrix} x(k+1) \\ s(k+1) \end{bmatrix} = \begin{bmatrix} A & 0 \\ P(A-I) & I \end{bmatrix} \begin{bmatrix} x(k) \\ s(k) \end{bmatrix} + \begin{bmatrix} B \\ PB \end{bmatrix} q(k) \tag{20}$$

with initial conditions

$$\begin{bmatrix} x(0) \\ s(0) \end{bmatrix} = \begin{bmatrix} x_0 \\ Px_0 - R \end{bmatrix} \tag{21}$$

The extra states introduced by the recursion on the slack variables are called "monitors," and a one-step ahead prediction requires that $q(k)$ satisfy the relation $s(k+1) \geq 0$, i.e.,

$$P(A-I)x(k) + s(k) + PBq(k) \geq 0 \tag{22}$$

The results of Yamalidou, et al. can be obtained for regular Petri nets (i.e. Petri nets without self-loops, c.f. Lafortune and Yoo (1990)). In the case of Petri nets,

$$A \equiv I, \; P \equiv \begin{bmatrix} -I \\ P_2 \end{bmatrix}, \text{ and } R \equiv \begin{bmatrix} 0 \\ R_2 \end{bmatrix} \tag{23}$$

In this case, the slack variables $s(k)$ represent controller memory, and the one step ahead constraint $s(k+1) \geq 0$ yields the relation

$$x(k+1) \geq 0 \tag{24}$$
$$s(k) + P_2 Bq(k) \geq 0 \tag{25}$$

Under the no self-loop condition of regular Petri nets, the first inequality is the standard Petri net firing rule, and the second consists of problem specific exclusion constraints. Yamalidou, et al. (1994b) demonstrated the utility of this approach when applied to several sample problems, and Guia, et al. (1992, 1993) proved several analytical results.

We introduce the notation $\hat{x}(k+j \mid k)$, $\hat{s}(k+j \mid k)$, and $\hat{q}(k+j \mid k)$ to denote projected values the state, slack, and input variables at time $k+j$ given information available up to time k. The input vector is partitioned into *controllable* and *uncontrollable* inputs, $\hat{q}^c(k+j \mid k)$ and $\hat{q}^u(k+j \mid k)$, respectively, as

$$\hat{q}(k+j \mid k) = \begin{bmatrix} \hat{q}^c(k+j \mid k) \\ \hat{q}^u(k+j \mid k) \end{bmatrix} \tag{26}$$

Likewise, B is partitioned conformably with $q(k)$ as $B = \begin{bmatrix} B^c & B^u \end{bmatrix}$. For purposes of this development, we make the following assumptions:

Definition 3 (One Event Supervision) *Let one event occur on the horizon k to k+N, and that the event occurs at k. That is,*

$$\hat{q}^c(k+j \mid k) = 0 \tag{27}$$

for $1 \leq j \leq N$.

Assuming that no uncontrolled events occur, then the operating constraints $\hat{s}(k+j \mid k) \geq 0$ yield

$$s(k) + P(A^j - I)x(k) \geq -PA^{j-1}B^c q^c(k \mid k) \qquad (28)$$

for $j = 1, 2, \ldots, N$. In contrast to the case of Petri nets, the enabling conditions of $q^c(k \mid k)$ depends on the current state $x(k)$ in addition to the current slack $s(k)$.

Supervision of a Fuel/Air Combustor

The safe ignition of a fuel/air combustor (shown in Fig. 2) requires careful monitoring of the fame state, and avoiding unsafe situations where an excess of unburnt fuel is in the combustion chamber. This example has been frequently used in literature to illustrate synthesis and validation of logic based supervisors (Moon, et al., 1992; del Sagrado Corázon Sanchez Carmona, 1994; Lamport, 1993). Here we develop an LHM model, and demonstrate one-event ahead supervision.

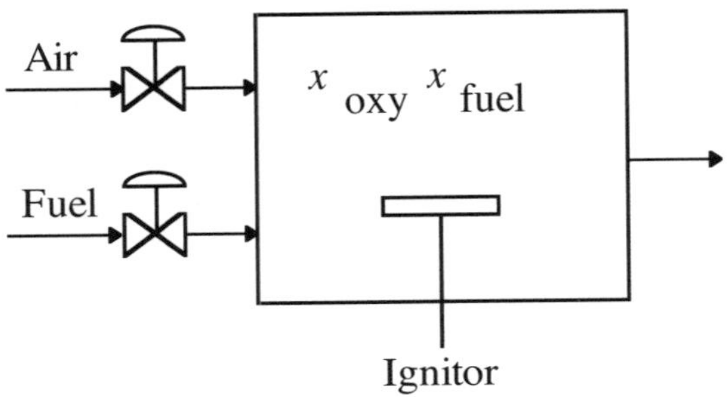

Figure 2. A fuel/air combustor.

The continuous state variables in the process correspond to the oxygen and fuel composition inside the combustion chamber. A qualitative model for the mixing dynamics can be written

$$x_{oxy}(k+1) = (1 - a_1)x_{oxy}(k) + a_1 v_{air}(k) \qquad (29)$$

$$x_{fuel}(k+1) = (1 - a_2)x_{fuel}(k) + a_2 v_{fuel}(k) \qquad (30)$$

where a_1 and a_2 are dilution coefficients for the mixing chamber, and $v_{air}(k)$ and $v_{fuel}(k)$ are discrete variables that denote valve state. The valve and ignitor states, in turn, are modeled by the condition-event Petri net

$$v_{air}(k+1) = q^+_{air}(k) - q^-_{air}(k) \qquad (31)$$

$$v_{fuel}(k+1) = v^+_{fuel}(k) - v^-_{fuel}(k) \qquad (32)$$

$$v_{ign}(k+1) = q^+_{ign}(k) - q^-_{ign}(k) \qquad (33)$$

where $q^{\pm}_j(k)$ are discrete controllable points.

There are several distinct classes of operating constraints. The first class are physical constraints which state that the valves and ignitors are either open or closed, and must be in one of those two states. These types of constraints are routine and not explicitly written here. The more significant constraints are posed by safety issues. For this illustration, we consider two constraints

$$x_{fuel} \leq 0.02x_{oxy}(k) + 0.98v_{ign} \qquad (34)$$

$$x_{oxy} \geq 0.8v_{ign}(k) \qquad (35)$$

The first of these constraints requires that the ignitor be on before the fuel composition reaches a flammability threshold. The second requires there to be a flow of air before the ignitor is turned on.

The one event ahead hybrid supervisor constructed above was applied to this sample. As shown in Fig. 3, the supervisor was used to screen commands that were passed down from a higher-level operations planning module. The supervisor monitors the process state, and applies a requested command if enabled. If not enabled within a time-out window, the command is rejected and presumably an error recovery procedure is invoked.

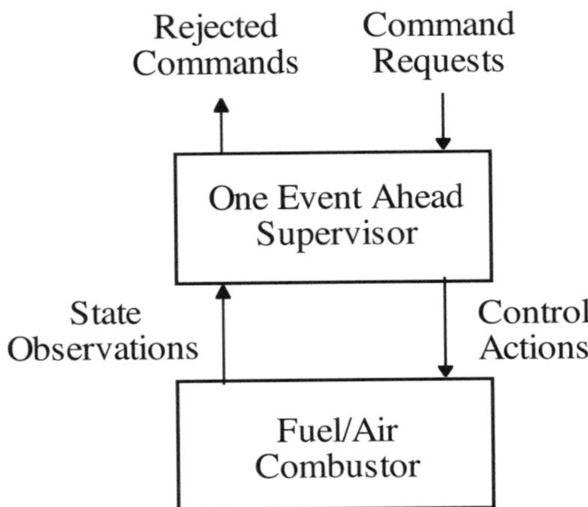

Figure 3. Hybrid supervision of the fuel/air combustor.

Figure 4 shows the result of applying the one event ahead hybrid supervisor to a series of commands. For this simulation, the model parameters were set to $a_1 = 0.01$, $a_2 = 0.02$, and the supervisor look ahead horizon was set to $N = 5$. A series of commands were sent to the supervisor to turn on air, turn on ignitor, turn on fuel, and then execute a safe shutdown. The supervisor accepted these commands, waiting until it identified a horizon for safe operation.

Conclusions

Intelligent methods in control are becoming part of mainstream control approaches. They are application driven for the most part and they represent our hope to meet the challenges of tomorrow.

Acknowledgment

This work was partially supported by NSF grants MSS-9216559, IRI91-09298, and CYS92-08567.

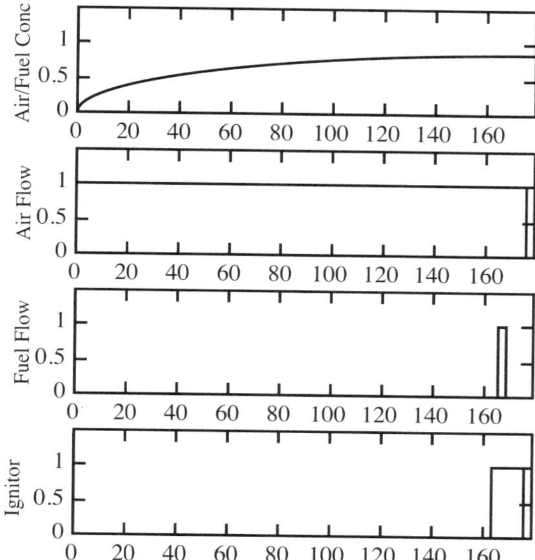

Figure 4. One event ahead hybrid supervisor applied to the fuel/air combustor example.

References

Alur, R., C. Courcoubetis, T.A. Henzinger and P.H. Ho (1993). Hybrid automata: an algorithmic approach to the specification and verification of hybrid systems. In R. Grossman, A. Nerode, A. Ravn and H. Rishel (Eds.), *Hybrid Systems*. Springer-Verlag. *Lecture Notes in Computer Science*, **736**, pp. 207-229.

Antsaklis, P.J. (1994). Defining intelligent control. *IEEE Control Systems Magazine*, pp. 4-5 and 58-66. Report to the Task Force on Intelligent Control, Chair: P.J. Antsaklis.

Antsaklis, P.J. and J.C. Kantor (1995). Intelligent control for high autonomy process control systems. ISIS Laboratory Technical Report ISIS-95-004, University of Notre Dame, Dept. of Electrical Eng.

Antsaklis, P.J., M.D. Lemmon and J.A. Stiver (1994). Modeling and design of hybrid control systems. *2nd IEEE Mediterranean Symposium on New Directions in Control and Automation*. Chania, Crete, Greece. pp. 440-447.

Antsaklis, P.J., M.D. Lemmon and J.A. Stiver (1995). Learning to be autonomous: intelligent supervisory control. In M.M. Gupta and N.K. Sinha (Eds.), *Intelligent Control: Theory and Practice*. IEEE Press, Piscataway, NJ.

Antsaklis, P.J. and K.M. Passino (1993a). Introduction to intelligent control systems with high degree of autonomy. In P.J. Antsaklis and K.M. Passino (Eds.), *An Introduction to Intelligent and Autonomous Control*, Ch. 1. Kluwer Academic Publishers. pp. 1-26.

Antsaklis, P.J. and K.M. Passino (Eds.) (1993b). *An Introduction to Intelligent and Autonomous Control*. Kluwer Academic Publishers.

Antsaklis, P.J., J.A. Stiver and M.D. Lemmon (1993). Hybrid system modeling and autonomous control systems. In R.L. Grossman, A. Nerode, A.P. Ravn and H. Rishel (Eds.), *Hybrid Systems*. Springer-Verlag. *Lecture Notes in Computer Science*, **736**, pp. 366-392.

Back, A., J. Guckenheimer and M. Myers (1993). A dynamical simulation facility for hybrid systems. In R.L. Grossman, A. Nerode, A.P. Ravn and H. Rishel (Eds.), *Hybrid Systems*. Springer-Verlag. *Lecture Notes in Computer Science*, **736**, pp. 253-267.

Balemi, S., P. Kozák and R. Smedinga (Eds.) (1993). *Discrete Event Systems: Modeling and Control*. Vol. **13** of *Progress in Systems and Control Theory*. Birkhäuser.

Benveniste, A. and P.L. Guernic (1990). Hybrid dynamical systems and the signal language. *IEEE Trans. on Automatic Control*, **5**, 535-546.

Cassandras, C. and P. Ramadge (1990). Towards a control theory for discrete event systems. *IEEE Control Systems Magazine*, **10**, 66-68.

Chaochen, Z., A.P. Ravn and M.R. Hansen (1993). An extended duration calculus for hybrid real-time systems. In R.L. Grossman, A. Nerode, A.P. Ravn and H. Rishel (Eds.), *Hybrid Systems*. Springer-Verlag. *Lecture Notes in Computer Science*, **736**, pp. 36-59.

Chung, S.-L., S. Lafortune and F. Lin (1992). Limited lookahead policies in supervisory control of discrete event systems. *IEEE Trans. on Automatic Control*, **37**, 1921-1935.

David, R. and H. Alla (1994). Petri nets for modeling of dynamic systems — a survey. *Automatica*, **30**, 175-202.

del Sagrado Corazón Sanchez Carmona, A. (1994). *Formal Specification and Synthesis of Sequential/Logic Controllers for Process Systems*. Ph.D. Thesis, Imperial College.

Delchamps, D.F. (1990). Stabilizing a linear system with quantized state feedback. *IEEE Trans. on Automatic Control*, **35**, 916-924.

Guia, A., F. DeCesare and M. Silva (1992). Generalized mutual exclusion constraints on nets with uncontrollable transitions. *Proc. 1992 IEEE Intl. Conf. on Systems, Man and Cybernetics*. Chicago, IL. pp. 974-979.

Guia, A., F. DeCesare and M. Silva (1993). Petri net supervisors for generalized mutual exclusion constraints. *Proc. 12th IFAC World Congress*, Vol. 1. Sydney, Australia. pp. 267-270.

Guckenheimer, J. (1995). A robust hybrid stabilization strategy for equilibria. *IEEE Trans. on Automatic Control*. **40**, 321-326.

Hadj-Alouane, N.B., S. Lafortune and F. Lin (1994). Variable lookahead supervisory control with state information. *IEEE Trans. on Automatic Control*, **39**, 2398-2410.

Holloway, L.E. and B. Krogh (1990). Synthesis of feedback control logic for a class of controlled petri nets. *IEEE Trans. on Automatic Control*, **35**, 514-523.

Lafortune, S. and H. Yoo (1990). Some results on petri net languages. *IEEE Trans. on Automatic Control*, **35**, 484-485.

Lamport, L. (1993). Hybrid systems in TLA+. In R.L. Grossman, A. Nerode, A.P. Ravn and H. Rishel

(Eds.), *Hybrid Systems*. Springer-Verlag. *Lecture Notes in Computer Science*, **736**.

Lemmon, M.D. and P.J. Antsaklis (1993). Inductively inferring valid logical models of continuous-state dynamical systems. *J. Theoretical Computer Sci.*, **137**.

Lemmon, M., J. Stiver and P. Antsaklis (1993). Event identification and intelligent hybrid control. In R.L. Grossman, A. Nerode, A.P. Ravn and H. Rishel (Eds.), *Hybrid Systems*. Springer-Verlag. *Lecture Notes in Computer Science*, **736**, pp. 265-296.

Li, Y. and W.M. Wonham (1993). Control of vector discrete-event systems. I. the basic model. *IEEE Trans. on Automatic Control*, **38**, 1214-1227.

Li, Y. and W.M. Wonham (1994). Control of vector discrete-event systems. II. controller synthesis. *IEEE Trans. on Automatic Control*, **39**, 512-531.

Moon, I., G.J. Powers, J.R. Burch and E.M. Clarke (1992). Automatic verification of sequential control systems using temporal logic. *AIChE J.*, **38**, 67-75.

Pantelides, C.C. (1995). Modelling, simulation, and control of hybrid processes. *Workshop on Analysis and Design of Event-Driven Operations in Process Systems*. Centre for Process Engineering, Imperial College of Science and Technology, London.

Passino, K.M. and P.J. Antsaklis (1989). On the optimal control of discrete event system. *Proc. 28th IEEE Conf. on Decision and Control*. Tampa, FL. pp. 2713-2718.

Passino, K.M. and P.J. Antsaklis (1992). Event rates and aggregation in hierarchical discrete event systems. *J. of Discrete Event Dynamical Systems*, **1**, 271-288.

Passino, K.M. and P.J. Antsaklis (1994). A metric space approach to the specification of the heuristic function for the A* algorithm. *IEEE Trans. on Systems, Man and Cybernetics*, **24**, 159-166.

Passino, K.M., A.N. Michel and P.J. Antsaklis (1994). Lyapunov stability of a class of discrete event systems. *IEEE Trans. on Automatic Control*, **39**, 269-279. Correction, July 1994, p. 1531.

Peleties, P. and R. DeCarlo (1994). Analysis of a hybrid system using symbolic dynamics and petri nets. *Automatica*, **30**, 1421-1427.

Raisch, J. and S. O'Young (1994). A discrete-time framework for control of hybrid systems. *Proc. 1994 Hong Kong Intl. Workshop on New Directions of Control and Manufacturing*. pp. 34-40.

Ramadge, P. and W.M. Wonham (1989). The control of discrete event systems. *Proc. of the IEEE*, **77**, 81-89.

Scheuring, R. and H. Wehlan (1993). Control of discrete event systems by means of boolean differential calculus. In S. Balemi, P. Kozák and R. Smedinga (Eds.), *Discrete Event Systems: Modeling and Control*. Vol. 13 of *Progress in Systems and Control Theory*. Birkhäuser. pp. 79-93.

Stiver, J.A. and P.J. Antsaklis (1992). Modeling and analysis of hybrid control systems. *Proc. 31st Conf. on Decision and Control*. Tucson, AZ. pp. 3748-3751.

Stiver, J.A. and P.J. Antsaklis (1993). On the controllability of hybrid control systems. *Proc. 32nd Conf. on Decision and Control*. San Antonio, TX. pp. 3748-3751.

Stiver, J.A., P.J. Antsaklis and M.D. Lemmon (1994). Digital control from a hybrid perspective. *Proc. 33rd IEEE Conf. on Decision and Control*. Lake Buena Vista, FL. pp. 4241-4246.

Stiver, J.A., P.J. Antsaklis and M.D. Lemmon (1995a). Interface design for hybrid control systems. Technical Report of the ISIS Group ISIS-95-001, University of Notre Dame, Dept. of Electrical Eng.

Stiver, J.A., P.J. Antsaklis and M.D. Lemmon (1995b). A logical DES approach to the design of hybrid control systems. *Mathematical and Computer Modeling*, to appear.

Tittus, M. and B. Egardt (1994). Control-law synthesis for linear hybrid systems. *Proc. 33rd IEEE Conf. on Decision and Control*. Lake Buena Vista, FL. pp. 961-966.

Yamalidou, K., L. Moody, M. Lemmon and P. Antsaklis (1994a). Feedback control of petri nets based on place invariants. *Proc. 33rd IEEE Conf. on Decision and Control*. Lake Buena Vista, FL. pp. 3104-3109.

Yamalidou, K., L. Moody, M. Lemmon and P. Antsaklis (1994b). Feedback control of petri nets based on place invariants. Technical Report of the ISIS Group ISIS-94-002, University of Notre Dame, Dept. of Electrical Eng.

Yang, X., P.J. Antsaklis and M. Lemmon (1994). On the supremal controllable sublanguage in the discrete event model of nondeterministic hybrid control systems. *IEEE Trans. on Automatic Control*, to appear.

EXPERT CONTROL AND FUZZY CONTROL

Karl-Erik Årzén and Karl Johan Åström
Department of Automatic Control
Lund Institute of Technology
Box 118, S-221 00 Lund, Sweden

Abstract

Fuzzy control and expert control are two paradigms in intelligent control. A major aim of fuzzy control is to model the control actions of the operator. Expert control, instead, focuses on the generic control knowledge possessed by an experienced control engineer. Both approaches are discussed and the conceptual similarities and differences are pointed out.

Keywords

Intelligent control, Fuzzy control, Expert control, Heuristics, Rule-based systems.

Introduction

Fuzzy control has its roots in manual control. A strong motivation for the approach is the desire to mimic the control actions of an experienced process operator. This approach is possible when it is not technically or economically justified to develop a physical or mathematical process model. Fuzzy sets, the foundation of fuzzy control, were introduced thirty years ago (Zadeh, 1965) as a way of expressing non-probabilistic uncertainties. Since then, fuzzy set theory has developed and found applications in database management, decision support systems, signal processing, data classifications, computer vision, etc. In 1974, the first successful application of fuzzy logic to control was reported (Mamdani, 1974). Control of cement kilns was an early industrial application (Holmblad and Ostergaard, 1982). Since the first consumer product using fuzzy logic was marketed in 1987, fuzzy control has received enormous attention. A number of CAD environments for fuzzy control design have emerged together with VLSI hardware for fast execution. Fuzzy control is being applied industrially in an increasing number of cases, e.g., Froese (1993).

The field of automatic control has for a long time focused on algorithms. To obtain flexible systems it is useful to add other elements like logic, sequencing, reasoning and heuristics. Such features are found in many conventional control systems. A knowledge-based system is one way to describe heuristics. Such a description leads to the notion of an expert control system of the type proposed in Åström, et al. (1986), which is a flexible architecture for combining real-time algorithms and logic. An expert controller can be characterized as a hierarchical system where a number of algorithms are juggled by a knowledge-based system.

An overview of fuzzy control is given in Section 2. Section 3 describes expert control and gives examples of the type of knowledge contained in an expert controller. Some conceptual similarities and differences between the two approaches are pointed out in Section 4.

Fuzzy Control

The early work in fuzzy control was motivated by a desire to

- directly express the control actions of an experienced human operator in the controller, i.e., to mimic his behavior, and
- to obtain smooth interpolation between discrete and controller outputs.

Since then the application range of fuzzy control has widened substantially. In most cases a fuzzy controller is used for direct feedback control. This is the approach that will be discussed in the following. However, it can also be used on the supervisory level as, e.g., a self-tuning device in a conventional PID controller. Also, fuzzy control is no longer only used to directly express a priori process knowledge. For example, a fuzzy controller can be derived from fuzzy model obtained through system identification. Therefore, it is difficult to define what a fuzzy controller is. A very general definition is:

> A *Fuzzy Controller* is a controller that contains a mapping, often linear, that has been defined using fuzzy logic-based rules.

The key issues in this definition are the non-linear mapping and the fuzzy logic-based rules.

Fuzzy Sets, Rules and Inference Systems

A fuzzy controller consists of a set of rules, each stating the control action to be taken in a certain process state. The process states and control actions are expressed on linguistic form, e.g.,

IF Error IS Large THEN
Control Action IS Large

Each linguistic relation is represented by a *fuzzy set*. A classical set is a set with a crisp boundary. An element either belongs to the set or does not belong to the set. A fuzzy set is a set with a non-crisp boundary. An element is a member of a fuzzy set to a degree between 0 and 1. A fuzzy set A is characterized by its *membership function*, $\mu_A : x \to [0,1]$.

The main idea of fuzzy sets is that statements like `Error IS Large` are not just true of false, but can be fulfilled to any degree in [0,1]. For a given observation x, the membership function determines to what extent the corresponding linguistic relation applies. Common types of membership functions used in fuzzy control are triangular functions and trapezoidal functions, or the smoother Gaussian functions and bell-shaped functions.

Fuzzy logic generalizes the Boolean set operators to operate on fuzzy sets. The most common operator definitions are:

Operation	Definition	Name
$\mu_{A\ AND\ B}(x)$	$\mu_A(x) \cdot \mu_B(x)$	Product
	$\min[\mu_A(x), \mu_B(x)]$	Minimum
$\mu_{A\ OR\ B}(x)$	$\min[1, \mu_A(x) + \mu_B(x)]$	Bounded Sum
	$\max[\mu_A(x), \mu_B(x)]$	Maximum
$\mu_{NOT\ A}(x)$	$1 - \mu_A(x)$	-

The process of reasoning with information given by fuzzy sets is called *approximate reasoning*. A detailed presentation of this is outside the scope of this paper. The interested reader is referred to Jang and Sun (1995). A key issue is the fact that a fuzzy rule, e.g. `if x is A then y is B` where A and B are linguistic values defined by fuzzy sets on X and Y, respectively, is defined as a binary fuzzy relation R on the space $X \times Y$. The reasoning principle applied is called *generalized modus ponens*, an approximation of ordinary modus ponens. Using this, if we have the implication rule `if the pressure is high then the volume is small` and we know that `the pressure is slightly high` then we may infer that `the volume is slightly small`. Other issues that make a detailed presentation of general fuzzy reasoning quite technical are the facts that the antecedents in a fuzzy control rule are defined on different domains, e.g., `if the pressure is high and the temperature is low then ...` and the fact that the inputs to the fuzzy rules are fuzzy sets.

In most fuzzy controllers the inputs are regarded as exact an represented as crisp values. This simplifies the computations substantially and the so called *Mamdani fuzzy inference system* is obtained. Consider the ith rule in a fuzzy controller rule base consisting of m rules:

IF x_1 IS A_1^i AND ... AND x_n IS A_n^i THEN u IS B^i

Each rule defines a fuzzy implication

$$\mu_{G^i}(u) = \mu_{A_1 \times \ldots \times A_n \to B}(x,u) \tag{1}$$

The most common implication rules are:

Definition	Name
$\mu_{G^i}(u) = \min\left[\mu_{A_1 \times \ldots \times A_n}(x), \mu_B(u)\right]$	Minimum
$\mu_{G^i}(u) = \mu_{A_1 \times \ldots \times A_n}(x) \cdot \mu_B(u)$	Product

Here, $\mu_{A_1 \times \ldots \times A_n}(x)$ is called the degree of fulfillment, d, of the rule.

All m rules are evaluated in parallel and the total control action is computed by aggregating the control recommendation from each rule using the union operator, i.e., OR.

$$G = \bigcup_{i=1}^{m} G^i \tag{2}$$

The aggregation-OR is typically defined as pointwise maximization or as summation, i.e.,

$$\mu_G(u) = \sum_{i=1}^{m} \mu_{G_i}(u) \tag{3}$$

During *defuzzification* the fuzzy set is transformed to a numerical value. The most common defuzzification strategy is center of gravity.

$$u = \frac{\int_{-\infty}^{\infty} \mu_G(w) \cdot w \, dw}{\int_{-\infty}^{\infty} \mu_G(w) \, dw} \tag{4}$$

Thus, the control signal is calculated as the ratio between the moment and the area of its fuzzy set.

In a fuzzy controller the above formulae become equivalent to the following calculations:

1. The fuzzy sets of all inputs are evaluated.
2. The degree of fulfillment for each rule is determined by applying the fuzzy set operators AND, OR and NOT.
3. The contribution of each rule to the control signal is determined by fuzzy implication.
4. The output fuzzy set of the controller is formed by aggregating the individual contributions.
5. The output of the controller is obtained by defuzzification of the output fuzzy set.

The calculations are either performed on-line at each sampling instant or using a precomputed look-up table. The input and output signals are usually normalized to a common range, typically [-1,1].

In principle any combination of implication method and fuzzy set operators can be used (Mizumoto, 1994). However, in practice only a few combinations are used. Common choices are to use *min* and *max* as fuzzy AND and OR operators and *min* as implication method, or to use product and bounded sum as fuzzy AND and OR operators and product as implication method. The above steps are summarized in Fig. 1 for the latter case.

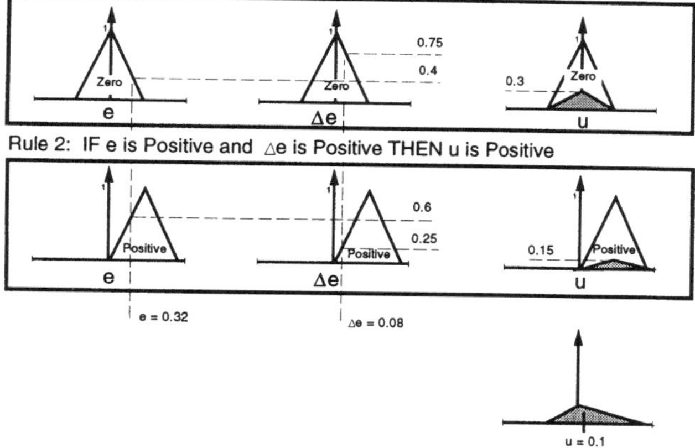

Figure 1. The calculations involved in fuzzy control.

An internal block diagram view of a Mamdani fuzzy inference system is shown in Fig. 2.

The *Takagi-Sugeno fuzzy inference system* is another common fuzzy inference system (Takagi and Sugeno, 1985). Here, a rule has the form

```
if x is A and y is B then u=f(x,y)
```

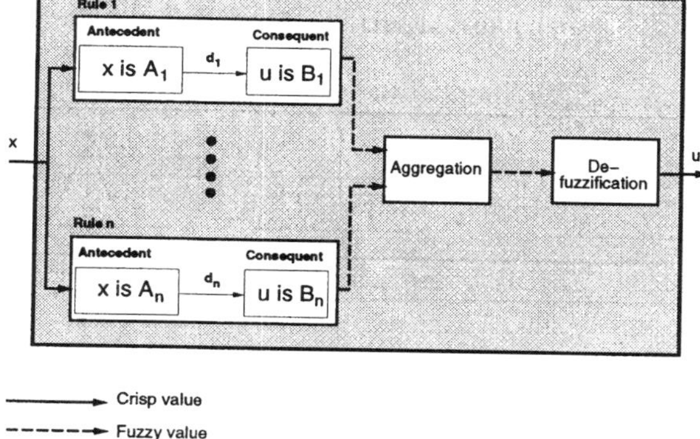

Figure 2. The internal view of a Mamdani fuzzy system with crisp inputs and output. Note that x is vector-valued.

where A and B are fuzzy sets and $u = f(x,y)$ is a crisp function that describes the output of the controller in the range defined by the fuzzy sets of the antecedent of the rule. A zero-order Takagi-Sugeno fuzzy model is obtained if f is a constant. This is a special case of the Mamdani model if the output of each Mamdani-rule is defined by a crisp number (i.e., a singleton fuzzy set). The aggregation and defuzzification of Fig. 2 is now replaced with the more efficient weighted average.

Nonlinearities

Characteristic for fuzzy controllers are that they contain a static nonlinear mapping defined by a fuzzy inference system. The external input-output view of a fuzzy inference system is given by Fig. 3.

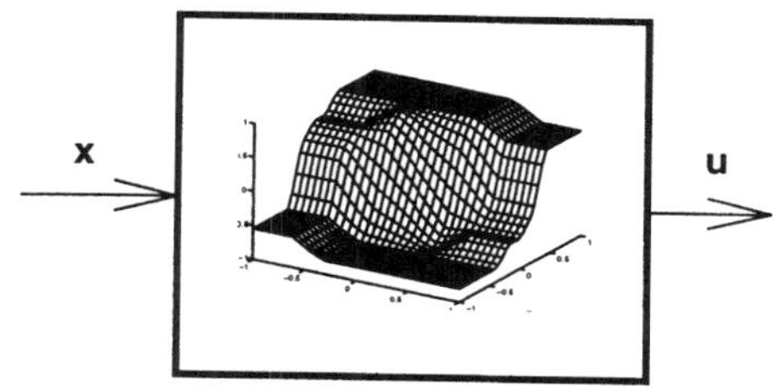

Figure 3. The external view of a fuzzy inference system with crisp inputs and output.

The way fuzzy logic is applied in most commercial fuzzy controllers, each rule defines one point in the mapping. The interpolation between these points is taken care of by the fuzzy logic. Thus, interpolation is the main reason for using fuzzy logic instead of classical Aristotelian logic. In order to fully understand fuzzy control it is important to understand the nature of this interpolation.

A particular question is under which conditions the mapping is affine or piecewise affine. If one knows how to design a controller that is affine in a certain region it is, e.g., possible to design a fuzzy controller with linear behavior for small control errors and nonlinear behavior for large errors. It is then possible to make use of the tuning and design rules for linear controllers. Affine fuzzy controllers are easiest to obtain when product and summation is used as AND and OR and product is used as implication method. However, it can be obtained also for other choices. For a more complete treatment, the reader is referred to Meyer-Gramann (1993) or Johansson (1993).

Three types of nonlinearities that can easily be defined are shown in Fig. 4. Nonlinearities that contain areas with constant output can be used to implement dead-zones and to take actuator saturations into account. Utilizing this, fuzzy controllers are often designed to approximate time-optimal "bang-bang" controllers (Kawaji and Matsunga, 1994). Smooth nonlinearities can be used to compensate for plant nonlinearities similar to a feedback linearization scheme. Finally, jump discontinuities can be used to implement controllers that behave similarly to multi-level relay controllers.

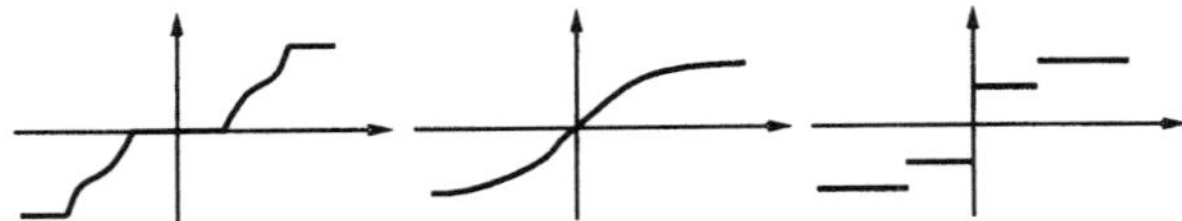

Figure 4. Many types of smooth and discontinuous nonlinearities can be described by fuzzy logic.

Fuzzy inference systems have a strong relationship to analytical non-linear function approximation methods such as splines or wavelets, and to neural networks (Benveniste, et al., 1994). It has been proved that fuzzy models are universal approximants (Wang, 1992), similar to neural networks. The zero-order Takagi-Sugeno fuzzy model is very similar to a radial basis function network. A fuzzy inference system can be expressed in network form (Jang, 1993), and can be trained with gradient methods equivalent to the back-propagation procedure. This is utilized in so called "neuro-fuzzy" models where the prior knowledge available about the function to be approximated is expressed as fuzzy rules and used to initialize the structure and weights of the network. After training, the resulting net can be reinterpreted in terms of fuzzy sets and rules.

Fuzzy Controller Structures

The fuzzy inference system in a fuzzy controller defines a non-linear mapping. The inputs and outputs to this mapping determines the structure of the fuzzy outputs to this mapping determines the structure of the fuzzy controller. The dynamics of the controller are outside the fuzzy system, as shown in Fig. 5. The linear filters on the input are used to generate the inputs to the fuzzy system. These are typically the process output, controller error, error integral or error derivative. The output linear filter consists of an integrator if the fuzzy system output is the control signal increment. Depending on which input and output signals that are used different controller structures can be implemented. For example, if the fuzzy block is chosen as $u = \mathcal{F}(e,\dot{e})$ the controller is structurally equivalent to a PD-controller. If the fuzzy mapping is designed to be linear, exact equivalence is obtained. Similarly, it is possible to define fuzzy controllers that are structurally equivalent to all forms of conventional controllers, e.g, PID controllers on position or velocity form, state feedback controllers or general polynomial controllers.

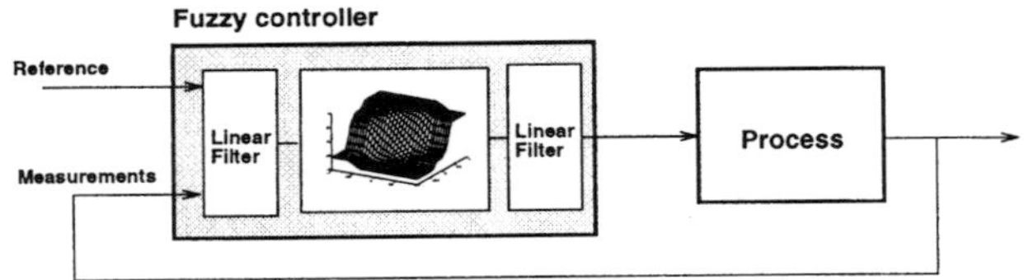

Figure 5. The fuzzy controller structure.

Several lessons can be learned from conventional PID control (Åström and Hägglund, 1995). To reduce the sensitivity to measurement noise it is important to limit the high frequency gain in the derivative part. In process control it is common practice to take derivatives on the process output instead of the error. This avoids large overshoots at set point changes. If the fuzzy controller uses integrators, an anti-windup scheme should be included.

Unfortunately, much fuzzy logic control literature concentrates on the fuzzy logic system part of the fuzzy controller. The dynamic part of the controller is largely ignored.

The Paradoxical Success

Fuzzy control has always been a controversial subject (IEEE, 1993a; IEEE, 1993b). This owes partly to lack of mutual understanding between the fuzzy control community and the traditional control community and partly to exaggerated claims in certain papers on fuzzy control. Many people active in fuzzy control have no classical control background. This typically leads to reinventions of results that are well-known in classical control. At the same time many classical control engineers have a very "fuzzy" idea of what fuzzy control really is. The empirical nature of fuzzy control where the importance of mathematical models is de-emphasized is sometimes regarded as "un-scientific."

In Elkan (1993) the success of fuzzy logic and particularly fuzzy control is presented as a paradox. Still, fuzzy control has had an indisputable success. There are several reasons for this. Fuzzy control is a direct approach to nonlinear control design. The rule-based formalism is intuitive and easy to understand for non-control engineers. Each rule represents local process knowledge about how the control signal should be selected for certain input signals. The local nature of the rule makes it possible to build up a controller in a step-wise fashion. Fuzzy control makes it easy to implement nonlinear control elements. Nonlinear controllers can, potentially, give better control performance than linear controllers both for nonlinear and linear processes. Therefore it is not surprising that fuzzy control has outperformed, e.g., PID control in different comparative studies. However, several of the comparative simulation studies are also unfair because they only compare the control system performance at one point, e.g., for one size of the change in reference value. Another reason for the success is the way the technique is packaged. CAD environments for fuzzy control have user-friendly, graphical environments. They are available on industrially accepted hardware and they can automatically generate C code. All this makes fuzzy control straight-forward to apply in industry. So straight-forward, that it is possible for people without any classical control background to put together a working control system. However, if these persons had some basic knowledge of control and non-linear systems, the result would have been even better. The user-friendly packaging of fuzzy control is an area where the conventional control community has a lot to learn.

Expert Control

The notion of expert control was introduced as an attempt to structure systems that process both signals and symbols. The idea was to obtain a strict separation of signal processing and logic where the logic was implemented in an expert system. See Åström, et al. (1986). It was found that such an approach gave many advantages. It gave a system with a much nicer structure. The knowledge representation became transparent and debugging was simplified. Several prototypes were implemented in this framework such as automatic tuners, safety networks for adaptive control and systems for diagnosis. A summary of several experiences is given in Åström and Årzén (1992).

Signal and Symbol Processing

A mixture of signal processing and symbol processing is found at many levels of process control systems. Traditional systems for process control had cabinets of analog controllers and cabinets with relays for sequencing and interlocks. The functions of signal processing and symbolic processing became more intertwined with the advent of distributed control systems (DDCs) and programmable logic controllers (PLCs) when the different functions were performed in the same equipment.

A mixture of signal processing and control is also found internally in many controllers. PIC controllers with schemes for protection against integral windup is a typical example, see Åström and Hägglund (1995). In this case logic and signal processing are often mixed in the code. The logic appears as **if-then-else** statements in the controller code. Systems with selectors and systems with gain scheduling are other examples where different controller parameters are used in different operating conditions. More complicated mixtures of logic and signal processing are found in automatic tuners and adaptive controllers (Åström and Wittenmark, 1995). The safety logic which assures safe operation of the adaptive system under a variety of operating conditions can be quite extensive, see Åström (1993). Other examples are found in control of nuclear reactors and military aircraft. Such systems have many different models and complicated mode switching mechanisms. The so-called configurable flight control systems represents an extreme case where it is attempted to construct control systems that can operate in spite of severe system damages.

The systems mentioned above are often designed heuristically with support from extensive simulations and implemented in an ad hoc manner. Code for logic and signal processing are often mixed. This leads to systems that are difficult to program, modify and test. The systems are also very difficult to analyze and design. The number of possibilities increases very rapidly with the number of logic variables. This means that only a small number of possible cases are investigated in simulations and tests.

Knowledge-Based Systems

Expert systems, or knowledge-based systems as they are called, are a spin-off of research in AI, see Barr and Feigenbaum (1982), Hayes-Roth, et al. (1983) and Waterman and Hayes-Roth (1983). They were designed to solve heuristic problems in an organized way. The key idea is to represent knowledge in terms of rules of the type

If {premises} **then** {conclusions or actions}

The system consists of a database with rules and an inference engine. The inference engine is an algorithm that draws conclusions based on the data and the rules. Several strategies can be used for this purpose. The forward chaining strategy is data driven. Starting with premises in the database, it generates conclusions by applying the rules until all possibilities are exhausted. Simultaneously it executes the corresponding actions. This can also generate new conclusions. Backward chaining is another strategy which is hypothesis driven. Starting with a statement like, Reduce variations in the process output of loop 5, the strategy finds rules that have this conclusion. It then chains all rules backwards from conclusions until it finds premises that support the desired conclusion or finds a contradiction.

It is natural to group the rules into classes that are associated with different algorithms and different tasks to be performed. It is very convenient to have generic rules, i.e., rules that apply to classes of objects. This was pursued in Årzén (1987) and Årzén (1989) where the usefulness of a black board architecture was demonstrated.

Expert systems have been applied to a wide variety of problems with varying success. Some commonly given criteria for success are that the problem is nontrivial and sufficiently complex, that the problem can be solved by human experts and that experts are available. The control problems we are considering satisfy all of these criteria.

Expert systems were originally developed to solve static problems. The control problems we are considering are not static. A statement may, e.g., suddenly switch from true to false because of a change in the physical system being controlled. Reasoning with time is an area where many theoretical problems are unresolved (Laffy, et al., 1988). Some pragmatic approaches are taken to deal with these issues. One method is to replace the dynamic problem with a static problem by assuming that all premises hold over a small sliding time-window. It is also important that conclusions are reached in a reasonable time. Since the time increases rapidly with the number of rules, it is useful to structure the rules into groups and to focus the reasoning to a given group.

Commercial programming environments for knowledge-based real-time systems have been available for around 10 years now. A good example is G2 from Gensym Corporation (Moore, et al., 1990), which is aimed at supervisory control applications. In G2 the developer can combine a number of programming paradigms, e.g., object-oriented programming, procedural programming, rule-based programming, and event-driven programming. It is also possible to implement Petri Nets, state machines, and neural networks in G2, e.g., Årzén (1994).

The Expert Controller

An example of an expert controller is shown in Fig. 6. The system has a collection of signal processing algorithms for control, control design, parameter estimation, excitation, diagnosis and logging. There is also a knowledge-based system that coordinates the operation of the algorithms. The block labeled expert system in Fig. 6 has a database, a rule base and an inference engine. It communicates with the operator by a user interface. The database could simply be a list of strings with statements like: *"The crossover frequency is 0.5 Hz," "The pressure is above normal"* or *"The gain varies substantially over the operating range."* Alternatively, the database could be represented as objects with attributes defined in class definitions. Natural objects in the expert control domain are control loops, numerical algorithms, models derived by the expert system, etc. The system shown in Fig. 6 can be used to implement an advanced controller with facilities for automatic tuning, gain scheduling and adaptation. The system also has features to assist diagnosis, loop assessment and performance assessment.

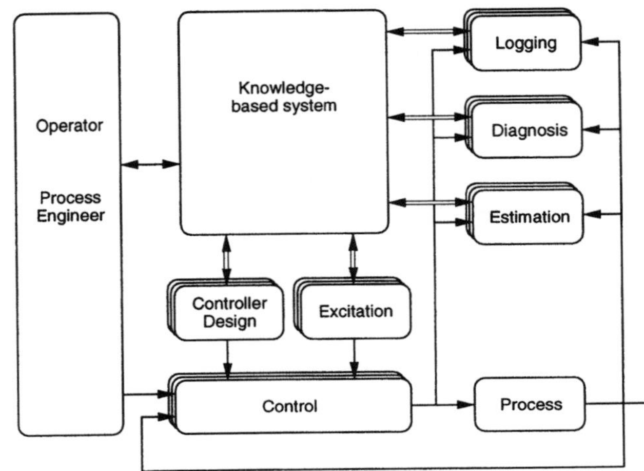

Figure 6. Expert control architecture.

The system has an ordinary feedback loop with a process and a controller. There are also a number of auxiliary functions for parameter estimation, control design, supervision, fault detection and diagnosis. There may be several alternative algorithms for the same task. For example there may be several different controllers. This is indicated by the different layers in the figure. The controller may be a simple PI controller or a more complicated algorithm based on an observer and state feedback. There are also algorithms for generating perturbation signals to excite the process. The fault detection and diagnosis tasks are aimed at finding faults that are local to the control loop that the expert controller is part of. This differs from the plant-wide approach to diagnosis taken by the majority of the work in diagnosis, e.g., Prasad and Davis (1992). The signal processing algorithms indicated by boxes can communicate with the expert system by sending and receiving data. The parameters of the algorithms can be changed and algorithms can be replaced. These are coordinated by an expert system, or a knowledge-based system, which decides what algorithm to use when. The knowledge-based system also interacts with the operator or the process engineer.

The system in Fig. 6 is very general. Many different systems can be implemented in this framework. A gain scheduling controller can, e.g., be obtained by having one controller whose parameters are changed based on a measured signal. The table for the different parameters can be represented by a collection of rules. A controller with automatic tuning, of the type discussed in Åström and Hägglund (1984) can be implemented by using two control algorithms, a PID controller and a relay feedback with sequencing and logic. Such a controller can be represented very conveniently as is discussed in Årzén (1993). An adaptive controller based on recursive parameter estimation and control design, see Åström and Wittenmark (1995), is obtained by combining the functions of estimation, excitation, controller design and controller selection. The supervision of the adaptive controller can also very conveniently be implemented in the system. The robot control system described in Brockett (1990) and Brockett (1993) is a system of very different character which also fits Fig. 6. In this case the control algorithms represent the different motion command and the language and its interpreter are represented by rules.

Expert control matches well the trend toward distributed intelligence in modern control systems shown, e.g., by smart sensors and actuators and fieldbus technology. An expert controller can be as a smart controller that is responsible for control and diagnosis at the local feedback loop level. The controller communicates with supervisory level control and diagnosis functions.

Knowledge Representation

Domain knowledge is a key issue in expert control. In this section we will illustrate acquisition of knowledge and reasoning by discussing a single loop controller. The discussion will be restricted to control of a single loop. Many issues can be dealt with in this way. There are however other questions that require a global view of the system, where the interaction of many loops is important.

Automation of control system design and operation should consider the tasks of design, commissioning, normal operation and emergencies. Control system design involves issues like control performance, modeling and choice of control laws. Commissioning involves initialization, tuning, trouble shooting and loop auditing. Normal operation involves supervision, diagnosis and fault detection. To perform these tasks we have to represent knowledge about *process dynamics*, *level and rate saturation in actuators*, *disturbances*, and *specifications*.

There is an interplay between several of these factors. Dynamics is, in principle, no limitation for linear systems that are strictly positive real (SPR) or with first- and second-order dynamics. Here, the speed of response is limited by measurement noise and actuator saturations. Large pole excess and non-minimum phase dynamics impose limitations of the achievable performance. It is thus essential to find methods to determine whether the performance is limited by the dynamics of other factors. It is also essential to characterize the complexity of the dynamics, e.g., the presence of oscillatory modes, the order of the dynamics, etc. For systems with difficult dynamics an attempt can be made to change the system so that the dynamics become simpler. Time delays can be reduced by repositioning sensors and actuators. Dynamics can be improved by replacing sensors and actuators with devices having faster responses. An attempt to use local feedback to make the dynamics simpler and more reproducible can be made.

The disturbances include set point changes, load disturbances and measurement noise. It is essential to find the ranges and the character of these disturbances. The range of set point changes the required precision in the controlled variable and the maximum loop gain indicate whether proportional control is sufficient or integral action is needed. The magnitude of the error due to load disturbances depends on the amplitude and frequency characteristics of the disturbance and of the loop gain.

Several actions could be contemplated with respect to the disturbances. They can be reduced at the source. Feedforward control can be considered if there is a measurable signal, which is correlated with the disturbance and appropriately located. Filtering can also be used to reduce disturbances and possibly to reconstruct signals that can be modeled.

Measurement noise results in variations in the control signal. Together with actuator saturation this limits the achievable regulator gain and thus also the achievable bandwidth. If an actuator saturates because of measurement noise and high gain, an attempt can be made to reduce the gain, to reduce the disturbance level by filtering or to replace the actuator with a more powerful device.

Model uncertainty is another limiting factor. It can be minimized to some extent by having a high loop gain at those frequencies where the uncertainty is large. To maintain a high loop gain, however, it is necessary to know the phase reasonably well around the cross-over frequency. Uncertainties in the time delay, which give very large phase uncertainties at high frequencies, is a severe limitation on the achievable bandwidth.

Several of the issues discussed above pertain to selection and positioning of sensor and actuators, particularly their sizing and resolution. An important task of an expert control system is also to assess if good design choices have been made. Capabilities to help in auditing control systems can therefore be very valuable. Useful knowledge for this purpose can be derived by observing the operation of a control system. Investigation of static process characteristics gives important information for this purpose. It is also useful to have diagnosis systems that indicate if some component of the control loop is degrading.

The Need for More Structure

Rules are the standard knowledge representation formalism in expert systems. Rules are also a natural way to describe much of the logic that is built around conventional control algorithms. Rules are however not very well suited for problems that have a strong sequential element. Although expert control is not dominated by sequential elements, some parts, e.g. control design, are clearly sequential.

The sequential parts of the problem can be represented in different ways. One approach is to combine the rules with a procedural programming language. This solution is adopted in the G2 expert system shell. Finite state machines can also be used as described in Bencze and Franklin (1995). Petri Nets is another alternative, see David and Alla (1992). Sequential function charts, Grafcet, is another possibility to structure the activation and deactivation of rule groups. A rule group can be seen as a knowledge source specialized on one specific subproblem. A discussion of this is given in Årzén (1989), which also gives a powerful extension of Grafcet.

Many methods for representing sequences have the drawback that the sequential parts are fixed. Planning is the automatic generation of a sequence of actions that lead to a desired goal. One example is to find a method to bring an oscillating system to a stable operation, another is to move a system from one operating condition to another in a smooth way. Planning has received a lot of attention from researchers in AI and robotics. See, e.g., Sacerdoti (1977) or Passino and Antsaklis (1992). One possibility is to characterize each action by preconditions and postconditions. The preconditions tell what is required to perform an action and the postconditions describe possible situations after the action. Many of the tasks required in expert control can be described as planning problems.

Several tasks performed in an expert control system are broadly speaking an attempt to automate what a control engineer does in designing a good system. The notion of scripts, which was originally introduced in Schank and Abelson (1977) to deal with natural language, is a powerful way of describing the much simpler task of characterizing a control system design. In Larsson and Persson (1991) scripts are used to describe system identification procedures.

Hybrid Systems

One possibility is to consider system structuring as a high-level control problem. A number of low-level functions is first grouped into primary tasks. Each task should perform a given function but it should also generate signals that can be used for higher level decision making and it should also be able to receive symbolic commands to modify its behavior. This is illustrated in Fig. 7. This figure only shows two levels, but the approach can be extended to an arbitrary number of levels.

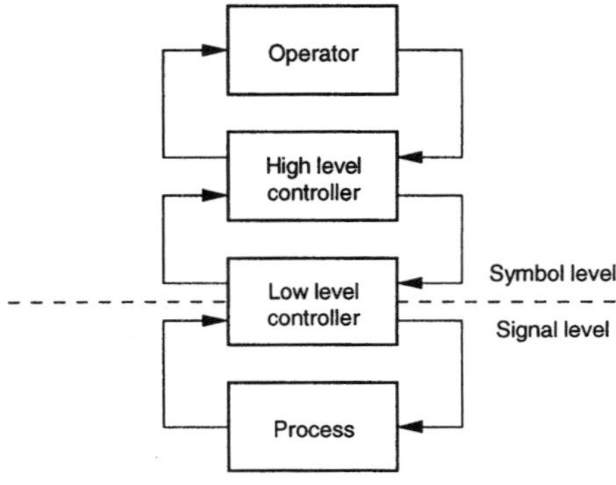

Figure 7. A structuring mechanism.

To illustrate the idea let us consider a basic control task. The box denoted controller could thus contain a PID algorithm and various diagnostic algorithms. The control task provides the basic low-level control, but it also delivers the following symbolic signals: *Normal*, *Emergency* and *Uncertain*. The control task should also be able to receive symbolic commands such as: *Resume normal operation*, *Determine stability margins*, *Retune immediately*, *Determine plant transfer function* or *Investigate disturbance characteristics*.

From the high-level controller the system below the dotted line in Fig. 7 can be viewed as a discrete control system. Control of this discrete system can be regarded as the problem of controlling an automata. If the response of the low level system to the control actions is not purely deterministic the problem may be regarded as control of a Markov chain. A system which is a combination of continuous time systems and a finite state machine is called a hybrid system. There is some progress in understanding the behavior of such systems theoretically, see Brooks (1990), Antsaklis, et al. (1991) and Åström (1992).

Differences and Similarities

Fuzzy control, when viewed as a way of representing manual control actions, and expert control are similar in the sense that both represent knowledge in terms of rules. Both systems also contain ordinary signal processing elements. In fuzzy control the signal processing is represented by the fuzzification. In expert control there is a much larger collection of signal processing algorithms. In fuzzy control, see Fig. 5, the rule processing is in the direct feedback path while the expert control system is organized hierarchically with an ordinary feedback loop at the lowest level, see Fig. 6. The presence of a signal/symbol interface is another similarity. In fuzzy control the numerical measurement and control signals are converted into symbolic values on linguistic form. The reasoning is performed on the symbolical level. During defuzzification, the linguistics values are converted back to numerical form. In expert control the numerical/symbolical interface consists of the numeral identification and monitoring algorithms that extract information in symbolical and numerical form. One may also view fuzzy control as a technique for incorporating uncertainties in control rules. From that point of view fuzzy logic can be seen as a subset of expert control.

Expert control and fuzzy control differ significantly with respect to the type of expertise that they encompass. Fuzzy control in its original meaning aims at representing knowledge about how a specific process should be controlled. The knowledge typically emanates from an experienced process operator or from the process designers. Expert control attempts to represent generic knowledge about feedback control as well as specific knowledge about the particular process, i.e. the knowledge of experienced control and process engineers. The knowledge in an expert control system includes both theoretical control knowledge, empirical heuristics or "rules-of-thumb" knowledge and knowledge acquired during the operation of the process. This knowledge consists of both procedural and declarative elements.

Conclusions

There are many reasons why approaches such as expert control and fuzzy control appear. One reason is an intellectual curiosity about new methods. Another is a dissatisfaction with current approaches. The primary control functions can be improved. There are situations where nonlinear controllers can give a good improvement, which is a case where fuzzy control may be considered. A major criticism against current distributed control systems is that they generate a lot of data but very little information. It would undoubtedly be highly valuable to have systems that could use the data generated during normal operation to induce information about the process and the disturbances. This would typically include information about dynamics, nonlinearities and time variations. To implement such a function we need conventional control algorithms as well as logic and reasoning. An expert controller is very well suited to implement such a task.

The expert controller is very well suited to implement a good scheme for performance assessment, which would tell how well a control system is performing in relation to historical data and fundamental performance limits, see Kozub and Garcia (1993) and Åström (1991). Integrated diagnosis and control is another promising area. By exploiting the signals in the primary control loop it is possible to obtain information that is much more useful for diagnosis than simple bound on the error signal. The information processing required for this, which can be conveniently expressed in terms of algorithms and logic, is also very well suited for expert control.

Yet another area where the expert control paradigm is useful is to design intelligent user interfaces. It would be highly desirable to have user interfaces that can adapt to the knowledge and the working habits of the different users. This is again a task that can be conveniently expressed in terms of algorithms and rules. A simple experiment in this direction is reported in Larsson and Persson (1991).

The properties of systems obtained when using fuzzy and expert control are quite complex. Techniques for analysis and design of such systems are badly needed.

Acknowledgments

This work has been supported by research contracts from the Swedish Research Council for Engineering Sciences (TFR) under contract 92-956, the Swedish National Board for Industrial and Technical Development (NUTEK) under contract 93-00734 and the Swedish Natural Science Research Council (NFR) who has supported our participation in the Espirit-III Basic Research Working Group FALCON (Fuzzy Algorithms for Control).

References

Antsaklis, P.J., Passino, K.M. and Wang, S.J. (1991). An introduction to autonomous control systems. *IEEE Control Systems Magazine*. **11**(4), 5-13.

Årzén, K.-E. (1987). *Realization of Expert System Based Feedback Control*. Ph.D. thesis. Dept. of Automatic Control, Lund Institute of Technology. Lund, Sweden.

Årzén, K.-E. (1989). An architecture for expert system based feedback control. *Automatica*. **25**(6), 813-827.

Årzén, K.-E. (1993). Expert control — Intelligent tuning of PID controllers. *Applied Control: Current Trends and Modern Methodologies*. S. Tzafestas, ed. Marcel Dekker Inc.

Årzén, K.-E. (1994). Grafcet for intelligent supervisory control applications. *Automatica*. **30**(10).

Åström, K.J. (1991). Assessment of achievable performance of simple feedback loops. *International J. of Adaptive Control and Signal Processing*. **5**, 3-19.

Åström, K.J. (1992). Autonomous control. *Future Tendencies in Computer Science, Control and Applied Mathematics*. A. Bensoussan and J.-P. Verjus, eds. Vol. 653 of *Lecture Notes in Computer Science*. Springer-Verlag. pp. 267-278.

Åström, K.J. (1993). Intelligent tuning. *Proc. of IFAC 4th International Symposium on Adaptive Systems in Control and Signal Processing, Grenoble 1992*. Pergamon Press. Invited Plenary Paper.

Åström, K.J., Anton, J.J. and Årzén, K.-E. (1986). Expert control. *Automatica*. **22**(3), 277-286.

Åström, K.J. and Årzén, K.-E. (1992). Expert control. *An Introduction to Intelligent and Autonomous Control*. K.M. Passino and P.J. Antsaklis, eds. Kluwer Academic Publishers.

Åström, K.J. and Hägglund, T. (1984). Automatic tuning of simple regulators with specifications on phase and amplitude margins. *Automatica*. **20**, 645-651.

Åström, K.J. and Hägglund, T. (1995). *PID Controllers: Theory, Design and Tuning*. 2nd ed. Instrument Society of America, Research Triangle Park, NC.

Åström, K.J. and Wittenmark, B. (1995). *Adaptive Control*. 2nd ed. Addison-Wesley, Reading, Massachusetts.

Barr, A. and Feigenbaum, E.A., eds. (1982). *The Handbook of Artificial Intelligence*. William Kaufmann. Los Altos, California.

Bencze, W. and Franklin, G.F. (1995). A separation principle for hybrid control system design. *IEEE Control Systems*. **15**(2), 80-85.

Benveniste, A., Juditsky, A., Delyon, B., Zhang, Q. and Glorenned, P.-Y. (1994). Wavelets in identification. *Proc. of 10th IFAC Symposium on System Identification, SYSID'94*. Vol. 2, pp. 27-48.

Brockett, R.W. (1990). Formal languages for motion description and map making. *Robotics*. R.W. Brockett, ed. American Math Society, Providence, Rhode Island.

Brockett, R.W. (1993). Hybrid models for motion control systems. *Essays on Control: Perspectives in the Theory and its Applications*. H.L. Trentelman and J.C. Willems, eds. Birkhäuser.

Brooks, R.A. (1990). Elephants don't play chess. *Designing Autonomous Agents*. P. Maes, ed. MIT Press, Cambridge, Massachusetts.

David, R. and Alla, H. (1992). *Petri Nets and Grafcet: Tools for modeling discrete events systems*. Prentice-Hall.

Elkan, C. (1993). The paradoxical success of fuzzy logic. *Proc. of AAAI*.

Froese, T. (1993). Applying of fuzzy control and neuronal networks to modern process control systems. *Proc. of the EUFIT'93*. Aachen. Vol. 2, pp. 559-568.

Hayes-Roth, F., Watermann, D. and Lenat, D. (1983). *Building Expert Systems*. Addison-Wesley, Reading, Massachusetts.

Holmblad, L. and Ostergaard, J. (1982). Control of a cement kiln by fuzzy logic. *Fuzzy Information and Decision Processes*. M. Gupta and E. Sanchez, eds. North-Holland, Amsterdam.

IEEE (1993a). Reader's forum. *IEEE Control Systems Magazine*.

IEEE (1993b). Reader's forum. *IEEE Control Systems Magazine*.

Jang, J.-S.R. (1993). ANFIS: Adaptive network-based fuzzy inference systems. *IEEE Trans. on Systems, Man and Cybernetics*. **23**(3), 665-685.

Jang, J.-S.R. and Sun, C.-T. (1995). Neuro-fuzzy modeling and control. *Proc. of the IEEE*.

Johansson, M. (1993). *Nonlinearities and Interpolation in Fuzzy Control*. Master thesis. Dept. of Automatic Control, Lund Institute of Technology. Lund, Sweden.

Kawaji, S. and Matsunga, N. (1994). Fuzzy control of VSS type and its robustness. *Fuzzy Control Systems*. A. Kandel and G. Langholz, eds. CRC Press.

Kozub, D.J. and Garcia, C.E. (1993). Monitoring and diagnosis of automated controllers in the chemical process industries. *Technical Report*. Shell Development Company. Houston, Texas.

Laffey, T.J., Cox, P.A, Schmidt, J.L., Kao, S.M. and Read, J.Y. (1988). Real-time knowledge-based systems. *AI Magazine*. **9**(1), 27-45.

Larsson, J.E. and Persson, P. (1991). An expert system interface for an identification program. *Automatica*. **27**, 919-930.

Mamdani, E. (1974). Application of fuzzy algorithm for control of simple dynamic plant. *Proc. IEEE*. **121**, 1585-1588.

Meyer-Gramann, K. (1993). Easy implementation of fuzzy controller with a smooth control surface. *Proc. of the EUFIT'93*, Aachen. Vol. 1, pp. 117-123.

Mizumoto, M. (1994). Fuzzy controls under product-sum-gravity methods and new fuzzy control methods. *Fuzzy Control Systems*. A. Kandel and G. Langholz, eds. CRC Press. pp. 275-294.

Moore, R., Rosenof, H. and Stanley, G. (1990). Process control using a real-time expert system. *Preprints 11th IFAC World Congress*. Tallinn, Estonia.

Passino, K.M. and Antsaklis, P.J. (1992). Modeling and analysis of artificially intelligent planning systems. *An Introduction to Intelligent and Autonomous Control*. K.M. Passino and P.J. Antsaklis, eds. Kluwer Academic Publishers.

Prasad, P.R. and Davis, J.F. (1992). A framework for knowledge-based diagnosis in process operations. *An Introduction to Intelligent and Autonomous Control*. K.M. Passino and P.J. Antsaklis, eds. Kluwer Academic Publishers.

Sacerdoti, E. (1977). *The Structure of Plans and Behavior*. Elsevier, New York.

Schank, R.C. and Abelson, R.P. (1977). *Scripts, Plans, Goals and Understanding*. Lawrence Erlbaum Associates. Hillsdale, New Jersey.

Takagi, T. and Sugeno, M. (1985). Fuzzy identification of systems and its application to modeling and control. *IEEE Trans. on Systems, Man and Cybernetics*. **15**, 135-156.

Wang, L. (1992). Fuzzy systems are universal approximators. *Proc. First IEEE Conference on Fuzzy Systems*. San Diego, pp. 1163-1169.

Waterman, D.A. and Hayes-Roth, F., eds. (1983). *Pattern-Directed Inference Systems*. Academic Press, New York.

Zadeh, L. (1965). Fuzzy sets. *Information and Control*. **8**, 338-353.

Zadeh, L. (1994). Foreword. *Fuzzy Control Systems*. A. Kandel and G. Langholz, eds. CRC Press.

PROCESS MODELING AND CONTROL USING NEURAL NETWORKS

Lyle H. Ungar[1]
Department of Chemical Engineering
University of Pennsylvania
Philadelphia, PA 19104-6393

Eric J. Hartman, James D. Keeler[2] and Greg D. Martin
Pavilion Technologies Inc.
Austin, TX 78727

Abstract

Neural networks are proving valuable for use in process modeling, optimization, virtual sensing and control. This article provides an introduction to how these universal multivariable function approximators can be used as virtual sensors and as process models in control applications, and presents a set of case studies in which neural networks are used to replace continuous emission monitoring systems to meet Federal and State emissions monitoring requirements at a fraction of the cost of hardware monitors. Neural network plant models also provide a basis for control and optimization.

Keywords

Neural networks, Process control, Virtual sensors.

Introduction

Neural networks have been extensively studied in academia as process models and controllers, and are increasingly being using in industry. This paper provides a selective overview of the state of research on neural networks for process control and a detailed case study of a current industrial application.

Neural networks have been used in a variety of different control structures and applications, serving as controllers and as process models or parts of process models (e.g. as virtual sensors). They have been used to recognize and forecast disturbances, to detect and diagnose faults, to combine data from partially redundant sensors, to perform statistical quality control, and to adaptively tune conventional controllers such as PIDs. Many different structures of neural networks have been used, with feedforward networks, radial basis functions, and recurrent networks being the most popular. Most of the examples cited here use standard feedforward networks; radial basis functions (RBFs) are more popular in adaptive control, since they are linear in the coefficients of the basis functions (Sanner and Slotine, 1991).

What are Neural Networks and Why Use Them?

Neural networks can be viewed as multivariate nonlinear nonparametric estimation methods: they are typically used to approximate a function $\boldsymbol{y} = f(\boldsymbol{x})$, where the functional form of f is unknown. This viewpoint then raises the obvious question: why use this particular approximation method as opposed to the many statistical methods that are widely used? In summary, the answer is that neural networks are universal function approximators (Cybenko, 1989; Hornik, et al., 1989; Hartman, et al., 1990) that typically work much better in practical applications than more traditional (polynomial) function approximations methods.

Neural network methods share a loose inspiration from biology in that they are represented as networks of simple neuron-like processors. A typical "neuron" takes in a set of inputs, sums them together, takes some function of them, and passes the output through a weighted connection to

[1,2] *ungar@central.cis.upenn.edu* and *keeler@pav.mcc.com*.

another neuron. The neuron is thus just a predictor variable, or a function of a nonlinear combination of predictor variables. The connection weights serve as adjustable parameters which are set by a "training method," that is, they are estimated from part of the data. Actual biological neural networks are of course incomparably more complex than artificial networks.

Many different network architectures are used, typically with hundreds or thousands of parameters. The resulting equation forms are general enough to solve a large class of nonlinear classification and estimation problems and complex enough to hide a multitude of sins. The most widely used of these is the multilayer network with sigmoidal activation functions or, as it is often called, the "backpropagation network." (Werbos, 1974; Rumelhart, et al., 1986).

Feedforward sigmoidal networks can be represented graphically as a set of nodes, representing the neurons, connected by links. Each node takes as its input, I, the weighted sum of the outputs of the nodes which feed into it plus a "bias" term, θ_i, corresponding to the constant in a linear regression. It produces as its output, o_i, a function of this total input. Many different functions can be used; the most common is the sigmoidal function

$$\sigma(x) = 1/(1 + e^{-x}). \tag{1}$$

The entire network can be written as

$$\hat{y}_k = \sigma\left(\Sigma_{j=1} w_{o[j,k]}\, \sigma\left(\Sigma_{i=1} w_{i[i,j]} x_i + \theta_{i[j]} \right) + \theta_{o[k]}\right) \tag{2}$$

where x_i represents the ith input, $w_{i[i,j]}$ represents the weight on the link between input node i and hidden node j, and $w_{o[j,k]}$ represents the weight on the link between hidden node j and output k. More layers of the network can also be used. Two hidden layers can be beneficial, but theory indicates that more than two hidden layers provides only marginal benefit in spite of the major increase in training time. Such networks are called "feedforward" because information is passed only in a "forward direction" from the inputs to the outputs. There are no connections from neurons backward to neurons located closer to the input.

The above equation form has several advantages. One can show that with enough neurons, neural networks of the above form can - not surprisingly - approximate any well-behaved function arbitrarily well (Cybenko, 1989; Hornik, et al., 1989).

For problems with many inputs there is a significant advantage in using representations such as neural networks. One can distinguish between methods in which a fixed set of basis functions $\phi_i(x)$ is used, such as:

$$y = \Sigma w_i \phi_i(x) \tag{3}$$

where the $\phi_i(x)$ may be sines, cosines, or polynomials as in a Fourier or a Taylor series, and those in which the basis functions are adjusted on the basis of the data

$$y = \Sigma w_i \phi_i(x; w_j) \tag{4}$$

Neural networks fall in this latter category, with the "basis functions" ϕ_i being the sigmoids containing the weights on the first layer of links.

When variable basis functions are used, fewer basis functions are needed, and so fewer data points are needed to achieve a given accuracy. More specifically, the approximation error scales as $n^{-1/d}$ for fixed basis function methods like linear and polynomial regression (where n is the number of data points and d is the input dimension), and as $n^{-1/2}$ for neural networks (Barron, 1994). When the input dimension, d, is much greater than two, the neural networks give far greater accuracy for a given number of data points. Computation times for training scale similarly.

Neural networks are often viewed as "black-box" estimators where there is no attempt to interpret the model structure. This approach is fine for many applications, but for process optimization and control it is important to ensure that the model has the proper physical relationships. This can be accomplished through sensitivity and response-surface analyses, which allow one to numerically and graphically determine the behavior of the model.

Network Training

The parameters in the model — the weights — are chosen to minimize the residual sum of squares error

$$E = \Sigma \|\hat{y}_k - y_k\|^2 \tag{5}$$

where $\|\hat{y}_k - y_k\|$ is the norm of the difference between the outputs of the network $\hat{y}_k$ and the response variables being predicted and the sum is over all data points in the training set.

Once the network structure has been chosen, this is simply a nonlinear least squares problem, and can be solved using any of the standard nonlinear least squares methods. For example, the weights can be estimated by any optimization method such as sequential quadratic programming. Neural network researchers often speak of the "backpropagation" algorithm or "delta learning rule," which is simply gradient descent.

Two more related problems must be addressed: how to determine the architecture (how many neurons and layers) and how to avoid overfitting. Since neural networks are highly redundant and overparameterized, it is easy to fit the noise in the data as well as the signal. Put differently, there is a tradeoff between reducing bias (fitting the training data well) and variance, which effects how well future data will be fit.

To avoid this problem, the standard neural network modeling practice is to separate the data into a "training set," which is used to train the model, and a "testing set," which is used to determine when to stop training. Only the training set is used to adjust the network weights. During training, the model is evaluated periodically using the testing set. A decrease in testing performance signals that overfitting is beginning. In addition, a "validation set" may be set aside for evaluating the model on data that was in no way involved in the training process. One can also use Bayesian techniques to select network structures and to penalize large weights or reduce overfitting.

Neural Networks for Control

The different uses of neural networks covered in this paper can be characterized by a set of mappings as shown in Table 1. Neural networks may be used in direct control, where they provide a mapping from the current state or, more commonly, current and recent sensor readings as in an ARMA model (Ungar, 1995b) and the desired next state of the plant to the control action to be taken. In indirect control, the networks are used as a nonlinear model of the plant: given the current state of the plant and a proposed control action, they predict the next state of the plant. This model can then be used in a variety of controllers such as MPC (described below).

Table 1. Input-Output Mappings for Different Uses of Neural Networks.

direct control current state + desired next state -> control action

indirect control current state + control action -> next state

virtual sensors easy-to-measure properties -> hard-to-measure properties

fault diagnosis sensor readings -> faults

offline optimization current state + desired state -> optimization settings

Neural networks can also be used as a piece of a larger model of the plant. Often one wishes to control variables such as concentration or viscosity, which are relatively hard to measure accurately. Other variables may be much easier to measure and be related to the variables of interest in a deterministic but highly nonlinear way. For example, a variety of spectral techniques such as infrared (IR) measurements can be used to predict concentrations of chemical species and hence quality properties such as viscosities. Neural networks can be trained on the basis of data collected in the laboratory to learn the mapping between the easier-to-measure and the harder-to-measure variables and then used as "virtual analyzers." For example, it is often inconvenient and expensive to measure the exact chemical composition of the gases going up a stack, yet it is important to be able to monitor and control them to meet environmental regulations. Cheaper and more reliable sensors can be used to measure related variables and use neural networks to estimate the emissions.

Finally, neural networks are being used to detect and diagnose faults and to compensate for sensor failures or drifts. These applications are closely related to control, serving either as preprocessors (filters) for controller inputs or as components in supervisory control schemes running above conventional process controllers.

This article focuses on the most important uses of neural networks in chemical process control: model predictive control and virtual analyzers; for work on fault detection and diagnosis, see Venkatasubramanian and Chan (1989), Kramer (1993), etc.

Model Predictive Control with Neural Networks

As mentioned above, there are a number of ways that neural networks can be used in process controllers. In almost all cases, the strengths and weaknesses of the control scheme remain unchanged when a neural network is used in place of ARMA or mechanistic models. Many different neural network architectures can be used in neuro-controllers. The most popular are backpropagation networks and radial basis functions. In both of these cases, an ARMA style model is typically used in which the inputs to the neural network include lagged values of the relevant measured variables. Alternatively, when the process has variable time delays, it may be more advantageous to use an internally recurrent network and input just the current measured variable (Qin, et al., 1992; Karjala and Himmelblau, 1994).

The Internal Model Control (IMC) framework (Garcia and Morari, 1982; Nahas, et al., 1992) provides a good example of how neural networks can be incorporated into controllers (see Fig. 1). In conventional IMC, a model of the plant, typically linear, is partially inverted to determine a control action. Neural networks can be used as the controller (for direct control) or as the plant model (to be "inverted" for indirect control), or as both (Psichogios and Ungar, 1991).

For indirect control the model of the plant is learned by training a neural network so that its output $\hat{y}_t$ approximates the function $y_{t+1} = f(y_t, u_t)$ where y_t is a set of plant measurements at time t and u_t contains (independent) control actions. Lagged values of y and u may also be given as inputs to the network. Finding a control action requires partially inverting this model to find the control action u_t to minimize the difference $\left\|\hat{y}_{t+1} - y_{t+1}^{sp}\right\|$ between the plant state predicted at time $t+1$, $\hat{y}_{t+1}$, and that specified by the set point y_{t+1}^{sp}. If the function f were linear, the minimum would be easy to find analytically. However, neural networks are nonlinear, and require using Newton's method or similar nonlinear equation solving methods to find the optimal control action. Fortunately, the Jacobian (the derivative of f with respect to each of the inputs or, equivalently, the linearization of f) required for such methods is generally available since it is required when using the backpropagation or conjugate gradient algorithms to determine the network weights.

As an alternative to the computationally burdensome partial inversion of the plant model, a neural network can be trained to directly approximate the desired control function:

$$u_t = g(y_t, y_{t+1}^{sp}) \tag{6}$$

This relationship provides a control law for a direct feed forward controller. In general, the model (or inverse model) learned will be imperfect, and feedback must be used to stabilize the system. IMC provides one framework for doing so, as shown in Fig. 1. Consider the case where the model gives an estimate $\hat{y}_{t+1}$ of the plant output which is higher than the actual output y_{t+1}. The difference $y_{t+1} - \hat{y}_{t+1}$, which is the estimation error, is subtracted from the set point y_{t+1}^{sp} so that the controller tries to achieve a lower target y. If the error in the model is roughly constant, this will cancel the error due to model inaccuracy.

Under certain conditions, one can train the controller network directly from past plant data, but if the controller is used in an IMC framework, one must be careful to be sure it is an accurate inverse of the plant model, since although IMC compensates for approximate plant models, it does not compensate for inaccuracies in inverting the model. (If the plant model is not invertable this will, of course, be hard to do.) One technique often used is to propagate errors back through the plant model to the controller (Jordan, 1990).

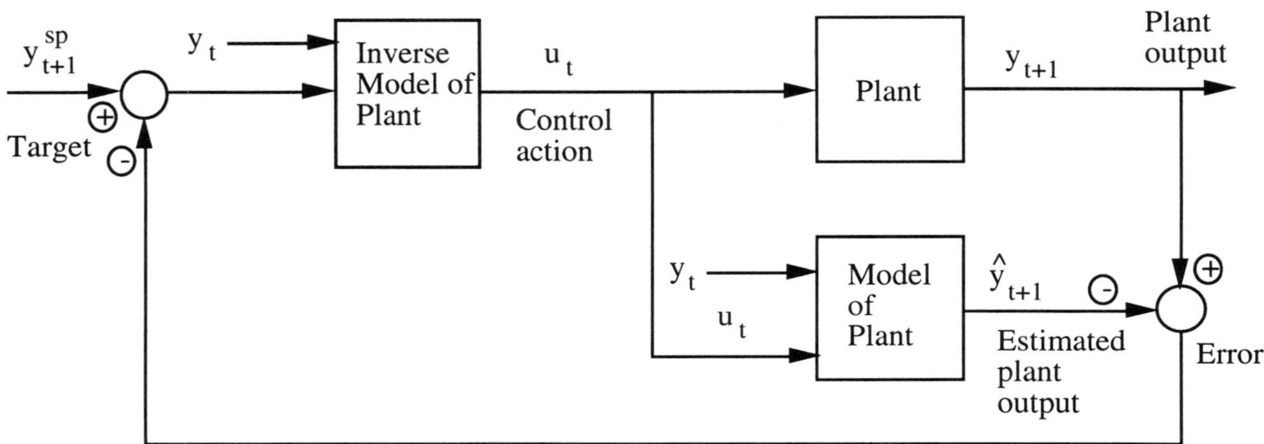

Figure 1. IMC Control Structure.

IMC, although relatively easy to implement and widely used, has several limitations. It is typically used as a one step-ahead predictor, and as such will not work on unstable plants or when there is a non-minimum phase response which requires looking more than one time step in to the future to determine the correct answer. (For example, increasing the concentration of a reactant will often initially decrease the temperature and then later increase it; one step ahead controllers fare badly on such problems.) However, IMC is useful for quasi-steady-state supervisory control where one step ahead control is adequate.

For plants where control actions have significant constraints, or for plants which are unstable or exhibit inverse response, it is more advisable to use neural networks in model predictive controllers.

The most widely used method of incorporating neural networks into controllers is within the framework of model predictive control (MPC), where the neural network is used as a process model (Psichogios and Ungar, 1991; Hunt, et al., 1992). In this case, the method of MPC and all of its advantages and disadvantages remain unchanged; it simply becomes easier to obtain an accurate, nonlinear model.

In a typical MPC architecture, one uses an optimizer to pick a sequence of control actions, u, to minimize the difference between the target, y^{sp}, and the actual value $\hat{y}$ over the next N time steps. More formally, one wishes to pick u to find the minimum:

$$\min \sum_{i=1}^{N} a_i \left[y_i^{sp} - (\hat{y}_i - d) \right]^2 \tag{7}$$

where d is an estimate of persistent disturbances, and is used to provide a form of feedback to account for modeling errors, measurement errors, and process disturbances and a_i

is a weighting, typically giving more weight to errors in the near future.

Usually, one sets the control action constant over the last portion of the time horizon ($j = M$ to N).

$$u_j = u_{j+1} \ , \ j = M,\ldots,N-1 \text{ for } N > M \tag{8}$$

As mentioned above, MPC allows one to set bounds on the output y, e.g., to keep a temperature in some range:

$$y_{\min} \le y_i \le y_{\max} \ , \ i = 1,2,\ldots,N \tag{9}$$

and on the control action, u_i,

$$u_{\min} \le u_i \le u_{\max} \ , \ i = 1,2,\ldots,N \tag{10}$$

and on the rate of change of the control action

$$|u_i - u_{i+1}| \le \Delta u_{\max} \ , \ i = 1,2,\ldots,M-1 \tag{11}$$

More advanced systems include the multivariate, or "combinatorial" constraints, such as $\Sigma c_i \, u_i \ = k_1 \ u_i \ u_j \ = k_2$ and so on.

One then uses an optimization technique such as Sequential Quadratic Programming (SQP) to minimize Eqn. (7) subject to the constraints, Eqns. (8-11), and the neural network model of the plant,

$$\hat{y}_{t+1} = f(y_t, y_{t-1}, \ldots u_t) \tag{12}$$

Such optimization methods typically use gradient descent using the derivatives of the objective function, Eqn. (7), with respect to the control variables u_i, with Lagrange multipliers added to handle constraints. Note that if one is going to optimize over many steps, it is important to train the neural network to minimize the multistep prediction error rather than the one step ahead prediction error.

Limitations of Neural Networks for Control

Since neural networks are universal approximators, it is tempting to mistakenly view these systems as a panacea. Neural networks represent, in some sense, a revolution in our capability to do high-dimensional nonlinear regression. However, they share the limitations generic to any empirical or regression modeling technique such as PLS or PCA.

For example, a standard system identification problem is that careless use of closed-loop data can tend to result in a model of the controller instead of the plant. Another example is that non-causal correlations must be avoided; one simple solution for neural networks is to exclude non-causal connections from the architecture.

An issue of importance when using any optimization or control scheme is the ability to ensure the satisfaction of any required physical constraints, such as mass or energy balance and fixed ratios in ratio controllers. Again, these issues are generic to regression-based technologies, including the more conventional DMC or IDCOM-type controllers, and these issues can be solved in various ways. Such constraints can certainly be applied in optimization routines using neural network models.

It is also important to ensure that the representation of the model in the time domain (where most neural network models are built and evaluated) does not cause problems in the frequency domain. With insufficient data, regression models can learn to give a good representation of the time series, but the model may contain high-frequency response components which can cause stability problems for control. Again, there are a variety of approaches to solving this problem, and they parallel the conventional model predictive control methods.

Much progress has been made in understanding the limitations generic to regression modeling, and advanced modeling packages contain techniques for dealing with them. Nevertheless, one must exercise care in using neural networks, as one must with any system identification and control techniques (Eaton, et al., 1994).

It bears repeating that although neural networks ease the task of building accurate multivariable nonlinear models, they do not eliminate the problems and difficulties involved in developing accurate and robust controllers. Although neural networks give good nonlinear models, their use is subject to all the usual limitations and dangers of conventional models. One must still determine appropriate control structures based on the plant and disturbance structure, including time delays and frequency response characteristics. IMC structures are not appropriate for controlling unstable systems, and accurate plant models are of no use if the disturbances dominate process behavior or if the plant is inherently hard to control due to poor design in which case no control scheme can help.

Hybrid Networks and Virtual Analyzers

Neural networks can also be used in conjunction with first principles models (Psichogios and Ungar, 1992) or as virtual analyzers (Piovioso and Owens, 1991). Often one has a reasonable partial theoretical model of a plant, but lacks certain parts of the model such as the dependence of viscosity or reaction kinetics on temperature and concentration. Neural networks can be used to learn that portion of the model. The overall hybrid model contains both the theoretical equations such as mass and energy balances and the neural networks, and can be used in an MPC scheme as described above.

More formally, plants can often be described in the form

$$d\boldsymbol{x}/dt = \boldsymbol{f}(\boldsymbol{x},\boldsymbol{u},\boldsymbol{p}) \tag{13}$$

$$p = g(x,u) \quad (14)$$

where the function f is known - e.g., kinematics or mass and energy conservation but contains "parameters" p which depend on the plant state x and possibly the control action u in unknown ways. Observations of x and u can then be used to train a neural network to approximate the unknown function g using an extension of the backpropagation algorithm even though the parameters p cannot be measured. The resulting plant model consisting of Eqns. (13) and (14) can be trained more accurately, using less data, and extrapolates better than pure neural network plant models which do not incorporate the known form of f (Psichogios and Ungar, 1992).

A similar situation arises when one cannot easily measure some quality variables (e.g. viscosity, photodegradation, or even concentrations of certain chemical species). For example, consider the control of the viscosity of a polymer product, where viscosity cannot be measured online. Temperatures, y, in the reactor can easily be measured, and perhaps supplemented with measurement of near-IR absorption. Online measurement of viscosity is difficult. Often companies take samples, which are then sent to a laboratory for analysis. It can take from a half hour up to several hours to get an analysis from the laboratory. A neural network trained with product samples whose viscosity, z, has been measured in the lab can be trained to predict z as a function of variables y such as temperature and pressure which are measured on line:

$$z = f(y) \quad (15)$$

Such neural networks can be used as *virtual analyzers* to give online estimations of viscosity for a controller.

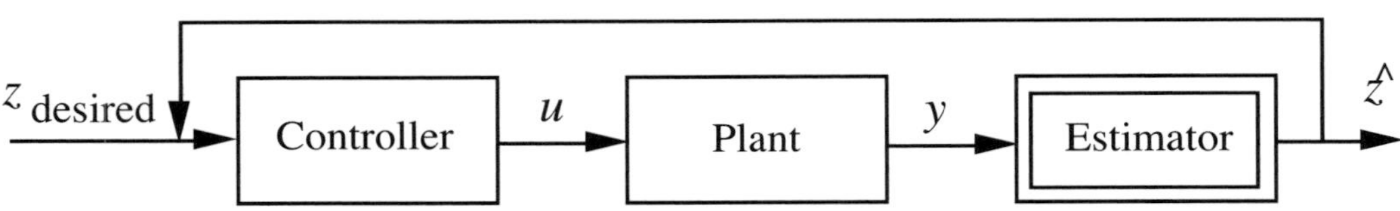

Figure 2. Plant and controller using estimator as virtual sensor for z.

Willis, et al. (1992) give a nice example of the use of virtual analyzers for inferential control of penicillin production. To optimally control penicillin production, one would like to have accurate online estimates of biomass concentration during fermentation. Such measurements are unfortunately not available. Willis, et al. use a neural network to estimate biomass from carbon dioxide evolution rate, batch age, and the rates that the two primary substrates are fed to the reactor. These plant measurements (y in Fig. 2) are then used to estimate the biomass (z), which is the controlled variable.

Neural networks have also been used to minimize the effect of sensor noise and drift (data reconciliation), as elements of nonlinear statistical process controllers (SPC), and for fault diagnosis, sometimes as a prelude to reconfiguring control loops to cope with faults such as sensor or actuator failure. Neural networks can be trained which take a set of sensor readings as inputs and predict the same sensor readings as outputs (Kramer, 1992). When fewer hidden nodes are used than inputs, this has the effect of nonlinearly projecting the sensor readings onto a smaller dimensional space and then reconstructing the sensor readings, thus reducing the noise in the signal. Such rectified sensor readings take advantage of the correlation structure of the sensor readings and, like other forms of filtered signals, can reduce the effect of sensor noise on control.

PEMS Virtual Analyzer: A Case Study

The 1990 Federal Clean Air Act Amendments require manufacturers to install continuous emissions monitoring systems (CEMS) to monitor emissions such as sulfur dioxide (SO2), nitrogen oxides (NOx), carbon monoxide (CO), and more. A CEMS typically relies on delicate instruments that require frequent recalibration and incur significant maintenance costs. A virtual analyzer called a predictive emissions monitoring system (PEMS) can serve as an inexpensive alternative to a hardware CEMS. Pavilion Technologies, Inc. has found that its PEMS product, the Software CEM, is typically more accurate than a CEMS (see below) and is 1/3 to 1/2 as expensive to install and certify.

Since a PEMS does not require daily recalibration and is not exposed to harsh environments, maintenance costs are also small in comparison. In addition, the virtual analyzer model provides the ability to reduce emissions. Sensitivity calculations yield insight into the causes of emissions, and constrained optimization routines yield input settings to reduce emissions while maintaining process performance and cost criteria.

Achieving the 95% up-time mandated by the EPA for demonstration of continuous compliance for a PEMS requires the ability to continue to operate in the event of sensor failure. Therefore, detection and correction of sensor failures is an essential component of a PEMS. This "sensor validation" task can be accomplished with an essentially all-sensors-to-all-sensors neural network model, along with error tolerances for failure identification and a reconstruction algorithm to replace failures with corrected values.

The Software CEM is currently installed on a wide variety of operating units, including steam boilers, gas turbines, furnaces, and reciprocating engines. Over 30 installations have passed one or more relative accuracy test audits (RATA), and are certified, licensed, and operating continuously. Over 100 installations are currently in progress.

Models and Algorithms

A Software CEM consists of two models: an Emissions Model and a Sensor Model. The Emissions Model is a feedforward neural network, trained using emissions readings from a temporary monitor as output data and available process measurements as input data. An adjustable model bias allows correction for constant-offset process drift.

The Sensor Model is typically an autoassociative neural network model, trained using the Emissions Model input variables as both input and output data, and possibly with additional input variables.

A sensor failure alarm is set whenever the difference between a sensors's value and the Sensor Models' output value for that sensor exceeds a threshold. The threshold for each variable is set by a thorough perturbation study on the model, which is parameterized by the maximum acceptable false-positive alarm rate specified by the user. The output of this exercise includes the average number of failures that can occur without failing RATA.

When one or more alarms occur, the reconstruction algorithm is triggered. This algorithm identifies which sensor(s) have failed, replaces the failed values with corrected ones, and notifies the operator of the failure(s). The task of the reconstruction algorithm is complicated by the fact that an alarmed sensor is not necessarily the failed sensor, since a failed sensor can cause variables other than or in addition to itself to alarm. This complication is overcome by identifying which single variable is most likely to have failed, estimating its true value, and iterating these steps until there are no more alarms. This iterative procedure is similar to the iterative measurement tests common in Data Reconciliation algorithms (Heenan and Serth, 1986). The reconstruction algorithm operates by minimizing the input-to-output error in the Sensor Model.

Building and Installing a Software CEM

The steps to building and installing a Software CEM are:

1. Connect a certified CEM to the stack to gather emissions data.
2. Operate the plant over the range of anticipated operations for 3 to 5 days.
3. Create the Emissions model and the Sensor Model.
4. Install the Software CEM system in a plant computer receiving process data.
5. Perform a Relative Accuracy Test Audit (RATA) for PEMS certification.

Case Studies

<u>Virtual Emissions Sensor For a Gas-fired Boiler</u>

A Software CEM PEMS was installed in May of 1993 to monitor NOx emissions from a 221 mmBTU/hr natural gas-fired boiler at the Arkansas Eastman Company (Collins, et al., 1993). Performance of the system is shown if Fig. 3. The system passed the RATA test one month later, in June of 1993, and was certified. It operates continuously with the approval of the EPA and under permit with the Arkansas Department of Pollution Control and Ecology. Now in operation for over two years, the system has passed all RATA tests, has had 100% limit compliance and better than 99% uptime.

<u>Virtual Analyzer Exceeds Accuracy of a Hardware Sensor</u>

NOx emissions data from a gas turbine at a chemical company in Texas were collected in one-minute samples using two well-calibrated hardware CEMS operated by skilled personnel from a emissions measurement company. Data from the CEMS considered by the operators to be the more accurate of the two was used to train a PEMS Software CEM model. The relative accuracy of the second CEMS to the first was 3%, and significant differences of up to 15% of the range of the data occurred over periods of hours several times during the five day data collection period. The relative accuracy of the PEMS model to the first CEMS was 1%, and differences never exceeded a few percent of the range of the data.

Currently this company has installed and certified eight PEMS systems. The longest running has been in operation for nearly two years and has passed all RATA tests.

<u>The Sensor Validation Component</u>

As emphasized above, sensor validation is an essential component of a virtual analyzer. Figure 4 displays the failures and reconstructed values for a temperature sensor used as an input to a PEMS Software CEM. The temperature transducer was on cylinder 3 of a 16-cylinder

8,000HP Cooper-Bessimer Reciprocating Engine. The Emissions Model used seven inputs to calculate two outputs, NOx and CO. The inputs to the Sensor and Emissions Models were eight temperature transducers plus the intake and exhaust manifold pressures and temperatures.

As can be seen in Fig. 4, the temperature transducer experienced gross failure in early July and intermittently throughout the summer. Whenever the error in the Sensor Model exceeded the tolerance factor for the transducer, the model was used to estimate a corrected value for the transducer (see Models and Algorithm Section).

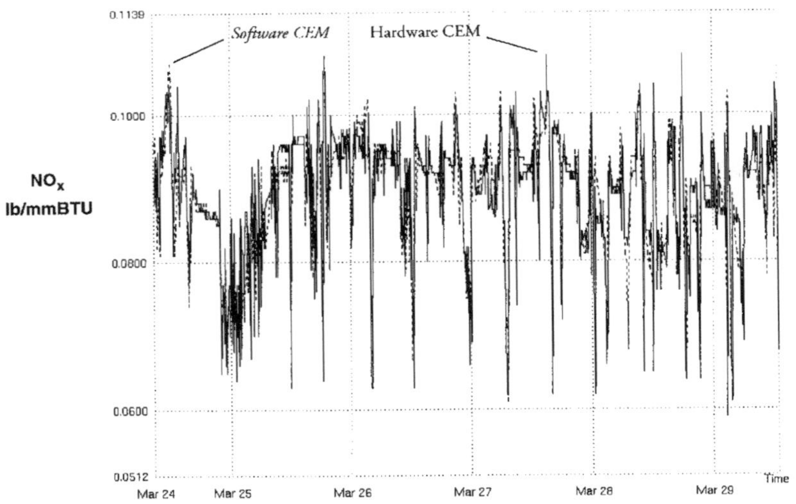

Figure 3. NOx emissions from the Software CEM versus hardware CEM values on the test set. Each point represents a one-minute sample.

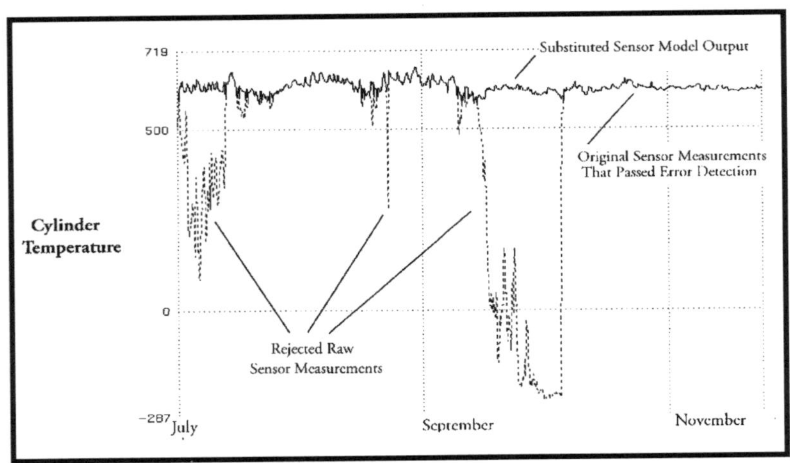

Figure 4. This figure shows the final, validated values of the sensor validation system. The dotted lines indicate raw measured values that were rejected via the error detection threshold and the solid line indicates the final reconstructed values. Note that in most circumstances (e.g. from October to November), there is no reconstruction; the raw values pass error detection and are used as the final output of the sensor validation system.

Optimization Using a Virtual Analyzer

A virtual analyzer neural network model can be used not only to monitor a process, but to improve it as well. Nonlinear constrained optimization codes (e.g. SQP) can be integrated with virtual analyzer models to compute setpoints for process inputs that improve the values of the sensed quantities (e.g. reduce emissions) while satisfying process performance and/or cost constraints (e.g. yield, quality, profit).

For example, in some PEMS applications both NOx and CO emissions are to be monitored, and it is often possible to reduce both emissions simultaneously. This can be valuable, since in addition to avoiding fines for noncompliance, emissions reduction can sometimes earn tradable credits. In combustion, NOx and CO production are typically inversely related (see Fig. 5). Using a virtual analyzer model of NOx and CO, one can run the optimization routine whenever external influences or operating goals change and keep operation at or near the "sweet spot," where both emissions are minimized.

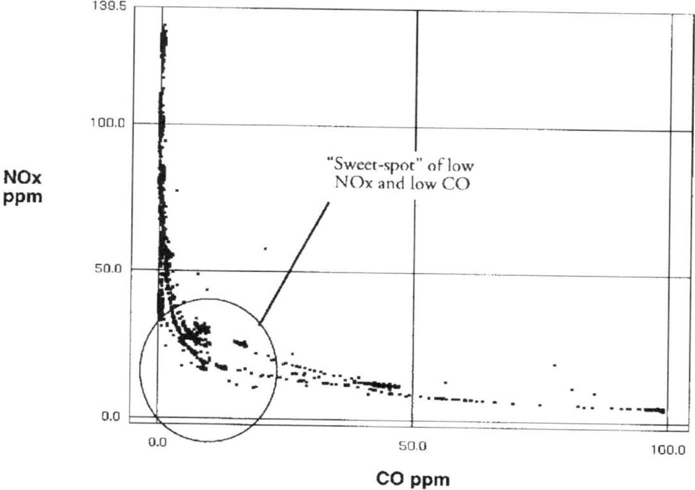

Figure 5. Measured values of NOx and CO during one week operation of a gas turbine. Note the inverse relationship between the two variables — decreasing one increases the other. At the "sweet-spot" region of operation, both NOx and CO are relatively low. Optimization using the Software CEM can find the optimal operating condition to reduce emissions under changing operating constraints or loads, thereby increasing plant efficiency and lowering pollution levels.

Virtual Analyzers, Optimization, and Control

Virtual analyzers constitute one of several broad classes of neural network applications for process industries. In the table below, the applications lower in the table are generally more complex. Not surprisingly, the number of applications fielded by Pavilion's customers is larger for the simpler types of applications. Economic benefit varies by specific application; many of Pavilion's customers are realizing millions of dollars of documented annual savings with the first four types of applications. The references after each of the five application areas below describe real-world projects.

(1) Optimization Off-line QSS

Although often the least demanding type of application, off-line optimization frequently yields the greatest economic benefit. The model created and analyzed off-line is "quasi-steady-state" (QSS) — steady state is assumed, but the incorporation of dead-times improves the model. Sensitivity analysis often yields valuable process information, such as identifying key variables previously thought to be unimportant. Constrained nonlinear optimization improves the mean of the process outputs by computing static setpoints for the process inputs. (See Spangler, 1994).

(2) Virtual Analyzer On-line QSS

A virtual analyzer model can be used to perform optimization or control. (See this Section and Collins, et al., 1993; Keeler, et al., 1993; Keeler, 1995; Schweiger and Rudd, 1994).

(3) Supervisory Control On-line QSS Open-loop

Process readings are automatically fed to the neural network model and the optimization routine. One step-ahead quasi-steady-state optimizations to determine setpoints for process inputs are computed on an ongoing basis. The operator can accept or reject the computed setpoints. (See Schweiger and Rudd, 1994).

(4) Supervisory Control On-line QSS Closed-loop

Like (3), but the computed setpoints are downloaded directly to the controllers. (See Spangler, 1994).

(5) Trajectory Control On-line Dynamic Closed-loop

A dynamic model is created and used for full Model Predictive Control. (See Section on Model Predictive Control with Neural Networks).

5. Discussion

The field of neural networks for process control is much too extensive to be fully summarized here: hundreds of papers are published each year in the area, and several commercial vendors offer control systems using neural networks. Neural networks have been shown to be are attractive models to use when processes are nonlinear and good first principles (mechanistic) models are not available, either for entire processes or for parts of the process. The most common commercial applications of neural networks are as virtual sensors or virtual analyzers.

Most control schemes using neural networks are identical to those that do not use neural networks; the "neuro-controllers" just use neural networks as process models (e.g. in MPC), as virtual sensors, or as controllers. Neural networks can also be used to forecast disturbances (see Ungar, 1995) as part of a feedforward control system. Often, neural controllers are used as a feedforward component of a more complex control architecture in which a standard feedback controller rejects disturbances and gives adequate control while the neural network is being trained. (See, e.g., Lee and Park (1992) and citations there to Kawato and others)

Many other uses of neural networks are possible, including their use as data preprocessors, as fault detectors (possibly coupled to a system which reconfigures the control loops when faults are discovered), and as automatic tuners for conventional controllers. Looking farther into the future, neural networks will be used to provide continuous optimization of plants, perhaps using a reinforcement learning scheme (Barto, 1990).

References

Barron, A.R. (1994). Approximation and estimation bounds for artificial neural networks. *Machine Learning*, **14**, 115-133.

Barto, A.G. (1990). Connectionist learning for control. In W.T. Miller, R.S. Sutton and P.J.Werbos (Eds.), *Neural Networks for Control*. MIT Press. pp. 5-58.

Bhat, N. and T.J. McAvoy (1990). Use of neural nets for dynamic modeling and control of chemical processes. *Comp. Chem. Engng.*, **14**, 573-583.

Collins, M., W. Jackson, K. Terhune and J. Havener (1993). NOx emissions using process insights and the software CEM. Technical Report, Pavilion Technologies, Inc. (to be published).

Eaton, J.W., J.B. Rawlings and L.H. Ungar (1994). Stability of neural net based model predictive control. *Proceedings of the ACC*. pp. 2481-85.

Garcia, C.E. and M. Morari (1982). Internal model control. 1. a unifying review and some new results. *Ind. Eng. Chem. Proc. Des. Dev.*, **21**, 308-323.

Hartman, E., J.D. Keeler and J. Kowalski (1990). Layered neural networks with gaussian hidden units as universal approximators. *Neural Computation*, **2**, 210-215.

Haykin, S. (1994). *Neural Networks: A Comprehensive Foundation*. Macmillan, NY.

Heenan,W. and R. Serth (1986). Detecting errors in process data. *Chemical Engineering*, Nov. 10, 99-103.

Hornik,K., M. Stinchcombe and H. White (1989). Multilayer feedforward networks are univeral approximators. *Neural Networks*, **2**, 359-366.

Hunt, K.J., D. Sbarbaro, R. Zbikowski and P.J. Gawthrop, (1992). Neural networks for control systems — a survey. *Automatica*, **28**, 1083-1112.

Jordan, M.I. (1990). Learning inverse mappings using forward models. *Proceedings of the Sixth Yale Workshop on Adaptive and Learning Systems*, pp. 146-151.

Karjala, T.W. and D.M. Himmelblau (1994). Dynamic data rectification by recurrent neural networks vs. traditional methods. *AIChE J.*, **40**, 1865-1875.

Keeler, J., J. Havener, E. Hartman and T. Magnuson (1993). Achieving compliance and profits with a predictive emissions monitoring system: Pavilion's Software CEM. Technical Report, Pavilion Technologies, Inc.

Keeler, J. (1995). Pavilion's Software CEM: a low-cost, reliable alternative for enhanced monitoring. Technical Report, Pavilion Technologies, Inc., in preparation.

Kramer, M.A. (1992). Autoassociative neural networks. *Comp. Chem. Engng.*, **16**, 313-328.

Lee, M. and S. Park (1992). A new scheme combining neural feedforward control with model-predictive control. *AIChE J.*, **38**, 193-200.

Kramer, M.A. (1993). Diagnosing dynamic faults using modular neural nets. *IEEE Expert*, **8**(2), 44-53.

Nahas, E.P., M.A. Henson and D.E. Seborg (1992). Nonlinear internal model control strategy for neural network models. *Comp. Chem. Engng.*, **16**, 1039-1057.

Narendra, K.S. and K. Parthasarathy (1990). Identification and control of dynamical systems using neural networks. *IEEE Trans. on Neural Networks*, **1**, 4-27.

Piovoso, M.J. and A.J. Owens (1991). Sensor data analysis using neural networks. *Proceedings of the Fourth International Conference on Chemical Process Control (CPC IV)*. South Padre Island, TX.

Psichogios, D.C. and L.H. Ungar (1991). Direct and indirect model based control using artificial neural networks. *I&EC Res.*, **30**, 2564-2573.

Psichogios, D.C. and L.H. Ungar (1992). A hybrid neural network — first principles approach to process modeling. *AIChE J.*, **38**, 1499-1511.

Qin, S-Z, H-T Su and T.J. McAvoy (1992). Comparison of four nerual ent learning methods for dynamic system identification. *IEEE Trans. Neural Networks*, **3**, 122-130.

Rumelhart,D.E., G.E. Hinton and R.J. Williams (1986). Learning internal representations by error propagation. In D.E.Rumelhart, J.L.McClelland and the PDP Research Group (Eds.), *Parallel Distributed Processing: Explorations in the Microstructure of Cognition, Volume 1: Foundations*. MIT Press/Bradford, Cambridge, MA.

Sanner, R.M. and J.-J.E. Slotine (1991). Direct adaptive control with gaussian networks. *Proc. 1991 Automatic Control Conference*. pp. 2153-2159.

Schweiger, C.A. and J.B. Rudd (1994). Prediction and control of paper machine parameters using adaptive technologies in process modeling. *The TAPPI 1994 Process Control Symposium Proceedings*.

Spangler, M.V. (1994). Uses of process insights at a refractory gold plant. *Pavilion Users Conference*. October 17.

Ungar. L.H. (1995a). Forecasting. In Arbib (Ed.), *Handbook of Neural Networks*. MIT Press.

Ungar. L.H. (1995b). Process control. In Arbib (Ed.), *Handbook of Neural Networks*. MIT Press.

Venkatasubramanian, V. and K. Chan (1989). A neural network methodology for process fault diagnosis. *AIChE J.*, **35**, 1993.

Werbos, P.J. (1974). Beyond regression: new tools for prediction and analysis in the behavioral sciences. Ph.D. thesis, Harvard University

Willis, M.J., G.A. Montague, C. Di Massimo, M.T. Tham and A.J. Morris (1992). Artificial neural networks in process estimation and control. *Automatica*, **28**, 1181-1187.

SESSION SUMMARY: INTELLIGENT CONTROL

Michael Nikolaou[1]
Chemical Engineering Department
Texas A&M University
College Station, TX 77843-3122

Babu Joseph[2]
Department of Chemical Engineering
Washington University
St. Louis, MO 63130-4899

Introduction

As the title suggests, the main theme of this session is Intelligent Control (IC). Basic issues that are addressed include:

1. What is IC?
2. What are the capabilities of IC?
3. What are typical paradigms of IC?
4. What theoretical and computational tools are available for the design and operation of IC systems?
5. Some applications of IC to industrial processes.

The definition of Intelligence is an interesting philosophical problem, that has often led to heated debates. While the resolution of this problem per se is certainly a goal worth pursuing, the target aimed at by the authors of these three papers is more modest, albeit no less important for control engineering. All three authors are discussing methods for using computers to automate a number of process control tasks, that would not be practical to automate by using "traditional" hardware and software tools. New theoretical and computational tools could provide this automation capability. Therefore, the incorporation of these new tools would endow the characteristic of Intelligence to control systems. While this interpretation of Intelligence has the drawback of evolving according to the day's state of the art in automation technology, it has the distinct advantage of providing a clear research direction: Relegate to the computer an increasing number of tasks currently handled by human operators. This is accomplished by employing a broad array of multi-disciplinary computer-based tools. By not merely focusing on human aspects of intelligence, IC has the potential to even transcend human intelligence.

The Papers

The first paper, titled "Intelligent Control for High Autonomy Process Control Systems," focuses on two issues: (1) the precise definition of intelligent control and, (2) the autonomy of process control systems.

Intelligent Control means different things to different workers in this field. The authors describe an effort by various researchers who have made significant contributions to this field to come up with a consensus. A rather broad definition is arrived at in an effort to include as many differing points of view as possible. In the second part of the paper the authors focus on the autonomy of control systems. The level of autonomy of a control system is the degree of its ability to handle a wide range of situations with minimal human intervention. While classical control theory has stressed controller analysis and design methods in the form of mathematical theorems that guarantee certain facts if the theorem's assumptions are satisfied, the quest for enhanced controller autonomy focuses on the assumptions themselves and aims at understanding (1) How to make less restrictive the assumptions needed during controller design (2) What happens when these assumptions are violated. The paper discusses issues such as adaptation and learning, planning under large uncertainty, and coping with large amounts of data. Specific emphasis is placed on hybrid control systems, learning control, and control of discrete event systems.

[1,2] *m0n2431@acs.tamu.edu* and *joseph@wuche.wustl.edu*.

The second paper, titled "Expert Control and Fuzzy Control: Differences and Similarities," examines the incorporation of expert systems and fuzzy logic into process control systems. Combined with numerical algorithms, expert systems and fuzzy logic can be used to endow control systems with additional functionality, such as logic, sequencing, reasoning, and heuristics. These capabilities can be used for the development of a hierarchical control structure. In that structure, a simple controller functions at the lowest level, supervised by a cascade of controllers at higher levels. The first higher-level controller could, for example, dictate how the low-level controller might adapt. Then, the second higher-level controller could select the method, according to which the first higher-level controller would dictate the low-level controller's adaptation, and so on. Expert and fuzzy systems can be used at all levels. At the lowest level the combination of traditional controllers with the additional features provided by expert and fuzzy controllers can lead to much greater autonomy. The authors stress that, in addition to providing control capabilities that are hardly feasible by algorithmic methods, fuzzy and expert controllers can be useful in providing convenient means for man-machine interaction. They can accomplish that through capturing human knowledge and heuristics (expert systems) or by replicating imprecise human reasoning (fuzzy logic). The authors alert readers to cautiously interpret comparison test results in literature, after careful examination of the conditions and assumptions under which these tests were performed. This approach will help select relevant results and identify applications for which fuzzy/expert control is indeed advantageous. Many IC concepts are now being integrated into the software provided with commercial DCS systems. Features particularly useful are self testing and diagnosis and adaptation.

The final paper in this session, titled "Process Modeling and Control Using Neural Networks," is an effort to elucidate the capabilities of neural networks that can be fruitfully applied to process control. The two key issues associated with neural networks, namely nonlinear approximation and parallel computations, are thoroughly examined and their relevance to process control is elaborated on. The relationship between ANN and Statistics is discussed at length and the similarities between the two subjects often ignored in the ANN literature are highlighted. The paper closes with a detailed discussion of a successful application of neural networks as soft sensors in a series of industrial processes.

Conclusions

Intelligence can manifest itself in a number of ways in control systems. While IC paradigms can endow control systems with functionality that traditional methods could hardly make available, they can also improve the analysis, design, operation, and maintenance characteristics of control systems. Of course, IC methods are by no means a panacea. They should be simply viewed as tools for process control.

Research and development on IC is an active area. In fact, IC methods are moving from the realm of academic curiosity to the field of industrial reality.

At times, the enthusiasm with which IC methods were greeted led to overly optimistic expectations that did not fully materialize. Separation of exaggeration from fact is an issue addressed by all three authors of this session. For successful applications, related literature should be consulted with care, and thorough scrutiny should be exercised before the right method is selected for the right application.

ROLE AND PROSPECTS FOR INTELLIGENT SYSTEMS IN INTEGRATED PROCESS OPERATIONS

G.V. Rex Reklaitis
School of Chemical Engineering
Purdue University
West Lafayette, IN 47907

Lowell B. Koppel
Setpoint, Inc.
Houston, TX 77079-2905

Abstract

Integration of process operations has been stimulated by global competition, focus on customer needs, cost pressures, societal expectations, as well as rapid technological changes and has been facilitated by remarkable performance advances in computing technology. Intelligent systems (IS) technologies have received mush attention in a wide range of process engineering applications, including process operations. This paper examines the role of intelligent systems technologies in the realization of integrated process operations and assesses the impact of IS on this domain to date. The review and assessment show that IS tools and methods have indeed had a significant role in process operations. However, as yet the impact has principally been attained at the individual operational task level rather than as an integrating technology per se. Future IS contributions relevant to the integration goal include: development of hybrid architectures for collaborative use of traditional and IS technologies, creation of intelligent bottleneck diagnosis systems for process improvement, development of intelligent process information management systems, and design of intelligent operator tutoring systems. Specific problem areas identified for future IS application include IS roles in the execution of hierarchical modular process optimization, the integration of planning to operations, and in decision support for integrated supply and transport systems. These problem areas are discussed in the context of petroleum/petrochemicals operations.

Keywords

Process operations, Integration, Data acquisition, Monitoring, Diagnosis, Process control, Scheduling, Planning, Intelligent systems.

Introduction

The dual objectives of this paper are to assess the role that intelligent systems developments have played in the realization of integrated process operations and to project future roles for these technologies in the process operations domain. To achieve these objectives, the paper will first summarize the current understanding of what is computer integrated process operations (CIPO), next will review the state of development of process integration efforts in the batch and continuous processing domains, and then will present an assessment of the impact of intelligent systems technologies on process operations to date. The paper will conclude with a projection of future needs and opportunities and a discussion of some concrete operational problems whose solution requires at root an integrated approach and appears fertile ground for the application of IS tools.

Computer-Integrated Process Operations

A short definition of this domain of process systems engineering is that it entails the exploitation of computing technologies to link, support, and execute the operational tasks and decision processes which arise in chemical manufacturing. In short, it may be characterized as computer integrated manufacturing (CIM) for the process industries (Pekny, et al., 1991). It is a relatively new domain of organized research and development: the first international conference on the subject was held just eight years ago (Reklaitis and Spriggs, 1987) and the second in 1993 (Rippin, et al., 1994). The interested reader is invited to consult these volumes for more details on the component technologies involved in CIPO.

The impetus for integration of operations has come from multiple sources:

- Strong business focus on meeting the needs of the customer through timely delivery of consistent quality products
- Global competition which is exposing all CPI sectors to both international and domestic challenges and requiring that competition be met in all markets through global coordination of resources
- Intense cost pressures which motivate market leaders to be low cost producers
- Increasing societal expectations and legally enforced demands that chemical manufacturing will be environmentally benign and inherently safe
- Rapidly accelerating technological changes which require manufacturing to be nimble in exploiting computing and process developments.

As might be expected, the enabling developments for integration have principally arisen from computing technology, a sector which has achieved remarkable performance advances in the last decade. The costs of memory, computing cycles, and high speed communications have been dramatically declining allowing ever larger applications to be addressed. For instance, since 1991 cost per MIPS has declined by a factor of four, cost to store a gigabyte of data by a factor of three, and cost to send a megabyte of data coast to coast via modem by a factor of five (Stewart, 1995). The strides in user interface design concepts and implementation tools and boom in networking utilities such as the world-wide web have made electronic information and applications accessible to a broad spectrum of users without the burden of extensive formal training. Moreover, software engineering methodology and tools have progressed significantly so that the software infrastructure for integrated operations has become less costly to create, more readily maintainable, and much more amenable for adaptation to new, unanticipated situations.

The functional task set which computer aided process operations encompasses corresponds to the strategic, tactical, execution, and evaluation activities which arise in any organized enterprise. Most commonly, the tasks are partitioned into the following six plant operational levels:

- Planning: the allocation of production resources and assignment of production targets for the plant averaged over a suitable time scale, often months or quarters
- Scheduling/Optimization: the determination of the timing and sequence in the execution of manufacturing tasks or the selection of operating variable values so as to achieve production targets in a feasible and possibly optimal fashion
- Execution: the performance of schedules and coordination of control actions so as to correct faults and achieve unit level requirements
- Monitoring and Diagnosis: the tracking of process execution, detection of departures from normal operation, and identification of causes
- Regulatory Control: the application of corrective actions and phases in response to disturbances
- Data Acquisition: the collection and recording of process information and associated data validation, smoothing, and corrective measures.

Increasingly this core set of tasks is expanded to include higher level decisions such as:

- Process design/retrofitting: an activity which can be viewed as capacity generation
- Multi-site coordination: the joint execution of the planning tasks of multiple production sites and their associated logistics subsystems
- Supply chain management: the integration of multiple enterprises (producers, suppliers, and customers) and their common logistics functions.

In traditional organizational environments, these various tasks were to a large degree isolated from each other and executed in a hierarchical manner, with upper level decisions imposed on lower levels with minimal feedback up the chain to influence decisions except indirectly through performance reporting. The coordination of these tasks in a consistent and efficient manner so as to allow evaluation and readjustment of decisions is the essential purpose of manufacturing integration efforts.

Scope of Integration

Two categories of integration efforts normally are involved in CIPO: task integration and systems integration. Task integration refers to the linkage of operational tasks described above, while systems integration refers to the linkage of the system components: data, applications, and people.

Task Integration

For implementation purposes as well as organizational reasons, the various operational tasks are best modularized. However, the design of the integration framework must recognize that the operational tasks are in fact not isolated but instead do interact strongly and thus the framework must incorporate active feedback links. The execution of the various tasks is normally supported by one or more applications and the enterprise data base.

In order to capture alternative views of the rich interrelationships between tasks, a number of integration models have been proposed in the literature (Benson, 1995). These can roughly be categorized into wheel and hub, hierarchical, and hierarchical with feedback models. The wheel and hub models reflect the philosophical orientation of the authors of the model. For instance, the knowledge-centered model, shown in Fig. 1, places the data and models which represent the core knowledge of the manufacturing enterprise at the center of the hub and the various tasks and their associated applications communicating with the hub and with each other. Alternative wheel and hub models emphasize the people dimension of the enterprise by placing people at the hub of the financial dimension be centering on the financial base of the enterprise. While these models are useful for conveying the management philosophy or values of the enterprise, they are of limited utility in guiding implementation activities.

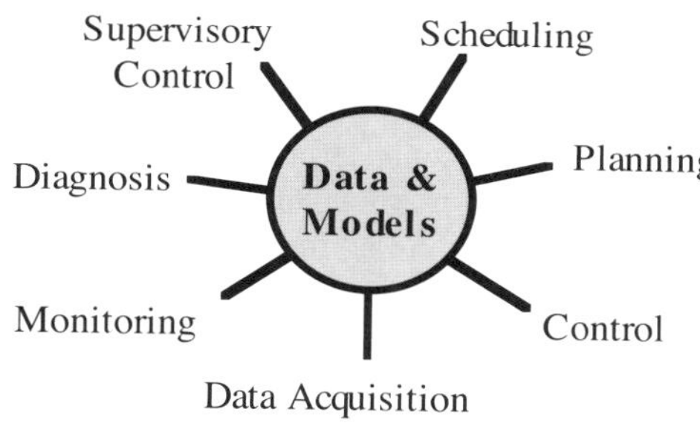

Figure 1. Knowledge-centered integrated model.

The traditional hierarchical pyramid, shown in Fig. 2, is a useful functional decomposition because it exploits the inverse relationship which exists between the scope of the decisions made at each level and the degree of detail which is considered at that level to obtain a partitioning of the decision space in which the decision problem at each level is of comparable difficulty.

Thus the upper levels involve decisions which have broad impact on the enterprise but employ a coarse grain view of the production components. Conversely the lower levels involve decisions of localized impact but which require a fine grained view of the entities being acted upon. The hierarchy also is intuitive in that it reflects natural human organizational tendencies. However, the top down decision flow, task level decision isolation, and limited feedback to upper levels constrain the attainable performance of the enterprise. Specifically, since actual operational constraints are aggregated at the higher levels, decisions made at these levels must be conservative to insure feasibility at the lower execution levels. Thus, in effect the manufacturing system is forced to operate at interior local optima.

Figure 2. Traditional functional hierarchy.

The preferred framework is the hierarchical model with feedback, as shown in Fig. 3. This model features communication between hierarchical tasks, with targets, allowable ranges, and prices passed down the hierarchy and constraints, actual performance results, and profit figures passed up the hierarchy. Communication of information such as flows and quality constraints between peers at the task level occurs in the direction set by the selected operational strategy.

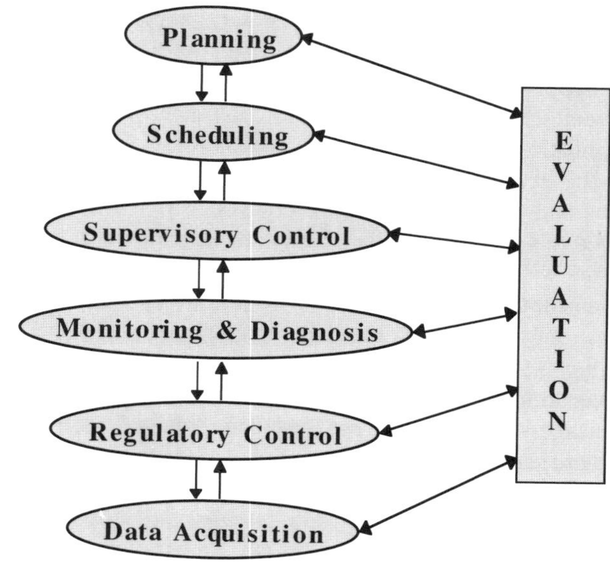

Figure 3. Hierarchical structure with feedback.

A key feature of this framework is the evaluation block in which human decision makers play an essential role. The evaluation function receives as input feedback from all levels in the form of exception and transaction reports and uses this information to evaluate system performance and initiate appropriate corrective actions. The evaluation block completes the loop in the plan-do-see cycle of effective decision making.

System Integration

This aspect of integration refers to the linkages and interfaces between the information, applications, and human users of the composite system. *Data integration* is achieved in large part through the definition of an enterprise data base which contains all of the information which is archived (published) for enterprise-wide use (subscription). This cache of information is not all of the data generated in the enterprise, rather, it is the data which has archival rather than merely transitory value. The contents of the data base includes:

- business (costs, prices, orders, etc.)
- technical (equipment, specifications, P&ID's, physical properties, etc.)
- process transactional (daily production summaries, inventory status, etc.)
- process operational information (temperatures, pressures, flows, etc.)

In general the physical implementation of the enterprise data base need not be monolithic: indeed it is typically distributed.

Application integration is critical to CIPO since multiple applications must normally be orchestrated to achieve desired task outcomes. Thus, the output of one application must be readily transferable to another. For instance, the planning application must communicate with the scheduling application; the planner to the optimizer, the optimizer to the controller, or the process flowsheet drawing application to the flowsheet simulator. Applications typically have their own local data base to which data may be downloaded from the enterprise data base and will generate information which must be selectively published in the enterprise data base. In general, not all information that is transferred between applications is transmitted through the enterprise data base since not all of the output generated by an application is of subscription quality. Considerable progress has been made in application integration through the development of data exchange standards, such as those promulgated by PDXI (Motard, et al., 1994), and the use of object-oriented data structures in the implementation of integrated systems (Kramer, 1995).

A key aspect of application integration is the integration of the software tools and solution methods used by the various applications and tasks and the rationalization of these tools into a manageable and coherent tool set. Most operational tasks do not lend themselves to attack via a single numerical/analytical tool. For instance, fault diagnosis requires the coordinated use of fault trees, rules, neural nets, and mathematical models (Venkatasubramanian, 1994). Scheduling problem solution may require both mathematical programming approaches and rule based systems (Baker, 1994). The encapsulation of solution technologies within each individual application leads to proliferation of different solution methodologies, with attendant maintenance, support, and training problems.

Finally, effective system integration requires that the human users of these systems be treated as an integral system component. *People integration* thus requires that interfaces be designed with the user and not the developer in mind. Users must be provided with consistent access to data, menus and commands, applications, documents, and reports. CIPO systems must be designed with the human as an active agent in the evaluation and decision processes at each task level.

Status of Integration Efforts

In reviewing process integration efforts, it is convenient for discussion purposes to differentiate between batch processing operations and continuous operations. Thus, before reviewing these efforts, we digress briefly to contrast the features of these two modes of operation.

Batch vs. Continuous Operations

While at the aggregate task level the differences between batch and continuous operations are small, at the task detail level and in the relative importance of the different tasks of the functional hierarchy there are substantive differences which definitely have an impact at the CIPO implementation stage. The key distinguishing features of these two operational modes arise from:

- the nature of the production facilities: dedicated versus multipurpose
- the temporal characteristics: continuous flow versus discrete material handling and limited versus wide range of process dynamics
- the level of available process engineering information and models.

The consequences of these differences are quite important and have a direct impact on system design.

In the continuous case, process step interaction occurs through flows and quality/composition of streams. However, reporting of process performance is normally transactionalized. Typically, rigorous engineering models are available for key steps and thus model based control and set-point optimization have an important role. Since plant information is collected essentially synchronously via multiple redundant sensors, reconciliation of measurements to establish the most likely plant status is important. By contrast, in batch operations, process step interactions occur through the timing of the events which are associated with the steps and through quality values of step intermediates. Reporting of process performance occurs on a batch-wise or lot-wise basis. Since step models are usually not available, fault detection and control occur through profile matching. Since equipment is usually shared across processing steps, the recipe description and the equipment network description are distinct and separate. Thus, recipe management is required in addition to equipment management, resulting in separate information structures that must be maintained and coordinated (see Fig. 4). Moreover, since equipment must be assigned to specific recipe tasks in specific time intervals, scheduling becomes an important function.

Process Integration — Continuous Processes

General overviews of the status of process integration efforts are available in Benson (1995), Rippin, et al. (1993), and Pekny, et al. (1991). Efforts in the continuous domain are reviewed by Venkatasubramanian and Stanley (1993) and Venkatasubramanian (1994). As indicated there, the most intense level of activity has been directed at the integration of the lower levels of the operational hierarchy from data acquisition through supervisory control levels. A significant number of these projects have built upon the G2 platform. Some initial efforts have been reported on incorporating the dynamic scheduling tasks as part of the supervisory control level, as described, for instance, by Setpoint at this conference (Bezanson and Fusillo, 1995). However, the planning levels are generally not integrated with the lower levels: rather, the planning tasks is still largely executed as an isolated application. Efforts are underway in several organizations to link enterprise resource planning applications with manufacturing planning/scheduling applications. For instance, DuPont reports progress in integration of SAP and MIMI applications (Beadling, 1995). However, there efforts are yet at the level of assessment of functional requirements,

Process Integration — Batch Processes

The status of integration efforts in the batch domain have been reviewed from the research perspective by Macchietto (1993), Puigjaner (1995), and from the implementation perspective at the World Batch Forum (Craig, 1995). A general architecture for integrated batch operations has been adopted as the SP88 standard (Fisher, 1995). The standard defines the terminology, models, and functionality for the control requirements of batch manufacturing plants as well as the data structures and language guidelines for implementation. The basic batch manufacturing hierarchy which SP88 advances is shown in Fig. 4. Vendors are beginning to offer a limited degree of integration capabilities up through the supervisory level (Benner, 1995; Young, 1995). However, integration of the scheduling level remains at the exploratory and development levels (Puigjaner, et al., 1994; Macchietto, 1995). Moreover, full integration of planning and scheduling functions is yet at the research stage (Pantelides, 1993; Bassett, et al., 1994; Subrahmanyam, et al., 1995) while the integration with enterprise resource planning remains at the conceptual discussion phase. While there appears to be wide recognition of the merits and benefits of integration of batch operations, vendor offerings are as yet fragmented (Brown and Nelson, 1995; Puigjaner, 1995). Moreover, at present, such integration projects are often viewed by manufacturers to be costly, requiring unacceptably high levels of customized implementation, and considerable efforts in re-engineering of manufacturing processes (Adler, 1995).

Based on available literature sources, one must conclude that developments in integration of process operations are being actively pursued by academic research groups, by technology vendors, and by CPI manufacturers. The situation has certainly progressed from the "islands of automation" state which was the rule a decade ago. The islands are now much larger, but fully integrated operations are as yet a rarity.

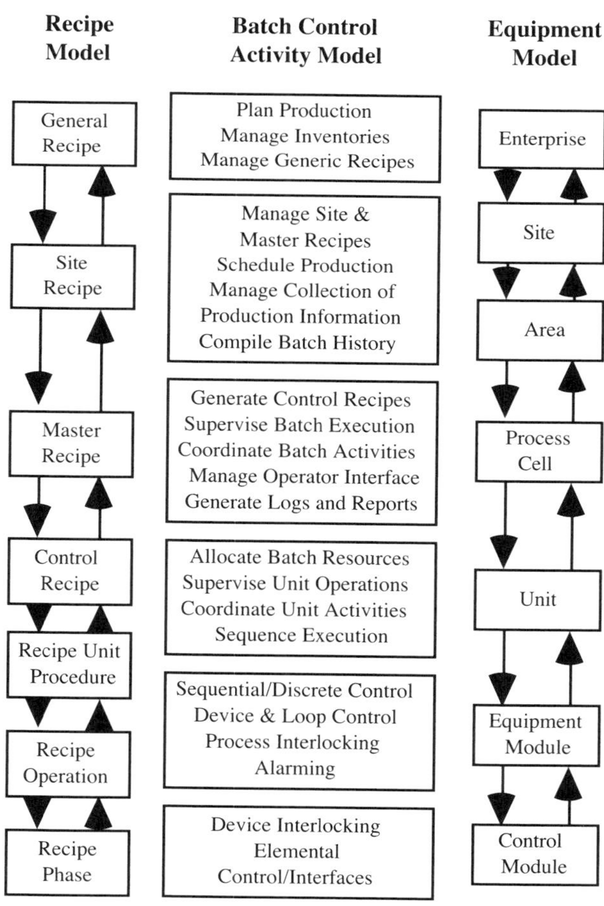

Figure 4. SP88 batch control activity model.

Intelligent Systems in Operations

In order to assess the impact of intelligent systems technologies on process operations, in general, and integration efforts, in particular, it is appropriate to establish what these systems actually are. Any such definition is a moving target as the field has been changing very rapidly, with successive waves of new technologies first fete-ed but then either discarded or else rapidly absorbed into the inventory of conventional computing tools. For present purposes, we paraphrase a definition offered by Stephanopoulos and Han (1994):

> *Intelligent systems: software which automate engineering tasks that are usually carried out by humans and, therefore, expand automation to tasks not covered by traditional numerical algorithms.*

Perhaps a more prosaic, short-lived, but functionally practical approach is to define IS in terms of a specific collection of current tools and methodologies. Following Kramer (1995), such a list of IS components should contain:

- Rule-based systems: including expert systems and fuzzy logic strategies
- Induction and learning methods: encompassing neural networks, wavelets, inductive decision trees, and hybrid systems
- Search methods: for instance, genetic algorithms and constraint directed search
- Representation and implementation methods: including object-oriented technologies, natural language processing, graphical languages, and qualitative simulation.

In addition, the IS domain can be credited with introducing or at least reinforcing software engineering design practices such as the separation of representation from models, procedures, and algorithms as well as the concept of open software systems. Thus, these concepts can be appended to the IS component inventory.

Role of IS in Integrated Operations

Recognizing that fully integrated process operation remains a goal rather than being an achieved reality, the main role of IS technologies in facilitating integration to data has been through the increasing adoption of IS representation and software engineering paradigms. Object-oriented implementation, using languages such as C++ or higher level object oriented programming tools, is now almost a foregone conclusion for any new CIPO development (e.g. Shah, et al., 1995). Object-oriented software environments for linking tools and applications have been introduced in the process systems domain (Ballinger, et al., 1994) and will in the near term see rapid adoption. Graphical object modeling to represent information rich structures such as P&ID's and recipes and equipment networks is used in most contemporary implementations (e.g. Kissling, et al., 1995). Although open systems standards for process operations applications have not been adopted in any subdomain of the process industries, continuous or batch, the need for such standards is widely accepted and their formulation at least at the discussion stage. Thus, the role of IS technologies in integration of process operations has principally been to facilitate the efficient creation and maintenance of large software systems and the development of "smarter" user interfaces.

Specific applications of IS tools have been almost exclusively confined to individual task domains. The use of expert systems and neural networks, separately or in an integrated fashion, for monitoring and diagnosis has proliferated (for a review of diagnosis developments, see Venkatasubramanian, et al. (1994)). The exploitation of neural networks, wavelets, and decision trees for trend analysis of real-time operational data is an area of vigorous investigation (see for instance, Bakshi and Stephanopoulos (1994)). Expert systems and fuzzy logic applications for supervisory control are proliferating in the process control literature and increasingly in practice, addressing issues such as controller tuning, emergency handling, and management of complex control schemes (see Årzén and Åström, 1995). The application of neural networks for model predictive and adaptive control, rightly or wrongly, has spread surprisingly quickly into contemporary control practice (Stephanopoulos and Han, 1994), with many apparently successful studies reported at virtually every control forum held in the last two or three years. Rule-based scheduling has been in use for quite some time, usually in the form of analyst defined and coded problem specific heuristics (Baker, 1993). More recently several commercial products employing expert system tools and constraint derived search-based methods which support the scheduling function have been reported (e.g. Rosenof, 1995; Bezanson and Fusillo, 1995). The use of IS technologies clearly will have a role in the scheduling domain, but the precise role is not yet well defined.

Impact of IS on Operations

The applications literature indicates quite clearly that the impact of IS is strongest in the trend analysis, monitoring, and fault diagnosis tasks. Conventional statistical and numerical techniques alone clearly have not sufficed to advance these task domains: hybrid approaches in which IS methods are combined with more traditional techniques seem to be leading to significant advances. While the volume of process control applications which employ IS, especially neural network, methods is certainly large and still growing, it is not clear whether IS techniques have necessarily impacted advances in that task area to an extent commensurate with the level of activity. Here too it seems that the role of IS will be to complement mathematical models and statistically based methods rather than to displace them.

The scheduling and planning tasks have been an application domain for mathematical programming techniques for quite some time, especially in the petroleum refining business. At the planning level, linear and mixed integer programming models and solution methods continue to hold sway, supported by data base technologies and graphically oriented user interfaces. Because of excessive CPU time requirements, scheduling applications, while perhaps initially posed as mixed integer mathematical programs, have in the past typically been solved using heuristic rules (e.g. Kudva, et al., 1991). Thus, it is not surprising that commercial scheduling tools based on IS methodologies (rule- and constraint-directed search based) have appeared in recent years. However, major parallel progress has also taken place on the mathematical programming side in terms of expressive problem representations, effective formulations, highly efficient solution methods, and sophisticated user interfaces (Zentner, et al., 1994, 1995; Shah, et al., 1995). Moreover, the mathematical programming framework offers the prospects for rigorous linkage between the master scheduling and planning levels which inherently deal with different levels of time and resource aggregation (Bassett, et al., 1994; Subrahmanyam, et al., 1995). However, IS-based techniques such as logic-modeling (Raman and Grossmann, 1993), intelligent branching strategies, and learning mechanisms (Realff, 1992) show considerable promise in enhancing the solution of large MILP-based applications. Furthermore, it would appear that IS-based scheduling approaches are likely to be effective in those short term dynamic (or reactive) scheduling cases where schedule feasibility is the primary concern. Thus, although IS impact on scheduling and planning is as yet quite modest, long-range it will be significant in partnership with mathematical model-based, combinatorial optimization approaches.

Although the role and the impact of IS on the actual development of computer integrated operations systems is modest to date, the potential is quite considerable as indicated by recent innovative research. For instance, Venkatasubramanian (1994) reports on the development of a prototype knowledge-based framework for continuous operations which integrates the lower level process operational tasks. The integration framework, shown in Fig. 5, encompasses the tasks of data acquisition, regulatory control, intelligent monitoring, data reconciliation, fault diagnosis, supervisory control, and communication with the operator. The framework is hybrid in nature: drawing upon a variety of IS tools, first principles dynamic models, statistical analysis methods, empirical knowledge, and operator decision inputs. It uses a blackboard-style architecture, implemented in G2. While exploratory in nature, this development indicates very promising directions for future commercial implementations.

Future IS Roles in Integrated Operations

In this section of the paper, we briefly sketch-out some general issues arising in integrated operations in which IS will have an important role to play and will describe in some detail several specific operational problems in which IS technology could profitably be engaged.

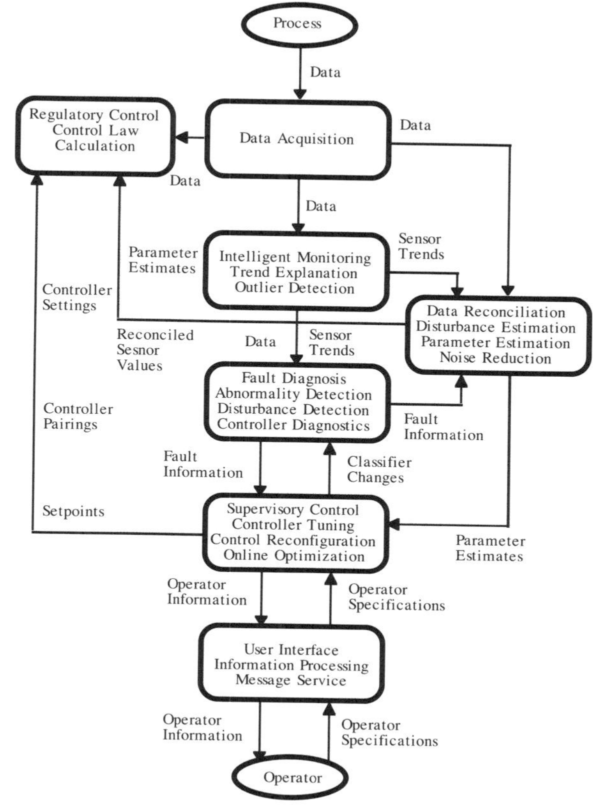

Figure 5. Integration framework.

General Issues

Developments to date have clearly shown that IS technology has had an important role to play in addressing key operational tasks individually. Repeatedly we have seen that process operational tasks do require multiple perspectives and tools. Typically, no single problem view, representation, methodology, or tool has been able to accommodate the complex features of the domain of chemical operations. The key to the future of CIPO is thus the development of hybrid architectures, along the lines proposed by Venkatasubramanian (1994), which allow alternative problem views and solution technologies to be collaboratively engaged and exploited.

A number of additional issues exist which IS technology may be enlisted to address. For instance, IS-based approaches could have potential as intelligent diagnostic agents for identifying processing and economic bottlenecks in the operation and for suggesting corrective actions leading to overall operational improvements. The very interactions which an integrated system seeks to capture and exploit, make identification and tracking of potential process improvements much more challenging.

IS tools also have an important role to play in the development of hypermedia-based and context dependent information management systems, such as smart P&ID's which allow convenient branching to equipment design specs, equipment maintenance history, operating history, shift and operator performance results, and so forth. Management of the enormous oceans of information which support integrated operations will require much effort and creativity in the design of intelligent navigation systems for traversing the information seas.

Considerable scope exists for using IS to help make integrated systems more understandable and valuable to the operator/engineer users. IS-based diagnostic tools must be developed which can extract diagnostic information and suggest corrective actions if infeasibilities or failures are encountered within or across CIPO applications. Furthermore, IS technology will be key to the design of intelligent systems for training and tutoring operators and new engineers. Context dependent computer-aided training is particularly important in integrated operations in order to adequately convey to the trainee the scope and impact of operational decisions on the entire process system.

Examples

We conclude this paper with examples. These examples represent areas under currently active development or scrutiny in the hydrocarbon processing industry (HPI). They are driven by the substantial economic rewards for integration of operations from the supply chain to the delivery of products. Figure 6 shows the location of the examples on the operations hierarchy. The examples are in three broad areas:

- Integrated optimization
- Integration of planning to operations
- Integrated supply and transport.

The discussion of each example area begins with a description of the business issues involved. The discussion then presents some prospective roles for IS in the solution. All examples share the characteristic of a likely preference or requirement for some human interaction in the decision process. The prospective roles visualize IS as a piece of the integrated solution, rather than as the complete solution.

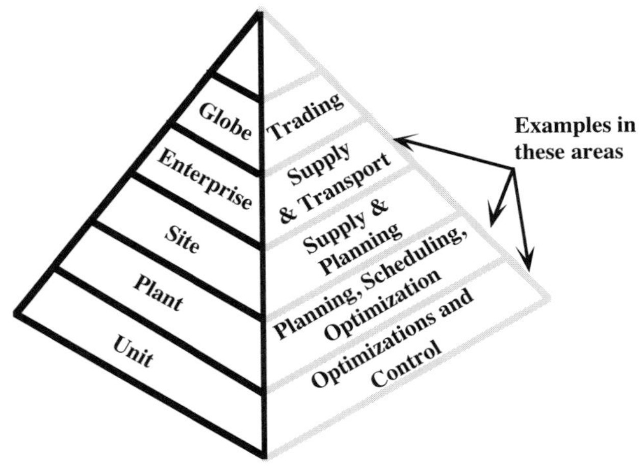

Figure 6. Location of examples.

Integrated Optimization

Issues

The HPI has a substantial and expanding installed base of modular optimizers. These optimize a plant envelope whose boundaries are crossed by one or more streams not traded in external markets. Rather, these streams are used within the plant boundaries.

Figure 7 illustrates modular optimization of a catalytic reforming unit (CRU). In this illustration, neither the naphtha feed nor the reformate product are traded externally. A key question is how to set integrated prices for these streams. We define *integrated prices* as those that will lead the modular optimizer to the same CRU operating conditions as would be found by a business-wide optimizer.

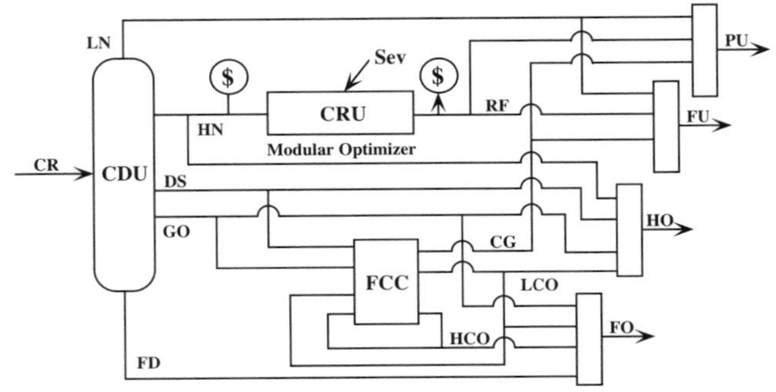

Figure 7. Modular optimization of CRU.

The relative advantages and disadvantages of integrated modular optimization, as opposed to a single business-wide optimization, are outside the scope of this paper. However, Fig. 8 shows that the CRU module may be quite complex all by itself. Crowe, et al. (1995) report that current technology was taxed by the successful real-time optimization of this parallel train of Penex and Platformer, using rigorous process models and detailed constraints.

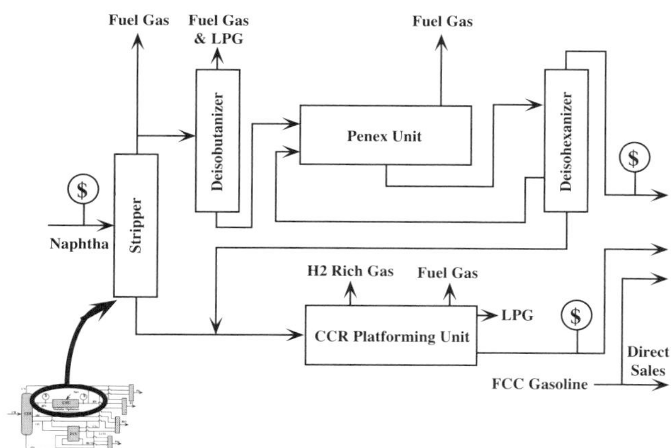

Figure 8. Detailed flowsheet of CRU.

Figure 9 shows an integrating optimizer using the refinery external or gate prices to calculate integrated prices for the CRU modular optimizer. In this case, the integrated prices will lead the modular optimizer to CRU operating conditions that are optimum for the given *refinery-wide* optimization. Figure 10 illustrates a situation where *refinery-wide* optimization must integrate with a higher-level *site-wide* optimization. This optimization hierarchy can extend in both directions. Figure 11 suggests a general structure. A module can be as small as a single piece of process equipment, and as large as several sites or an entire business.

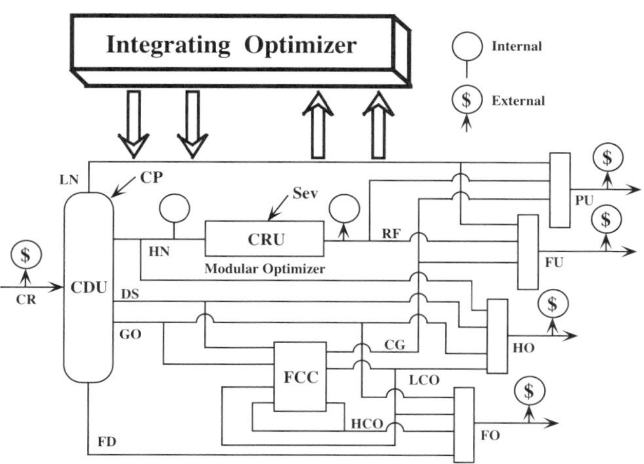

Figure 9. Integrated optimization of CRU.

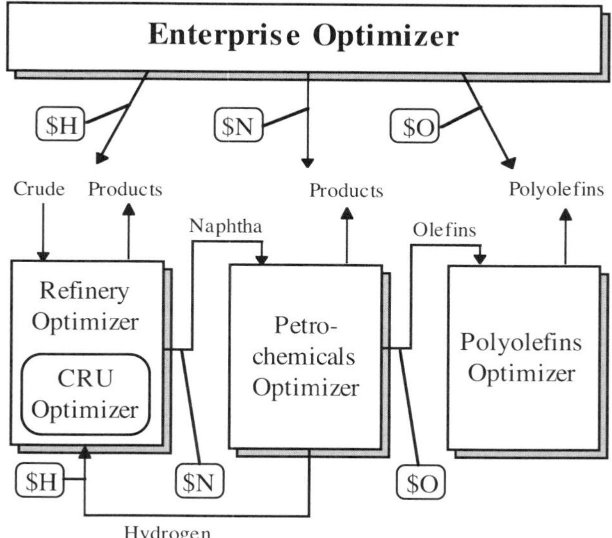

Figure 10. Enterprise optimization.

At the level of Fig. 10, many sites use human business processes to estimate integrated prices for naphtha, olefins, and hydrogen. For example, the refinery may have two broad alternatives for its naphtha pool: process it further to gasoline blending components or supply it to the sister petrochemical plant. The site may use published naphtha spot market prices to guide the allocation of naphtha between these alternatives, and to account for profitability. It may do this even though it never actually sells any of its naphtha on the open market.

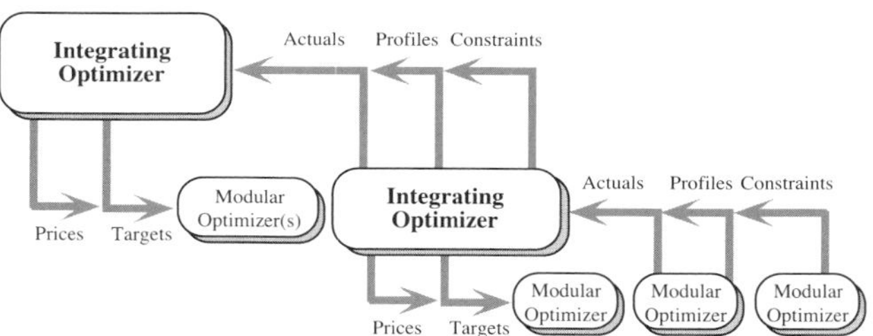

Figure 11. General integrating optimization.

Alternately, plants sometimes use approximation methods to estimate integrated prices for modular optimizers. One method tracks each stream that crosses the module boundary to all its ultimate sources or destinations as products that are bought or sold in external markets. The method then sets the stream price according to the external market prices of these products, allowing for intervening processing costs.

Prices set by these human business processes, or calculated by approximate methods, normally do not integrate. They do not lead the individual module to the correct optimum. The main advantage of the methods is simplicity. The main disadvantage is the potential loss of substantial profit opportunities, particularly because of the large cash value represented by flow rates of typical HPI streams.

True integrated prices can be calculated by existing mathematical programming techniques. The techniques fall into two general classes: hierarchical decomposition and autonomous agents. As with many mathematical programming techniques, the prices represent necessary conditions for an integrated optimization. Hierarchical decomposition is more widely used, and will be discussed here.

Hierarchical decomposition assembles marginal values from the integrating optimizer into integrated prices for the modular optimizer(s). The assembled marginal values come from three main sources: inventory constraints; operating constraints on flows and qualities; and throughput sensitivity of market prices, yields and operating costs. The integrated prices reflect effects of both stream flow and stream qualities.

IS Roles and Prospects

An obvious role for IS in hierarchical modular optimization is solution analysis. The immediate issues are convergence, feasibility, and optimality of the algorithm and solution. The number of variables is large. People are generally involved at the upper portions of the hierarchy, and can benefit from IS assistance in assuring that the optimizer has converged, and that its solution is feasible and at least locally optimal. IS can also assist be recommending remedies where appropriate. Operating companies are already looking at these prospects, not only for integrated optimization, but also for unintegrated production planning tools, typically linear programs (LP's).

Two additional issue that arise in integrated optimization hold prospects for important IS roles in implementation. These are "stability" and pivoting constraints.

We are calling a mathematical program "unstable" if it has locally optimal solutions with radically different structures, but with practically equal objective functions. For example, if not correctly managed by people, an LP can attempt to "turn a refinery upside down" to make a few extra pennies on its profit function. The phenomenon occurs because of the multiple alternatives for disposition of streams, as suggested in Figs. 7-10. These illustrations are in fact much simpler than real plants.

IS can assist by mode recognition. A mode is one specific combination of dispositions active in the optimal solution. Examples of dispositions are: reformate to regular gasoline, light cycle oil to fuel oil, naphtha to spot market, cat cracker feed from spot market, fuel gas to flare, etc. The number of different dispositions in a typical refinery might be on the order of 10 to 50. The potential number of different modes is therefore high. However, it is not nearly in the combinatorial range of 2^{10} to 2^{50}, because most combinations of dispositions are not practically important. The IS can inform the human when the solution mode has changed. Conceivably, it could also provide advice on effective controls for undesirable changes.

Pivoting constraints refers to a change in the set of active constraints in the optimal solution. For example, a small change in crude composition may shift the refinery optimum from a condition where fluid cat cracking (FCC) capacity is limiting to one where heating oil sales are limiting. As plants continue to balance capacities via de-bottlenecking activities, this phenomenon occurs more frequently.

The actual differences between the two operating points may be quite small. Nevertheless, changes in active constraints have large effects on marginal values. The marginal values determine the integrated prices for modular optimizers. IS can assist by recognizing the constraint pivot and analyzing whether it is necessary or worthwhile to adjust the integrated prices.

Finally, there is a potential role for expert systems to advise on the transfer prices in situations such as Fig. 10. Inputs to this system can include spot prices and their elasticity, plant maintenance schedules and history, plant capacity plans and operating costs, spot futures and options markets, distressed cargo information, seasonal variations, etc. Rather than using a mathematical program, IS can make qualitative recommendations on changes from the current situation, such as "small increase in naphtha price and moderate decrease in hydrogen price."

Integration of Planning to Operations

Issues

Figure 12 illustrates the scope area. Operating guidelines are passed down from a planning and scheduling level to the operations level. The planning and scheduling level normally contains an LP, and may also contain a scheduling function. The output from this level is often in natural language. Some examples follow:

1. "heavy reformate 99 RON min"
2. "max feed" or "min vent"
3. "max kero"
4. "feed at 42 Mbbl/d; all LCGO, fill out with CDU2 HSGO"
5. "makeup H_2 as necessary"
6. "draw down resid at 2 Mbbl/d"
7. "build jet to max, T-206/207"

To integrate the operations without computer support, people translate the guidelines to detailed operating targets and orders. Later, people compare actual operations with the planning guidelines. These comparisons may generate much discussion. Are the planning models unrealistic? Or, did we not operate the plant according to the guidelines?

Some operating companies are currently installing information system support for the comparison step. The authors are not aware of any available technology for the translation step.

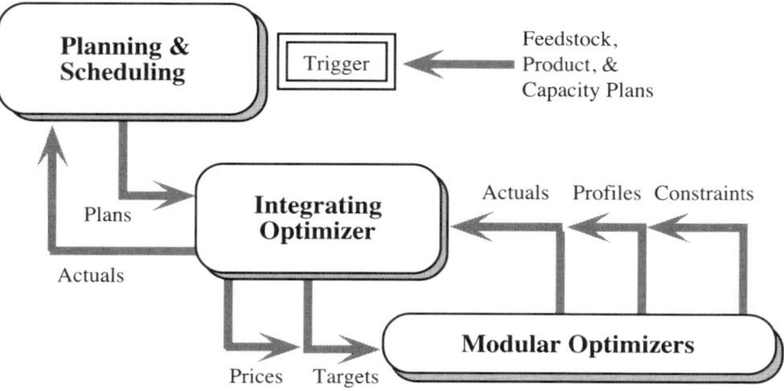

Figure 12. Integration of planning in operations.

IS Roles and Prospects

The translation step is a natural role for IS. Accomplishing this translation will more tightly integrate the operations. The integration will assure that actual operations conform more closely to planned operations, and can produce audit trails and store explanations of exceptions.

Translation of example (1) is straightforward. It is a lower bound on a stream quality.

Example (2) is different. The maximum feed or minimum vent from a process unit may be governed by one, or by a combination, of several secondary detailed constraints, such as column flooding, or reboiler or condenser heat exchange capacity. The planners cannot know in advance which of the many detailed constraints will be active, and do not greatly care so long as the unit feed or vent is always extremized. Therefore, example (2) does not translate to a direct upper or lower bound, but rather to a mode instruction for the supervisory control layer.

Example (3) is similar to example (2) but differs in the complexity of the plan versus actual follow-up comparison. The challenge is to list all the possible detailed operating constraints that could combine to actively limit the kero production from a fractionator. The IS can then assist the follow-up comparison by analyzing the actual operational data and verifying that at least one of the constraints or constraint combinations was active. If not, there is a disconnect between the planning and operational layers. One possibility is that the planning models are inaccurate, and that maximum kero does not produce maximum profit. In this case, the lower level optimizers with their more detailed models may be finding a better operation. A complementary possibility is that the operations control layers are not correctly organized to force the fractionator to maximum kero. IS can help analyze the data, and lead to corrections in the business processes.

Example (4) translates to a mathematical equation relating the setpoints of two flow controllers: the total feed rate to the unit, in this case an FCC; and the flow split of the high-sulfur gasoil from one of the crude fractionators. Supporting this type of translation via IS could be accomplished in a number of rule-based ways.

Example (5) illustrates an operating guideline that may be too vague for IS translation. One alternative is to exclude such guidelines from the IS scope, and continue to process them via operator translation. Another is to revise the business processes so that the guidelines are more specific.

Example (6) can be simple or complex. If there is only one inflow and one outflow to the resid inventory, example (6) is similar to example (4). Otherwise, translation will require sending directives to an operational control strategy that monitors the resid inventory and manipulates flows in some preferred order.

Example (7) is a case where translation requires writing orders for field devices that switch a rundown tank. In this case, the switch is between tanks T206 and T207, and is ordered when T206 has reached its maximum inventory.

To summarize this example area, the suggested roles for IS in integrated operations are:

- translation of planning guidelines to operating orders
- comparison of planned with actuals
- detection of mismatches in models, constraints, or operating strategies.

Integrated Supply and Transport

Issues

As enterprises become more global, the integration of supply and transport with manufacturing and marketing activities becomes more crucial to profitability. Figure 13 diagrams the business issues.

An HPI enterprise has *resources* available to commit to its customers or to its manufacturing facilities. These resources are normally inventories. They may be stationary in tanks at depots or plant sites. They may be mobile, such as in pipelines or carried on ships currently at sea. They may be contemplated, such as product campaigns or batches currently in production, or even currently scheduled for production.

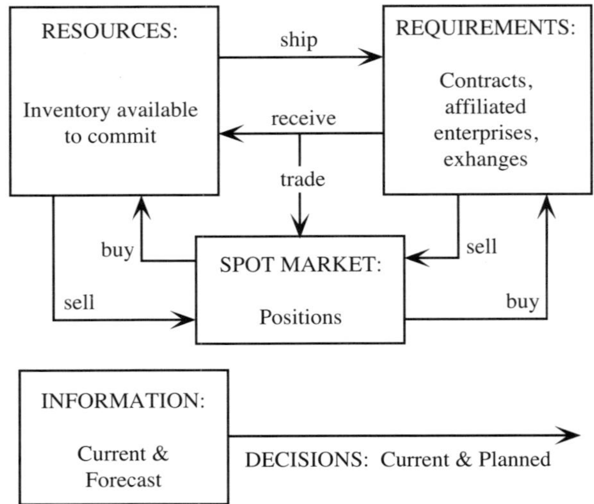

Figure 13. Integrated supply and transport.

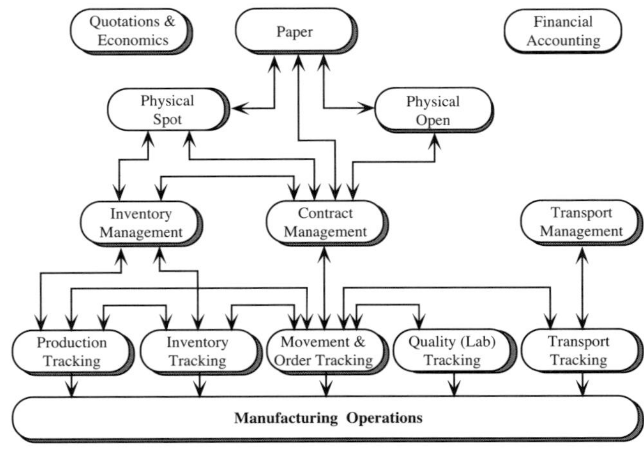

Figure 14. Information system for supply and transport.

The enterprise has commercial *requirements* to meet. These usually result from contracts it has signed to deliver products or to receive feedstocks. Other commercial requirements arise through negotiated exchanges, agreements to partners on processing capacity, etc.

One business objective to supply and transport is to match resources and requirements in a way that minimizes costs, including transport, inventory, and manufacturing. However, the fungibility of HPI products and feedstocks creates additional profit opportunities in the *spot market*. The enterprise can adjust its resources and/or meet its requirements by taking future positions or options on feedstocks or products.

The profit opportunities from well-executed supply and transport are large. Figure 13 suggests that integrated supply and transport has two facets: timely and accurate information about current resources, requirements and positions; and intelligent planning. Planning means defining all the activities shown by arrows in Fig. 13. These activities determine what is to be shipped and received, bought and sold, and from where to where.

IS Roles and Prospects

Figure 14 diagrams an information system that might meet the requirement for complete, timely, and accurate information to the supply and transport activity. The main point of the diagram is that the information requirements are complex, both in terms of scope and detail. Even if such a system were available, it would be too demanding for humans to absorb and comprehend everything in it.

A prospective role for IS is to filter the information. Ideally, humans want to be presented with important changes that have occurred since the last planning activity, and with forecasted problems that arise by straightforward projection of the current plans. They can then focus their examination of the rest of the information on those items that are most likely to be relevant to formulating their supply and transport plans.

Another role for IS is to support the planning activity. Figure 15 diagrams a mathematical model. A *forecast* is an anticipated commercial requirement, and can be either in or out (receive or ship). A *plan* is an anticipated activity that wholly or partly satisfies one or two forecasts, and may involve one or two facilities. A *facility* is a location where inventory is physically stored. A *movement* is an actual physical activity, such as a shipment, that wholly or partly fulfills a plan.

The main point of the diagram is that the business activity is relatively easy to model, and that it is relatively easy to formulate rules for the planning activity. Example rules are:

- A forecast must be satisfied by the sum of all related plans.
- A plan must help fulfill either two forecasts (for example an exchange), or one forecast via one facility (for example a purchase or sale), or no forecasts via two facilities (for example a supply movement).
- Updates to a plan must account for existing movements.

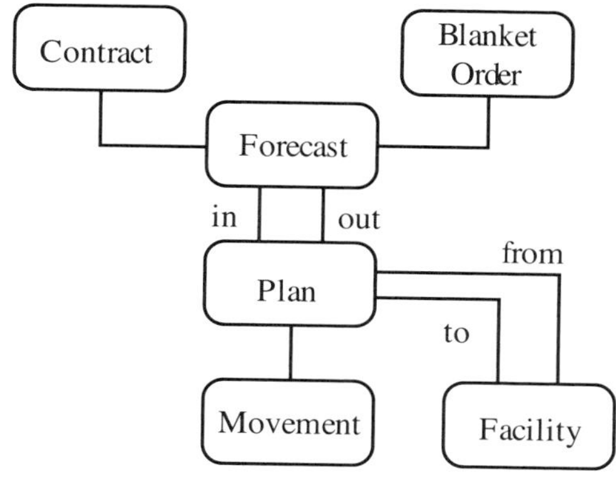

Figure 15. Business activity model.

A role for IS is to assist the supply and trading activity by recommending plans that:

- meet commercial requirements
- reduce cost (sum of transport, inventory, and manufacturing)
- buy low and sell high on the spot market
- manage risk (via hedges).

The IS requirements are similar to the programmed trading rules applied daily in the financial markets. However, the problem is more complex. The manufacturing component is one complicating factor. The reduced level of fungibility of HPI feedstocks and products relative to the fungibility of cash and securities is another. Nevertheless, the economic rewards for good decisions are high owing to the large cash volumes of feedstocks and products handled daily. Small differences in unit prices or costs lead to large differences in profitability. In fact, the level of entrepreneurial activity in pure HPI product trading, without ownership in physical storage and manufacturing facilities, is growing. This activity could reduce HPI profit margins, unless the industry takes advantage of its physical facilities to capitalize on the trading aspects of supply and transport.

Conclusions

The preceding assessment very clearly points to the conclusion that IS tools and methods have had a significant role in advancing process integration. The impact has to date principally been at the task level rather than as an integrating technology per se. If computer integrated operation is compared to a modern composite material, then IS technology can be viewed as the fiber filler which is present in relatively small mass amounts but which has a large synergistic effect both locally and globally in strengthening the entire composite. We anticipate with eagerness future contributions from the IS domain which will help to make CIPO a pervasive reality in the chemical process industries.

References

Adler, D. (1995). Does a MES reduce the cost of production for batch pharmaceuticals? In L. Craig (Ed.), *Proc. World Batch Forum*. Philadelphia, PA.

Årzén, K.E. and K.J. Åström (1995). Expert control and fuzzy control: differences and similarities. *Proc. ISPE'95*.

Bakshi, B.R. and G. Stephanopoulos (1994). Representation of process trends. part III: multi-scale extraction of trends from process data. part IV: induction of real-time patterns from operating data for diagnosis and supervisory control. *Comp. Chem. Eng.*, **18**, 267-302, 303-332.

Ballinger, G.H., R. Bañares-Alcántara, D. Costello, E.S. Fraga, J. Krabbe, H. Lababidi, D.M. Laing, R.C. McKinnel, J.W. Ponton, N. Skilling and M.W. Spenceley (1994). épée: a process engineering software environment. *Comp. Chem. Eng.*, **18**, S283-S287.

Baker, T.E. (1994). An integrated approach to planning and scheduling. In D.W.T. Rippin, J.C. Hale and J.F. Davis (Eds.), *Proc. 2nd Intl. Conf. on Foundations in Computer Aided Process Operations*. CACHE, Austin, TX.

Bassett, M.H., D. Dave, F.J. Doyle III, G.K. Kudva, J.F. Pekny, G.V. Reklaitis, S. Subrahmanyam, D.L. Miller and M.G. Zentner (1995). Perspectives on model based integration of process operations. *Proc. PSE'94*. Kyongju, Korea.

Beadling, W. (1995). Process manufacturers integrate planning and scheduling with business systems. *Chesapeake Tide*, **7**, July issue, 3.

Benner, S. (1995). MES in batch manufacturing: an MES vendor view. In L. Craig (Ed.), *Proc. World Batch Forum*. Philadelphia, PA.

Benson, R.S. (1995). Computer integrated management: an industrial perspective on the future. *Comp. Chem. Eng.*, **19**, S543-S552.

Bezanson, L. and R. Fusillo (1995). Pro-Sked — a technology for building intelligent transportation and production scheduling systems. *Proc. ISPE'95*.

Brown, J. and R. Nelson (1995). Meeting the challenge of automation technology. In L. Craig (Ed.), *Proc. World Batch Forum*. Philadelphia, PA.

Craig, L. (Ed.) (1995). *Proceedings of the World Batch Forum*. Philadelphia, PA.

Crowe, T.J., S.S. Milner, P. Eriksson, T. Wallberg and A.A. Mason (1995). Refinery multiunit optimization: a case study. *1995 UOP Technology Conf.*, San Diego, CA.

Fisher, T.G. (1995). SP88 update — now and the future. In L. Craig (Ed.), *Proc. World Batch Forum*. Philadelphia, PA.

Kissling, J.L. (1995). Flexible software structures and change management. In L. Craig (Ed.), *Proc. World Batch Forum*. Philadelphia, PA.

Kramer, M. (1995). What happened to AI? *Acta Chim. Slov.*, **42**, 101-113.

Kudva, G., A. Elkamel, J.F. Pekny and G.V. Reklaitis (1994). A heuristic algorithm for scheduling multiproduct plants with production deadlines, intermediate storage limitations, and equipment changeover costs. *Proc. 4th Intl. Conf. on Process Systems Engineering*. Montebello, Canada. See also *Comp. Chem. Eng.*, **18**, 859-876.

Macchietto, S. (1994). Bridging the gap — integration of design, operations scheduling and control. In D.W.T. Rippin, J.C. Hale and J.F. Davis (Eds.),

Proc. 2nd Intl. Conf. on Foundations in Computer Aided Process Operations. CACHE, Austin, TX.

Macchietto, S. (1995). Integrating control, planning, and scheduling in batch operations management. Presented at *ISPE'95*.

Motard, R.L., M.R. Blaha, N.L. Book and J.J. Fielding (1995). Process engineering databases — from the PDXI perspective. *4th Intl. Conf. on Foundations of Computer Aided Process Design. AIChE Symposium Series*, **91**, 304.

Pantelides, C.C. (1994). Unified frameworks for optimal process planning and scheduling. In D.W.T. Rippin, J.C. Hale and J.F. Davis (Eds.), *Proc. 2nd Intl. Conf. on Foundations in Computer Aided Process Operations*. CACHE, Austin, TX.

Pekny, J., V. Venkatasubramanian and G.V. Reklaitis (1991). Prospects for computer-aided process operations in the process industries. In L. Puigjaner and A. Espuna (Eds.), *Proc. Computer-Oriented Process Engineering - 91*. Barcelona, Spain.

Puigjaner, L., A. Huercio and A. Espuna (1994). Batch production control in a computer integrated manufacturing environment. *J. Proc. Control*, **4**, 281-290.

Puigjaner, L. and A. Espuna (1995). Prospects for integrated management and control of total sites in the batch manufacturing industry. *Proc. ESCAPE V*. Bled, Slovenia.

Raman, R. and I.E. Grossmann (1993). Symbolic integration of logic in mixed integer linear programming techniques for process synthesis. *Comp. Chem. Eng.*, **17**, 909-917.

Realff, M.J. (1992). *Machine Learning for the Improvement of Combinatorial Optimization Algorithms: A Case Study in Batch Scheduling*. PhD Thesis, MIT.

Reklaitis, G.V. and H.D. Springs (Eds.) (1987). *Proc. 1st Intl. Conf. on Foundations of Computer Aided Process Operations*. CACHE-Elsevier, New York.

Rippin, D.W.T., J.C. Hale and J.F. Davis (Eds.) (1994). *Proc. 2nd Intl. Conf. on Foundations of Computer-Aided Process Operations*. CACHE, Austin, TX.

Rosenof, H. (1995). Dynamic scheduling for a brewery. In. L. Craig (Ed.), *Proc. World Batch Forum*. Philadelphia, PA.

Shah, N., K. Kuriyan, L. Liberis, C.C. Pantelides, L.G. Papageorgiou and P. Riminucci (1995). User interfaces for mathematical programming based multipurpose plant optimization system. *Comp. Chem. Eng.*, **19**, S765-S772.

Stephanopoulos, G. and C.-H. Han (1994). Intelligent systems in process engineering: a review. *Proc. PSE'94*. Kyongju, Korea.

Stewart, T.A. (1995). What information costs. *Fortune*, July 10, 119-121.

Subrahmanyam, S., M.H. Bassett and J.F. Pekny (1995). Issues in solving large scale planning, design, and scheduling problems in batch chemical plants. *Comp. Chem. Eng.*, **19**, S577-S582.

Venkatasubramanian, V. and G.M. Stanley (1994). Integration of process monitoring, diagnosis, and control: issue and emerging trends. In D.W.T. Rippin, J.C. Hale and J.F. Davis (Eds.), *Proc. 2nd Intl. Conf. on Foundations in Computer Aided Process Operations*. CACHE, Austin, TX.

Venkatasubramanian, V. (1994). Towards integrated process supervision: current status and future directions. *Proc. IFAC Workshop on Integrating AI/KBS for Process Control*. Lund, Sweden.

Venkatasubramanian, V., S.N. Kavuri and R. Rengaswamy (1994). Process fault diagnosis — a review. *AIChE J.*, submitted.

Young, S. (1995). Technology: the enabler for tomorrow's agile enterprise. In L. Craig (Ed.), *Proc. World Batch Forum*. Philadelphia, PA.

Zentner, M.G., J.F. Pekny, D.L. Miller and G.V. Reklaitis (1994). RCSP++: a scheduling system for the chemical process industry. *Proc. PSE'94*. Kyongju, Korea.

Zentner, M.G. A. Elkamel, J.F. Pekny and G.V. Reklaitis (1995). A language for describing process scheduling problems. *Comp. Chem. Eng.*, **19**, in press.

AUTONOMOUS DECENTRALIZED SYSTEMS FOR FUTURE PROCESSING/MANUFACTURING

Ichiro Koshijima
CIM Technology Center
Chiyoda Corporation
2-12-1 Tsurumi-chuo, Tsurumi-ku, Yokohama, Japan

Kazuo Niida and Tomio Umeda
Department of Industrial Management
Chiba Institute of Technology
2-17-1 Tsudanuma, Narashino, Chiba 275, Japan

Abstract

From the socioeconomic viewpoint, the concept of the mass-production system is said to be no longer panacea for manufacturing systems in the post-industrial era. A new concept is instead emerging now as the Autonomous Decentralized System (ADS) which both internally manages itself and maintains its order by interactive coordinations among its subsystems, responding to the externally changing conditions. This concept, though immature and its functionalities not fully realized yet, is attracting many researchers with interdisciplinary character. In this paper, upon reviewing systems of the conventional and autonomous plants, the authors present this concept and a new framework of designing and operating an Autonomous Decentralized Chemical Plant (ADChP) under the new paradigm.

Keywords

Autonomous decentralized system, Situation awareness, Swarm intelligence, Reflex based control.

Introduction

Process and manufacturing industries have been a basis for economic growth. After the Ford method was established, the mass-production system has been a life-line of industrial existence. Since the 1950s, new technologies, from pneumatic control devices to digital computers, have been developed to increase the productivity. This system has caused not only remarkable social and economic changes but also concomitant problems common in the industrialized naions as decapitulated below:

- Changes in the market requirements from mass production, to mass customization (various productions in small quantity).
  Shorter lead time in production and diversified demands require higher flexibility and robustness in the manufacturing system.
- Changes in social conditions from the secondary to the tertiary industry.
  Because of shortage of skilled labor and reluctance of younger generations to work in the process and manufacturing industries, sophisticated and automated manufacturing systems are required.
- Changes in environmental requirements from the industry-oriented to the nature-oriented.
  Depleted natural resources and abandoned industrial wastes require the preservation of natural resources and the recycling of industrial wastes. More emphasis should be placed on "function performance index (= life-cycle cost / performance)" than cost performance index (= hardware cost / performance). The reusability of production equipment is essential and would also lead to the increase of the function performance index.

When a certain system becomes more complex, as indicated in the preceding paragraph, it is all the more difficult to function under a single centralized supervisory system. Here is borne a new concept whose functional order is generated only by coopeative interactions among its subsystems without a centralized supervisory system.

In Japan, to solve the above problems, the Intelligent Manufacturing Systems (IMS for short) program was started in 1992 under the commission of the Ministry of International Trade and Industry in Japan. In the IMS program, several new manufacturing systems including the modularized systems and the bionic (holonic) systems have been taken into consideration based on the "Autonomous Decentralized System (ADS)" concept.

The priority research project named "the Decentralized Autonomous Systems" was organized by Ichikawa and conducted by Ito and others in 1990 under the sponsorship of Ministry of Education, Science and Culture. Though 114 researchers joined this project from various research fields ranging from biology to computer science (Ito, et al., 1990), there is no report on the ADS in the chemical processing field as far as the authors know.

In this paper, the authors present a new framework of processing systems' design and operations, applying the ADS concept.

Bachground of New Concepts

Before embarking on the main subject of "ADS in Chemical Industries," allow us to explain a few concepts for clear understanding of the paper. The authors would also refer to architecture in robotics which you are already familiar with and show how it can be applied to chemical industries.

Definitions of ADS

The biological system shown in Fig. 1 is a good example of the ADS. The system has a hierarchical structure of organization. After 4 billion years of the natural selection, a cell acquires an autonomy, such multiple homogeneous cells are coordinating under the

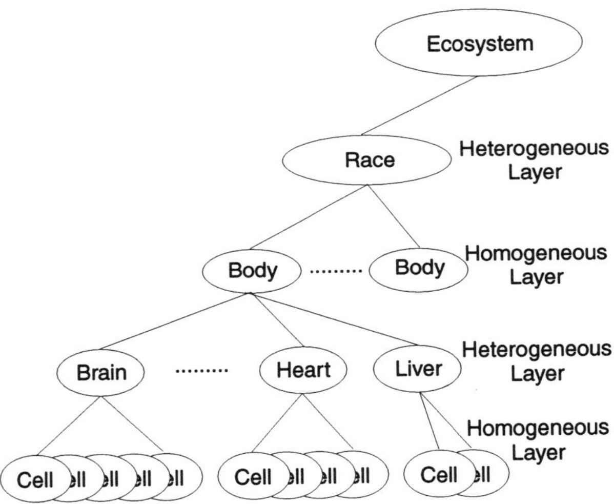

Figure 1. Hierarchy in biological system.

same objective and form a heterogeneous organ. Because of a small deviation in its coordination, every individual has the same kind of organs with different characteristics. It makes a diversity in the race and enhances a flexibility to survive the struggle for existence. In this biological system, homogeneous layers characterize the autonomous decentralized system.

From the viewpoint of systems engineering, Ihara, et al. (1984) proposed the following two properties that characterize the ADS.

Autonomous Controllability:

> This is a property to describe the independence of a subsystem from its surroundings. Each autonomous decentralized subsystem can continue to manage itself and perform its own responsible functions, even though their operating conditions and/or structure changed.

Autonomous Coordinability:

> This property shows the interaction among subsystems for forming a global order in the entire system. Each autonomous decentralized

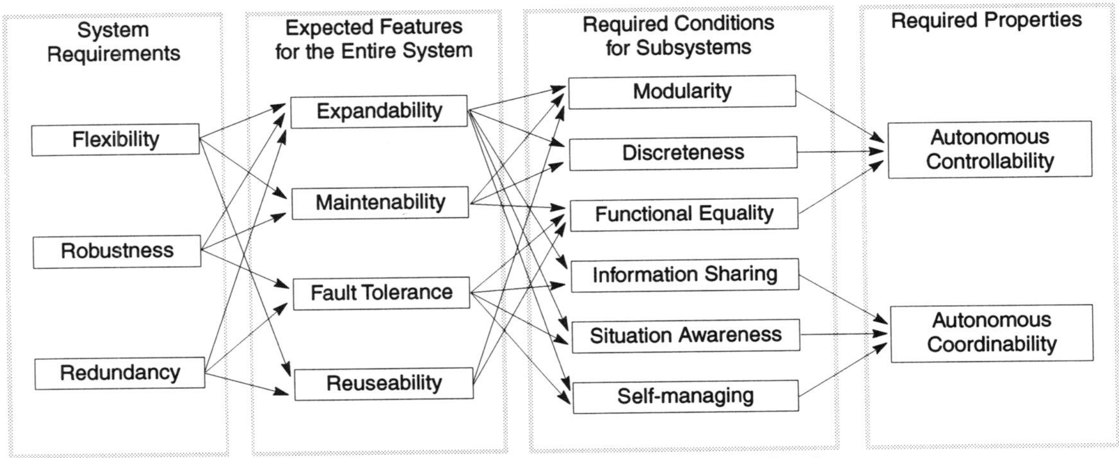

Figure 2. Characteristics of autonomous decentralized system.

subsystem can coordinate its objective among its surroundings, even though their operating conditions and/or structure changed.

Therefore each subsystem in the ADS is designed and operated under the following conditions as shown in Fig. 2.

Modularity: A subsystem is modularized as a self-containing package to contain sufficient mechanisms for self-recognition and self-organization.

Discreteness: To connect and/or disconnect internal and/or external linkages, each subsystem has an ability for handling discrete events and their perturbations.

Functional Equality: Subsystems have the equivalent functions and performance so that they can be replaceable.

Information sharing: Each subsystem sends its information and receives required information without any direction from the information sender. Information exchange for generating the functional order of the entire system is limited among local subsystems, because of the reduction of the communication load.

Situation awareness: Each subsystem contains an interpreter and decision-maker to initiate proper action according to the situation.

Self-managing: Each subsystem contains intelligence to manage itself and to coordinate its role among local subsystems.

Architecture in Robotics

In robotics there are two approaches to realizing an autonomous decentralized system, as shown in Fig. 3.

Multi Robot System (MRS):

This system is based on multiple self-organizing robots. Each robot has sufficient intelligence to execute a given task. Through cooperative work, multiple robots execute an advanced task efficiently. Customized robots for a special task are the key point to enhance the total efficiency.

Distributed Robotics System (DRS):

This system is based on a group of simple robots. Every robot is equipped with the same hardware and software. Though it does not have any specialities, it can be applied to every task. Each robot has situation awareness about the neighbor's behavior, and initiates situated actions. A group of robots makes cooperative work possible (swarm intelligence), though none of the robots has sufficient intelligence to execute a given task alone. The group generates diverse and complex systems to meet the given tasks without a centralized robot. Unlimited number of homogeneous robots is the key point to enhance the flexibility, robustness and redundancy of the entire system.

The DRS is a new concept to overcome the complicated self-organizing mechanism. Characteristics of each concept are shown below.

Item	MRS	DRS
Efficiency	High	Low
Capability	High	Low
Mechanics	Complex	Simple
No. of Units	Small	Large
Control	Complex	Simple
Flexibility	Low	High
Robustness	Low	High
Redundancy	Low	High

Research works related to both concepts are still in progress. Currently, there are no industrial applications under the DRS concept.

Multi Robot System

Intelligent CPU

Intelligent CPU

Communication

- Multiple Self-Organizing Robots
- Sufficient Performance

Distributed Robotics System

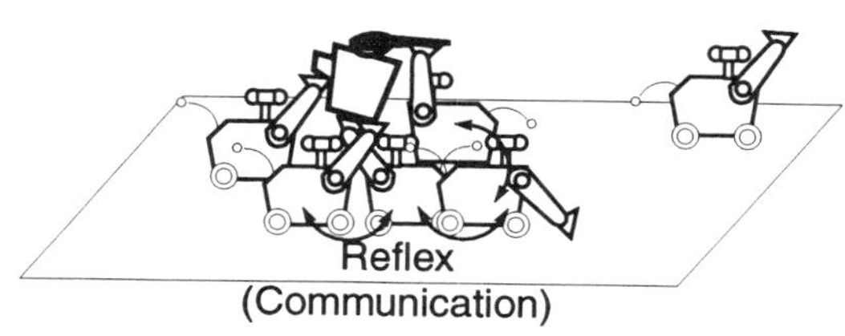

- Group of Simple Robots
- Insufficient Performance
- Reflex Based Swarm Intelligence

Figure 3. Autonomous decentralized systems in robotics.

Autonomous Decentralized Chemical Plant

Architecture in Chemical Industries

In chemical industries, continuous-process plants are designed and operated under the unit operation concept. The unit operation concept is defined to simplify the designing of complicated equipment by applying a single and fixed functionality. It reduces the interaction among unit operations with local and distributed control systems. This independence of the unit operation is a key point to the design and control of a complicated chemical plant, though its heterogeneous, and fixed system structure restricts flexibility, robustness and redundancy.

Though a batch-process plant is still designed under the unit operation concept, it can satisfy some measure of expandability with auxiliary batch-units. It also can adapt to external changes by changing its feedstocks and batch schedule. For example, several plants install a multiple of small package boilers for their utility system, instead of a single large boiler. By starting up or shutting down package boilers, they can control the energy utilization flexibly. The batch process achieves a certain level of decentralized autonomy. Furthermore a pipeless batch-process plant (Tanaka, et al., 1985) introduces mobile (or replaceable) reactors to reduce restrictions of fixed system structure. The pipeless plant maximizes flexibility for mass customization by changing permutations and combinations of prepared subsystems.

The batch-process concept derives such flexibility from the introduction of auxiliary units and discrete events. From the viewpoint of systems analysis, this introduction to the unit operation concept extend the degrees of freedom in time and space. The batch-process requests the following facilities:

- Multi-in-single batch-units for self-containment during batch operation
- Multi-purpose fixed-units for supporting every batch operation
- Centralized control system which makes a hierarchy in the control structure

It is, however, difficult to divide required functions between both units for unplanned changes in external conditions. It is also difficult to allocate required tasks among lower level control subject to the upper level control of hierarchy .

The unit operation concept, and the batch-process concept are similar to the MRS concept in robotics. As shown in Table 1, they have the same weakness for extending their concepts to an autonomous decentralized chemical plant (ADChP).

Architecture of ADChP

In the batch and pipeless plant, the physical size of each batch unit is so large that it enhances the discreteness in time and space. To manage enhanced discreteness, each plant requires a complicated control system. The authors attempt to increase the continuousness in the total structure and its control sequence by reducing the size of

Table 1. Comparison between the Conventional Plants and the Autonomous Plant.

		Continuous Plant	Batch Plant	Pipeless Batch Plant	ADChP
Processing System	Function of Subsystem	Fixed	Fixed	Fixed	Changeable by switching internal structure
	Structure of Entire System	Fixed	Changeable by rescheduling the batch sequence	Changeable by rerouting mobile reactors	Changeable by adding or removing subsystems
Control System	Function	Keep independency	Keep schedule	Keep schedule	Keep autonomy
	Structure	Fixed to each unit operation under the hierarchical control system	Fixed to each batch unit under the hierarchical control system	Distributed with each mobile reactor under a centralized scheduling system	Distributed with each subsystem with an autonomous control system
System Requirements	Expansion	Limited	Possible with auxiliary batch units	Possible with auxiliary mobile reactors	High with homogeneous subsystem
	Maintenance	Offline	Online at batch units	Online at mobile reactors	Real time at each subsystem
	Fault Tolerance	Limited, excepting hot backup units	Possible with auxiliary batch units	Possible with auxiliary mobile reactors	High with homogeneous subsystem
	Reusability of Subsystem	Low	Low	High	High

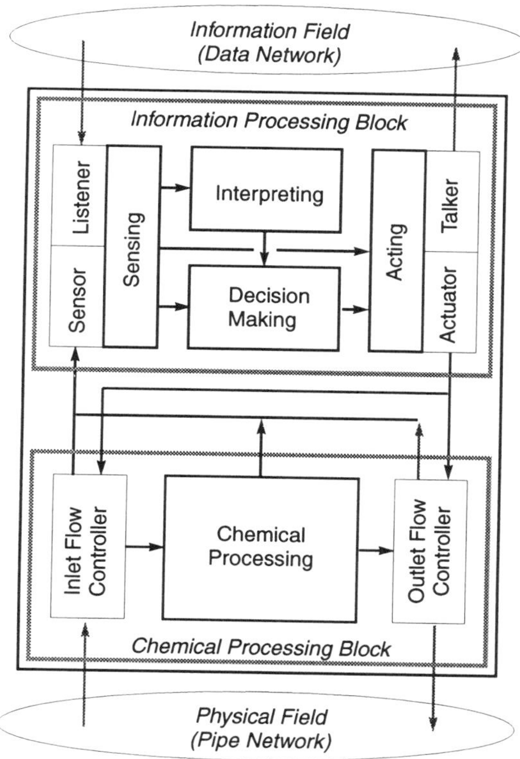

Figure 4. Conceptual structure of autonomous module for chemical processing.

the unit, while keeping the degrees of freedom.

It is, therefore, considered that the DRS is a key concept in the design and control of ADChP. To apply the DRS concept to chemical plants, a conceptual structure of autonomous module for chemical processing (ChPM: Chemical Processing Module) is shown in Fig. 4. The module consists of the following two blocks:

- An information processing block (IPB) which manages itself by mean of internal data and external information.
- A chemical processing block (CPB) where mass transfer and heat exchange operations take place

The IPB has four functions: sensing, interpreting, decision-making and acting. Through so-called listeners and talkers, the IPB is connected to the information field where whole operational information is exchanged. The sensor and the actuator link the IPB to the CPB. The CPB has a multi-purpose chemical processing mechanism with inlet and outlet flow controllers. The CPB is connected to the physical field where other ChPMs are physically coupled to the pipe network.

The general structure of the ADChP implemented with the ChPMs is illustrated in Fig. 5. In the ADChP, pipe connections specify process flows and utility flows among the ChPMs.

There are two types of information fields; the local information field and the generic information field. Through data connections provided along the pipe connections, the local information fields are spontaneously formed to share the local information. Through data connections to the data network, each ChPM reaches the generic information field.

A chemical plant consists of two subsystems, i.e., a process subsystem and a utility subsystem. Four different types of structure are applicable to construct a chemical plant as shown in Table 2. The Type 2 structure is already applied by advanced chemical plants where automatic package boilers act as autonomous modules. In the area of process synthesis, Nishio, et al. (1984) proposed an analytical solution method to figure out structural changes under the optimal condition of heat and power supply systems. In their formulation, a synthesis problem is transformed into a fuel minimization problem. It is

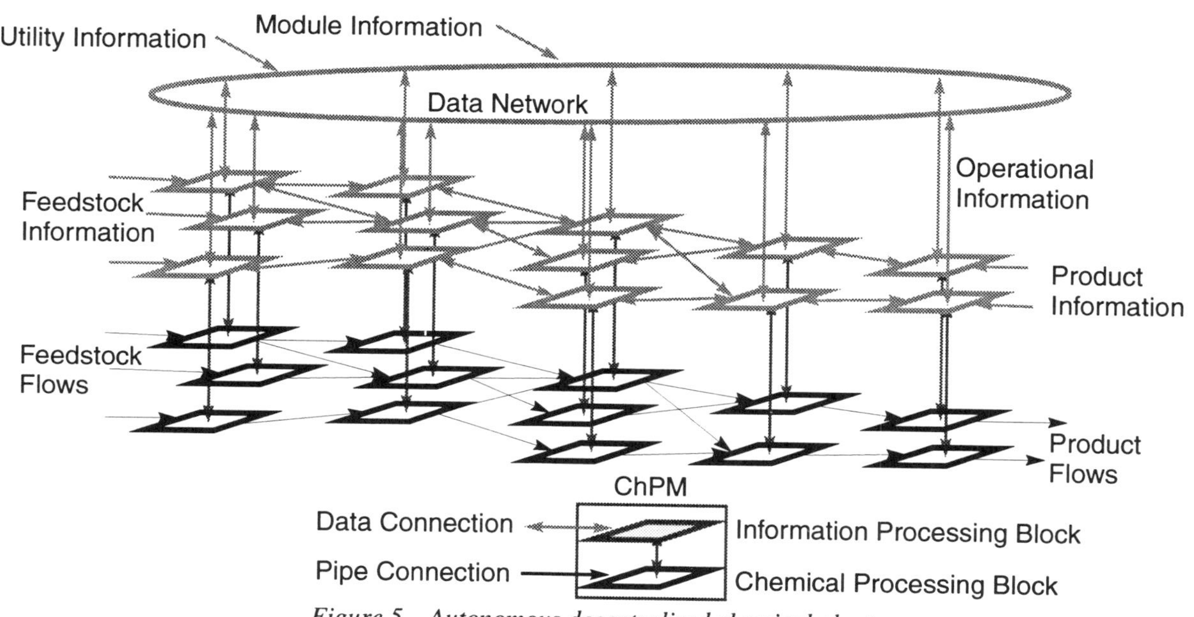

Figure 5. Autonomous decentralized chemical plant.

possible to interpret their approach from the autonomous systems viewpoint. Extending their formulation to optimize the number of the utility modules each having a fixed performance and capacity, a utility subsystem can be designed as an autonomous system.

Assuming above ChPM and the ADChP, it is necessary to clarify the following items;

- How to design and control a chemical plant as a homogeneous system which consists of ChPMs
- How to define the situation awareness and situated actions in the ChPM

Design of ADChP

The equations for designing a ADChP are shown below.

$$\underset{\{M,\ \delta_{ij}^{kn}\}}{Optimize}\ \varphi(\Phi_1,\ldots,\Phi_M) \tag{1}$$

$$\Phi_n = \phi_n(X_{n1}^p,\ldots,X_{nk}^p,Z_{n1}^p,\ldots,Z_{nm}^p,\delta p_{n1}^{11},\ldots,\delta p_{ni}^{mj} \\ X_{n1}^u,\ldots,X_{nk}^u,Z_{n1}^u,\ldots,Z_{nm}^u,\delta u_{n1}^{11},\ldots,\delta u_{ni}^{mj},D_n,P_n) \tag{2}$$

subject to

$$Z_{nm}^p = f_n^p(X_{n1}^p,\ldots,X_{nk}^p,X_{n1}^u,\ldots,X_{nk}^u, \\ Z_{n1}^u,\ldots,Z_{nm}^u,D_n,P_n) \tag{3}$$

$$Z_{nm}^u = f_n^u(X_{n1}^p,\ldots,X_{nk}^p,Z_{n1}^p,\ldots,Z_{nm}^p, \\ X_{n1}^u,\ldots,X_{nk}^u,D_n,P_n) \tag{4}$$

$$X_{nk}^p = \sum_{m=1}^{M}\sum_{j=1}^{J}\delta p_{ni}^{mj} Z_{mj}^p \tag{5}$$

$$X_{nk}^u = \sum_{m=1}^{M}\sum_{j=1}^{J}\delta u_{ni}^{mj} Z_{mj}^u \tag{6}$$

$$g_n(X_{n1}^p,\ldots,X_{nk}^p,Z_{n1}^p,\ldots,Z_{nm}^p, \\ X_{n1}^u,\ldots,X_{nk}^u,Z_{n1}^u,\ldots,Z_{nm}^u,D_n,P_n) \geq 0 \tag{7}$$

where , the sub/superscript p and u show a process module and a utility module, respectively. f_n is a system equation and g_n is a constraints equation for the nth ChPM. δ_{ni}^{mj} shows a connection among the ChPMs as follows;

$$\delta_{ni}^{mj} = \begin{cases} 1 & (j\text{th stream of } m\text{th ChPM is} \\ & \text{connected to } i\text{th stream of } n\text{th ChPM}) \\ 0 & (j\text{th stream of } m\text{th ChPM is not} \\ & \text{connected to } i\text{th stream of } n\text{th ChPM}) \end{cases} \tag{8}$$

The number of modules M and connections among ChPMs δ_{ni}^{mj} are unknown variables and must be determined by an optimization under fixed f_n, g_n and D_n to meet the plant specifications including a range of the product capacity and the utility consumption. In the above equations, if allowable ranges of X_{ni} and Z_{nm} are specified with fixed connections δ_{ni}^{mj}, the modules related to the specified variables are designed as conventional unit operation blocks. In this approach, required specifications are not only the external specifications, such as the feedstock, product and utility specifications, but also the functional specifications of the plant. Under the traditional plant design method, the functional specifications are determined by a process flow diagram with unit operations. It is possible to solve the above equations without the functional specifications by a mixed integer nonlinear optimization technique, such as the genetic algorithm and Hopfield neural networks. To reduce convergence time, it is necessary to supply the functional specifications explicitly.

Table 2. Available System Configuration.

Type	Process Subsystem	Utility Subsystem	System Description
1	Conventional	Conventional	Conventional plant
2	Conventional	Autonomous	Utility fluctuations caused by process operations to meet product specifications and demands are absorbed by the flexible utility subsystem
3	Autonomous	Conventional	Though autonomous reconfiguration of the process subsystem takes place, cumulative utilities are still in certain ranges.
4	Autonomous	Autonomous	Full autonomous plant

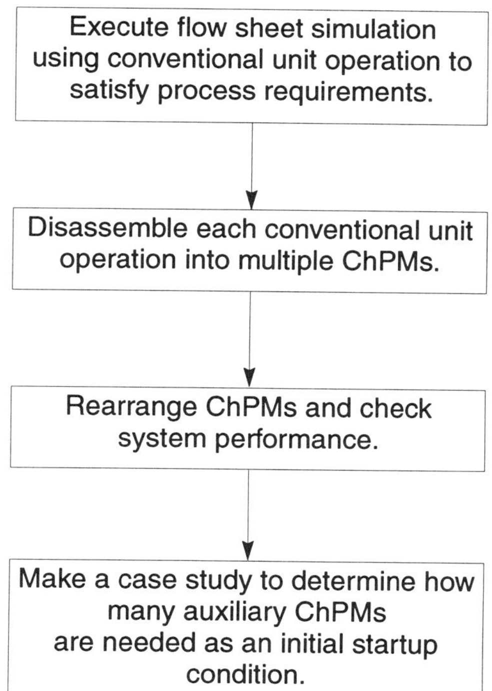

Figure 6. Design logic of ADChP.

Considering a possibility of on-line optimization which creates a global order in an ADChP, it is sufficient to define an initial structure (skeleton) as the result of the design activity. In PSE'94, the authors proposed a design procedure shown in Fig. 6 to define the skeleton structure of the ADChP (Koshijima, et al., 1994). In the first step of this procedure, the system requires a heterogeneous structure using conventional unit operations, and each unit operation is locally decomposed into a set of homogeneous ChPMs. This step defines one of the local minimum solutions, $\delta_{ni}^{mj}|_{init}$ in the solution space. This operation does not affect the system boundary, and it increases the degrees of freedom by the potential application of an unlimited number of ChPMs.

In this logic, $\delta_{ni}^{mj}|_{init}$ which defines a skeleton structure of ADChP and its functional specifications are implicitly set through a conventional flowsheet simulation by a human designer. Starting from the skeleton structure, the entire system structure is autonomously formed as the result of the global order. This is the significant point of the proposed concept. From the viewpoint of system analysis, the ADChP with ChPMs paradigm transforms a global non-linear problem into a number-of-units problem by using a homogenous multiple module. It also integrates the design optimization and the online optimization into the autonomous behavior.

Control of ADChP

As shown in Fig. 4, there are the following three signal paths between the sensing and the acting functions.

Reflex Control Path:
: This path directly connects the sensing function to the acting function. Without any interpreting and decision-making processes, the ChPM immediately generates hard-wired actions against the sensed data.

Conditioned Reflex Control Path:
: According to the decision table, sensed data is transformed into specific action. There are no interpretations for recognizing and forecasting actions among local ChPMs.

Situated Action Control Path:
: Receiving local information including its own information, the ChPM is aware of its situation by the interpreting function. After specifying its situation, the ChPM selects available strategies by the decision making function.

Though the authors (Koshijima, et al., 1992) developed a fault recognition method by learning normal operation patterns, there exist unsolved problems for interpreting process conditions. It is, therefore, difficult to install situation awareness, which includes a fault diagnosis function.

Shibao, et al. (1994) proposed a distributed maximum principle method to realize the optimum collaborative control of the distributed chemical plant. Though their method does not require situation awareness for controlling each subsystem, a whole system structure including the information network is fixed and does not change its configuration autonomously.

To realize the situated control, it is necessary to define a strategy of how to keep the global order under the optimal condition shown in Eqn. (1). The authors wish to propose a load sharing strategy as basic awareness. Under this strategy, each ChPM tries to absorb a deviation caused by internal and/or external changes, and to minimize the propagation of deviation among ChPMs. The strategy has the following stages according to the situation.

First stage: Change Internal Manipulated Parameter
: Each ChPM accepts input changes and keeps the process output states within the allowable or specified range by tuning process variables and utility consumption.

Second stage: Change the External Structure
After the ChPM recognizes that it cannot keep the output states by itself, it changes local structure by adding new ChPM.

The situated control in the first stage is given by the follows equations:

$$\underset{\{P_n\}}{Min} \sum_{n=1}^{M} \sum_{m=1}^{M} \Delta Z_{nm}^{p} \tag{9}$$

subject to

$$\left.\begin{array}{l} \Delta X_{ni}^{p} \neq 0 \\ Pl_n \leq P_n < Ph_n \end{array}\right\} \text{for First Stage} \tag{10}$$

When the process variable P_n reaches to the upper or the lower band, Ph_n or Pl_n, the situated control enters the second stage. To install a new module, there are two types of proliferation.

Proliferation	Violated Outlet State
Parallel	Quantitative (Flow Rate)
Serial	Qualitative (Temperature, Compositions)

After the installation, the control recursively applies the strategy to the newly added ChPM. To minimize Eqn. (9), the outlet deviations are locally back-propagated to upstream ChPMs through the local information field. The detailed explanations will be given elsewhere. (Koshijima, 1995b)

The basic awareness for this control is summarized as follows;

ΔX	ΔZ	Situation	Action
$=0$	$\neq 0$	Internal fault	Send a warning to the data network and add new ChPM
$\neq 0$	$=0$	Success to absorb	Keep operation and send the allowance of P_n to the local information field as its margin
$\neq 0$	$\neq 0$	Fail to absorb	Add new ChPM and send the new structure to the local information field

$$@\left(\begin{array}{l} \text{Inlet deviation: } \Delta X = X_{(t)} - X_{(t+1)} \\ \text{Outlet deviation: } \Delta Z = Z_{(t)} - Z_{(t+1)} \end{array}\right)$$

Example of the ADChP

In PSE'94, the authors proposed a multi-purpose processing module for the constituent processing unit for the ADChP (Koshijima, et al., 1994). To simplify the problem, its basic functions are restricted to the mass/heat transfer capability under a continuous operating condition.

The micro ChPM (mChPM) shown in Fig. 7 consists of two chambers and several flow control units. In each chamber, mixing or separation of streams takes place. Both streams in the chamber can exchange heat. The flow control unit (FCU) has both a transport function and a sensor one acting simultaneously. Unlike traditional control valves, the FCU has only two states, such as an On or Off position. This makes the mChPM a discrete device. Turning FCUs on or off, the mChPM has a simple reflex mechanism and defines its specific function by a combination of internal reflexes. External reflex reaction among mChPMs is the self-organizing mechanism of the ADChP. (Koshijima, et al., 1995a)

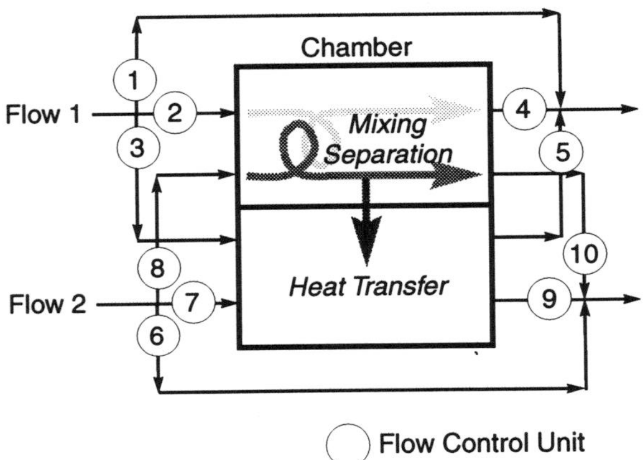

Figure 7. Micro chemical processing module.

Table 3 shows a function chart of the mChPM demonstrating the internal reflexes. One mChPM can cover heat and mass transfer, separation or mixing functions. Eleven pipe lines and four mixers are also formed by the combination of FCUs, though they are omitted in Table 3. It is not feasible to assign an mChPM as a pipe line or a mixer. This possibility, however, may support redundant behavior and heterogeneous growth of the ADChP.

Though a single mChPM cannot execute a given process requirement because of its small capacity, the proliferation mechanism makes a swarm of mChPMs which satisfy the process requirements. Because most mChPMs are operated at maximum efficiency under the proliferation mechanism, it is supposed that the total efficiency of the ADChP at any moment is higher than

Table 3. Function Chart of the mChPM.

mChPM Function	Flow Control Unit No.									
	Flow 1					Flow 2				
	1	2	3	4	5	6	7	8	9	10
Heat Exchanger 1		On		On			On		On	
Heat Exchanger 2		On			On		On			On
Heat Exchanger 3			On	On				On	On	
Heat Exchanger 4			On		On			On		On
Flash Drum 1		On		On						On
Flash Drum 2			On		On				On	
Flash Drum 3					On		On		On	
Flash Drum 4				On				On		On
Absorber 1		On		On				On		On
Absorber 2			On		On		On		On	

that of the traditional chemical plant which is operated under the non-optimal condition.

The mChPM has no self-transportation means. Automatic guided vehicles (AGV) are installed to support a mobility for dynamic structure changes. An image of the ADChP is illustrated by a computer graphics system and is shown in Fig. 8. In this figure, complicated pipe lines are omitted to clearly show the stacked structure of the ADChP. The size of each mChPM shown in this image has 1/4 of 12ft-container to apply petrochemical industries. To meet the DRS concept, the size of mChPM should be determined so that hundreds of modules are utilized as one group.

In the pharmaceutical and the fine chemical industry, the ADChP can be applied, because of its batch-oriented process and essentially less consumption of energy and waste. Comparing to conventional chemical processes, initial investments and operating costs are not a primary objective, and the quality of products and the reliability of production process are more important in these industries. The ADChP approach, therefore, become more feasible in this field.

Conclusions

In this paper, the authors introduced the ADS concept and proposed an Autonomously Decentralized Chemical Plant for a next generation plant. The authors pointed out

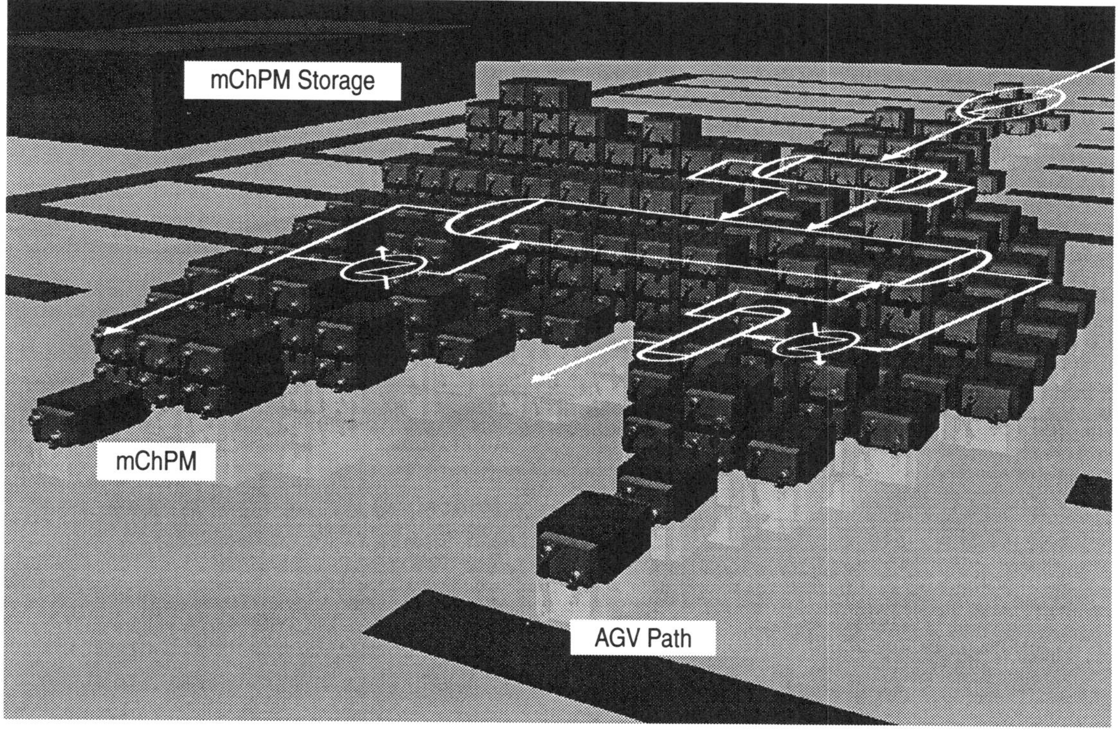

Figure 8. Image of Autonomous Decentralized Plant with mChPMs

that the design problem of the ADChP results in the operation problem. The load sharing strategy for operation is also proposed to minimizing the propagation of deviation among ChPMs. Further studies will be needed in taking other sharing strategies for cooperative work.

In considering recent socioeconomic trends, it is the authors' belief that mass-production system and any plants of large scale now prevailing in the industrial world could not assure their bright future including the chemical plants. It is necessary to change such an old paradigm of designing and operating chemical plants. One of the successful methods to shift this paradigm is to seek the breakthrough of the traditional technology as shown in this paper. Slicing the time and the space of the chemical plant, the authors reached the mChPM-based ADChP.

Though the ADS concept is not fully established and more research is necessary, the authors attempt to disclose our present status of the research so as to stimulate members of the process systems engineering community to pay attention to this new and challenging research area.

Acknowledgments

The authors wish to express their thanks to Chiyoda Corporation for supporting the present study and for permitting its publication. They also express their appreciation to Messrs. I. Harase and T. Higuchi for their contributions in the early stages of this work.

Nomenclature

X_{ni} = the *i*th input vector to the *n*th ChPM
Z_{nm} = the *m*th output vector from the *n*th ChPM
D_n = a decision vector
P_n = a parameter vector

A sub/superscript p and u show a process module and an utility module respectively.

References

Asama H. (1994). Trends of distributed autonomous robotic systems. In H. Asama, et al. (Eds.), *Distributed Autonomous Robotic Systems*. Springer-Verlag, Tokyo. pp. 3-8.

Ihara H. and K. Mori (1984). Autonomous decentralized computer control systems. *IEEE Computer*, **17**(8), 57-66.

Ito M., et al. (1990). Special Issue on Decentralized Autonomous Systems. *J. of the Society of Instrument and Control Engineers (SICE in Japan)*, **26**(1), 1-81.

Koshijima I. and K. Niida (1992). Neural network approach to fault detection under unsteady state operation. *Proc. of IFAC Symp. On Line Fault Detection in the Chemical Process Industries*. Newark, U.S.A. pp. 174-179.

Koshijima I., K. Niida and T. Umeda (1994). A new approach to the design and control of autonomous decentralized chemical plant. *Proceedings of Fifth Inter. Symp. on Process Systems Engineering (PSE'94)*. Kyongju, Korea. pp. 889-892.

Koshijima I., K. Niida and T. Umeda (1996). A micro module approach to the design and control of autonomous decentralized chemical plant. *J. Proc Cont.*, **6**(2/3), 169-179.

Nishio M., I. Koshijima, K. Shiroko and T. Umeda (1984). Structuring of optimal solution space in a certain class of system synthesis. a case study of heat and power supply systems. *Ind. Eng. Chem. Process Des. Dev.*, **23**, 450-456.

Shibao K. and Y. Naka (1994). Optimization of the decentralized systems by autonomous cooperation. In H. Asama, et al. (Eds.), *Distributed Autonomous Robotic Systems*. Springer-Verlag, Tokyo. pp. 29-40.

Tanaka K. and M. Hirayama (1985). Multi-production batch plant oriented control system by KAYAKU "M-POCK-K1." *Chem. Economy Eng. Review*, **17**(4), 32-39.

SESSION SUMMARY: INTELLIGENCE IN INTEGRATED MANUFACTURING

Joseph F. Pekny
Purdue University

Iori Hashimoto
Kyoto University

Introduction

The pace and intensity of change in the international manufacturing environment has provided an enormous impetus for the utilization of new approaches and technology. In particular, the need to simultaneously improve manufacturing efficiency, product quality, and customer responsiveness has motivated an integrated view of manufacturing that encompasses suppliers through customers and a spectrum of time scales from data acquisition through long range corporate planning. The need to develop and deliver tools that support an integrated view of manufacturing has resulted in the use of a variety of intelligent systems (IS) technologies that have now come to be considered mainstream techniques. For example and although mundane, the notion of object oriented programming has become an indispensable paradigm for developing, debugging, and supporting the ever more complex software systems necessary to maintaining an integrated view of manufacturing. Indeed, without object oriented development tools the engineering of manufacturing data systems and higher level software would be prohibitively expensive, if not intractable. This and other examples given in the papers from this session show IS technologies to be an indispensable part of the existing tools for supporting manufacturing. However, the future promises to provide an even larger role for IS technologies.

Consider that fully achieving integrated manufacturing requires a tight coordination among low level activities such as data acquisition, monitoring, and diagnosis; intermediate level activities such as process control and scheduling; and high level activities such as long range planning and supply chain coordination. The paper by Dr. Macchietto provides a discussion of the issues associated with making detailed operations match high level schedules and plans. As his paper shows, one of the key issues is that of data consistency so that information used in higher level models accurately aggregates the many subtasks required in plant operations. In fact, one of the principal lessons of IS technology is the importance of representation to solving problems. Future tools for supporting manufacturing activities will have to use representations that readily provide performance metrics and still quickly respond to changing manufacturing conditions and large volumes of data. The challenge will be developing representations that allow decomposition so that solutions to low level problems contribute to the global optimum. As Dr. Macchietto's work shows, plant activity must be consistent with production schedules and both must be consistent with business planning, otherwise the opportunity exists for overly aggressive plans which can lead to customer service difficulties or excess production capacity which hurts profitability.

The paper by Drs. Reklaitis and Koppel asserts that the ultimate measure of the effectiveness of IS technologies is going to be achievement oriented in the sense of how much capital costs are reduced in both equipment and inventory, the degree of responsiveness in meeting more stringent quality and order to delivery times, and the amount of flexibility inherent in processes. In these terms, existing processes exhibit considerable opportunity for improvement. While the integration of low level manufacturing activities has benefited from the type of automated support provided by IS technologies, the integration of intermediate and high level manufacturing activities has barely begun. A necessary condition to achieving further integration is a greater degree of automation of high level manufacturing activities, for example in the scheduling and planning areas. This follows because the current reliance on manual methods is not well suited for large volumes of information, the short time scales required to implement changes, and the combinatorial complexity inherent in managing highly integrated operations. Largely missing in existing methodology are the frameworks necessary to directly tie manufacturing activities to corporate profits and other gross measures of effectiveness. For example, process scheduling is largely driven to satisfy customer orders but rarely currently considers the various cost factors so that the notion of cost optimal schedules is not yet widely achievable. As the paper by Drs. Reklaitis and Koppel suggests, part of the problem is that companies do not yet have sufficient data to cost individual manufacturing activities and the tools for integrating such data to achieve overall corporate objectives must be still be developed.

Perhaps the greatest contribution of research on IS technologies has been the introduction of new and very different approaches to solving problems. In this regard, the paper by Drs. Koshijima, Niida, and Umeda upholds this tradition by proposing the development of autonomous decentralized systems as an alternative to existing means of designing and operating manufacturing processes. The goal is to create small fully capable and adaptable modules that interconnect using a data network and a reconfigurable material flow network. By analogy to complex biological systems composed of relatively simple and autonomous organisms, the underlying power of the proposed approach lies in the capabilities of large collections of such modules interacting using the material flow and information networks to collectively accomplish manufacturing objectives. The appeal of such an approach stems from the promise of biotechnology and nanotechnology to fabricate highly capable and miniature modules that can be easily combined. Like other proposals in the IS area, many of the ideas of Koshijima, Niida, and Umeda may readily transfer over to conventional manufacturing regardless of the long term disposition of the proposed approach.

With respect to integrated manufacturing, the ultimate goal of computer based support tools must be the ability to promote learning about a process. In particular, tools must unambiguously indicate the level of performance against a rigorous standard and suggest the limitations to improvements. Given that a high level of integration will amplify the effect of changes in market demand, IS technology will be essential to helping human engineers and business managers adapt manufacturing capability accordingly. For the present, the papers of this session show the broad and significant impact that IS technologies are already having on the development of integrated manufacturing systems. With respect to the future, competition in the world economy promises greatly increased pressure on manufacturing systems. Such pressure suggests that the need for advances in IS technologies and how they are applied to support manufacturing will continue for the foreseeable future.

DESIGN AND EVALUATION OF COMPUTER-AIDED PROCESS MODELING TOOLS

Matthias Jarke
Information Systems Dept., RWTH Aachen
Ahornstr. 55, 52072 Aachen, Germany

Wolfgang Marquardt
Process Engineering Dept., RWTH Aachen
Turmstr. 46, 52064 Aachen, Germany

Abstract

The modeling of chemical processes (called c-processes in this paper) is itself a complex process (m-process). C-process knowledge lies largely in the methodology and experience how this m-process is conducted, and cannot be adequately captured by looking just at the resulting c-process models. Traceability, selective guidance of the modeler, and experience-based improvement of the m-process become central issues of c-process modeling environments. From recent work in information systems requirements engineering, we adapt a characterization of m-processes along the three dimensions of representation, domain understanding, and social agreement in the modeling team. This characterization has implications for the software architecture of process modeling environments which can be highlighted by comparing the interaction patterns of the agents in c-processes and m-processes. An environment based on this software architecture has been coupled with the VeDa process engineering model to experiment with a process-centered approach to c-process model development.

Keywords

Chemical process, Computer-aided modeling, Process-centered environment, Modeling process.

Introduction

The application of advanced process technology to the continuous improvement of existing chemical processes and for the development of novel ones is one of the prerequisites for the success or even survival of a business in the chemical process industries which face increasing pressure from environmental and safety regulations, growing demands on product quality and availability, and increasingly competitive markets. A major enabler of advances in process technology is mathematical modeling of the process under consideration, its analysis preferably by simulation, and its integration in model-based methodologies to support the whole process life cycle.

Advances in process engineering demand models of adequate complexity tailored to the requirements of a variety of application areas. A multi-faceted family of models of varying degree of detail is required to adequately support problem solving in its entirety.

The requirements on the sophistication of the models are also increasing. This trend is caused by the steadily growing knowledge on c-process modeling which allows a more accurate prediction of process behavior and hence better solutions to the process engineering problem.

Finally, increasing complexity and variety of the design of process units, which frequently comprise more than one unit operation, contribute to the growing effort of mathematical modeling. Since modular (or block oriented) modeling approaches commonly offered by established process simulators are not applicable without adaptation and modification of the available model libraries in many cases, model development and implementation is a time-consuming and demanding engineering task. In order to spread state-of-the-art process engineering methodologies into routine industrial practice, the effort for development and validation of c-process models needs to be reduced.

Shortcomings of Current Technology

Information technology support for c-process modeling is seen as a major factor in achieving this goal. However, the functionality of modeling tools in established simulation systems is limited in several respects (Marquardt, 1994; Pantelides and Britt, 1994). Recent discussions with industrial modeling practitioners brought up major shortcomings of the modeling technology routinely used in chemical and process industries (Malik, et al., 1994):

1. Model representation should not only include equations but also operations, model assumptions and limitations, information on feasible degrees-of-freedom specifications and on model initialization.
2. Most engineers have problems in formulating non-standard process models by writing equations. The interaction between a modeler and the modeling tool needs to be lifted from the equation level to the knowledge level.
3. Reuse and modification of existing models to derive similar models in an efficient and traceable manner is not adequately supported by any of the established process simulators.
4. The different versions of a model being built during a modeling project for model initialization, model refinement, or different application contexts need to be documented together with their interrelationships.
5. A modeling tool should adopt, store and retrieve available and proven explicit modeling knowledge to be used in guiding the modeling process. In addition, the modeling experience gathered over time by the modeler, forming part of the implicit modeling knowledge, should be captured gradually by the tool.
6. The libraries of predefined c-process models are unsatisfactory in current simulators. The development and storage of families of models for the same process need to be supported to effectively assist different model-based process engineering applications, e.g. by distributed parameter systems in up to three spatial dimensions, population balance models, or ill-defined mixtures of chemicals (polymers, crude oil, proteins, electrolytes, etc.).

Summarizing, current c-process modeling tools do not adequately support the m-process, the reuse and version management of non-standard c-process models such as detailed steady-state or dynamic models of standard equipment nor models for non-standard equipment. Hence, unnecessarily high engineering effort is spent during c-process modeling. As a consequence, the application of model-based process engineering techniques is impeded and, ultimately, competitive advantage is hindered.

Some New Approaches

Several advanced tools of varying maturity are being developed at universities to overcome at least some of the problems with established and routinely used modeling software as identified above. A detailed overview is given in (Marquardt, 1994); below, we provide a brief summary.

The tools may be roughly classified into modeling languages, modeling expert systems, and interactive knowledge based modeling environments.

General *modeling languages* such as `Dymola` (Elmqvist, et al., 1993), `Omola` (Mattson and Andersson, 1993), `Ascend` (Piela, et al., 1992), or `gProms` (Barton and Pantelides, 1994) focus on generic structured means for the declarative representation and hierarchical decomposition of complex mathematical models. Other languages such as `Model.la` (Stephanopoulos, et al., 1990) or `VeDa` (Marquardt, et al., 1993; Bogusch and Marquardt, 1995) include domain-specific concepts on top of application independent language concepts. Though basic methods such as type checking or initialization procedures can be associated with a model formulation, these languages do not support inference methods that actively guide the modeler and critique his decisions.

Modeling expert systems such as `Modex` (Meyssami and Asbjörnsen, 1989) or `Profit` (Telnes, 1992) aim at automatically producing an adequate solution from a formal specification of the modeling problem with minimum user interaction. The modeler needs to state all requirements on the envisioned model and code them in some specification language. Since modeling problems are typically ill-defined as any other design problem, requirements and solution can only evolve gradually during the problem solving process. Hence, in principle, a complete requirements specification cannot be stated a priori. Therefore, the idea of expert systems (at least in its purest form) is unsuited for supporting the m-process.

In contrast to autonomously acting expert systems, *interactive knowledge-based modeling environments* or construction kits (Fischer and Lemke, 1988) support the aggregation of elementary concepts to build the envisioned model. Guidance and critiquing components may be included to support the m-process. Automatic generation of the mathematical models is not intended; rather, an interactive user interface supporting direct manipulation involves the modeler actively into the problem solving process. At the moment, there is no modeling software tool following this paradigm, though prototypical systems show some of the basic features of a construction kit approach (e.g. `Modass` (Sörlie, 1992)).

Overview of the Paper

Though most of the above shortcomings with existing technology are partially addressed by one or the other academic prototype environment, there is yet no consensus on the way modeling support should be offered to an industrial modeling team. This paper therefore presents a novel approach to eventually support all phases of c-process modeling, resulting from an interdisciplinary collaboration of the Process Engineering and Information Systems groups at the Technical University of Aachen.

In contrast to previous work, we emphasize that support should focus on the *process of modeling* (the m-process) as carried out by a team of modelers, rather than on the product of this activity, the mathematical model of the c-process. The next section analyzes the process of model development in chemical engineering building on recent results from software engineering. The derived understanding of the m-process leads to new implications for computer-aided c-process modeling environments in chemical engineering. Our ideas are illustrated by a first experimental implementation of such a modeling tool.

From Requirements Engineering to the Modeling of Chemical Processes

In this section, we use an analogy with software engineering to analyze the modeling process along three important dimensions. This interpretation is a relatively simple abstraction of the real multi-dimensional modeling problem emphasizing the most important coordinates with the goal of highlighting some demands on modeling process support. Nevertheless, it unifies and reconciles a number of crucial aspects of the modeling process.

The basic ideas may carry over to a broader domain of application such as c-process design, with its associated operating procedures. The modeling tool is intended to act as a model server in the sense of Pantelides and Britt for a variety of process engineering software tools in an open CAPE environment. Models can therefore be taken from one application to another, i.e. from simulation during design to control analysis, data reconciliation, or operator training. This is however beyond the scope of the paper.

Requirements Engineering Processes

Requirements engineering (RE) as a task within system development is charged with determining what an initial vague system vision entails and if it can be accomplished in a given technical, cognitive, and social context.

When an RE process begins, problem understanding and the knowledge about the system to be built are still vague, incomplete and opaque. A number of stakeholders (analysts, clients, domain experts, managers, etc.) with different skills and hence their own understanding are involved. Their views are typically represented informally by pictures, graphics or notes in natural languages.

The RE process should reduce these uncertainties to the point where stakeholders share an agreed requirements specification that is sufficiently complete and consistent to drive and evaluate the subsequent development stages. Despite the many open design issues, crucial parts of the specification should be expressed in a formal language to limit ambiguity and to enable computerized analysis.

The discussion indicates that the RE process must be described and supported in at least three dimensions (Pohl, 1994): specification, representation, and agreement.

The *specification dimension* deals with the degree of understanding achieved. Generally speaking, the requirements specification must comprise all major functional decisions to make sure that the system will serve its purpose for the clients. This understanding may be expressed by the aid of standards or guidelines which define a set of canonical concepts to be used to describe the system function and important nonfunctional aspects such as reliability or performance.

The *representation dimension* deals with the different informal, semiformal, and formal notations used to express requirements. In particular, it is important to link related statements expressed in different notations to each other and to informal statements.

The *agreement dimension* captures the degree to which ideas, tasks, and object information is shared in the design team. To accomplish agreement, individual information gathering must be accompanied by selective information sharing and channeling in the team. Information chunks are discussed and evaluated in argumentations. These are concluded by consensus or explicit management decisions.

The RE process is viewed as a trajectory in this three-dimensional space. It shows how the understanding of the system and hence the specification is refined, the system representation is transformed from informal to formal languages, and the personal views of individual team members are reconciled to form a common team view.

Obviously, the specification is apt to change during the whole system life cycle; it is therefore important to maintain the trace in all three dimensions.

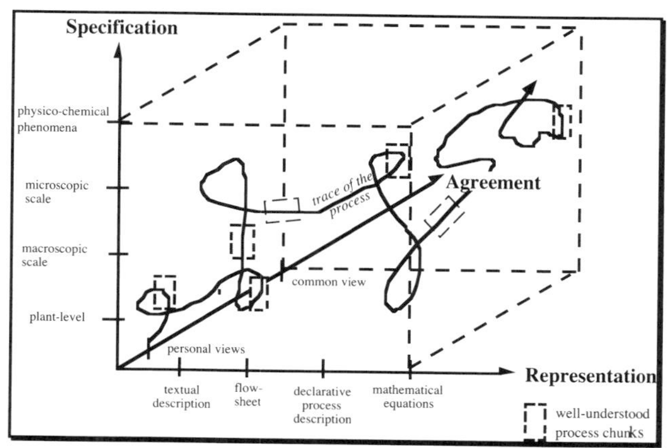

Figure 1. An example m-process for c-processes in three dimensions, adapted from (Pohl, 1994).

The Modeling Process in Three Dimensions

The process of developing c-process models has many similarities to the RE process. Figure 1 shows three dimensions of m-processes, with adapted coordinates for specification, representation, and agreement.

The *specification dimension* relates to the concepts used for c-process modeling on different granularity of c-process structure, behavior, and functionality. A coarse specification looks at the plant level with process units as the major conceptual entities. More detail is provided on the process unit level which for simplicity of presentation is split in two levels of scales. The macroscopic scale refers to a part of a process unit such as a tray of a distillation column or a tube of a catalytic fixed bed reactor. The microscopic scale refers to a volume element in a part of a process unit, such as a catalyst pellet, a gas bubble, or a volume element of a homogeneous fluid. Physico-chemical phenomena such as reaction and inter- or intra-phase transport detail the structural description on the appropriate scale. Concepts on a molecular scale are essential to express the physico-chemical properties of the fluid mixtures processed in the plant.

The *representation axis* shows various formalisms for c-process representations. In the early stages of c-process modeling only informal natural language representations are used, then flowsheets and other schematic drawings are added. Semi-formal and formal languages (e.g. conceptual information or data models) are employed to represent the c-process specification explicitly and completely. Finally, mathematical equations are used to code the behavior of the most detailed model. These equations are implemented and analyzed by some numerical solution engine (e.g. simulation or optimization packages etc.). Ideally, maps are included to transform knowledge coded in one type of formalism into another going either from less formal to more formal or vice versa. Consistency of the various formalisms and the development of partially automatic transformations are important and challenging issues.

The *agreement dimension* spans from the personal views and information resources of all involved people, to the view of the whole team on the final c-process model with all associated information. This view may be the result of an evolved argumentation process between modelers, domain experts, plant operators, etc. leading to an ideally completely agreed and commonly accepted c-process model, or the result of a management decision in case no agreement can be reached on certain issues.

The m-process can be visualized by a trajectory through this problem space. This trajectory may show different characteristics depending on the modeling problem as well as on the composition of the modeling team.

A simple top-down process is illustrated in Fig. 1. It flows from a coarse to a fine granularity of specification, from imperfect personal views to the shared view of the modeling team, and from an informal representation, to the set of equations ready for subsequent numerical analysis.

Obviously, many other starting points and processes are possible. For example, a project might start at the molecular level in order to clarify the physico-chemical properties of the chemicals involved. As another example, a set of equations might be available from a previous project but needs to be explained informally for reuse.

Another complication arises if different parts of the c-process are treated by different kinds of m-processes. In this case, the trajectory must capture a complex composite m-process comprising a large number of sub-processes; each can be understood to reside in a three-dimensional problem space (Lohmann and Marquardt, 1996).

Formal Representation of M-Processes

A formal language for the description of m-processes must be flexible enough to cope with all the above complications. The m-process is a goal-driven but ill-structured and not adequately understood engineering activity. Complete planning or automation is infeasible. However, certain steps in the workflow of modeling are well understood, indicated by filled boxes along the trajectory in Fig. 1. We call these *m-process chunks*.

One example of such a process chunk is the symbolic manipulation of mathematical equations to provide compact model formulation, partial analytical solutions, derivative information etc. (Cellier and Elmqvist, 1993). Other process chunks could support pruning rules for specifying modeling objects and their aggregation (Rozenblit and Huang, 1991) where the data in the if-part are requested from the user by means of dialogue, guidance and consulting.

The observation that m-processes can be supported only by local freely composable process chunks rather than by a complete workflow model has two major consequences when defining a formal *process meta model* defining the language in which process chunks can be specified. Figure 2 shows such a process meta model in the form of an entity-relationship diagram extended with inheritance (isA) relationships (Jarke, et al., 1994).

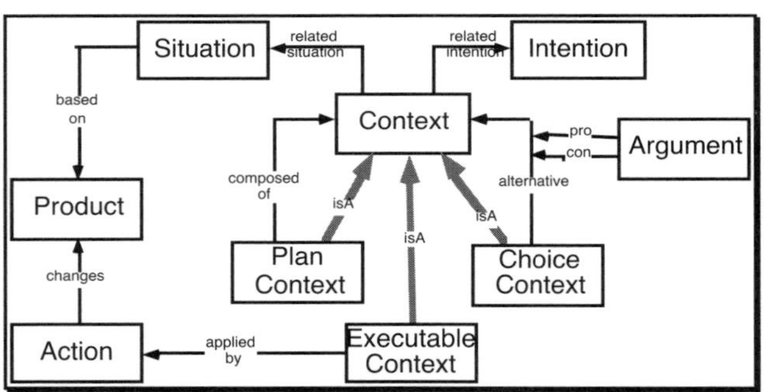

Figure 2. Process meta model for m-processes.

The parceling of the m-process in process chunks necessitates *contextual process modeling*. Not all of the m-process is well enough understood to be guided by computer. This is not even desirable because modeling creativity should be preserved. Therefore, the m-process model needs a means to recognize when such guidance can actually be made available. To achieve this goal, the context specification of a m-process chunk consists of an objective (situation) and a subjective (intention) part.

The objective part is called a *situation* and is based on the documented status of the m-process, i.e. mostly on what part of the product (c-process model) has been completed. The situation of a given m-process chunk thus determines if and on what part of the product the chunk could be invoked at any given moment. For example, the m-process chunk for a parameter estimation method could be linked to the situation that the necessary unit models have been adapted, aggregated to a flowsheet, and compiled.

The subjective part is called the *intention* and describes the modeling goal or task the user wants to achieve with a certain m-process chunk. In the example situation, one intention could be to actually run a simulation, another to change an earlier modeling decision. Both intentions lead to different plans among which the modeler can choose.

In addition to allowing contextual m-process support, the language must distinguish different types of process chunks because the *degree of support* varies. Some m-process chunks can be formalized to the extent of *executable* algorithmic procedures which automatically transform an input specification into the desired result. Others define a *plan* which guides the human modeler. Yet others just give argumentative advice on *choices* among modeling alternatives to be made by human modelers.

Formally, the distinction between these three different types of process chunks is expressed by the isA hierarchy of contexts shown in Fig. 2. In total, a m-process chunk is thus defined as a triple *<situation, intention, content>*. where *content* depends on the subclass of context : *executable contexts* are associated to actions on the product, *plan contexts* to decomposition into subcontexts, and *choice contexts* with alternatives to be considered.

Some First Conclusions

The above view of the m-process shows that knowledge about modeling is not confined to application-specific concepts represented by some formalism as addressed (at least implicitly) in previous work. These concepts form just one of the important ingredients to talk about, and finally to organize the modeling process. It is the m-process in its whole entirety which comprises the modeling knowledge. Only understanding of the m-process can result in improving the performance of human modelers, and in adequate computer-based assistance.

Understanding c-process modeling as an activity in the three-dimensional problem space of Fig. 1 suggests the further development of different kinds of concepts for adequate support and improved performance. These include an extensible set of concepts for the specification of a c-process and its associated mathematical model, a number of consistent representations of varying degree of formality together with transformation procedures, and support methodologies for collaborative decision making.

If concepts along all three coordinates of the cube would be available in sufficient quality, the route to a gradually improving understanding of the m-process and ultimately to better performance is via an inductive learning process: The performance of an experienced modeler, encoded in the series of decisions and steps (s)he has undertaken during a particular modeling project, needs to be recorded and subsequently analyzed. Over time, useful patterns may be extracted from these observations, leading to heuristics for better guidance or even novel algorithmic procedures for certain parts of the modeling process.

Implications for Computer-Aided Process Modeling Environments

Unsurprisingly, due to an inadequate understanding and technological limitations, current modeling software does not reflect the characteristics identified above. One may categorize current tools by means of the cube of Fig. 1 to point out some of the shortcomings.

All of the established tools and most of the recent explorative prototypical developments are confined to the specification-representation plane.

In modular flowsheeting systems like `Aspen+`, `ProII`, or `Hysys`, specification concepts are largely confined to the plant level (process unit models) and to the molecular level (physical property models). Equation-oriented modeling systems like `Speedup`, `Ascend` or `gProms` do not provide additional specification concepts; rather, they offer a richer set of representational constructs (i.e. mathematical equations). More elaborate specification concepts are provided in `Model.la`, `Modex`, `Modass` or `Profit`, further enriched representational capabilities in `ASCEND`, `gProms`, `Modass`, and `Profit`. Still, none of the known systems covers the problem space as defined in the last section to a significant extent.

Novel software tools for computer-aided modeling need to be founded on the understanding of c-process modeling as an ill-structured collection of activities proceeding in a three-dimensional space. This characterization suggests the design of process-centered rather than product-centered modeling environments, in order to focus on the dominating role of the activities performed by the modeler (the m-process) rather than on the model as the result of the activity. The implications on the design of such novel environments are discussed in detail in this section.

Modeling Agents and Patterns of their Interaction

In order to highlight the proposed process-centered view of model development, it is useful to compare the interaction patterns of the agents involved in m-processes to those involved in an operating c-process. C-processes and m-processes have in common that they involve three identifiable domains of agents (Dowson, 1993):

- the *process performance domain* in which the process is actually happening: In c-processes the agents are process units in the plant which are manipulated by a set of actuators on the basis of available sensor information. In m-processes, the agents are the modeling tools which are steered by their human users.

- The *process enactment domain* in which the process is managed with the assistance of some *process engine*: In c-processes the relevant agent is a distributed process control system with all implemented controllers steered by human operators. In m-processes it is a guidance tool steered by the project leader to suggest or enforce a modeling procedure.
- the *process modeling domain* in which the concepts underlying the process are maintained: In c-processes, the agents are the process engineers who define the design of the c-process and its control system as well as the way it is operated. In m-processes it is the systems engineers (or researchers) who prescribe the concepts for the structuring of models and procedures for their development.

Despite these apparent similarities, there are also striking differences between c-processes and m-processes which need to be understood to offer adequate modeling support. We claim that these differences concern mostly the interaction patterns between the three domains (Figs. 3 and 4).

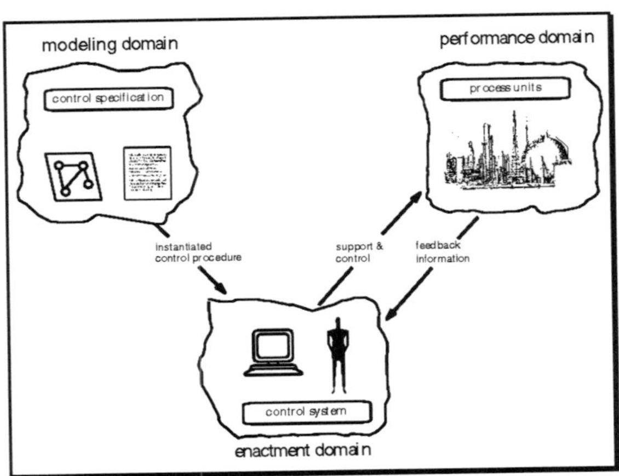

Figure 3. Interaction patterns in c-processes.

In c-processes (Fig. 3), the enactment domain starts a process by selecting a predefined start-up procedure from the modeling domain, and customizing it to the present application (e.g. by fixing set points, controller parameters, and the like). This instantiated process model is often compiled into process control software and drives the performance domain. After start-up, the normal mode procedure is selected and parameterized to drive the process. A feedback loop with the enactment domain is included.

This feedback is not sufficient for m-processes. In contrast to our characterization of the m-process (Fig. 1), it assumes implicitly that the m-process model covers all important aspects of the m-process and that the performance domain has a subservient role, not an active and creative one.

Furthermore, Fig. 3 does not offer enough feedback links to enable continuous learning in the m-process. This learning should be partly based on the experience of the enactment domain (how are m-process models customized? in which parts of a m-process model does frequent feedback indicate problems or opportunities for improvement?), but partly also on experience in the performance domain (are there recurring patterns of success or failure which could be generalized? do people work around existing processes ?).

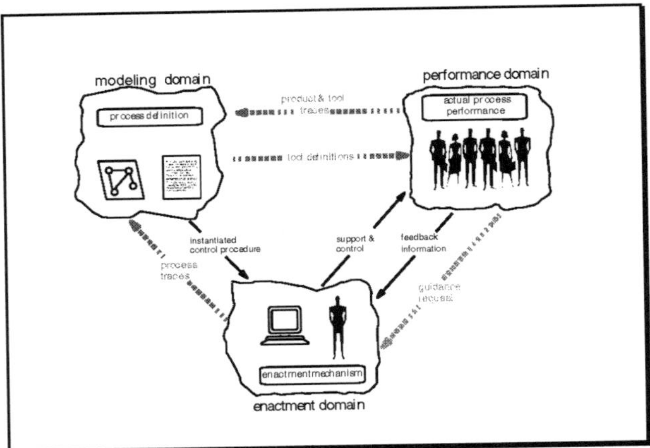

Figure 4. Enriched interaction patterns required for the support of m-processes.

Therefore, an environment for m-processes must offer a richer interaction pattern than one for c-processes. Figure 4 identifies four additional kinds of links (Jarke, et al., 1994):

- *product and tool traces:* Because much of the m-process is not supported by defined process guidance, the performance domain must use its local modeling tools to leave a trace of its actions, including the c-process models (product) it creates or changes, in the modeling domain to enable maintenance and reuse. This trace should record the process along all three dimensions of Fig. 1. In environments that currently support this kind of link, such as KBDS/épée (Banares-Alcantara, 1995; Banares-Alcantara and Lababidi, 1995; Costello, et al., 1995), trace coverage is limited to a fixed dependency model of the m-process.
- *tool definitions:* To write the trace in the modeling domain, tools have to register with the modeling domain and they have to know what to write. Ideally, the kind of trace written should be problem-dependent. This can be accomplished when the trace is structured according to a process model of the services the tool offers. By sending changed tool definitions to the performance domain, the

modeling domain can adapt tool behavior to changing process definitions.

- *process traces:* Process enactment is only active during the execution of some process chunk. Then, the process engine should leave a trace of its activity in the modeling domain. This trace determines the history and status of the overall modeling process. Comparison of the process trace (planned action) and the product and tool traces (actual action) is an additional means of controlling and continuously improving the m-process.
- *guidance request:* Following the idea of situated action and intention in Fig. 2, the modeler should be in the driver´s seat, rather than just giving feedback to a controlling enactment domain. If modelers believe they are in a situation where a process chunk is available, they can ask the process engine for advice which will then retrieve, offer, and on demand enact the selected process chunk.

Process-Centered Environments

So-called process-centered environments are currently being investigated in a number of fields, including software engineering, workflow management, mathematical method and model management, concurrent engineering, and, of course, chemical process design. In terms of supporting software technology, the performance domain is supported by modeling tools such as editors or model transformers, the enactment domain by project management tools or so-called process engines which run process scripts or operational procedures, and the modeling domain by model repositories and associated browsing and editing facilities.

These environments tend to address some of the aspects listed above but still lack a broad coverage. A detailed overview is given in (Pohl, 1995; Pohl, et al., 1996); here, we just summarize the major approaches.

Process-free performance support: Most modeling, CAD, and CASE environments are product- rather than process-centered. They offer modeling tools without a defined way-of-working. Transformation tools may assist in the development of program code from specifications. Design artifacts are stored in a repository. In terms of Fig. 4, only product traces are offered. No process steps can be documented since the enactment domain is not technically linked to the modeling domain or the performance domain.

Document flow: This class includes typical workflow products but also some design environments, such as the requirements traceability product `RDD-100` and in part épée. In addition to the design product, they also trace coarse-grained linkages between design documents. These linkages can concern the workflow in the performance domain, or the dependencies created by transformations.

Process programming The office automation and software process communities pursue solutions very close to the one shown in Fig. 3. The enactment domain fully determines what happens — it "programs" the human modelers (Osterweil, 1987). A few environments, including `EPOS` and `SLANG` (IEEE, 1993), include feedback links from enactment to modeling domain to enable process evolution but still lack the direct link between performance and modeling domain we identified as necessary for capturing creative, ill-defined processes.

Modeling and design as search: Based on design research in the artificial intelligence community (Stallman and Sussman, 1977; Dhar and Jarke, 1988), the m-process is understood as a search process in a large information space, driven by decisions among modeling alternatives (Marquardt, 1992). Dependency-directed backtracking saves rework when these decisions turn out to be wrong . This offers product and tool traces, and often some knowledge about decision alternatives in choice contexts (cf. Fig. 2).

Human-centered process support: The concurrent engineering and human factors communities complement product model capture with performance support for the collaboration aspects of modeling (co-decision, co-ordination, co-operation). The n-dim project at CMU (Levy, et al., 1994; n-dim Group, 1995) is a representative of this type.

The `PROART` *System*

The Information Systems group at RWTH Aachen has developed an experimental requirements engineering environment called `PROART` (Jarke, et al., 1994; Pohl, 1995) which covers all the links in Fig. 4 and offers some support for all three dimensions of modeling in Fig. 1.

`PROART` offers data and m-process integration of tools in the performance domain as shown in Fig. 5. The individual concepts employed during human-computer interaction are supported by graphical or textual tools. In the background, all the information is represented in the deductive and object-oriented conceptual modeling language `Telos` (Mylopoulos, et al., 1990) which is supported by the `ConceptBase` system (Jarke, et al., 1995).

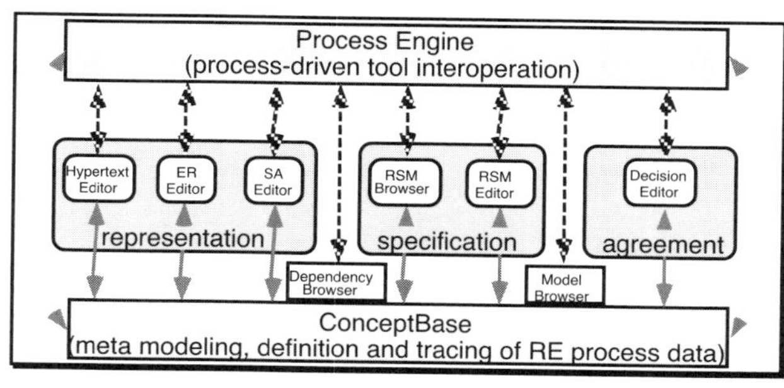

Figure 5. The PROART modeling environment.

The *process modeling domain* implemented in ConceptBase (Jarke, et al., 1995) maintains process definitions, tool definitions, product, process, and tool traces under the process meta model shown in Fig. 2.

The *process enactment domain* is, for simple process steps, handled within the tools themselves. Complex plan contexts which need the contextual interoperation of multiple tools are controlled by a separate process engine.

The modeling and transformation tools of the *performance domain* are sandwiched between the repository and the process engine. Their behavior — the objects shown on the screen and the operations applicable to them — can be adapted to the currently relevant process situation and intention, as defined in active process chunks. Traces are automatically written by the tools but can be additionally annotated by human developers. The tools developed for PROART, thus also the product and tool traces, are grouped according to the three dimensions shown in Fig. 1. They support a range of representations, a structure of pre-defined domain knowledge (e.g. system or organization specific standards), and conceptual models for co-operation, co-ordination, and co-decision. To offer a more global view of the m-process, the repository is equipped with its own monitoring, editing, and browsing tools.

An Experimental M-Process-Centered C-Process Modeling Environment

The flexibility of the PROART architecture allows its rapid adaptation to other domains. Since mid-1994, a version called PROART/CE is being developed to support model generation for c-process simulation.

PROART/CE is enriched by typical representations of c-process engineering, such as flowsheets and equations. The specification dimension is covered by a subset of the VeDa concepts for c-process modeling, proposed by the Process Engineering group at RWTH Aachen.

The discussion of the functionality of PROART/CE is organized around the modeling process in the three dimensional space introduced above.

An initial case study is reported for illustration. The example problem comprises a controlled stirred tank reactor and an evaporator as graphically represented in Fig. 6, using the notation of (Marquardt, 1995) in two of the windows of a PROART/CE screendump. Instead of the development of a c-process model from scratch, the case study focuses on the revision of an already existing model to stress from the beginning the requirements on reuse, modification and documentation.

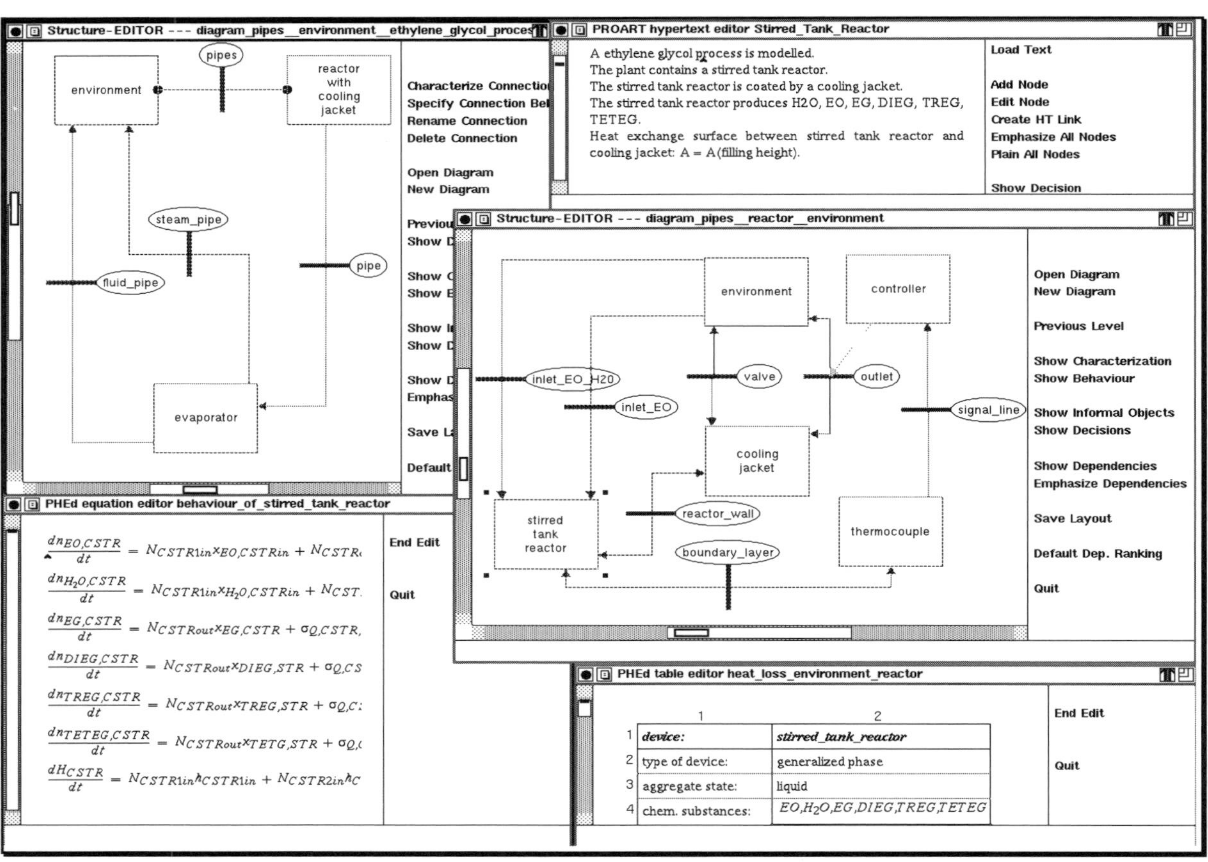

Figure 6. Informal, semi-formal, and formal representations of a stirred tank reactor.

The Specification Dimension

The VeDa conceptualization of c-processes and their associated models has been developed in (Marquardt, et al., 1993; Marquardt, 1994, 1995; Bogusch and Marquardt, 1995). VeDa comprises a canonical type hierarchy of concepts for a structured representation of the c-process with varying degree of detail and granularity.

The structure of the c-process is modeled by two different classes of material entities, devices and connections, distinguished according to the functionality assigned to them during the modeling process. By definition, a *device* stores extensive quantities; its role is the transformation of its internal (intensive) state variables (temperature, pressure, concentrations etc.) according to known sources and fluxes acting from the environment on the device. In contrast, a *connection* by definition never stores any extensive quantity; it provides the fluxes to the adjacent devices according to driving forces determined by the states of those devices. Major specializations of device are the generalized phase capturing all kinds of physico-chemical phenomena explicitly, and the signal transformer which like a black box just maps input quantities to output quantities. The connections are refined accordingly. Devices and connections can form aggregates of arbitrary complexity : complete c-process models can be built up by composition from a small set of elementary concepts.

Part of the structural concept hierarchy is shown as a PROART/CE screendump in Fig. 7 on the lower right together with some general PROART types.

In VeDa, the behavior of every structural concept is distinguished by the phenomena occurring and by their abstraction. The phenomena are encoded by means of process quantities whose values are restricted by some physico-chemical law, i.e. by a set of governing equations. The behavior description of a model is also organized in a hierarchical concept structure. Aggregation is used to compose complex equation sets from elementary concepts. This part of the hierarchy is not yet implemented in PROART/CE. Rather, like in equation-oriented modeling languages, behavior is abstracted by a flat set of equations and associated with each of the devices and connections.

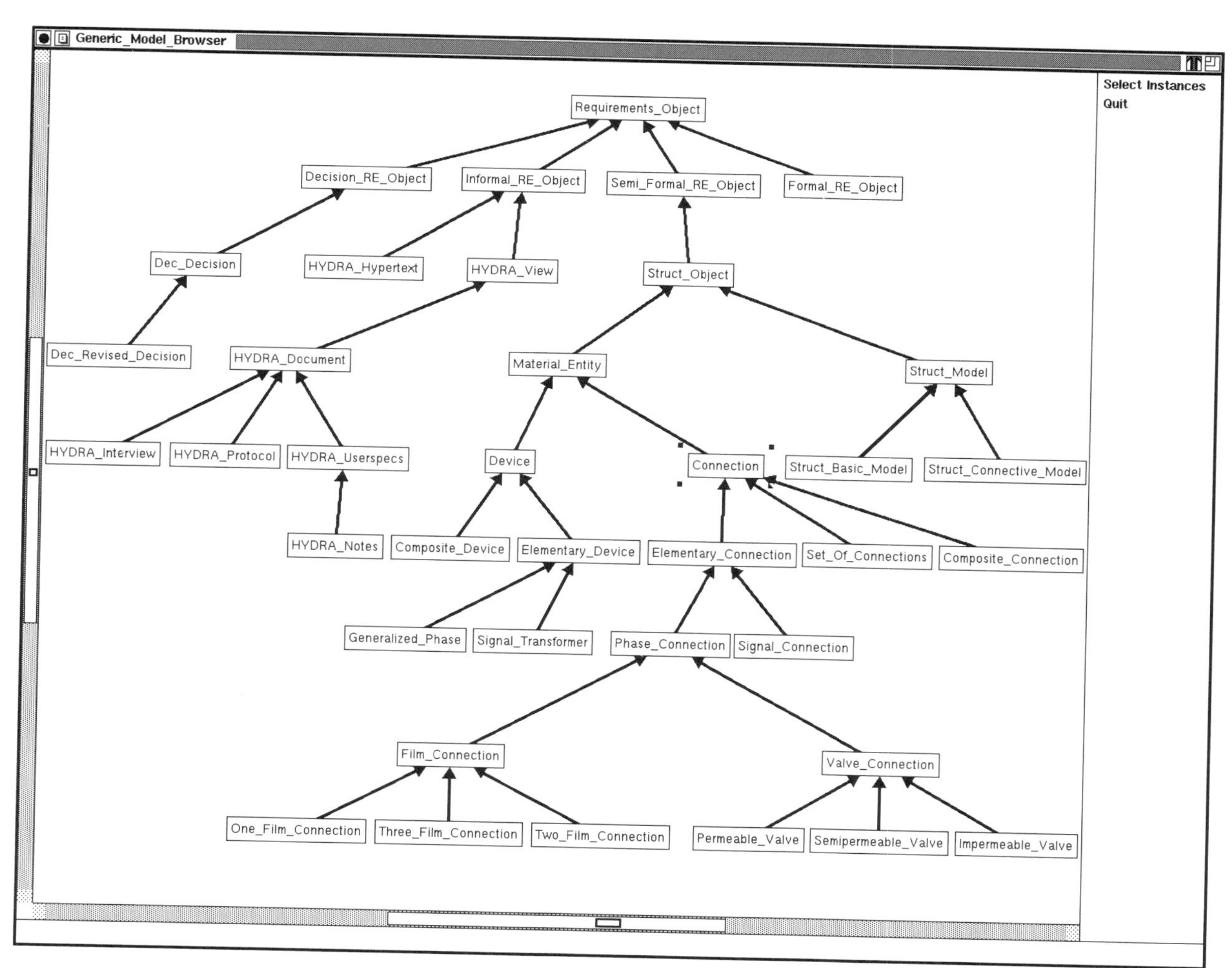

Figure 7. Process engineering domain knowledge captured in a VeDa-based class hierarchy of the repository.

The Representation Dimension

As outlined above, established process simulators use either predefined building blocks (e.g. process units) with a one-to-one mapping to a set of equations and simulation code, or just equations to represent the model with fine granularity without using any domain-specific concepts. In both cases, the documentation of the m-process is insufficient because only certain aspects are captured. Therefore, the m-process lacks traceability. This impedes maintenance and later reuse and modification.

In order to capture also informal knowledge, a hypertext tool is used to maintain statements in natural language as the least formal representation type (Haumer, 1994). As an example, the screendump of Fig. 6 shows in the upper right corner a window displaying a number of statements which have been entered in an early modeling phase during a brainstorming session.

The structure of the process and its associated model is depicted by a flowsheet-like graphical representation of which two hierarchy layers are also shown in Fig. 6. Devices are shown as rectangular boxes whereas connections are given by filled bars and an associated ellipse. The lower right representation is a hierarchical refinement of the upper left. An unlimited number of decomposition and aggregation levels are supported by the tool, in order to graphically build up complex and nonstandard process unit models from the canonical set of basic building blocks. Further information about a flowsheet element can be entered via a form, shown for the *stirred-tank_reactor* at the lower right corner of Fig. 6.

The operations offered in the menu bars on the right of all windows are chosen by PROART/CE automatically depending on the m-process situation. Based on the active m-process chunk, the m-process engine lets the tools indicate only operations which are currently applicable.

In the current version of the prototype, the behavior description is represented by a largely unstructured set of equations for simplicity. As an example, we refer to the lower left window of Fig. 6 which shows some of the equations describing the stirred tank reactor contents abstracted by a device in the flowsheet representation.

All the examples shown illustrate informal or semi-formal conceptual representations. The Telos representations underlying the graphics or texts are outside the scope of this paper but, of course, crucial to define the links between the different representations.

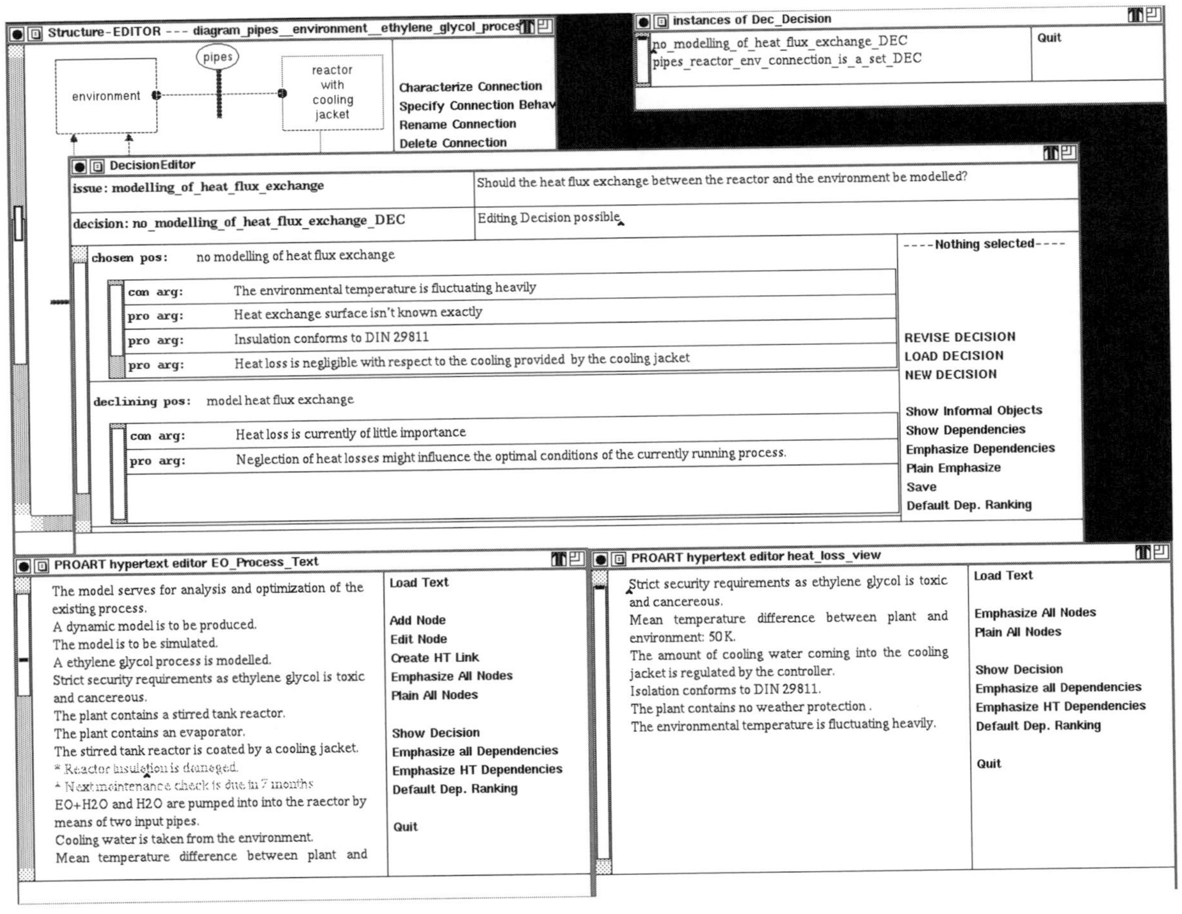

Figure 8. Changing a modeling decision based on traceability back to informal representations.

The Agreement Dimension

In the agreement dimension, a team of modelers typically starts with individual vague ideas which need to be first collected and then structured and agreed on to define the scope of the modeling task. In PROART (Jacobs, 1995), this is supported by tools for collaborative idea collection (brainwriting) and for collaborative idea structuring (following the Japanese KJ technique). The output of such a process is deposited in an informal hypertext as shown in Figs. 6 and 8 (the structuring of the text in nodes is shown by different colors on the screen and does not reproduce well in the black and white of the screendumps). Later in the m-process, this is augmented by coordinating parallel work on shared tasks or shared artifacts (Jarke, et al., 1992), and cooperative decision making when conflicts arise (Jarke, et al., 1987; Pohl and Jacobs, 1994).

From this family of tools, our experiment has used the Decision Editor. This tool, shown in the middle of Fig. 8, documents design decisions using a variant of the Issue-Based Information Systems (IBIS) model, similar to those used in the gIBIS (Conklin and Begeman, 1988) and REMAP systems (Ramesh and Dhar, 1992).

The major advance over these systems is the automatically supported linking of design decisions to their informal rationale (represented in the hypertext editor), and to their consequences in terms of semi-formal and formal product definitions (represented e.g. by flowsheets or equations). This way, backward and forward traceability can be supported in a decision-oriented manner.

Model Revision — The Use of Traces

Using Fig. 8, we continue the example to illustrate the use of decision-oriented traceability for model revision.

Assume that the reactor model in Fig. 6 is used for online optimization of a production reactor. During model development, a modeler decided not to consider heat losses from the cooling jacket to the environment. Figure 8 shows the decision editor which documents the pro and contra arguments to back up the decision. One of the pro arguments refers to an adequate insulation according to the standard DIN 29811. Therefore, heat losses were considered small as compared to the reactor cooling duty. There are dependency links (not shown) between these arguments and the relevant statements represented in the hypertext tool.

The design rationale is formally defined as a structured view on the hypertext which groups the various informal statements related to the heat loss decision. This view — created when making the original decision by simply marking the relevant sentences — is shown in the lower left window. It traces the arguments back to the textual facts entered during the brainstorming session.

Assume that a revision of the model is required since additional facts on the plant status become available. This new situation is defined by entering the new facts into the original hypertext, displayed in the lower left of Fig. 8. The new facts — marked by asterisks (*) — refer to a damage in the insulation of the cooling jacket and a projected maintenance check of the reactor scheduled in only several months time.

The modeler clicks the connection between the cooled reactor and the environment in the graphical representation as shown in the small window in the upper left corner. The decisions related to the specification of the properties of this connection are retrieved automatically and listed in the decision instances window on the top right. The modeler selects the decision related to heat flux and the system prompts with the large window of the decision editor and the heat loss view in the hypertext tool as explained above.

Now, the previous and the current situations are completely transparent to the modeler. A sound decision can be taken on the information available regarding a revision of the model. Obviously the modifications in the decision editor will be propagated by the tool automatically to request additional information and decisions from the modeler to consider heat losses in the model representation.

Conclusions

At the moment, there is limited experience with the novel approach suggested. Its first prototype implementation can be considered a test of feasibility of transferring not only requirements engineering ideas but also corresponding software to the domain of c-process modeling. The simple case study illustrates the potential of a process-centered approach to computerized modeling support. Further development of PROART/CE into an industrial-strength prototype and field studies with industrial modeling teams are mandatory prior to a final evaluation of the approach.

Despite the limited coverage of the current prototype, it shows that the m-process can be captured much more broadly than in previous efforts, considering the three dimensions of specification, representation, and agreement.

The resulting option of complete traceability provides the key to maintenance of existing models in ever changing environments, the reuse of existing models as a starting point of model development in a similar project as well as to quality control of the m-process as well as of the c-process model. Hence, this approach may not only lead to reduced model development times but also to a prolonged usage of a model or model-based application.

The more efficient use of capital investment may significantly contribute to higher competitiveness of a business. Since we aim at a general approach applicable to all kinds of ill-defined design processes, the same ideas and software framework should carry over to other process engineering tasks such as process and control system design, process operations, scheduling and planning etc. If such an extension were feasible, a quite large group of users would be able to take advantage of such technology.

Despite the effort of automating trace capture as much as reasonable, using a process-centered environment does not come for free. Compared to traditional approaches, the modeler will initially need more time to solve a problem due to some "extra" information (s)he has to supply for the trace, such as the design decisions taken and their rationale. Disregarding certain situations in which such traceability may simply be mandated by law (safety-critical systems, environmental protection issues), this extra effort must be viewed as an investment in the future.

The return on this investment is hard to determine as it is characterized by all kinds of external influences: the type of c-process, the market situation, rates of personnel changeover in the modeling team or the management, and the degree of expected reuse. At this point we can only hypothesize, that the time integral of engineering effort over the life cycle of a plant can be drastically reduced by process-centered technology to support the mathematical modeling of c-processes in particular, or all kinds of process engineering activities in general.

Acknowledgments

The authors appreciate the development of the case study and the implementation of the PROART-based prototypical process-centered environment to support chemical process modeling by a cross-disciplinary team consisting of M. Baumeister, R. Bogusch, R. Dömges, B. Lohmann, and K. Pohl. Partial financial support by Deutsche Forschungsgemeinschaft (Postgraduate College "Informatics and Technology" and grant Ma 1188/5-1) and by the European Commission (ESPRIT 6353 NATURE) is gratefully acknowledged.

References

Banares-Alcantara, R. (1995). Design support systems for process engineering I: requirements and proposed solutions for design process representation. *Computers Chem. Engg.*, **19**, 267-277.

Banares-Alcantara, R. and H.M.S. Lababidi (1995). Design support systems for process engineering II: KBDS : an experimental prototype. *Computers Chem. Engg.*, **19**, 279-301.

Barton, P.I. and C.C. Pantelides (1994). The modeling of combined discrete/continuous processes. *AIChE J.*, **40**, 966-979.

Bogusch, R. and W. Marquardt (1995). A formal representation of process model equations. *Proc. ESCAPE-5*. Bled, Slovenia, June.

Cellier, F.E. and H. Elmqvist (1993). Automated formula manipulation support for object-oriented continuous-system modeling. *IEEE Control Systems*, April, 28-38.

Conklin, J. and M.L. Begeman (1988). A hypertext tool for exploratory policy discussion. *ACM Trans. Office Information Systems*, **6**, 303-331.

Costello, D.J., E.S. Fraga, G.H. Ballinger, R. Banares-Alcántara, J. Krabbe, D.M. Laing, R.C. McKinnel, J.W. Ponton, N. Skilling and M.W. Spenceley (1994). épée — a process engineering software environment. Technical Report 1994-19, Dept. Chemical Engineering, Univ. Edinburgh, UK.

Dhar, V. and M. Jarke (1988). Dependency-directed reasoning and learning in systems maintenance support. *IEEE Trans. Software Engineering*, **14**, 211-227.

Dowson, M. (1993). Software process themes and issues. *Proc. 2nd Intl. Conf. Software Process*. Berlin, Germany, pp. 54-62.

Elmqvist, H., F.E. Cellier and M. Otter (1993). Object-oriented modeling of hybrid systems. *European Simulation Symposium*. Delft, The Netherlands.

Fischer, G. and A. Lemke (1988). Construction kits and design environments: steps toward human problem-domain communication. *Human-Computer Interaction* **3**, 179-222.

Haumer, P. (1994). *Structuring and Integration of Informal Knowledge in Requirements Engineering by Means of Hypertext*. Diploma thesis, Information Systems Dept., RWTH Aachen, Germany.

IEEE (1993). *IEEE Transactions on Software Engineering*, Special Section on the Evolution of Software Processes, **19**, 12.

Jacobs, S. (1995). Teamwork support in requirements processes. In M. Jarke, C. Rolland, A. Sutcliffe (Eds.), *The NATURE of Requirements Engineering*. Springer-Verlag, to appear.

Jarke, M., R. Gallersdörfer, M. Jeusfeld, M. Staudt and S. Eherer (1995). ConceptBase — a deductive object base for meta data management. *Journal of Intelligent Information Systems*, **4**, 167-192.

Jarke, M., M.T. Jelassi and M.F. Shakun (1987). MEDIATOR: towards a negotiation support system. *European Journal of Operations Research* , **31**, 314-334.

Jarke, M., C.G. v. Maltzahn and T. Rose (1992). Sharing processes: team support in design repositories. *Intl. J. Intelligent and Cooperative Information Systems*, **1**, 145-167.

Jarke, M., K. Pohl, R. Dömges, S. Jacobs and H.W. Nissen (1994). Requirements information management: the NATURE approach. *Engineering of Information Systems*, **2**, 609-638.

Levy, S., E. Subrahmanian S. Konda, R. Coyne, A. Westerberg and Y. Reich (1993). An overview on the n-dim environment. Report 05-65-93, Engineering Design Research Center, Carnegie-Mellon University .

Lohmann, B. and W. Marquardt (1996). On the systematization of the process of model development. *Proc. ESCAPE-6*. Rhodes/Greece, to appear.

Malik, T., S. Nagel, M. Pons and W. R. Johns (1994). Personal communication. *CAPE Meeting*, Heidelberg, Germany, November 1994.

Marquardt, W. (1992). Rechnergestützte Erstellung verfahrenstechnischer Prozeßmodelle. *Chem.-Ing.-Tech.*, **64**, 25-40. English translation in Int. Chem. Engg., **34**, (1994), 28-46.

Marquardt, W. (1994). Trends in computer-aided process modeling. *Proc. 5th Intl. Conf. Process Systems Engineering*. Kyongju, Korea (to be published in *Comput. Chem. Engg.)*.

Marquardt, W. (1995). Towards a process modeling methodology. In R. Berber (Ed.), *Model-based Process Control*. Kluver.

Marquardt, W., A. Gerstlauer and E.D. Gilles (1993). Modeling and representation of complex objects: a chemical engineering perspective. *Proc. IEA/AIE ´93*. Edinburgh, UK, pp. 219-228.

Mattson, S.E. and M. Andersson (1993). Omola — an object-oriented modeling language. In M. Jamshidi and C.J.

Herget (Eds.), *Recent Advances in Computer Aided Control Engineering*. Elsevier, pp. 291-310.

Meyssami, B. and O.A. Asbjörnsen (1989). Process modeling from first principles — method and automation. *Proc. Summer Computer Simulation Conf.* Austin, TX, pp. 292-299.

Mylopoulos, J., A. Borgida, M. Jarke and M. Koubarakis (1990). Telos: a language for representing knowledge about information systems. *ACM Trans. Information Systems*, **8**, 325-362.

n-dim Group (1995). n-dim — an environment for realizing computer-supported collaboration in design work. Report 05-93-95, Engineering Design Research Center, Carnegie-Mellon University, Pittsburgh, Pa.

Osterweil, L. (1987). Software processes are software, too. *Proc. 7th Intl. Conf. Software Engineering*. San Francisco, CA.

Pantelides, C.C. and H.I. Britt (1994). Multipurpose process modeling environments. *Proc. Foundations of Computer-Aided Process Design Conf. (FOCAPD ´94)*. Snowmass, CO.

Perkins, J.D., R. W. H. Sargent and R. Vàsquez-Romàn (1994). Computer generation of process models. *Proc. 5th Intl. Conf. Process Systems Engineering*, Vol. 1. Kyongju, Korea, pp. 123-125.

Piela, P.C., R.D. McKelvey and A.W. Westerberg (1992). An introduction to ASCEND: its language and inter–active environment. *J. Man. Info. Sci.*, **9**, 91-121.

Pohl, K. (1994). The three dimensions of requirements engineering: a framework and its applications. *Information Systems,* **19**, 243-258.

Pohl, K. (1995). *Process-Centered Requirements Engineering*. John Wiley Research Studies Press, to appear.

Pohl, K., R. Dömges and M. Jarke (1996). Tool integration in open CAPE environments. *Proc. ESCAPE-6*. Rhodes/Greece, to appear.

Pohl, K. and S. Jacobs (1994). Concurrent engineering: enabling traceability and mutual understanding. *Concurrent Engg. - Research and Applications,* **2**, 4.

Ramesh, B. and V. Dhar (1992). Supporting systems development by capturing deliberations during requirements engineering. *IEEE Trans. Software Eng.*, **18**, 498-510.

Rozenblit, J.W. and Y.M. Huang (1991). Rule-based generation of model structures inmultifaceted modeling and system design. *ORSA J. Computing*, **3**, 330-344.

Sörlie, C.F. (1990). *A Computer Environment for Process Modeling*. Doctoral dissertation. Laboratory of Chemical Engineering, Norwegian Institute of Technology, Trondheim, Norway.

Stallman, R. and G. Sussman (1977). Forward reasoning and dependency-cirected backtracking in a system for computer-aided circuit analysis. *Artificial Intell.*, **9**, 135-196.

Stephanopoulos, G., G. Henning and H. Leone (1990). Model.la: a language for process engineering. Part I and II. *Comput. Chem. Engg.*, **14**, 813-869.

Telnes, K. (1992). *Computer-aided Modeling of Dynamic Processes based on Elementary Physics*. Doctoral dissertation. Division of Engineering Cybernetics, Norwegian Institute of Technology, Trondheim.

COMBINING ARTIFICIAL INTELLIGENCE AND OPTIMIZATION IN ENGINEERING DESIGN: A BRIEF SURVEY

Jonathan Cagan, Ignacio E. Grossmann and John Hooker
Engineering Design Research Center
Carnegie Mellon University
Pittsburgh, PA 15213

Abstract

The objective of this paper is to propose generalized representations of engineering design models that involve quantitative and qualitative aspects. Solution methods are surveyed, and AI and OR models are classified by their positions along generality, structured/unstructured and numeric/qualitative axes. Research directions are identified that combine AI and OR by moving models away from the poles of the axes. Finally, some design problems are presented to illustrate various ways in which AI and optimization can be combined for tackling engineering design problems. This is a condensation of a longer paper.

Introduction

Over the last few years there has been great interest in developing solution approaches to design problems that rely on combining AI and Optimization. These efforts have been motivated by the fact that most work in engineering design and synthesis has involved the development of methods that either rely only on AI techniques (commonly expert systems), or only on optimization techniques (commonly NLP and MINLP models). It has become evident that there are several shortcomings in approaching synthesis problems with only one methodology (e.g. see Fenves and Grossmann, 1991). The major problems for the AI-based methods are difficulties in integrating qualitative knowledge with analysis models that are used in engineering design, and accounting for interactions of design decisions. The major problems for the optimization-based methods are difficulties in making use of engineering knowledge that is not expressible in the form of equations, and limitations in solving large-scale problems.

It is the objective of this paper to discuss some of the insights and concepts that have emerged in the last few years in the combination of AI and Optimization techniques for dealing with qualitative and quantitative issues in engineering design problems. We propose generalized representations of engineering design models as well as a taxonomy of solution methods. We then explore how AI and OR models may be combined by analyzing how they differ. We arrange models in a three-dimensional cube, where the axes correspond to specific-general, numeric-symbolic and structured-unstructured polarities. OR and AI models tend to lie at opposite corners of the cube. We indicate several strategies for combining AI and OR by moving models toward the center of the cube. Finally, we illustrate how AI and optimization can be combined for tackling some specific engineering design problems.

This paper is a condensed version of a much longer paper (Cagan, et al., 1995).

Models and Solution Methods for Engineering Design

Conceptually, we can represent computational models for design as follows. Let the *continuous variables* be x and the *discrete variables* be y. The physical or cost parameters which are normally specified as fixed values are represented by ϑ.

Equations and inequality constraints can be represented as the vectors of functions h and g, that must satisfy,

$$\begin{aligned} h(x,y,\vartheta) &= 0 \qquad (1) \\ g(x,y,\vartheta) &\leq 0 \end{aligned}$$

It should be noted that the representation in Eqn. (1) could range from a relatively simple model to a very complex model that tries to capture in as much detail as possible the underlying physics of the system. The solution to the equations in (1) are often not uniquely defined since they commonly involve several degrees of freedom.

The logical relations that define symbolic relations will be given by a set of propositions that must hold true; that is,

$$L(x,y,\vartheta) = \text{TRUE} \tag{2}$$

Finally, the design goal (or goals) can be expressed as the objective function $F(x,y,\vartheta)$. This function is a scalar for a single-criterion optimization, and a vector of functions for a multiobjective optimization.

With the above definitions for a design model, the general computational problem for a design can be formulated as follows:

Given ϑ, possibly with a description of its fluctuations (e.g. distribution functions), find values for x and y in order to satisfy:

$$\begin{aligned} h\ (x,y,\vartheta) &= 0 \\ g\ (x,y,\vartheta) &\leq 0 \\ L\ (x,y,\vartheta) &= \text{TRUE} \end{aligned} \tag{3}$$

while possibly optimizing the goal or goals as given by the objective function $F(x,y,\vartheta)$.

In Cagan, et al. (1995) we classify design problems.

Overview of Solution Techniques

In this section we will briefly overview some of the major solution techniques that can be used to solve particular cases of problem (Eqn. (3)). The intent is not to provide a comprehensive review, but rather give a brief sketch of algorithmic tools that are available to designers.

Two major problems that arise in design are analysis or evaluation and synthesis or optimization. The solution techniques for analysis can be generally classified as:

a. Equation solving
b. Symbolic analysis

For synthesis and optimization they can be generally classified as:

a. Mathematical programming
b. Heuristic search techniques

Equation solving techniques are widely used and implemented in most of the design tools. Here, the particular model:

$$\begin{gathered} \text{Given } \vartheta \text{ and y, find x to satisfy} \\ h(x,y,\vartheta) = 0 \end{gathered} \tag{4}$$

gives rise to a system of algebraic equations for the static case. Linear models can be effectively solved with matrix factorization methods (e.g. L/U decomposition) which can exploit the inherent sparsity in large scale design problems (Carnahan and Wilkes, 1980; Dahlquist and Anderson, 1974; Pissanetzky, 1984). Nonlinear models are considerably more difficult to solve, and in general require the iterative solution of linearized equations. This is for instance the case in Newton's method and its variants known as Quasi-Newton methods. The former requires analytical derivatives for the jacobian matrix, while the latter will predict jacobian approximations based on function values (Dahlquist and Anderson, 1974). First order methods, such as successive substitution, can also be used. They have the advantage of not requiring derivative information, but are slower to converge (Carnahan and Wilkes, 1980). Methods for solving algebraic equations are available in many computer codes (e.g. IMSL, 1987; Piela, et al., 1991; Rice, 1983). It should also be noted that for the case of large and complex nonlinear equations, which for instance arise in simulators (e.g. ASPEN in chemical engineering (Aspen-Technology, 1991), SPICE in electrical engineering (Banzhaf, 1989)), most of the equations are treated as a "black box" routine which is converged externally as an implicit function.

For the dynamic case, the problem in Eqn. (4) gives rise to differential/algebraic systems of equations. In the simplest case (only non-stiff ODE's) explicit methods such as Euler and linear multistep methods can be used (Carnahan and Wilkes, 1980; Dahlquist and Anderson, 1974). For stiff ODE's implicit methods such as backward differences and collocation methods are required. These can be extended to the case when algebraic equations are included, but care must be exercised to handle the so called "index problem" which may introduce large errors even if small integration steps are used with implicit methods. Codes implementing methods for integrating differential equations include IMSL (1987) and DASSL (Petzold, 1982).

When spatial equations are also included in a model this will often give rise to partial differential equations which are commonly solved by finite element methods (Becker, et al., 1982). These have become a major tool for design analysis and are implemented in codes such as ANSYS (1992).

Symbolic analysis tools are used mostly on design problems that are expressed in qualitative terms. Here the particular model has the form:

$$\begin{gathered} \text{Given } \vartheta\text{, find x and y to satisfy} \\ L(x,y,\vartheta) = \text{TRUE} \end{gathered} \tag{5}$$

Most of the computational techniques for logic are restricted to the case when the variables x are absent or prespecified. Symbolic methods for solving these problems include theorem proving and resolution techniques such as the ones implemented in PROLOG (Bratko, 1986; Sterling and Shapiro, 1986; Dodd, 1990) and PROLOG III (Colmerauer, 1990). For simpler cases (e.g. Horn clauses), forward and backward chaining methods can be used (Barr and Feigenbaum, 1981). The latter are implemented in a number of expert systems shells (e.g. VP-Expert (1989), EXSYS (1990)). Quantitative approaches to solving this problem have also been developed (e.g. see Hooker, 1988). They rely on the idea that it is possible to systematically transform Eqn. (5) into a linear programming problem with which the inference problem is solved as an optimization problem.

As for synthesis and optimization problems mathematical programming models have the general form:

$$\begin{aligned}&\text{Given } \vartheta\text{, find x and y to}\\&\text{minimize } F(x,y,\vartheta)\\&\text{subject to}\\&\qquad h(x,y,\vartheta) = 0\\&\qquad g(x,y,\vartheta) \leq 0\end{aligned} \tag{6}$$

Here $F(x,y,\vartheta)$ is assumed to be a scalar function. The case when $F(x,y,\vartheta)$ is a vector of functions gives rise to multiobjective optimization problems, which as opposed to the scalar case, have in general an infinite number of solutions. These are given by trade-off or pareto-optimal curves in which the various objectives cannot be improved simultaneously.

The mathematical programming problem may be a linear program (LP) in which the discrete variables y are not present; a mixed-integer linear program (MILP) in which both x and y are present; a nonlinear program (NLP) in which the functions may be nonlinear and only x is present; or a mixed-integer nonlinear program (MINLP) in which both x and y are present.

For LP problems the most common method is the simplex algorithm (Hillier and Lieberman, 1986) which is implemented in many computer codes (SCICONIC (1986), OSL (IBM, 1991), LINDO (Schrage, 1986), ZOOM (Marsten, 1986), MINTO (Nemhauser, et al., 1994), CPLEX (Bixby, 1992, 1994). Interior point methods have been recently developed which have shown to outperform simplex in problems involving many thousand of constraints (Marsten, et al., 1990; Lustig, Marsten and Shanno, 1994). MILP techniques rely commonly on branch and bound methods that use the simplex algorithm for LP as subproblems in each node of the tree (Nemhauser, et al., 1989). Therefore, these are commonly available in the same simplex codes. It should also be noted that the structure in special cases of LP and MILP problems can be greatly exploited, such as is the case in network flow problems. Here specialized algorithms can efficiently solve very large scale versions of these problems as opposed to the general purpose LP and MILP methods.

The most common techniques for NLP are the reduced gradient method (Reklaitis, et al., 1983), which is available in the computer codes MINOS (Murtagh and Saunders, 1985), GINO (Liebman, et al., 1986), CONOPT (Drud, 1991), and the successive quadratic programming (SQP) method (Gill, et al., 1989), which is available in the codes NPSOL (Gill, et al., 1983) and OPT (Vasantharajan, et al., 1990). The former method tends to be better suited for models with explicit equations (mostly linear) and the latter for "black box" models. Finally, MINLP techniques which are only recently starting to be applied in design problems, include the Generalized Benders decomposition (Geoffrion, 1972) and the Outer-Approximation method (Grossmann, 1990) which is implemented in DICOPT++ (Viswanathan and Grossmann, 1990). All these nonlinear optimization methods can only find a local optimum solution, unless the problem is convex in which case a local optimum is also a global optimum (Bazaraa and Shetty, 1979).

It should be noted that modeling tools such as GAMS (Brooke, et al., 1988) and ASCEND (Piela, et al., 1991) interface automatically with several of the computer codes cited above, greatly facilitating the formulation and solution of optimization problems.

Heuristic search techniques are methods that do not necessarily have a rigorous mathematical basis, but have the advantage of being simple to implement. Furthermore, for most cases they do not make any special assumption on the form of the equations or functions which can make them useful for complex or poorly understood design problems. Specific heuristic techniques are discussed below.

AI vs. OR Models

Having presented a classification of solution methods from an engineering design perspective, we now address the more general question as to what makes a model an OR model or an AI model? One way to answer this question is to classify models along three dimensions, each of which has a pole associated with OR and one associated with AI. This analysis clarifies the various senses in which OR and AI can be combined. It also helps one to say more precisely what sort of research thrusts could profitably combine AI and OR styles of modeling. The latter will be addressed in the next section.

The three axes along which models can be classified are:

1. numeric/symbolic
2. specific/general
3. structured/unstructured

In each case, the first attribute is normally associated with OR models, and the second with AI models. We begin with a brief description of these polarities.

Quantitative vs. Symbolic Models

In the schema presented earlier, quantitative models have only constraints of the form $h(x,y,\theta) = 0$, $g(x,y,\theta) \leq 0$, and symbolic models have only constraints of the form $L(x,y,\theta)$ = TRUE. A symbolic model might also be called a *logic model*, particularly if the propositions in $L(x,y,\theta)$ belong to a formal logical language.

Structured vs. Unstructured Models

Although it is hard to define what is meant by a structured model, it is a concept widely employed by modelers. Consider, for instance, an assignment model, which is a very special case of a linear programming problem (Nemhauser and Wolsey, 1988). All instances of an assignment model are very similar. They all exhibit the same, fairly simple pattern. This makes an assignment model highly structured.

Structured models tend to have two advantages. First, their relative simplicity is more conducive to understanding. If one can fit a structured model to a situation, one's grasp of it is likely to improve. A reason for the popularity of network flow models is that they are easy to understand and seem to illuminate the subject matter.

Another advantage of structured models is that they *tend* to be easier to solve. Network flow models, for instance, are much easier to solve than general linear programming problems (Bazaara, et al., 1990).

Specific vs. General Models

A specific model is one that applies only to a narrow range of problems, whereas a general model fits a wide variety of problems. A specific model "presupposes structure in the problem." An assignment model is very specific; problems are rarely so neat. Logic programming, on the other hand, presupposes little structure in the problem. It applies to any problem that is expressible in its logical formalism. This results in considerable generality, particularly if one considers W.V. Quine's assertion that first order predicate logic is adequate to formulate all of science (Quine, 1961).

A highly structured *model* need not presuppose structure in the *problem*; i.e., it need not be specific. A neural network model, for instance, is structured because it is a particular type of nonlinear regression model in which the gradient of the error function can be computed recursively via back propagation (Rumelhart, et al., 1986). A math programmer would regard this as a highly structured nonlinear programming model. But it is very general because it can formulate problems ranging from traveling salesman problems to visual recognition problems.

Fig. 1 attempts to classify some models along the specific/general and numeric/symbolic axes. One can imagine a structured/unstructured vertical axis, with the structured pole at the upper end. OR models should occur on the left side of the diagram, with preference for the upper left; they should also have a high elevation on the vertical axis. AI models should gravitate toward the right, with a preference for the lower right (although neural networks, genetic algorithms, etc., are moving the center of gravity toward the upper right). They should occur mainly at low elevations on the vertical axis.

	Specific	General
Numeric	ODE,PDE LP NLP Multi-objective Stochastic models MILP ILP MINLP DP Nonserial DP	Neural networks Constraint logic programming Randomized local search models
Symbolic	Combinatorial optimization (special structure) Monotonicity analysis Discrete models Classical combinatorics	Constraint programming Expert systems Logic programming

Figure 1. Classification of models.

What We Want In a Model

We want general models because they presuppose less structure in the problem. They are more likely to fit messy, real-world situations. One software package will have wide application.

We want structured models because they reveal structure that helps us to understand the problem, and because they are more likely to be tractable.

The ideal would seem to be models that are as general as possible while being as structured as possible. But there are two additional factors.

a) A model that *mirrors* the problem structure is superior to one that does not. A neural network, for instance, does not model a problem in the same way that an assignment model does. An assignment model mirrors the structure of an assignment problem, whereas the neural network model mirrors a neural network! A neural network models by reflecting the structure of a *problem-solving device*. This sort of model sacrifices one type of explanatory power.

b) An ideal model is easily calibrated, where 'calibration' is used in the general sense of tailoring a model to a specific problem. It might involve adjusting parameters and coefficients, formulating constraints for a mathematical programming model, or designing a solution space for a local search algorithm (more on this later). In the case of the assignment model, both calibration and solution are easy. But calibration of the neural network is hard, since it must be "trained" on a large test set that one never knows quite how to design.

Just as less structured models *tend* to be harder to solve, more general models *tend* to be harder to calibrate, or at least harder to calibrate in a way that makes a good solution possible. In a genetic algorithm, for example, it is relatively easy to define a problem representation, a crossover operation, mutations, etc. But it is harder to define these so that good solutions evolve (Goldberg, 1989).

So we want models that are a) general, b) structured, and c) easy to calibrate, all the while acknowledging that a structured model may fail to be tractable and may fail to have explanatory value.

In Cagan, et al. (1995) we evaluate several specific models in the light of the foregoing analysis.

Research Directions

The specific research strategies that could synthesize OR and AI seem to depend on which quadrant of Fig. 1 in which one begins. We consider one of the quadrants here; the remainder are discussed in Cagan, et al. (1995).

Specific Numeric Models

These typically OR models can move toward AI by becoming more general, and one route to greater generality is to become more symbolic.

One way to make a quantitative model more symbolic is to mix symbolic constraints with quantitative ones. Constraint programming does this (Van Hentenryck, 1989; Tseng, 1993). In this field the goal is generally to find a feasible solution, one of which may be a bound on the objective function. A job shop scheduling problem, for instance, can be described by a mixture of logical rules and numerical inequalities, plus a lower bound on the makespan. The object is to find a feasible schedule that meets this bound.

A number of constraint programming packages have been developed, particularly in Europe. These include CHIP (Dincbas, et al., 1988; Simonis and Dincbas, 1993), CHARME (Bull Corp., 1990; Drexl and Jordan, 1995; Sciamma, et al., 1990), PECOS, CHLEO, ConstraintLisp and other packages (Banel, et al., 1992; Remy, 1990). Constraint programming is less accepted in the U.S. and until very recently was largely ignored by the OR community. Perhaps one reason for this is that, except to the extent it is undergirded by the theory of logic programming, constraint programming does not generally have the kind of theoretical grounding and deep analysis enjoyed by mathematical programming. But mixed logical/numerical constraint sets can be systematically analyzed in the way that numerical constraint sets have been.

A related thrust is the use of logical methods in the solution of mathematical models, particularly those with combinatorial complexity. PROLOG III and CHIP takes some steps in this direction, but the potential is much greater. Propositional logic, for instance, is useful in the solution of mixed integer programming problems. In fact one can develop logic-based concepts and theory, parallel to those of branch-and-bound, cutting planes, facet-defining cuts, etc. (Hooker, 1994).

Another related idea is to replace some quantities with qualitative descriptions and to try to deduce properties of the solution. This occurs in symbolic optimization, monotonicity analysis, activity analysis, etc. To date these approaches have been used primarily to guide numerical algorithms, but they may have potential as more general models.

As already noted, there is an interesting way to make quantitative models more general without making them more symbolic: they can be used to model (in the mirroring sense) the problem solving process, rather than the problem itself. The influence of AI is clear here, particularly in the case of neural networks, which crudely model a problem-solving brain. Genetic algorithms (which are somewhat less numeric) model a problem-solving evolutionary process. Simulated annealing algorithms (Aarts and Korst, 1989) and some neural networks model nature's way of minimizing energy. Tabu search (Glover, 1989, 1990) models a search process with short-term memory; the early literature even suggested 7 as the ideal length of a tabu list, because human short-term memory generally has room for 7 chunks of information. An obvious research direction is to continue to model problem-solving processes, and it is being actively pursued. Such new search strategies as scatter search (Glover, 1994) are being invented, such social problem-solving processes as asynchronous teams (Talukdar, et al., 1993) and ant colonies (Colorni, et al., 1994) are being investigated, and so on.

Examples

In this section we examine two examples of recent research which combines AI and Optimization. They are intended to give the reader a flavor of the merging of the two approaches. We first examine a qualitative abstraction of the Karush-Kuhn-Tucker (KKT) conditions. We then describe an approach to generate the optimal topology of a network flow problem is discussed.

In Cagan, et al. (1995) we also discuss logic-based methods for optimal design of chemical processing networks (Raman and Grossmann, 1993) as well as design space expansion (Cagan and Agogino, 1987, 1991a, 1991b; Aelion, et al., 1991a, 1991b, 1992).

Qualitative Optimization: QKKT

Williams and Cagan (1994) introduced a qualitative abstraction of the Karush-Kuhn-Tucker (KKT) conditions of optimality (Karush, 1939; Kuhn and Tucker, 1952) called the Qualitative KKT (QKKT) conditions. By combining a hybrid algebra combining signs and reals (Williams, 1991), QKKT uses signs (direction of monotonicities) of the objective function and constraints to determine whether a point can be stationary (called *qstationary*) or whether a point is non-stationary. By using a square bracket to denote "sign"[1], QKKT states that a feasible point $\mathbf{x}^*$ is stationary only if:

$$[\nabla f(\mathbf{x}^*)] + [\lambda]^T [\nabla \mathbf{h}(\mathbf{x}^*)] + [\mu]^T [\nabla \mathbf{g}(\mathbf{x}^*)] \supseteq 0^T, \qquad \text{(QKKT1)}$$

subject to

$$[\mu]^T [\mathbf{g}(\mathbf{x}^*)] = 0^T, \text{ and} \qquad \text{(QKKT2)}$$

$$[\mu_i] \neq \hat{-}, \qquad \text{(QKKT3)}$$

where f is the objective function, $\mathbf{x}$ are the variables, and μ and λ are the inequality and equality Lagrange Multipliers, respectively.

KKT says that to be stationary there must exist a weighted sum ($\vec{\mathbf{w}}$) of $\nabla \mathbf{g}$ and $\nabla \mathbf{h}$ that exactly cancels ∇f (note $\vec{\mathbf{w}}$ is a row vector). QKKT says a point is *nonstationary* unless there exists a $\vec{\mathbf{w}}$ that lies in the quadrant diagonal from that which contains ∇f. QKKT1 results in a set of equations (one for each variable) consisting of the signs of the Lagrange multipliers and other sign terms. QKKT, by combining AI and optimization, gives a much simpler condition than KKT to rule out non-stationary points.

This condition and the techniques that follow from it fall in the symbolic/specific (lower left) quadrant of Fig. 1. By abstracting the quantitative KKT condition to the qualitative QKKT condition and using powerful symbolic reasoning techniques, a potentially powerful merging between the AI and optimization fields emerges. This is exploited in activity analysis (Cagan and Williams, 1994) and monotonicity analysis (Papalambros and Wilde, 1988; Choy and Agogino, 1986; Agogino and Almgren, 1987). Both techniques are discussed in Cagan, et al. (1995).

Shape Annealing

As an example of how optimization can generate design topologies by breaking away from local minima, we examine the design technique of *shape annealing* introduced by Cagan and Mitchell (1993). Here a design problem is again formulated as an optimization problem. However, the problem knowledge is modeled as a *shape grammar* with the properties of shapes described by Stiny (1980). Concepts of simulated annealing are used to create a technique which generates optimally directed solution shapes.

The shape annealing algorithm executes by applying a shape rule to an initial design. If the modification improves the design based on an objective it is accepted as a new state. If it generates an inferior design then it can still be accepted with a certain probability which is a function of the number of iterations executed and the progress of the annealer; toward the beginning of the process almost all inferior solutions are accepted and as the algorithm progresses, only those solutions which improve the objective are accepted. In shape annealing, previous designs can be re-gained if they are superior; for every rule which modifies a shape topology, there is a complementary rule which removes that modification, while for rules that modify variable values, the value can continuously be modified to reach any feasible value. Shape grammars are a concise, formal and computable representation of a design space while simulated annealing is a powerful exploratory search mechanism. Thus, shape annealing effectively explores the combinatorial number of feasible design configurations. Inferior solutions are pursued to get out of local minima, and optimally directed design topologies are derived.

Applications of the shape annealing technique include solutions to constrained geometric knapsack problems - knapsack problems constrained by geometry and relative component orientations - (Cagan, 1994), generation of truss structures (Reddy and Cagan, 1994) and component layout (Szykman and Cagan, 1994). As an illustration of shape annealing, Fig. 2 shows a shape grammar that manipulates triangles. With this grammar, topologies can change, shapes can change, and sizing of the triangle lines can change. Reddy and Cagan use this grammar to modify the topology of truss structures. As these rules are applied, a finite element analysis evaluates the constraint violations, and the simulated annealing algorithm determines whether to accept the change. The result is an optimally directed topology and optimal shape of the truss structure. A variety of constraints can be incorporated through the finite element analysis. Fig. 3 shows two structures generated from shape annealing which are analyzed for stress and Euler buckling constraints; Fig. 3b includes a geometric obstacle.

Shape annealing incorporates randomized local search models (simulated annealing) and expert systems/logic models (categorizing shape grammars). In Fig. 1 this method would be classified in the right quadrants (general) merging the top (numerical) and bottom (symbolic) categories.

---

[1]A sign is one of four values, namely $\hat{+}$ if the quantity is positive, $\hat{-}$ if the quantity is negative, 0 if the quantity is zero, and $\hat{?}$ if the quantity is unknown ([x] for $x > 0$ is $\hat{+}$, while [x - y] is $\hat{?}$ unless additional information is known).

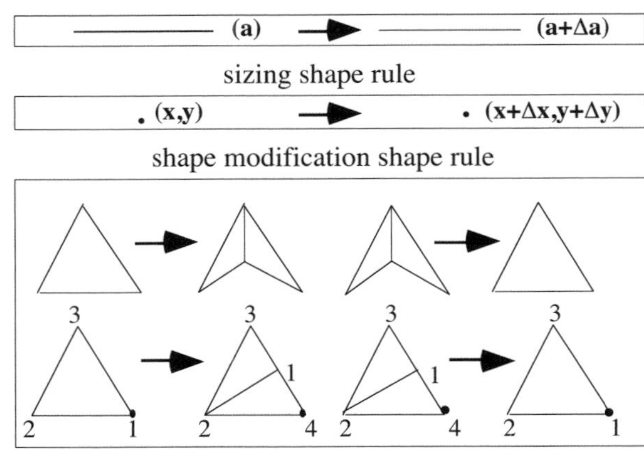

Figure 2. Shape grammar for truss generation (from Reddy and Cagan, 1994). The "•" represents a fixed point.

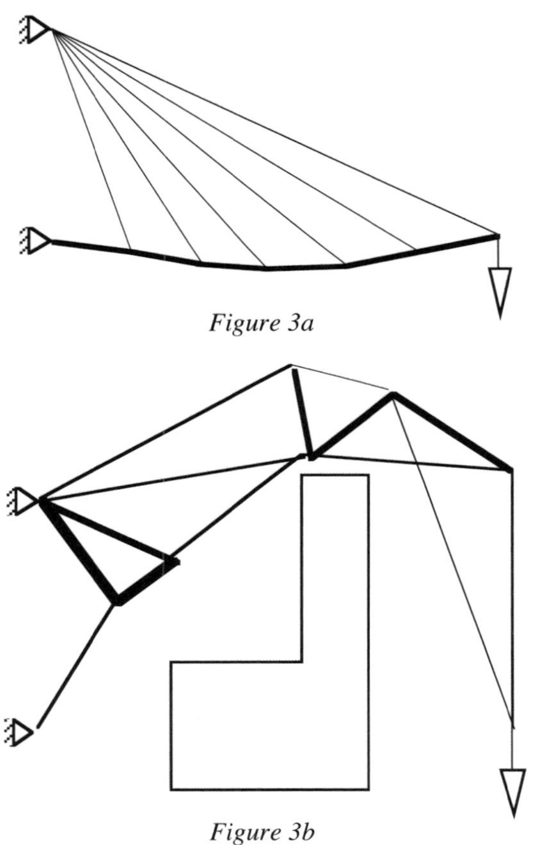

Figure 3a

Figure 3b

Figure 3. Truss structures generated by shape annealing (from Reddy and Cagan, 1994).

Conclusions

This paper provides a conceptual overview of the problem of combining artificial intelligence and optimization, and motivates the need for and suggest directions of future research. AI and optimization together make sense: few problems can be solved by structured optimization methods alone, and AI techniques are limited in their search for general designs. Rather, the combination of AI and optimization can provide problem specific reasoning, symbolic representations, and powerful numerical optimizing search. The framework is here; future work must continue to merge and expand the focus of combined AI and optimization problem solving.

Acknowledgments

The authors would like to acknowledge financial support from the Engineering Design Research Center, an NSF sponsored research center.

References

Aarts, E. and J. Korst (1989). *Simulated Annealing and Boltzmann Machines: A Stochastic Approach to Combinatorial Optimization and Neural Computing*. Wiley, New York.

Aelion, V., J. Cagan and G. Powers (1991). Inducing optimally directed innovative designs from chemical engineering first principles. *Comput. Chem. Engng.*, **15**, 619-627.

Aelion, V., J. Cagan and G. Powers (1992). Input variable expansion - an algorithmic design generation technique. *Res. Engng. Des.*, **4**, 101-113.

Agogino, A.M. and A.S. Almgren (1987). Techniques for integrating qualitative reasoning and symbolic computation in engineering optimization. *Engng. Optimization*, **12**, 117-135.

ANSYS (1992). Version 5.0. Swanson Analysis Systems, Inc.

Aspen Technology (1991). *Aspen-Plus User Guide*. Release 8.3. Cambridge, Mass.

Banzhaf, W. (1989). *Computer-aided Circuit Analysis using SPICE*. Prentice Hall, Englewood Cliffs, NJ .

Barr, A. and E.A. Feigenbaum (Eds.) (1981). *Handbook of Artificial Intelligence*. William Kaufmann, Inc., Los Altos, CA.

Bazaraa, M.S. and C.M. Shetty (1979). *Nonlinear Programming*. John Wiley.

Bazaara, M. S., J. J. Jarvis and H. D. Sherali (1990). *Linear Programming and Network Flows*. Wiley, New York.

Becker, E., J.T. Oden and G.F. Carey (1982). *Finite Elements: An Introduction*. Prentice Hall, Englewoods Cliffs, NJ.

Brooke, A., D. Kendrick and A. Meeraus (1988). *A GAMS User's Guide*. Scientific Press, Palo Alto.

Bull Corporation (1990). *CHARME VI User's Guide and Reference Manual*. Artificial Intelligence Development Centre, Bull S.A., France.

Cagan, J. (1994). A shape annealing solution to the constrained geometric knapsack problem. *Computer-Aided Design* , **28** , 763-769.

Cagan, J. and A.M. Agogino (1987). Innovative design of mechanical structures from first principles. *Artificial Intelligence in Engineering Design, Analysis, and Manufacturing*, **1**, 169-189.

Cagan, J. and A.M. Agogino (1991a). Dimensional variable expansion - a formal approach to innovative design. *Research in Engineering Design*, **3**, 75-85.

Cagan, J. and A.M. Agogino (1991b). Inducing constraint activity in innovative design. *Artificial Intelligence in Engineering Design, Analysis, and Manufacturing*, **5**, 47-61.

Cagan, J., I. Grossmann and J. N. Hooker (1995). Combining artificial intelligence and optimization in engineering design. Engineering Design Research Center, Carnegie Mellon University, Pittsburgh, PA 15213 USA.

Cagan, J. and W.J. Mitchell (1993). Optimally directed shape generation by shape annealing. *Environment and Planning B*, **20**, 5-12.

Carnahan, B. and J.O. Wilkes (1980). *Applied Numerical Methods*. Wiley, New York.

Choy, J.K. and A.M. Agogino (1986). SYMON: Automated SYMbolic MONotonicity Analysis for Qualitative Deisgn Optimization. In G. Gupta (Ed.), *ASME Computers in Engineering*, Vol. 1. pp. 207-212.

Colmerauer, A. (1990). An introduction to Prolog III. *Communications of the ACM*, **33**, 52-68.

Colorni, A., M. Dorigo and V. Maniezzo (1994). Distributed optimization by ant colonies. In F. Varela and P. Bourgine (Eds.), *Proceedings of ECAL91—Eurpoean Conference on Artificial Life*. Elsevier. pp. 134-142.

Dahlquist, A.B. and N. Anderson (1974). *Numerical Methods*. Prentice Hall, Englewood Cliffs, NJ.

Dincbas, M., P. van Hentenryck, H. Simonis, A. Aggoun, T. Graf and F. Bertier (1988). The constraint programming language CHIP. *Proceedings of the International Conference on Fifth Generation Computer Systems FGCS-88*. Tokyo.

Dodd, A. (1990). *PROLOG: A Logical Approach*. Oxford University Press, New York.

Drexl, A. and C. Jordan (1995). A comparison of logic and mixed-integer programming solvers for batch sequencing with sequence-dependent setups. *INFORMS Journal on Computing* (to appear).

EXSYS Inc. (1990). *EXSYS User's Manual*. Albuquerque, NM.

Geoffrion, A.M. (1972). Generalized benders decomposition. *J. of Optimization Theory and Applications*, **10**, 237-260.

Fenves, S.J. and I.E. Grossmann (1992). An interdisciplinary course in engineering synthesis. *Res. Eng. Des.*, **3**, 223-231.

Gill, P.E., W. Murray, M.A. Saunders and M.A. Wright (1983). *User's Guide for SOL/NPSOL: A FORTRAN Package for Nonlinear Programming*. Dept. Optns. Res., Stanford Univ., Technical Report SOL 83-12.

Glover, F. (1989). Tabu search—Part I. *ORSA Journal on Computing*, **1**, 190-206.

Glover, F. (1990). Tabu search—Part II. *ORSA Journal on Computing*, **2**, 4-32.

Glover, F. (1994). Genetic algorithms and scatter search: unsuspected potentials. *Statistics and Computing*, **4**, 131-140.

Goldberg, D. E. (1989). *Genetic Algorithms in Search, Optimization and Machine Learning*. Addison-Wesley, Reading, MA.

Grossmann, I.E. (1990). Mixed-integer nonlinear programming techniques for the synthesis of engineering systems. *Res. Eng. Des.*, **1**, 205-228.

Hooker, J.N. (1988). A quantitative approach to logical inference. *Decision Support Systems*, **4**, 45-69.

Hooker, J.N. (1994). Logic-based methods for optimization. In A. Borning (Ed.), *Principles and Practice of Constraint Programming, Lecture Notes in Computer Science* , **874**, 336-349.

IBM (1991). *Optimization Subroutine Library. Guide and Reference-Release 2*. Kingston, NY.

IMSL Math. Library (1987). *FORTRAN Subroutines for Mathematical Applications*. Houston, TX.

Karush, W. (1939). *Minima of Functions of Several Variables with Inequalities as Side Conditions*. MS Thesis, Dept of Mathematics, Univ. of Chicago, Chicago, IL.

Kuhn, H.W. and A.W. Tucker (1951). Nonlinear programming. In J. Neyman (Ed.), *Proceedings of the Second Berkeley Symposium on Mathematical Statistics and Probability*. Berkeley, CA: University of California Press.

Liebman, J., L. Lasdon, L. Schrage and A. Warren (1986). *Modelling and Optimization with GINO*. Scientific Press, Palo Alto.

Marsten, R. (1986) *User's Manual for ZOOM/XMP*. Dept. of Management Information Systems, Univ. of Arizona.

Marsten, R., M. Saltzman, J. Lustig and D. Shanno (1990). Interior point methods for linear programming: Just call Newton, Lagrange and Fiacco and McCormick! *Interfaces*, **20**, 105-116.

McGeoch, C. (1995). Toward an experimental method for algorithm simulation. *INFORMS Journal on Computing* (to appear).

Murtagh, B.A. and M.A. Saunders (1985). *MINOS User's Guide*. Systems Optimization Laboratory, Dept. of Operations Research, Stanford Univ.

Nemhauser, G.L., A.H.G. Rinnory Kan and M.J. Todd (Eds.) (1989). Optimization. *Handbooks in Operations Research and Management Science*, Vol. 1. North Holland, Amsterdam.

Nemhauser, G. L., M. W. P. Savelsbergh and G. C. Sigismondi (1994). MINTO: a mixed integer optimizer. *Operations Research Letters*, **15**, 47-58.

Nemhauser, G. L. and L. A. Wolsey (1988). *Integer and Combinatorial Optimization* . Wiley, New York.

Papalambros, P. and D.J. Wilde (1988). *Principles of Optimal Design*. Cambridge University Press, Cambridge.

Petzold, L.R. (1982). A description of DASSL: a differential/algebraic system solver. Sandia Tech. Rep. 82-8637.

Piela, P.C., T.G. Epperly, K.M. Westerberg and A.W. Westerberg (1991). ASCEND: an object-oriented computer environment for modeling and analysis. *Comput. Chem. Eng.*, **15**, 53-72.

Pissanetzky, S. (1984). *Sparse Matrix Technology*. Academic Press, London.

Quine, W. V. (1961). *From a Logical Point of View: Nine Logico-Philosophical Essays*. Harvard Univ. Press, Cambridge, MA.

Reddy, G. and J. Cagan (1994). An improved shape annealing algorithm for truss topology generation. *ASME J. of Mechanical Design*, (to appear).

Raman, R. and I.E. Grossmann (1993) Symbolic integration of logic in mixed integer linear programming techniques for process synthesis. *Comput. Chem. Eng.*, **17**, 909.

Reklaitis, G.V., A. Ravindran and K.M. Ragsdell (1983). *Engineering Optimization — Methods and Applications*. Wiley.

Rice, J.R. (1983). *Numerical Methods, Software, and Analysis: IMSL Reference Edition*. McGraw Hill, New York.

Rumelhart, D. E., G. E. Hinton and R. J. Williams (1986). Learning internal representation. In D. E. Rumelhart and J. L. McClelland (Eds.), *Parallel Distributed Processing: Exploration in the Microstructure of Cognition. Vol. 1: Foundations*. MIT Press, Cambridge, MA. pp. 318-362.

Schrage, L. (1986). *Linear, Integer and Quadratic Programming with LINDO*. ScientificPress, Palo Alto.

Sciamma, D., J. Gay and A. Guillard (1990). CHARME: a constraint oriented approach to scheduling and resource allocation. *Artificial Intelligence in the Pacific Rim, Proceedings of the Pacific Rim International Conference on Artificial Intelligence*. Nagoya, Japan. pp. 71-76.

SCICONIC/VM User Guide (Version 1.4) (1986). Scicon Ltd. Milton Keynes.

Simonis, H. and M. Dincbas (1993). Propositional calculus problems in CHIP. In F. Benhamou and A. Colmerauer (Eds.), *Constraint Logic Programming: Selected Research*. MIT Press, Cambridge, MA. pp. 269-285.

Stiny, G. (1980). Introduction to shape and shape grammars. *Environment and Planning B*, **7**, 343-351.

Szykman, S. and J. Cagan (1993). Automated generation of optimally directed three dimensional component layouts. *Advances in Design Automation, Proceedings of the ASME Design Automation Conference*, Vol. 1. Albuquerque, NM, September 19-22. pp. 527-537.

Talukdar, S., P. DeSouza and S. Murthy (1993). Organizations for computer-based agents. *International J. of Eng. Intelligent Systems for Elec. Eng. and Comm.*, **1**, 75-87.

Tseng, E. (1993). *Foundations of Constraint Programming*. Academic Press.

Van Hentenryck, P. (1989). *Constraint Satisfaction in Logic Programming*. MIT Press, Cambridge, MA.

Vasantharajan S., J Viswanathan and L.T. Biegler (1990). Reduced SQP implementation for large-scale optimization problems. *Comput. Chem. Eng.*, **14**, in press.

Viswanathan, J. and I.E. Grossmann (1990). A combined penalty function and outer-approximation method for MINLP optimization. *Comput. Chem. Eng.*, **14**, 769-782.

VP-Expert (1989). Paperback Software, Berkeley, CA.

Williams, B.C. (1991). A theory of interactions: unifying qualitative and quantitative algebraic reasoning. *Artificial Intell., Special Volume on Qualitative Reasoning About Physical Systems II*, **51**, 39-94.

Williams, B. C. and J. Cagan (1994). Activity analysis: the qualitative analysis of stationary points for optimal reasoning. *Proceedings, 12th National Conference on Artificial Intelligence*, Vol. 2. AAAI Press/MIT Press. pp. 1217-1223.

TOWARDS INTEGRATED FRAMEWORKS OF REASONING AND COMPUTATION IN CHEMICAL ENGINEERING

Matthew J. Realff
School of Chemical Engineering
Georgia Institute of Technology
Atlanta, Georgia 30332-0100

Ioannis Kevrekidis
Department of Chemical Engineering and
Program in Applied and Computational Mathematics
Princeton University
Princeton, NJ 08544

Abstract

We discuss two distinct selected cases of "multi-tiered" computation: analysis and interpretation of first-level numerical results leads to a second computational level through gradual adaptation or through radical change of the algorithms used. We briefly discuss other realizations of such hierarchical computational approaches in modern chemical engineering modeling practice.

Keywords

Automated reasoning, Optimization, Dynamical systems, Numerical bifurcation, Machine learning, Branch and bound algorithms, Continuation, Branch switching.

Introduction

When an algorithm has been implemented, it is difficult to conceptually distinguish its parts in "reasoning" as opposed to calculation, since all computations are ultimately implemented in the same set of logic gates. For the purposes of this paper, "calculation" will be taken to represent computations used to directly find the answer to an engineering problem that has been posed, while reasoning will represent computations *about* the problem or problem-solving algorithm. We will use selected examples to illustrate the combination of these two types of computation, and the resulting benefits in expanding the classes of engineering problems that we can use computers to solve. Our two paradigms of modeling progress made through the use of reasoning come from (a) the use of learning within design and optimization algorithms, and (b) the use of non-linear systems theory to guide the simulation of systems with complicated dynamics. We will also briefly mention a number of other cases where "hierarchical" computation appears in modern chemical engineering modeling. The direction towards effective implementation of reasoning rests on engaging complex representations of knowledge. This knowledge goes beyond that necessary for solving the problem and addresses the theory behind the problem solving itself.

Case I: Implementing "Reasoning" in Optimization Computations

Optimization problems are in general hard computational problems, both in the theoretical sense of worse case complexity bounds (Garey and Johnson, 1979), and in practical algorithm performance. Most successful algorithms have relied on exploiting special structures present in a particular formulation of an optimization problem, e.g. the asymmetric Traveling Salesman Problem (Miller and Pekny, 1991), or have relied on clever encodings of the original problem space that enable general purpose algorithms to achieve good computational performance (Shah, et al., 1993). The special structures have meant that techniques constructed to solve one problem have been difficult to transfer systematically to other problems, although frequently the abstract principles behind the specialized approaches are the same. The encodings are often problem specific and it has been hard to generalize the features of a good encoding and automatically apply them in new formulations.

From this we draw the implication that the *a priori* synthesis of general optimization algorithms that will be effective across a wide range of problems with many discrete alternatives is unlikely. Furthermore, optimization techniques are being applied to a wider range of problems as decentralized computational resources spread to all levels of an organization. This leads to a knowledge acquisition bottleneck. Thus, the need to have algorithms that will customize themselves based on observing their own behavior is growing. It is possible that this need can be met by the application of reasoning and learning techniques intertwined with algorithm execution.

A Description of Search Problems

To enable more concrete discussions of reasoning and learning within search algorithms we need to be more explicit about the components of the algorithm. We will assume that a search algorithm can be characterized by the following, general, elements.

1. A representation of the search space. We will assume that this is comprised of:
   (a) A set of features that characterize the space and from which all information about solutions within the space can be derived.
   (b) A feasibility predicate that computes from the features of a particular element of the search space its membership in the feasible set.
   (c) An objective function that computes the value of an element of the search space.
2. A function that estimates the quality of the search space. The function has one general feature that we must define, its admissibility, (Pearl, 1984). A function is admissible if it provides a lower bound on the value of the solution, if we are minimizing, or an upper bound if we are maximizing. Many enumeration methods compare the objective function value of a feasible solution to that of a bound on a subspace solution value. If the bound is worse then the subspace is eliminated from further consideration. An optimal solution will not be eliminated by a method that uses an admissible bound.
3. Control knowledge that enables us to eliminate sets of solutions from the search space without evaluating all their members. The bound elimination described in (2) is the most common form of control knowledge.
4. A decomposition operator that splits, or specializes, the space into subspaces that can be successively searched.
5. A search function that chooses the next subspace to decompose from the available subspaces.

The Role of Reasoning In Optimization

The major roles that reasoning will play, or is already playing, in optimization problem solution synthesis are:

1. In tools to support the correct translation of real problems to particular problem representations such as linear programs.
2. The *a priori* detection of special structures within a formulated problem, such as constraint subsets that form network flow problems, and the subsequent application of algorithms to exploit the structure.
3. Reasoning within the algorithm to exploit the results of the search itself to improve performance, or understand problem structure, either within the same problem or in future similar problems.

In addition to the roles that reasoning plays in problem solution synthesis, it is also playing a major role in automated understanding and analysis of the solution output (Greenberg, 1993).

The first two roles have been extensively researched and a catalogue of research appears in (Greenberg, 1993). We will not discuss them further, continuing with the theme of this paper, the combination of reasoning and calculation within an algorithm. We will focus specifically on how machine learning methods can be applied to achieve the goal of exploiting search results to improve future search performance.

What is Machine Learning ?

Machine Learning is a catch all term for attempts to build "computer programs able to construct new knowledge or to improve already possessed knowledge by using input information" (Michalski, 1990). Approaches to machine learning have been categorized in many different ways; we will use the classification scheme of (Michalski, 1990) to elucidate the connections between reasoning and learning and their use in optimization and search. The key distinction is between synthetic and analytic learning. Synthetic learning attempts to construct new or better knowledge by an inductive process, whilst analytic learning aims to reformulate existing knowledge into a better form using deduction. The key difference between these types of learning is the distinction between induction, which hypothesizes the premises of an argument entailed by established conclusions, and deduction, which uses established premises to find conclusions. These two methods use background knowledge that the learner may have in different ways. In analytic learning the background knowledge is combined with observations to deduce new conclusions, in synthetic learning the background knowledge is used to constrain the premises that the learner is allowed to consider.

To apply machine learning to search algorithms we can use several dimensions to classify the learning problem. First, we can distinguish between two types of learning, intra-trial and inter-trial. In intra-trial learning information is gathered about search subspaces, as they are generated, and when the same, or similar, subspaces are encountered in another part of the search space the results found in the previously encountered spaces can be used to influence their evaluation. This type of learning is most appropriate in situations where constraint interactions are complex and finding feasible solutions is hard, such that implicit unfeasible subspaces are frequently encountered and difficult to detect by *a priori* reasoning. We will equate this behavior with non-persistent learning, or remembering. Once the trial is over none of the information remembered during the trial is retained for future problem-solving activity, and the state of the problem solver itself is not altered from one problem-solving episode to the next.

Inter-trial learning uses the solution of one or more problems within a problem class to learn something that can be applied in solving problems drawn from that class in the future. For example, if we have to find the schedule of a batch plant on a repeated basis then it is useful to try to customize the search algorithm to work for that specific application. The key issues in learning over multiple trials are:

1. How far the learning is able to generalize the problem class in which the learnt result will apply.
2. How many trials are required to learn something.
3. The correctness of the resulting learnt concept — will the search algorithm still continue to find the optimum result, or can we predict how its behavior will be modified ?

The key difference between synthetic and analytic learning is the guarantee that each places on its result, and the number of examples that are typically used in learning, addressing issues (2) and (3) above. In synthetic learning the best that can be achieved is that the acquired knowledge is consistent with the examples. Work in PAC (Probably approximately correct) learning has provided guarantees that, provided enough examples are seen, the learning will almost always generate a concept that will misclassify only a small fraction of examples (Valiant, 1984). However, this guarantees that the concepts that it learns are probably correct most of the time, and not provably correct. As such, this places certain types of modifications to search algorithms off-limits to synthetic learning, if we wish to guarantee that the algorithm will continue to find optimal solutions. The review of the various approaches to learning in optimization will follow the general divisions outlined above, intra- versus inter-trial, and analytical versus synthetic learning.

Intra-trial Learning

The overall goal of intra-trial learning, or remembering, is to improve the efficiency of a search strategy in which no learning takes place on a single instance. It should be clear that all successful search algorithms employ some form of "remembering." For example, a typical branch and bound algorithm caches the objective function of the best solution found so far and uses this, along with a lower bound function, to eliminate other partial solutions. It also remembers which tree branches have been explored. What distinguishes the different techniques is the type, complexity, and level of generalization, of the knowledge that is retained to influence the future path of the search technique.

Analytic Intra-trial Learning

Dechter (1990) explores the use of learning in backtracking search, particularly within the constraint satisfaction problem (CSP) (Nudel, 1983). Backtracking search tries to find an assignment of a value to a variable that is consistent with the assignments to variables that have previously been made, and with the constraints that exist between subsets of the variables. This type of search is very similar to depth-first branch and bound algorithms, except that in CSP and many other applications of backtracking search, a consistent set of values for the variables is sought rather than an optimal set of values.

The goal of the learning in backtracking search is to find hidden constraints, that is a set of values of subsets of the variables which result in a dead end and therefore require a backtracking step. The actual assignment of values to variables that causes the dead end will not be encountered again in the branching process. The reasoning problem is to identify subsets of the variables which are causing the unfeasibility, and which could therefore reoccur on other branches. This can be thought of as deducing the reason for failure, and it is based on graph theory and equation solving. Finding the minimum subset of variables can require exponential computation, and thus various weaker methods for finding subsets are used. This results in a tradeoff between the effort spent in finding smaller subsets and the extra time spent in the search. Techniques based on these ideas have been applied in problems with specialized constraint structures (Bayardo and Miranker, 1994; Dechter, 1990), and shown to improve computational performance. Richards, et al. (1995) extends this approach to more general constraint structures and Maruyama, et al. (1991) uses it in optimization problems by iteratively solving the constraint satisfaction problem.

Synthetic Intra-trial Learning

The analytical framework above can be relaxed to generate a synthetic learning strategy, where the frequency of particular assignments to variables that lead to dead ends can be assessed and then this frequency information used to bias the selection algorithm away from instantiating the combination of those variables at those values (Glover, 1986). Glover (1986) also describes a method where common features of solutions constructed by heuristics are extracted and then a constrained search performed such that these features are preserved in the final solution.

Inter-trial Learning

The goal of inter-trial learning is to perceive some regularities in one or more searches performed for a given class of problems and to exploit these regularities in future problem-solving activity. In analytical learning there is an explicit use of the theory that surrounds the particular search method and problem class. Deductive reasoning takes place after a single example has been solved applying background knowledge to the example. In synthetic learning this theory, or background knowledge, is used to pick features that are likely to be good predictors of search performance. The learning methods then inductively classify many examples and use the induced rules to change the problem-solving behavior from its default state.

Analytical Inter-trial Learning

The learning methodology used in analytical inter-trial learning has been explanation-based. The overall goal is to specialize general principles of search control that apply across the general problem representation to the specific problem class. This specialization is carried out by analyzing examples using proven knowledge about the problem domain and the problem formulation. Thus, the components of this methodology are, (1) A training example, (2) A target concept (3) A theory of the domain (4) An operationality criterion. For an introduction to explanation-based learning see Minton, et al. (1990). The key difference between an algorithm that uses explanation-based learning and a traditional computation is that the explanation-based component employs knowledge about the search method itself to analyze its performance. It detects examples either of good aspects or imperfections in the way it is working. It then combines these examples with its knowledge about how it is searching to reorganize its knowledge or specialize some general notion to the particular problem domain.

To illustrate the use of explanation-based learning in combined reasoning and optimization, we will use the work of Realff (1992) in learning dominance and equivalence rules for branch and bound algorithms. First, we define dominance and equivalence rules. These rules are a form of search control knowledge, and are used to eliminate one search subspace y, by using another one x. x is said to dominate y if it satisfies a set of logical conditions, the principal one of which is that the search space y cannot contain a solution with a better objective function that the search space x. This means that we need not explore y to find the optimal solution. These conditions supplement the traditional lower bound elimination of branch and bound algorithms, since a feasible solution need not be generated to use them, and they can eliminate solution spaces whose lower bound is better than the current best objective function value. Ibaraki (1978) shows that the stronger the control knowledge of a branch and bound algorithm the fewer the nodes enumerated during the search, which motivates acquiring dominance and equivalence rules. For more details of the definitions of dominance and equivalence and proofs of the reduction in enumeration see Ibaraki (1978). The components of the explanation-based learning method in Realff (1992) are detailed below.

1. The training examples are partial solution pairs (x,y) which have been expanded during the search and which can be shown to satisfy some or all of the conditions for x to dominate y. For example, the partial solutions might represent incomplete batch chemical plant schedules. The general problem representation could be as a mixed integer linear program (MILP) (Nemhauser and Wolsey, 1988). Thus the examples would be represented by the set of values that binary variables have taken in the formulation, and the corresponding state of the constraints in the MILP;
2. The target concept is the set of pairs (x, y) such that for the specific problem class x dominates y. For the above example, the problem class is defined by the constraints of the MILP formulation. As long as the types of the constraints and variables remain the same and the number of each is simply expanded, the problem is considered to be in the same class;
3. The domain theory is a sufficient theory that enables the proof of the dominance of y by x. For example, if the feasible region of the MILP in the solution x is larger than that in y, then the objective function value in y can be no better than that in x, and hence x dominates y. This theory is expressed in terms of comparing the relative sizes of the hyperspaces of each constraint. This theory in turn leads to simple comparisons between variable values in the two partial solutions. In order to reason about the dominance conditions the theory must be explicitly represented. The form of the representation is

as horn clauses, or antecedent consequent rules in expert system terminology; and

4. The explanation of why a training example is part of a concept must terminate with terms that are considered "primitive" to the domain and hence can be easily evaluated. The operationality criterion expresses which terms can be considered primitive. For MILP's we consider simple comparisons between the values that variables take in different subproblems, and classifications of different constraints, to be operational.

The components of the system are then orchestrated to produce an explanation. In our example, the explanation is the transformation of the complex notion of dominance to simpler, specific, terms. It is a deductive process which uses the domain theory to connect the concept to the facts of the example, and hence it generates a provably correct reason as to why the two subproblems, x and y are an example of a new dominance rule in the particular domain. The explanation is generated by backward chaining, as in a rule-based system, from the goal to the facts. At this point nothing has been learned, since the proof is only for the specific example. The next step is to generalize the explanation, and in the process generate a rule that can apply beyond the specific instance in which it was proved.

The generalization is effected by replacing the specific facts of the example by variables. The necessary bindings of the variables are propagated from the root of the explanation through the antecedent consequent rules to the leaves. A new rule can then be generated that cuts out all the intermediate reasoning that connects the root to the leaves. This generalization procedure has been shown to be sound (Mooney, 1990) and hence any rule generated will be guaranteed to be consistent with the domain theory. Thus, if the domain theory is correct, the rule will represent a valid dominance rule. The generalization performed by this process is limited to replacing constants with variables, and as such generates a rule that can only be applied to problems that have the same number of constraints and variables. This would be very limiting in most applications, for example in scheduling where the number of jobs and units would have to be constant, and potentially the scheduling horizon as well. To overcome this limitation an extension of the original generalization algorithm was used (Shavlik, 1990), which transformed the explanation structure by analyzing recurring consequents. This algorithm enabled the generalization of the number of constraints in the problem and hence the number of variables.

The above method has been applied to learning dominance rules in several scheduling applications, single machine scheduling and flowshop scheduling (Baker 1974), and to a simplified version of the MILP formulation for general chemical engineering scheduling problems (Shah, et al. ,1993). For more details see Realff (1992).

Synthetic Inter-trial Learning

In contrast to analytical learning methods which can modify elements of the control knowledge to eliminate solution subspaces, synthetic methods cannot do this without potentially violating the optimal solution seeking behavior of the algorithm. Those methods that do not violate this property cannot directly eliminate parts of the search space. Most often the target of synthetic learning is the fifth component of the search algorithm, the search function. Other methods are free to modify the other components but can lose the guarantee that the optimal solution will be found.

There have been several methods devised to modify the search function. The main differences between them are that they employ different learning algorithms, and have different representations of the search function. In the operations research literature this approach is represented by Target Analysis (Glover, 1986; Glover and Greenberg, 1989). The search function is represented by a set of criteria which are weighted and combined linearly. In implicit enumeration techniques, the output of the search function is an assignment of a value to a variable which will be used to specify the subspace which will be searched next. The learning takes place after a problem is solved and takes the form of adjusting the weights on the criteria so that, with hindsight, the choice at each branching point is the one that will lead to directly to the optimum solution. The weights can be adjusted by any technique that "learns" a discriminant or classification function between good and bad branches, where the good branch is defined as the one on the path to the solution and a bad branch is a competing feasible choice. The target analysis approach can be applied in other enumeration algorithms, such as tabu search (Glover, 1989). Laguna and Glover (1993) evaluate the methodology for scheduling problems where target analysis is used to synthesize a rule to select the local neighborhood search operator to apply in the tabu search.

In the Artificial Intelligence literature, Rendell (1990) applied a similar technique to A^* search solving the NxN puzzles problem. Six learning methods were applied, and the search space features were determined by the existing knowledge of effective heuristics for this problem, such as the Manhattan distance. The most successful method was a hybrid-genetic algorithm which simultaneously learnt a partition of the feature space and the weight vector. The partitioning was based on the utility of underlying search state in finding the solution. The weight vector was found by regression. This type of learning in game playing has a long history; Samuel (1959) developed a checkers playing program that selected moves based on a search function that was a weighted combination of features of the board position where the weights were adjusted by analyzing wins and losses.

Other approaches modify different parts of the search algorithm. Cerbone and Dietterich (1992) modifies a combination of the quality, decomposition and search functions. Inductive learning is applied to gradient descent search applied to optimizing truss design. The objective of the learning is to map from a description of the problem to a small set of stress states in which an optimum, or very near optimum, solution will lie. These are the only states that will be evaluated. The training examples consist of instances within the problem class, described by the appropriate feature vector, and the optimal "action," in this case, stress state. The learning method is decision trees (Quinlan, 1986) combined with a Min Cover algorithm (Garey and Johnson, 1979). The output of the algorithm is a decision tree that enables the mapping from the features describing the problem to a set of stress states to be evaluated. The algorithm is shown to find solutions that are within 5% of the optimum, with about a fourfold decrease in computation time compared to exhaustive search of the stress state. The generalization achieved by the learning algorithm is limited to the continuous parameters of the problem. The number of truss elements is fixed, thus the decision tree for a three member truss is not generalized to any other number of members. In addition, the method used a large number of training examples, 500, for a problem with 64 stress states (6 members). These training examples have to be solved to optimality, and it is not clear that for larger problems the number of examples required will not be prohibitively large.

Bramanti-Gregor, et al. (1992) present techniques for the improvement of heuristics used in A* search. Two improvements based on learning are presented. The first transforms an admissible heuristic to a "near admissible" one in an attempt to combat the slowing of search towards the leaves of the tree as the bounding function becomes unable to distinguish between search states. The learning algorithm outputs a multiplier, greater than unity, that raises the value of the admissible estimate. The level of sub-optimality is fixed by the value of this multiplier, in the same way the tolerance parameter in branch and bound algorithms controls their convergence. As has been seen in branch and bound algorithms, the tolerance of a small gap between the lower bound and best feasible solution can dramatically alter search performance. The "self-tuning" of this gap is a natural extension of current practice. This latter performs the inverse of making an admissible heuristic inadmissible, outputting a multiplier less than unity that transforms an inadmissible heuristic to an admissible one. The value of the multiplier is learned via a post problem-solving analysis which plots the true objective function value and the heuristic value as a function of the depth of the tree.

Another approach to learning bounds on solution values is presented in Modi, et al. (1993). This uses the Soar architecture (Laird, et al., 1986) to solve distillation column sequencing problems. The objective is to find the appropriate multiplier on the value of the output of a heuristic evaluation of a sequence choice, such as marginal vapor rate or marginal cost, that will generate good solutions that are unlikely to have too great an error in cost compared to the optimal solution. The search architecture allows the solution method to try different heuristics that are increasingly difficult to evaluate. The different sequence options are tried and the multiplier required to make the best solution cost equal to the next best computed. If this multiplier is less than the one currently being used by the system the next level of heuristics in invoked, until the default strategy of full search is implemented. If the heuristics fail to be able to make choices based on the tolerated error then the result of the full search enables the multipliers to be computed at different errors between the best solution and the one that would be found using the heuristic.

In complex domains where finding good, or feasible, solutions is an important precursor to full-scale optimization, learning has been employed to synthesize problem-solving heuristics. In terms of our search framework, such heuristics correspond to inadmissible control knowledge, since they either entirely eliminate search by specifying the path through the search space or severely curtail the search to a few states. The basic approach is to find some successful search path through the search space and to cache the result to use the path again in the future without having to search for it again. Generalization methods are used to transform this simple caching into sequences or subsequences that are simpler and apply in a greater range of situations. In AI planning methodologies this has been described as macro-operator construction (Fikes, et al., 1972). Macro-operator techniques have also been combined with explanation-based learning to construct operator sequences (Mitchell, et al., 1983; Minton, 1988). The extent to which these operators can be applied in new situations is a function of the level of generalization. As in analytical inter-trial learning, the generalization is governed both by the theory about the problem that can be represented and employed, and the generalization method itself.

An Integrated Approach

It should be clear that the four different ways of learning to improve search algorithm performance are not mutually exclusive and could be combined into a multi-strategy approach to the problem (Michalski and Tecuci, 1994). In intra-trial learning the dynamic memory of the problem-solver is added to by reasoning over the results of the search performed so far. The results of this reasoning need not be discarded at the end of the search, but could be archived for an inter-trial learning method to explore after several problems have been solved. Analytical methods based on strong domain knowledge can be used to reason over the regularities that are "discovered" by the synthetic approaches to learning. The results of the analytical learning can be relaxed by replacing conditions with simpler ones or dropping them altogether. These empirical adjustments can then be tested and a synthetic learning algorithm used to learn which lead to the most successful problem-solving results.

All of the learning methods are predicated on having an explicit representation of the search algorithm itself available to be modified. We have laid out components of a search problem and each of these can be the target of experimentation, not just the search function and control knowledge. This implies that a much richer set of representations are needed for algorithms, and that monolithic programs will present a barrier to future experimentation in this area.

Case II: Implementing "Reasoning" in Nonlinear Dynamics Computations

Most chemical engineering systems are strongly nonlinear, whether due to the kinetics of chemical reactions and their strong temperature dependence, or from nonlinearities in the transport equations, like the inertial terms in the Navier-Stokes. Consider a model in the form of a "dynamical system," in this case a set of coupled nonlinear ordinary differential equations (say the CSTR with a single exothermic irreversible reaction, as considered in Uppal, et al. (1974)) depending on a number of parameters (some representing physical properties, like heat capacities, and some representing operating conditions, like flow rates). Let us state, as our goal, the *complete* classification of the dynamic behavior of such a model. Let us also assume that we have at our disposal a number of basic general purpose scientific computing tools: linear equation solvers, eigensolvers and integrators - these are the "building blocks" of our algorithms and we consider them available.

There are two conceptual ways of "going to the computer" with such a dynamical system. The first is to use the computer as a pure simulator, the equivalent of an experiment that one performs in the laboratory: set parameter values (design the experiment), set initial conditions (start up the reactor), integrate and observe. The result of this simple simulation experiment is th*e entire transient* time series, which, say, eventually approaches a steady state. If the question we want answered is not knowledge of the transient, but knowledge of the final steady state, the second way of "going to the computer" becomes apparent: we should write an algorithm that finds a steady state directly. Simulation *does* of course find the steady state, in the same sense that the true experiment does: when the simulation goes on but the variables do not (appreciably) change in time, we are there. But an algebraic equation solver, with a good initial guess, can find the steady state very fast (quadratic convergence) and accurately. Furthermore, the floating point work involved in this single "steady state" finding is essentially the same as the work in a *single* integration step for an implicit integrator simulating the initial value problem. The "reasoning" in this case is just the realization that a steady state finding algorithm is the correct tool to answer the desired question. It is straightforward to implement as a call to a root-finding subroutine at the "tail end" of an integration that is slowing down as it approaches a (possible) steady state.

Our illustrative example (the Kuramoto-Sivashinsky equation, KSE (Brown, et al., 1991))

$$V_t + \frac{\alpha}{2}(V_x)^2 + \alpha V_{xx} + 4V_{xxxx} = 0. \tag{1}$$

is at once more complex and easier than the CSTR: we will use a dynamical system that comes from discretizing a partial differential equation (sixteen as opposed to two ODEs), but there will only be a single operating parameter in this set of ODEs.

This equation has been derived as a model of spatiotemporal instabilities in a number of physical settings. In the context of thin film flow down an inclined plane, the instability parameter α contains various physical property values of the fluid and is inversely proportional to the square of the length scale over which periodic boundary conditions are applied. We discretize this equation through a spectral scheme representing the solution *U(x,t)* as a truncated Fourier series:

$$U(x,t) \approx a_0 + \sum_{n=1}^{N} a_n(t)\cos(nx) + \sum_{n=1}^{N} a_n + N(t)\sin(nx) \tag{2}$$

then, after suitable projection (we use Galerkin), a set of $2N+1 \equiv M$ ordinary differential equations (ODEs) arises from the KSE:

$$\dot{\mathbf{a}} = \vec{g}(\vec{\mathbf{a}};\alpha). \tag{3}$$

Steady states for this problem (corresponding to spatially nonuniform solutions of the partial differential equation) can be found by solving $\vec{\mathbf{g}}(\vec{\mathbf{a}};\alpha) = \vec{0}$.

We have already, however, encountered a different type of question: not what the dynamics are, but what the *long-term* dynamics are, states of the system that are *invariant*, i.e. in some sense stationary or converged, states that do not change with time. There are several types of such states: steady states, *standing waves* - oscillatory solutions that change shape in space and time but repeat themselves exactly after a finite period, traveling waves (which look like steady states in a frame moving with the appropriate constant speed), *modulated traveling waves* (which look like standing waves in a frame moving with the appropriate constant speed), quasiperiodic and spatiotemporally chaotic states etc. How to find *all* these states for *each* parameter value?

While no complete answer is in general possible, the implementation of "reasoning" in available scientific computing packages (like AUTO (Doedel, et al., 1991)) does assist in such a search.

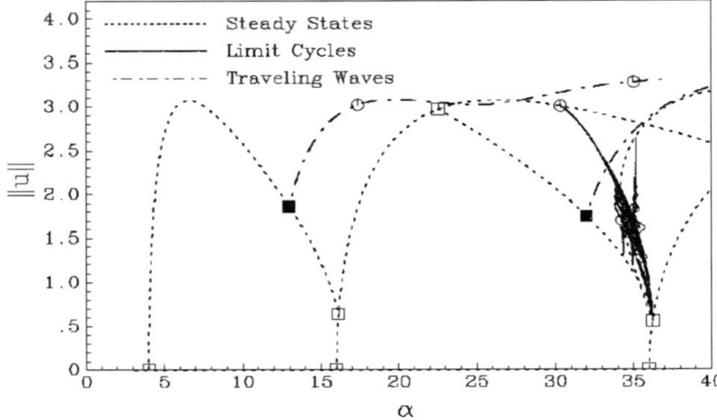

Figure 1. Bifurcation diagram for the KSE. □ marks steady state and traveling wave bifurcations. o marks Hopf bifurcation points.

To begin with, once a steady state has been found, what happens to it in parameter space is followed via *continuation*. This widely used technique uses a "basic scientific computing building block" (a nonlinear equation solver based on iterative solution of linear equations) but on an *augmented* set of nonlinear equations, built partially on the model, and partially on the Implicit Function Theorem. The specific continuation algorithm implemented in AUTO, which was used to construct the bifurcation diagram in Fig. 1, parameterizes the solution branch with a pseudo-arclength s solving

$$\vec{\mathbf{g}}(\vec{\mathbf{a}}_1; \alpha_1) = \vec{0} \tag{4}$$

$$(\vec{\mathbf{a}}_1 - \vec{\mathbf{a}}_0)^T \frac{d\vec{\mathbf{a}}_0}{ds} + (\alpha_1 - \alpha_0)\frac{d\alpha_0}{ds} - \Delta s = 0. \tag{5}$$

Here $\vec{g}(\vec{\mathbf{a}}_0;\alpha_0) = \vec{0}$, meaning that we assume we know a regular solution point, and the slopes with respect to the pseudo-arclength s are calculated using the implicit function theorem and standard linear algebra "scientific computing building blocks."

Reasoning here consists once more of the realization that a continuation code must be used, and calling it as a subroutine after the appropriate initialization (finding the first steady state). As the parameters change one encounters "problems": a branch of solutions may turn backwards at a so-called "turning point" or "fold" (two of these make up the usual S-shaped curve of multiple steady states in CSTR bifurcation diagrams). Another possibility is that a *bifurcation point* is reached, where a new branch of steady states bifurcates from the first one. There are several such points in Fig. 1: new branches bifurcate from the zero solution at α=4, 16 and 36. Turning points and bifurcation points are important because they signal a change in the number of steady states. They have a common characteristic (the determinant of the Jacobian of the M steady state equations is zero) but they can be discriminated based on the rank of the *augmented MxM+1* matrix with the additional column $\frac{\partial \vec{\mathbf{g}}}{\partial \alpha}$. Furthermore, the *slope* of the new branch can be locally approximated and computed based on the implicit function theorem and basic scientific computing.

Reasoning in this case (as implemented in AUTO) corresponds to

1. upon convergence of steady state continuation step, check for the vicinity of fold or bifurcation points (has the sign of the determinant of the Jacobian changed since the last step ?);
2. discriminate between the two (based on the rank criterion above);
3. locate them exactly (call a subroutine for that purpose);
4. in the case of a bifurcation point, when done with the current branch of steady states come back to it, calculate the slope of the new bifurcation branch and start continuation on it.

In addition to the "old" reasoning theme of which algorithm to write (based on the system model) to best answer the correct question, we clearly have a new theme: once a "higher" organizing center for the behavior is found (here, a bifurcation point that separates different types of behavior in parameter space), write a *new*, higher level algorithm, based on the system model again, planned to locate and characterize this new type of center. In addition, there is a "supervisory" level of reasoning: check the ongoing solution in order to decide when to activate this higher level of computation. Everything we discussed so far in essence has only to do with the solution of algebraic equations, and does not drastically involve dynamical systems theory: indeed, the basic mathematical tool is the implicit function theorem. A new, interesting twist comes

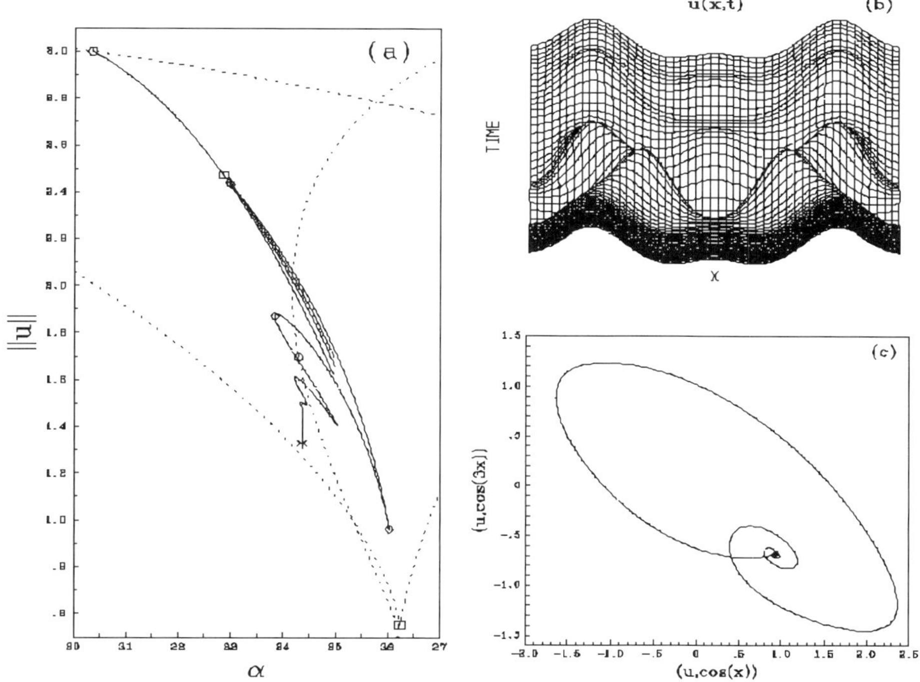

Figure 2. (a) Partial bifurcation diagram showing a limit cycle branch in detail, (b) Space-time plot, and (c) an a_1-a_3 phase space projection of a limit cycle approaching homoclinicity ($\alpha \sim 34.37$, marked with a $*$).

from Hopf bifurcation points, where a pair of complex conjugate eigenvalues of the Jacobian cross the imaginary axis during continuation. Then, the Hopf bifurcation theorem tells us that a branch of periodic solutions (*limit cycles*) is born (supercritically or subcritically, the direction depending on some higher order terms in a local Taylor expansion). Here the supervisory reasoning (check for the crossing of a complex eigenvalue pair) during continuation is again present, but now a new level of algorithmic complication needs to be added: we must solve two-point boundary value problems in time, looking for solutions $\vec{\mathbf{a}}(t)$ and a period T of the type

$$\begin{aligned} \dot{\vec{\mathbf{a}}} &= \vec{\mathbf{g}}(\vec{\mathbf{a}};\alpha) \\ \vec{\mathbf{a}}(t=0) &= \vec{\mathbf{a}}(t=T). \end{aligned} \tag{6}$$

This is essentially a simple subroutine call to another almost standard piece of scientific software (a boundary value problem solver for ODEs, even though there is a small difficulty with the time-translation invariance of the solution). We have changed, however, "in mid-stride" the type of object we are looking for: from steady states we are now looking for limit cycles (oscillatory solutions).

Our last illustration of yet another level in this hierarchical structure of computational dynamics exploration comes from the *continuation* of limit cycle branches as the parameter varies. It is interesting that, since boundary value problem solvers also eventually end up in a set of nonlinear coupled algebraic equations, the same machinery used for steady state continuation (when given the steady state equations) is used to perform limit cycle continuation, given the *discretized in time* equations for the BVP. Continuation of limit cycles may yield a picture like the one in Fig. 2a: a branch of limit cycle solutions (plotted through its norm) takes a "corkscrew" appearance by going through repeated turning point bifurcations, and apparently asymptotes to some final state. A "dynamical systems" person would immediately recognize this construction as characteristic of a so-called "Shil'nikov loop": a homoclinic connection of a saddle-type unstable *steady state* with itself, illustrated in Figs. 2b and 2c. As the limit cycle approaches this object, its period tends to infinity (because a lot of time is spent close to the steady state when movement slows down); this causes the continuation algorithm for limit cycles to eventually fail. Once more, a supervisory "reasoning" can be built in the limit cycle continuation code (is the period tending to infinity ? are there large periods of time over which the solution changes very slowly ? is there a saddle-type steady state in the vicinity of this slow-down ?). AUTO has not taken this step yet. On the other hand, it *has* taken the next step: write an algorithm to find this new type of "object" (in this case a homoclinic loop) which in a sense underlies and organizes the behavior of the limit cycle branches in its neighborhood. Algorithms for this purpose are built in AUTO94 (see also (Friedman and Doedel, 1993; Champneys and Kuznetsov, 1994)). These algorithms (involving approximations of saddle-type stable and unstable manifolds and forming appropriate infinite-time BVPs) are as "alien" to limit cycle continuation as limit cycles are to steady state continuation.

A simpler type of "new algorithm," that would not involve the formulation of a completely different problem, but only a simple modification of the existing one, is the case of finding traveling wave and modulated traveling wave solutions. Travelling waves are solutions of the original equation of the type $V(x,t) = V(x - Ct)$ where C is a (constant) wave speed. A little thought shows that these are *limit cycle* solutions of the original equation, but they are *steady state* solutions of

$$V_t + CV_x + \frac{\alpha}{2}(V_x)^2 + \alpha V_{xx} + 4V_{xxxx} = 0. \tag{7}$$

Here the speed C is unknown, but this can be remedied if one observes that infinitely many solutions exist *for the correct C*: they constitute shifts of each other in space. The addition of a pinning condition does therefore allow one to find the additional unknown speed C. Realization of the fact that moving in a traveling frame would simplify the solution process comes from visualization of the results, but can also reasonably be automated. We actually have two options here: one is to again "augment" the steady state equations with a pinning condition that will allow us to find the unknown C; the other is to make a transformation to a *single* variable $\chi = x - Ct$ turn the problem into a fourth order ODE and look for *limit cycles* of that problem. Similarly, a pinning condition can be used to turn *modulated traveling waves* of the original equation (quasiperiodic solutions) into *standing waves* (limit cycles of the above equation) thus allowing us to use a simpler algorithm.

A final comment concerns spatiotemporally chaotic solutions: there is no fixed point algorithm that will locate them exactly. One needs to find them by simulation, and decide on how to discriminate between transient approaches to them and deterministic chaotic wandering on them (an admittedly difficult task). Again, visualization in phase space is a convenient tool for "supervisory reasoning" here. The "new algorithms" that will be called when convergence to a chaotic solution is decided will be algorithms to describe its statistics: dimension, Lyapunov exponents and various correlation functions.

The issues raised by this example range from the "traditional" supervisory and hierarchical computation of certain system solutions (like steady states and the detection of bifurcating branches) to new issues, like switching to the computation of new types of objects (limit cycles, homoclinic orbits) when distinct quantitative and/or only qualitative supervisory criteria are met.

Other Examples of Reasoning in Non-linear Dynamical Systems

The reasoning built in the current algorithms for the study of the dynamics of chemical engineering models (and nonlinear dynamical systems more generally) tends to fall more under the category of "intra-trial" learning discussed above: forms of remembering as well as the ability to switch to new types of solution-finding methods are built in all "good" codes, and what differs between alternatives is again the type and level of complexity of the knowledge retained. There have been great strides towards incorporating these techniques in a computational environment that encompasses modeling, simulation, stability and operability analysis and possibly optimization of large scale transport processes (Brown, et al., 1980; Christodoulou, et al., 1995).

More recently, however, as the scope of modeling techniques expands across space scales from the molecular to the macroscopic and in time scales from picoseconds to real-time, there is much more effort in intelligently (or rather, correctly) interfacing a hierarchy of algorithms addressing the individual scales. This is very well and concisely discussed by Professor D. Theodorou in the editorial of a recent *Chem. Eng. Science* "symposium in print" on molecular modeling (Theodorou, 1994). In traditional modeling, microscopic effects were included in macroscopic conservation equations in the form of empirical correlations, or theories with experimentally fitted coefficients (like closure models for turbulent flows, or constitutive equations for the rheology of suspensions or multiphase flows). Today the energetics and kinetics of chemical reactions, e.g. (vanSanten, 1990; Klein, et al., 1991) as well as the rheology of complex suspensions, e.g. (Brady and Bossis, 1988) can be calculated from more fundamental models.

What is in some sense more computationally exciting, however, is that most recently there have been efforts of actively combining algorithms working at different scales in the same computational experiment. Illustrations of this approach include, among others, the combination of Particle-in-Cell and Dynamic Monte Carlo methods in simulations of electron dynamics in RF glow discharges (Economou and Lymberopoulos, 1995), hybrid Monte Carlo — continuum computations in cases where the continuum descriptions are inadequate (O'Connell and Thompson, 1995) as well as hybrid continuum—stochastic simulations for viscoelastic flows (CONNFFESSIT (Laso and Oettinger, 1993)). One of the major challenges in the use of reasoning in chemical engineering computations is to study and establish a rational (if not rigorous) interface between knowledge obtained computationally at such different scales and levels of phenomenology, whether in isolated stages or in simultaneous computation.

Conclusions

The goal of this paper was to illustrate the role that reasoning can play in enriching the class of chemical engineering problems that can be addressed computationally. In chemical engineering optimization we are driven to solve larger problems and to find global optima of processes where models can be very sophisticated or very rudimentary, with corresponding levels of accuracy. Reasoning about the solution of those models presents us with the opportunity to change the representation and dimensions of our search, compressing it to look first for qualitatively different behaviours and then to engage in parameter optimization. Reasoning within the optimization algorithm not only serves the purpose of improving efficiency, and therefore increase the size of problems that can be solved, but also reveals regularities and dependencies within the search space that using a black box technique would render unobservable. This can lead to a refinement in our understanding of the problem itself and our own discovery of new ways to synthesize chemical engineering systems that exploit new classes of phenomena.

Acknowledgments

Ioannis Kevrekidis was partially supported through the Exxon Education Foundation, and gratefully acknowledges the hospitality of the Center for Nonlinear Studies at Los Alamos National Laboratory. Matthew Realff was partially supported by a grant from Molten Metal Technology Inc.

References

Baker, K.R. (1974). *Introduction to Sequencing and Scheduling*. Wiley, New York.

Bayardo, R.J. and D.P. Miranker (1994). An optimal backtrack algorithm for tree-structured constraint satisfaction problems. *Artif. Intell.*, **71**, 159-181.

Brady, J.F. and G. Bossis (1988). Stokesian dynamics. *Ann. Rev. Fluid Mech.*, **20**, 111-157.

Bramanti-Gregor, A., H.W. Davis and F.G. Ganschow (1992). Strengthening heuristics for lower cost optimal and near optimal solutions in A* search. *10th European Conference on Artificial Intelligence*. Vienna, Austria. John Wiley & Sons.

Brown, H.S., I.G. Kevrekidis and M.S. Jolly (1991). A minimal model for spatiotemporal patterns in thin film flow. In R.Aris, D.G. Aronson and H.L. Swinney (Eds.), *Patterns and Dynamics in Reactive Media*. Springer Verlag.

Brown, R.A., L.E. Scriven and W.J. Silliman (1980). Computer-aided analysis of nonlinear problems in transport phenomena. In P.J. Holmes (Ed.), *New Approaches to Nonlinear Problems in Dynamics*. SIAM Publications, Philadelphia.

Cerbone, G. and T. G. Dietterich (1992). Inductive learning in engineering: a case study. *Proc. of Conference on Adaptive and Learning Systems*. Orlando, Florida.

Champneys, A.R. and Y.A. Kuznetsov (1994). Numerical detection and continuation of co-dimension two homoclinic bifurcations. *Int. J. Bifurcations and Chaos*, **4**, 785-822.

Christodoulou, K.N., S.F. Kistler, and R.P. Schunk (1995). Advances in computational methods. In T. M. Schweizer and S.F. Kistler (Eds.), *Liquid Film Coating: Scientific Principles and Technological Implications*. Chapman and Hall, New York. (In press)

Dechter, R. (1990). Enhancement schemes for constraint processing: backjumping, learning, and cutset decomposition. *Artif. Intell.*, **41**, 273-312.

Doedel, E., H.B. Keller and J.P. Kernevez (1991). Numerical analysis and control of bifurcation problems, (I) bifurcations in finite dimensions. *Int. J. Bifurcations and Chaos*, **1**, 493-520.

Economou, D.J. and D.P. Lymberopoulos (1995). Spatiotemporal electron dynamics in RF glow discharges: fluid vs. dynamic monte carlo simulations. *J. Phys. D.*, (in press).

Fikes, R., P. Hart and N. Nilsson (1972). Learning and execution of generalized robot plans. *Artif. Intell.*, **3**, 251-288.

Friedman, M. and E. Doedel (1993). Computational methods for global analysis of homoclinic and heteroclinic orbits: a case study. *J. Dyn. Diff. Equs.*, **5**, 37-57.

Garey, M.R. and D.S. Johnson (1979). *Computers and Intractability: A Guide to the Theory of NP-Completeness*. Freeman, New York.

Glover, F. (1986). Future paths for integer programming and links to artificial intelligence. *Comp. Opns. Res.*, **13**, 533-549.

Glover, F. (1989). Tabu search — Part I. *ORSA J. Comp.*, **1**, 190-206.

Glover, F. and H.J. Greenberg (1989). New approaches for heuristic search: a bilateral linkage with artificial intelligence. *Eur. J. Opns. Res.*, **39**, 119-130.

Greenberg, H.J. (1993). *A Computer-Assisted Analysis System for Mathematical Programming Models and Solutions: A User's Guide for ANALYZE*. Kluwer, Boston, MA.

Greenberg, H.J. (1993). *An IMPS Bibliography*. University of Colorado at Denver.

Ibaraki, T. (1978). Branch and bound procedure and state-space representation of combinatorial optimization problems. *Inf.* Control, **36**, 1-27.

Klein, M.T., M. Neurock, A. Nigam and C. Libanati (1991). Monte carlo modeling of complex reaction systems: an asphaltene example. In A. Sapre and F. Krambeck (Eds.), *The Mobil Workshop*. Van Nostrand Reinhold, New York.

Laguna, M. and F. Glover (1993). Integrating target analysis and tabu search for improved scheduling systems. *Expert Sys. Appl.* , **8**, 287-297.

Laird, J.E., P.S. Rosenbloom and A. Newell (1986). Chunking in Soar: the anatomy of a general learning mechanism. *Machine Learning*, **1**, 11-46.

Laso, M. and H.C. Oettinger (1993). Calculation of viscoelastic flow using molecular models: the CONNFFESSIT approach. *J. Non-Newtonian Fluid Mech.*, **47**, 1-20.

Maruyama, F., Y. Minoda, S. Sawada, Y. Takizawa and N. Kawato (1991). Solving combinatorial constraint satisfaction and optimization problems using sufficient conditions for constraint violation. *International Symposium on Artificial Intelligence*, Cancun, Mexico.

Michalski, R. (1990). Research in machine learning: recent progress, classification of methods, and future directions. In Y. Kodratoff and R. Michalski (Eds.), *Machine Learning An Artificial Intelligence Approach*. Morgan Kaufmann, San Mateo, CA.

Michalski, R. and G. Tecuci (1994). *Machine Learning A Multistrategy Approach*. Morgan Kaufmann, San Francisco, CA.

Miller, D.L. and J.F. Pekny (1991). Exact solution of large asymmetric travelling salesman problems. *Science*, **251**, 754-761.

Minton, S. (1988). *Learning Search Control Knowledge: An Explanation-Based Approach*. Kluwer Academic Publishers, Boston, MA.

Minton, S., J.G. Carbonell, C.A. Knoblock, D.R. Kuokka, O. Etzioni and Y. Gil (1990). Explanation-based learning: a problem solving perspective. In J.G. Carbonell (Ed.), *Machine Learning Paradigm and Methods*. MIT Press, Boston. pp. 64-118.

Mitchell, T.M., P. Utgoff and R. Banerji (1983). Learning by experimentation: acquiring and refining problem-solving heuristics. In R.S. Michalski (Ed.), *Machine Learning An Artificial Intelligence Approach*. Morgan Kaufmann, Los Altos, CA.

Modi, A.K., A. Newell, D.M. Steier and A.W. Westerberg (1993). Building a chemical process design system within SOAR — 2. learning issues. *Comp. Chem. Eng.*, **19**, 345-361.

Mooney, R.J. (1990). *A General Explanation-Based Learning Mechanism and its Application to Narrative Understanding*. Morgan Kaufmann, San Mateo, CA.

Nemhauser, G.L. and L.A. Wolsey (1988). *Integer and Combinatorial Optimization*. Wiley Interscience, New York.

Nudel, B. (1983). Consistent-labeling problems and their algorithms: expected-complexities and theory-based heuristics. *Artif. Intell.*, **21**, 135-178.

O'Connell, S.T. and P.A. Thompson (1995). A new tool for studying complex fluid flows. *Phys. Rev. Letters*, (submitted).

Pearl, J. (1984). *Heuristics*. Addison-Wesley, Boston.

Quinlan, R. . (1986). Induction of decision trees. *Machine Learning*, **1**(1), 81-106.

Realff, M.J. (1992). *Machine Learning for the Improvement of Combinatorial Optimization Algorithms: A Case Study in Batch Scheduling*. Ph.D., MIT.

Rendell, L. (1990). Induction as optimization. *IEEE Trans. Sys. Man. Cyber.*, **20**, 326-338.

Richards, T., Y. Jiang and B. Richards (1995). Ng-backmarking an algorithm for constraint satisfaction. *BT Tech. J.*, **13**(1), 102-109.

Samuel, A.L. (1959). Some studies in machine learning using checkers. *IBM J. Res. Develop.*, **3**, 210-220.

Shah, N., C.C. Pantelides and R.W.H. Sargent (1993). A general algorithm for short-term scheduling of batch operations-II. computational issues. *Comp. Chem. Eng.*, **17**, 229-244.

Shavlik, J.D. (1990). *Extending Explanation-Based Learning by Generalizing the Structure of Explanations*. Morgan Kaufmann, San Mateo, California.

Theodorou, D. (1994). Editorial note. *Chem. Eng. Sci.*, **49**, 2715-2716.

Uppal, A., W.H. Ray and A.B. Poore (1974). On the dynamic behavior of continuous stirred tank reactors. *Chem. Eng. Sci.*, **29**, 967-985.

Valiant, L.G. (1984). A theory of the learnable. *Comm. ACM*, **27**, 1134-1142.

vanSanten, R.A. (1990). Computational advances in catalyst modeling. *Chem. Eng. Sci.*, **45**, 2001-2011.

SESSION SUMMARY: INTELLIGENCE IN MODELING AND NUMERICAL COMPUTING

Angelo Lucia
Department of Chemical Engineering
Clarkson University
P.O. Box 5705, Potsdam, NY 13699-5705

Bernt Nilsson
Department of Automatic Control
Lund Institute of Technology
P.O. Box 118, S-221 00 Lund, Sweden

Introduction

This session is concerned with the use of intelligence in modeling and numerical computing. When we think of modeling we think of simulation tools built around a presumed engineering language, graphical interfaces and the necessary software for describing the model in the computer. This is the current state of the art. Only a small amount of effort has gone into the human component associated with modeling within the design task. Often times, users are required to accept a particular representation of a model, irrespective of whether they agree with it or not. Furthermore, different users of the same model may not require the same detail and frequently only certain "chunks" of the process are well understood.

In the first paper, Jarke and Marquardt address these and many other related issues in building their information system-based model of process modeling. In particular, they address directly the issues of specification (i.e., the degree of understanding of the process), representation (i.e., formalism used to express knowledge about the process) and agreement (i.e., an archive of agreement or disagreement of all aspects of the model reached by the design team).

A completely separate but equally important aspect of the design task is that associated with solving the model. For this, as we know, we have a variety of accepted engines for solving the nonlinear algebraic, differential, differential/algebraic, and optimization models that arise in describing chemical processes. However, a great many of the tools presently in use in simulation and design exploit system knowledge within or to guide these more algorithmic engines.

In the second paper, Cagan, et al. show how combined use of techniques from operations research (OR) and AI can be used to successfully attach engineering design problems. Two examples are presented, one in shape annealing and the other in qualitative optimization, that show that the combined approach can lead to algorithms that provide problem-specific reasoning, symbolic representations and powerful optimizing searches and thus overcome the individual shortcomings of OR and AI techniques alone.

In the final paper in this session, Realff and Kevrekidis discuss two cases of "multi-tiered" computation that combine analysis and interpretation at a first level and adaptation or radical change at a second level in illustrating a hierarchical computational approach. They discuss what they mean by "reasoning" in computations and address issues related to intra- and inter-trial computational learning in both optimization and nonlinear dynamics.

Conclusions

Artificial intelligence and knowledge-based techniques are becoming important in both modeling and computation today. Even though modeling and computation are very different tasks, both research areas use and exploit ideas from AI.

In this session, we have seen one example of how research in information systems can be applied to process modeling. The paper by Jarke and Marquardt represents the development of a process model as a path in three-dimensional space, those dimensions being representation, agreement and specification. As a consequence, future modeling software must be able to support the actual modeling process in a way that goes beyond the present descriptive languages and graphical tools.

The other two papers, one by Cagan, et al. on combined use of optimization and AI and the other by Realff and Kevrekidis on reasoning in equation-solving, discuss the use of qualitative knowledge and reasoning inside and as a supervisor to numerical solution algorithms of various types (i.e., optimization, algebraic and differential equation-solvers, etc.). These papers show that, while there are many methods available to solve well-defined mathematical problems, there are few tools that also take into account qualitative knowledge during the solution procedure. Both papers clearly show that the use of AI in numerical computing leads to algorithms that are capable of solving problems that were previously intractable and that this approach possesses the potential for refining our understanding of the problem itself.

PRODUCT AND PROCESS DESIGN WITH MOLECULAR-LEVEL KNOWLEDGE

Michael L. Mavrovouniotis
Chemical Engineering Department
Northwestern University
Evanston, IL 60208-3120

Abstract

We present an overview of intelligent-system methods and applications that rely on the representation, generation, and manipulation of molecular structures. Most existing methods are deductive; inductive computational methods are a promising approach for taming complexity, development of heuristics, and navigation of databases or long results produced by synthesis algorithms. Chemical *product design* follows the paradigm established by the identification of molecular structures from spectra. An initial selection of acceptable building blocks (usually functional groups) is followed by construction of feasible chemical structures; the search space is pruned by restricting the numbers of groups from various categories. Genetic algorithms carry out an opportunistic search, recombining the structures of good candidates to derive better ones. Detailed three-dimensional conformations are essential in drug design but have not been used in chemical product design. *Process design* with molecular-level knowledge is a descendant of computer-assisted organic synthesis, which played a major role in the development of molecular representation methods and the growth of chemical databases. The generation of reaction networks and the construction of pathways from a given reaction database are discussed.

Keywords

Product design, Process design, Solvent design, Group contributions, Polymer design, Reaction systems, Reaction paths, Reaction generation, Pathway synthesis, Reaction processes.

Introduction

Design activities in chemical engineering range from the design of products (including pure components, solvents, mixtures, and polymers) to the design of processes (such as reactor systems, separation trains, or heat exchanger networks). Equally diverse are the types of knowledge, tools, and techniques that are brought to bear on these tasks. General mathematical techniques (such as optimization methods) often distance themselves from the details of the chemical process or product in question; other methods offer intermediate physical insight (e.g., thermodynamic relations or separation diagrams). In this paper we focus on the use of molecular-level information: representation and manipulation of molecular structures. We will not discuss methods that deal only with derivative information, such as the equilibrium diagrams used in separation design by Malone and Doherty (1995) or Wahnschafft and Westerberg (1993). We will be interested in methods in which molecular-structure knowledge is not a priori lumped but rather an integral part of the task.

The molecular-level information is often compressed into a simple numerical form. Beyond very coarse molecular features (such as the molecular weight), it sometimes turns out that problems which appear difficult on the surface can be resolved through simple linear correlations on the most basic structural features (atoms and bonds), at least for a good initial approximation; this turns out to be the case, for example, in gas chromatographic retention indices of drugs (Rohrbaugh and Jurs, 1988). In other cases, an elaborate codification is needed: In the simulation of Carbon-13 Nuclear Magnetic Resonance spectra of norbornan-2-ols, torsional descriptors were developed to encode the effect of strain on chemical shift resonance (Rohrbaugh, et al., 1988). It is perhaps more appropriate to consider the use of molecular-knowledge not as a stationary target but rather as a *direction* for computer applications. In this paper, we provide a sampling of methods and applications of molecular information in product design and in process modeling and design.

The earliest computational efforts in this field were the identification of molecular structures from spectra and the computer-assisted planning of organic synthesis. In the DENDRAL project (Lederberg, 1965; Lindsay, et al., 1980) molecular features were determined from spectroscopic data and then assembled into molecular structures consistent with the observed mass spectra. This problem is quite analogous to product design, with the spectrum as the target property and additional constraints stemming from elemental analysis and knowledge of certain structural groups.

The Computer-Assisted Organic Synthesis (CAOS) work initiated by Corey and Wipke resulted in the LHASA system (Corey, 1967; Corey, et al., 1985a, 1985b). Given the general similarity between laboratory synthesis and batch reaction processes, CAOS is analogous to the synthesis of batch processes. CAOS is of interest for two additional reasons: (a) Its molecular structure representation and perception modules formed the foundation of chemical databases; similar modules are essential for any molecular-information based system. (b) The generation and application of reactions is of interest in *modeling*, analyzing, and simulating complex chemical reaction processes.

Organic Synthesis by Computer

Computer-Assisted Organic Synthesis (CAOS) has been an area of intense research activity in computational chemistry, beginning in the late 1960's (Corey, 1967). In CAOS, the objective is to devise a path for synthesizing a target product in the laboratory, given a list of starting materials which are available (primary source materials) or can be synthesized through standard known paths (secondary source materials). As was mentioned earlier, much of its influence has been indirect: Its basic representations and algorithms (such as canonicalization and substructure perception) for molecules and reactions facilitated the creation of chemical databases and served as the foundation for other computational efforts.

The overall structure of CAOS procedures involves detailed analysis of molecular structures which are generated by the application of reactions. A generic retrosynthetic (i.e., searching from target products towards precursors) procedure is shown in Fig. 1. The key iterative step in CAOS is the application of a reaction operator to a target compound to derive its simpler precursors (following the most common retrosynthetic mode), or the application of a reaction to precursors to derive a product (in the less common forward synthetic mode). The selection, encoding, and application of reactions in CAOS can be made in two ways, exemplifying a trade-off which is common in the application of molecular-level knowledge:

a. In a heuristic associative manner, one may seek and apply known patterns of synthesis. The drawback in this approach is that it is unlikely to uncover particularly innovative solutions. Also, the definition and perception of the appropriate pattern in the molecular structure is often complicated (e.g., for compounds with ring systems). The primary advantage of the approach is that only a small number of possibilities are considered at a time, and only modest pruning of the search space is necessary.
b. In a logic-driven manner, one seeks all reactions theoretically possible in any given situation. This approach is characterized by higher potential for discovery, easier perception, smaller search space, and more challenging pruning.

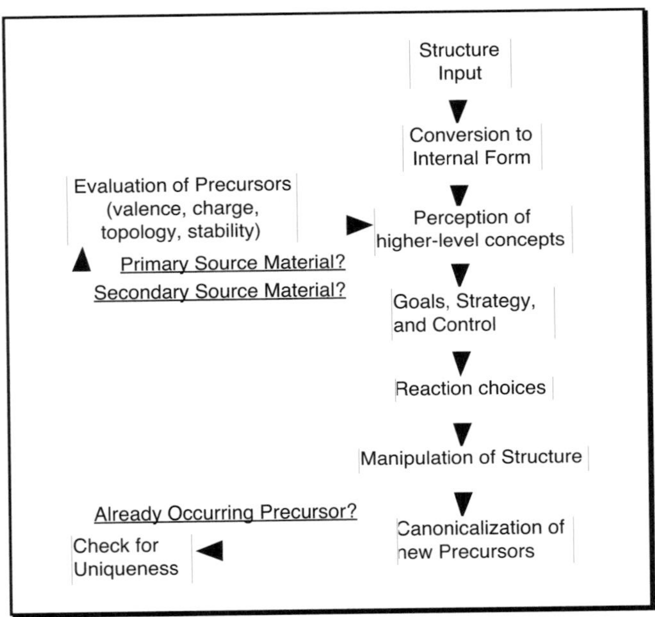

Figure 1. A retrosynthetic approach to Computer-Assisted Organic Synthesis.

CAOS requires many underlying basic methodologies for a system (even an experimental one) to be assembled. It is no accident that each research effort in CAOS involved many researchers and lasted over a decade. We will discuss CAOS in some detail, precisely because it represents a prototype of a high-complexity multi-faceted problem that is intimately dependent on molecular-level knowledge. The modules implied in Fig. 1 are required in any system that must apply reactions, reason about reactions, or recognize reactivity; the peculiarities of each system show up primarily in the strategy and control of the search.

Molecular Representation

Two general kinds of representation can be used in the computational representation of molecules:

a. Line-based extensions of the classical nomenclature
b. Graph or object-oriented representations

The first type has the advantage of being extremely compact and therefore suitable for fast input of molecules and efficient storage. The second type is normally more convenient for the manipulation of structures. A common approach is thus to accept input as line-based and convert it automatically to a graph or object-oriented form for internal use.

The structure of Fig. 2 would be represented in Wisswesser Line Notation as QY1&VY1&U1, where Q represents the hydroxyl group, Y the carbons on the main chain, 1 the methyl side chains, and & serves as a fragment delimiter. Another line notation, used by Prickett et al. (1993), would represent the same compound as (C (O) C C (db O) C (C) db C), where side chains are enclosed in parentheses and db denotes a double bond; as is common with line notations, hydrogens are omitted from these notations since they can be easily inferred from the valence. In graph or object-oriented methods, each bond and atom is described by detailed features (Barone and Chanon, 1986), as shown in Fig. 2 and Tables 1-2. We note that this representation is *conceptually* object-oriented, but in many cases it is implemented in the form of numerical arrays.

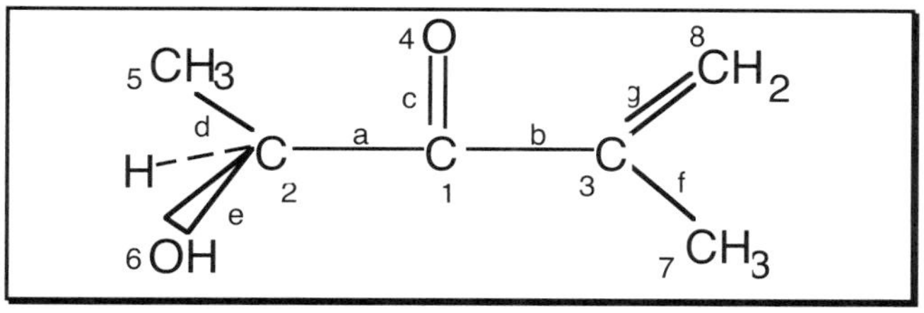

Figure 2. A chemical structure represented in Tables 1 and 2.

The matrix descriptions given in Tables 1-2 include qualitative stereochemical information, although they obviously do not entail numerical coordinates. Omission of some of the columns would give a topological (non-stereochemical) representation of the connectivity of the atoms and bonds, sufficient for many applications. We also note that there is considerable redundancy in the representation. For each atom, for example, there is a list of bonds that are attached to it, as well as list of neighboring atoms. Clearly, this representation is not intended to be minimal or compact, and it opts to cache information that may be frequently needed in analyzing the molecular structure. A drawback of the redundancy (in addition to the storage requirements) is that alteration of the structure requires more effort to maintain consistency.

Table 1. Atom Table for the Chemical Structure of Fig. 2.

Atom #	Type (a)	Stereo (b)	NBS (c)	NATCH (d)	ATBD (e)	BD (f)
1	1	0	4	3	234	abc
2	1	1	3	3	165	aed
3	1	0	4	3	178	bfg
4	4	0	2	1	1	c
5	1	0	1	1	2	d
6	4	0	1	1	2	e
7	1	0	1	1	3	f
8	1	0	2	1	3	g

(a) C=1, O=4; (b) stereocenter=1 no stereocenter=0;
(c) number of valence used; (d) number of bonds other than to H; (c) and (d) differ when there are double or triple bonds;
(e) number of atoms connected to the atom considered;
(f) bonds involved.

Table 2. Bond Table for the Chemical Structure of Fig. 2.

Bond #	B type (a)	B stereo (b)	Atom 1	Atom 2
a	1	0	1	2
b	1	0	1	3
c	2	0	1	4
d	1	0	2	5
e	1	1	2	6
f	1	0	3	7
g	1	0	3	8

(a) 1=single bond, 2=double bond, 3=triple;
(b) stereo 0=none, 1=Atom 2 up with respect to atom 1, 3=Atom 2 down with respect to atom 1.

Canonicalization and Feature Perception

Molecules are converted to a canonical form, i.e., a form which is guaranteed to be the same regardless of the arbitrary way in which the molecule may have been initially presented by the human user or derived by computational application of a reaction. Canonicalization is essential because (Bersohn, 1977; Esack and Bersohn, 1974):

- It facilitates comparison of structures, so that a new compound can be compared to the available source materials, target products, and other precursors.
- It permits the identification of classes of equivalent atoms. For example, 2,3-dimethyl butane has 6 carbons which form only two classes of chemically equivalent atoms.

Subsequent reactions need to be applied only once for each class of equivalent atoms.

- It provides a basis for stereochemical description. For example, 2,3-dimethyl butane does not have any optical isomers, because of the equivalence of atoms mentioned above, while 2,3-dimethyl pentane has an stereo center in position 3.

Canonicalization algorithms rely on the iterative refinement of a ranking of atoms. Initially, an index is assigned to each atom based on topology (e.g., number of neighbors) and/or the element (heteroatoms usually receiving higher priority than carbon atoms). This gives an initial ranking of the atoms, with many ties. This ranking is updated by examining the scores of the neighbors of each atom iteratively (and in concentric rings around each atom), until no more changes occur. Atoms that continue to tie are equivalent.

After canonicalization (and comparison to other compounds), important structural features of the molecule are extracted. These may be specific to the reaction strategies employed, or generic features, such as the rings in the molecule. In complex ring systems, a few rings suffice to form a *basis* for the ring system: these are usually the smallest rings (Barone and Chanon, 1986), such as rings (1) and (10) in the system shown in Fig. 3. These are chemically significant because they will normally play the most important role in determining the 3-dimensional shape of the molecule. Other rings may be significant from the viewpoint of synthetic strategies or reactivity in bridged polycyclic structures (Corey and Petersson, 1972; Corey, et al., 1975). All rings can be constructed as algebraic superpositions of basic rings (e.g., ring (2) as a superposition of (1) and (10)), if we use a sign to signify the direction to each ring.

As mentioned earlier, the molecular representation, canonicalization, and feature perception methods have had a significant impact on chemical structure databases.

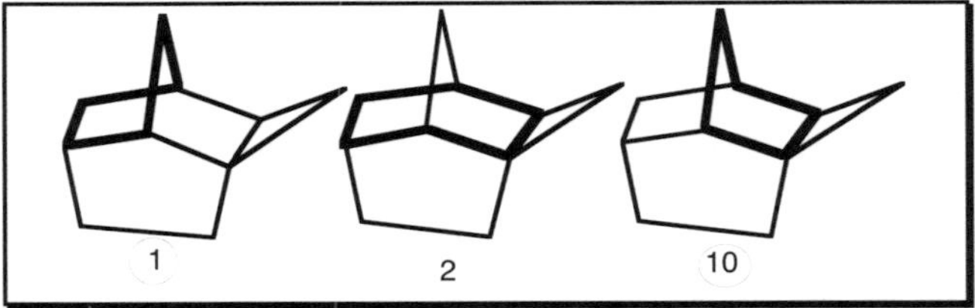

Figure 3. Perception of rings in a complex multi-ring system.

Reaction Representation

The classical approach (Corey, 1967; Corey, et al., 1985a, 1985b) followed the associative/heuristic method of representing a repertoire of known generic synthetic paths, which modify specific searchable sub-structures.

The constitutional approach (Ugi, et al, 1972; Gasteiger and Jochum, 1978) on the other hand seeks a fundamental logic-based definition of all feasible reactions. In the constitutional approach, chemical reactions are viewed as interconversions of isomeric ensembles of molecules. Each ensemble can be represented as a bond-electron matrix (atoms × atoms), with the off-diagonal elements representing the bond order between the respective atoms, and the diagonal elements representing the numbers of valence electrons that do not participate in a bond. In the case of a multi-molecular ensemble, the bond-electron matrix is formed for the whole ensemble. A reaction matrix describes the change caused by the occurrence of the reaction; it is equal to the difference between the two bond-electron matrices. The smallest principal submatrix of the reaction matrix is the characteristic R-matrix of the reactive site; it defines a generic reaction that is applicable to a variety of systems.

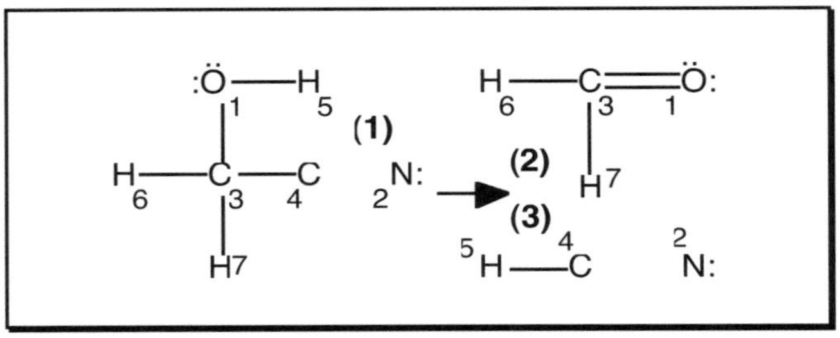

Figure 4. A chemical reaction represented in Tables 3-5.

In the example shown in Fig. 4, the reactant is a unimolecular ensemble, but the product is an ensemble of two molecules. The consistent representation of the ensembles as Bond-Electron matrices (Table 3) requires that we identify a uniform numbering of the atoms for both ensembles; the appropriate numbering should reflect the true mechanism of the reaction (i.e., the correct destination of each atom in the products). The reaction matrix is then formed as the difference of the two (Table 4). The principal submatrix in this example involves atoms 1, 3, 4, and 5, giving the R-matrix (Table 5).

The constitutional approach has the advantage of a compact, fundamental description of reactions with significant potential for discovery of reactions. A set of 1900 carbon-carbon bond formation reactions has been compactly described by only 30 R-matrices; in another example, among 100 irreducible R-matrices considered, only 12 corresponded to known reactions (Bauer and Ugi, 1982), while the remaining were fundamentally valid, i.e., they satisfied basic chemical conditions. The constraints imposed on R matrices enforce conservation of electrons and valence restrictions. However, this is not enough to ensure that an R-matrix is realistic or useful; furthermore, a useful R-matrix will appear to be applicable to a great number of situations for which the reaction turns out to be infeasible. In other words, the R-matrix description is too localized too abstract to capture the richness and complexity of organic reaction behavior. The result is a dramatic growth in the size of the search space, requiring the superposition of pruning heuristics which compromise the neatness of the constitutional approach.

Other constitutional approaches have attempted to introduce additional context, beyond the bonding changes encoded in the R-matrix. One such approach involves the separation of each reaction into two half-reactions, each involving no more than 3 linear carbon atoms (Hendrickson, 1975; Ghose, 1981).

Table 3. The Bond-Electron Matrices for Compound 1 (Fig. 4) and the Ensemble of Compounds 2 and 3 (Fig. 4).

(1)	1	2	3	4	5	6	7
1	4	0	1	0	1	0	0
2		2	0	3	0	0	0
3			0	1	0	1	1
4				0	0	0	0
5					0	0	0
6						0	0
7							0

(2)U(3)	1	2	3	4	5	6	7
1	4	0	2	0	0	0	0
2		2	0	3	0	0	0
3			0	0	0	1	1
4				0	1	0	0
5					0	0	0
6						0	0
7							0

Table 4. The Difference between the Bond-Electron Matrices of Reactants and Products for the Reaction of Fig. 4.

R	1	2	3	4	5	6	7
1	0	0	+1	0	-1	0	0
2		0	0	0	0	0	0
3			0	-1	0	0	0
4				0	+1	0	0
5					0	0	0
6						0	0
7							0

Table 5. The R-Matrix for the Reaction of Fig. 4.

R	1	3	4	5
1	0	+1	0	-1
3		0	-1	0
4			0	+1
5				0

Search Strategy and Pruning the Reaction Space

While much of the previous discussion was generically applicable to the computational representation of reactions, the particular strategies used for navigating the space of potential reactions are strictly specific to CAOS systems and are discussed here only briefly.

Effectiveness is gained if a program plans both synthetically (from precursors towards targets) and retrosynthetically (from products towards precursors) (Corey, et al., 1985b). The primary CAOS focus has historically being retrosynthetic, with only occasional bilateral approaches (Ugi, et al., 1979). The forward (synthetic) work of Varkony, et al. (1975) targeted the planning and interpretation of isotope labeling experiments for the elucidation of pathways and mechanisms.

The heuristic use of "chemical islands" (Johnson, et al., 1989) is similar to planning islands in AI planning. The idea is to postulate intermediates that divide the ultimate path into two segments (which are then pursued independently), so that the depth of the search tree is reduced. This strategy was used by Corey and Petersson (1972) and Corey, et al. (1975) for strategic bond disconnections in polycyclic structures.

The combinatorial explosion of answers is the main difficulty (Hendrickson, et al., 1981; Bersohn, et al., 1978), and efforts to overcome it have led to increased specialization of programs (Barone and Chanon, 1986).

We reiterate the trade-offs of reaction representation in CAOS: If reactions are treated as molecular rearrangements subject only to the most fundamental (low level of abstraction) restrictions, then the number or possibilities explodes, and only a small fraction of the generated paths are realistic. If reactions are given more structure to conform to our higher-level notions of how reactions work, then the chance of any kind of surprising result diminishes, i.e., the program will incorporate not only our knowledge but also our bias and ignorance. This trade-off between detailed experiential knowledge vs. application of fundamental principles is common in intelligent systems.

Chemical Reaction Processes

A pathway operating in a chemical reaction process has several characteristics which are not taken into account by CAOS. Many intermediates in a pathway are continuously produced and consumed at substantial rates without leaving (or entering) the process in only negligible amounts. A designed pathway should provide for this type of balance of the intermediates; in many cases, this means recycling of intermediates internally in the reaction system, i.e., an intermediate consumed by one reaction in the pathway is produced by another in the same amount. This use of the intermediates is directly related to the yield of the desired product, which is also ignored by CAOS. These issues arise regardless of whether the reactions are to be generated or are a priori known and stored in a database.

The synthesis of a desirable pathway from known reactions was addressed by Mavrovouniotis, et al. (1990) for biochemical pathways, and subsequently by Mavrovouniotis and Stephanopoulos (1992) for other reaction systems; a unified view of the two is given by Mavrovouniotis (1995). For example, in the oxidation of methane to methanol, which involves the 15 steps listed in Table 6, one of the pathways synthesized leads to methanol and to ethane as a stoichiometric byproduct, as shown in Fig. 5; other pathways involve no byproduct. Such distinctions, which are outside the CAOS domain, are crucial in analyzing the behavior of an existing reaction system or designing a pathway.

Table 6. Steps in the Mechanism of Oxidation of Methane to Methanol.

s_1: $CH_4 + O_2 \rightarrow CH_3 + HO_2$
s_2: $CH_3 + O_2 \rightarrow CH_3O_2$
s_3: $CH_3O_2 \rightarrow CH_2O + OH$
s_4: $CH_3O_2 + CH_4 \rightarrow CH_3O_2H + CH_3$
s_5: $CH_3O_2H \rightarrow CH_3O + OH$
s_6: $CH_3O \rightarrow CH_2O + H$
s_7: $CH_3O + CH_4 \rightarrow CH_3OH + CH_3$
s_8: $OH + CH_4 \rightarrow CH_3 + H_2O$
s_9: $CH_3 + CH_3 \rightarrow C_2H_6$
s_{10}: $CH_3 + OH \rightarrow CH_3OH$
s_{11}: $CH_3 + CH_3O \rightarrow CH_3OCH_3$
s_{12}: $CH_2O + CH_3 \rightarrow CH_4 + CHO$
s_{13}: $CHO + O_2 \rightarrow CO + HO_2$
s_{14}: $CH_2O + CH_3O \rightarrow CH_3OH + CHO$
s_{15}: $CHO + CH_3 \rightarrow CO + CH_4$

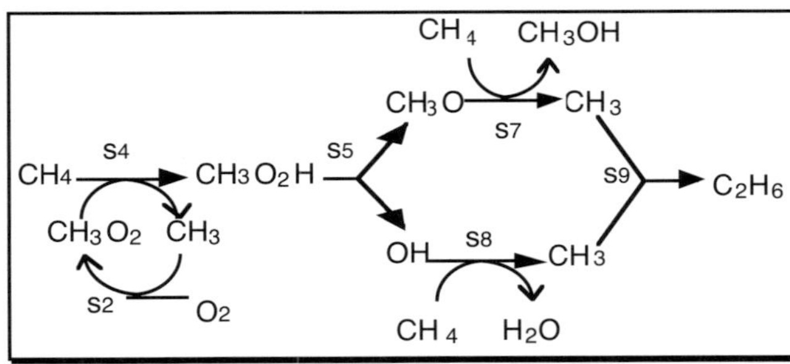

Figure 5. A pathway for the system of Table 6; methanol and ethane are produced in stoichiometric proportions.

Fujiwara, et al. (1994) developed a system for synthesizing industrial chemical reaction processes from a database of reactions, based on stoichiometry, heuristic weights on the reactions, and reactivity (kinetic) information extracted from the literature. Such heuristics facilitate the synthesis of the most promising pathways.

The above methods synthesize pathways from given sets of reactions. A task which has more similarities to CAOS is the generation of reactions occurring in a process, in order to model and simulate the process. In this case, we seek the construction of an entire reaction network not just a good path or set of paths (which is the aim of CAOS). Pruning a portion of the network should be based on an assessment that the path will occur at negligible rates and is insignificant for the behavior of the reaction network. Note also that the generation of the reaction network can only be achieved in the forward direction, from the components fed into the process towards intermediates and products, in contrast to the usual CAOS retrosynthetic approach.

It is, of course, assumed that the generic types of reactions are known. The objective is to infer the specific reactions occurring in a process, given a set of initial components.

One category of approaches for this task relies on the recursive application of R-matrices or boolean matrices (Agnihotri and Motard, 1980; Froment, 1991) – a logic-based or constitutional view. Such methods can be quite successful for very narrow classes of compounds and reactions, where constitutional representation of the reactions will not lead to large numbers of implausible yet formally correct reactions. Another set of methods follows a wholly procedural approach (Broadbelt, et al., 1994; Liguras and Allen, 1989), in which reaction application and pruning are encoded into subroutines written in an ordinary programming language (usually C). These methods do not have a reaction representation in the usual sense, i.e., there is no recognizable *description* of the reaction that is in any way distinct from its eventual use. Because of this lack of explicit representation, it is difficult to make adjustments to the specificity and pruning of the reactions.

Table 7. Description of β-Scission.

```
(Label C1+ (Find positive-carbon))
(Label C2 (Find carbon attached-to C1+))
(Label C3 (Find carbon attached-to C2))
(Forbid (Primary C3))
(Disconnect C2 C3)
(Increase-bond-order (Find bond connecting C1+ C2))
(Subtract-charge C1+)
(Add-charge C3)
```

The combination of empirical and constitutional views in an explicit representation was sought by Prickett and Mavrovouniotis (1995) in a reaction generation language. In this approach, generic reactions are described by code in the custom language. This code is compiled and the reactions are automatically applied to generate the network. The generic reaction of β-scission is represented by an early version of the language as shown in Table 7 and Fig. 6.

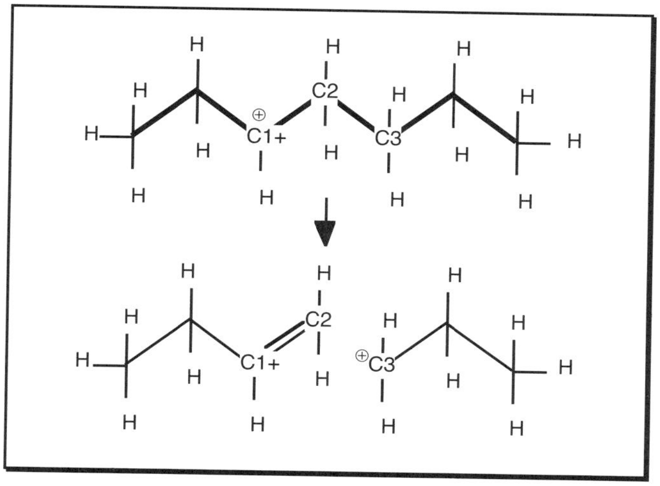

Figure 6. An example of β-scission, with the atoms labeled to conform to Table 7.

The advantage of a reaction-description language of this type is that it allows easy formulation and modification of reaction types and pruning rules. The overall goal of this approach is to accommodate a continuum of heuristic and constitutional approaches, with the user free to select the appropriate level of detail and pruning. The language also allows compile-time optimization of the reactions. For example, search statements can be re-ordered by the compiler so that the most rare part of the reaction site (a charged atom, for example) is considered first; if it cannot be located on the candidate compound, then other (more common) portions of the reaction site will not be considered at all.

In the direction of less detailed representations, a group-based lumped view of reaction systems was presented by Quann and Jaffe (1992). Each compound is represented as a vector based on 22 functional groups; naturally (as in group-based product design) the same combination of groups may correspond to a variety of distinct chemical compounds, which are lumped together. Generic reactions are likewise represented in this group vector space.

We will not consider the design of distillation and related vapor-liquid equilibrium processes. Many of the newer techniques for complex distillation design (e.g., Malone and Doherty, 1995; Bernot, et al., 1991, 1990, 1993; Foucher, et al., 1991; Wahnschafft and Westerberg, 1993; Wahnschafft, et al., 1991, 1992, and 1993) rely on very detailed information on vapor-liquid equilibrium behavior, directly dependent on the molecular structures of the compounds. Nevertheless, these methods are outside the scope of this paper, because they do not explicitly consider molecular structures.

Product Design

The products of interest in the context of chemical engineering design are pure chemical compounds, mixtures, or macromolecules. Product design is the synthesis or selection of chemical structures; the need for molecular-level knowledge in this task is self-evident. Much of the work in product design has been devoted to the requisite analysis problem: the prediction of properties from structure. Here, we focus on the synthesis problem.

Spectra

As discussed in the introduction, the identification of molecular structures from spectra can serve as a model for product design. The generic approach to the problem is based on identifying the distinct peaks of the spectrum and matching them against compounds or functional groups; this leads to a list of groups that are required to be present in the structure and a list of groups that cannot be present. Candidate structures are generated based on this information and their (simulated) spectra are matched against the target. This was DENDRAL's approach for mass spectra (Lindsay, et al., 1980), as shown in Fig. 7. Cross, et al. (1986) followed an essentially similar approach for MS-MS spectra (a secondary mass spectrum taken in tandem with the primary spectrum), and Harner, et al. (1986) for NMR spectra of biological interest. Schrader, et al. (1979), in their work on infrared and Raman spectra, combined a basic pattern-matching approach for generating initial candidate compounds with simulation of the expected spectra of each candidate through force field computations.

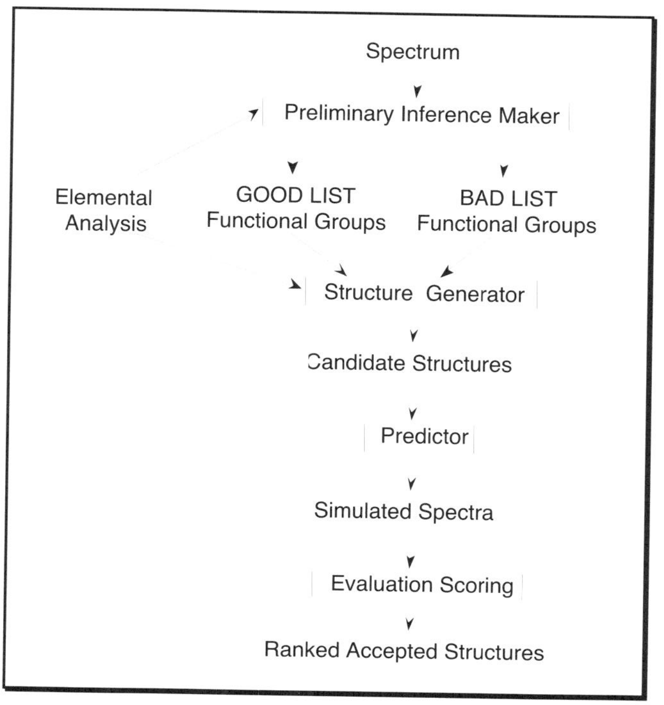

Figure 7. The DENDRAL methodology for identifying structures from spectra.

DENDRAL's success is largely due to its ability to search the entire space of structures (through its isomer-generation algorithms) while using prior knowledge about functional groups that are known to be present in the compound (or known to be absent from it). In one study, involving a compound with molecular formula $C_9H_{18}O$, chemical instability arguments allowed 1936 plausible structures; knowledge (from chemical properties) that the compound is a ketone narrowed the space down to 82 structures; and final comparison of the spectra led to only 2 structures (Lindsay, et al., 1980).

Some of the difficulties in spectra interpretation cited by Zupan (1989) have been overcome in the last few years. Spectra databases are extensively available electronically. Computational handling of molecular structures is within the computational capabilities of almost any chemical laboratory. Other core difficulties in spectroscopic analysis by intelligent systems remain (Zupan, 1989), including the combinatorial explosion in theoretically possible chemical structures; irreproducibilities of spectra (variations) creating imperfect matches; and non-locality and diversity of rules linking spectra to structure. It is particularly hard to derive confidence measures and similarity judgments for spectra or between spectra and structures. Even if some confidence measures exist for individual fragments or features, the overall confidence is hard to derive because features are not independent.

The identification of molecular structures from spectra with programs such as DENDRAL is analogous to product design. The (observed) spectrum is analogous to a design objective, with other sources of information about the compound (elemental analysis, results of chemical tests on presence or absence of groups) serving as design constraints.

Group-Contribution Based Design for Physical Properties

The objective here is the synthesis of chemical structures that satisfy a set of physical property specifications (Gani, et al., 1991; Macchietto, et al., 1990; Constantinou, et al., 1995; Joback and Stephanopoulos, 1989). The properties are related to the structure through group-contribution techniques; for pure component properties, only a linear combination of contributions is involved, while for mixtures the relation between structure and properties may involve more elaborate expressions, such as those of the UNIFAC method.

The product design procedures usually consist of the following general steps (reminiscent of DENDRAL):

- Property specifications
- Selection of acceptable groups
- Formation of feasible compounds
- Prediction of property values
- Screening and selection

In the first step, target values or ranges are selected for the desired properties. Some properties may be chosen as primary goals to be used during the search while others may be delegated to ranking acceptable structures. Naturally, a set of specifications may or may not be reachable. This issue of consistency has received attention primarily in the work of Joback and Stephanopoulos (1989), where an entire set of physical properties is reduced to only 3 abstract properties called *factors*. By reducing the property targets to specifications on the factors, one can assess the intrinsic consistency and satisfiability of the specifications and make the necessary trade-offs altering the target values) before even considering specific chemical groups or structures.

Joback and Stephanopoulos (1989) provided both an automatic design procedure and an interactive software system which displayed visually the effects of adding particular groups to the partial evolving compound structure. Their automatic procedure and more recent methods (Constantinou, et al., 1995; Gani, et al., 1991) generate feasible chemical structures from groups by relying primarily on the *valence* of groups. Examples of groups of valence 1, 2, and 3 are shown in Table 8.

Table 8. Groups of Different Valence.

Valence	Groups (> or < signify two single bonds available)
1	$-CH_3$, CH_3COO-, $-CONH_2$, $-OH$, $-CHO$, ...
2	$-CH_2-$, $>CHNH_2$, $-CH_2NH-$, $-CH{=}CH-$, ...
3	$>CH-$, $-CH_2N<$, $>CH{-}O-$, ...

Table 9. Topologies of a Compound Consisting of 6 Groups. Each topology involves a specific number of groups from each valence category.

Groups of Valence 1	Groups of Valence 2	Groups of Valence 3	Groups of Valence 4	Topology of compound	
2	4	0	0	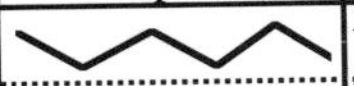	A
3	2	1	0	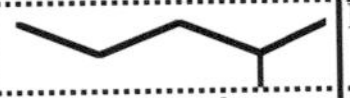	B
4	0	2	0	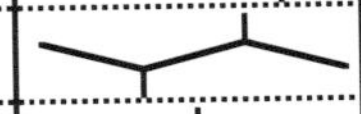	C
3	2	1	0	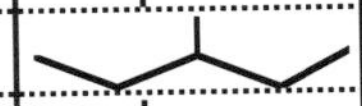	D
4	1	0	1	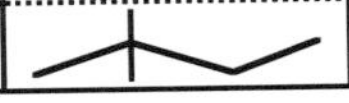	E

The initial choice of groups for a particular problem is usually heuristic and may serve to satisfy external requirements (such as avoidance of unsaturated groups to guard against polymerization) or to direct the search towards promising structures (by focusing on groups which are known to lead to the desired behavior). Given the acceptable groups, one may construct the feasible topologies of the compound, based solely on valence. For example, for an acyclic compound with 6 groups, the possible topologies correspond to the possible isomers of hexane (Table 9). For considerably larger structures, the number increases rapidly; for 40 groups, we would have

6×10^{13} structures (Bolcer and Hermann, 1994). All the methods explore this space of structures, attempting to cover all feasible topologies and all feasible choices of groups for any one topology. The methods differ in the details of their search strategy: They may rely either on screening as the compounds as they are (partially) constructed or on testing for feasibility after complete compounds are generated.

We emphasize that the topological complexity is inherent in the problem; no degree of pruning based on chemical feasibility can eliminate these topological structures — in contrast to the actual choice of groups, where chemical stability and feasibility can play a significant role, especially when a comprehensive set of groups (including unusual ones) is used. Topological simplifications are possible (Pretel, et al., 1994), but always at the expense of neglecting a portion of the search space — and potential good solutions within it.

The simplicity of the property estimation methods does provide indirect reduction of the complexity: Because in a group-contribution scheme the contributions are independent of the neighbors of the group, many isomers may have the same property value (for example, structures B and D in Table 9). Furthermore, the independence of the contributions makes it unnecessary to consider the exact configurations of groups within the topological structure. Newer property-estimation methods, such a conjugation-based approach (Mavrovouniotis, 1990; Constantinou, et al., 1993, 1994) eliminate this artificial equivalence of isomers.

In the selection of specific groups, in order to avoid infeasible or chemically implausible structures, Gani, et al. (1991) further subdivide the groups into *categories* (Table 10). Groups of category 1 are not subjected to any special restrictions. However, groups of higher categories are limited in their total number in a molecule and in their interconnections. For example, groups of category 2 may not be joined with each other or with any group of a higher category. Four groups of categories 2-5 may not be attached to the same group. Other detailed restrictions depend on the total number of groups: For a compound with 8 groups, no more than 4 may belong to categories 3-5 and, of these, no more than 3 may belong to category 4.

We should point out that these heuristics combine several classes of arguments:

- Heavily substituted compounds are likely to be chemically unstable (for example, two hydroxyl groups attached to the same carbon).
- In many applications, heavily substituted compounds are undesirable because they are likely to have higher costs.
- Limitations on the accuracy of group-contribution methods are primarily due to group interactions, making the property estimates especially unreliable for heavily substituted compounds.

In the development of intelligent systems, we usually prefer to make each argument explicit, rather than combine several (entirely different) arguments into a single implicit procedure, as in this case. However, in the absence of detailed models for each separate issue (stability, desirability/cost, and prediction accuracy), their combination is unavoidable.

With the pruning rules limiting the search space, feasible compounds can be formed, and their property values are easily computed from group contributions. A final screening and selection step is largely driven by external heuristics.

A similar approach of pruning the space of chemical structures was followed by Pretel, et al. (1994) in solvent design, with a classification of attachments (groups) for acyclic saturated compounds into:

- K= severely restricted (such as –OH)
- L= partially restricted (such as $-CH_2Cl$)
- M= unrestricted carbon attachments; single valence, such as $-CH_3$, or dual valence on main chain
- J= unrestricted carbon attachment; dual valence ($-CH_2-$) on side chain

If the number of unrestricted attachments is equal to or greater than the number of type K attachments then no restriction is placed on the remaining attachments. Otherwise, the pruning rules that accompany the classification are:

1. Type K attachments can only be combined with unrestricted carbon attachments (M, J)
2. Type L attachments can be combined with L, M, or J
3. Type J attachments change to L after a combination with a type K attachment

A family of more mathematical optimization-oriented techniques have also been proposed (Naser and Fournier, 1991; Macchietto, et al., 1990). In particular, the solvent-design procedure of Macchietto, et al. (1990) uses an MINLP formulation which includes constraints on severely restricted attachments, i.e., a group classification conceptually similar to that used by Gani, et al. (1991) and Pretel, et al. (1994).

All methods must make their choice of basis groups, based on the abilities of the property estimation method they employ. Larger, more detailed groups offer the promise of better accuracy, but they increase the size of the basis set and require more experimental data for the regression. An element introduced by Constantinou, et al. (1995) is the use of second-order groups which are substructures larger than (ordinary) first-order groups; the careful selection of these groups may improve accuracy without excessive increase in adjustable parameters. Conjugation-based methods (Mavrovouniotis, 1990; Constantinou, et al., 1993, 1994) have similar goals.

Table 10. Subdivision of Groups into Categories. Higher categories are subjected to more restrictions.

Valence	Category 1	Category 2	Category 3
Class 1	$-CH_3$	$-CH_2NH_2$, $-CH_2CN$, ...	CH_3CO-, $-CONH_2$, ...
Class 2	$-CH_2-$	$>CHNO_2$	$-CH_2CO-$, $-CH_2COO-$, ...
Class 3	$>CH-$		$-CON(CH_2-)_2$

Valence	Category 4	Category 5
Class 1	$-OH$, $-CHO$, ...	$-CH_2NH_2$, CH_3NH-, ...
Class 2	$>CHNH_2$, $-CH_2NH-$, ...	$CH_2=C<$, $-CH=CH-$, ...
Class 3	$-CH_2N<$, ...	$-CH=C<$, ...

For mixture design (Klein, et al., 1992; Constantinou, et al., 1995), progress has been made primarily for mixtures of a small number of components defined a priori. The objective is to derive a low-cost blend of solvents by combining a set of available mixtures, subject to constraints on the properties of the final blend. The difficulty associated with extending this problem to arbitrary synthesis of solvents is the complexity increase: For a mixture of three compounds, the cardinality of the set of feasible solutions is obviously raised to the cubic power. Since, in any estimation method for the properties of mixtures, pairwise interactions of groups must be considered, the complexity of each calculation for the mixture also increases. Finally, heuristics are more difficult to derive, because the effect of any one group is related to the groups in the rest of the compounds in the mixture.

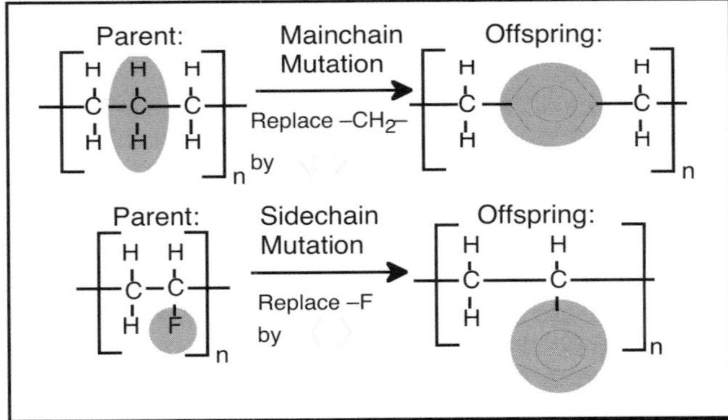

Figure 8. Genetic operators for polymer design (Venkatasubramanian, et al., 1994). The mutation operator carries out a local alteration of the chemical structure, through replacement of an atom or group.

Genetic Algorithms for Product Design

A genetic algorithm attempts to derive a good solution by generating a large set of mediocre ones and then combining and altering them iteratively; in each iteration, a much larger set of candidates is produced, and the set is reduced back to its initial size by keeping the best candidates. The chief elements of a genetic algorithm are: A representation language for the candidate solutions; a fitness measure that assesses the quality of each candidate solution; genetic operators that modify a candidate or create a hybrid of two or more candidates; and a reproduction plan that selects the candidates that will be genetically altered.

The design of molecular structures is a natural domain for genetic algorithms. Venkatasubramanian, et al. (1994) presented a comprehensive approach for molecular design, and in particular polymer design, through genetic algorithms. The genetic operators (Figs. 8-9) correspond to local alterations of the chemical structures or combination of features (substructures) from a number of candidate molecules.

Brown, et al. (1994a, 1994b) applied genetic algorithms for matching chemical graphs, for substructure searching. Clark, et al. (1994) applied genetic algorithms in the 3D domain for drug design, as will be discussed in the next subsection.

Drug Design

The basis of modern drug-receptor theory (Wilson, 1986) is that the drug fits its biological receptor in much the same way as a key fits a lock; this applies to ordinary enzymatic activity (Karp and Mavrovouniotis, 1994), although in drug activity the recognition between the receptor and the drug causes changes in regulation rather than catalysis of a chemical reaction. The traditional approach to drug discovery has been large-scale empirical testing for a desirable pharmacological activity, followed by a systematic local search (limited modification of a lead compound to optimize its pharmacological effect). The theoretical and computational pharmacologist instead aims to design a drug to interact with some given target receptor by investigating drug-receptor interactions at the molecular level. The analysis step is thus the determination of the geometric arrangement of atoms and other properties necessary for activity. The synthesis step is the construction of alternative structures incorporating the arrangement.

This exemplifies the shift from process-level experiments to molecular-level knowledge. The structure of the receptor is often unknown, and it must be inferred from the structures of compounds that bind to it. The first task is thus the identification of the structure-activity relation for existing drugs. This allows one to predict structure-activity relation for new drugs of the same general type. The ultimate goal is to predict (design) entirely new classes of drugs. In matching drugs to receptors in increasing detail, one considers not only the topology and structure of drug and receptor but also their local electronic properties. The physicochemical properties of drug must also be analyzed, because they determine whether the drug will have an opportunity to interact with the receptor.

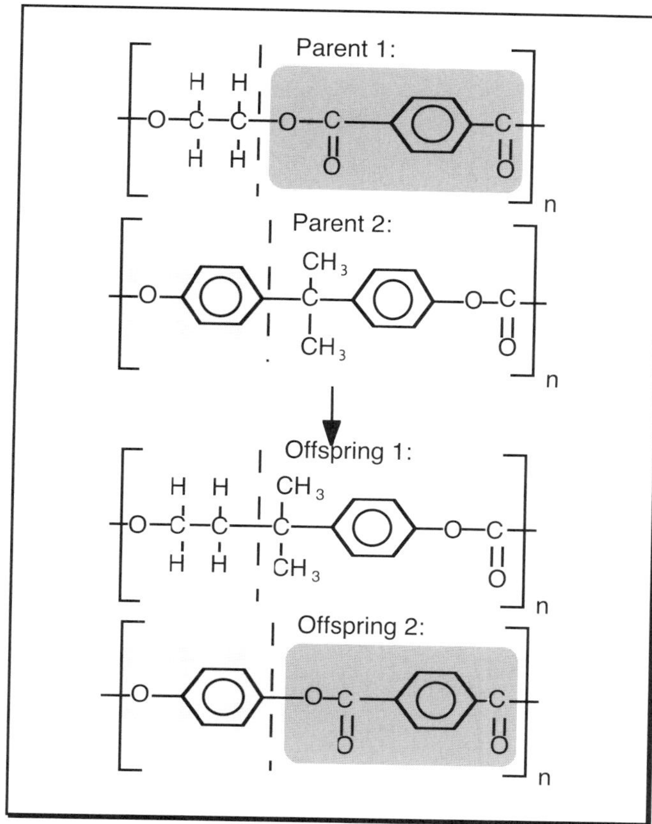

Figure 9. Genetic operators for polymer design (Venkatasubramanian, et al., 1994). The crossover operator replaces a substructure (along the backbone) in one candidate molecule by a substructure taken from another candidate molecule.

What makes drug design especially difficult, compared to design of other chemical product classes (such as solvents or refrigerants), is the fact that three-dimensional analysis of the chemical structure is essential. Atoms which are separated by several bonds may act together to bind to the receptor. Topologically disparate molecular structures may place a small set of atoms in the same three-dimensional positions — and thus bind to the same receptor. For example, in Fig. 10, the two structures bind to the nicotinic receptor (Blaney and Dixon, 1994), because they place the shaded carbon, oxygen, and nitrogen atoms in the appropriate positions in their three-dimensional conformations.

In difficult cases, a static three-dimensional structure is not enough. In Fig. 11, the key feature required for activity is flexibility of rotation of a particular nitrogen atom from an angle of 90° above the plane of the aromatic ring to an angle of 30° below the plane (Wilson, 1986). Structure (b) allows this flexibility while structure (c) does not.

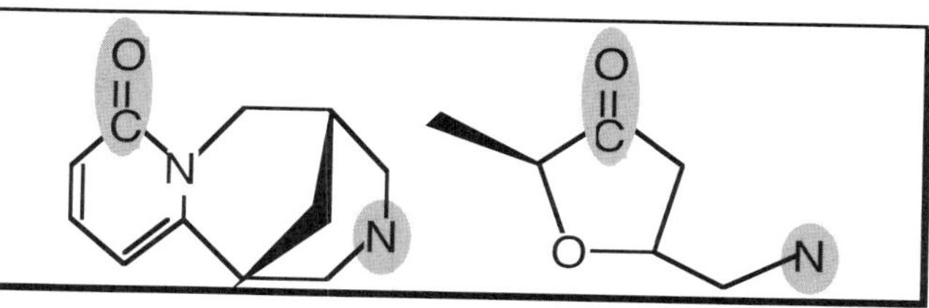

Figure 10. Nicotinic receptor agonists (Blaney and Dixon, 1994).

Clark, et al. (1994) take genetic algorithms to the three-dimensional drug-design domain: They seek conformations of 3-dimensional molecular structures that place certain atoms in the appropriate positions to display pharmacological activity. The primary information that is encoded and then genetically searched consists of the torsion angles of rotatable bonds. Their fitness measure includes two components: The Van der Waals energy component to favor lower-energy conformations; and a pharmacophore-matching score which examines the pairwise distances.

Martin, et al. (1990) point out several difficult open issues in drug design. There are too many hits in response to a query. This indicates that we have only limited understanding of the pharmacophores; the definition of the target is not sufficiently discriminating. Conformational flexibility is difficult to take into account; structures are assumed either entirely rigid or in certain cases entirely flexible (DesJarlais, et al., 1986), without intermediate states. The flexibility of the target (the receptor) in response to the specific ligand is not considered, even though the receptor does undergo conformational changes in response to the drug.

Inductive Methods

Our discussion so far has focused on deductive methods. Inductive methods that postulate the general patterns from the specific instances (including machine learning, clustering, and automated classification) are equally valuable but have not been developed to the same extent. There are many methods for induction on numerical properties that are derived from the molecular structure; these include correlations among physical properties, such as the use of the factor space (Joback and Stephanopoulos, 1989), or correlations of physical properties to structure, such as group-contribution or conjugation-based methods (Mavrovouniotis, 1990; Constantinou, et al., 1993, 1994). None of these methods, however, involves direct induction on the molecular structure itself.

Inductive methods on molecular structures are very useful for navigating through the flood of information in databases: Any automated organization, clustering, or classification of molecular database entries facilitates both the maintenance and the use of the database. In database queries or synthesis of molecular structures, induction is useful for navigating through long hit lists of results.

Often, a large set of results can be partitioned into subsets such that the crucial distinctions are accentuated and separated from minor variations. Design methods often follow a cycle: A weak generation (synthesis) method gives too few solutions; if it is strengthened, it yields too many solutions, which must be analyzed to derive pruning rules and isolate the essential features; such analysis leads to better synthesis methods which eliminate bad solutions and describe other solutions more succinctly, so that the solution space (or at least its description) is again reduced. But if we are unable to analyze the results, because of their sheer volume, we do not have the insights from which to improve the design method. In other words, we may need heuristic rules to tame the complexity, but it is hard to derive such rules manually precisely because of the complexity. Thus, we need automated inductive methods.

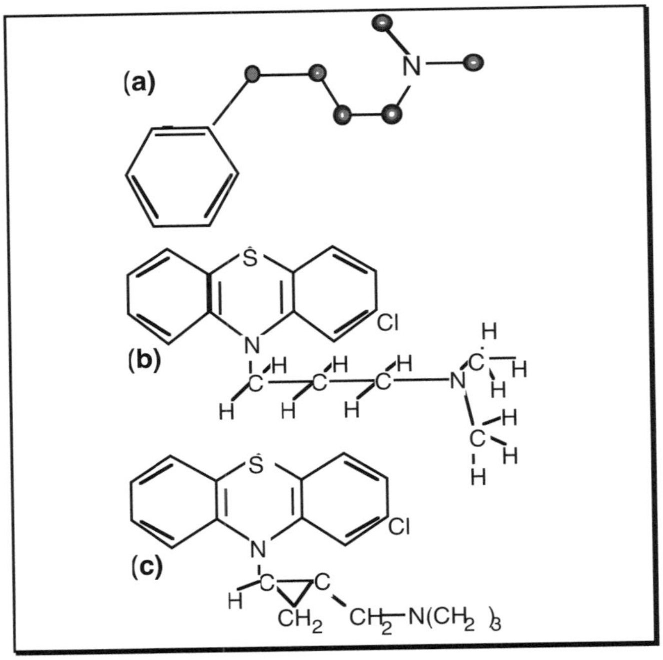

Figure 11. (a) The general structure required by neuroleptics (Wilson, 1986). The nitrogen must have flexibility of rotation from +90° to −30° with respect to the plane of the ring. (b) An example of a sufficiently flexible structure. (c) An inflexible structure which fails the test.

Induction of molecular structural features has been used to derive rules or criteria for predicting biological activity. Carcinogenesis (Bahler and Bristol, 1993) has been studied through the decision-tree induction method of Quinlan (1986, 1987): Molecular features (which may be related to the structure of the molecule but also with various macroscopic physical, chemical, and biological properties) are recursively selected to achieve classification. From known data on substrates and inhibitors of proteins, Lewis, et al. (1991) and Lewis (1992) induced structural criteria for toxicity with respect to these proteins. King, et al. (1994) carried out induction of topological constraints in protein structures. Woodruff, et al. (1986) developed a structure-elucidation program for infrared spectra, which examines large sets of known spectra to create its own correlations of peaks to functional groups. The ensemble distance geometry method (Sheridan, et al., 1986; Leach, 1991; Blaney and Dixon, 1994) operates on a list of biologically active molecules, containing a set of preserved atoms (or groups) believed to be responsible for the biological activity, clustering them based on geometric patterns in their conformations.

Induction methods have also been used to define reaction types. Rose and Gasteiger (1994) presented a system for hierarchically classifying chemical reactions, based on topological and physicochemical features. As illustrated in Fig. 12, a reaction class involves molecular structures with only partially specified substituents. Given a new (specific) individual reaction, the reaction class can be extended so that its description includes the new instance.

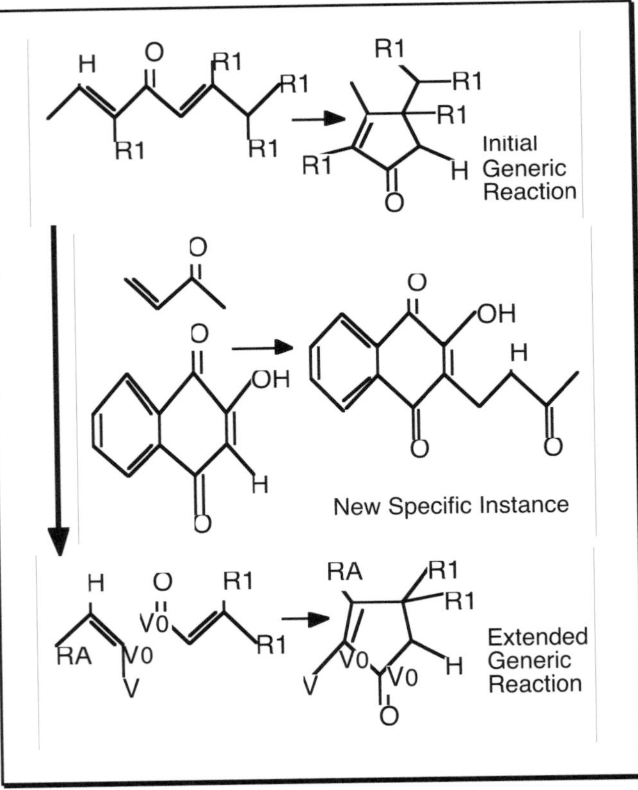

Figure 12. Induction on reactions, to form generic reaction descriptions that subsume individual reaction instances.

Conclusions

A few common themes emerge from the specific methods discussed in this article. The first is the selection of the level of detail and level of empiricism in describing molecular structures and their manipulation. Molecular structures can be defined through:

- Indices or properties which abstract the structure into a numerical vector
- Sets or vectors of functional groups
- Topological description of the connectivity of atoms and bonds
- Topological description with additional stereochemical labels on atoms or bonds
- Three-dimensional static description
- Three-dimensional flexible or dynamic description

Each of these levels of details has natural applications and limitations. Increasing computational power facilitates the inclusion of more detail.

The second theme is the role of the search strategies. Some methods employ an opportunistic search approach; genetic algorithms are a prime example in this category. Other methods aim for a systematic search that will explore, in some fashion, the entire space. We can view this distinction as a choice of (ever important) pruning strategies; opportunistic search corresponds to a form of aggressive pruning in which large portions of the space are neglected unexamined. In the case of computer-assisted organic synthesis, the same distinction manifests itself (Fig. 13) in the contrast between associative methods, which attempt to encode only well-established types of reactions based on the chemist's experience, and logic-centered or constitutional methods, which aim to encode all reactions that are in principle feasible. Ideally, one would like the flexibility to include both associative (opportunistic) and constitutional elements, in adaptation to each specific case, perhaps through domain-specific computer languages (Prickett and Mavrovouniotis, 1995).

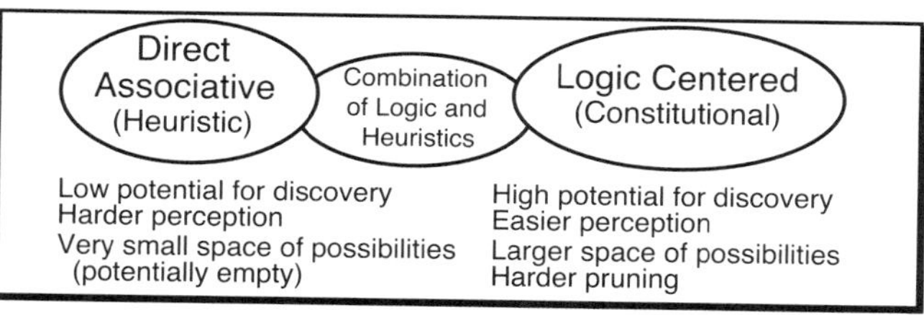

Figure 13. Heuristic-associative methods vs. logic-centered constitutional methods.

The role of molecular-level knowledge in product and process design, and in process systems engineering in general, will continue to expand. A gradual move to more detailed molecular descriptions is likely in product design. In process design, the application of molecular-level knowledge has been quite limited, but there are favorable conditions for its expansion: There is increasing need for tighter design, analysis, and control of complex chemical processes, in response to economic pressures and environmental restrictions, as cited by other papers in this volume. There is also an expanding capability for detailed process measurements. High resolution measurement devices include tandem or hyphenated analytical methods such as GC-MS (Gas Chromatography-Mass Spectroscopy), which provide a wealth of information on the chemical details of the behavior of the process. This will allow process measurements to be exploited in combination with a priori molecular-level knowledge (Mavrovouniotis and Bonvin, 1995).

Barcza (1994) discussed the generalization and future of chemical information with a number of conceptual diagrams; two of them, focusing on structure (emphasizing supra-molecular structures) and transformations/reactions (beyond simple chemical reactions) provide intriguing extensions of molecular-level information and its role (see Figs. 3 and 5 in Barcza, 1994). Process systems engineering is likely to not only consider molecular-level knowledge but also, in the future, play a role in engineering supramolecular and generalized conversion and transformation systems.

References

Agnihotri, R.B. and R.L. Motard (1980). In R.G. Squires and G.V. Reklaitis (Eds.), *Computer Applications to Chemical Engineering*. *ACS Symposium Series*, **124**, 193-206.

Bahler, D. and D. Bristol (1993). The induction of rules for predicting chemical carcinogenesis in rodents. In L. Hunter, D. Searls and J. Shavlik (Eds.), *Proceedings of the First International Conference on Intelligent Systems in Molecular Biology (ISMB-93)*. AAAI Press, Menlo Park, California. pp. 29-37.

Barcza, S. (1994). Far beyond the structure diagram. increasing the dimensionality of chemical information retrieval: structure → transport → transformation → energetics → logic and control [networks]. *J. Chem. Inf. Comput. Sci.*, **34**, 22-31.

Barone, R. and M. Chanon (1986). Computer-aided organic synthesis (CAOS). In G. Vernin and M. Chanon (Eds.), *Computer Aids to Chemistry*. Chapter I. John Wiley & Sons, New York. pp. 19-102.

Bauer, J. and I. Ugi (1982). *J. Chem. Res.*, **11,** 3101.

Bernot, C., M.F. Doherty and M.F. Malone (1990). Patterns of composition change in multicomponent batch distillation. *Chem. Engng. Sci.*, **45,** 1207-1221.

Bernot, C., M.F. Doherty and M.F. Malone (1991). Feasibility and separation sequencing in multicomponent batch distillation. *Chem. Engng. Sci.*, **46,** 1311-1326.

Bernot, C., M.F. Doherty and M.F. Malone (1993). Design and operating targets for nonideal multicomponent batch distillation. *Ind. Eng. Chem. Res.*, **32,** 293-301.

Bersohn, M. (1977). In T.E. Wipke and W.J. Howe (Eds.), *Computer Assisted Organic Synthesis, ACS Symposium Series, Volume 61*. American Chemical Society, Washington, DC.

Bersohn, M., A. Esack and J. Luchini (1978). *Comput. Chem.*, **2,** 105-111.

Blaney, J.M. and J.S. Dixon (1994). Distance geometry in molecular modeling. In K.B. Lipkowitz and D.B.

Boyd (Eds.), *Reviews in Computational Chemistry, Volume 5*. Chapter 6. VCH Publishers, New York. pp. 299-335.

Bolcer, J.D. and R.B. Hermann (1994). The development of computational chemistry in the United States. *Reviews in Computational Chemistry*, **5,** 15-63.

Broadbelt, L.J., S.M. Stark and M.T. Klein (1994). Computer generated pyrolysis modeling: on-the-fly generation of species, reactions, and rates. *Ind. Eng. Chem. Res.* **33**, 790-799.

Brown, R.D., G.M. Downs, G. Jones and P. Willett (1994a). Hyperstructure model for chemical structure handling: techniques for substructure searching. *J. Chem. Inf. Comput. Sci.*, **34,** 47-53.

Brown, R.D., G. Jones, P. Willett and R.C. Glen (1994b). Matching two-dimensional chemical graphs using genetic algorithms. *J. Chem. Inf. Comput. Sci.*, **34**, 63-70.

Clark, D.E., G. Jones, P. Willett, P.W. Kenny and R.C. Glen (1994). Pharmacophoric pattern matching in files of three-dimensional chemical structures: comparison of conformational-searching algorithms for flexible searching. *J. Chem. Inf. Comput. Sci.*, **34,** 197-206.

Constantinou, L., K. Bagherpour, R. Gani, J.A. Klein and D.T. Wu (1995). Computer-aided product design: problem formulations, methodology, and applications. *Computers chem. Engng.*, in press.

Constantinou, L., S.E. Prickett and M.L. Mavrovouniotis (1993). Estimation of thermodynamic and physical properties of acyclic hydrocarbons using the ABC approach and conjugation operators. *Ind. Engng. Chem. Res.*, **32**, 1734-1746.

Constantinou, L., S.E. Prickett and M.L. Mavrovouniotis (1994). Estimation of properties of acyclic organic compounds using conjugation operators. *Ind. Engng. Chem. Res.*, **33**, 395-403. (with supplemental material in the microfiche edition)

Corey, E.J. (1967). General methods for the construction of complex molecules. *Pure Appl. Chem.*, **14,** 19-37.

Corey, E.J., W.J. Howe, H.W. Orf, D.A. Pensak and G. Petersson (1975). General methods of synthetic analysis. strategic bond disconnections for bridged polycyclic structures. *J. Am. Chem. Soc.*, **97**, 6116-6124.

Corey, E.J., A.K. Long and S.D. Rubenstein (1985a) *Science*, **228,** 408-418.

Corey, E.J., A.K. Long, T.W. Greene and J.W. Miller (1985b) *J. Org. Chem.*, **50,** 1920-1927.

Corey, E.J. and G. Petersson (1972). An algorithm for machine perception of synthetically significant rings in complex cyclic organic structures. *J. Am. Chem. Soc.*, **94,** 460-465.

Cross, K.P., P.T. Palmer, C.F. Beckner, A.B. Giordani, H.G. Gregg, P.A. Hoffman and C.G. Enke (1986). Automation of structure elucidation from mass spectrometry-mass spectrometry data. In T.H. Pierce and B.A. Hohne (Eds.), *Artificial Intelligence Applications in Chemistry. ACS Symposium Series* **306**. American Chemical Society, Washington, DC. pp. 321-336.

DesJarlais, R.L., R.P. Sheridan, J.S. Dixon, I.D. Kuntz and R. Venkataraghavan (1986). Docking flexible ligands to macromolecular receptors by molecular shape. *J. Med. Chem.*, **29,** 2149.

Esack, A. and M. Bersohn (1974). *JCS Perkin I*, 2463-2470.

Foucher, E.R., M.F. Doherty and M.F. Malone (1991). Automatic screening of entrainers in homogeneous azeotropic distillation. *Ind. Eng. Chem. Res.*, **30**, 760-772.

Froment, G.F. (1991). Fundamental kinetic modeling of complex processes. In A.V. Sapre and F.J. Krambeck (Eds.), *Chemical Reactions in Complex Systems: The Mobil Workshop*. Van Nostrand Reinhold: New York.

Fujiwara, I., M. Sato, E. Kunugita, N. Kurita and M. Mitsuhashi (1994). Expres: an expert system for synthesizing chemical reaction cycles. *Computers chem. Engng.*, **18,** 469-480.

Gani, R., B. Nielsen and A. Fredenslund (1991). A group contribution approach to computer-aided molecular design. *AIChE J.*, **37,** 1318-1332.

Gasteiger, J. and C. Jochum (1978). *Topics Curr. Chem.*, **74**, 93.

Ghose, A.K. (1981). *J. Scientific Indus. Res.*, **40,** 423-431.

Harner, T.J., G.C. Levy, E.J. Dudewicz, F. Delaglio and A. Kunar (1986). Artificial intelligence, logic programming, and statistics in magnetic resonance imaging and spectroscopic analysis. In T.H. Peirce and B.A. Hohne (Eds.), *Artificial Intelligence Applications in Chemistry. ACS Symposium Series* **306**. American Chemical Society, Washington, DC. pp. 337-349.

Hendrickson, J.B. (1975). *J. Am. Chem. Soc.*, **97**, 5784-5800.

Hendrickson, J.B., E. Braun-Keller and G.A. Toczko (1981). *Tetrahedron*, **37**, Suppl. 1, 359-370.

Joback, K.G. and G. Stephanopoulos (1989). Designing molecules possessing desired physical property values. *Proc. FOCAPD'89*. CACHE Corporation, Austin, Texas. pp. 363-387.

Johnson, P.Y., I. Burnstein, J. Crary, M. Evans and T. Wang (1989). Designing an expert system for organic synthesis: the need for strategic planning. In B.A. Hohne and T.H. Pierce (Eds.), *Expert System Applications in Chemistry*. Chapter 9. American Chemical Society, Washington, DC. pp. 102-123.

Karp, P.D. and M.L. Mavrovouniotis (1994). Representing, analyzing, and synthesizing biochemical pathways. *IEEE Expert*, **9**(2), 11-21.

King, D., D.A. Clark, J. Shirazi and M.J.E. Sternberg(1994). Inductive logic programming used to discover topological constraints in protein structures. In R. Altman, D. Brutlag, D. Karp. R. Lathrop and D. Searls (Eds.), *Proc. of the Second International Conference on Intelligent Systems in Molecular Biology (ISMB-94)*. AAAI Press, Menlo Park, California. pp. 219-226.

Klein, J.A., D.T. Wu and R. Gani (1992). Computer aided mixture design with specified property constraints. *Computers chem. Engng.*, **16,** S229-S236.

Leach, A.R. (1991). A survey of methods for searching the conformational space of small and medium-size molecules. In K.B. Lipkowitz and D.R. Boyd (Eds.), *Reviews in Computational Chemistry II*. Chapter 1. VCH Publishers, New York. pp. 1-55.

Lederberg, J. (1965) Topological mapping of organic molecules. *Proc. Natl. Acad. Sci. USA*, **53,** 134-139.

Lewis, D.F.V. (1992). Computer-assisted methods in the evaluation of chemical toxicity. In K.B. Lipkowitz and D.B. Boyd (Eds.), *Reviews in Computational Chemistry*. Volume 3. VCH Publishers, New York. pp. 173-222.

Lewis, D.F.V., C. Ioannides and D.V. Parke (1991). COMPACT: a form of discriminant analysis for the identification of potential carcinogens. In C. Silip and A. Vittoria (Eds.), *QSAR: Rational Approaches to the Design of Bioactive Compounds*. Elsevier, New York. pp. 525-527.

Liguras, D.K. and D.T. Allen (1989). Structural models for catalytic cracking. 1. model compound reactions. *Ind. Eng. Chem. Res.*, **28**, 665-673.

Lindsay, R.K., B.G. Buchanan, E.A. Feigenbaum and J. Lederberg (1980). *Applications of Artificial Intelligence for Organic Chemistry: The DENDRAL Project*. McGraw-Hill, New York.

Macchietto, S., O. Odele and O. Omatsone (1990). Design of optimal solvents for liquid-liquid extraction and gas absorption processes. *Trans. IChemE Part A*, **68**, 429-433.

Malone, M.F. and M.F. Doherty (1995). Separation system synthesis for nonideal liquid mixtures. *AIChE Symposium Series*, **91**(304), 9-18.

Martin, Y.C., M.G. Bures and P. Willett (1990). Searching databases of three-dimensional structures. In K.B. Lipkowitz and D.B. Boyd (Eds.), *Reviews in Computational Chemistry*. VCH Publishers, New York. pp. 213-263.

Mavrovouniotis, M.L. (1990). Estimation of properties from conjugate forms of molecular structures: the ABC approach. *Ind. Engng. Chem. Res.*, **29**, 1943-1953.

Mavrovouniotis, M.L. (1995). Symbolic and quantitative reasoning; design of reaction pathways through recursive satisfaction of constraints. In G. Stephanopoulos (Ed.), *Artificial Intelligence Approaches in Product and Process Engineering*. Academic Press.

Mavrovouniotis, M.L. (1992). Synthesis of reaction mechanisms consisting of reversible and irreversible steps: II. formalization and analysis of the synthesis algorithm. *Ind. Engng. Chem. Res.*, **31**, 1637-1653.

Mavrovouniotis, M.L. and D. Bonvin (1995). Towards design of reaction paths. *AIChE Symposium Series*, **91**(304), 41-51.

Mavrovouniotis, M.L. and G. Stephanopoulos (1992). Synthesis of reaction mechanisms consisting of reversible and irreversible steps: I. a synthesis approach in the context of simple examples. *Ind. Engng. Chem. Res.*, **31**, 1625-1637.

Mavrovouniotis, M.L., G. Stephanopoulos and G. Stephanopoulos (1990). Computer aided synthesis of biochemical pathways. *Biotech. Bioeng.*, **36**, 1119-1132.

Naser, S.F. and R.L. Fournier (1991). A system for the design of an optimum liquid-liquid extractant molecule. *Computers Chem. Engng.*, **15**, 397-414.

Pretel, E.J., P.A. Lopez, S.B. Bottini and E.A. Brignole (1994). Computer-aided molecular design of solvents for separation processes. *A.I.Ch.E. J.*, **40**, 1349-1360.

Prickett, S.E., L. Constantinou and M.L. Mavrovouniotis (1993). Computational identification of conjugate paths for estimation of properties of organic compounds. *Molecular Simulation*, **11**, 205-228.

Prickett, S.E. and M.L. Mavrovouniotis (1995). A language for describing and generating complex reaction systems. (Manuscript in preparation).

Quann, R.J. and S.B. Jaffe (1992). Structure-oriented lumping: describing the chemistry of complex hydrocarbon mixtures. *Ind. Eng. Chem. Res.*, **31**, 2483-2497.

Quinlan, J.R. (1986). Induction of decision trees. *Machine Learning*, **1**, 86-106.

Quinlan, J.R. (1987). Simplifying decision trees. *Int. J. Man-Machine Studies*, **27**, 221-234.

Rohrbaugh, R.H. P.C. and Jurs (1988). Prediction of gas chromatographic retention indices for diverse drug compounds. *Anal. Chem.*, **60**, 2249.

Rose, J.R. and J. Gasteiger (1994). HORACE: an automatic system for the hierarchical classification of chemical reactions. *J. Chem. Inf. Comput. Sci.*, **34**, 74-90.

Schrader, B., D. Bougeard and W. Niggemann (1979). Determination of the structures of organic molecules by computer evaluation and simulation of infrared and raman spectra. In J. Bargon (Ed.), *Computational Methods in Chemistry*. Plenum Press, New York. pp. 37-64.

Sheridan, R.P., R. Nilakantan, J.S. Dixon and R. Venkataraghavan (1986). The ensemble approach to distance geometry: application to the nicotinic pharmacophore. *J. Med. Chem.*, **29**, 899.

Ugi, I., J. Bauer, J. Brandt, J. Friedrich, J. Gasteiger, C. Jochum, W. Schubert and J. Dugundi, J. (1979). Computer programs for the deductive solution of chemical problems on the basis of mathematical models – a systematic bilateral approach to reaction pathways. In J. Bargon (Ed.), *Computational Methods in Chemistry*. Plenum Press, New York. pp. 275-300.

Ugi, I., P. Gillespie and C. Gillespie (1972). *Trans. NY Acad. Sci.*, **34**, 416-422.

Varkony, T.H., D.H. Smith and C. Djerassi (1978). *Tetrahedron*, **34**, 841-852.

Venkatasubramanian, V., K. Chan and J.M. Caruthers (1994). Computer-aided molecular design using genetic algorithms. *Comput. Chem. Engng.*, **18**, 833-844.

Wahnschafft, O.M. and A.W. Westerberg (1993). The product composition regions of azeotropic distillation columns. 2. separability in two-feed columns and entrainer selection. *Ind. Eng. Chem. Res.*, **32**, 1108-1120.

Wahnschafft, O.M., T.P. Jurain and A.W. Westerberg (1991). SPLIT: a separation process designer. *Computers Chem. Engng.*, **15**, 565-581.

Wahnschafft, O.M., J.P. Le Rudulier and A.W. Westerberg (1993). A problem decomposition approach for the synthesis of complex separation processes with recycles. *Ind. Eng. Chem. Res.*, **32**, 1121-1141.

Wahnschafft, O.M., J.P. Le Rudulier, P. Blania and A.W. Westerberg (1992). SPLIT: II. automated synthesis of hybrid liquid separation systems. *Comput. Chem. Engng.*, **16**, S305-S312.

Wilson, S. (1986). *Chemistry by Computer*. Plenum Publishing, New York.

Woodruff, H.B., S.A. Tomellini and G.M. Smith (1986). Elucidation of structural fragments by computer-assisted interpretation of IR spectra. In T.H. Pierce and B.A. Hohne (Eds.), *Artificial Intelligence Applications in Chemistry*. *ACS Symposium Series* **306**. American Chemical Society, Washington, DC. pp. 312-320.

Zupan, J. (1989). *Algorithms for Chemists*. John Wiley & Sons, New York.

KNOWLEDGE-BASED APPROACHES IN PROCESS SYNTHESIS

Chonghun Han and George Stephanopoulos
Massachusetts Institute of Technology
Cambridge, MA 02139

Y.A. Liu
Virginia Polytechnic Institute and State University
Blacksburg, Virginia 24061

Abstract

This paper reviews the recent developments and applications of knowledge-based systems (KBS) for process synthesis, emphasizing the modeling of the design process, synthesis methodologies and applications, and innovative approaches. It describes the recent trends of KBS for process synthesis, including the increasing integration of multiple knowledge representations and problem-solving methodologies and the expanding range of industrial applications to narrowly defined classes of process synthesis problems.

Keywords

Design, Process synthesis, Artificial intelligence, Knowledge-based systems.

Introduction

Process synthesis has been an active research area since Rudd (reference 1) wrote a series of papers on a system synthesis principle that has made a profound impact on the subsequent design research. Nishida, et al. (2) define process system synthesis as an act of determining the optimal interconnection of processing units as well as the optimal type and design of the units within a process system. When the performance of the system is specified, the structure of the system and the performance of process units are not determined uniquely. The task of synthesis is to select a particular system out of the large number of alternatives that meet the specified performance.

Many review papers (2-7), books (8-12) and more than 500 papers have appeared in the area of process synthesis. The impressive progress made in the artificial intelligence and mathematical programming has opened new paradigm in process synthesis. The deeper understanding of the design process has resulted in many powerful systematic design procedures for the preliminary and detail designs. The thermodynamic insight into the process (e.g., pinch analysis) has led to the unparalleled contribution to the industry in terms of cost savings and productivity.

This paper focuses on the knowledge-based approach to process synthesis. We emphasize the development of knowledge-based systems (KBS), modeling of the engineering design process, design methodologies and applications, and innovative approaches. We do not discuss the mathematical formulation of the design process and its solution through mathematical programming techniques. For interested readers, there are review papers available on the applications of mathematical programming to process synthesis (13,14). In addition, several books on artificial intelligence in chemical engineering have appeared recently (15-18).

KBS Approach in Design

KBS: Definition and Key Elements

Definition and Structure of KBS

Definition. A knowledge-based system (KBS) is a computer program that has a specialized knowledge about a specific area and solves a specific class of problems using the knowledge. The better the knowledge, the better the performance of the system.

Stephanopoulos and Han (18) suggest that knowledge is today's and future's competitive advantage in industrial manufacturing and have listed the roles of knowledge-based approach to gain such advantages as: (a) to represent and preserve knowledge using various knowledge representation schemes; (b) to clone knowledge once it has been represented within a KBS, (c) to make knowledge more precise by identifying inconsistencies and conflicts during the construction and testing of knowledge representations, (d) to centralize knowledge by incorporating various forms of knowledge into a single KBS, and (e) to port knowledge by copying and sending a KBS to other places.

Structure of KBS. A KBS contains: (1) a knowledge base; (2) an inference engine; and (3) a user interface. The *knowledge base* contains specific, in-depth information about the problem at hand. The knowledge consists of facts, rules, and heuristics. To utilize the knowledge, a KBS relies on its *inference engine*. The engine uses inference mechanisms to process the knowledge and draw conclusions. The *user interface* provides smooth communication between the program and the user (16).

Knowledge Representation

A *representation* is a set of conventions about how to describe a class of things. Winston (19) suggests that a good representation should a) make the importance of objects and relations explicit, b) expose natural constraints, c) bring objects and relations together, d) suppress irrelevant detail, e) be transparent, f) be complete, g) be concise, h) be fast, and i) be computable. Furthermore, he has listed various powerful representations: a) semantic nets, b) state space, c) goal tree, d) rules and rule chaining, e) frames, classes, instances, slots, slot values, and inheritance and f) logic and resolution proof.

Chemical engineers have applied many representation schemes that bring important objects and relations together, while suppressing irrelevant details. A process flowsheet is one of the most successful representation schemes. It is routinely used in process simulation systems. In separation processes involving; (a)vapor/liquid mixtures, (e.g., azeotropic, extractive, and reactive distillations), (b) liquid/liquid mixtures (e.g., extraction), and (c) solid/liquid mixtures (e.g., extractive crystallization), ternary composition diagram and residue curve map (RCM) (20) are significant design tools. In process integration, the composite curve, grand composite curve, and grid diagram (21) based on the idea of pinch point have been key representations of evolution and deployment of pinch analysis. In batch processing, Reklaitis (22, 23), Reklaitis, et al. (24), and Rippin (25) have reviewed many efficient tools (e.g., Gantt chart, recipe network, etc.) to represent networks of all types of continuous, semicontinuous or batch processing tasks. We have found the concepts of state-transition network (26), material-based representation (27), and data-flow diagram (28) particularly useful.

Problem-Solving Methods

After a good representation scheme has been established for the problem, the problem-solving methods should be designed. For a small-scale problem, a purely algorithmic scheme can be used. However, for a fairly large-scale problem, it is inevitable to use some combinations of heuristic, evolutionary and algorithmic schemes. Sometimes, the interactive user intervention is indispensable. Some examples of problem-solving methods (16, 19) are listed as follows.

Generate-and-test. Similar to trial-and-error or total enumeration method. Heavily used in many knowledge-based approaches, i.e. design alternative management in hierarchical decision procedure (29).

Means-end analysis. We try to find the difference between the initial state and the goal state and apply appropriate operator to reduce the difference. For the generated smaller differences, the same strategy is applied in an iterative manner until all the differences are resolved. This has been applied to an early process synthesis program, *AIDES* (30) and a learning program, *CDP-Soar* (31).

Problem decomposition. Sometimes, it is possible to convert difficult goals into one or more easier-to-achieve subgoals. Each subgoal, in turn, may be further divided into one or more lower-level subgoals. Problem decomposition is sometimes called *goal reduction*. For example, the hierarchical decision procedure (32) employs this strategy to reduce the scope of design from the overall plant to input/output structure to recycle structure, and Han, et al. (29) implement the decision procedure using goal reduction.

Search methods. (depth-first search, breadth-first, best-first, branch-and-bound, discrete dynamic programming, A*). Efficient search methods are one of key methods to the success of KBS in process synthesis because of combinatorial nature of synthesis. These search methods have been used in many applications (16).

Constraint satisfaction and conflict resolution. This is a pruning or filtering technique that eliminates certain potential solutions within the state space when they do not satisfy a specific set of constraints. This technique is particularly useful in handling constraints during a design process (29, 33).

We believe that the key ideas of these methods can be far more advanced by utilizing the structure and characteristics of the design problem and by recognizing the nature of the engineering design process. Therefore, we shall describe additional "problem-solving methods" under the section "model of design process for implementation."

User Interfaces

Given that the design process is complex, the human-computer interface is an essential element for the success of KBS. The interface allows a user to monitor the progress of the evolving design, provide decisions, and guide the direction of design in a productive way.

There are several criteria for a user-interface design (34). A good user-interface design (a) needs a thorough understanding of the problems that will be addressed and the ways in which users will approach the problems, (b) provides flexible navigation — how and from what point the user accesses the different functions, (c) can be customized by the user to accommodate different working styles, (d) provide prompts and context-sensitive help to guide the user, (e) anticipate and prevent errors, (f) use color, shape, size, and texture purposefully to provide useful information to the user, and (g) should be consistent throughout the system.

KBS Development Process

In this section, we briefly describe a procedure based on the software engineering principle to develop a KBS.

Step 1. Requirement Analysis

When is a KBS development warranted? In particular, when is a KBS development possible, appropriate and justifiable? We refer the reader to Quantrille and Liu (16, pp. 397-401) for specific answers.

Step 2. Knowledge Acquisition

When the KBS development is warranted, the developer should acquire knowledge on the problem from the *domain experts*. To define a design task, we must acquire knowledge defining: (a) the class of *problems* that can be solved, (b) the class of *candidate solutions* that contains a set of acceptable solutions to the problem, and (c) the *domain theory*, the body of domain-specific knowledge that is accessed in solving such problems, and constraints. How can such design knowledge be either easily acquired from domain experts, or otherwise automatically added to the knowledge base?

Knowledge acquisition using graphic interfaces. Tong and Sriram (35) suggest that using graphical interfaces facilitate acquiring design knowledge from experts.

Model-based knowledge acquisition. Tailoring a particular knowledge acquisition method to a specific design model is another way to simplify knowledge acquisition. As the overall structure of the design process model is already known, the knowledge engineer tries to find the decisions and design heuristics that can fill up the framework. *ConceptDesigner* (29) and *PIP* (36) are examples of this approach , where a hierarchical decision procedure is used as a framework to be filled up with the knowledge of domain expert.

Case-based knowledge acquisition. Given a problem together with a solution for a similar problem, we use the solution to evaluate new solution to the given problem. This strategy of *case-based knowledge acquisition* is useful to ensure whether there is any missing information about design decisions or design process.

Step 3. Object Modeling

A model is an abstraction of a process or an object for the purpose of understanding it before building it. Thus, modeling helps us deal with systems that are too complex to understand directly. A useful object modeling technique (OMT) is that developed by Rumbaugh, et al. (28) at General Electric. In OMT, a system is modeled from three related but different viewpoints, each capturing important aspects of the system, but all required for a complete description: (a) an object model for the static, structural, "data" aspects of a system, (b) the dynamic model for the temporal behavioral, "control" aspects of a system, and (c) the functional model for the transformational, "function" aspects of a system. A typical software procedure incorporates all three aspects: it uses data structures (object model), it sequences operations in time (dynamic model), and it transforms values (functional model).

Step 4. Logic Inference Modeling

Rules are declarations of policy or condition that must be satisfied. Rules, then, allow the user to specify policies or conditions in small, stand-alone units using explicit statements. Forward- and backward-chaining are two reasoning mechanisms. When more than one rule is triggered, we generally want to perform only one of the possible actions, thus requiring a *conflict-resolution strategy* to decide which rule actually fires. The conflict resolution strategies (19) are: (a) rule ordering, (b) context limiting, (c) specificity ordering, (d) data ordering, (e) size ordering, and (f) recency ordering. Note that the proper choice of a conflict resolution strategy depends on the situation, making it difficult or impossible to rely on a fixed conflict resolution strategy or combination of strategies. When a set of rules shows a common behavior which can be abstracted into a generic rule, the generic rule, also called *meta rule*, enables more robust and efficient reasoning process.

Step 5. Verification

Once a KBS has been developed, it should not be used without thorough verification. There exist many techniques for the verification of KBS. Several issues for process synthesis in particular are as follows:

1. The synthesized flowsheet should be tested using a rigorous process simulation system. The result will show whether the flowsheet is thermodynamically feasible and all design

decisions have been made in a logically consistent manner (37).

2. The production rules should be verified for logical completeness and consistency using the available verification methods (38)
3. The synthesized design alternatives for the same process should be evaluated under their optimal conditions. This will keep our focus on the overall view of the flowsheet, rather than a small part.

The KBS development process is an iterative process, not a sequential process. During the process, the interaction with the domain experts such as design engineer or process chemists is extremely important.

Key Elements of KBS Approach to Process Synthesis: Model of Design Process

Why do we need a model for the design process?

If the computer is to understand how the design is done, the knowledge base of a KBS must have an explicit description of the following: (a) the *design steps* that it must go through from the problem specification to the final solution; (b) the decision-making and numerical *design tasks* that it must carry out at each design step; and (c) the elements that are needed by each design task, i.e. specifications, units, materials, relationships, constraints, etc. The first and the second requirements necessitate the availability of *a model of the design process*, while the third implies the need for design-oriented languages. The model of the design process represents the methodology that the human designer has instructed the computer to carry out.

Requirements for the Model

Stephanopoulos, et al. (39) have provided an extensive discussion on what is needed for the creation of an intelligent computer aid for design that can carry out specific design activities. Also, Mostow (40) provides a superb tour in modeling the process of design through the various issues as follows: the state of the design, the goal structure of the design process, design decision, rationales for design decisions, control of the design process, and the role of learning in design.

Design Methodologies for Process Synthesis

We summarize below four fundamental approaches for the synthesis of chemical process flowsheets.

1. Systematic Generation of Base-Case Design. The systematic generation builds the flowsheet from smaller, more basic components strung together in such a way that raw materials become eventually transformed into the desired products. Some examples are *AIDES* (30), *BALTAZAR* (41, 42), and the hierarchical decision procedure (32). All the above works are based on the top-down design. Lien (43) proposes a bottom-up opportunistic design.

2. Evolutionary Synthesis. The evolutionary synthesis (42, 44-45) starts with an existing flowsheet from the same or a similar product, and then make the objectives of the specific case at hand. Most process synthesis works in industry use this approach. The evolutionary synthesis is often combined with elementary decomposition theory, heuristic strategy and branch-and-bound strategy.

3. Superstructure Optimization. This approach views synthesis as an optimization over structure, and starts with a larger superstructure that contains, embedded within it, many alternatives and redundant interconnections and then strips parts of the superstructure away, while simultaneously optimizing other design parameters. *Task assignment* (46) and all MINLP (Mixed-Integer Nonlinear Programming)-based synthesis methodologies belong to this category. In particular, MINLP-based synthesis has been extensively applied to many process synthesis areas. See the latest review by Grossmann and Kravanja (14).

4. Targeting. Strong bounds on performance of system are derived to reduce the enormously huge search space. Generally, the targeting information gives useful insights about the global solution (although not complete information) and is much easier to obtain. For example, we can readily establish the design targets for minimum utility requirements prior to the actual synthesis of a heat-exchanger network. Linnhoff (21) gives a latest review of the various targeting approaches in process integration, while Morgan (47) convincingly describes the importance of using design targets to improve process designs and the design process.

Model of Design Process for Implementation

Having identified a design methodology for a given class of design problems, we can adopt a variety of artificial intelligence techniques to map the methodology into a computer model (48). For example, three models of the design process that have led to practical KBS implementation are as follows.

1. Planner-Generator-Tester. Quantrille and Liu (16) and Brunet and Liu (37) have developed an Expert System for SEParation synthesis (*EXSEP*) that uses planner-generator-tester. *Planner* trims the state space to a manageable level, *generator* identifies and creates feasible separations from the existing state space, and *tester* evaluates different feasible separations based on rank-ordered heuristics and chooses an attractive separation from the ones generated.

2. Task-oriented approach. Many KBS applications in process synthesis involve design strategies and knowledge that are well-structured. Task-oriented approach recognizes this structure in the design task and exploits it by describing the design task in terms of identifiable types of knowledge and a specific problem-solving strategy. Myers, et al. (49) have developed *STILL*, an expert system for the design of sieve-tray distillation columns. Han, et al. (29) have improved the concept of design task to a design agent so that the generic function of a high-level agent can be inherited to a low-level agent.

3. *Planner-Scheduler-Designer* (29). *Planner* defines the hierarchical planning of design activities, by prescribing the characteristics of the intermediate design milestones that the design process must go through. *Scheduler* determines the sequence of the design steps to be taken as one attempts to advance the current state of a processing scheme to the next milestone, defined by the Planner. *Designer* maintains the representation of the processing scheme being synthesized and other domain-specific knowledge to do the following: acquire necessary data, reason with a specific set of rules, execute a design algorithm, carry out an optimization procedure, update the state of the evolving design, etc.

KBS Environments for Implementation

Advances in the area of graphic user interface, database, and object-oriented programming and the available knowledge-based development tools such as KEE, Nexpert, G2, made it possible to prove the design process model by applying the developed system to various design cases (27, 29, 36). The concept of *bench-marking*, applying the system to a set of pre-selected design case studies, looks very promising to compare different design models from diverse perspectives.

Applications

Total Process Synthesis

AIDES (30) decomposes the synthesis problem into three levels: a) selection of raw material and chemical reactions, b) selection of products, and c) selection of mixing, splitting and separation. It also employs a means-end analysis. Lu and Motard (50) present a heuristic-evolutionary approach to total flowsheet synthesis. Hierarchical decision procedure proposed by Douglas (10, 32) is a practical approach for synthesis from scratch. However, it depends heavily on heuristics and does not deal with all circumstances. To overcome this weakness, Douglas has extended the scope of the original decision procedure from continuous, single-product processes to multistep reaction processes to the design of solid processes, to gas-liquid-solid separation processes, to the design of solid processes and polymer processes, and to the design for waste minimization (see literature cited in Han, et al. (29)).

PIP (36) is a system that implements Douglas' hierarchical procedure for conceptual process flowsheets. *ConceptDesigner* (29) is another implementation of Douglas' hierarchical synthesis of process flowsheets. In contrast to *PIP*, the system has incorporated an explicit and formal model of the design methodology and represented the model with a network of design agents. The agent representation allows the rule-driven selection of a design and an easy generation of design alternatives for the selection of the best one through a combinatorial optimization.

Mizsey and Fonyo (51, 52) combine the hierarchical procedure with the algorithmic methods by incorporating a user-driven synthesis technique and an efficient bounding strategy for cross-level interactions.

Reaction Pathway Synthesis

Mavrovouniotis (53) gives a latest review on how to discover improved reaction sequences for producing large organic and biochemical molecules. Recognizing that earlier works employed the thermodynamic feasibility as a preliminary screening which reduces the number of chemical production schemes, Fornari and Stephanopoulos (54) show that addition of a few constraints related to role specification, gross added value and demand/supply requirements results in a dramatic decrease of alternative reaction paths. Knight and McRae (55) present an approach to process integration based on the choice of the system chemistry. Fujiwara, et al. (56) create a system called *EXPRES* for the synthesis of chemical reaction cycles.

Reactor Network Synthesis

Reactor network synthesis is the problem of deciding the optimal type(s), arrangement(s), and size(s) of the chemical reactors when the reaction mechanism and kinetics are given. Because of the complexity and highly nonlinear characteristics of chemical reactions, reactor network synthesis has received relatively limited attention until recently.

Superstructure approach (57, 58) where a structural optimization is performed on postulated network of idealized reactors, has led to an effective synthesis strategy. However, the approach has also drawbacks: a) nonlinear nature of reaction processes making it difficult to get a "rich enough" superstructure, and b) numerous nonconvexities with the possibility of local maxima. *Reactor targeting approach* (59, 60) is one intuitively analogous to targets employed in heat exchanger networks. As strong bounds on network performance can be derived in terms of concentrations without explicit construction of a network, the targeting information gives useful insights about the global solution and is much easier to obtain. Due to the characteristics of reactor network synthesis, i.e. nonlinearity of reactions, there are very few KBS, if at least they exist.

Separation System Synthesis

Separation system synthesis involves the selection of the methods and sequences for separating a multicomponent mixture into desired product streams. In sharp separations, each component being separated appears almost completely in one and only one product; in sloppy or nonsharp separations, some components in the feed appear simultaneously in two or more product streams. There have been over 100 papers on separation system synthesis, focusing mostly on sharp separations. A review of the literature appears elsewhere (61).

A number of expert systems for separation system synthesis have been reported (16, pp. 420-26). We believe, however, that some expert systems (particularly those for sharp separation sequencing) have been mis-guided in addressing a design task for which the available knowledge (e.g., heuristics) lends itself to a not-so-different optimization problem. Separation system synthesis involving sloppy separations, especially solvent-based separations, is relatively more complex, since a thermodynamic feasibility analysis of potential separations must be performed (62).

There have been very few reports on the development and applications of KBS for solvent-based separation system synthesis. Some examples are: (a) *EXSEP* (16, 37) applies the plan-generate-test approach to heuristically synthesize separation systems involving ordinary distillation, absorption, extraction and stripping; (b) *SSAD* (Separation Synthesis ADvisor) (63) employs a task-oriented approach to heuristically select the methods and sequences for separating liquid and gas/vapor mixtures; and (c) *SPLIT* (64, 65) combines multiple knowledge sources into an integrated system (i.e., a blackboard) with a mathematical optimization software for azeotropic separations. Significantly, SPLIT has now been implemented as a synthesis module of the commercial software, *ADVENT* (66). This commercial implementation demonstrates the growing industrial importance of azeotropic, extractive and reactive distillations (20, 67-68). Other significant studies include: (a) KBS for reverse-osmosis desalination (69) and for membrane permeators (70); and (b) KBS for heat-integrated distillation processes (71-72).

Batch Process Synthesis

Reklaitis (22) defines the batch process synthesis problem as follows: Given a) a set of product specifications, such as the amount, selling price, and the production horizon, b)a set of feasible equipment items, c) the recipe information for each product, d) the status and transfer rules for the intermediates resulting from each recipe step, e) the set of equipment types, f) resource utilization requirements associated with each step of the recipe, g) the lost production time and costs associated with change-overs between products, h) inventory charges for each product and intermediate per unit time, i) a performance function, determine a feasible design that optimizes the selected performance measure, including specification of the operating strategy and equipment configuration and the number of each type of processing unit and intermediate storage vessel and their sizes or capacities.

Reklaitis (22-23), Reklaitis, et al. (24) and Rippin (25) give updated review of the literature on batch process synthesis. Three papers that present heuristic and hierarchical synthesis methods of particular interest are: Yeh and Reklaitis (73), Patel and Mah (74), and Iribarren, et al. (75).

Considering the diverse classes of problems involved in batch process, there are a variety of research needs for KBS as follows: a) a systematic procedure for the efficient decomposition of the synthesis problem, b) the unit sizing and costing procedures, c) the characterization of dominant effects in different situations to aid in identification of appropriate measures of improvement, and d) most importantly, the development of a framework to deal simultaneously with both process design and operations.

PROVAL (76) is developed at Merck as a computer-aided process evaluation tool to produce a preliminary batch design. *BATCHES* (77) is a simulator that combines both discrete and continuous simulation methodologies to model the start-up, execution and shut-down of batch operations. *BioSep-Designer* (78) encompasses the optimal design of the configuration of protein recovery and purification processes. *BatchDesign-Kit* (27) is a software system to support the development and design of batch processes for manufacturing pharmaceuticals and specialty chemicals, by integrating economic and ecological considerations. It consists of three subsystems: Process_Synthesizer for batch process synthesis, Process_Assessor for the rapid generation and evaluation of alternative flowsheets, and Solvent_Selector for *optimal selection of solvent*.

Process Synthesis for Environment

Most of the process synthesis work presented in the previous sections can be adopted for environmental applications, e.g., source reduction, resource recovery and recycling, waste minimization and treatment. Here, we focus primarily on process wastes rather than main products. Thus, only research works directly related to the environment are reviewed.

Source reduction. Douglas (79), and Smith and Petela (80), among others, have demonstrated how process synthesis and operations greatly affect the production of wastes in a chemical process. Ciric and Huchette (81) introduce multiple objective optimization approach to sensitivity analysis of process costs. Flower, et al. (82) examine a quantitative approach to establishing targets for the amount of waste that a process will produce in relation to process economics. Wang and Smith (83-84) propose a general methodology based on targeting for the design of distributed effluent treatment systems. Targets were used to determine quickly the scope of and to screen alternatives in water reuse and regeneration before design.

Process subsystem synthesis. This area has been an active area of research and has a lot to offer to waste minimization. On the separation sub-area, (85) develop the theory of *mass-exchanger networks* (MENs), based on the pinch concept for heat-exchanger networks. Their work has been extended to cases involving simultaneous mass-exchange and regeneration, and reactive mass exchange, and applied to waste minimization problems in petroleum refineries and synthetic fuel plants. See recent references cited in Srinivas and El-Halwagi (86). Sung, et al. (87) and Rossiter (88) report the industrial experience on applying process integration to environmental problems at the Southern California Edison Co. and other sites. Ahmad and Barton (89) present a methodology for automatic targeting of maximum feasible solvent recovery of an arbitrary number of components by batch distillation. On the reactor subarea, the research on reactor network synthesis described in the previous section can be made useful. Efforts to formally combine reaction and separation system synthesis have also appeared in the literature (e.g., Balakrishna and Biegler (59)).

Economic profitability analysis. There is a need for analytical techniques that can efficiently determine the sensitivity of economic profitability to changing disposal costs and regulations. Ciric and Jia (90) present a method for computing the sensitivity of net profits to waste treatment costs using multiobjective optimization.

KBS for the environment. *MIN-CYANIDE* (91) is an expert system for cyanide waste minimization in electroplating plants. The system resorts to the fuzzy logic for knowledge representation and manipulation of imprecise information. The system evaluates and recommends process changes to minimize cyanide production. *EASY* (92) is an environmental assessment system developed by Merck to help quantify potential environmental impact and implications of manufacturing processes during the early stages of development. *ENVIROCAD* (93) is a design tool for efficient synthesis and evaluation of integrated waste recovery, treatment and disposal processes.

Selection of Physical Property Prediction Models, Materials, Equipment and Catalyst

Selection is a classification problem where a knowledge-based approach is extremely powerful. Many expert systems have been developed and being used. Some examples are as follows.

Selection or design of catalysts. Bañares-Alcántara, et al. (94) for catalyst selection, Hu, et al. (95) for the design of alcohol synthesis catalysts and Kito, et al. (96) for the design of multicomponent catalysts.

Selection of process equipment. Bakker, et al. (97) for mixer, Chuang, et al. (98) for a vapor-liquid contactor, Hanratty and Joseph (99) for a laboratory reactor, Lahdenperä, et al. (100) for solid-liquid separation equipment, Yang, et al. (101) for process equipment and Venkatasubramanian (102) for pump selection.

Selection of materials. Bieker and Simmrock (103) for the selection of solvents for extractive and azeotropic distillation, Modi, et al. (104) for plant-wide identification of solvents for pharmaceutical and specialty-chemical batch processes, Venkatasubramanian (105) for plastics selection and Shacham, et al. (106) for material selection and corrosion control.

Selection of physical property models. Bañares-Alcántara, et al. (107), and Gani and O'Connell (108).

Design Environment

Intelligent design environments greatly facilitate the development of KBS for flowsheet synthesis. An example is *Design-Kit* (109), a completely object-oriented environment with a modeling language, i.e. MODEL.LA. (39), integrated expert-system shell, and with external databases and programs. We refer the reader to Han and Stephanopoulos (29) and three other papers being presented at ISPE'95 for further discussions on design environments.

Innovative Approaches in Process Synthesis

What kind of innovation to process synthesis can computer bring to us? We describe a number of new developments below.

Innovating through Search

Can computer create innovative designs? Aelion, et al. (110) introduce a new paradigm: create innovative designs by searching through the space of artifacts generated from first principles. Lien (43), proposes that for creative designs, the process should be seen as opportunistic rather than goal-driven (as in the routine design). Both studies point to a challenging direction to design innovation through computer search.

Cooperating Expert Systems and Concurrent Engineering

The objective of cooperative design and concurrent engineering is to facilitate effective coordination and communication in various disciplines involved in engineering, and consequently to produce a robust design within a shorter time frame. Within the cooperative design environment, each user interacts with other users through a global database where (i) solutions are posted by a user, (ii) critiques are articulated by other users, (iii) conflicts are identified, (iv) mechanisms for negotiation and resolution of the conflicts are invoked (48).

Wang, et al. (111) apply concurrent engineering to fluid catalytic cracking process design. The work of Simmrock and his students has advanced from an expert-system development for a specific design task, to cooperating knowledge-integrating systems, i.e. PROSYN-M (Process Synthesis Manager) (71, 112). These studies represent exciting areas for continuing research and development.

Learning

Modi, et al. (31) explore the potential to include automatic learning in a design agent. They implement a simple distillation sequencing system, *CPD-Soar*, within Soar (113), which is an integrated software architecture for both learning and problem solving. When a similar distillation sequencing problem is given, *CPD-Soar* is able to accelerate problem solving by applying the learned heuristic knowledge. Obviously, automating the learning process in KBS is an important area for research.

Combining Symbolic Processing and Algorithmic Optimization

Grossmann and his students are active in integrating symbolic processing (e.g., logic-based reasoning techniques) with algorithmic optimization (e.g., MINLP) for process synthesis applications. See, for example, Floudas and Grossmann (114) and Viswanathan and Grossmann (115). Their work clearly indicates the increasing importance of integrating different methodologies for solving complex process synthesis problems.

Combining Expert Systems and Neural Networks

Baughman and Liu (116) illustrate how to combine the qualitative reasoning skills of an expert system with the quantitative modeling capabilities of a neural network to develop an *expert network* for bioseparation process synthesis. The integrating approach represents a significant tool for process synthesis and operations (17).

Summary

1. The value of KBS is in their ability to solve narrowly defined, specific industrial problems.
2. To keep up with the expanding scope of process synthesis and its related problems, we need to integrate an array of specific KBS, numerical computations, high-level databases (e.g. environmental regulations), and graphic interfaces.
3. More work is needed to understand the design processes for various process synthesis problems (e.g., batch processes, processes for environment, and total processes). This includes the systematic procedures, more powerful design heuristics, abstraction and learning from similarity between different classes of synthesis problems, the interactions between process subsystems, the systematic generation of design alternatives and their management, search techniques, theoretical bounding for the overall system and its subsystems.
4. The design of integrative but still open computer-aided design environments to support process synthesis will continue to be an important research issue. Of particular significance will be the architecture and integration of different computer environments including both software (e.g. object-oriented database) and hardware (e. g. massively-parallel computers).

References

Introduction

1. Rudd, D.F. (1968). The synthesis of system designs, I. elementary decomposition strategy. *AIChE J.*, **14**, 343-349.
2. Nishida, N., G. Stephanopoulos and A.W. Westerberg (1981). A review of process synthesis. *AIChE J.*, **27**, 321-351.
3. Hendry, J.E., D.F. Rudd and J.D. Seader (1973). Synthesis in the design of chemical processes. *AIChE J.*, **19**, 1-15.
4. Hlavacek, V. (1978). Synthesis in the design of chemical processes. *Comput. Chem. Eng.*, **2**, 67-75.
5. Westerberg, A.W. (1980). A review of process synthesis. In R.G. Squires and G.V. Reklaitis (Eds.), *Computer Applications to Chemical Engineering . ACS Symp. Ser,* **124**. American Chemical Society, Washington D.C. pp. 53-87.
6. Westerberg, A.W. (1987). Process synthesis: a morphological view. In Y.A. Liu, et al. (Eds.), *Recent Developments in Chemical Process and Plant Design*. Wiley, New York. pp. 127-145.
7. Stephanopoulos, G. and C. Han (1994). Intelligent systems in process engineering: a review. In E.S. Yoon (Ed.), *Proc. 5th Int. Symp. Process Systems Engineering*. Kyongju, Korea. pp. 1319-1336.
8. Rudd, D.F., G.J. Powers and J.J. Siirola (1973). *Process Synthesis*. Prentice-Hall, Englewood Cliffs, NJ.
9. Kumar, A. (1981). *Chemical Process Synthesis and Engineering Design*. McGraw-Hill, New Delhi, India.
10. Douglas, J.M. (1988). *Conceptual Design of Chemical Processes*. McGraw-Hill, New York.
11. Hartmann, K. and K. Kapick (1990). *Analysis and Synthesis of Chemical Process Systems*. Elsevier, New York.
12. Smith, R. (1995). *Chemical Process Design*. McGraw-Hill, New York.
13. Grossmann, I.E. and M.M. Daichendt (1994). New trends in optimization-based approaches to process synthesis. In E.S. Yoon (Ed.), *Proc. 5th Int. Symp. Process Systems Engineering*. Kyongju, Korea. pp. 94-109.
14. Grossmann, I.E. and Z. Kravanja (1995). Mixed-integer nonlinear programming techniques for process systems engineering. Report EDRC 06-184-95, Carnegie Mellon Univ., Pittsburgh, PA.
15. Mavrovouniotis, M.E. (Ed.) (1990). *Artificial Intelligence in Process Engineering*. Academic Press, San Diego, CA.

16. Quantrille, T.E. and Y.A. Liu (1991). *Artificial Intelligence in Chemical Engineering*. Academic Press, San Diego, CA.
17. Baughman, D.R. and Y.A. Liu (1995). *Neural Networks in Bioprocessing and Chemical Engineering*. Academic Press, San Diego, CA.
18. Stephanopoulos, G. and C. Han (Eds.) (1995). *Intelligent Systems in Process Engineering. Adv. Chem. Eng.*, **21**. Academic Press, San Diego.

KBS Approach in Design

Definition , Structure and Development of KBS

19. Winston, P. (1992). *Artificial Intelligence*, 3rd ed. Addison-Wesley, Reading, MA.
20. Fien, G.J.A.F. and Y.A. Liu (1994). Heuristic synthesis and shortcut design of separation processes using residue curve maps: a review. *Ind. Eng. Chem. Res.*, **33**, 2505-2522.
21. Linnhoff, N. (1994). Use pinch analysis to knock down capital costs and emissions. *Chem. Eng. Prog.*, **90**(8), 32-57.
22. Reklaitis, G.V. (1990). Progress and issues in computer-aided batch process design. In J.J. Siirola, I.E. Grossmann, and G. Stephanopoulos (Eds.), *Foundations of Computer-Aided Process Design*. CACHE-Elsevier, Amsterdam. pp. 241-276.
23. Reklaitis, G.V. (1995). Award lecture: Computer-aided design and operation of batch processes. *Chem. Eng. Edu.*, **29**, Spring, 76-85.
24. Reklaitis, G.V., D.W. T. Rippin and A. Sunol (Eds.) (1994). *NATO ASI Series F., Batch Processing Systems Engineering*. Springer Verlag, New York.
25. Rippin, D.W.T. (1993). Batch process system engineering: a retrospective and prospective review. *Comput. Chem. Eng.*, **17**, 1-13.
26. Kondili, E., C.C. Pantelides and R.W.H. Sargent (1988). A general algorithm for scheduling of batch operations. In *Proc. Third Int. Symp. Process Systems Engineering*. Sydney. pp. 62-75.
27. Linninger, A.A., S.A. Ali, E. Stephanopoulos, C. Han and G. Stephanopoulos (1995). Synthesis and assessment of batch processes for pollution prevention. In *Pollution Prevention via Process and Product Modifications*, *AIChE Symp. Series*, **90**(303). pp. 46-58.
28. Rumbaugh, J., M. Blaha, W. Premerlani, F. Eddy, and W. Lorensen (1991). *Object-oriented Modeling and Design*. Prentice Hall, Englewood Cliffs, NJ.
29. Han, C., G. Stephanopoulos and J.M. Douglas (1995). Automation in design: the conceptual synthesis of chemical processing schemes. In G. Stephanopoulos and C. Han (Eds.), *Intelligent Systems in Process Engineering*, Ch. 2. Academic Press, San Diego, CA.
30. Siirola, J.J., G.J. Powers and D.F. Rudd (1971). Synthesis of system designs. III. toward a process concept generator. *AIChE J.*, **17**, 677-682.
31. Modi, A.K., A. Newell, D.M. Steier, and A.W. Westerberg (1995a,b). Building a chemical process design system within soar — 1. design Issues, and 2. learning issues. *Comput. Chem. Eng.*, **19**, 75-89 and 345-361.
32. Douglas, J.M. (1985). A hierarchical decision procedure for process synthesis. *AIChE J.*, **31**, 353-362.
33. Sheppard, C.M., L.T. Beltrami and R.L. Motard (1991). Constrained-directed nonsharp separation sequence design. *Chem. Eng. Comm.*, **106**, 1-32.
34. Britt, H.J., J.A. Smith and J.S. Warek (1989). Computer-aided process synthesis and analysis environment. In J.J. Siirola, I.E. Grossmann and G. Stephanopoulos (Eds.), *Foundations of Computer-Aided Process Design*. CACHE-Elsevier, Amsterdam. pp. 281-307.
35. Tong, C. and D. Sriram (Eds.) (1990). *Artificial Intelligence in Engineering Design*, Vol. III. Academic Press, San Diego, CA.
36. Kirkwood, R.L. (1987). PIP — process invention procedure. a prototype expert system for synthesizing chemical process flowsheets. Ph.D. thesis, Dept. of Chem. Eng., Univ. of Massachusetts, Amherst, MA.
37. Brunet, J.C. and Y.A. Liu (1993). Studies in chemical process design and synthesis. 10. an expert system for solvent-based separation process synthesis. *Ind. Eng. Chem. Res.*, **32**, 315-334.
38. Renard, F.X., L. Sterling and C. Brosilow (1993). Knowledge verification in expert systems combining declarative and procedural representations. *Comput. Chem. Eng.*, **17**, 1067-1090.

Key Elements of KBS Approach to Process Synthesis

39. Stephanopoulos, G., G. Henning and H. Leone (1990a,b). MODEL.LA. a modeling language for process engineering-I. the formal framework; and II. multifaceted modeling of processing systems. *Comput. Chem. Eng.*, **14**, 813-846 and 847-869.
40. Mostow, J. (1985). Toward better models of the design process. *AI Magazine*, Spring, 44-57.
41. Mahalec, V. and R.L. Motard (1977a). Procedures for the initial design of chemical processing systems. *Comput. Chem. Eng.*, **1**, 57-68.
42. Mahalec, V. and R.L. Motard (1977b). Evolutionary search for an optimal limiting process flowsheet. *Comput. Chem. Eng.*, **1**, 149-160.
43. Lien, K.M.(1989). A framework for opportunistic problem solving. *Comput. Chem. Eng.*, **13**, 331-342.
44. Nath, R. and R.L. Motard (1981). Evolutionary synthesis of separation processes. *AIChE J.*, **27**, 578-587.
45. Stephanopoulos, G. and A.W. Westerberg (1976). Studies in process synthesis, II: evolutionary synthesis of optimal process flowsheets. *Chem. Eng. Sci.*, **31**, 195-204.
46. Umeda, T. and A. Ichikawa (1975). A rational approach to process synthesis: an extensive use of task assignment concept. *Chem. Eng. Sci.*, **30**, 699-707.
47. Morgan, S.W. (1992). Use process integration to improve process designs and the design process. *Chem. Eng. Prog.*, **88**(9), 62-68.
48. Stephanopoulos, G. (1990). Artificial intelligence and symbolic computing in process engineering design. In J.J. Siirola, I.E. Grossmann and G. Stephanopoulos (Eds.), *Foundations of Computer-Aided Process Design*. CACHE-Elsevier, Amsterdam. pp. 21-47.
49. Myers D.R., J.F. Davis and D.J. Herman (1988). A task-oriented approach to knowledge-based systems for process engineering design. *Comput. Chem. Eng.*, **12**, 959-971.

Application of KBS Approach to Process Synthesis

Total Process Synthesis

50. Lu, M. D. and R. L. Motard (1985). Computer-aided total flowsheet synthesis. *Comput. Chem. Eng.*, **9**, 431-445.
51. Mizsey, P. and Z. Fonyo (1990a). Toward a more realistic overall process synthesis — the combined approach. *Comput. Chem. Eng.*, **14**, 1213-1236.
52. Mizsey, P. and Z. Fonyo (1990b). A predictor-based bounding strategy for synthesizing energy integrated total flowsheets. *Comput. Chem. Eng.*, **14**, 1303-1310.

Reaction Pathway Synthesis

53. Mavrovouniotis, M. L. (1995). Symbolic and quantitative reasoning: design of reaction pathways through recursive satisfaction of constraints. In G. Stephanopoulos and C. Han (Eds.), *Intelligent Systems in Process Engineering*, *Adv. Chem. Eng.*, **Vol. 21**, Ch. 3. Academic Press, San Diego, CA.
54. Fornari, T. and G. Stephanopoulos (1994a,b). Synthesis of chemical reaction paths: 1.the scope of group contribution methods; and 2. economic and separation constraints. *Chem. Eng. Comm.*, **129**, 135-157 and 159-192.
55. Knight, J. P. and G. J. McRae (1993). An approach to process integration based on the choice of the system chemistry. Paper no. 153d, *AIChE Mtg.*, St. Louis, MO.
56. Fujiwara, I., M. Sato, E. Kunugita, N. Kurita and M. Mitsuhashi (1994). EXPRES: an expert system for synthesizing chemical reaction cycles. *Comput. Chem. Eng.*, **18**, 469-480.

Reactor Network Synthesis

57. Achenie, L.E.K. and L.T. Biegler (1990). A superstructure-based approach to chemical reactor network synthesis. *Comput. Chem. Eng.*, **14**, 23-40.
58. Kokossis, A.C. and C.A. Floudas (1991). Synthesis of isothermal reactor-separator-recycle systems. *Chem. Eng. Sci.*, **46**, 1361-1383.
59. Balakrishna, S. and L.T. Biegler (1993). A unified approach for the simultaneous synthesis of reaction, energy and separation systems. *Ind. Eng. Chem. Res.*, **32**, 1372-1382.
60. Lakshmanan, A. and L.T. Biegler (1994). Isothermal and nonisothermal targeting of reactor networks for process synthesis: a more comprehensive approach. Paper 221c, *AIChE Mtg.,* San Francisco, CA, Nov.

Separation System Synthesis

61. Liu, Y.A. (1995). Separation system synthesis. In M. Howe-Grant and J. I. Kroschwitz (Eds.), *Kirk-Othmer Encyclopedia of Chemical Technology*, 4th ed., Vol. 21. Wiley, NY, in press.
62. Liu, Y.A., T.E. Quantrille and S.H. Cheng (1990). Studies in chemical process design and synthesis. 9. a unifying method for the synthesis of multicomponent separation sequences with sloppy product streams. *Ind. Eng. Chem. Res.*, **29**, 2227-2241.
63. Barnicki, S. D. and J. R. Fair (1990,1992). Separation system synthesis: a knowledge-based approach. 1. liquid-mixture separations, and 2. gas/vapor mixtures. *Ind. Eng. Chem. Res.*, **29**, 421-432, and **31**, 1679-1694.
64. Wahnschafft, O. M., T. P. Jurian and A. W. Westerberg (1991). SPLIT: a separation process designer. *Comput. Chem. Eng.*, **15**, 565-581.
65. Wahnschafft, O. M., J. P. Le Rudulier, P. Blania and A. W. Westerberg (1992). SPLIT II. automated synthesis of hybrid liquid separation systems. *Comput. Chem. Eng.*, **16**, S305-S312.
66. Aspen Technology, Inc., (1995). Meet SPLIT: Aspen Tech's new distillation synthesis technology. *The Aspen Leaf*, February, p. 3.
67. Doherty, M. F. and G. Buzad (1992). Reactive distillation by design. *Trans. IChemE., Series A*, **70**, 448-458.
68. Doherty, M. F. and G. Buzad (1994). New tools for the design of kinetically controlled reactive distillation columns. *Comput. Chem. Eng.*, **18**, S1-S13.
69. Papafotious, K., D. Assimacopoulos and D. Marinos-Kouris (1992). Synthesis of a reverse-osmosis desalination plant: an objected-oriented approach. *Trans. IChemE., Series A*, **70**, 304-312.
70. Pettersen, T. and K. M. Lien (1994). Synthesis of separation systems using membrane permeators. In E. S. Yoon (Ed.), *Proc. 5th Int. Conf. Process Systems Engineering*. Kyongju, Korea. pp. 835-842.
71. Gerhard, S., W. Schuttenhelm and K. H. Simmrock (1994). Cooperating knowledge integrating systems for the synthesis of energy-integrated distillation processes. *Comput. Chem. Eng.*, **18**, S131-S135.
72. Wahl, P. E. (1991). Synthesis of heat-integrated distillation sequences: approaches combining knowledge bases and operations research techniques. Ph.D. thesis. Norwegian Inst. of Tech., Trondheim, Norway.

Batch Process Synthesis

73. Yeh, N. C. and G.V. Reklaitis (1987). Synthesis and sizing of batch semicontinuous process: single product plants. *Comput. Chem. Eng*, **11**, 639-654.
74. Patel, A.N. and R.S.H. Mah (1993). Heuristic synthesis in the design of noncontinuous multiproduct plants. *Ind. Eng. Chem. Res.*, **32**, 1383-1395.
75. Iribarren, O.A., M.F. Malone and H.E. Salomone (1994). A heuristic approach for the design of hybrid batch-continuous processes. *Trans. IChemE.*, **72**, Part A, 295-306.
76. Bamopoulos, G., E. Hsu, and S. Bacher (1986). PROVAL — a tool for batch process design and evaluation. *AIChE Mtg.*, New Orleans, LA, April.
77. Joglekar, G.S., S.M. Clark, D.S. Carmichael and G.V. Reklaitis (1987). BATCHES and the simulation of multipurpose plants. *AIChE Spring Mtg.*, Houston.
78. Siletti, C.A. and G. Stephanopoulos (1990). BioSep Designer: a knowledge-based process synthesizer for bioseparations. In C. Tong and D. Sriram (Eds.), *Artificial Intelligence Approaches in Engineering Design*. Academic Press, San Diego, CA.. pp. 295-316.

Process Synthesis for Environment

79. Douglas, J. M. (1992). Process synthesis for waste minimization. *Ind. Eng. Chem. Res.*, **31**, 238-243.
80. Smith, R. and E. A. Petela (1991-92). Waste minimization in the process industries: 1. the problem, 2: reactors, 3: separation and recycle systems, 4: process operations, 5: utility waste, *The Chemical Engineer*, 24-25 (Oct., '91), 17-23, (Dec. '91), 24-28 (Feb. '92), 21-23 (Apr. '92), 32-35 (Jul. '92).
81. Ciric, A. R. and S. G. Huchette (1993). Multiobjective optimization approach to sensitivity analysis: waste treatment costs in discrete process synthesis and optimization problems. *Ind. Eng. Chem. Res.*, **32**, 2636-2646.
82. Flower, J. R., S. C. Bikos and S. W. Johnson (1994). Process synthesis for pollution targeting. *AIChE Mtg.*, San Francisco, CA, November.
83. Wang, Y. P. and R. Smith (1994a). Wastewater minimization. *Chem. Eng. Sci.*, **49**, 981-1006.
84. Wang, Y. P. and R. Smith (1994b). Design of distributed effluent treatment systems. *Chem. Eng. Sci.*, **49**, 3127-3145.
85. El-Halwagi, M. M. and V. Manousiouthakis (1989). Synthesis of mass-exchange networks. *AIChE J.*, **35**, 1233-1244.
86. Srinivas, B. K. and M. El-Halwagi. (1994). Synthesis of reactive mass-exchange networks with general nonlinear equilibrium functions. *AIChE J.*, **40**, 463-472.
87. Sung, R. D., A. P. Rossiter and H. Klee, Jr. (1993). Process integration and the environment. In T. Berntsson (Ed.), *Proc. IEA Workshop on Process Integration*. IEACADDET, Sittard, Sweden. pp. 263-279.
88. Rossiter, A. P. (1995). Process integration and pollution prevention. In *Pollution Prevention via Process and Product Modifications. AIChE Symp. Series*, **90 (303)**, 12-22.
89. Ahmad, B. S. and P. I. Barton (1994). Solvent recovery targeting for pollution prevention in pharmaceutical and specialty chemical manufacturing. *AIChE Mtg.*, San Francisco, CA, Nov.
90. Ciric, A. R. and T. Jia (1994). Economic sensitivity analysis of waste treatment costs in source reduction projects: continuous optimization problems. *Comput. Chem. Eng.*, **18**, 481-495.
91. Huang, Y. L., G. Sundar and L. T. Fan (1991). Min-Cyanide: an expert system for cyanide waste minimization in electroplating plants. *AIChE Mtg.*, Houston, TX, Apr.
92. Venkataramani, E. S., M. J. House and S. Bacher (1992). An expert-system-based environmental assessment system (EASY). *Workshop on pollution prevention in the process industries*, September 10-11, Newark, NJ.
93. Petrides, D. P., K. G. Abeliotis, and S. K. Mallick (1994). ENVIROCAD: a design tool for efficient synthesis and evaluation of integrated waste recovery, treatment and disposal processes. *Comput. Chem. Eng.*, **18**, S603-S607.

Selection of Physical Property Prediction Models, Materials, Equipment, and Catalyst

Selection or design of catalyst.

94. Bañares-Alcántara, R., A.W. Westerberg, E. I. Ko, R. and M. D. Rychener (1987, 1988). DECADE — a hybrid expert system for catalyst selection. I: expert system considerations; and II. final architecture and results. *Comput. Chem. Eng.*, **11**, 265-278, and **12**, 923-938.
95. Hu, X. D., H. C. Foley and A. B. Stiles (1991). Design of alcohol synthesis catalysts assisted by a knowledge-based expert system. *Ind. Eng. Chem. Res.*, **30**, 1419-1427.
96. Kito, S., T. Hattori and Y. Murakami (1990). An expert system approach to computer-aided design of multi-component catalysts. *Chem. Eng. Sci.*, **45**, 2661-2667.

Selection of process equipment

97. Bakker, A., J. R. Morton and G. M. Berg (1994). Computerizing the steps of mixer selection. *Chem Eng.*, **101**, March, 120-129.
98. Chuang K. T., G. X. Chen and M. Rao (1992). An expert system for selecting a vapor-liquid contactor. *Can. J. Chem. Eng.*, **70**, 794-799.
99. Hanratty, P. J. and B. Joseph (1992b). Decision-making in chemical engineering and expert systems: application of the analytic hierarchy process to reactor selection. *Comput. Chem. Eng.*, **16**, 849-860.
100. Lahdenperä, E., E. Korhonen, and L. Nyström (1989). An expert system for selection of solid-liquid separation equipment. *Comput. Chem. Eng.*, **13**, 467-474.
101. Yang, J., T. Koiranen, A. Kraslawski and L. Nyström (1993). Object-oriented knowledge-based systems for process equipment selection. *Comput. Chem. Eng.*, **17**, 1181-1189.
102. Venkatasubramanian, V. (1988). PASS: an expert system for pump selection. *Knowledge-Based System in Chemical Engineering. CACHE Case Study Series*, Vol. II. CACHE Corp., Austin, TX.

Selection of materials

103. Bieker, T. and K. H. Simmrock (1994). Knowledge integrating system for the selection of solvents for extractive and azeotropic distillation. *Comput. Chem. Eng.*, **18**, S25-S29.
104. Modi, A., J. P. Aumond and G. Stephanopoulos (1994). Plant-wide identification of solvents for pharmaceutical and specialty-chemical batch processes. Paper no. 220b, *AIChE Mtg.*, San Francisco, CA, Nov.
105. Venkatasubramanian, V. (1988) CAPS: an expert system for plastics selection. *Knowledge-Based System in Chemical Engineering. CACHE Case Study Series*, Vol. III. CACHE Corp., Austin, TX.
106. Shacham, M., O. Shacham and K. Amaral (1987). Applications of expert systems in chemical engineering and corrosion control. *Corros. Rev.*, **7**, 151.

Selection of physical property models.

107. Bañares-Alcántara, R., A. W. Westerberg and M. D. Rychener (1985). Development of an expert system for physical property predictions. *Comput. Chem. Eng.*, **9**, 127-142.
108. Gani, R. and J. P. O'Connell (1989). A knowledge-based system for the selection of thermodynamic models. *Comput. Chem. Eng.*, **13**, 397-403.

Design Environment

109. Stephanopoulos, G., J. Johnston, T. Kriticos, R. Lakshmanan, M. Mavrovouniotis and C. Siletti (1987). DESIGN-KIT: an object-oriented environment for process engineering. *Comput. Chem. Eng.*, **11**, 655-674.

Innovative Approaches in Process Synthesis

110. Aelion, V., J. Cagan and G. Powers (1991). Inducing optimally directed innovative designs from chemical engineering first principles. *Comput. Chem. Eng.*, **15**, 619-627.
111. Wang, X. Z., M. L. Lu, E. K. T. Kam, and C. McGreavy (1994). Concurrent engineering application in fluid catalytic cracking process design. In E. S. Yoon (Ed.), *Proc. 5th Int. Symp. Process Systems Engineering*. Kyongju, Korea. pp. 381-386.
112. Schembecker, G., K. H. Simmrock and A. Wolff (1994). Synthesis of chemical process flowsheets by means of cooperating knowledge integrating systems. *I ChemE Symp. Ser.*, **133**, 333-341.
113. Laird, J. E., A. Newell and P. S. Rosenbloom (1987). Soar: an architecture for general intelligence. *Art. Intell.*, **33**, 1-64.
114. Floudas, C. A. and I. E. Grossmann (1994). Algorithmic approaches to process synthesis: logic and global optimization. Report EDRC 06-174-94, Carnegie Mellon University, Pittsburgh, PA.
115. Viswanathan, J. and I. E. Grossmann (1994). Symbolic logic, optimization and process synthesis. Report EDRC 06-172-94, Carnegie Mellon University, Pittsburgh, PA.
116. Baughman, D. R. and Y. A. Liu (1994). An expert network for predictive modeling and optimal design of extractive bioseparations in aqueous two-phase systems. *Ind. Eng. Chem. Res.*, **33**, 2668-2687.

A PERSPECTIVE ON INTELLIGENT SYSTEMS FOR PROCESS HAZARDS ANALYSIS

Venkat Venkatasubramanian[1]
Laboratory for Intelligent Process Systems
School of Chemical Engineering
Purdue University
West Lafayette, IN 47907

Malcolm L. Preston
ICI Engineering Technology
The Heath
Runcorn, Cheshire WA7 4QD UK

Abstract

Process safety, occupational health and environmental issues are ever increasing in importance in response to heightening public concerns and the resultant tightening of regulations. The process industries are addressing these concerns with a systematic and thorough process hazards analysis (PHA) of their existing as well as new facilities. Given the enormous amounts of time, effort and money involved in performing the PHA reviews, there exists considerable incentive for automating the process hazards analysis of chemical process plants. In this paper, we review the progress towards such automation through the development of intelligent systems. The substantial progress that has been made in the *HAZOPExpert* project in the US and the *STOPHAZ* project in Europe suggests that the technology is now well beyond proof of concept and is ready for industrial applications and commercial exploitation. This progress has promising implications for inherently safer design, operator training and real-time fault diagnosis.

Keywords

Process hazards analysis, Process safety analysis, Hazard identification, HAZOP, Intelligent systems.

Introduction

As modern chemical plants have become very complex, it has become quite difficult to analyze and assess in detail the inherent hazards in these systems, thus raising environmental, occupational safety and health related concerns. The plants are often operated at extremes of pressure and temperature to achieve optimal performance, making them more vulnerable to equipment failures. Furthermore, since we are also dealing more and more with toxic substances and genetic engineering technology, the results of a major industrial accident can be quite devastating, as seen in the Bhopal, India accident. Industrial statistics show that even though major catastrophes and disasters from chemical plant failures may be infrequent, minor accidents are very common, occurring on a day to day basis, resulting in many occupational injuries, illnesses, and costing the society billions of dollars every year (Bureau of Labor Statistics, 1992; McGraw-Hill Economics, 1985; National Safety Council, 1994).

Hence, industrial practitioners view safety as an important design objective in process engineering in order to prevent accidents. Engineers involved in the design and operation of the chemical plants systematically ask questions such as, "What can go wrong?," "How likely is it to happen?," "What range of consequences might there be?," "How could they be averted or mitigated?," "How safe is safe enough?," and so on in order to evaluate and improve the safety of the plant. The answers to these and

[1] Author to whom all correspondence should be addressed.

other related questions are sought in what is known as Process Hazards Analysis (PHA) or Process Safety Analysis (PSA) of a chemical process plant. Process Hazards Analysis is the systematic identification, evaluation and mitigation of potential process hazards which could endanger the health and safety of humans and cause serious economic losses. This is an important activity in Process Safety Management (PSM) which requires a significant amount of time, effort and specialized expertise. The importance of this activity was recently underscored by the Occupational Safety and Health Administration's PSM standard Title 29 CFR 1910.119 in the United States, which requires that initial PHAs of all the processes covered by the standard to be completed by no later than May 26, 1997 (OSHA, 1992).

A wide range of methods such as Checklist, What-If Analysis, Failure Modes and Effects Analysis (FMEA), Fault Tree Analysis (FTA) and Hazard and Operability (HAZOP) Analysis are available for performing PHA (CCPS, 1985). Whatever method is chosen, the PHA, typically performed by teams of experts, is a laborious, time-consuming and expensive activity which requires specialized knowledge and expertise. For PHAs to be thorough and complete, the team can not afford to overlook even "routine" causes and consequences which will commonly occur in many plants. The importance of performing a comprehensive PHA is illustrated by Kletz (1986, 1988, 1991) with examples of industrial accidents that could have been prevented if only a thorough PHA had been performed earlier on that plant.

A typical PHA can take 1-8 weeks to complete, costing over $13,000 per week. By an OSHA estimate, approximately 25,000 plant sites in the United States require a PHA (Freeman, et al., 1992). It is further estimated that a typical plant site might require five to ten PHA studies over the five year OSHA compliance schedule. Thus, an estimated $1-3 billion will be spent by the US process industry collectively between 1995 and 1997, on PHAs alone. This cost excludes getting the process safety information updated before the PHA can be performed.

Given the enormous amounts of time, effort and money involved in performing PHA reviews, there exists considerable incentive to develop intelligent systems for automating the process hazards analysis of chemical process plants. An intelligent system that can reduce the time, effort and expense involved in a PHA review, make the review more thorough and detailed, minimize human errors, and free the team to concentrate on the more complex aspects of the analysis which are unique and difficult to automate is needed. Also, an intelligent PHA system can be integrated with CAD systems and used during early stages of design, to identify and decrease the potential for hazardous configurations in later design phases where making changes could be economically prohibitive. It would facilitate automatic documentation of the results of the analysis for regulatory compliance. Also these PHA results can be made available online to assist plant operators during diagnosis of process disturbances.

Despite the obvious importance of this area, there has only been limited work on developing intelligent systems for automating PHA of process plants. In this paper, we will review the important approaches towards the automation of PHA from the perspective of intelligent systems. Of the various methods, HAZOP analysis is the most widely used and recognized as a preferred PHA approach by the chemical process industries. Hence, the main focus of this paper would be on intelligent systems for HAZOP analysis of chemical process plants. In this context, we will review the two ongoing major efforts in developing intelligent systems for HAZOP analysis, namely, the *HAZOPExpert* project at Purdue University in the US and the *STOPHAZ* project in Europe.

Approaches to Process Hazards Analysis: A Brief Summary

As noted earlier, a number of approaches such as Checklist, What-If Analysis, Failure Modes and Effects Analysis (FMEA), Fault Tree Analysis (FTA) and Hazard and Operability (HAZOP) Analysis are available for performing PHA (CCPS, 1985). We will briefly review these different PHA approaches, their advantages and limitations in this section

Checklist

The Checklist approach to PHA involves evaluating the safety of equipment, materials and operating procedures by asking a list of questions. This list of questions is often prepared by experienced engineers. One can envision Checklists for performing PHA during the different phases of process development - design, construction, startup, operation and shutdown. A Checklist approach is easy to use for compliance with standard procedures. But the Checklists are limited by the experience of its authors and hence they should be audited and updated frequently. Each company maintains its own in-house Checklists in order to comply with their standard safety requirements and procedures.

What-If Analysis

What-If analysis involves the systematic examination of unexpected events and their consequences by asking questions such as, "What if the coolant valve fails closed?" or "What if the operator pumps the wrong material?" and so on. The What-If procedure can be used during the design, construction, operation or modification of a plant. The What-If procedure is not as systematic and logical compared to the other PHA approaches. In addition to a thorough understanding of the process intent, the What-If analysis requires the users to synthesize possible events which cause process deviations leading to undesired consequences in the plant. It is a useful method if the persons performing the analysis are experienced and ask the right questions; otherwise, the results are likely to be incomplete.

Failure Modes and Effects Analysis

The Failure Modes and Effects Analysis (FMEA) method is used for identifying all the ways in which the equipment can fail and the corresponding potential effects on the process. FMEA identifies single failure modes of process equipment that results in an accident. The FMEA can be used during the design, construction and operation phases of a plant to qualitatively rank the effects of the various failure modes of the process equipment on the plant operation. FMEA is not good at identifying interactive combinations of equipment failures that lead to accidents. Also, operator errors and human interaction are generally not examined in FMEA.

Fault Tree Analysis

Fault trees are used in analyzing system reliability and safety. Fault tree analysis was originally developed by the Bell Telephone Laboratories in 1961 for aerospace applications. A fault-tree is a logic tree that propagates primary events or faults to the top level event or a hazard. The tree usually has layers of nodes. At each node different logic operations like AND, OR are performed for propagation. A general fault tree analysis consists of the following four steps: (i) system definition, (ii) fault tree construction, (iii) qualitative evaluation and (iv) quantitative evaluation as discussed by Fussell, et al. (1974). The major problem with fault trees is that the development is prone to mistakes at different stages. The fault tree constructed is only as good as the developer's mental model of the system. The algorithm of Lapp and Powers (1977) for the automated synthesis of fault treed addresses some aspects of this concern.

Hazard and Operability Analysis

The Hazard and Operability (HAZOP) analysis is used for the identification of the hazards and operability problems in a plant that could compromise the plant's safety and productivity. In HAZOP analysis, a team of four to six people familiar with the different aspects of the plant under consideration systematically identify every conceivable deviation from design intent in the plant, determine all the possible abnormal causes and the adverse hazardous consequences of those deviations. The experts in the HAZOP study team are chosen to provide the knowledge and experience in different disciplines for all aspects of the study to be covered comprehensively. The HAZOP procedure involves examining the process Piping and Instrumentation Diagram (P&ID) systematically, line by line or section by section (depending on the level of detail required), by generating deviations of the process variables from their normal state using the "guide words" approach. Since HAZOP analysis is the most widely used and recognized as the preferred PHA approach, we will focus our attention on intelligent systems for automating HAZOP analysis.

Intelligent Systems for Automating HAZOP Analysis

HAZOP analysis was developed in the late 1960s at ICI in the UK. The basic principle of HAZOP analysis is that hazards arise in a plant due to deviations from normal behavior. A group of experts systematically identify every conceivable deviation from design intent in the plant, find all the possible abnormal causes, and the adverse hazardous consequences of that deviation. The possible causes and consequences of each deviation so generated are then considered, and potential problems are identified. In order to cover all possible malfunctions in the plant, the process deviations to be considered are generated systematically by applying a set of "guide words," namely, NONE, MORE OF, LESS OF, PART OF, REVERSE, AS WELL AS and OTHER THAN, which correspond to qualitative deviations of process variables. The flow chart of the guide word HAZOP analysis procedure is given in Fig. 1.

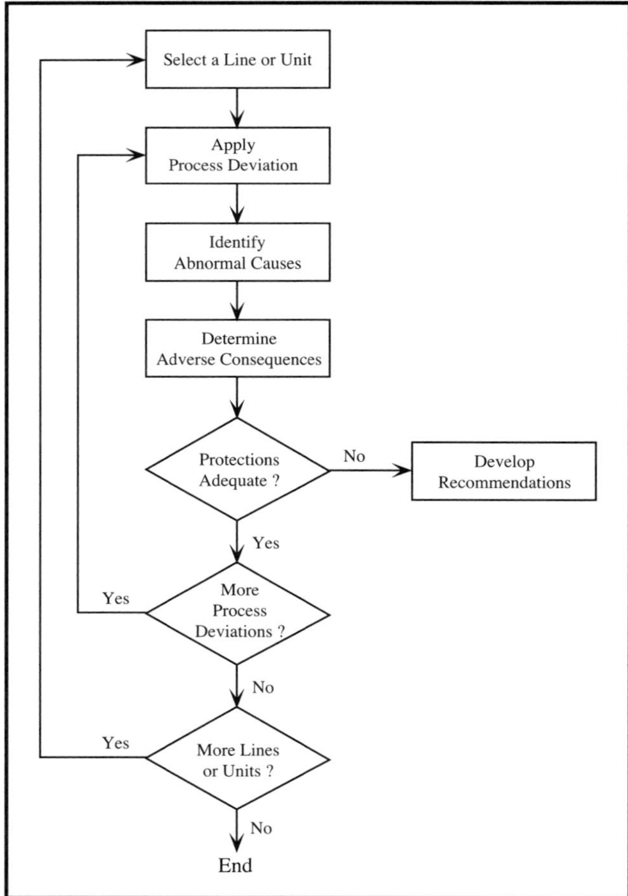

Figure 1. The HAZOP analysis procedure.

The meaning of the guide words could be understood from the process variable deviations they imply:

NONE	No flow, Zero composition of a particular component, etc.
MORE OF	Higher flow, temperature, pressure, etc.
LESS OF	Lower flow, temperature, pressure, etc.
REVERSE	Reverse flow
PART OF	Change in composition
MORE THAN	Impurities present, e.g. ingress of air, water, acids Extra phase present, e.g. vapor, solids
OTHER THAN	Complete substitution, of one material for other

The guide words and process variables should be combined in such a way that they lead to meaningful process variable deviations. Therefore, all the guide words can not be applied to all process variables. For example, when the process variable under consideration is temperature, only the guide words MORE OF and LESS OF lead to meaningful process variable deviations. In addition to identifying the hazards in a process plant, the HAZOP study also identifies operability problems which prevent efficient operation of the plant. Detailed descriptions of the HAZOP analysis procedure with illustrative examples are given in Lawley (1974), CCPS (1985), Knowlton (1989), and Kletz (1986).

Variants on this basic structure of HAZOP analysis have been developed to make the approach more thorough. For example, ICI has adopted a six stage Hazard Study methodology which not only embraces HAZOP within its Hazard Study 3, but also the appropriate elements of the other PHA techniques (Preston and Turney, 1991). Hazard Study 1 is the key SHE (Safety, Health and Environment) study during process conception, including inherent Safety and Environmental Impact. Hazard Study 2 is carried out on the process flow diagrams to identify "top events" and the need for further quantification, including QRA (frequency/consequence) design modification and hazard elimination/minimization. Hazard Study 3 relates to the engineering phase as the classic line by line critical examination of the Engineering Line Diagram (ELD) or P&ID prompted by guide words. Hazard Studies 4 and 5 relate to construction and commissioning and Hazard Study 6 is a final audit after the plant is in beneficial operation. Although developed for new plants, the 6-stage methodology is equally applicable to modifications and on-going plant safety reviews. Indeed an adaptation of Hazard Study 2 has been specifically targeted to SHE Assurance requirements of existing plants as "Process Hazard Review (PHR)" (Turney and Roff, 1995) and has been successfully applied to a range of technologies.

One of the important challenges in automating HAZOP analysis is handling the huge amount of process specific information which is required as the input for performing HAZOP. It is desirable to develop a system that is context-independent so that it can be used for the HAZOP analysis of a wide variety of processes and will also be able to find the process-specific hazards for the various processes.

Hazard Identification by Propagation of Faults

As one of the first attempts, Parmar and Lees (1987a, 1987b) used a rule-based approach to automate HAZOP analysis and showed its application for the hazard identification of a water separator system. They represented the knowledge required for propagating faults in each process unit using qualitative propagation equations and event statements for initiation and termination of faults. The initiation event statements described the initiation of a fault in a malfunctioned unit, the propagation equations described the propagation of faults through a normal operating unit, and the termination event statements described the termination of a fault in a unit causing the unit to malfunction. The initiation and the termination statements are the cause and consequences, respectively, the determination of which is the goal of the HAZOP analysis procedure.

The decomposition of the P&ID into process units was done at a coarser level. The P&ID of the plant was divided into lines consisting of pipes and other units (such as pumps, control loops, valves, etc.) through which a process stream passes, and vessels. And the control loop which consists of sensor, controller and control valve and its bypass was represented as a single process unit. The starting point of HAZOP analysis is a process variable deviation in a line. The causes are generated by searching for the initial events and the consequences by searching the terminal events. But the causes and consequences generated for a process variable deviation are confined to the line under consideration and the vessel connected to it. Thus, this method finds only the immediate causes and immediate consequences, unlike the actual HAZOP analysis in which the causes and consequences are propagated to the end of the process section under consideration to find all the adverse consequences due to every abnormal cause. This automated hazard identification system was implemented using FORTRAN 77 and Prolog. The computer method generates a more exhaustive list of results compared to the conventional analysis performed by the HAZOP team and hence the need for development of pruning rules and further enhancement using the expert systems approach is discussed.

HAZOPEX Expert System

A rule-based expert system prototype called HAZOPEX was developed using the KEE shell by Karvonen and co-workers (1990). The HAZOPEX system's knowledge base consisted of the structure of the process system and rules for searching causes and consequences. The rules for the search of potential causes are of the type, "IF deviation type AND process structure/conditions THEN potential cause." One important drawback of these rules is that the condition part of the rules depends on the process structure. For example, for a process variable deviation in a line, the rules for finding the causes has to check what unit is connected upstream and will find the appropriate cause. This increases the number of rules required as the number of process units increases, thereby limiting the generality of the system. Also in HAZOPEX, the identification of abnormal causes was more emphasized and less was said about the adverse consequences, though in actual HAZOP analysis the identification of adverse consequences is given priority. In that sense, this work had more of a diagnostic flavor. HAZOPEX's performance was evaluated on a small part of an ammonia system case study for which HAZOPEX was found to include useful knowledge about the potential causes of deviations which can be used as a check-list for the user.

Hazard Identification using Qualitative Simulation

Waters and Ponton (1989) developed a quasi-steady-state qualitative simulation approach based on the earlier work of De Kleer and Brown (1984), and attempted to automate HAZOP analysis using this approach. They found the approach to be highly combinatorial, thus restricting its practical usefulness. Their qualitative simulation approach will explore all behaviors that are consistent with the initial state of the system and the input deviations. But, many of these explored behaviors will not be useful from a safety perspective. Their qualitative simulation program was written in Prolog and implemented on a Sun 3/50 workstation. They reported that the time taken to perform qualitative simulation substantially exceeded that required for a numerical simulation involving considerable detail for the simple systems tested.

A prototype hazard identification system, called Qualitative Hazard Identifier (QHI), was developed by Catino and Ungar (1995). QHI works by exhaustively positing possible faults from a library, automatically building qualitative process models, simulating them and checking hazards. One main difference between the QHI algorithm and the HAZOP analysis procedure is that the QHI's analysis starts with the faults whereas the HAZOP analysis starts with process deviations. The QHI approach requires that the library of faults be complete. The model of a given physical system is built by choosing the physics and chemistry from a pre-defined library of fundamental physical and chemical phenomena such as heat and mass transfer, chemical reaction and phase equilibrium. This modeling approach was implemented using the Qualitative Process Compiler (QPC) developed by Crawford, et al. (1990) and Farquhar (1993). The qualitative simulation of these models thus generated are performed using the QSIM algorithm of Kuipers (1986) and possible behaviors are determined. Perfect controller and pseudo-steady-state assumptions were used in order to simplify the complexity of the qualitative simulation.

The QHI algorithm was applied for the hazard identification of a reactor section of a nitric acid plant. For this case study, QHI evaluated the effects of the faults: leaks, broken or partially and completely blocked filters and pipes, controller failures leading to valve failures, etc. The number of behaviors generated by qualitative simulation of a fault ranges between 10 and 1600, with 40 to 100 behaviors for most of the faults. For some of these faults the simulation and hazard identification were completed in a matter of seconds whereas for many others the QHI took hours to days. Also, for some faults the memory on a Sun SparcStation was exhausted and the QHI could not identify the hazards. This is an important drawback of this approach as it is too fine-grained for industrial-scale process hazards reviews, even though the deeper-level modeling aspect of this approach is quite appealing.

HAZOPExpert: A Model-Based Expert System

HAZOPExpert is a model-based expert system for automating HAZOP analysis developed by Venkatasubramanian and Vaidhyanathan (1994). In their approach, they recognized that while the results of a HAZOP study may vary from plant to plant, the approach itself is systematic and logical, with many aspects of the analysis being the same and "routine" for different process flowsheets. It turns out that a substantial amount of time and effort is spent on analyzing these "routine" process deviations, their causes, and consequences. Hence, they focused on these "routine" cause-and-effect analyses by developing generic models which can be used in a wide variety of flowsheets, thus making the expert system process-independent. They also recognized that the process-specific components of knowledge, such as the process material properties and process P&IDs, have to be flexibly integrated with the generic models in an appropriate manner. To address this integration, they developed a two-tier knowledge-based framework by decomposing the knowledge base into "process specific" and "process general" knowledge, represented in an object-oriented architecture.

Process-specific knowledge consists of information about the materials used in the process, their properties (such as corrosiveness, flammability, volatility, toxicity, etc.) and the P&ID of the plant. The process-specific knowledge is likely to change from plant to plant and is provided by the user. Process-general knowledge comprises the process unit HAZOP models that are developed in a context-independent manner, which remain the same irrespective of the process plant under consideration. The HAZOP model of a process unit consists of its class definition and generic qualitative causal model-based methods for identifying and propagating abnormal causes and adverse consequences of process variable deviations. Based on this framework, an expert system called *HAZOPExpert* has been implemented using Gensym's G2 real-time expert system shell. *HAZOPExpert's* inference engine allows for the interaction of the process-general knowledge with the process-specific knowledge to identify the valid abnormal causes and adverse consequences for the given process variable deviations for the particular HAZOP study section of the plant under consideration.

HAZOPExpert is not meant to replace the HAZOP team. It's objective is to automate the "routine" aspects of the analysis as much as possible, thereby freeing the team to focus on more complex aspects of the analysis that can not be automated. It can be used in an interactive mode or fully automated batch mode.

HAZOP-Digraph (HDG) Models

The initial version of *HAZOPExpert* was recently modified into a HAZOP-Digraph (HDG) model-based framework to facilitate model development and refinement by the users as well as to tackle more complex process configurations (Vaidhyanathan and Venkatasubramanian, 1995). The overall architecture of the HDG model-based *HAZOPExpert* system is shown in Fig. 2. The HDG models are modified signed directed graphs (SDG) developed for the purpose of hazard identification. SDG is a graph with nodes connected by directed arcs. The nodes of the SDG represent events or variable deviations from their steady state value and the directed arcs of the SDG represent the causal relationship between the nodes. The nodes of the SDG can have the values + and - for the high and low deviation of the process variable they represent, and 0 for normal value. The directed arcs lead from the "cause" nodes to the "effect" nodes and they can have + or - arc-gains indicating the direction of the causal influence between the nodes. The SDGs are more compact than truth tables, decision tables, or finite state models for representing qualitative causal models. A framework for automated development of SDG models of chemical process units has been proposed by Mylaraswamy, et al. (1995).

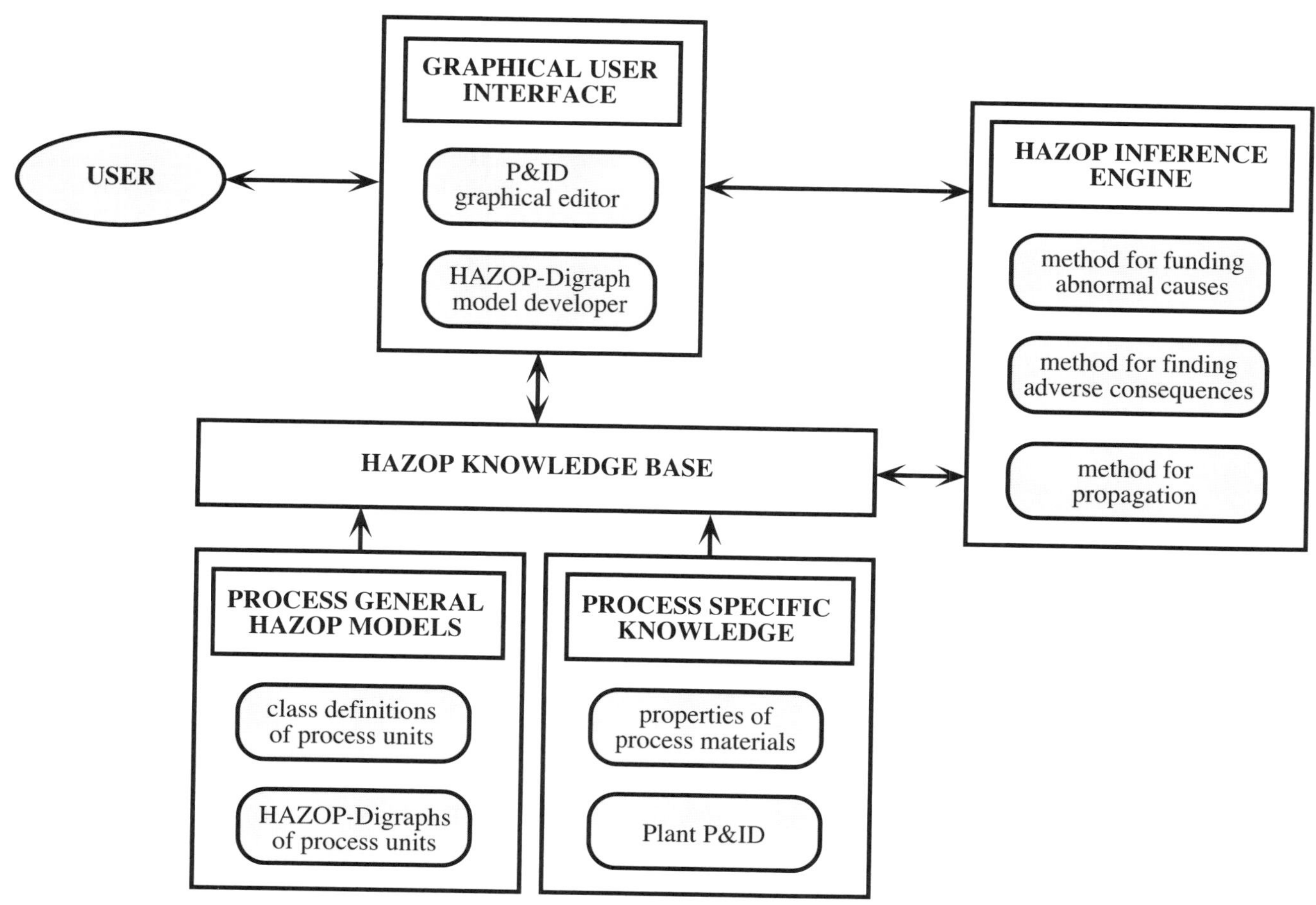

Figure 2. Architecture of HAZOPExpert.

Hazard-Digraphs provide the infrastructure for graphically representing the causal models of chemical process systems in a transparent manner to the user. The knowledge about finding the abnormal causes and the adverse consequences are incorporated into these digraphs. The HDG models of the process units are used for propagating the process variable deviations and for finding abnormal causes and adverse consequences by interacting with the process-specific knowledge. The HDG models are developed in a context-independent manner so that they are applicable to a wide variety of flowsheets. The user can build a new HDG model or add more knowledge to the existing HDG model using the graphical HDG model developer. A graphical HAZOP-Digraph model building tool is provided for this purpose.

The graphical user interface (GUI) of *HAZOPExpert* is shown in Fig. 3. This figure displays the essential features of the GUI, namely, the process unit HAZOP model library, the P&ID graphical editor and the HAZOP results windows. *HAZOPExpert's* model library currently has generic models for process units such as pump, tank, surge drum, heat exchanger, condenser, accumulator, reboiler, stripper, controller, valve, pipe, etc. The P&IDs of the process and the process materials properties are the process-specific information that are to be supplied by the user. If the P&IDs of the process are available in CAD format, they can be automatically imported into *HAZOPExpert*. Otherwise, the user can easily draw the P&IDs of the process using the P&ID graphical editor in *HAZOPExpert*. Similarly, if the process material property data are available in any database format they can be imported automatically into *HAZOPExpert*. Once the P&IDs and the process materials property data are input into the system, the corresponding HAZOP models of the process units in the P&IDs get connected automatically internally in the appropriate manner. This greatly simplifies knowledge acquisition.

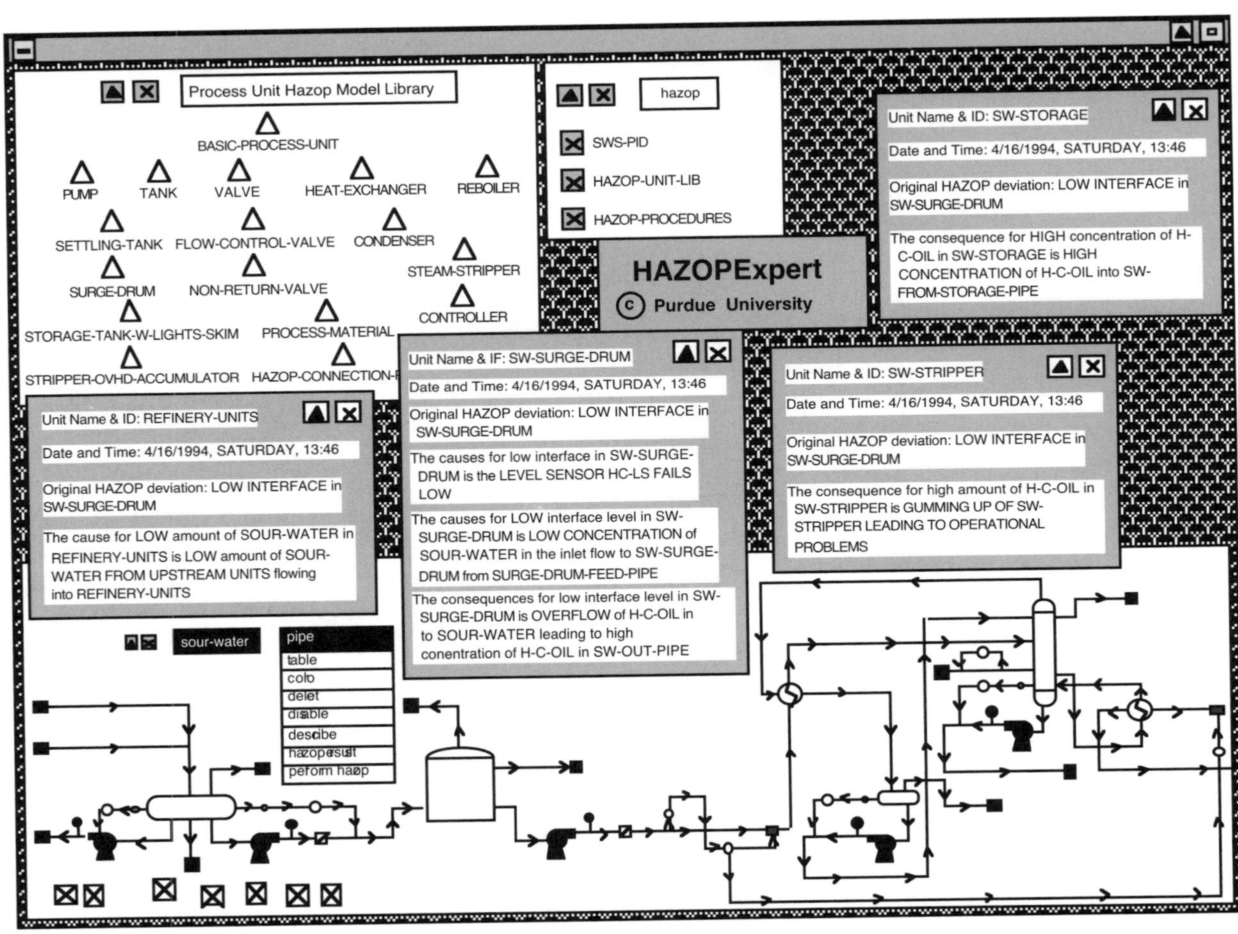

Figure 3. The graphical user interface of HAZOPExpert.

The user can then initiate any process variable deviation in any pipeline or process unit. *HAZOPExpert* will systematically perform HAZOP analysis for the process variable deviation both upstream and downstream until the end points of the P&ID are reached. The HAZOP results are also stored in user-defined files which can be automatically imported into spreadsheet software such as Microsoft Excel, or relational databases such as Oracle, or word-processing software such as Microsoft Word, to generate the standard HAZOP review tables. These can be further manipulated and formatted in a user-defined manner for meeting regulatory compliance and other requirements.

Performance of HAZOPExpert on Industrial Case Studies

HAZOPExpert is the first intelligent system, and to date, the only system, in the published literature that has been found to be successful on industrial-scale HAZOP case studies by HAZOP professionals. It has been tested on a number of actual process systems of varying degrees of complexity by HAZOP consultants at the Arthur D. Little Company, Cambridge, MA, USA. In this section, we will present results from one such case study, namely, the sour water stripper plant case study that has been reported by Vaidhyanathan and Venkatasubramanian (1995). The consultants from A.D. Little had performed a HAZOP review for this process earlier and their results were used to compare with the expert system's performance. The P&ID of the sour water stripper plant as drawn in the GUI of *HAZOPExpert* is shown in Fig. 3. In this process, there are 26 pipes, 5 flow control valves, 5 non-return valves, 5 pumps, 1 surge drum, 1 storage tank, 1 stripper, 1 condenser, 1 stripper overhead accumulator and 6 controllers. This process contains a refinery sour water stream that is separated in a surge drum to remove slop oil from the sour water. The sour water is pumped into a storage tank where any carried over slop oil can be skimmed off. From the storage tank the sour water is pumped through a heat exchanger to a steam stripper where ammonia and hydrogen sulfide are stripped from the water. Hydrocarbon oil is a flammability hazard and hydrogen sulfide and ammonia are toxic hazards. The release of these materials is a safety concern for the plant. Also, if there is poor separation of hydrocarbon oil from the sour water, the oil will escape into the stripper. This can gum-up the stripper which will cause operational problems.

HAZOP analysis was performed using *HAZOPExpert* for all conceivable process variable deviations in all the units and pipes in the sour water stripper plant. In any process unit, there are 3 deviations for flow (high, low, zero), 2 deviations for temperature (high, low), 2 deviations for pressure (high, low), 3 deviations for level (high, low, zero), and 3 deviations for the process materials concentration (high, low, zero). Thus, HAZOP analysis was performed for a total of 734 deviations in all the process units in the sour water stripper plant. It identified 100 abnormal causes and 90 adverse consequences for these deviations. As an example, the HAZOP results found by *HAZOPExpert* for the process variable deviation, "low interface level in sw-surge-drum" are shown in the HAZOP results windows of the process units in Fig. 3. *HAZOPExpert* has propagated the deviation all the way downstream to the stripper and has found the ultimate consequence, "gumming up the stripper due to high amounts of hydrocarbon oil entering the stripper." For a human team, it is often difficult for the team members to propagate the consequences through all the downstream units as far as *HAZOPExpert* does.

When a team of personnel perform the conventional HAZOP analysis it is not possible for them to consider all process variable deviations (734 in the sour water stripper case study) in each of the pipes, valves and pumps separately. So, they group a number of connected pipes, valves, and pumps and other units into study nodes and perform HAZOP for these study nodes. This way, the total number of deviations considered by the HAZOP team will be far less. In the sour water stripper case study performed by the HAZOP team, 135 total process variable deviations were considered for HAZOP analysis, and 32 abnormal causes and 32 adverse consequences were reported. All these causes and consequences were identified by *HAZOPExpert*. In addition, it had identified a number of other causes can consequences. Many of these were judged to be less important by the human team. The generation of an excessive number of causes and consequences is due to the qualitative nature of the HAZOP analysis procedure implemented in *HAZOPExpert*. In conventional HAZOP analysis, experts filter their initial HAZOP results using additional quantitative information in the form of the design specifications and normal operating conditions of the process units, and the quantitative properties of the process materials. These aspects of expert's reasoning have not been incorporated in *HAZOPExpert* yet and efforts are underway to include additional quantitative information and semi-quantitative reasoning for filtering the HAZOP results, and for ranking of the various adverse consequences identified by *HAZOPExpert*.

The performance on other case studies were similar, except in some cases the system missed certain causes and consequences as they had not been modeled as a part of the HDG models library.

From a computational perspective, *HAZOPExpert* is quite fast. On a Sparc 1 class machine, for the sour water stripper case study, it generated all the causes and consequences for a given deviation in a matter of seconds. The entire HAZOP analysis considering all 734 deviations can be performed in about two hours in an automated batch mode of the *HAZOPExpert*. *HAZOPExpert* has been implemented in G2, the real-time expert system shell marketed by Gensym, Inc. Hence, *HAZOPExpert* will run on any hardware platform that can run G2, including PCs supporting Windows NT. Efforts are currently underway to make *HAZOPExpert* available as a commercial product.

STOPHAZ Project

STOPHAZ is a 3 year multi-team, multi-country collaborative project supported by the European Commission under the ESPRIT III program. ESPRIT is a research program targeted at Information Technology including Intelligent Systems. STOPHAZ aims to improve the performance of the design process, particularly SHE (Safety, Health and Environment) and economics, through the provision of a set of support tools for the design and safety engineer.

The project is being undertaken by a consortium of the following 10 European organizations with ICI as the overall project manager:

Imperial Chemical Industries PLC	UK
SFK-Software for knowledge	UK
Snamprogetti SpA	Italy
Intrasoft SA	Greece
Loughborough University	UK
TXT SpA	Italy
VTT Safety Laboratory	Finland
Bureau Veritas SA	France
Aspen Tech Europe SA	Belgium
Hyprotech SL	Spain

In order to ensure the solutions which STOPHAZ will deliver satisfy the urgent needs of the users, an industry-wide survey was completed. 85 people from 27 separate organizations in six European countries were interviewed, making it unique in the Process Safety/Process Systems field. Coverage included operating companies, consultants, contractors and safety specialists in R&D, Offshore/oil and gas, petrochemicals and fine chemicals.

Following analysis of the survey results the STOPHAZ partners identified three key areas where computer tools will provide great benefit to those involved in safety-related design. The three areas cover many of the issues identified in the survey and are to:

- Offer context sensitive design advice, calculations and supporting data to those developing, checking, modifying or auditing an ELD
- Check the ELD, at various stages of development, for inherent hazards
- Assist in the efficient development of effective operating instructions for plant personnel.

STOPHAZ will address these three areas through three functional modules:

ELDER	(ELD helpER)
HAZID	(automatic HAZard IDentification)
CHOPIN	(Computerized Helper for OPerating INstructions).

The functional modules will form part of an integrated tool, supported by a Common Software Environment. Through this environment the main modules will have access to a variety of external data sources which may include incident data bases, physical properties data, legislation and calculations. The external data will be used to support the advice algorithms provided by the main modules. Figure 4 shows a representation of the STOPHAZ architecture. The central "box" represents the common software environment, the functional modules being shown on the right. External data sources are shown at the bottom of the figure. To the left are shown the tool which will generate the plant description, which is accessed by HAZID and ELDER, and an interface to a CAD system.

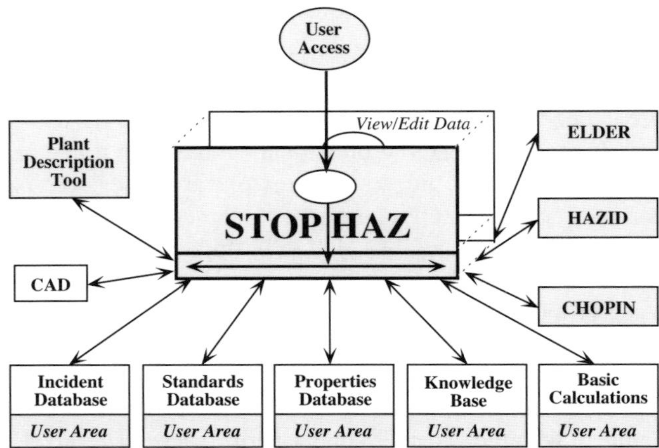

Figure 4. Representation of the STOPHAZ architecture.

The first prototype of STOPHAZ, including the ELDER and HAZID modules, will be completed during the summer of 1995. The second prototype will be completed a year after, with the current project ending in December 1996. The prototypes will be tested and evaluated within the consortium and by other selected process industry companies.

STOPHAZ has a variety of target users including process engineers (carrying out process designs and modifications), safety specialists (carrying out safety studies or specifying appropriate SHE standards), commissioning teams (completing operating instructions or understanding the design philosophy), and functional engineering designers (developing the process engineering package into detailed design). Figure 5 illustrates the tasks which STOPHAZ will support throughout the design lifecycle, be they applied to new plant or plant modifications, and shows at which stage of the design process the support is most appropriate. The left hand column shows the milestones in the design process. The design phases which progress the design are shown in the second column and the next three columns detail the tasks which will be supported by STOPHAZ.

Design Status	Design Task	ELDER Tasks	AutoHazid Tasks	OIH Tasks
Firm PFD Firm ELD HAZOP	ELD Development	ELD Development	Pre-Screen ELDs Full ELD Check	Draft OIs for HAZOP
Firm Plant Data Package Commissioning	Detailed Design & Operating Instruction Development	Reports Used by design/ops/ commissioning teams		Detail OIs
Operations	Use of Operating Instructions	Auditing operating plants		Use of OIs

TIME ↓

Figure 5. The use of STOPHAZ against the design life cycle.

ELDER Module

The purpose of ELDER is primarily to offer to the developer, or checker, of an ELD advice which assists them in determining the suitability of process equipment configurations by advising on appropriate SHE considerations. The advice offered will represent current best process design practice and the system will provide back up information, when requested, to justify the advice. Incident data and standards will be related to the advice as will back up calculations which will be required to determine leak rates, etc. The system will also be of help to those auditing existing operating plants.

ELDER will provide advice appropriate to the task which the user is carrying out and takes into account decisions which the user has taken. This context sensitivity will result in the user being offered less irrelevant advice, e.g. no information of explosions will be offered to a user who is considering the pumping of pure water. ELDER will assist the user in taking decisions based on best practice advice and will not automatically check that the user has adopted best practice, i.e. it is an advisor, not an auditor. The advice offered by the system will be fully adaptable to a particular user's needs. Access to common information such as standards and legislation will be provided as well as a framework to allow company specific codes, incidents and experience to be added.

An example of the sort of advice which ELDER will offer is that when considering tanker loading the user will be prompted to consider a number of potential problems, for example "protection against impact." Should the user not have considered this problem fully then ELDER will provide some possible arrangements to provide the protection and detail some incidents which have occurred when impact protection was not adequate. Another example would be, if the user has not considered the effects of leaking connections then the system will prompt the user by detailing the problems which may occur because of leaks, provide some information on the likely frequency of leaks, detail incidents which have occurred due to leaks and provide a simple calculation to determine leak rates.

HAZID Module

The purpose of the automatic hazard identification module (HAZID) of STOPHAZ is the identification and reporting of feasible and important hazards inherent in the design of a chemical processing plant. The hazard identification is to be based on a plant description and to be achieved through the use of generic knowledge of process streams and process equipment and through the application of an inferencing procedure. It is assumed that the problem is tackled no earlier than when draft EDLs or P&IDs have been prepared. Draft layout and operating instructions may also be available.

The style of tackling the problem is an emulation of the HAZOP technique (Jefferson, et al., 1995). Access to similar data to that used by a human HAZOP team is therefore an important requirement. The earlier, cheaper and more certain (in some respects) identification of hazards is the primary goal. Elimination of the conventional use of hazard and operability studies is not the primary goal. On the other hand, significant savings in time spent in HAZOP meetings, and particularly in following up actions arising from HAZOP meetings, are anticipated. Benefits should also follow from making any necessary changes earlier in the design life-cycle. In particular, earlier attention to hazard analysis (leading to earlier conclusions on necessary protections) and fewer late modifications will shorten the critical path of the design. Another benefit will be in the capture of expertise relevant to hazard and operability study and the access to this expertise for younger engineers.

The visible output envisaged is primarily in the form of tables similar to those produced in HAZOP. Typically a table has columns for location ("Item/Line/Stage"), deviation, cause, consequence ("effect or hazard"), protections ("Preventive or corrective measures") and some columns related to actions arising. The invisible output will be a data base of cause-to-consequence chains which can be operated upon or manipulated. One such manipulation will generate the HAZOP forms but other manipulations are to be considered and may prove attractive.

Systems Issues

In order to maximize the potential user base for STOPHAZ it will be targeted for the PC/Windows platform. In order to maximize functionality and flexibility the design and implementation of the modules and common software will follow the object-oriented paradigm using C++. STOPHAZ will appear to the user as an integrated set of tools operating within a common, open software environment. It also introduces a number of integration issues including data exchange protocol standards (e.g. STEP and PDXI) and interfacing to process plant data and CAD information.

Conclusions

Process safety, occupational health and environmental issues are ever increasing in importance in response to heightening public concerns and the resultant tightening of regulations. The process industries have responded by confirming safety, occupational health and environmental issues as some of the prime performance measures.

This paper has reviewed how manual, human dominated hazards analysis techniques and methodologies are being leveraged by an intelligent systems approach to reduce the time, effort and cost involved, and to improve the thoroughness of the analysis. Of the various PHA methodologies, the HAZOP analysis is the prime one for automation using intelligent systems due to its widespread usage in the process industry and its systematic and comprehensive nature. In both the US and Europe, synergistic developments, namely the *HAZOPExpert* and STOPHAZ projects, are now well beyond proof of concept and ready for industrial applications and commercial exploitation.

The next steps on this safety, health and environmental improvement journey will be testing and benchmarking against the proven human methodologies. Initially this will be by pre-screening studies or use on lesser risk processes. Once assured, the longer term aim may well be to move towards process conception and synthesis to create inherently safer designs and operating plants that tend towards zero defects. A more immediate development could be the use of online hazard reviews for the training of operators for abnormal situation management. The online hazard models can also be adapted for fault diagnosis applications.

Acknowledgments

The lead author gratefully acknowledges the National Institute for Occupational Safety and Health which supported this work in part through the grant R01 OH03056. He also acknowledges Messrs. Fred Dyke, Scott Stricoff and David Webb of Arthur D. Little, Inc., Cambridge, MA, for their assistance with the HAZOP case studies and the evaluation of *HAZOPExpert* as well as for some financial support.

Reference

Bureau of Labor Statistics (1992). *Occupational Injuries and Illnesses in the United States by Industry (1983).* Government Printing Office, Washington DC.

Center for Chemical Process Safety (CCPS) (1985). *Guidelines for Hazard Evaluation Procedures.* AIChE, New York.

Crawford, J., A. Farquhar and B. Kuipers (1990). QPC: a compiler from physical models into qualitative differential equations. *Proc. Amer. Assoc. Artificial Intelligence (AAAI-90).* pp. 365-372.

Catino, C.A. and L.H. Ungar (1995). A model-based approach to automated hazard identification of chemical plants. *AIChE Journal*, **41**(3).

De Kleer, J. and J.S. Brown (1984). A qualitative physics based on confluences. *Artif. Intell.*, **24**, 7-83.

Farquhar, A. (1993). PhD Dissertation, University of Texas at Austin, Dept. of Computer Science.

Freeman, R.A., R. Lee and T.P. McNamara (1992). Plan HAZOP studies with an expert system. *Chem. Eng. Prog.*, August, 28-32.

Fussell, J.B., G.J. Powers and R.G. Bennetts (1974). Fault treed — a state of the art discussion. *IEEE Trans. Reliability*, **23**, 51-55.

Jefferson, M., P. Cheung and A.G. Rushton (1995). Automated hazard identification by emulation of HAZOP studies. *Proc. 8th Intl. Conf. on Industrial and Engineering Applications of AI and Expert Systems.* Melbourne, Australia, June.

Karvonen, I., P. Heino and J. Suokas (1990). Knowledge based approach to support HAZOP studies. Research Report, Technical Research Centre of Finland.

Kletz, T.A. (1986). *HAZOP & HAZAN Notes on the Identification and Assessment of Hazards.* The Institute of Chemical Engineers, Rugby, England.

Kletz, T.A. (1988). *What Went Wrong? – Case Histories of Process Plant Disasters*, Ch. 18, 2nd ed. Guif Publishing, Houston, TX. pp. 189-196.

Kletz, T.A. (1991). Incidents that could have been prevented by HAZOP. *Journal of Loss Prevention Process Ind.*, **4**, 128-129.

Knowlton, R.E. (1989). *Hazard and Operability Studies: The Guide Word Approach.* Chematics International Company, Vancouver.

Kuipers, B.J. (1986). Qualitative simulation. *Artif. Intell.*, **29**, 289-338.

Lapp, S.A. and G.J. Powers (1977). Computer-aided synthesis of fault trees. *IEEE Trans. Reliability*, **26**, 2-13.

Lawley, H.G. (1974). Operability studies and hazards analysis. *Chem. Eng. Prog.*, **70**, April, 105-116.

McGraw-Hill Economics (1985). *Survey of Investment in Employee Safety and Health.* McGraw-Hill Publishing Co., New York.

Mylaraswamy, D., S.N. Kavuri and V. Venkatasubramanian (1994). A framework for automated development of causal models for fault diagnosis. *AIChE Annual Meeting.* San Francisco.

National Safety Council (1994). *Accident Facts (1994).* National Safety Council, Chicago.

OSHA (1992). Process safety management of highly hazardous chemicals; explosives and blasting agents; final rule 29 CFR 1910.119. *Federal Register*, February 24, 6356-6417.

Parmar, J.C. and F.P. Lees (1987a). The propagation of faults in process plants: hazard identification. *Reliability Eng.*, **17**, 277-302.

Parmar, J.C. and F.P. Lees (1987b). The propagation of faults in process plants: hazard identification for a water separator system. *Reliability Eng.*, **17**, 303-314.

Preston, M.L. and R.D. Turney (1991). The process systems contribution to reliability engineering and risk assessment. *Proc. COPE'91.* Elsevier, Barcelona, Spain. pp. 249-257.

Turney, R.D. and M.F. Roff (1995). Improving safety, health and environmental protection on existing plants: process hazards review. *Proc. 8th Intl. Symp. on Loss Prevention and Safety Promotion in the Process Industries.* Antwerp, Belgium. pp. 93-105.

Vaidhyanathan, R. and V. Venkatasubramanian (1995). Digraph-based models for automated HAZOP analysis. *Reliability Eng. and System Safety*, to appear.

Venkatasubramanian, V. and R. Vaidhyanathan (1994). A knowledge-based framework for automating HAZOP analysis. *AIChE Journal*, **40**, 496-505.

Venkatasubramanian, V. and R. Vaidhyanathan (1994). HAZOPExpert: a model-based expert system for HAZOP analysis. *AIChE Spring National Meeting*. Atlanta.

Waters, A. and J.W. Ponton (1989). Qualitative simulation and fault propagation in process plants. *Chem. Eng. Res. Des.*, **67**, 407-422.

SESSION SUMMARY: KNOWLEDGE-BASED PRODUCT AND PROCESS DESIGN

Kristian M. Lien
The Norwegian Institute of Technology

Jeffrey J. Siirola
Eastman Chemical Company

Introduction

This session contains three papers, addressing Product Design, Process Synthesis and Process Safety. Before giving a brief outline of the three papers, we will address three fundamental questions:

1. *Why* have people developed knowledge based design systems?
2. *What* have such systems contributed?
3. *How* have these contributions been achieved?

Finally, we will ask whether this can be expected to be the same way in the future.

Why?

There are two diametrically opposite reasons for the desire to develop and use knowledge based systems in design. First, there is the expectation that intelligent systems may ultimately outperform human design experts, and thus replace expensive personnel. Second, and opposite, there is the view that intelligent assistance systems may allow the human design expert to use his experience and problem solving capabilities more effectively.

These two views are extremes on a continuous scale, and where on this scale the majority of intelligent design systems are most likely going to be, is still an unresolved issue. This is no surprise. The two extremes reflect fundamentally different views on *knowledge*, *people*, and *machines*, questions which themselves are far from being resolved.

Design knowledge may be understood as a *state description*, a set of true, unambiguous statements about process design, and with this understanding, knowledge may be transferred to formal procedures, and finally automated, without significant losses of information in the transfer process. But it may also be understood as *complex, dynamic process*, where competence, insights and skills are not only *retrieved and used*, but also *created*, challenged and further developed along and intertwined with the developing process design.

People may be viewed, staticly, as *design knowledge operators* who need to be guided by somebody more competent than themselves, or they may be viewed as creative and competent decision makers in need of information systems giving them easy access to the information they need to make qualified decisions in the most efficient way.

Finally, *machines* may be viewed in many different ways, and this is in particular the case for information processing machines. It is safe to say that we have not yet by far explored the full potential of computers and our interaction with them.

What?

On an overall level, we may distinguish three different kinds of results that have come out of the work on knowledge based design systems during the last decade.

First, new *strategies* for process and product design have been developed. The design process itself has been studied, and procedures like Douglas' Hierarchical Decomposition procedure have been proposed as generic approaches to large classes of design problems.

Second, new *insights* into the nature of specific design subtasks have been gained, e.g. classification schemes for selection of alternative separation technologies.

Third, *computerized tools* have been created, e.g. for classification and selection of equipment, or as "expert interfaces" to complex simulation systems.

It is probably true that knowledge based systems have not had the same rate of successful implementation into practical use in design as in areas like diagnosis. This should not be interpreted as signs of failure, it should rather be interpreted as a sign of the tremendous complexity of the area of process design.

How?

How have the contributions in this area been produced? In the systems that have been developed, various kinds of powerful data structure and search mechanisms have been exploited, and thus, on the implementation level, these systems have frequently embedded in them quite advanced enabling technologies. But the added value of most of these systems is not found on the implementation level, but rather on the conceptual level.

It is clear that there has been a substantial structuring and formalization of design knowledge in the areas that have been studied. This has created valuable knowledge for tutoring purposes. *Created* is a key concept here. The notion of "Knowledge Mining" or "Knowledge Elicitation" that was so popular in the mid-80s does not seem to have had much impact in the area of design. Rather than "digging out" the knowledge from design experts, the focus in design more seems to have followed a *constructive* pattern — the design knowledge has been actively created. This may have been substantially influenced by the fact that much design knowledge is *visually oriented* and not easily expressed in words alone. (Imagine e.g. the reasoning and problem solving involved in the manipulation of McCabe-Thiele diagrams or ternary diagrams in distillation). Thus, creation of new human-oriented representations (graphics) enabling visual reasoning over geometric objects seems to be a key feature in many design tasks that the knowledge based systems need to map into. The transition from "picture space" to "verbal space," which is needed in order to let computers, which are "language machines," reason over transformations of graphical representations, is intrinsically hard.

Brief Overview of the Papers of This Session

The first paper in this session, by M. Mavrovouniotis, addresses *product and process design with molecular level knowledge*. This paper clearly shows that research in this area is often based on deep theoretical theories, rather than on the more shallow "expert knowledge." The paper presents a thorough overview of theories, methods and systems developed. With respect to product design, the paper addresses design and synthesis of pure components with desired properties, of solvents, of mixtures, of macromolecules, as well as of microstructure materials. With respect to process design, the focus is on reaction pathway synthesis.

The second paper, by C. Han, G. Stephanopoulos and Y. Liu, addresses the area of *knowledge based approaches in process synthesis*. This paper is an extensive review of a large number of contributions from the area, with applications from a diverse variety of subdomains. Total process synthesis, reaction pathways, reactor networks, separation systems, batch processing systems, synthesis for environmental purposes, selection of physical property calculation methods, of materials, of equipment and of catalysts. The paper also addresses the area of design support environments, and has a section on innovative approaches.

The third and final paper of the session, by V. Venkatasubramanian and M. Preston, addresses *intelligent systems for process hazards analysis*. The paper presents various approaches to HAZOP analysis, it analyses the nature of HAZOP studies, and presents some systems and tools that are used or in the process of being developed.

The Future: Needs and Challenges

It appears that during the last decade, substantial efforts have been invested in the creation of procedures and systems reflecting existing design practices. For application domains with a well-defined and limited scope, such as selection among alternative technologies or types of equipment, it is expected that useful and productivity-enhancing tools and systems will continue to be built. Within domains that have matured and where the development is predictable, the structuring, formalization and conservation of past experience is important.

However, in domains where there is rapid change, or where substantial changes in technology are needed, past experience will be less focused, and instead there will be a need for new and more fundamental design methods. This will e.g. be the case in the area of *multifunctional units*, where it is attempted to reduce investment and operating costs by combination of more than one traditional unit operation into one physical piece of equipment. In such areas, exemplified by technologies such as membrane reactors, or reactive distillation, it appears that the focus on configuration of *units*, i.e. Equipment, is not sufficient. In addition there is a need for methods that are able to determine how one should configure combinations of the basic phenomena, such as reaction, mass and heat transfer, in a synergistic manner, i.e. how one should "team up with the natural coupling" between various phenomena that conventionally are allocated to separate pieces of equipment.

Presently, the optimal configuration of process unit operations is a hard combinatorial problem, and this suggested kind of configuration from even smaller building blocks, the basic phenomena, will make the combinatorial problem even harder. In addition to the tools of combinatorial optimization, there is clearly a need in this area for the development of new methods, new overall design approaches and guidelines that make it possible to predict potentially interesting combinations of phenomena, as well as to give hints on the desired *macro-level characteristics* that these new systems should exhibit. Doherty and coworkers' and Westerberg and coworkers' methods for design of reactive or highly non-ideal distillation systems and Glasser and coworkers' Attainable Region approach to complex reaction systems are well known examples of such new methods that are based on a thorough understanding of the underlying physiochemical phenomena, rather than on the behavior of equipment. The phenomena based state vector representation suggested by Hauan and Lien (to appear in the Proceedings of ESCAPE-6, Rhodes, Greece, 1996), and Tondeur and Kvaalen's Equipartition principle (Ind. Eng. Chem. Res. Vol. 26, p. 50, 1987) and our own extension of this principle — the Isoforce Theory — (Ind. Eng. Chem. Res. Vol. 34, p. 3001, 1995) are other examples of this new line of methods that is presently being developed.

AI IN DESIGN: REVIEW AND PROSPECTS

B. Chandrasekaran [1]
Laboratory for AI Research
The Ohio State University
Columbus, OH 43210

Abstract

In this paper, I first discuss how to characterize AI as a set of ideas and technology for application to design. The goal is to try to understand what kinds of tasks in design is AI technology likely to be useful in. Then I give a brief *task analysis* of design. The result is a task structure which shows possible methods for design, the subtasks for the methods, and methods for the subtasks, etc., recursively. We can then characterize the methods in terms of the knowledge that they need and other properties, so that an informed decision can be made about when AI-based methods and what kind might be used for a design problem. I then proceed briefly to describe some work on Functional Representation, an approach for representing the structure and function of devices. I briefly mention the possible role of FR in design, including the construction of device libraries and design rationale.

Keywords

Design problem solving, Task structure of design, Functional representation, Device libraries, Design rationale.

Introduction

What do people have in mind when they talk about applying AI ideas to design? A pragmatic view might simply say that an AI technique is one that arose in the "AI community" and using any such technique, say, rule-based systems, a knowledge-representation scheme (such as frames or semantic nets), A* algorithm or neural nets, to solve some design problem is applying AI to design. This way of thinking is not entirely satisfactory. So-called AI techniques can often be used as general purpose programming languages. For example, one can use a rule-based language to implement a multiplication algorithm. For many problems for which neural net techniques are used, other statistical techniques can be used with similar results. See Chandrasekaran (1994a) for a detailed analysis of the problems arising from an excess of fascination with mechanisms *per se*, and not enough on the properties of the task. A technique-based view does not come to grips with, let alone make explicit, the relation between the task (in this case design), its subtasks, and under what conditions what techniques are appropriate. For this we need a detailed task analysis of design.

Another common way of characterizing AI is in terms of a couple of dichotomous distinctions on one side of which lies AI and on the other side lies traditional computing:

- heuristic vs. algorithmic
- qualitative vs. quantitative

Heuristic vs. Algorithmic. AI is normally said to focus on the former. The term "heuristic" is generally used to refer to knowledge that is used to control search: the heuristic helps to suggest alternatives that that might lead to correct or satisfactory solution, but is not guaranteed to do so. In this sense heuristic is contrasted with algorithmic, the latter being a procedure that is guaranteed to lead to the solution of interest, but might take a longer time than search that uses heuristics. Heuristic search is certainly part of the repertoire of design behavior — e.g., Westerberg (1989) points to its importance in process design — but there are many other

[1] *chandra@cis.ohio-state.edu*

aspects of the design activity, such as problem representations, that have to be done appropriately before heuristic search can play a role. In any case, many design problems involve little, if any, heuristic search if the problem is properly represented.

Qualitative vs. Quantitative. Computation in traditional engineering is numerical in character — typically some parameters in scientifically-based equations are instantiated with numerical values and the values of other parameters are then calculated. Examples are the use of finite-element methods in civil engineering and mass-action formulas in chemical engineering. Numerical computation is often contrasted with human behavior which is often said to be "qualitative" in nature. Hence the emphasis that AI approaches are qualitative (Stephanopoulos (1990a) points to the numerical-qualitative distinction in his overview of AI in design). The term "qualitative" is interpreted to mean slightly different things, however. In the sense in which researchers in qualitative physics use the term, "qualitative" refers to the human tendency in commonsense reasoning to use terms like "increase" or "positive change" in describing behavior as opposed to actual values by which some value increased: e.g., "if the valve is turned to the right that flow will increase." Since engineers often have to deal with systems whose inputs and parameters are only qualitatively known, it is useful to have techniques to simulate physical systems qualitatively. For example, during conceptual design, an engineer may not have committed to numerical values for all the parameters, but might still want to reason qualitatively (in this sense) about the partial design.

In another sense, "qualitative" is often used as equivalent to the use of symbolic representations. When people describe the world, we do not talk in numbers or equations, but string predicates and relations together in some natural language. Symbol structures are the natural way of representing such descriptions. For example, the structure of an artifact may be represented as a symbolic representation that might be

device D; parts {d1, d2}, Relation {R(d12)}
component d1: attributes (a11, a12); values (v11, v12)
component d2: attributes (a21, a22): values (v21, v22)

to say that it consists of parts d1 and d2, each with certain properties, connected together in some relation R. To the extent that reasoning involves operating on descriptions like this, cognitive behavior is often characterized as qualitative while traditional computation is said to be "number crunching."

Whichever sense "qualitative" is used, the qualitative-quantitative distinction does not fully characterize the role of AI in applications. Intelligent behavior involves dealing with numerical information just as much other types of symbolic information. Humans handle numbers and other symbols equally well in reasoning, subject of course to short term memory limitations, which in any case can be surmounted by memory aids such as paper. Of course it is useful to generate techniques, such as those pursued by qualitative physics, by which we can reason about behavior in an approximate way, but that kind of qualitativeness is not the only or essential feature of human reasoning in design or any other problem solving activity. The other sense of "qualitative," i.e., manipulating descriptions in the form of symbol structures would, without additional qualifications, include all kinds of computational processes, not especially the processes that AI is concerned with elucidating and supporting.

Characterizing Intelligent Behavior Computationally

AI's subject matter is computational modeling of intelligence and construction of computational artifacts that display intelligence. The proposals in AI and cognitive science can be broadly categorized into two classes (Chandrasekaran and Josephson, 1994), which I will dub "deliberative" and "subdeliberative," both inspired by aspects of human cognition.

Deliberative architecture. Deliberation is what we do when we engage in some thought process over time, typically in the pursuit of some problem solving goal: we have a series of thoughts about something such that, if we are successful, the final thought corresponds to or contains the solution. The content of a thought is often an assertion about some world state under some conditions. Some of the thoughts come to deliberation from memory or a "knowledge base," while other thoughts are derived in some manner by means of the thought process.

A substantial part of the research in AI can be divided into two sets of questions:

- *Ontology*: What kinds of states and relations should be included in the representation? What is the relation between representations and problem solving goals? This is broadly the Knowledge Representation problem. Good progress has been made on this question for classes of problems: for example, we know a lot about what kinds of things we would be interested in talking about if our goal is diagnosis, design or simulation.
- *Engine*: What kinds of computational, or inference, engine can model how deliberation moves from thought to thought, in such a way that a solution to a problem is achieved?

There have been many proposals for the deliberative thought engine. One set of ideas is inspired by logic, or the idea that intelligence is correct reasoning. Another set of ideas is inspired by the idea that intelligence is characterized by goal-directed problem solving, which in turn is search for a solution. The logic view of deliberation proposes that the relationship between thoughts is one of logical inference making. This view of intelligence then has focused on the kinds of inferences that characterize human intelligent behavior. In this view, the knowledge base is a model of the world of interest (things that are true), and the goal of thinking is to make inferences from this knowledge base such that one can conclude what actions will lead to the goal of making the world go to a desired state.

Problem solving as search is a long-standing idea in AI. The idea is that the agent is searching in a problem space for a path from the current state to a goal state, where at each state many actions may be available, and the task is to come up with a series of appropriate actions so that the goal state is reached. The emphasis in this approach has been on search strategies, use of heuristics to control search and so on, and less on the notion of inference-making that dominates the logic approach. Specifically, the Soar framework (Laird, Newell and Rosenbloom, 1987) proposes that the search is conducted using a form of subgoaling in a recursive way.

As it turns out, the two views of deliberation are not only mutually compatible, they actually complement each other. Each state generated in the problem space by applying an operator in the search approach can be viewed as an inference, and each operator (or action) as a possible rule of inference. Conversely, the reasoning view still has a search component since, given many different inferences that can be made from a given world state, choices have to be made about which inferences are likely to lead to the inference about the goal state of interest. Of course, there are differences in focus: the logic view as commonly practiced emphasizes very general domain-independent inference rules such as modus ponens or default inference rules, while the operators in the search view, when interpreted as inferences, are not so general. In fact, they are ideally suited to very context-dependent inferences, since complex preconditions can be attached to them. But broadly speaking, an engine based on the logic view still has to come up with goal-directed search regimes, and an engine based on the search view still has to come up with a good ontology for the states and state transitions, a concern traditionally associated with logic approaches.

Perhaps a simple illustration involving design might make the ideas clearer. Each state in the design process may be thought of as a partial description of design: components and connections at some level description and detail. An operator then is one of the following possible actions:

- instantiation of some value for some component or connection.
- addition or deletion of some new component in some connection relation to existing partial structure.
- refinement of the detailed structure of one of the components.

In this way, the design process can be thought of as a search in the space of structural descriptions, the initial state as the null state, and the goal state as a state in which the structural description is completely specified. I will elaborate on this picture later on, but for now, the above example should serve to clarify the idea of design as search in a problem space.

The above proposal for the deliberative engine, one that unifies the logic and search views, still requires a "memory": one which contains the knowledge base of the problem solver and which can retrieve the relevant pieces of knowledge and operators to move the search forward in the deliberative engine. Different types of theories of memory have been proposed. In the traditional logicist view, memory or the knowledge base consists of abstract propositions. Both in the Soar view and in the approach that has been called case-based reasoning, memory is largely composed of earlier problem solving experience: production rules that capture successes in problem space for Soar, and cases as frame-like entities for (Schank, 1982). The search view of deliberation is compatible with these different views about memory. For example, in case-based design, memory supplies a case that is closest to the current problem specifications, and the task is to modify it so that the specifications of the new design task are satisfied. Modifying a design consists of the operations of addition or deletion of components, and changing the values of any of the components or connections. This still involves deliberative search.

Subdeliberative architecture. There is an alternative view of human cognition in which either much or all of the interesting activity takes place below awareness and not subject to deliberative control. There are problems that we solve, such as being reminded or visual perception, in which the solution occurs to us within a few hundred milliseconds, and there is no explicit awareness of intermediate states as in deliberative problem solving. In the alternate view of intelligence, not only memory retrieval, but many other tasks, perception and even problem solving, are performed by such subdeliberative architectures. Both symbolic and non-symbolic models have been proposed for this kind of behavior. Schank's work on dynamic memory is a proposal within the traditional symbolic AI paradigm for how memory might be organized to do not only simple memory retrieval, but problem solving as well. Neural network models are the most well-known proposal for subdeliberative architectures in the non-symbolic framework. The latter have typically been successful for problems that can be couched as classification (visual perception) or associative retrieval. I have not done an extensive literature survey of the uses of neural network ideas for design, but I would suspect that they have been used to solve those parts of the design problem that can be couched in a classification framework. In any case, problems involving ontology will still have to be solved before we can use neural networks effectively. See Chandrasekaran, Goel and Allemang (1989) for a discussion of this issue.

Integrating the Two Architectures. There are a number of positions on whether and how to integrate the deliberative and subdeliberative architectures. We have already described one proposal: deliberation is where problem solving takes place, and memory is simply a repository of knowledge and experience. Another proposal is to deny any role for a deliberative architecture, and simply model all thought and intelligent behavior as a repeated use of one subdeliberative architecture. In this view, given a problem, the subdeliberative architecture simply generates a series of states that lead to the solution, without any role played by deliberation. Intermediate positions are possible in which both deliberation and subdeliberation solve problems and the results are coordinated sequentially. See Chandrasekaran and Josephson (1995) for a discussion on integrating architectures.

Using AI in Design

Based on the above picture of AI, we can see how AI might help. When we want the design process to explore alternatives at run time for any of the subtasks of design (to be described shortly), we can envision a knowledge-based search engine in the deliberative mode. In order to do this effectively for design, we need to have ideas for the following:

- *Task-specific ontology for design.* Representations for devices, components, connections, functions, behavior, structure can be used to represent design artifacts, specifications and their properties. Also, design knowledge in the form of design plans may also be part of the issues in knowledge representation. Substantial work has been done in the process engineering community in coming to grips with the representation issue. See, for example, Stephanopoulos (1990b), Lind (1994), and Saelid, Mjaavatten and Fjalestad (?). Later in this paper, I will describe a device representation framework that also contributes to this issue.
- *Operations in the design space.* Strategies for setting up and exploring design problem spaces can be supported. A task analysis of design can help in identifying these strategies and specification of the problem spaces. The design task includes as subtasks tasks of analysis, simulation, etc., so computational support for invoking, conducting, and organizing all the activities can be significant contribution of technologies from AI.
- *Memory (knowledge base) organization.* Strategies for indexing and retrieving relevant knowledge, including past experience, can be valuable.
- *Learning.* Successful experiences in the design problem space may be abstracted and cached in memory, so that future design can be faster.

Search-based techniques might be integrated with traditional numerical algorithms which may be appropriate for any subtask. For example, at various points, design verification might involve calculating some formula to make sure that the maximum temperature is within allowed limits. Thus qualitative and quantitative methods can be integrated, just as it happens in humans. Some of the subtasks might also use qualitative techniques in the sense of approximate solutions as well, e.g., in simulation as part of design verification.

Other types of problem solving architectures, such as neural networks, can be used to supply the solutions to any subtask of design.

Design Problem Solving

I have written elsewhere (e.g., Chandrasekaran, 1990) analyzing the structure of the design task, and what follows is largely extracted from it. A task structure is constructed by identifying alternative methods for each task. For each method, the knowledge needed and the subtasks that it sets up are identified. This recursive style of analysis provides a framework in which we can understand a number of particular proposals for design problem solving as specific combinations of tasks, methods and subtasks. Most of the subtasks are not really specific to design as such. The analysis shows that there is no one ideal method for design, and good design problem solving is a result of recursively selecting methods based on a number of criteria including knowledge availability.

A common theme for design as a process is this: it involves mappings from the space of design specifications to the space of devices or components (often referred to as mapping from behavior to structure), typically conducted by means of a search or exploration in the space of possible subassemblies of components. The design problem is formally a search problem in a very large space for objects that satisfy multiple constraints. The search also often takes place at multiple levels of abstraction (Westerberg, 1989). The notion of design as search is an oft-repeated theme in the process design community (Myers, Davis and Herman, 1990; Westerberg, 1989). What is needed to make design practical are strategies that radically shrink the search space.

A designer is charged with specifying an artifact that delivers some functions and satisfies some constraints. A design task is characterized by a (possibly large and generally only implicitly specified) set of primitive components and a repertoire of primitive relations or connections between components. Functions may be explicitly stated or implicit (such as safety). Design specifications may additionally contain constraints, on the properties of the artifact (e. g., weight limit), on the manufacturing process, or on the design process itself (e. g., time limit).

A Task Structure for Design

A method can be defined for our purposes here as some way of organizing the computation to achieve the task. Methods are typically described in terms of computations some of which may be explicitly set up as subtasks. For example, the "logarithmic method" for multiplying two numbers may be described as "compute the logarithms of the two numbers, add them, and compute the antilogarithm of the sum." Computing the logarithms and antilogarithms may be set up as subtasks for which different methods may be described. A task analysis of this type can be continued recursively until methods whose operators are all directly achievable (within the analysis framework) are reached.

The most common top level family of methods for design can be characterized as *Propose-Critique-Modify* (PCM) methods. That is, a proposal for some part of the design problem is made, which is then critiqued and modified, and this process is repeated until the design problem is solved. The PCM method sets up four subtasks: *Propose, Verify, Identify Cause of Failure (if any), Modify*.

The Propose subtask. Design proposal methods use domain knowledge to map part or all of the specifications to partial or complete design proposals. The proposal may be partial in the sense that the design for a component might be suggested, or it may be a high level solution, needing additional refinement. Three families of methods can be identified:

- Problem decomposition/solution composition. In this class of methods, domain knowledge is used to map subsets of design specifications into a set of smaller design problems. Use of design plans (Brown and Chandrasekaran, 1989; Myers, Davis and Herman, 1990) is a special case of decomposition methods.
- Retrieval of cases from memory which correspond to solutions for design problems which are similar or "close" to the current problem.
- Family of methods that solve the design problem as a constraint satisfaction problem and use a variety of quantitative and qualitative optimization or constraint satisfaction techniques. This method is applicable only in special circumstances, and we will say no more about it here.

Decomposition and case-based methods help reduce the size of the search spaces, since the knowledge they use can be viewed as the compilation or chunking of earlier (individual or community) search in the design space.

For the decomposition method, the knowledge needed is of the form, $D \rightarrow \{D_1, D_2, \ldots D_n\}$, where D is a given design problem, and Di's are "smaller" subproblems. A number of alternate decompositions for a problem may be available, in which case a selection needs to be made, with the attendant possibility of backtracking and making another choice. Repeated applications of the decomposition knowledge produce *system-subsystem hierarchies*. In well-trodden domains, effective decompositions are known and little search for decompositions needs to be conducted as part of routine design activity. Decomposition knowledge in design generally arises when the functional specifications can be decomposed into a set of subfunctions, or, even better, in part-subpart decompositions. Design commitments made by decomposition may be to intermediate level design abstractions which need to be further refined at the level of primitive objects. For example, in designing a process, the designer might commit to a particular process flowsheet — which is a sequence of basic process units — but the design of the equipment to support the flowsheet is posed as a subtask to be solved by any of the available methods.

The decomposition method requires solving the subtasks of (1) generating specifications for subproblems, and (2) gluing the subproblem solutions into a solution to the original design problem. In most routine design, solutions to these subtasks are precompiled and no run-time problem solving is required. Design plans are a special case of decomposition knowledge, representing a precompiled partial solution to a design goal (Friedland, 1979; Mittal, Dym and Morjaria, 1986; Brown and Chandrasekaran, 1989), where gluing the solutions together is explicitly included in the plan.

Decomposition is an often-used techniques in AI approaches to process design. For example, Banares-Alcantara, et al. (1987) describes a design system which decomposes the problem into a sequence of tasks, some of which are design tasks, while others corresponding to the preprocessing and recomposition subtasks discussed in the previous paragraph. Another example is Wahnschafft, Jurain and Westerberg (1991).

In configuration tasks (Mittal and Frayman, 1989) subproblem solutions are given as part of the problem (i.e., a set of key components is already available), and the subtasks of specification generation and solution recomposition dominate at problem-solving time.

A major source of design proposal knowledge is design cases, instances of successful past design problem solving. Cases can arise from an individual's problem solving experience or that of a design firm or community. Cases can be episodic (i. e., represent one problem solving episode) or can represent the result of abstraction and generalization over several episodes. Design plans can be considered to be fairly abstracted versions of numerous cases. Indexing of cases with a rich vocabulary of features of the case and the goals it satisfies is a key idea in case-based reasoning. Matching and retrieval can be driven by associative processes on these indices.

The Verify Subtask. In addition to direct calculation using prespecified formulae, the most common method for this task is some form of simulation. Simulation takes as input a description of the structure of the system and generates as output the behaviors of interest. There are domain-specific quantitative simulation methods which use equations that directly describe the results of this composition. From an AI point view, interesting techniques are: qualitative simulation (see Forbus (1988) for a survey), consolidation (Bylander, 1988), functional simulation (Sticklen, Kamel and Bond, 1991) and visual simulation are of interest. Little AI research has been done so far on visual representations and simulations (but see, Narayanan and Chandrasekaran (1990) and Glasgow, Narayanan and Chandrasekaran (1995)). The role of simulation in design for the purpose of verification can be seen in (Beltramini and Morato, 1988).

Critiquing. Critiquing is really a generalized version of the diagnostic problem, i.e., a problem of mapping from undesirable behavior to parts of the structure responsible for the behavior. Modification of design can be directed to these candidates. Of course localization of responsibility for failure will not always work, and the entire approach to the design may need to be changed.

Criticism requires information about how the structure of the device contributes to (or is intended to contribute to) the desired overall behavior. Dependency analysis (Stallman and Sussman, 1977) may be useful here. Proposals for critiquing in the case-based reasoning literature typically use domain-specific, pre-compiled patterns. (Goel, 1989) uses a functional analysis of the proposed design for critiquing.

Modification. This subtask takes information about failure of a candidate design as its input and then changes the design so as to get closer to the specifications, by changing a functional subpart or adding components to the proposed design. Modification will be aided by knowledge that relates the desired changes in behavior to possible structural changes (Goel, 1989).

A related search approach is one where modification is done by some form of hill-climbing: parameters are changed, direction of improvement noted, and additional changes are made in the direction of maximal improvement. This is applicable if the design problem is one of choosing parameters for a predetermined structure.

If modification is for adding new functions modularly, i.e., by the creation and integration of separate substructures that deliver the functions, the design of the additional structures can be viewed simply as a new design problem.

Use of the Task Structure. The task analysis provides a clear road map for knowledge acquisition. Methods can be chosen based upon the types of answers we want and the knowledge available to support them. Different types of methods may be used for different subtasks: a design system may use a knowledge-based problem solving method for the subtask of creating a design, but use a quantitative method such as a finite element method for the subtask of evaluating the design. The choice of methods can be precompiled as in first generation knowledge systems, or can be selected at run time as in TIPS (Punch, 1989), DSPL++ (Herman, 1990) and Soar-based approaches. Methods can be directly supported by high-level shells as in the generic task methodology (Chandrasekaran, 1986) or more flexible architectures can be employed to integrate the methods as in SOAR.

Device Representations

An important issue in knowledge representation for design is how to represent devices, their structure, functions and behavior. A framework called the Functional Representation (FR) has been developed over the last decade and applied to device representation in many domains including process engineering. For a review, see Chandrasekaran (1994b, 1994c). In this section, I give a brief account off this work and discuss its importance to design. The discussion owes much to the description of FR in Josephson (1993).

FR is complementary to the more common bottom-up device representations in which the behavioral characteristics of each component in isolation are represented, and the behavior as a whole is inferred, given information about how components or processes are combined. FR differs in taking a more top-down view in which device functions are explicitly represented, along with the roles that components and processes play in achieving those functions. FR is especially appropriate for representing a designer's intent, but is also suitable for representing unintended functions and behaviors. Recent work has made progress in combining the strengths of FR and component-centered representations. Iwasaki and Chandrasekaran (1992) have demonstrated a way to do design verification by using FR in conjunction with qualitative simulation.

FR allows one to represent the following:

Devices. Components of devices are devices in their own right. A device may have associated *Ports.*

A device will usually have a description of its structure — given as a set of components, their associated ports, and a set of *Connections* between ports. Connections allow the passage of *Substances*, which may be material (e.g., water), or abstract (e.g., heat or information).

States. A state is a partial description of the device or its environment at a moment or over an interval of time. This is usually given as a Boolean combination of predicates over *State-variables* (which may be discrete or continuous).

Functions. A device has a set of functions associated with it. (Keuneke, 1991) distinguished four types of functions: to *Achieve*, to *Prevent*, to *Maintain*, and to *Control*. Iwasaki and Chandrasekaran (1992) added *Allow*. Each function type is specified somewhat differently. We will focus here on Achieve functions, since the representation for them is the oldest and best developed.

An Achieve function is associated with an ordered pair of states, called the IF state and the TO-MAKE state. The idea is that the TO-MAKE state is achieved, starting from the IF state, by using the particular function of the device. To explain how the TO-MAKE state is achieved, one points to the responsible device and function.

Sometimes an additional state is associated with a function called the PROVIDED state. It is used to specify conditions (other than those specified by the IF state) under which the function can be expected to achieve its TO-MAKE state, e.g., standard operating conditions. Sometimes a TRIGGERING state is also associated with a function, giving conditions under which the device is expected to immediately begin a process to achieve the function.

Causal-Process Descriptions. Functions are accomplished by way of causal processes. The CPD is a "causal story" describing how the function is accomplished, and if available, is attached to the function of a device. A CPD is a directed graph, where the nodes are states and the edges are *Annotated State Transitions*.

Annotated State-Transitions. Transition between states in the CPD may be annotated in a number of ways so as to explain the transition. The annotations may be of the following kinds.

- Inference. A change of state description, rather than an actual change of underlying state. The inference shows how the second state description follows from the first one, given the requisite knowledge.
- Function of a Component. A particular component, by performing one of its functions, is responsible for the state transition. FR thus provides a kind of recursive decomposition: functions are explained by the causal processes by which they are achieved, and causal processes are explained by the functions of the components that are responsible for state transitions that make up the causal process. This way the FR represents how the functions of a device arise from the functions of its components.
- General Knowledge Principle. The state transition can be understood to occur as the result of some general principle, for example falling as a result of gravity, or increasing in temperature due to friction. A convenient way to represent general knowledge principles, especially scientific laws, is in the form of analytical equations or functions relating parameters of the antecedent state to parameters of the subsequent one. When this is done, the FR can be used to guide numerical simulations. This use of FR has been explored extensively by Sticklen and his colleagues.

More than one annotation may be given, corresponding to knowledge of more than one "actor" capable of playing the role. (This way FR can encode "multiple realizability.")

As we said, a CPD is a directed graph whose edges are annotated state transitions. Inference procedures may traverse CPDs in either direction, forwards or backwards. Traversing in the forwards direction, "consequence finding," moves from cause to effect, and supports predictive inference. Traversing backwards, "antecedent finding," moves from effect to cause, and supports abductive inference. Design and planning are logically dependent on prediction. Diagnosis and process monitoring (any that goes beyond directly-observables) are logically dependent on abduction.

Goel (1989) showed how to use FR representations for organizing case libraries of designs, and how the cases can be indexed by function, so a designer can retrieve candidate designs for components or whole devices. Chandrasekaran, Goel and Iwasaki (1993) describes the use of FR for capturing design rationale. Iwasaki and Chandrasekaran (1992) demonstrated the use of FR for design verification. In (Chandrasekaran, 1994c) uses of FR for diagnosis and monitoring are cited and described.

Decision Support for Design Using FR

FR-based representations can support computer assistance in several ways for the various subtasks of design identified in the task analysis. We will only mention a few of them here. Most obviously a good browsing facility for a well-stocked device library can assist a designer in generating good design proposals to begin with. Beyond this, if the library is indexed by the functions that the devices can achieve, then retrieval by function can aid a designer by suggesting ways of achieving functions that she would not have thought of otherwise.

FR can support several forms of design criticism. Designs retrieved based on similarity of function can be criticized to bring out remaining functional discrepancies. Iwasaki and Chandrasekaran (1992) demonstrated design verification based on matching the causal state trajectories envisioned by the designer (and encoded in FR using CPDs) with those predicted by qualitative simulation. It is common for modern engineers to test designs with simulations, but typically this is a computationally intensive process of brute-force number crunching at the lowest level of detail. Toth (1993) demonstrated how an

FR representation can be used to constrain a simulation to follow just those causal paths relevant to answering a particular design question, thereby potentially saving an enormous number of computations, and making it feasible to use more simulation runs.

In behavior-oriented approaches for modeling devices, the behavioral characteristics of each conceptually distinct physical phenomenon is represented as an independent piece of knowledge, often called a "model fragment." Each model fragment is a context-independent description of a physical phenomenon. A model fragment typically consists of the conditions for the phenomenon to take place, including the objects that must exist and the conditions they must satisfy, and the consequences, including the functional relations among quantities that will hold, and other effects on quantities. Given a description of the physical structure of a device and the initial condition, a reasoning system is to select the applicable model fragments and to predict the behavior of the entire device by composing the knowledge of the behavior of individual components and the physical processes through which they interact. The behavior of each component or physical process should be represented in a context-independent manner so that each description does not make any implicit assumptions about the physical context in which the component is placed, or about the function of the whole device of which it is a part. This has been articulated as the no-function-in-structure principle and the locality principle (de Kleer and Brown, 1984). The purpose of these principles is to ensure that a prediction of behavior will be objective in the sense that it will not be biased by any hidden assumptions about the functions of the whole device. Though most of the model-based reasoning systems that have been built to date are based on either the behavior- or function-oriented approaches, the two are complementary, and there is much to be gained from combining them. Generating and understanding design rationale, especially design rationale that concerns behavior, requires both types of knowledge: knowledge of intended functions, and knowledge of the underlying structures and general physical principles. In engineering design, a designer typically first has in mind the function to be achieved by the artifact. Then she formulates a conceptual causal mechanism to achieve the function. Finally, she produces a particular design of the physical mechanism to implement the conceptual causal mechanism. A capacity for simulating the behavior of the designed device based solely on the design, and knowledge of general physical principles, regardless of the intended function, is valuable in foreseeing the consequences of design decisions. At the same time, knowledge of the intended function of the device is indispensable in order to evaluate the design based on such predictions.

Conclusions

I felt I could be most useful to the process engineering community interested in use of AI ideas by trying to characterize what AI is and how its products — ideas as well as technology — can be most usefully applied for the design task in engineering. I have avoided a detailed review of substantial work that has already gone on in the process engineering community in this area, mainly because much of this work, distinguished as it is, is still in what Newell (1981) has called the Symbol Level of AI, i.e., with respect to specific representational formalisms and algorithms to use them. My goal in this paper has been to review the issues in design problem solving at the Knowledge Level, i.e., at a level where we can see the task and its decomposition independent of the specific representational formalisms, and delineate the kinds of knowledge needed for getting the task accomplished. The overall framework that I described also allows for integration of qualitative and quantitative techniques.

I have also given a view of how to characterize human cognitive behavior in problem solving as an interaction between deliberative and subdeliberative architectures. It is my belief that this characterization helps in understanding what AI can do to support problem solving in general, and design in particular.

I described the general problem as one of identifying knowledge representation primitives and engines to use the knowledge representation. Goal-directed search and indexing memory and retrieval from it provide the substance for the basic inference mechanisms. For design, both the task structure and the general problem of device representation provide the basic material for knowledge representation primitives. I briefly described an emerging device representation framework called Functional Representation and pointed to its role in design.

Acknowledgments

I acknowledge the support by ARPA, Order No. A714, monitored by USAF Materiel Command, Rome Laboratories, Contract F30602-93-C-0243 in the preparation of this paper. I also thank John Josephson whose formulation of FR and its role in design I have used in this paper. My thanks are due to Jim Davis who put me in touch with process design literature.

References

Beltramini, L. and R. L. Morato (1988). KNOD — a knowledge-based approach for process design. *Computers Chem. Eng.*, **12**, 939-958.

Banares-Alcantara, R., A. W. Westerberg, E. I. Ko and M. D. Rychener (1987). Decade — a hybrid expert system for catalyst selection — I. expert system consideration. *Computers Chem. Eng.*, **11**, 265-277.

Brown, D. C. and B. Chandrasekaran (1989). *Design Problem Solving: Knowledge Structures and Control Strategies*. San Mateo, CA: Morgan Kaufmann.

Bylander, T. (1988). A critique of qualitative simulation from a consolidated viewpoint. *IEEE Trans. Systems, Man and Cybernetics*, 252-263.

Chandrasekaran, B. (1986). Generic tasks in knowledge-based reasoning: high-level building for expert system design. *IEEE Expert*, 23-30.

Chandrasekaran, B. (1990). Design problem solving: a task analysis. *AI Magazine*, 59-71.

Chandrasekaran, B. (1994a). AI, knowledge and the quest for smart systems. *IEEE Expert*, **9**(6), 2-6.

Chandrasekaran, B. (1994b). Functional representations: a brief historical perspective. *Applied Artificial Intelligence*, **8**, 173-197.

Chandrasekaran, B. (1994c). Functional representation and causal processes. *Advances in Computers*, 38, 73-143.

Chandrasekaran, B., A. Goel and D. Allemang (1989). Connectionism and information processing abstractions: the message still counts more than the medium. *AI Magazine*, **9**(4), 24-34.

Chandrasekaran, B., A. Goel and Y. Iwasaki (1993). Functional representation as design rationale. *IEEE Computer*, 48-56.

Chandrasekaran, B. and S. G. Josephson (1994). Architecture of intelligence: the problem and current approaches to solutions. In V. Honovar and L. Uhr (Eds.), *Artificial Intelligence and Neural Networks: Steps toward Principled Integration*. Academic Press. pp. 21-50.

de Kleer, J. and J. S. Brown (1984). A qualitative physics based on confluences. *Artificial Intelligence*, **24**, 7-83.

Forbus, K. D. (1988). Qualitative physics: past, present and future. In H. Shrobe (Ed.), *Exploring Artificial Intelligence*. San Mateo, CA: Morgan Kaufmann. pp. 239-296.

Friedland, P. (1979). Knowledge-based experimental design in molecular genetics. *Proc. of the 6th International Joint Conference in Artificial Intelligence, (IJCAI)*. Tokyo. pp. 285-287.

Glasgow, J., N. H. Narayanan and B. Chandrasekaran (Eds.) (1995). *Diagrammatic Reasoning: Cognitive and Computational Perspectives*. Cambridge, MA: MIT Press.

Goel, A. K. (1989). *Integration of Case-Based Reasoning and Model-Based Reasoning for Adaptive Design Problem Solving*. Ph.D. dissertation, The Ohio State University.

Herman, D. J. (1990). DSPL++: A High-Level Language for Building Design Systems with Flexible Use of Multiple Methods (tentative title). Ph.D. dissertation, The Ohio State University.

Iwasaki, Y. and B. Chandrasekaran (1992). Design verification through function- and behavior-oriented representation: bridging the gap between function and behavior. In J. S. Gero (Ed.), *Artificial Intelligence in Design '92*. Kluwer Academic Publishers. pp. 597-616.

Josephson, J. R. (1993). The functional representation language FR as a family of datatypes. Technical Report, Laboratory for AI Research, The Ohio State University.

Keuneke, A. (1991). Device representation: the significance of functional knowledge. *IEEE Expert*, April, 22-25.

Laird, J. E., A. Newell and P. S. Rosenbloom (1987). SOAR: an architecture for general intelligence. *Artificial Intelligence*, **33**, 1-64.

Lind, M. (1994). Modeling goals and functions of complex industrial Ppant. *Applied Artificial Intelligence*, **8**, 259-283.

Mittal, J., C. Dym, C. and M. Morjaria (1986). PRIDE: an expert system for the design of paper handling systems. *IEEE Computer*, 102-114.

Mittal, S. and F. Frayman (1989). Towards a generic model of configuration tasks. *Proc. of the 8th International Joint Conference on Artificial Intelligence*. Detroit, MI.

Myers, D. R., J. F. Davis and D. J. Herman (1990). A task-oriented approach to knowledge-based systems for process engineering design. *Computers Chem. Eng.*, **12**, 959-971.

Narayanan, N. H. and B. Chandrasekaran (1990). Qualitative simulation of spatial mechanisms: a preliminary report. *Proc. of the AAAI 1990 Workshop on Artificial Intelligence and Simulation*.

Newell, A. (1981). The knowledge level. *AI Magazine*, 1-19.

Punch, W. (1989). *A Diagnostic System Using Task-Integrated Problem Solver Architecture (TIPS), Including Causal Reasoning*. Ph. D. dissertation, The Ohio State University.

Saelid, S., A. Mjaavatten and K. Fjalestad (?). An object-oriented operator support system based on process models and an expert system shell. *European Symposium on Computer-Aided Process Engineering*.

Schank, R. (1982). *Dynamic Memory: A Theory of Learning in Computers and People*. New York, NY: Cambridge University Press.

Stallman, R. and G. Sussman (1977). Forward reasoning and dependency-directed backtracking in a system for computer-aided circuit analysis. *Artificial Intelligence*, **9**, 135-196.

Stephanopoulos, G. (1990a). Artificial intelligence in process engineering — current state and future trends. *Computers Chem. Eng.*, **14**, 1259-1270.

Stephanopoulos, G. (1990b). Model.LA. a modeling language for process engineering — I. the formal framework. *Computers Chem. Eng.*, **14**, 813-846.

Sticklen, J., A. Kamel and W. E. Bond (1991). Integrating quantitative and qualitative computations in a functional framework. *Engineering Applications of Artificial Intelligence*, **4**(1), 1-10.

Toth, S. (1993). *Using Functional Representation for Smart Simulation of Devices*. Ph.D. Dissertation, The Ohio State University.

Wahnschafft, O. M., T. P. Jurain and A. W. Westerberg (1991). Split: a separation process designer. *Computers Chem. Eng.*, **15**, 565-581.

Westerberg, A. W. (1989). Synthesis in engineering design. *Computers Chem. Eng.*, **13**, 365-376.

DISTRIBUTED AND COLLABORATIVE COMPUTER-AIDED ENVIRONMENTS IN PROCESS ENGINEERING DESIGN

Arthur W. Westerberg
Department of Chemical Engineering
and Engineering Design Research Center
Carnegie Mellon
Pittsburgh, PA 15213

The *n*-dim group[1]
Engineering Design Research Center
Carnegie Mellon
Pittsburgh, PA 15213

Abstract

n-dim is a computer environment to support collaborative design. It is also a history capturing mechanism for complex corporate activities such as design. We first discuss the use of information modeling to aid these processes and then outline relevant attributes of *n*-dim. The major part of this paper is to describe some of the applications we have built or are building on top of *n*-dim, where our long-term goal for each is to test our hypothesis that this system will support collaboration and will capture a usable history better than previous approaches.

Introduction

With the marked improvement in network and computer technology, we see the opportunity to create a new type of computer-based system to support collaboration among distributed teams of persons carrying out a complex process such as design. We also see the opportunity to support the capturing of these processes for subsequent reuse and analysis with a much higher "fidelity" than possible in the past. We base our approach on a system we call *n*-dim for capturing, organizing and sharing of data and information through the use of network-wide information modeling.

Any time we place a structure over data, we are organizing it (Robertson, et al., 1994). One example is when we organize files in directories and subdirectories, another is when we put data into databases, and another is when we use document management systems to develop and exchange documents for sign-off during a design process.

The Computer Supported Collaborative Work (CSCW) literature discusses issues on group collaboration and is rapidly developing, with many conferences (for example, see Turner and Kraut (1992)) occurring each year in this area. The formidable database literature is relevant to information modeling, with much current interest in the creation of distributed object-oriented databases (Oszu, et al., 1994; Stonebraker, 1994; Kim, 1995). Within chemical engineering, J. Ponton's group at the University of Edinburgh is developing the épée and KBDS systems (Ballinger, et al., 1994; Bañares-Alcántara, 1995) to record and organize information developed while designing chemical processes. These systems also capture design intent and track whether designs are meeting stated objectives.

This paper summarizes several design features we have previously described for *n*-dim (Westerberg, et al., 1989; Robertson, et al., 1994; see also summary description in Krieger, 1995). The major part of the paper then describes

[1] The work reported here is the product of the *n*-dim group comprising the following persons: Robert Coyne, Douglas Cunningham, Allen Dutoit, Eric Gardner, Suresh Knoda, Sean Levy, Ira Monarch, Robert Patrick, Yoram Reich, Eswaran Subrahmanian, Michael Terk, Mark Thomas and Arthur Westerberg. All are part of the Engineering Design Research Center except Y. Reich who is now at the Department of Soil Mechanics at Tel Aviv University, Israel.

several of the applications we have created or are creating on top of *n*-dim.

Information Modeling

To understand our approach to information modeling, imagine the existence of two distinctly different worlds in which we would like to organize information. Our first world is an empty one. In this world we first define a number of object types with which we wish to populate it. Our object type definitions are themselves constructed of parts which may be instances of previously defined types, giving us a powerful part/whole capability in creating new type definitions. To each such object type definition we can send a message to make an instance of itself. These instances contain slots into which we place values that give each instance its individual character. For example, we may create an instance of type "car" and indicate in its roof color slot that it has a blue roof. This world soon becomes filled with numerous instances of these object types. They are strongly organized through recording the parent-child and part/whole relationships among them. Over time we augment our world by defining new types with which we can populate it. We might describe our mind set in creating this first world as one of "patterns first, then instances."

In contrast to starting with no objects, our second world starts with a population of millions of objects of various types already existing. Some of these are text objects created by Word and others by FrameMaker; still others are simple e-mail messages. We find numerous "gif" files representing graphic objects. Some are objects defined within databases, such as an entry describing a reflux pump belonging to a column in our plant south of Houston. It is our goal to organize these already existing objects so that we can better understand them. We organize them by establishing a variety of as yet undefined relationships among them. One relationship we might devise is to indicate that this file is the output resulting from having run this other input file through that simulation program, labeling the grouping "Simulation 3 of new methanol process design."

We can further subdivide our first world. There are at least two ways an object can make an instance of itself. In a *class-based* object world, the type definition object contains rules on how to construct an instance of itself. If we later alter any of these rules, the parent type object will seek out and propagate these changes to all instances of it, i.e., to its children. In a *prototype-based* object system, the parent simply makes a copy of itself when creating an instance. The user of this system then edits in the ways this instance is to differ from its parent. If one later edits changes into the parent object, the parent will not seek out and propagate these changes to its children. The more recent literature on object-oriented systems argues that this latter approach more nearly mimics what we wish for design. To see the distinction, imagine we wish to design a new control system for our column. We start by going to our files and copying a blueprint for an earlier design we deem to be close to the one we want. We red-line this copy to create our new control system. In a class-based inheritance system, if someone subsequently alters the blueprint we copied, the system would seek out our copy and alter it to reflect these changes. We might find ourselves enormously annoyed when this happened.

We can augment our second world with the ability to add pattern definitions much like the first when, with experience, we find patterns that should be useful in the construction of future objects. We might describe our mind set for our augmented second world as "instances first, then patterns." Of course, once one has these patterns, we see a completion of the cycle. We can use these to create future objects.

We suggest that the two mind sets, "patterns first then instances" and "instances first then patterns," are profoundly different. This difference is as subtle and important as the distinction between class-based and prototype-based inheritance in object oriented systems. The first world characterizes the mind set behind the creation of object-oriented information systems. It is the mind set behind the creation of centralized database systems. The second world, without the ability to extract patterns, is the mind set behind the creation of the World Wide Web (WWW). The augmented second world is the mind set behind the information modeling system we are creating called *n*-dim.

We view our system as a base on which we and others can add applications. This base provides important functionality to all systems built on top of it, including history maintenance, access control and revision management. As *n*-dim tolerates diverse information objects and any kind of relationship between two objects, the diverse applications we describe below can co-exist if we were willing to have them do so.

n-dim

As we have described in earlier papers on this topic, our work in information modeling is based on experiences that we and others have had on collaborative design projects (Westerberg, et al., 1989; Robertson, et al., 1994). One goal is to *support collaborative work* by allowing the timely sharing of information, allowing users to develop personal and likely very diverse views of that information. A second goal is to *capture the real processes occurring* when an organization carries out a complex activity such as design. Such information can serve as an excellent prototype for those who wish to repeat the same type of activity later and as data to find patterns and establish understanding about the processes really occurring (in other words to develop theories about the design of this type of artifact). We suggest that it is only with this understanding that one can hope to make significant design process improvements.

The underlying premise for our approach is that we believe designers are always developing and using a diverse set of models. They sketch organization charts (a model interrelating the people on the project and in the company), they write quantitative models for the performance of physical artifacts (a simulation model, for example), they draw activity charts to establish critical paths and allocate resources vs. time and so forth. Our aid to support collaborative activities such as design is, therefore, in the form of a general modeling environment, an environment we call *n*-dim (***n*** **-dimensional i**nformation **m**odeling). One can place relationships over files and their contents, over objects stored in databases, over pages in the World Wide Web (WWW), and so forth, by creating *n*-dim models built of nodes and labeled links. A model contains a list of pointers, displayed as nodes, to anything one can see on the network. Between any two nodes one can place a labeled link that indicates a relationship between the objects to which the nodes point.

Figure 1 is an example of a typical model in *n*-dim. A simulation requiring two input files has produced an output file. Notes in the form of a text file describe this simulation. Each node is actually a pointer not unlike a URL (universal resource locator) on the WWW to a data file, text file or simulation package. Labeled links describe how these objects are related. The simple act of aggregating them implies these particular objects are related to each other, a relationship of type Universal that we call model1.

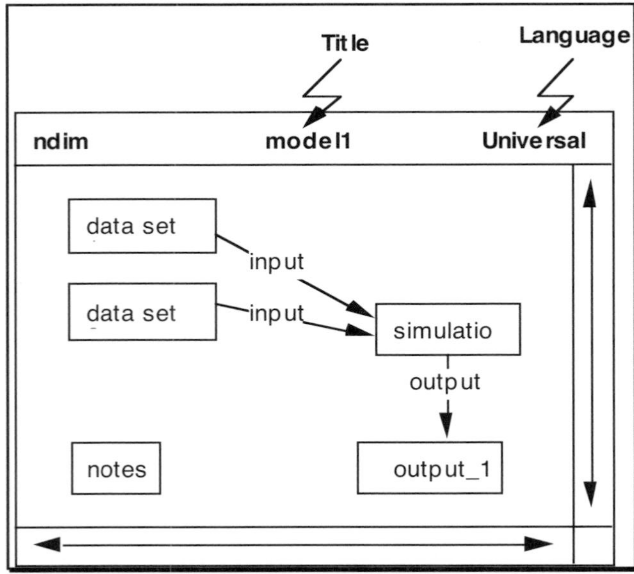

Figure 1. A typical n-dim model.

We summarize several properties/capabilities we have within *n*-dim.

- **Flat space**: We view all the objects anywhere in the network as existing in a flat space. The WWW adopts a similar viewpoint for all the files in it. It is over these objects that we place structure using *n*-dim models, which themselves become objects stored in the same flat space.
- **History keeping**: We want this system to record a history of any activity with a better fidelity than we have seen done in the past, keeping track of all communications, all the paths taken, all the mistakes made and so forth. We accomplish our history capture by keeping all the computer-based information the participants find, develop, organize and share while carrying out the project.
- **Private models**: A user prepares new models in his private workspace. Known as private models, the user may alter them at any time with no record kept of such alterations.
- **Published models**: We call the mechanism to keep information permanently in the system *publishing;* the name reflects what happens when one puts an article in the literature. In *n*-dim, to share an object one must publish it. A published object is immutable. We are free to make copies of it knowing it will not change, ever. In principle it will not go away either.
- **Revision management**: Published items will typically contain errors. We need a mechanism to correct errors. Any object can be designated as a revision of another. We have special operations in the system that allow us to trace revisions quickly.
- **Public models**: We have a specialized form of publishing/revision management that allows several users to share and revise a model. It is similar to what others have referred to as a "sticky" model. A user can create a public model. By controlling access to the operation (also an object) that can alter this model, the user controls who is allowed to alter it. Changes are by adding descriptions of the changes to the end of the internal model description. A user can alter a public model by adding parts to it and/or by apparently removing parts from it. He may do this while someone else is also modifying it. Rules dictate the revision sequence if two changes occur at the same time. Since we record every change made in sequence, we are able to reconstruct and display the revision history for it. Removed parts are in effect placed and then removed whenever the model is recreated for display. The default display for a public model is its latest revision.
- **Access and ownership control**: We are implementing access control mechanisms that allow users to restrict access to and alter

ownership of objects. If a user has access, he can see and copy the object. Otherwise he can neither see it nor copy it. The "owner" of an object can revise its access at any time; he can also transfer ownership to someone else (and later retrieve it if he desires). We are implementing access so that the system maintains a revision history of access and ownership for each published object (e.g., who has had access and when). Our access mechanisms will not defend against malicious behavior any more than the access mechanisms of the UNIX operating systems can.

- **Pedigree management**: If someone uses the *n*-dim "copy" command to copy an object and its associated operations (prototype-based inheritance), then *n*-dim automatically adds a link labeled "copy of" between the original and the copy in a special model that keeps the pedigree of that object. Anyone can trace both forward and backward on the pedigree model for any object; however, the system suppresses the display of parts to which a person has no access, appropriately displaying and linking those parts that he can see.
- **Inductive learning**: We want this system to aid users to learn how they should organize their information. This is in contrast to having a team of experts tell the user how he or she may organize information. We introduced the concept of a *modeling language* (ML) to support this learning. A designer may create an instance of a new kind of model he labels an activity model. Another team member may observe this model and like the way it organizes activities. She may start creating such models, agreeing verbally with the originator on the "rules" by which they construct these models. After a while, they have a very good understanding of rules for this model structure. One of them may wish to establish the construction rules more formally to pass them to others. To do this that person can write a modeling language describing how one is to construct an activities model.
  A modeling language is itself an *n*-dim model built of nodes and labeled links. The nodes are pointers to other languages in the system; call them modeling languages ML1, ML2, etc. We may use such a model, call it OurML, as a modeling language to construct a new model which we call OurNewModel. OurNewModel may contain a pointer to any object in the system provided it was constructed using one of the languages ML1, ML2, etc. We may place a labeled link between two objects provided that in OurML such a labeled link connects their modeling languages (e.g., ML1 —(linklabel)—> ML2). We have added mechanisms to state cardinality, "parentheses," and the "or" operator to augment our ability to create modeling languages. An example of cardinality is that we might wish to have precisely one object only constructed using the activities modeling language in OurNewModel, and it may have to have exactly one link emanating from it with the label "DoneBy."
- **Tool integration**: We want *n*-dim to be an environment within which one can integrate diverse tools. We shall talk more about tool integration later in this paper.
- **Scalability**: We expect there to be billions of objects stored all over the network. Searches for information have to be fast in spite of the amount of the information stored.

We now have a third generation version of *n*-dim with most of these properties implemented. It is built on top of a prototype-based object system BOS (Levy and Dutoit, 1994a) with an interactive interpreted language to create and manipulate objects called Stitch (Levy and Dutoit, 1994b). The BOS/Stitch system allows us to implement and test new ideas rapidly.

We are using this version of *n*-dim in a variety of ways to test our hypotheses that this form of environment can enhance collaborative activities in significant ways and that it will capture enough information to allow us to understand complex activities such as design better. We shall report on the outcome of these tests in a later paper. Here we describe briefly several of the tests that are underway.

Example 1: The Information WEB[2]

Motivation and Approach

The goal of the Information Web (IWEB) project is to develop an information modeling and management tool specifically designed to meet the needs of software engineers working in a group environment on large projects.

As more people work on a project, the requirements placed on the communication infrastructure which supports their work grows rapidly. The more people there are working on a project, the more difficult it is to maintain a shared understanding of a project's goals, status, design, and implementation. This problem is heightened in a part-

[2] Members of *n*-dim group active on this project: R. Coyne, A. Dutoit.

time and/or distributed development environment where both the managers and the developers involved in a project often have many other projects or unrelated activities for which they are responsible.

Current environments available to support collaborative software design and implementation are insufficient for the task at hand. Typically, for each project, developers cobble together an ad hoc combination of communication means such as e-mail, electronic bulletin boards, word processing and document preparation tools and a set of disjointed CASE tools which they use to support their work and communications.

With these requirements in mind, we are experimenting primarily within the context of the project-based software engineering courses in the School of Computer Science at Carnegie Mellon University. These courses reflect much that is typical of the "state of the art" and outstanding issues in the current practice of software engineering. The classes are large (typically 25 to 60 participants), require the delivery of a running prototype to a real client, and generally attempt to provide a realistic software engineering experience to the participants who are generally junior or senior computer science students.

IWEB has been developed on top the *n*-dim information modeling environment. It is presented here as an example of how computer supported collaborative work (CSCW) applications can be developed using *n*-dim. In this paper, we have focused on presenting specific modeling examples illustrating a selected number *n*-dim concepts which IWEB uses; a more detailed examination of IWEB itself is available in Coyne and Dutoit (1994).

IWEB Overview

Given that we found everyday intra- and inter-group communication to be a major bottleneck in the cases we studied (Coyne, et al., 1995), most of the core functionality of IWEB is targeted at improving information exchange and capture. On one hand, IWEB provides a gIBIS-like (Yakemovic and Conklin, 1990) Issue Modeling Language for structuring discussions and maintaining lists of unresolved issues. On the other hand, IWEB provides a notification mechanism allowing users to send informal notices to a specific user or group of people. The IWEB notification mechanism is very similar to an e-mail system with the exception that notices may contain references to issue models and/or other *n*-dim models. Users usually create issue models within an Issue Forest, which contains a set of related issues. One can view an issue forest as a context for discussion. Users may create and reorganize issue forests as their communication needs evolve. Similarly, users can post notices in Notification Boxes, which represent a target audience (e.g. a person, a team or the whole project). Anyone in the project can create a notification box as the organizational structure of the project evolves.

By providing users with both structured and unstructured media for communication, together with means for creating and restructuring contexts in which discussions occur, we hope to increase the amount of information exchanged while decreasing the amount of information lost or misdirected.

Figure 2 illustrates the use of issue forests and notification boxes. The leftmost window shows an issue forest used by the development team of a project for discussion of development related issues. The rightmost window shows a notification box used for notifying management. The issue forest of the development team contains two issues: Team Formation and Project Deliverables. The description of the Team Formation issue appears at the bottom of the issue forest window, together with its author and the date it was posted. The nodes displayed at the right of the issue are proposals and arguments related to the issue that other members of the development team have posted in response to the issue. The management notification box displays a notice entitled Team Formation Issue which one of the members of the development team created. The notice includes the issue being discussed (displayed in the upper right hand corner of the notification box) and includes a description (displayed in the lower right corner of the notification box). At this point, a management member may select the issue in the notification box and decide to include it in the management issue forest. From there on, the issue will live in both issue forests, i.e. any proposals, arguments or resolutions related to this issue will be seen in both issue forests.

Modeling Examples

Modeling Languages (MLs) & Operations. We implemented the issue forest and notification box MLs using the ML concept of *n*-dim. For example, the issue forest language is a model containing only those languages which can be included in an issue forest: the issue, proposal, argument and resolution languages. The issue forest language also provides operations for posting new nodes in an issue forest, or including existing nodes from other issue forests. All legal operations on issue forests are similarly implemented as operations in the issue forest ML.

Public Models. Operations on the issue forest and notification box languages ensure that any new issue forest, notification box, notices and nodes created within an issue forest are public (see our description of *public* models earlier), therefore accessible to the rest of the project. Making these models public also ensures that they cannot be removed or modified (only new models can be added), and thus provides a simple history mechanism.

Flat Space. Anyone can refer to issues, issue forests and notification boxes (as with all *n*-dim models) from more than one model. This allowed us to be able to share an issue across multiple issue forests. This also allows users to build their own models containing references to issues of interest: for example, one user may want to maintain his/her own "hot list" of issues by adding references to them in a private unpublished "current hot list" model.

Access Rights. Access rights allows us to restrict access of certain models by certain users. In the above example, managers may want to allow only managers to access the management issue forest. Separate access control of the operation for a model allows us to control who can add to a forest.

information created by the members of previous classes. We characterized the amount and type of information each group posted vs. the time they posted it. We were able to correlate the type of information exchanged and when it was exchanged during the term to the assessed success of the group. If this correlation is real, it should stand up to future testing, and it would represent one way we can teach and support future groups to improve their performance.

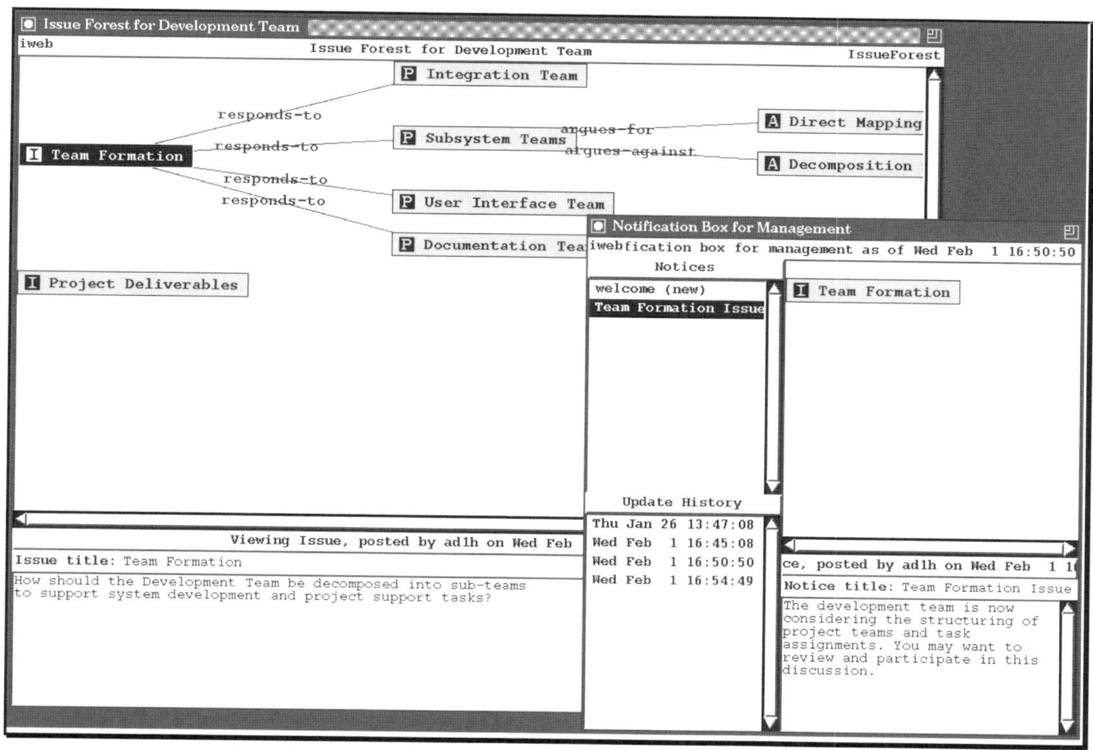

Figure 2. IWEB issue forest and notification box.

IWEB Status and Future Plans

The communication functions of IWEB have been implemented in *n*-dim. We are in the process of training about 20 students participating in software engineering classes to use the IWEB rhetorical model for capturing meeting minutes and structuring design rationale documents (weekly information exchange). Once the students are familiar with this mode of working, they will be introduced to the IWEB tool for daily information exchange. We plan to scale the use of IWEB to a full software engineering class by Fall '95 (roughly 60 participants) and hope that the data captured by the use of IWEB will enable us to improve our software engineering program, our software engineering process and the IWEB system itself.

To illustrate the type of information we expect to learn from this study, we analyzed the bulletin board

Example 2: An Industrial Engineering Design Support System[3]

We have worked on a project with ABB to support their process for designing a particular kind of electrical equipment. As a first step, two members of the *n*-dim group (Subrahmanian and Gardner) visited two of their equipment design groups for several days each, with the purpose of developing an information flow model and an understanding of roles played by all persons involved in their current design processes. They and personnel from ABB developed an information flow model, starting with the original customer order and ending with a final design. It showed several interesting features; for one, in each

[3] Members of *n*-dim group active on this project are: Eric Gardner, Sean Levy, Ira Monarch, Eswaran Subrahmanian.

group someone voluntarily assumed the role of organizing a catalogue of past designs. They saw points in the flow where significant delays occurred in passing information to those who really needed it. A particularly costly example was the delay in discovering a design flaw in the product and having a several month delay in feeding this information back to both the customer ordering personnel and the product designers. The mechanisms were in place; they simply did not function quickly.

For this project both ABB and the *n*-dim group selected a project to record the entire design process for this electrical product, a project we felt would test the hypothesis that they could significantly improve their design process if they improved the recording of it. At the present time the electrical product design involves establishing values for several tens of parameters. Many are discrete parameters indicating such things as the type of component to use; others are continuous parameters indicating such things as sizes. ABB captures each of its designs in a design database. A design starts with preliminary values for only a few of the parameters established by interpreting what the customer says he wants. The interpretation reflects the "tacit" information (e.g., colors for painting the product) available from previous contracts with the customer as well as taking numbers directly from correspondence between the customer and the ABB negotiation engineer. After successfully winning a bid to build a piece of electrical equipment for a customer, the negotiation engineer passes a preliminary design to the design manager with his best estimate for values for a few of the parameters, particularly those that specify what it must be capable of doing.

The design manager assigns a design engineer to complete and "optimize" the design. The assigned designer runs any of a number of design programs using input from the current design information. After running a tool the designer may incorporate some of the program output into the design data. The order in which to use the programs is up to the designer, reflecting his past experiences and preferences. Many times the designer has to set parameter values for the design that he thinks will lead to a reasonable design, which, with the running of subsequent tools, prove to be inappropriate. The designer may then reguess values for these parameters and repeat tool invocation following the previous sequence, or he may start down a different path to find a suitable design.

Some designer engineers are markedly more successful than others in creating good designs. Because only the latest design data is available in the design database, there was no record of the design steps taken to create it. Two deficiencies result. First, the process that created a design cannot be passed to someone else to use as a guide for a new design. Second, there is no data on which to form hypotheses relating the design process to the quality of the final design.

For this project we used the history capturing/publishing and language mechanisms of *n*-dim. The original correspondence between the order engineer and the customer is a record kept within *n*-dim. Being published it becomes immutable and can never be removed. We provided operators that could dissect these memos, allowing the user to use a mouse to highlight any string of characters within a memo and give that string a label. The labeled part became a "string" or, if appropriate, a "number" object in *n*-dim. *n*-dim models automatically captured relationships between that string or number object and the memo from which it came, and, if appropriate, between that number and the data field in the design database where it may have been used as part of the original design specification for the product. The system unobtrusively encourages the order engineer to annotate any of these relationships. This mechanism creates a trail from the customer input to the information put in as preliminary design data. We formulated *n*-dim languages (which themselves are *n*-dim models) for each of the types of *n*-dim models in this process.

We also captured in a similar manner every step taken by the design engineer. Whenever he or a tool he invoked accessed the design data, we added pointers and relationships describing this access into an *n*-dim model. We did this by placing code in front of the database that sorted out such accesses to it and that passed along information so we could construct appropriate *n*-dim models. Output put back into the database led to more pointers and relationships to describe it. Thus we could track everything done by a designer in the course of creating a design. Again, in as unobtrusive a fashion as we could, we asked for annotations describing the steps he took. We also provided tools to display to the designer the steps he took. He can point anywhere in this process and ask to repeat the design from that point, leading to branches in the design process. His supervisor can also observe the process and offer suggestions.

The designer can revise any of these models at any time to alter the way he organizes and displays any of the captured information. What a designer cannot do is delete earlier versions of any of the data as *n*-dim captures it as a part of the history of the design and of the design process. His revisions supply information on how to improve what the system should capture and how it might best organize and display this information as these revisions reflect changes the designer finds useful. We believe this evolutionary improvement is crucial for these systems.

Example 3: A Prototype Engineering Design Support System[4]

In this example we shall look in more detail at the actual objects created for a small prototype design support system that we created to illustrate the capabilities of *n*-

---

4 Members of *n*-dim group active on this project are: Eric Gardner, Sean Levy, Ira Monarch, Eswaran Subrahmanian.

dim. Engineers generate and use large amounts of heterogeneous information during the course of designing an artifact. In this *n*-dim example, we first look at how a designer might organize all of this information using a design folder model. We then explore, in greater detail, one particular aspect of the design folder in order to demonstrate some of the virtues of the *n*-dim approach and representation. To give a context to this discussion we have chosen a relatively simple domain: the design of hydraulic cylinders.

Since we have modeled this design environment in *n*-dim, the ability for multiple designers to collaborate on the design is inherently supported. We require that all relevant models be public using the *n*-dim publishing mechanism.

We have implemented a "Design Folder" modeling language which allows one to organize many different kinds of information used throughout the design process. We describe the components of the Design Folder model below.

- **Physical Model**: The Physical Model represents a hierarchical physical decomposition of the components of the design artifact
- **Design Database**: This *n*-dim model provides an interface to the relational database in which one stores all of the design specifications, parameters, and other features.
- **Design and Analysis Toolbox**: This model contains references to the various design and analysis tools the designer can utilize. He can tailor the toolbox specifically for the current design problem. The designers can build a library of many different toolboxes, each for a specific engineering domain or task.
- **Design Constraints**: Constraints are an important part of any design environment. There are *n*-dim models that provide a grammar for defining constraints relevant to the current design artifact. There is also a ML defined for grouping these constraints into logical classes.
- **Design Rules**: The design environment also supports the creation and use of design rules. Rules can be defined to suggest what action might be appropriate when a designer violates a particular design constraint. We concentrate on this part of the Design Folder to elucidate some of the important capabilities designed into the *n*-dim system.

The rule ML permits one to write rules of the form if <condition> then <action>. We refer to <condition> as the left hand side of the rule (lhs) and <action> as the right hand side (rhs). The ML specifying the grammar of the lhs of rules allows the combination of constraints using the operators "and" and "or." The constraint ML, in turn, allows one to construct algebraic constraints using the design parameters, real and integer constants, and unary and binary operators.

The rhs in the rule ML is currently modeled simply as string. This string is what the user sees when the lhs of the rule evaluates to false. A logical improvement to this current representation would be to define the rhs as a block of code that gets executed rather than a string that gets displayed.

The rule ML defines several useful operations, one for displaying a particular rule's definition and another for evaluating a rule. The first of these is fairly straight forward and warrants little discussion. The latter operation evaluates the rule in the context of the current values of the design parameters. The system displays a dialog box indicating whether or not the parameters values violate the rule.

Figure 3 shows a view of what the designer's screen might look like. The design folder is the model in the upper left. Notice that it contains parts corresponding to all of the elements of the design folder that we discussed above (i.e., Rules, Constraints, Design DB, Physical Model and Design Tools). The window covering the lower left portion of the design folder shows the interface to the design rules.

The dialog box covering part of the lower portion of the rules display was the result of evaluating the "check stress" rule. Notice that the text in the dialog box indicates that the lhs of the rule is not satisfied. It also contains the suggestion that the designer try adjusting the inside diameter or the wall thickness of the cylinder and then re-invoke the tool that calculate the stress to see if the problem is corrected.

The remaining three large windows show the physical model and two of its components. The CAD drawing associated with one of these components is also shown. In addition to having references to CAD models, the component models contain references to the design parameters associated with that part of the artifact. These parameter references point to the same models referenced in the design database model in the design folder.

Notice that there is a second dialog box displayed. This dialog box is the result of the designer invoking the set value operation on the inside diameter parameter. This was in response to the suggestion given by the "check stress" rule.

Example 4: Design and Manufacturing[5]

ACORN (Advanced Collaborative Open Research Network) is an ARPA funded project within the MADE (Manufacturing Automation and Design Engineering)

---

[5] Members of *n*-dim group active on this project: S. Konda, E. Subrahmanian, M. Terk, A. Westerberg. Other active participants include: S, Fenves, S. Finger, both in Civil Engineering.

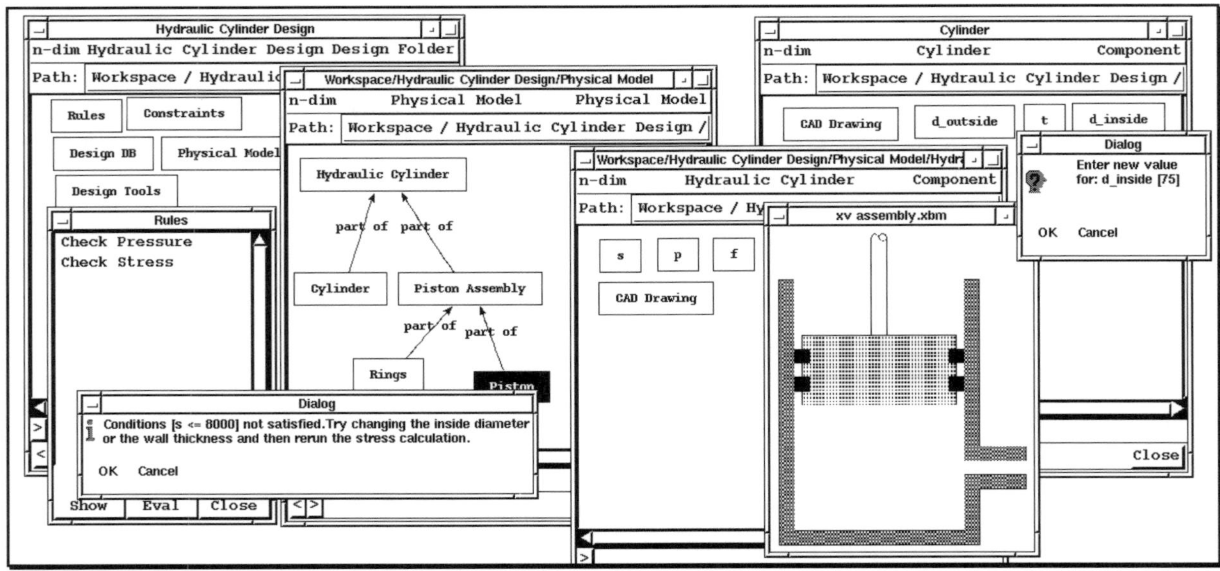

Figure 3. Engineering example screen shot.

program. ACORN is a testbed for using the World Wide Web to link together geographically dispersed team members who may reside in several different organizations and to link together diverse design and manufacturing resources. We currently are working with five other universities and eleven companies. Examples of services being provided are:

- University of Michigan: synthesize a computer board
- University of Pennsylvania: acquire the 3-D shape of a physical object
- University of Utah: supply a mechanical engineering parts catalogue
- MIT: advise on assemblability of a mechanical component
- Alcoa: create a part using stereolithography

We have a problem domain where, as users of this system, we wish to realize many of our *n*-dim objectives while cooperating with others not in our *n*-dim world; i.e., in this demonstration we show how we can use *n*-dim while interacting with external human and computational agents across the WWW. We have encapsulated the "driver" which NCSA Mosaic (a popular WWW browser) uses to exchange pages of information across the web. Our encapsulation identifies and saves WWW pages that pass to and from specific external agents. We have added additional operators to construct *n*-dim objects from the information in these pages. Thus we are able to record these transactions and convert them to objects we can organize and interrelate within *n*-dim as we please.

Example 5: Tool Integration[6]

We are embedding ASCEND (Piela, et al., 1993) in *n*-dim. ASCEND is an interactive, equation-based modeling environment. As a tool, it aids its users (called modelers) to create, write, debug, and execute code describing quantitative models based on algebraic and differential equations.

We saw four levels of embedding ASCEND which would be possible: (1) We could put an icon within *n*-dim which, when double clicked, would open the ASCEND system; control would then pass entirely to ASCEND. (2) We could allow the modeler to create a particular model instance in ASCEND, such as a model of a car. The modeler could create an *n*-dim "frame" (a data structure containing pairs of attribute names and corresponding values) in which he or she would place names and values for parameters for this model. We could create a translator to convert this to ASCEND input. The modeler could then run ASCEND and generate results which we would translate back into an *n*-dim frame. (3) We could attempt to support a modeler to create a new model. Input to ASCEND would be code; preliminary output would be diagnostic messages and instances of that code resulting from compiling it. Final output would be debugged code. (4) We could completely rewrite ASCEND to make it a part of *n*-dim.

The first is hardly a significant form of tool embedding. The second is a reasonable form, but again it is not much of test of embedding for a complex tool like ASCEND. The last approach is impractical as the ASCEND system involves over one hundred thousand

[6] Members of *n*-dim active on this project: M. Thomas, A. Westerberg.

lines of C-code. We, therefore, opted to embed ASCEND for its main purpose: to aid modelers in developing correct equational-based models.

Our original idea on how to do this was to allow a modeler to write ASCEND code visually using the node and link modeling paradigm of *n*-dim. After trying this approach on several models, we found that one could construct such a model rather easily, but it was virtually impossible for anyone else to understand the result. Even the person creating the code was lost when reviewing it a few hours later. ASCEND code in textual form, on the other hand, is fairly easy to understand. We concluded representing ASCEND code using nodes and links was not the way to embed ASCEND in *n*-dim, as nice as it sounded.

We further noted that *n*-dim is designed to support information handling when designing something, either individually or as a collaborating team. We therefore proposed to use *n*-dim to aid in writing ASCEND code (a design activity) line by line.

ASCEND supports a part/whole approach to modeling. A modeler starts by defining the types of the variables to be used, and creates instances of these types as parts of more complex type definitions. Type definitions themselves are arranged into hierarchies. As an example, a car could be put together by having a car body, an engine, wheels, and so forth, and each of these parts could be built of other parts. Wheels could be organized into generic_wheels which are then refined into spare_wheels and normal_wheels, where the last two each inherit the attributes of generic_wheels.

ASCEND also supports deferred binding of types. Deferred binding means a user can include an instance of a type in his or her model and can, at a later time, reach inside that instance and alter the type of one of its parts, provided the alteration is to a more refined type. For example, one could create an array of five generic_wheels for the car (including the spare). One generic_wheel could be altered to be a spare_wheel and the other four to be normal_wheels. The consequence is that one may not know the structure of a part simply by looking at its type definition.

To create executable code, one picks a type and asks ASCEND to compile an instance of it.

To support writing and debugging code representing a complex engineering object, the modeler must be able to see inside any of the type definitions in the system. We felt the modeler would be well-served if he or she could also see inside instances of the code he or she is preparing to catch any alterations caused by deferred bindings. Also, the car engine may have four cylinders, and the modeler may want to access attributes for the first one only. This again suggests a need to see inside the instance rather than just inside its type definition. We concluded, therefore, that to support a modeler for coding and debugging, we needed to provide him or her with access to both type definitions and to compiled instances. Our approach was to use the ASCEND system to provide any of these services that it already could, and to supply the rest as operations in *n*-dim.

To use the current embedding of ASCEND within *n*-dim, the modeler writes an ASCEND model by entering a few lines of code. Occasionally, he or she will trigger ASCEND to compile the current version of the model. We have created the ability for the support system to write a test model which contains an instance of the model being created as its only part. This test model is passed to ASCEND to compile, and ASCEND creates a partially compiled data structure (something it does nicely). If the compile step finds it is missing information to create certain instances (for example, the size of an array is unspecified), the support system adds a statement to the test model to prompt the user to put this specification into that model. In addition, *n*-dim models could exist which could contain all the ASCEND types organized into libraries. *n*-dim allows each model to be annotated separately; a modeler could use this annotation to aid in selecting a type.

Whatever part of the data structure the system could compile, the modeler can browse it within ASCEND. The modeler can then use the actual parts in the partially compiled ASCEND models to aid in programming. He or she can open a partially compiled instance to any depth (going into parts of parts of parts etc.) and visually pick a part to be a variable or object for the next line of code in the model being constructed. The system will bring that object name to that line of code, creating the right qualified name for it (from experience, incorrectly qualifying a name is the most common error made by modelers).

We believe this style to be very similar to programming on a spreadsheet — remembering where things are and not what they are called.

The modeler can, in addition, use *n*-dim to manage code revisions, to share models, to link in annotations, and so forth. A complete history of the development of the code is possible. A future addition will be to allow the modeler to use an already existing tool in ASCEND to define frame definitions for a particular model instance, allowing the attachment of ASCEND using the second level of tool integration posed earlier.

Example 6: PhD Notebook[7]

The last example, albeit one we shall only describe very briefly, is for our students to use the *n*-dim environment to keep their PhD notebooks. They are scanning in minutes from meetings, keeping a current bibliography of reference articles, and so forth. The environment allows them to organize the pages of the notebook using a variety of index schemes.

---

[7]Members of *n*-dim group active on this project: D. Cunningham, R. Patrick, and A. Westerberg.

Future Endeavors

There are a number of aspects of information modeling that we have considered or are considering in great depth and intend to incorporate into *n*-dim. Some of these features as they would be manifested in *n*-dim are outlined below.

Events and Rules

Information modeling goes beyond static representations of information. While operations (mentioned previously) give users the ability to impose behavior on models, they still fall short of providing a comprehensive environment which users can customize. In order to provide user customization, we intend to incorporate an event/rule mechanism into *n*-dim. This would most likely be done by using an existing rule-based system such as CLIPS (1994) or RAL (Forgy, 1994). The basic premise is that a user can establish rules that are invoked when particular events occur in the system. For example, a user may design a rule to have a particular operation invoked when a new issue is added to an issue forest. This gives a user the ability to configure *n*-dim in a radically different way.

Modeling User Interfaces

The nodes and links representation of *n*-dim is only one way to view information, and it is not always the most desirable — different views of information structures are often necessary.

In the case of IWEB, for instance, we wrote a separate user interface. We identified two important drawbacks to this exercise: an experienced programmer had to write the interface and someone attempting to use IWEB needs to explicitly load the IWEB user interface code. For this and other reasons we wish to model our user interface toolkit in *n*-dim such that users can build interfaces in the same manner that they design models. We could associate an interface with a particular model or modeling language. such that, when certain models are opened, the system uses the user-defined interface instead of the default interface.

Modeling Search Criteria

The ability to search for models in an extremely sophisticated matter is a crucial aspect of *n*-dim. Since one of our main requirements for *n*-dim is that it be useful in accumulating the information history of an entire organization over time, we anticipate rapidly accumulating a large number of published and public models. Finding specific models or models fitting certain criteria is therefore very critical. In addition, the criteria of a search is in itself useful information that a user may want to keep, publish, revise, etc. Again, the logical solution would be to model search criteria in *n*-dim.

References

Bañares-Alcántara, R. (1995) A process engineering design environment based on a model of the design process. *Intelligent Systems for Process Engineering (ISPE'95)*. Snowmass, CO July 9-14.

Ballinger, G.H., R. Bañares-Alcántara, D. Costello, E.S. Fraga, J. Krabbe, H. Lababidi, D. M. Laing, R.C. McKinnel, J.W. Ponton, N. Skilling and M.W. Spenceley (1994). épée: a process engineering software environment. *Comput. Chem. Engng.*, **18S**, S283-S287.

CLIPS (1994). *CLIPS Version 5 Reference Manual*. COSMIC Project, NASA.

Coyne, R. and A. Dutoit (1994). IWEB (Information WEB): information management for software. EDRC Technical Report EDRC-05-87-94, Engineering Design Research Center, Carnegie Mellon University.

Coyne, R., A. Dutoit, B. Bruegge and D. Rothenberger (1995). Teaching more comprehensive model-based software engineering: experience with objectory's use case approach. *Proc. Conf. Software Engng. Education*. New Orleans, January.

Forgy, C. (1994). RAL programming language. *Production Systems Technologies*. Pittsburgh, PA.

Kim, W. (1995) *Modern Database Systems*. Addison-Wesley, New York.

Krieger, J.H. (1995). Process simulation seen as pivotal in corporate information flow. *Chem. & Engng. News*, Mar. 27, 50-61.

Levy, S. and A. Dutoit (1994a). An overview of the basic object system. Unpublished Manuscript, *n*-dim Group, Engineering Design Research Center, Carnegie Mellon University, Pittsburgh, PA.

Levy, S. and A. Dutoit (1994b). BOS/Stitch reference manual. Unpublished Manuscript, *n*-dim Group, Engineering Design Research Center, Carnegie Mellon University, Pittsburgh, PA.

Oszu, M.T., D. Umeshwar and P. Valduriez (1994). *Distributed Object Management*. Morgan Kaufman, San Mateo, CA.

Piela, P., R. McKelvey and A. Westerberg (1993). An introduction to the ASCEND modeling system: its language and interactive environment. *J. Management Information Systems*, **9**, 91-121.

Robertson, J.L., E. Subrahmanian, M.E. Thomas and A.W. Westerberg (1994). Management of the design process: the impact of information modeling. *Proc. FOCAPD*. Snowmass, CO.

Stonebraker, M., P.M. Aoki, R. Devine, W. Litwin and M. Olson (1994). Mariposa: a new architecture for distributed data. *Proc. 10th Int. Conf. Data Engng*. Houston, TX, Feb.

Turner, J. and R. Kraut (Eds.) (1992). *CSCW '92*. ACM Press, New York.

Westerberg, A.W., P.C. Piela, E. Subrahmanian, G.W. Podnar and W. Elm (1989). A future computer environment for preliminary design. *Proceedings of FOCAPD*. Snowmass, CO, July 9-14.

Yakemovic, K. C. B. and J. Conklin (1990). Report on a development project use of an issue-based information system. *CSCW '90*. pp. 105-118.

DESIGN SUPPORT SYSTEMS FOR CONCEPTUAL PROCESS DESIGN

René Bañares-Alcántara and Jack W. Ponton
Department of Chemical Engineering
University of Edinburgh
EH9 3JL, Scotland, UK

Abstract

The conceptual stage of process design is arguably the most critical component of the whole design process, and so the development of systems to support this activity is of great importance. We present a list of functional and representational requirements for such a design support system. Criteria for the classification of design support systems are proposed with a view to evaluating current systems and setting guidelines for the future. A brief introduction to the KBDS design support system is presented as a showcase for these ideas. Extensions to improve "understanding" of the design artifact and process by a design support system are also described. These developments are key ones for the next generation of systems. We conclude with a description of a session using a design support system of the future in the year 2010.

Keywords

Design support systems, Conceptual process design, Design process, Design history, Knowledge representation, Design rationale.

Introduction

It may be argued that conceptual design is the most critical component of the overall design process because of its large impact on later design stages. It has been estimated that up to 85% of the life-cycle costs of a product can be committed at the end of the conceptual design phase, while only 5% of the actual life-cycle costs have been spent at that point (Nicholas, 1990). This clearly indicates that an improvement in the identification of better design alternatives during conceptual design may well result in a better quality product and lower life-cycle costs. Moreover, the cost penalty for the additional effort will be minimal.

In a survey among 200 UK engineering designers it was shown that 23% of their time is spent in paperwork (Court and Culley, 1993). Estimates also suggest that design engineers spend 50-80% of their time moving and organising data between stand-alone computer applications (Motard, 1989). Manual transfer of information is not only unchallenging but error-prone as well. Conceptual design can be improved by addressing problems such as maintenance of design history and its documentation for future retrieval and re-use, quality assurance and standardisation of documents, information sharing between design "agents," both human and computer, and information re-use for multiple design projects.

More significant than the volume of information involved in design is the issue of knowledge which it represents. A substantial amount of design documentation is recorded in personal diaries, memos, reports and logbooks which are very seldom structured or reused. Information on sheets of paper and computer files cannot be automatically related. Although the latter is available to the computer, the essence of its content, its "meaning" is not. A key role of a design environment, in our view, must be to capture, coordinate and organise the knowledge currently distributed in diverse and often inaccessible media.

These two observations emphasise the opportunity for a computer-based system to support conceptual design. Although the importance of conceptual design and current limitations of design practice have been recognised for some time, very few conceptual design support systems for process engineering have been developed. Perhaps the most important reason for this is the lack of understanding of how conceptual design is done. Everybody agrees, however, that conceptual design is an innovative and creative process. While this precludes the successful development of fully automatic design systems with currently available computer techniques, many possibilities remain open for the development of design support systems.

What characteristics must a design support system have? Can such a system be delivered with our current technology, and if so, what would it look like? Attempts to answer these questions constitute the bulk of this paper. Experience shows that a number of necessary requirements (Section 2.1) must be satisfied by a successful design support system.

Prior to the implementation of a design support system it is necessary to develop a representation of, and procedures to reason about, design knowledge. Section 2.2 presents a proposal of the types of knowledge that must be represented by a design support system. However, the development of a system that supports design is an open-ended problem. We attempt to organise the multitude of possibilities by proposing some classification criteria, Section 2.3.

In Section 3 we present a short history of design support systems in process engineering and classify each of the instances according to the classification proposed in Section 2.3. We then focus on KBDS, a design support system that has been developed focusing on the representation and use of the design process and with enough flexibility to allow the experimental application of diverse design models. Its representation accounts for the evolutionary, cooperative and exploratory nature of the design process in an explicit, prescriptive and integrated manner.

KBDS is undergoing a number of extensions resulting from feedback of internal and industrial users. We will focus on three of them: the improved "understanding," or representation at a finer grain of detail, of the design artifact by the design support system (Section 5.1), an example of the inclusion of domain-specific knowledge during the design process, via the incorporation of waste-minimisation criteria (Section 5.2), and the application of design process concepts to the area of modeling (Section 5.3).

Section 6 explores a possible design support system of the future. The last section summarises the document and lists the advantages we can expect from the development of design support systems.

Issues in Design Support Systems Development

Conceptual process design, as a subclass of engineering design, has a number of features in common with the latter. It is ill-structured, in the sense that its initial specifications are usually inconsistent, redundant or incomplete. It is an explorative and opportunistic task, spanning activities from the routine to the creative. It is evolutionary, generating a final design in an incremental fashion. It is performed by teams of engineers cooperating over long periods of time. Finally, it combines several sub-tasks, e.g. synthesis, simulation, and sometimes optimisation and control (Bañares-Alcántara, 1995).

Thus, a design support system must be embodied in a flexible and integrated computer-based environment. It should support the design process and enable the incremental development of a design product by a group of cooperating (human) designers and computer-based agents through the successive application of design tools, Fig. 1.

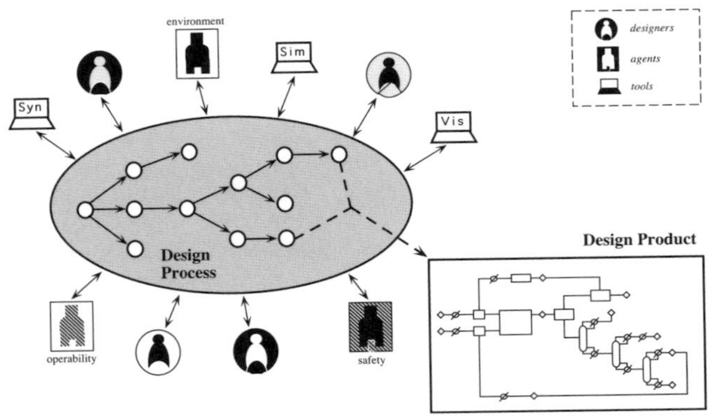

Figure 1. Designers, computer-based agents and tools operating upon the design process.

Based on the above, Sections 2.1 and 2.2 present what we believe are the necessary requirements that must be satisfied and the types of knowledge to be represented by a successful design support system. This leads into Section 2.3, which proposes classification criteria for the organisation of design support systems into different types. This classification aids in the evaluation of current achievements and, more important, sets some guidelines for future work.

Requirements of a Design Support System

Our work has focused on the understanding and improvement of the design process. This has implied a shift of focus from the traditional emphasis on the representation of the chemical plant, to the representation of the process by which the plant is created. A representation of the plant is still needed, but only as one component of the design process, as shown in Fig. 1. This new focus helps to identify a number of requirements for a design support system (Bañares-Alcántara and Lababidi, 1995):

- Exploration. Improved designs may be obtained if the designer is allowed to consider more alternatives while the system keeps their proliferation controllable. The design support system must assist in the management of those alternatives, i.e. their representation, control, evaluation and use.

- Evolution. A design support system must take into account the dynamic nature of the design process and the evolution, by incremental operator application, of the design proper or the design artifact. Each partial or alternative design is a cross-section of the design process, and is therefore related to other alternatives. Concepts used repeatedly should be represented as a single object, growing and evolving during the design process, and not as a series of unrelated items.
- Cooperation. A design support system must be able to support cooperation between multiple designers working concurrently on the same design artifact. This entails allowing the sharing of data and knowledge (communication), and the maintenance of consistency among sub-designs and concordance of intention (coordination).
- Integration. In order to carry out the process of conceptual design it must be possible to use within an environment all the standard tools of process design: physical property calculation routines, flowsheeting packages, equipment design, etc. This does not mean that the design support system should attempt to recreate tools that already exist, but rather, to integrate and provide access to them.
- Automation. For a design support system to become an active and useful assistant to the designer, it must interpret correctly the state of the design. Thus, it is necessary to represent explicitly the designers' intention, decisions, methods, and assumptions. This varied knowledge must be recorded, structured and related to the evolving representation of the design artifact description, constituting the design process history.

We believe that these requirements are necessary but may not be sufficient. Further experimentation and conceptualisation are needed. This must be based on testing and evaluating the application of a design support system on industrial problems.

Representation in a Design Support System

Representation of information is one of the key issues in the implementation of a computational system. A good representation must be complete, concise, transparent (easy to understand by the user) and simple to manipulate (create and access) by computer. Sections 2.2.1 and 2.2.2 present two complementary views of design representation.

Information and Representation

Ullman has proposed a *value of information scale* (Ullman and Herling, 1995) which recognises several degrees of information organisation and a corresponding relative value. According to this arrangement, information increases its value (usefulness) as it is organised and represented in higher levels of abstraction.

Data or parameters when combined with relationships become models; the interpretation of the behaviour of a model gives rise to knowledge; and using judgement based on knowledge leads to decisions, see Fig. 2.

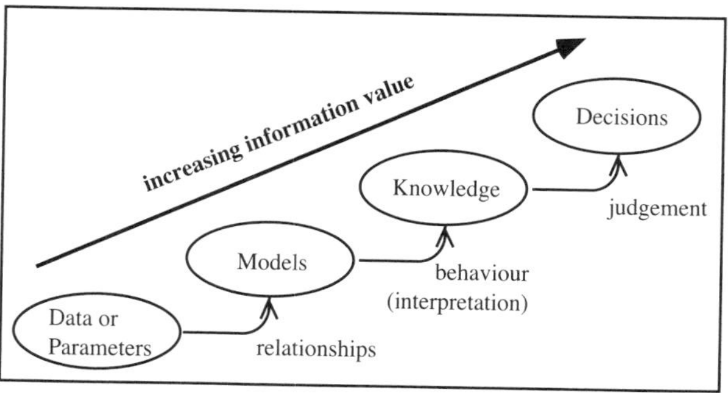

Figure 2. The value of information and its organisation.

This view is useful in understanding the role of design support systems. Current design tools maintain information at the Data and Model levels, but leave the maintenance and use of Knowledge and Decision levels to the designer. A design support system should be able to operate in the last two levels of the hierarchy as well.

Representation in a Comprehensive Model of Design

In order to address the requirements presented in Section 2.1, a comprehensive model of design should support the representation of all the items found in Table 1 (Mostow, 1985).

All of the design support systems that will be mentioned in Section 3 have a mechanism for the representation of the design state, item 1 above. One measure of their generality is the extent to which they include the rest of the items in the above categorisation.

Criteria for the Classification of Design Support Systems

The classification of design support systems in Table 2 is based on our own observations and on criteria presented in (Ullman, 1994) and (Peña-Mora, Sriram and Logcher, 1995). For each criterion in the table we list a number of possible options. Despite the format of the table, the number of possible options is not discrete, but is in reality a continuum. We have attempted to list the two extremes. Note that the criteria in Table 2 are applicable to the classification of design support systems operating on innovative or creative design problems, and not to automated design systems performing routine design.

Table 1. Representation required for a Comprehensive Model of Design.

1	Design state	Description of the elements of the design artifact, such as flowsheets, unit operations, variables, etc.
2	Design goals	These prescribe how the design artifact descriptions should be manipulated, and thus represent the purpose of the design activity
3	Design decisions	A statement of the choices between alternative design paths or satisfaction of design goals
4	Design rationale	Justifications for the selection of design goals, i.e. for the design decisions
5	Control of the design process	Concerned with the selection of the best goal to work on and the best plan with which to achieve it
6	Learning in design	This involves acquiring general knowledge about the application domain and specific knowledge about the particular design problem

The *focus* (1) of a design support system can either be the design artifact or the design process. Artifact-based systems can use first principles or higher levels of representation to record the design artifact. Process-based systems represent the design process as well, either by recording design steps by themselves or with their associated justifications or design rationale.

Support (2) to the designer can be limited to recording his or her actions, acting as a critic by providing analysis of design alternatives, or acting as a guide by generating design alternatives. This support is given through *interaction* with the designer (3), which can be passive (user-driven) or active. In the first case the designer may be constrained to a fixed order of actions or may be given the opportunity to interact with the system in a flexible manner. We term these two forms of interaction procedural and opportunistic respectively.

We identify two philosophies with respect to the *access to design tools* (4): monolithic, where the system is developed to provide a set of built-in design tools, and toolkit, where design tools are assumed as stand-alone and external. In turn, a system used as a toolkit may give access to a fixed number of design tools or may be built as an open-ended system, where the number and nature of design tools is left variable.

An important criterion is that of *maintainability by the designer* (5), i.e. the degree to which the scope of a system may be modified once it has been delivered to the final user. A system is said to be static when a designer is not able to modify its behaviour, tailorable when the modifications are limited to variations, and extensible when the behaviour can be expanded by the designer.

A system may document its interaction with a designer by generating a trace of the actions made by the designer or by itself. *Documentation* (6) may also include the result of organising, abstracting and explaining available information. This last step involves the manipulation and interpretation of information into knowledge as explained in Section 2.2.1.

A design support system may support the activity of only one designer or provide communication and coordination to team(s) of designers (7). The second situation engenders additional requirements for consistency maintenance and complexity management. In turn, and because of their very nature, process-based systems need to provide support for *multiple computer sessions* (8), while a single session may suffice for an artifact-based system.

Design support systems may constrain the design states to be always consistent. It can be argued that disallowing inconsistent states may stifle creativity, and thus the *co-existence of temporary inconsistencies* (9) during conceptual design may be desirable. However, it is important to recognise that some inconsistencies have to be resolved immediately while the resolution of others can be delayed. This decision could be based on a classification of inconsistencies according to their severity and urgency, whether they are local or global, and depending on the design process stage.

The *capture of design rationale* (10) in a process-based system may be passive or it may involve a degree of automatic extraction of rationale from the designer's actions. Alternatively, design rationale can be acquired on-line as the design process evolves or it may be recorded retrospectively, i.e. post-mortem.

Two other criteria are not listed in Table 2: generation and evaluation of design alternatives. In terms of design alternatives generation, a system can use a wide variety of methods, such as total enumeration or means-ends analysis. These alternatives can be evaluated with respect to cost, safety, controllability, environmental factors, etc. The evaluation may take into account different gradations of belief and uncertainty.

Design Support Systems in Process Engineering

In retrospect, it may be surprising that the earlier computer-based applications for design attempted to automate, rather than support, the design of complete chemical plants. Both AIDES (Siirola, et al., 1971), which was based on the General Problem Solver methodology, and BALTAZAR (Mahalec and Motard, 1977), which used a theorem-proving algorithm and evolutionary search, were developed in the 70's. See also Lu and Motard (1985). Their limited but interesting success focused research efforts in the 80's on the automation of design for simpler products using novel system architectures. This was the case of DECADE, a blackboard-based system for catalyst development (Bañares-Alcántara, et al., 1988) and other systems for the design of materials such as polymers (Venkatasubramanian, et al., 1992). Given the size and complexity of any creative design problem it was soon realised that automatic design could provide acceptable results only for the simplified or partial problems. Renewed efforts in the design automation resulted in two blackboard model systems: SPLIT (Wahnschafft, et al., 1991) and PROSYN (Schembecker, et al., 1994). However, it was also realised that there are interesting opportunities for the application of design support systems to creative design.

Table 2. Classification of Design Support Systems for Conceptual Process Design.

<table>
<tr><th>Criteria</th><th colspan="2">Options</th><th>Design Support System</th></tr>
<tr><td rowspan="3">1 Focus</td><td colspan="2">artifact-based</td><td>DESIGN-KIT,PIP</td></tr>
<tr><td rowspan="2">process-based</td><td>steps</td><td rowspan="2">n-DIM KBDS,IDIS</td></tr>
<tr><td>rationale</td></tr>
<tr><td rowspan="3">2 Type of support</td><td colspan="2">record alternatives history</td><td rowspan="3">PIP
KBDS,IDIS</td></tr>
<tr><td colspan="2">analysis of alternatives</td></tr>
<tr><td colspan="2">synthesis of alternatives</td></tr>
<tr><td rowspan="3">3 Interaction with designer</td><td colspan="2">passive, user-driven</td><td rowspan="3">PIP
n-DIM,KBDS,IDIS</td></tr>
<tr><td rowspan="2">active</td><td>procedural</td></tr>
<tr><td>opportunistic</td></tr>
<tr><td rowspan="3">4 Access to design tools</td><td colspan="2">monolithic or “wrapper”</td><td>PIP</td></tr>
<tr><td rowspan="2">toolkit</td><td>fixed</td><td>n-DIM,IDIS</td></tr>
<tr><td>open-ended</td><td>KBDS</td></tr>
<tr><td rowspan="3">5 Maintainability by designer</td><td colspan="2">static</td><td>PIP,IDIS</td></tr>
<tr><td colspan="2">tailorable</td><td>KBDS</td></tr>
<tr><td colspan="2">extensible</td><td>n-DIM</td></tr>
<tr><td rowspan="3">6 Documentation produced</td><td rowspan="2">trace</td><td>user’s actions</td><td rowspan="2">PIP,IDIS</td></tr>
<tr><td>computer actions</td></tr>
<tr><td colspan="2">manipulation</td><td>KBDS</td></tr>
<tr><td rowspan="2">7 Number of designers</td><td colspan="2">single</td><td rowspan="2">KBDS DES.-KIT,PIP,IDIS
n-DIM</td></tr>
<tr><td colspan="2">multiple (teams)</td></tr>
<tr><td rowspan="2">8 Computer sessions</td><td colspan="2">one computer session</td><td>DESIGN-KIT,PIP</td></tr>
<tr><td colspan="2">multiple computer sessions</td><td>n-DIM,KBDS,IDIS</td></tr>
<tr><td rowspan="2">9 Treatment of consistency</td><td colspan="2">design always consistent</td><td>PIP</td></tr>
<tr><td colspan="2">allows temporary inconsistency</td><td>KBDS</td></tr>
<tr><td rowspan="2">10 Rationale capture</td><td>passive</td><td>retrospective</td><td rowspan="2">KBDS,IDIS</td></tr>
<tr><td>active</td><td>on-line</td></tr>
</table>

More recent developments have advanced on two fronts: the maturity of ideas, methods and systems useful for design support, and the development of design support systems as such.

Examples of the first category are the ideas of Motard suggesting the use of object-oriented database technology and truth maintenance systems; the integration, rather than replication, of diverse design tools (Motard, 1989); and the implementation of KNOD, a knowledge-based front end to a process simulator (Beltramini and Motard, 1988). Above all we should mention the declarative modeling languages that facilitate the adequate implementation of models of the design artifact, item 1 in Table 1, e.g. ASCEND (Piela, et al., 1991). Modeling languages have been steadily evolving towards the inclusion of the representation of the design process, to the point where systems such as MODEL.LA (Stephanopoulos, et al., 1990) extend the representation to items 2, 3 and 4 of Table 1 and present some characteristic features of a design support system (see Marquardt (1994) for a review of modeling languages).

Two artifact-based systems were proposed in the second half of the 80’s: DESIGN-KIT (Stephanopoulos, et al., 1987), the first design support system in process engineering that allowed the specification of a design state through a graphical user interface and a set of analysis tools, and PIP (Kirkwood, Locke and Douglas, 1988), a system resulting from the implementation of Douglas' hierarchical design methodology (Douglas, 1988).

Three process-based systems are under development as of the time of writing this document: n-DIM, KBDS and IDIS. They satisfy the representation requirements listed in items 1 though 4 of Table 1 and provide a solid base for the last two items. n-DIM (Subrahmanian, et al., 1993; Westerberg, et al., 1989) is a system that supports the representation of the social and collaborative aspects of conceptual design. It emphasises that a design support system is a modeling environment, thus it records and organises models as they are developed and shared. This is the most generic system so far proposed, but for this reason requires substantial input from the designer. KBDS (Bañares-Alcántara, 1991) is a system to support the representation of the evolutionary, cooperative and exploratory nature of the design process. It takes advantage of the existence of other computer-based tools. KBDS is described in more detail in Section 4. IDIS (Goodwin and Chung, 1994) has been specifically focused on safety issues during design.

Where data is available, a classification according to the criteria in Section 2.3 has been attempted for DESIGN-KIT, PIP, n-DIM, KBDS and IDIS (see the third column of Table 2).

KBDS, an Example of a Design Support System for Conceptual Process Design

KBDS (Bañares-Alcántara and Lababidi, 1995) has been developed as a flexible tool to develop and test ideas about the representation of the design process, and to demonstrate how such a representation can be used to support the engineering design activity. As shown in Fig. 3 , the design process is represented in KBDS by means of four linked trees that evolve through time: one for design specifications and constraints, another for design alternatives, a third one for models of these alternatives, and finally, one for the design rationale supporting the decisions made during the evolution of the alternatives and the selection of the models, their specifications and constraints.

All the information in KBDS is represented in a prescriptive fashion, i.e. in a form that is amenable to computer processing. It is possible to superimpose on top of this structure a set of strategies and policies that reflect the design style used for a given application within a specific organisation.

KBDS supports features necessary for effective design support, i.e.:

- help in the control of design complexity by providing an incremental representation of the evolution of a chemical plant specification through the design process, by managing design alternatives, and by allowing the controlled propagation of values across the design history,
- monitor consistency between coupled partial designs,
- present information at different levels of detail and enables the designer to navigate through the resulting networks by the use of a graphical user interface (Lababidi and Bañares-Alcántara, 1993), and
- perform tedious and repetitive tasks within the design by interfacing transparently to external applications (Ballinger, et al., 1994), such as synthesis and simulation packages.

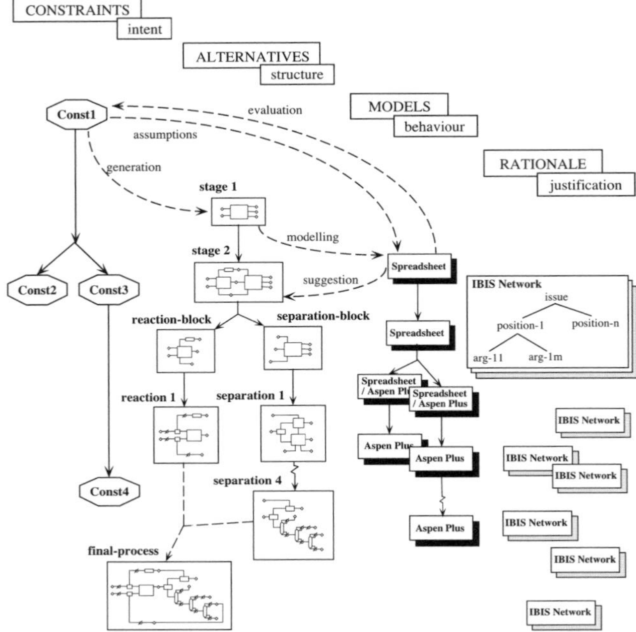

Figure 3. Design constraints, alternatives, models and rationale in KBDS.

The above features are provided by a set of internal tools to maintain, visualise, and manipulate design alternatives, constraints and rationale. The Flowsheet Tool, in the lower right of Fig. 4, is a flowsheet editor which enables the designer to create and edit the graphical representation of alternative designs.

The Design History tool, upper-center window in Fig. 4, supports a graphical representation of the design alternatives history by showing their interrelationship. Every symbol and icon in KBDS is bound to the underlying object-oriented knowledge representation.

Unit operations are defined in terms of variables, such as input-temperature and materials-of-construction. The number and nature of these variables can be set on-line by the designer through a menu (not shown). This makes the application of KBDS to other domains and its integration to diverse external packages possible.

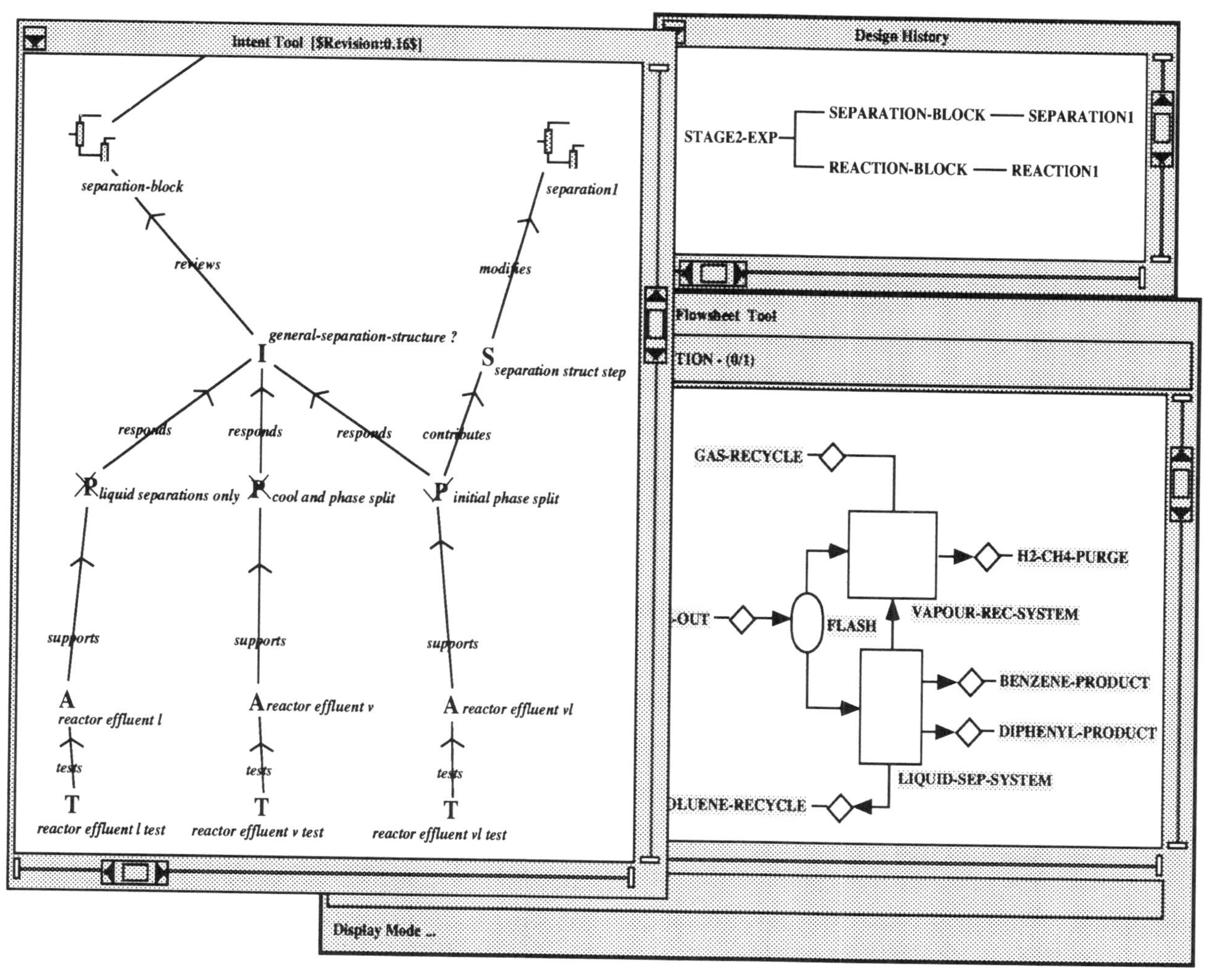

Figure 4. A design alternative, its context within the design history and its rationale.

KBDS can also maintain the designer's rationale. This is done by means of IBIS networks (Rittel and Webber, 1973), which store every decision along with its competing alternative decisions and the arguments used in the selection. IBIS structures can be used to keep track of the issues that have been discussed, which design alternative suggested them, and, if they are resolved, the options that were selected, the reasons for their selection and which alternative they affect directly. One such structure can be seen in the lower left window of Fig. 4. There are four ways in which design rationale can be used to the designer's advantage within KBDS (Bañares-Alcántara, King and Ballinger, 1995):

1. Dependency-directed backtracking. Storage of the design rationale in a form that the computer can process allows the design team to identify which parts of the plant must be re-designed when there has been a change in the internal assumptions or constraints of the plant, or any external factor affecting them.
2. Conflict detection and management. KBDS can identify inconsistencies between coupled design alternatives that were developed independently, e.g. the reaction and separation sections of a process.
3. Support for "what-if" studies. Weights can be assigned to the arguments supporting alternative decisions and these decisions can be automatically ranked in order of desirability.
4. Automatic report generation. KBDS automatically produces documents describing the evolution of the design alternatives and the argumentation that resulted in a given decision.

A design support system like KBDS can improve the design process by speeding re-design, improving the focus of re-use by providing a fuller understanding of its assumptions and implications, and act as a repository for the maintenance of "corporate memory."

Current Developments in KBDS

We now discuss three current developments pursued by the design group in our department: the representation of the design artifact, Section 5.1, the application of waste-minimisation criteria during the design process, Section 5.2, and the use of design process ideas in the area of modeling, Section 5.3.

The Design Artifact and its Representation

There are no unique or definitive answers to the questions of what the design artifact should be in the context of a specific design domain or how should it be represented. In general, the amount of information describing a design artifact should be a decision left to the designer.

In the case of a chemical process a current representation may consist of a set of flowsheets showing its structure, associated models, and a collection of textual documents (memos, manuals, etc.). In the future, as more computer tools are used during the process of design, other descriptions such as multimedia documents and virtual reality models will be part of the design artifact. Each one of the above representations is a partial description of the design artifact and it is used for a purpose, be it visualisation, evaluation, prediction, guidance, etc.

Having a deeper "understanding" of the underlying structure of each of these representations would put the computer in a position of recording argumentation, intent and justification at a finer level of detail, thus opening the possibility of having a more active role as a critic, guide or participant in the design process. By "understanding" we mean being able to manipulate and draw inferences from the structure, function, behaviour, purpose and rationale of the design artifact and their interrelationships.

The extraction of "understanding" from the complementary representations should be at least partially automated because we cannot rely upon the designer to input the necessary information, both, because it would be an enormous amount of work, and because we cannot be confident that the volunteered information will be complete, consistent and true.

We exemplify these ideas with two procedures: the derivation of function for a flowsheet and its substructures, and the extraction of keywords from text.

Function can be derived from the process structure, its behaviour and some problem-specific information, such as the roles of the chemical substances in the process, e.g. reactant, main product, waste; similarly to the approach taken in SPLIT (Wahnschafft, Jurain and Westerberg, 1991). Function can be later used for reverse engineering a process, as explained in Section 5.2.

We also envisage a system that would extract semi-automatically a list of keywords inferred from a text and its context. The designer would then edit the proposed keyword list before it is finally associated to the text. These keywords can help in the organisation and cross-referencing of design rationale nodes.

Application of Domain-specific Knowledge during the Design Process

The objective of this extension is to assist the engineer at any point during the design process in the identification and evaluation of alternatives for recycle, re-use, recovery and reclamation of waste streams. This will be done by improving the designer's understanding of a chemical process, its impact on the environment and the opportunities for waste reduction.

The major potential for waste minimisation exists through the identification of opportunities within an existing process, because this usually involves less capital expense and shorter implementation times than other choices within the EPA waste minimisation hierarchy (Stephan and Atcheson, 1989). Berglund and Snyder (1990) report that process structure changes or recycling accounted for 73% of 80% waste minimisation alternatives adopted at Union Carbide. Similarly, Fonyo, Kurum and Rippin (1994) find that 61% of 18% processes reported in the literature have eliminated or reduced the waste by recycling, re-use or reclamation techniques.

The approach consists of four sequential steps:

1. Derivation of the function of the constituent blocks in a chemical process, see Section 5.1.
2. Abstraction of the process structure. Abstraction will group a collection of functionally similar blocks into an "abstract" block. The new "abstract" process will have fewer blocks than the original process but will perform the same overall function. This abstraction step is the core of a reverse engineering task and can be seen as Douglas' hierarchical refinement step in reverse, as was indeed envisioned in Chapter 1 of Douglas (1988).
3. Suggestion of modifications to the process structure. Changes, such as recycling, may be proposed on the basis of the role of components and the function of the (abstracted) blocks. Waste-minimisation criteria will be used in this step.
4. Evaluation of the impact of a process structure modification. Prediction of the likely impact in the overall behaviour of the process.

Application of Design Support Techniques to Modeling

Modeling is one of the most important activities in engineering and, in particular, during the engineering design process. Its aim is to produce suitable mathematical models that will provide the main insights into the behaviour of a system. A model is understood in the rest of this section as the mapping of the relationship between physical variables onto mathematical structures such as differential-algebraic systems of equations.

Formulation of appropriate and consistent models is a knowledge-intensive, time consuming and error-prone task, and thus requires the intervention of experts. In addition, modeling presents similar requirements to the design task since various versions of a model are produced as it evolves alongside the specification of the chemical plant during design. Consistency must be maintained between models of different parts of the plant. These requirements justify the development of modeling support systems, or interactive knowledge-based modeling environments according to the terminology in (Marquardt, 1994).

A modeling support system is under development at our department to foster exploration of the model space, maintain the evolution and enhance the understanding of the model, improve cooperation between several modelers, and allow the integration with auxiliary tools such as physical property, flowsheeting or symbolic manipulation packages. The development is based on the requirements of design support systems (Bañares-Alcantara, 1995) and a process system representation which facilitates the generation of dynamic lumped parameter models (Vázquez-Román, 1992). This representation describes the physical properties and transfer mechanisms of the process system in terms of interconnected phases, reservoirs and vessels.

The modeling support approach can be seen as the adaptation of KBDS to the modeling task, Fig. 5.

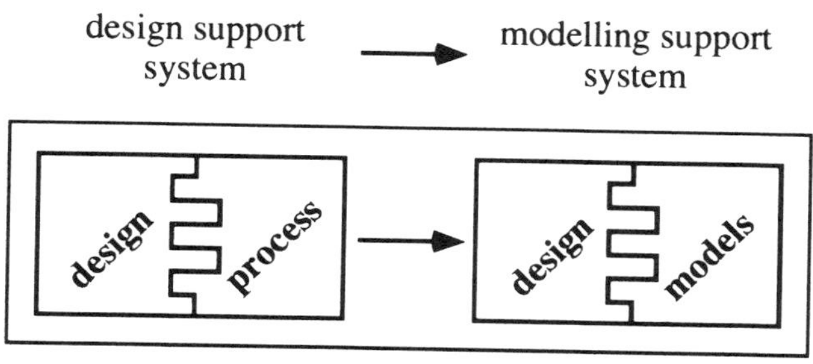

Figure 5. Application of design support techniques to modeling.

Potential benefits are the maintenance of consistency during modeling, enhanced documentation, support for the re-use of a model and its parts, as well as the provision of a graphical interface to models for their verification, modification and maintenance by the users. While this work is intended for modeling chemical process systems, the concept should be applicable in other engineering areas.

A Future Design Support System: an Imaginary Retrospective

[This section adapts an imaginary retrospective written as if the author were explaining, in the year 2010, how design is practiced (Ullman, 1991). In the original retrospective, designers interact with a DUDA (David Ullman's Design Assistant), a notebook sized combination of computer, information source, communication link and design process aide. DUDAs have a touch sensitive screen and a microphone that allows voice commands and external wireless communication. They have and use an understanding of the process of design, and the one we will refer to has a set of chemical engineering design templates. Zazil is a designer working in the same design team at another office].

After a simulation run on the latest design proposal and looking at the results in our respective DUDAs ...

René: "What should we do with this stream?"

Zazil hears me and, when I touch the stream with my pencil, the stream highlights on Zazil's DUDA, so she knows that "this stream" is the top of a distillation column. Additionally, both DUDAs have labeled the object that is the focus of the issue.

Zazil responds by touching the stream and dragging its downstream end to the fuel input of the furnace in the flowsheet while saying

"We could use it as fuel."

An extension of the stream now appears on both DUDAs as a blue dashed line labeled "option 1," indicating that it is a proposed change.

René: "But we might be burning potentially valuable material there," I object.

Zazil: "Let me check, I will get back to you in a moment," she says and starts searching for data such as fuel cost, combustion properties of the stream, potential market value of the stream, and trends in cost of the fuel and market values. Data are accessed through databases, either internal and renewed periodically by subscription, or external and accessed through cellular technology. Questions can also be asked from suppliers by transmission through cellular phone or translation of the voice message to e-mail.

Once all the data has been collected, she requests an economic evaluation of the options of burning vs. selling the stream. Given the relatively high cost of fuel and low value of the stream, her DUDA comes back with a recommendation to burn the stream. Zazil comes back to me with the results.

René: "OK, I see, but what about recycling it?" while I say this, I drag the downstream end of the stream to a special area in the screen labeled "Recycle." DUDA searches for all the possible places where the stream can be recycled and identifies two with a high certainty of success. Two new dashed line extensions to the stream are now displayed, one is violet, labeled "option 2" and going all the way back to the pretreatment section of the plant; the other is green, labeled "option 3" and goes to the reactor feed, but before reaching it, the stream crosses a green box labeled "separation." When I touch the violet line and say "evaluate" my DUDA highlights the quantities that would be substantially modified if I recycled the stream back to the pretreatment section of the plant, such as the sizes of the columns, conversion in the reactor and final product purity. When I touch the green box and say "show inside" my DUDA expands the green box into an the automatically synthesised structure of a separation sequence for the required purification of the stream. In both cases an estimate of the final cost is displayed for me and Zazil to see. None of these options is considered attractive at the point and we decide to use the stream as fuel. DUDA confirms that a similar solution is being successfully used in three other processes within the company and that the solution makes sense from the environmental point of view.

At the end of the interaction my DUDA has recorded the rationale for burning the light ends of the column. It noted that I was working with Zazil, that the issue was the use of a stream, and recorded the options we considered and the arguments we used to make the final decision. DUDA will successively refine design rationale from multi-media records to structured representations.

These records can be of use in the future. My DUDA can replay the history of a design or any of its parts by showing their evolution from sketches to their final state. I am also able to recall design decisions by touching an object and verbally requesting its decision history. I can also compare my decisions with similar cases stored in large design libraries, similarity being measured in terms of the process structure or abstracted structures obtained by reverse engineering. Finally, DUDA will flag a warning whenever some of the data that Zazil obtained from the databases changes and the modification may invalidate our decision.

Conclusions

There are many opportunities for the use of computer technology in the improvement of the conceptual design task in process engineering. However, the development of these systems is not straightforward, the main obstacle being the lack of understanding of how conceptual design is and should be done. While this precludes the successful development of fully automatic design systems with currently available computer techniques, many possibilities remain open for the development of design support systems.

We have proposed a set of necessary requirements for a successful design support system. Additionally, we have attempted to classify the few existing systems. KBDS, a flexible design support system focused on the representation and use of the design process, has been presented as a showcase for the above ideas. It supports the representation of the evolutionary, cooperative and exploratory nature of the design process and takes advantage of the existence of other computer-based tools.

Two types of extension to existing systems have been explored: the application of design support concepts to other domains, e.g. modeling support, and the maintenance of representation at a finer level of detail to improve the "understanding" of the design artifact and procedures by the system. This last extension will open for future designers the possibility of the computer having a more active role as an intelligent critic, guide or participant in the design process.

It is reasonable to expect in the future the following advantages from the application of design support systems:

- improved products from the design process.
- a better understanding of the design process.
- a more effective, better integrated and focused design process, and an improved understanding of its constituent components.

The last of these will lead to specific practical advantages in re-design and re-use of the parts of the design, and related activities, such as retrofit and reverse engineering.

References

Ballinger, G., R. Bañares-Alcántara, D. Costello, E. S. Fraga, J. Krabbe, J. King, M. Laing, R. McKinnel, J. W. Ponton, N. Skilling and M. Spenceley (1995). Developing an environment for creative process design. *Chemical Engineering Research and Design*, **72**(A), 316-324.

Bañares-Alcántara, R. (1991). Representing the engineering design process: two hypotheses. *Computer-Aided Design (CAD)*, **23**(9), 595-603.

Bañares-Alcántara, R. (1995). Design support systems for process engineering. I. requirements and proposed solutions for a design process representation. *Computers & Chemical Engineering*, **19**, 267-277.

Bañares-Alcántara, R. and H. M. S. Lababidi (1995). Design support systems for process engineering. II. KBDS: an experimental prototype. *Computers & Chemical Engineering*, **19**, 279-301.

Bañares-Alcántara, R., E. I. Ko, A. W. Westerberg and M. D. Rychener (1988). DECADE: A hybrid expert system for catalyst selection — II. final architecture and results. *Computers & Chemical Engineering*, **12**, 923-938.

Bañares-Alcántara, R., J. M. P. King and G. Ballinger (1995). Égide: a design support system for conceptual design support. In J.E.E. Sharpe, (Ed.), *AI System Support for Conceptual Design*. Springer-Verlag, Lancaster, UK.

Beltramini, L. and R. L. Motard (1988). KNOD — a knowledge based approach for process design. *Computers & Chemical Engineering*, **12**, 939-958.

Berglund, R. L. and G. E. Snyder (1990). Minimize waste during design. *Hydrocarbon Processing*, April, 39-42.

Court, A. W. and S. J. Culley (1993). A survey of information access and storage amongst engineering designers. University of Bath, UK.

Douglas, J. M. (1988). *Conceptual Design of Chemical Processes*. McGraw-Hill.

Fonyo, Z., S. Kurum and D. W. T. Rippin (1994). Process development for waste minimization: The retrofitting problem. *Computers & Chemical Engineering*, **18**, supplement, S591-S595.

Goodwin, R. and P. W. H. Chung (1994). An intelligent information system for design. In *ESCAPE4, 4th European Symposium on Computer Aided Process Engineering*. Symposium Series No. 133, 375-382.

Joback, K. G. and G. Stephanopoulos (1989). Designing molecules possessing desired physical property values. In J.J. Siirola, I.E. Grossmann and G. Stephanopoulos (Eds.), *Foundations of Computer-Aided Process Design*. CACHE, Elsevier, New York. pp. 363-387.

Kirkwood, R. L., M. H. Locke and J. M. Douglas (1988). A protype expert system for synthesizing chemical process flowsheets. *Computers & Chemical Engineering*, **12,** 329-343.

Lababidi, H. M. S. and R. Bañares-Alcántara (1993). An integrated graphical user interface for a chemical engineering design support system. *Chemical Engineering Research and Design*, **71**(A4), 429-436.

Lu, M. D. and R. L. Motard (1985). Computer-aided total flowsheet synthesis. *Computers & Chemical Engineering*, **9,** 431-445.

Mahalec, V. and R. L. Motard (1977). Procedures for the initial design of chemical processing systems. *Computers & Chemical Engineering*, **1**, 57-68.

Marquardt, W. (1994). Trends in computer-aided process modeling. In E.S. Yoon (Ed.), *The 5th International Symposium on Process Systems Engineering*. Korean Institute of Chemical Engineers, Kyongju, Korea. pp, 1-24.

Mostow, J. (1985). Towards better models of the design process. *AI Magazine*, **6**(1), 44-57.

Motard, R. L. (1989). Integrated computer-aided process engineering. *Computers & Chemical Engineering*, **13,** 1199-1206.

Nicholls, K. (1990). Getting engineering changes under control. *Journal of Engineering Design*, **1**(1), 5-15.

Peña-Mora, F., D. Sriram and R. Logcher (1995). Design rationale for computer supported conflict mitigation. *Journal of Computing in Civil Engineering*, **9**(1), 57-72.

Piela, P. C., T. G. Epperly, K. M. Westerberg and A. W. Westerberg (1991). ASCEND: an object oriented computer environment for modeling and analysis. *Computers & Chemical Engineering*, **15**, 53-72.

Rittel, H. W. J. and M. M. Webber (1973). Dilemmas in a general theory of planning. *Policy Sciences*, **4**, 155-169.

Schembecker, G., K. H. Simmrock and A. Wolff (1994). Synthesis of chemical process flowsheets by means of cooperating knowledge integrating systems. In *ESCAPE4. 4th European Symposium on Computer Aided Process Engineering.* Symposium Series No. 133, pp. 333-341.

Siirola, J. J., G. J. Powers and D. F. Rudd (1971). Synthesis of system designs: III. towards a process concept generator. *AIChE Journal*, **17**, 677-682.

Stephan, D. G. and J. Atcheson (1989). The EPA's approach to pollution prevention. *Chemical Engineering Progress*, **85**(6), 53-58.

Stephanopoulos, G., J. Johnston, T. Kriticos, R. Lakshmanan, M. L. Mavrovouniotis and C. Siletti (1987). Design-Kit: an object-oriented environment for process engineering. *Computers & Chemical Engineering*, **11,** 655-674.

Stephanopoulos, G., G. Henning and H. Leone (1990). MODEL.LA. a modeling language for process engineering — II. multifaceted modeling of processing systems. *Computers & Chemical Engineering*, **14,** 847-869.

Subrahmanian, E., et al. (1993). Equations aren't enough: Informal modeling in design. *AIEDAM*, **7**(4), 257-274.

Ullman, D. G. (1991). The foundations of the modern design environment: an imaginary retrospective. In K. N. Cross and N. Roozenburg (Eds.), *Research in Design Thinking*. Workshop meeting, May 29-31, Delft University Press, The Netherlands. pp. 61-73.

Ullman, D. G. (1994). Issues critical to the development of design history, design rationale and design intent systems. *DTM94 Conference*. Minneapolis, and in Journal review.

Ullman, D. G. and D. Herling (1995). Computer support for design team decisions. In J.E.E. Sharpe (Ed.), *AI System Support for Conceptual* Design. Springer-Verlag, Lancaster, UK.

Vázquez-Román, R. (1992). *Computer Aids for Process Model-Building*. PhD thesis, Imperial College, London.

Venkatasubramanian, V., K. Chan and J. M. Caruthers (1992). Designing engineering polymers: a case study in product design. *AIChE Annual Meeting*. Session 140 (d). Miami, FL.

Wahnschafft, O.M., T.P. Jurain and A.W. Westerberg (1991). SPLIT: a separation process designer. *Computers & Chemical Engineering*, **15,** 565-581.

Westerberg, A. W., P. Piela, E Subrahmanian, G. Podnar and W. Elm (1989). A future computer environment for preliminary design. In J. J. Siirola, I. E. Grossmann and G. Stephanopoulos (Eds.), *Foundations of Computer-Aided Process Design*. CACHE, Elsevier, New York. pp. 507-528.

SESSION SUMMARY: KNOWLEDGE AND CAD ENVIRONMENTS IN ENGINEERING DESIGN

Rudy Motard
Washington University
St. Louis, MO 63130

Sik Shum
UOP
Des Plaines, IL 60017

Introduction

These three papers introduce us to impressive developments in engineering information systems that imply new directions and extend our traditional view. The traditions are perhaps 15 years old, such as engineering databases, activity models, data models, and interactive simulation but, we appear to be facing a "paradigm shift."

Professor Chandrasekaran's review covers a vast area of research in the application of AI-based methods in design. A great deal of importance is awarded to representation. But, the principal dimensions in the art of AI are more on the cognitive level than in the real. For instance, a common intellectual view is to contrast heuristic search versus algorithmic operation or, again, qualitative reasoning versus quantitative analysis. We now realize that these distinctions are artificial, no pun intended, since an integrated approach uses all aspects of the problem ontology namely, all the states and relations that represent the problem. Similarly, the engine that fills in partial solutions may be based on logical inference or on search strategies. If the design process can be thought of as a search in the space of structural descriptions such as components and connections, further progress in the design is then possible by,

- instantiation of some value for a component or a connection,
- addition or deletion of some component,
- refinement (simulation?) of the detailed structure of some component.

Finally, we are led to consider a concept called Functional Representation as a framework to model devices, their structure, functions and behavior. It is a top-down view of a designer's intent what devices should do, along with the roles that other components and processes should play in achieving those functions. It is very reminiscent of the data modeling methodology which is sweeping the international standards community to achieve rigorous industrial product descriptions or process rationalizations. In the language of data modeling, devices may be regarded as entities, states as time- or context-dependent attributes and functions as associations.

Is there still a mystique to process design or can it be largely automated? Prof. Chandrasekaran opines that much of the work on design methodology in the process engineering community is still in what Allan Newell called the symbol level of AI. Thus, the problems are then said to be largely unstructured. Yet, many domains have yielded to highly structured operations research techniques and to new functional representations and associations. We are indeed in a period of development that one would call design engineering namely, the engineering of the design process itself. Prof. Chandrasekaran is helping us along the way.

One of us, working for a technology licenser, cannot overemphasize the importance of engineering design and the usefulness of an integrated computing environment (CAD, knowledge-base, simulation, etc.) that facilitates the engineering design process. Since the potential productivity gain and plant life-cycle rewards are tremendous, companies, such as UOP, have been investing substantial resources in this area. It is clearly an important application area.

Future operating problems can be caused by oversights during the design process. If at all possible, those problems should be recognized during the design phase. What better framework to confront difficult operating problems, even for operations people who have to deal with problems in plants that have already been built, than a thoroughly engineered and understood "map" of all the entities (static and dynamic) and relations that arise in the description of the plant? Thus, the applications to on-line systems are also worthwhile!

Sometimes engineers can be so narrowly focused on their specialty areas that they lose sight of the "big picture." Cross-fertilization between design and operations disciplines is important. Design decision support systems may actually provide that functionality. All of the different perspectives can be considered such as, is the process is easy to start up and control, does it have low environmental impact during operation, etc?

Tools used for design can be reused in operations. For example, models built to facilitate the design task can be reused for control, optimization, monitoring, diagnosis, etc. Resources invested in engineering the design process may be repaid in many different ways.

In this spirit, the second and third papers in the session address design support systems, an essential framework for archiving and understanding the historical evolution of a project.

Prof. Westerberg invites us to envision a design support environment which uses World Wide Web (WWW) cyberspace technology to enhance collaborative design. It is difficult to explain WWW other than to say it is a hypermedia environment rather than simply hypertext. It brings full documents, words and pictures, audio and video sequences into focus from any repository anywhere in the world. Embedded in the top-level document are links to other resources, again, anywhere. Of course, what is proposed is to use the technology within the enterprise to maintain confidentiality. For those who haven't encountered Mosaic and Netscape, the WWW browsers, Internet will catch up to you. New operating systems for personal computers are providing low-cost universal access to all these wonders.

Some of us have had the notion of putting university course notes on the Web. Our own experience is that it is not possible to enter scientific notation with an early version of an authoring language called HTML (hypertext markup language). There is a clumsy fix using graphics and storing images of the scientific equations as (.gif) files. In the near future we are promised SGML (standard generalized markup language) which has actually been standardized internationally (ISO 8879) as the medium for the world-wide exchange of full hypermedia documents across various platforms.

In any event, a Web document is a flat file with links to a wide variety of media and Prof. Westerberg proposes that we can then use this as an archive from which we build generalization by induction, bottom-up so to speak. That turns the data modeling process on its head; a good idea when analyzing what happened rather than describing what should happen as in a planning model. The links in the Westerberg model are relations between the objects generated during the collaborative design process. The objects include discursive issues and points of view among the team members. A generous sampling of application research on the properties of the support system is provided.

Prof. Rene Banares' computer becomes an active conceptual design assistant whose top-level entities are constraints, alternatives, models and rationale. The prototype harnesses knowledge-based processing on a higher level than simply an expert system. There is also induction built into that environment since the computer-based agents try to, 1) interpret the state of the design, 2) represent intention, decisions, methods and assumptions, 3) relate all of the latter and, 4) use them in a man-machine discourse. The interaction is from person to person as well, so that the KDBS design support system encourages collaborative design.

Both the prototypes are people-oriented, not merely automatons. Both are hierarchical and integrate many tools. n-dim can handle very generic design activities and appears to be more open-ended, whereas KDBS is more closely centered on process engineering. PROART (Jarke and Marquardt) described elsewhere in this volume adds life-cycle support and process operations to a KDBS-like environment.

Beyond ISPE'95 there is now a great wave of new technology growing out of the object-oriented paradigm, on which n-dim and KDBS are built, and called component software. There are a number of names for these methodologies (COM, OLE, SOM, etc.) but their goal is the same namely, a broad interoperablility of modules as distributed objects. The field presents opportunities in developing new generations of support systems that would span wide-areas or global enterprises of users, machines and data resources. We have seen some of these innovations already. Engineering software vendors like Intergraph and Autodesk have announced products and platforms for the computer-aided design field. Whether this development will lead to open systems software production remains to be seen. If it does, hang onto your hats.

Someone interested in these developments should also stay tuned to the international efforts to standardize product and process information modeling such as, ISO/STEP and ISO/EXPRESS. These, in some cases, form the basis for electronic data interchange and our professional societies are aware of it. AIChE-PDXI and ICHEME-EDI come to mind. The same data modeling process is useful in design support systems.

HUMAN-MACHINE INTERACTION IN PROCESS CONTROL

Hirokazu Nishitani
Graduate School of Information Science
Nara Institute of Science and Technology
8916-5 Takayama, Ikoma, Nara 630-01 Japan

Abstract

The diverse ideas related to human-machine interaction in process control have been brought together in this paper. Among the many topics concerning the interaction, design instruction of a human-machine interface and applications of interface technology are reviewed to realize better human-machine interaction aimed at the human-centered production system. The target of a system design that includes human and machine is to induce the machine system to be fitted for human ability and to optimize the total performance of the human-machine system. This helps lessen or helps avoid human errors and increases safety. The human behavior modeling is inevitable to consider how to fit the machine system to humans and vice versa. Some new approaches to the human interface are also presented for future needs.

Keywords

Human-machine system, CRT operation, Human factors, Human interface, Operator assistance, Human behavior modeling.

Introduction

As technologies to support artifacts have matured, people are mostly interested in ease of use. This depends on the relationship between human and artifact. The historical changes of the relationship can be summarized as follows: The first relationship between human and artifact began with simple instruments such as a stone ax and a bow and arrow for hunting at ancient times. Thereafter, instruments evolved into machines to help human activity. Machines have gradually increased in size and complexity by the development of new technologies. Human-machine interaction has been discussed in various areas as there exists a natural interface between every artifact and every human.

In the industrial system, automation has been promoted in order to increase productivity. But some tasks are not automated for reasons such as cost performance or expectation of human ability. There is certain evidence that human beings have superior ability which cannot be predicted by any analysis. As a result, however, there will be many isolated islands of automation. Additionally, automation often creates black-box processes with more sophisticated control techniques. This situation will impose a severe burden on the human operator.

There are two approaches to the above situation. One is to clarify the work of the human operator under specified circumstances and provide appropriate education and training; the other is to improve the system by operation support functions and intelligible interfaces. The role of operation support functions is to provide adequate information for the operator to exhibit ability. For this purpose, operator tasks under various situations of plant operation must be known. The operation support system with built-in intelligence will monitor the state of the plant, diagnose the status and give operation guidance to the operator in steady and unsteady operation.

Smooth interaction between the operation support system and human beings is a basic design specification of human-machine interaction. For this purpose, new sophisticated human-machine devices can be applied. A function in the operation support system to understand the operator's cognitive and thinking processes will also contribute to smooth interaction. To this end, both physiological and mental properties of human beings must be studied (Ogino, et al., 1988).

Human-machine interaction is a key factor in the process control system for safe and flexible production. In this paper, human-machine interaction in process control was focused upon and discussed by referring to sundry records, although most of them were published in Japan. In the first section, research areas related to human-machine interaction are surveyed. In the second section, the history of process control systems are reviewed and the historical changes of obstacles regarding the human-machine interface in process control is described. In the third section, design of the human-machine interface in process control is discussed. In the fourth section, human-machine interface technologies are reviewed. In the fifth section, new approaches to the human interface are introduced. In the final section, observations on human-machine interaction extracted from the preceding sections are summarized.

Related Areas Concerning Human-Machine Interaction

Generally system complexity brings difficulty in use and sometimes creates human errors. The research areas, called ergonomics and human engineering, appeared to cope with these situations. In the area of human-machine interaction, studies have been applied to optimize systems composed of humans and machines and to improve the labor environment. In this approach, the human is regarded as responsible physiological equipment.

On the other hand, mental responses in intellectual tasks, including pattern recognition, memory, problem solving, deduction and linguistic theory, were the focus of cognitive science and cognitive engineering. Ease of use and understanding and learning have been studied as user interface research in this area. Principles, guidelines and methodologies have been developed to support design and evaluation of various human interactive systems. Nowadays, the computer is the most common machine that supports human activity. Therefore, human-computer interaction is vigorously studied to design a better user interface for computer systems.

In the case of OA equipment such as computers and copying machines, the terminology of human interface aimed at human-centered systems is commonly used to represent the relationship between human and machine. But in industrial systems such as nuclear power plants and aviation, in which safety is a determinant and there are many factors of safety, the terminology of human factors has been used instead of human interface to represent the relationship between human and machine. In this case, the term ‘interface’ is usually used in a narrow sense as the interface hardware. The area of human factors is aimed at systemization of functional knowledge concerning the relationship between human and system. For example, it is defined as follows (Human Factors Dept., 1994): The technology concerned to optimize the relationships between people and their activities by systematic application of the human sciences, integrated within the framework of system engineering.

In Japan, the first research division on human interface was established in the Society of Instrument and Control Engineers (SICE) in 1984. The division has held a symposium every year since then. Papers in the SICE symposium on human interface are classified into two divisions of fundamentals and applications (Kawai, et al., 1992). The fundamentals division, often called useware, includes the following contents: ergonomics session for visual display terminal, mouse, keyboard, lighting, body temperature, one’s eyes, disabled person; usability session for physiological and psychological factors, learning, skills; manual session for documentation, technical writing, ease of reading; user model session for error factors, mental model, interactive model, guidance system, metaphor. On the other hand, the applications division includes the following human interface technologies: display technology session for text editor, desk-top model; pointing technology session for mouse, 3D measurement, input of location information, sensors; application systems session for evaluation of application systems including recognition or AI techniques.

In 1987, several research centers studying human interfaces were established in Japan. For example, the Institute of Human Factors at the Nuclear Engineering Research Center, the Human Factor Research Center at the Central Research Institute of Electric Power Industry, and the Human Interface Institute at NTT started this year. The last institute is composed of departments of language media, image media, voice information and visual information.

The difficulty of human-machine interface design is due to the uncertainties of human behavior; humans often do not behave logically. Necessity of human behavior modeling has been recognized for improved operation in nuclear power plants, chemical plants, aviation, and vehicles. AI technologies such as expert systems, fuzzy theory and neural networks have enabled descriptions of various forms of human behavior in the application field.

To prevent human errors in nuclear power plants, human behavior modeling based on mental abilities was started in the 1980’s (Yoshikawa, 1988). Since safety is a supreme order, a large amount of research money has been spent to study human-machine interaction in the nuclear plant. The needs of the times are reflected in the three-year research plan at a power company (Kawano, et al., 1993). In 1984-1986, experiments based on behaviorism such as measurement of human error rate were made. In 1987-1989, experiments based on cognitive science such as modeling of human cognitive process were attempted. In 1990-1992, experiments based on social psychology such as the behavior pattern of an operating team were executed. In 1993-1995, experiments of human-machine interfaces for human-centered systems are under way. However, it is said that observations and know-how obtained by experiments depend on the conditions of each subject and the environment, so it is difficult to generalize the result. This is due to ambiguous factors such as personality, motivation, and group dynamics and mental processes that cannot be determined externally.

Human-Machine Interaction in Process Control

History of Process Control Systems

At first, operation of valves and switches were done manually on the site of machines. At the beginning of process automation, pneumatic instrumentation control systems were installed to each processing unit. A small-size instrument panel was used to display the unit data to the operator. In the 1950's, electronic instruments were developed and brought with it, miniaturization of instruments. This realized the panel type with instrument displays arranged in rows. During the 1960's, the instrument panel and CRT of the operator console existed together in the control room. The panel was used as a back-up system, due to the inexperience of CRT. In the 1970's, the comprehensive instruments and control system called DCS were installed in the process industry. CRT operation based on the DCS enabled accurate and stable conditions in plants, using sophisticated computer control technology, which realized higher quality products, energy conservation, yield increase, and so on. It also saved space and labor in the control room. However, it also generated new problems in plant operation.

In the early days of automation, the operator had many tasks such as to keep the process running as closely as possible to a given condition, to preserve optimality, to detect failures, and to maintain safety (Crossman, 1974). These control tasks have not been changed even at the present time. Recently, the operator must have not only these control skills but also management skills required for process and equipment maintenance, facility management and production control. In this situation, computer aids in various forms are essential to support the operator. When considering computer aids, we must take into account the operator's cognition and behavior. Intelligible human-machine interface is also inevitable in order to realize smooth interaction between computer aided systems and human beings.

Given the situation where the responsibility of an operator is expanding, it is important that the operator be able to use a logical approach to understanding the given state at the time. The operator should logically assess the meaning, analyze the situation, and make a rational decision. This approach enables the operator not only to take over conventional control skills but also new skills via computer aids, for coping with unexpected situations and realizing a more flexible operation of the plant.

Obstacles in Human-Machine Interaction in the Early 1980's

Investigation by questionnaires in 1983 and 1992-1993 was critically reviewed to determine historical changes of obstacles in human-machine interaction in process control systems. The questionnaires are useful to grasp the aspects of human-machine interaction.

In 1983, the analog instrument and control system was still used in a fairly large number of plants with panel operation most common in process control. Under such conditions, questions of the five branches were asked to control system users in the process industry and replies were analyzed (Matsui, 1983):

As the first branch, questions on past records of digital instrumentation control systems, size of control systems, number of CRTs, purpose and accomplishment of CRT, obstacles and future needs for CRT operation were asked. Questions concerning combinations of panel and CRT were also included. As the second branch question, traditional panel operation, panel width, arrangement of semi-graphic parts and instruments, information display, alarm display and ease of use were asked. The third branch question concerned the human-machine interface in digital instrument and control systems, quality of keyboard input and display, display for complete understanding and information access, advantages and disadvantages of CRT operation and availability of CRT operation such as operator's load. The fourth branch question, concerned operator training, training methods and periods. The last branch question concerned new technologies for the human-machine interface, voice input and output, touch sensor input, screen manipulation, large-sized screen, combination of CRT and large-sized screen and other expectations.

As a result of factor analysis, 159 replies of control system users were characterized by the following six major viewpoints (Matsui, 1983):

1. Necessity of the panel for operation
2. Amount of information displayed at all times
3. Necessity of improvement of the human-machine interface in both panel operation and CRT operation
4. Necessity of an additional display form to CRT
5. Dissatisfaction with input form of CRT operation
6. Necessity of auxiliary functions for CRT operation

Obstacles in Human-Machine Interaction in the Early 1990's

In 1992 and 1993, investigations by questionnaire on reliability and security of control systems for extra safety in plant operation were applied to both users and makers of the process control systems, respectively (JEMIMA, 1993, 1994; Sone, et al., 1994). Figure 1 illustrates all factors of safety in the total plant operation control system, which was used for planning the survey. These factors were classified into the two levels of control system and plant operation. Questions for each level are summarized as follows:

(1) Reliability and security of control systems

Reliability and security of the hardware and software of DCSs was evaluated by both users and makers. Engineering and maintenance problems and future needs were also asked from reliability and security viewpoints.

(2) Improvement of safety in plant operation

Plant operation is a complex activity of control systems and human operators. In the investigation, the following five systems were examined to identify obstacles in human-machine interaction:

a. Advanced control systems to decrease operator burden
b. Human interface to prevent human errors
c. Operator assistance systems to aid decision making
d. Education and training systems
e. Preventive maintenance systems.

Figures 2-5 show the obstacles for advanced control, human interface, operator assistance, and education and training systems, respectively, regarding importance and settlement time expected. These obstacles are closely related each other. Therefore, comprehensive understanding of these relationships is required to improve human-machine interaction. Effective integration of these systems enables improved safety in plant operation.

Figure 6 shows a comparison of perspectives of human-machine technology by users and makers. Functions closely related to actual operation such as operation guidance, all-time display function, use of human senses, intelligent alarm are agreed by both users and makers. A gap is seen in multi-media applications, automatic screen invocation, use of human factors, improvement of resolution. The figure shows a tendency that the maker is primarily interested in active application of new technologies and the user is interested in its effectiveness.

The following issues are now under consideration to improve the human interface in recent CRT operation:

- To grasp the situation all over the plant
- To invoke the desired screen as soon as possible
- To be able to recognize screen information by others except the operating person
- To prevent misconception of important alarms by frequent occurrence
- To support the operator for emergency.

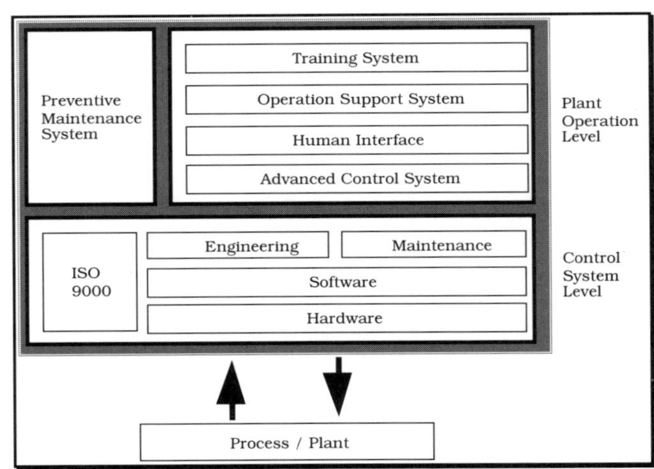

Figure 1. Reliability and security factors in plant operation.

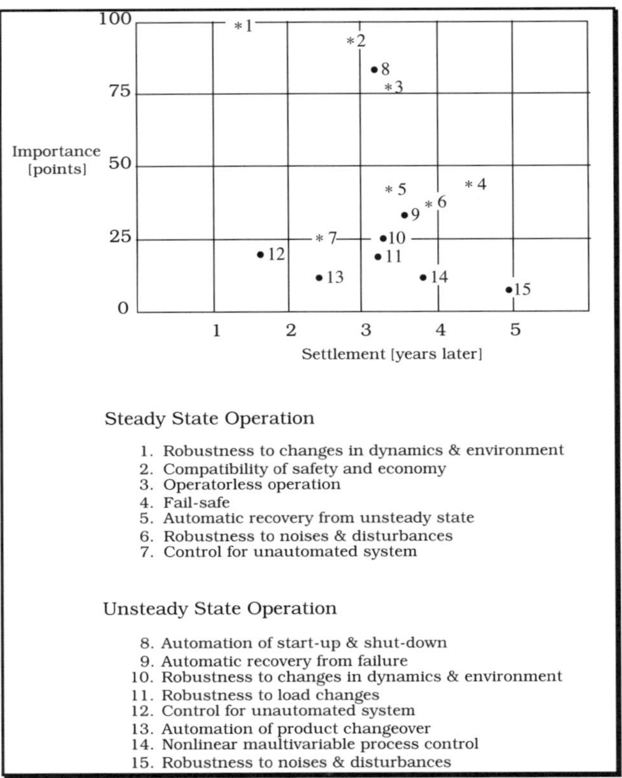

Figure 2. Settlement of obstacles for advanced control.

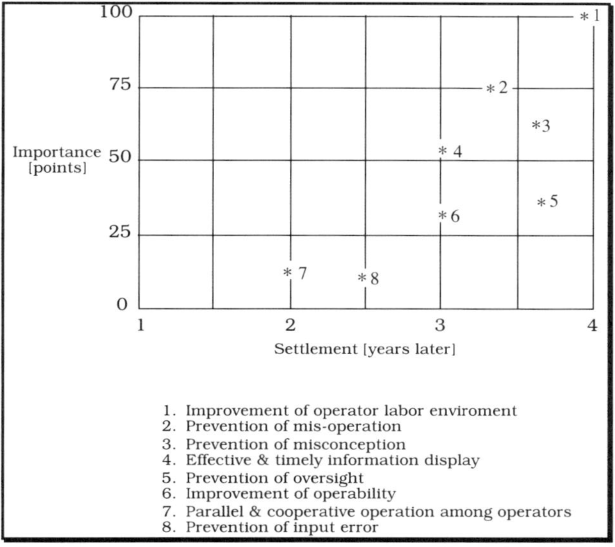

Figure 3. Settlement of obstacles for human interface.

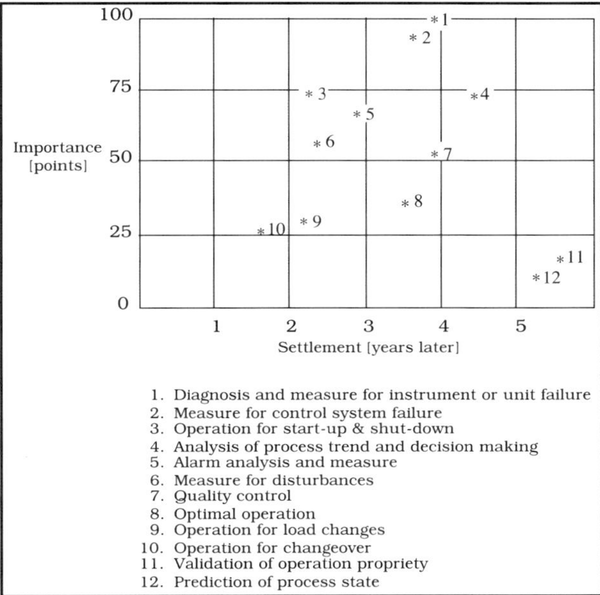

Figure 4. Settlement of obstacles for operator assistance.

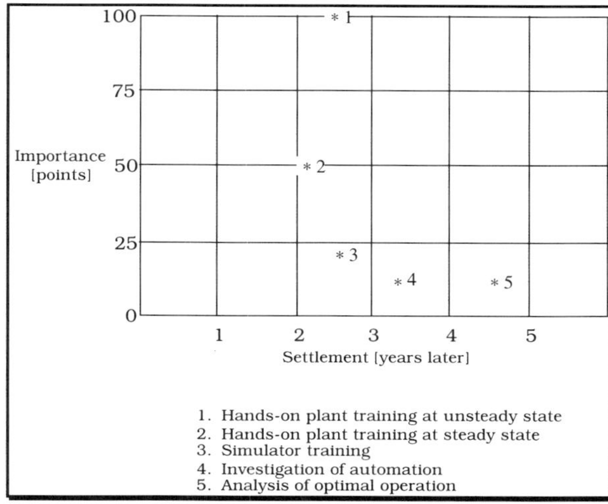

Figure 5. Settlement of obstacles for education and training systems.

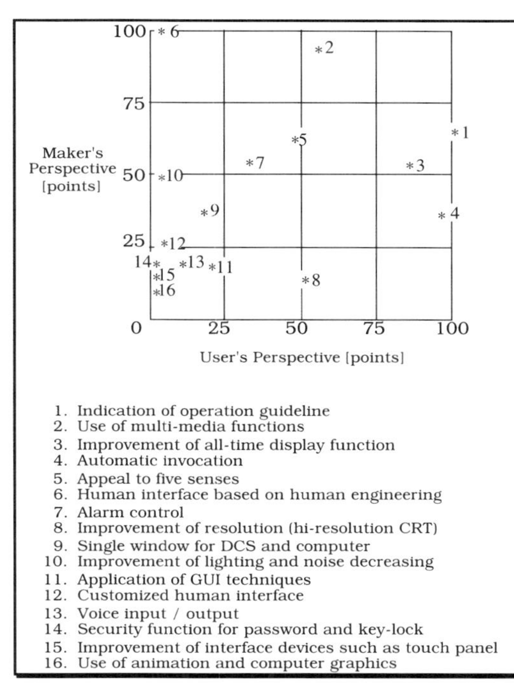

Figure 6. Comparison of user's perspective with maker's perspective for human interface.

Human-Machine Interface Design in Process Control

Model of the Human-Machine Interface

When we consider human-machine interaction in the plant operation control system, a basic model of interfaces is useful when identifying problem locations (Saeki and Nishida, 1989; Nishitani, 1994). There are several interfaces shown in Fig. 7. The interface between human and CRT is the major interest for the human-machine interface in the process control system. The interface between human and plant becomes an issue for maintenance. The human-human interface is important for the operating team in a large plant.

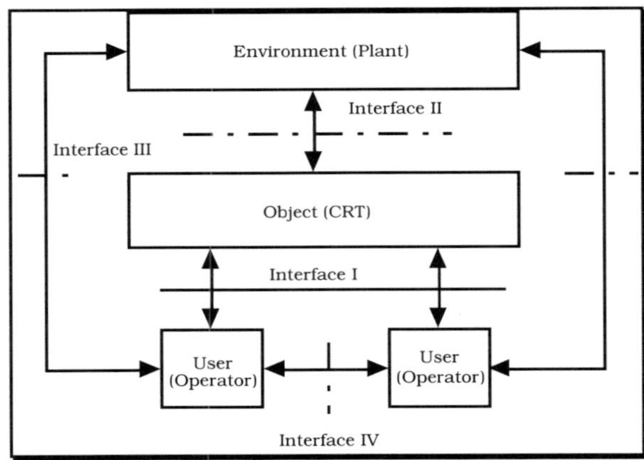

Figure 7. Interfaces of plant operation.

In CRT operation, the operator is an important element in the closed loop composed of display, operator, actuator, and controlled object. Figure 8 illustrates the closed loop model of the human-machine system in CRT operation. This model can also be used to represent manual control and sequential control functions by the operator. In this model, it is important to recognize that the human operator is an information processing element. Modeling this element requires an insight into human cognitive processes. The following are necessary to model the process:

a. How the operator observes the controlled processes?
b. How the operator knows the data displayed?
c. How the operator selects meaningful information?
d. How the operator uses information to predict a future process state?
e. How the operator makes the decision to take action?

The above observations are essential for human-machine interface design. The role of the human-machine interface is to display appropriate information and to help the operator to recognize the state and to make a rapid decision. For this purpose, intuitive indication of essential information on what is happening is inevitable. The operator cannot easily visualize what is happening in the processes. Computer aids for visualization of the process state are useful to give support to the operator's natural understanding. In particular, mass and energy balances are essential to understand the dynamic behavior of the plant. Such concentrated information is used for ecological interface design (Vincente and Rasmussen, 1992). Analogy in mass and energy is a powerful tool for understanding the phenomena in the chemical plant.

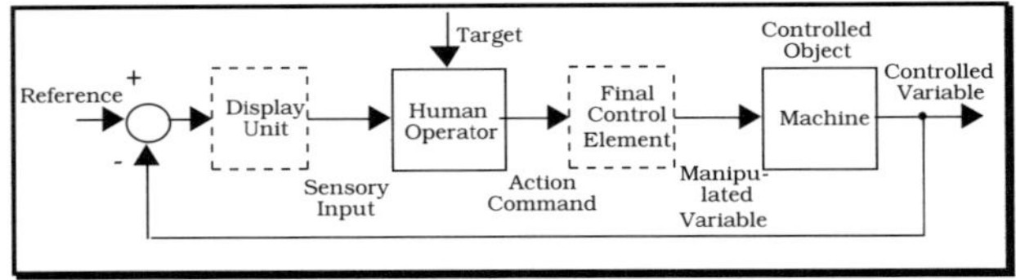

Figure 8. Closed-loop model of human-machine system.

Interface Design Instruction

The basic design targets of human-machine interface are improvement of supervision and control functions and decrease of operator's load. In particular, rapid detection of failure, automatic diagnosis of failure causes, and judgment support by operation guidance are essential to prevent spread of failure for emergency operation. The operator must understand well the controlled object through an intelligent human-machine interface. For this purpose, process engineers, operators and interface specialists such as human factor engineers must cooperate to improve the human-machine interface.

The main design factors of the human-machine interface in process control are summarized as follows (Dallimonti, et al., 1985):

1. The operator's role in the control system
2. Data required to accomplish the operator's role
3. Manageable database size by the operator
4. Display form of information
5. Facility provided for operator interaction to screens, databases, and controlled processes
6. Consistency among response times of human operator, interface, and controlled processes
7. Physical properties of the interface
8. Labor environment in the control room.

Although the first design factor is the most important one, there are few instructions. Usually the operator's role is not clearly defined. The following questions must be answered to clarify the operator's role and to give task assignment to the system and the human:

- Knowledge level of the operator
- Automation required to satisfy plant operation objectives
- Balancing of automation and the operator's role
- Direct manipulation or human supervisory control

With respect to the third design factor of manageable database size, an operator can handle 100-500 control loops around the steady state. But it is recommended that an operator should not handle simultaneously more than about 7 loops for unusual operation. The magical number of seven is verified in many psychological experiments (Miller, 1956; Human Factors Dept., 1994). Under a strained situation in emergency operation, an operator can handle one or two loops only (Takano, 1994). With respect to the fourth design factor of display, the operator tends to desire high density information on a display. Inversely however, an ergonomist tends to avoid complicated display. The display form of plant information to the operator, high fidelity of object on the display, and response times for interaction were important issues for the human-machine interface in the mid 1980's. With respect to the 7th and 8th design factors, there are many useful design instructions (Dallimonti, et al., 1985). Figure 9 shows the outline of human-machine interface design.

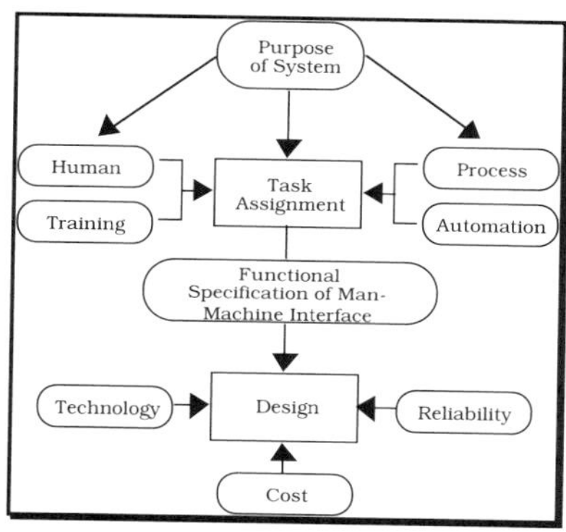

Figure 9. Outline of human-machine interface design.

There are many computer aided systems for flexible operation of the production system. In addition to the DCS, various advanced control systems are implemented for optimization and adaptation, to accommodate large variations in the production system environment. These functions are usually created in process computers connected to the DCS. Various computer systems for management are also connected to the DCS, to acquire actual operating data. Integration of basic control systems, advanced control systems and management systems is under way in the process industry. The operator is being concerned in these systems. Each system such as a DCS, a production control system and a quality control system has its own interface. The way the operator has to communicate with each of these systems is different. It is pointed out that operators tend to concentrate on just the DCS to run the plant, leaving the other systems mostly idle (Cox, 1991). A consistent operator interface to the various operation functions is desired to decrease CRT overload.

Human-Machine Interface Technologies

Multiple CRT

Short-term memory is required to synchronize sensory input in parallel to the operator and sequential information processing of humans. The state of the plant is recognized instantaneously in the short-term memory. However, when operator overload occurs at unusual situations such as alarm handling for emergency, the operator can easily get into trouble. Parallel display of a set of screens enables simultaneous viewing of different information and helps the operator recognize the situation. Because of this, multiple CRT is proposed in order to help human operators rapid recognition of emergency situations (Tomita, 1985). The layout of screens is also an important factor to support human recognition effectively.

Display

The CRT display has been improved with respect to resolution and color. Since the display screen on a CRT is too small to be viewed by multiple operators, a large-sized screen is desired for the operating team. It is useful to hold information in common for emergency operation. The large-sized screen has been used in power plants and town gas production plants. An integrated use of CRT and large-sized panel and integration of graphics and video image were proposed as a new human interface to integrate different media in process control systems (Tani, 1994).

Interactive Devices

Many invocation methods were introduced to improve the usability of CRT. There are touch target invocation, cursor target invocation, and user function-key invocation according to input devices. There are some functions such as display back/forward invocation, prior display invocation, associated display invocation for rapid operation. For the multiple CRT, cross screen invocation and display set invocation were proposed in order to enhance the superior human ability of pattern recognition (Tomita, 1985).

The touch screen installed as the input device on the most recent CRT has improved screen access and input operation. It enables zoom-in on the controlled object from the overview screen by tracing the hierarchical structure of graphic screen information and manipulating valves and switches on the screen. This releases the operator from operation while being able to see both the screen and keyboard (Sone, 1994).

Voice Input and Output

Voice is a natural and fast communication media for humans. Eyes and hands are free for voice input and output. This can increase a radius of operator action. Voice is also effective for calling attention to the operator. Voice alarm and voice guidance will help the operator in human supervisory control.

New Approaches to Human Interface

Usability Test by the Protocol Analysis

Evaluation of artifact from the viewpoint of ease of use is called the usability test. In the test, the experimenter observes reactions when a subject uses the artifact and discovers issues to be improved. After investigation, the experimenter will be able to find a solution to improve the quality of the user interface. A characteristic of recent information processing equipment is the existence of the complicated interaction between human and machine. Therefore, both sides must exchange information to be able to execute advanced tasks. Under this situation, it is useful to analyze the protocol for system evaluation. Since protocol is output from the cognitive process such as thinking and judgment, it cannot be analyzed by physical measurement only. Protocol analysis has, however, been applied to evaluate usability of machines (Asahi, 1993).

In our laboratory at NAIST, the protocol analysis is applied to evaluate the human interface of a recent DCS. The DCS is composed of an information command station with upper and lower CRTs and a field control station A training simulator of a boiler plant is installed in the field control station. Figures 10 and 11 illustrate the protocol analysis equipment and laboratory computer environment. Four video cameras make a film of synchronized four scenes of upper and lower CRTs, keyboard and operator. This video image with voice records is used for protocol analysis of DCS operation. We have been studying the following items using the training simulator in the DCS:

- Evaluation of display form on each graphic screen
- Evaluation of touch screen operation
- Evaluation of hierarchical screen manipulation
- Evaluation of multi-windows for emergency operation
- How to use multiple CRT
- Necessity of special functions to support the operator
- Data collection for operator behavior modeling
- Feedback to education and training.

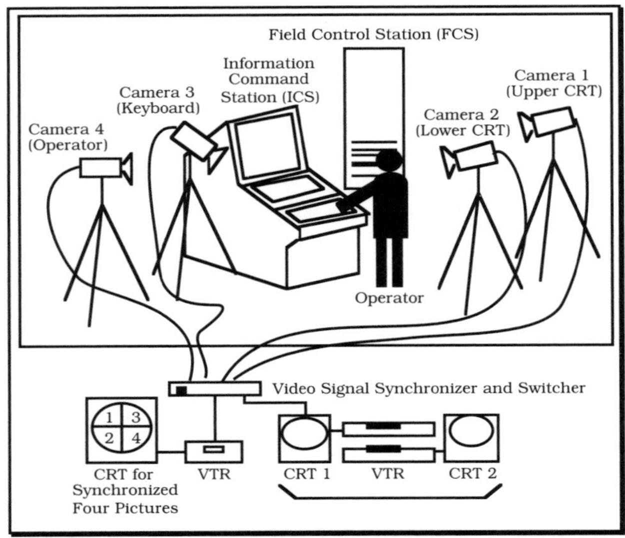

Figure 10. Protocol analysis equipment for usability test of DCS.

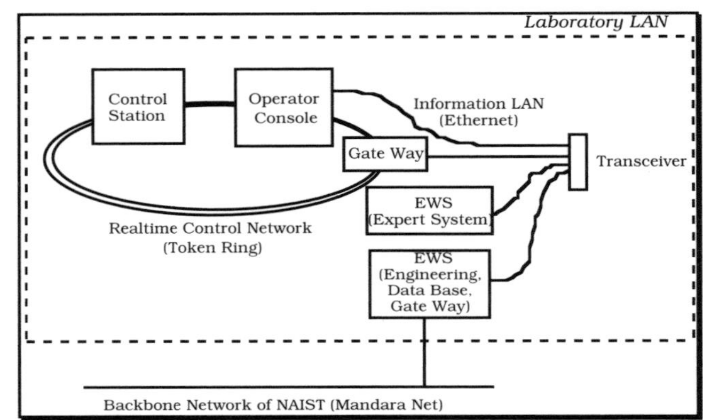

Figure 11. Experimental environment for human interface research at NAIST.

Virtual Environment for Design

Construction of guidelines and standardization of the human interface is slow in comparison with the development speed of multi-media technology. Therefore, how to design human interfaces using new media technology such as 3D computer graphics and animation is left to the designer. Under this situation, the virtual environment will be useful to evaluate proposed human interfaces for operation control systems.

A comprehensive virtual environment composed of a virtual prototype system, a virtual subject and a virtual training simulator has been developed for planning, conceptual design, detail design, human factor assessment and training (Doi, 1994). The prototyping system enables a designer to evaluate machine layout in the virtual space. The system provides high-quality image of 3D CAD data, high speed rendering using a simplified data architecture, interactive changes in layout and viewpoint, real-time physical simulation of lighting effect and natural operation under physical constraints. This system was applied to control room layout of a steam-power plant. The virtual subject can move in the virtual space instead of the human. A designer can evaluate human factors by observing and measuring motions of the virtual subject. A virtual training simulator can be realized by integrating the virtual control room and a dynamic simulator of an object plant. A trainee with DataGlove can push buttons, change switches and touch screen menus as he supervises trend graphs and annunciators. This system can be used as a training tool.

Human Behavior Modeling and Simulation

An approach to build a model under a set of fearless assumptions using advanced programming techniques such as object-oriented modeling techniques and to revise interactively the model through simulation became a practical means to understand the object and to identify the embedded problems. In nuclear power plants the central control room is rapidly changing by introduction of extensive supervision systems and various operational support systems. This also brings changes in personnel organization. Therefore, assessment of a new working environment should be evaluated in advance by some means, and desirable human-machine interfaces must be developed for the new environment. For this purpose, a behavior model of an operating team has been studied at a human factors research center. The research is aimed at the following (Takano, 1994):

- Assessment of the effect of a new panel of the control system
- Assessment of accident paths under mixed conditions of machine failure and human error
- Assessment of effects by personnel organization of the operating team
- Role assignment and communication method for effective team activity.

The operator behavior model shown in Fig. 12 is composed of four micro models: attention, thinking, action and utterance and three memories for: short term, medium term and long term. The attention micro model simulates whether the operator notices panel information and voice information. The thinking micro-model simulates the operator's understanding process of the current state with the mental model in medium-term memory. A mental model of event sequences is created at the moment of any alarm occurrence and it is modified when new information is obtained through the attention micro model. The thinking micro model also identifies causes of the alarm and decides on an action target. The mental model is a representation of predictable events over a time horizon. The action micro model actualizes the action intention decided upon by the thinking micro model and executes the action. The utterance micro model creates a verbal command via the utterance intention of the thinking micro model. The short-term, medium-term and long-term memory are for temporal records of information, working memory to manage mental models, and memory for knowledge bases such as alarms, plant parameters, actions and so on, respectively. The model to simulate group behavior of an operating team consisting of a leader and two assistants is shown in Fig. 13. The operator behavior model is used for each operator.

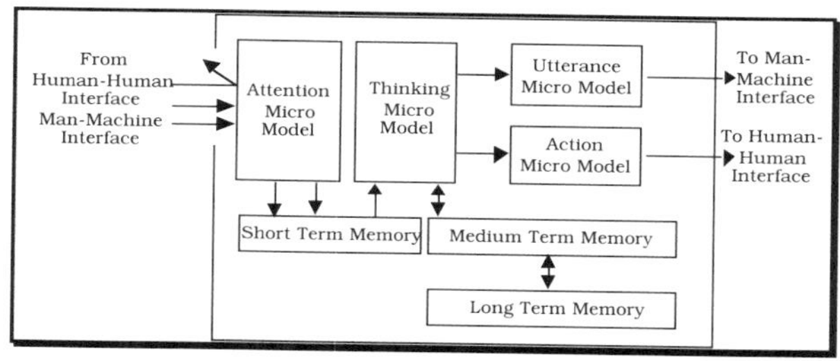

Figure 12. Operator behavior model.

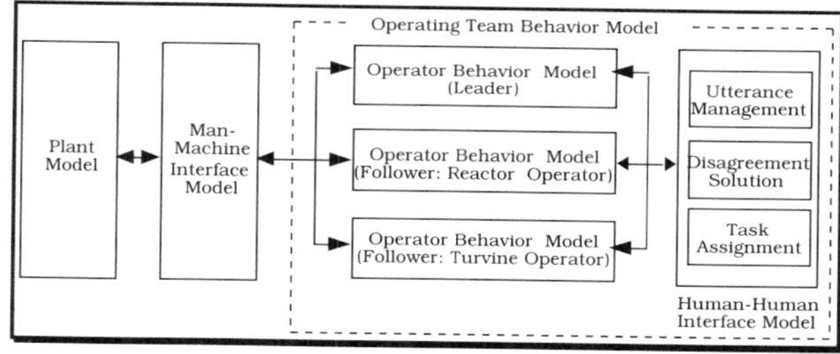

Figure 13. Behavior model of operating team.

Conclusions

In the preceding sections, diverse ideas related to human-machine interaction in process control were reviewed. What can we learn from these ideas? The following is a summary for consideration.

Firstly, the process control engineers are not always familiar with the fact that many experiments have been published by people who often don't know that others exist. This means that unfamiliar technical terms puzzle the process control engineers. The engineers must interpret observations on human-machine interaction when applying them to control system design.

Secondly, there are many pioneering experiments in the nuclear power industry. The results will be generally useful to chemical plants, but there are some obvious differences between nuclear power plants and chemical plants. The basic structure of a nuclear power plant is common and many operation procedures are standardized. On the other hand, there are various types of plants and a variety of plant operations in the chemical industry. This may give plants inherent obstacles in human-machine interaction. Knowledge of the controlled object should be effectively used when designing human-machine interaction in the chemical plant.

Thirdly, visualization of the process state is an essential support for the operator to understand what is happening in the process. In particular, the mass and energy balances are very important factors when predicting future trends. Visual display of the state predicted will help the operator's judgment.

Fourthly, many operation support systems with different functions are developed to improve human-machine interaction in the same plant. Since each system has its own human interface, there are many human interfaces in the operation control system. This will produce confusion in plant operation. To cope with this situation, a consistent human-machine interface to various operation support systems must be developed.

Lastly, human-machine interaction includes many aspects of studies in many different areas. This implies that know-how in human-machine interaction in process control is scattered over many areas. Therefore, interdisciplinary cooperation between these areas is advisable to realize the human-centered production system which provides synergistic production in a safe and comfortable manufacturing environment. High-tech should be applied not only to machines but also to humans so that they can give the best by using instinct and knowledge.

Acknowledgment

The author would like to thank Professor Shigehiko Yamamoto, of Kogakuin University for his valuable discussion.

References

Asahi, T. (1993). Product evaluation by protocol analysis. In H. Kaiho and E. Harada (Eds.), *Introduction to Protocol Analysis- What we can read from protocol data*, Ch. 8. Shinyosha, Tokyo. pp. 153-169.

Cox, R.K. (1991). The operations specialist's workstation. In Y. Arkun and W.H. Ray (Eds.), *Chemical Process Control IV*. CACHE, Austin. pp. 695-698.

Crossman, E.R.F.W. (1974). Automation and skill. In E. Edwards and F.P. Lee (Eds.), *The Human Operator in Process Control*. Taylor & Francis Ltd., London. pp. 1-24.

Dallimonti, R., M. Togo and Y. Suzuki (1985). Human-machine interface in systems control. *J. SICE*, **24**, 114-120.

Doi, M. (1994). Training simulator using a virtual operation room. *Proc. of International Symposium Info-Tech'94* . Kobe, Japan. pp. 85-90.

Human Factors Department (1994). Human Factors TOPICS. Tokyo Electric Power Co., Tokyo.

JEMIMA (Japan Electric Measuring Instrument Manufacturer's Association) (1993). *Investigation Report on Reliability and Security of Control Systems for Extra Safety in Plant Operation*. Tokyo.

JEMIMA (Japan Electric Measuring Instrument Manufacturer's Association) (1994). *Investigation Report on Reliability and Security of Control Systems for Extra Safety in Plant Operation*. Tokyo.

Kawai, K., T. Kawata and H. Tsunekawa (1992). Human interface. In T. Sumi and K. Hiroi (Eds.), *Theory and Application of Control System Technology*, Ch. 7. Denki-shoin, Tokyo. pp. 273-321.

Kawano R., J. Itoh and R. Kubota (1993). Human factors in nuclear power plant. *J. IEE Japan*, **113**, 821-827.

Matsui, J. (1984). Investigation of human-machine interface by questionnaire. *J. SICE*, **23**, 1028-1033.

Miller, G.A. (1956). The magical number seven, plus or minus two; some limits on our capacity for processing information. *Psychological Review*, **63**, 81-97.

Nishitani, H. (1994). Human-computer interaction in the new process technology. *Proc. of 5th International Symposium on Process Systems Engineering*. Kyongju. pp. 1367-1375.

Ogino T., T. Miki and M. Fujii (1988). Human interface in nuclear power plant. *J. SICE*, **27**, 71-74.

Saeki, Y. and S. Nishida (1989). Cognitive aspect of human interface. *J. IEE Japan*, **109**, 645-648.

Sone H. (1994). Human interface in the recent DCS. In *SICE Seminar on Advanced Supervision and Operation*. pp. 105-120.

Sone H., R. Mori, S. Yamamoto and K. Furuta (1994). A study on reliability and security of control systems and plant safety operation. *J. SICE*, **33**, 1054-1062.

Takano K. (1994). Simulation model for operating group behavior of nuclear power plant. In *SICE Seminar on Advanced Supervision and Operation*. pp. 29-73.

Tani, M. (1994). Human interface for multi-media integration. In *SICE Seminar on Advanced Supervision and Operation*. pp. 75-88.

Tomita Y. (1985). Human interface in process control systems and usability improvement. *J. SICE*, **29**, 459-466.

Vincente, K.J. and J. Rasmussen (1992). Ecological interface design: theoretical foundations. *IEEE Trans. Syst. Man. Cybern.*, **SMC-22**, 589-606.

Yoshikawa, H. (1988). Researches on human error prevention in nuclear power plant. *J. System and Control*, **32**, 168-175.

THEORY AND PRACTICE OF HUMAN-MACHINE INTERFACES

Philip J. Smith, Stephanie Guerlain, Jack W. Smith, Jr. and Rebecca Denning
Cognitive Systems Engineering Laboratory
The Ohio State University
Columbus, OH 43210

C. Elaine McCoy
Aviation Department
Ohio University
Athens, OH 45701

Charles F. Layton
Galaxy Scientific
Atlanta, GA 30345

Abstract

For a variety of reasons, many decision-support systems must be considered "brittle." Such brittleness can result from a failure of the designer to design for all of the potential scenarios, an intentional decision by the designer to use an oversimplified model of the decision task, a failure to correctly anticipate system behavior, or an error in the implementation of the design. Regardless of the cause, the net result is often a system failure without the graceful degradation often shown by human experts. To deal with such brittleness, it is often suggested that a person be kept "in the loop." There are, however, many different ways to do so. This paper examines the issue of the role of the computer system and its impact on human performance when brittleness is encountered in the use of a decision-support system.

Keywords

Aviation, Brittleness, Decision-support systems, Cognitive biases, Critiquing, Human error, Medicine, Planning, Cooperative problem-solving.

Introduction

The interface between a computer system and its user can be defined at many levels. Often, interface design is discussed in terms of the surface image presented to the user (i. e., the displays and controls presented to the user by the computer). At another level, however, the interface between the user and a computer system can be defined in terms of a number of other design features, such as the underlying conceptual model for the system and its intended role. This paper focuses on this latter emphasis, dealing with the design of cooperative problem-solving systems. In particular, three studies are reviewed, dealing with the effect of the user's role on her performance when using a decision-support system. The first study deals with a flight planning system for airline dispatchers; the second and third deal with the design of cognitive tools to assist blood bankers.

Study 1 — Flight Planning

Using flight planning for commercial airlines as a testbed, three alternative designs for a graphical flight planning tool were evaluated, using 27 dispatchers and 30 pilots as subjects. The results show that the generation of a suggestion or recommendation by the computer early in a person's own problem evaluation can have a significant impact on that person's decision processes, influencing situation assessment and the evaluation of alternative solutions. If the scenario is one where the computer's brittleness leads to a poor recommendation, this impact can strongly influence the person to make a poor decision.

The Application Area

Flight planning for commercial airlines involves dealing with multiple competing and complementary goals in an environment with substantial uncertainty in response to a variety of possible initiating events, including changes in weather, mechanical and other system failures, medical emergencies, runway closures, etc. (Hayes-Roth and Hayes-Roth, 1979; Hoc, 1988; Miller, Galanter and Pribram, 1960; Sacerdoti, 1974; Schank and Abelson, 1977; Suchman, 1987; Wilensky, 1983). In addition, flight planning involves coordination with the flight crew, air traffic controllers, and other company operations control staff.

The Flight Planning Testbed — An Evaluation

A prototype system, the Flight Planning Testbed (FPT), was developed to test several alternative design concepts in support of planning by human experts. FPT permits the user to view a variety of weather and flight data and to explore alternative flight plans. Users can sketch flight paths themselves, or they can ask the system to recommend a path generated by an optimization routine. The system then provides users with feedback about the implications of those flight paths in terms of time and fuel. (FPT simulates the behavior of a Boeing 737-200.)

Methods

This system was used as a testbed to study the effects of different design features on cooperative problem-solving performance. Each of the fifty-seven subjects (twenty-seven airline dispatchers and thirty airline pilots) was asked to think aloud (Ericsson and Simon, 1993) as he or she used one of three alternative system designs to solve four flight planning problems. These three designs represented variations on the levels and timing of support provided by the computer.

The 'Sketching Only' System

Using the 'sketching only' system, the human planner could sketch flight paths on a map display while the computer filled in lower-level details (such as fuel remaining, time of arrival, and recommended altitudes) using an optimization program. The user was responsible for proposing the flight paths (sequences of jet routes), while the computer computed an optimal altitude profile and estimates of time and fuel consumption. Routes were sketched by displaying the jet routes and navigational fixes and clicking on a sequence of the navigational fixes.

The 'Route Constraints and Sketching' System

The 'route constraints and sketching' system retained all of the capabilities of the 'sketching only' system and added one other feature: The user could specify higher-level constraints (maximum allowable turbulence, maximum allowable precipitation, and destination) on the solution she or he desired and then ask the computer to find the shortest distance route which satisfied those constraints.

The 'Automatic Route Constraints, Route Constraints, and Sketching' System

Using the 'automatic route constraints, route constraints, and sketching' version, the computer automatically suggested a deviation (based on default constraints of no turbulence, no precipitation, and the originally planned destination) as soon as it detected a problem with the original route. All of the tools available in the other two versions were also available in this version.

Of the four planning problems used in the experiment, the most important to this paper was Case 3, a scenario wherein the limitations of the computer's knowledge led to "brittle" performance so that the computer generated a very poor flight plan. The brittleness resulted from the fact that the computer treated forecasts as reality, leading to a poor solution. This case provided a critical test to see whether subjects would successfully detect and override the computer's suggestion in the 'route constraints' and 'automatic' versions, and is discussed below.

Results and Discussion

A total of twenty-seven dispatchers and thirty pilots from nine commercial airlines were studied. The dispatchers had from two to twenty-two years of experience as professional dispatchers, with an average of 9.2 years. The pilots represented eight major airlines and had an average of 9300 hours of flying experience as commercial pilots and 2800 hours of experience in military aircraft. There was no obvious relationship between any of the performance measures (reported in the following sections) and years of experience as either a dispatcher or a pilot.

Factors Influencing Route Selection

Given the nature of the data collected (concurrent verbal reports), it is impossible to identify all of the factors considered in selecting a particular alternative flight plan in one of FPT's scenarios. At various points in the transcripts, however, there is evidence of one or more factors being considered. Below is a composite list of all of the factors so identified. Many of these factors are interrelated:

1. Fuel consumption as a cost;
2. Fuel consumption as it relates to fuel reserves (a safety concern);
3. Arrival time as it relates to the published schedule and to passenger connections;
4. Turbulence (current; predicted; cause; levels of uncertainty associate with the forecast);
5. Thunderstorm activity (current; predicted; cause; levels of uncertainty associated with the forecast);
6. Passenger comfort;
7. The availability of alternatives or options to deal with an unexpected problem if it arises (including alternative routes to the planned destination and alternative destinations);
8. Characteristics of possible alternative destinations (weather; runway conditions; closings; air traffic activity; maintenance and support facilities);
9. Characteristics of the planned destination (weather; runway conditions; closings; air traffic activity);
10. Air traffic patterns enroute and on approach to the destination;
11. Preferred alternate routes by ATC;
12. Approval (or the likelihood of approval) of a reroute by ATC;
13. Expectations regarding the ability of the flight crew and ATC to detect and deal with minor problems on their own when they arise (without assistance from Dispatch);
14. Expectations regarding the likely air traffic along various routes due to rerouting to avoid the same storm;
15. Winds aloft and their effect on fuel consumption and arrival time;
16. Availability of jet routes or vectors.

It is clear that the complexity of this task is beyond currently feasible methods for designing an autonomous computerized problem-solver. This list also serves to point out that flight planning involves cooperation between several parties (ATC, the flight crew and Dispatch).

The Influence of System Design on Performance

Case 3 was a difficult planning problem that put the various system designs to a demanding test. The thunderstorms in Case 3 were not localized, and their tops were at varying altitudes. There were two likely directions for deviating, but neither was without potential problems. In particular, flight safety was a significant concern in this case.

To begin Case 3, a scenario was read to the subject:

> "It's summer and the aircraft is on a flight from Cheyenne to San Antonio. The aircraft got off the ground at 1900 Zulu and is now two minutes into the flight. Decide what you think the aircraft should do."

In the 'automatic route constraints, route constraints, and sketching' treatment condition, the computer automatically suggested a poor eastern route to the subjects. This eastern route passed between two large, severe thunderstorm cells in an area where such storms are notorious for their volatility. As a result, the eastern route was very undesirable, as it was very possible that the two cells would build and grow together. Furthermore, the eastern route passed extremely close to a forecast intense cell location. To come up with a western route, the subjects had to either modify the constraints on the computer or sketch their own routes.

Case 3 — Sample Subjects

Two representative subjects are described in detail below.

'Sketching Only' Subject 6

To begin, this subject looked at the current and forecast weather, checking the composite radar and clouds and the fronts. He concluded:

> "This stuff is pretty stationary, if anything drifting a little bit north. Some pretty hot cells, up to 43,000 right along the route. That's the wrong place to be. This is a pretty bad bunch of weather here."

He then sketched a far western deviation from DEN to HBU to FMN to ABQ to CNX to ROW to INK to ABI to SAT. While doing so he commented:

> "I'm just gonna circumnavigate this whole area. It's too nasty a weather system to be playing with. I'm gonna stay on the backside of this stuff."

He also noted that:

> "That route adds considerable time and burn, but it's too crummy to get foolish here. I just don't like that weather pattern. That thing is building and developing. This is definitely bad news. I just have to eat it, add an extra half hour to the flight time."

Noting that he "could be cutting it a little bit fine here [in terms of fuel]" he then modified his far western route to fly more directly from INK to SAT (flying from INK to JCT to SAT instead of from INK to ABI to SAT). He checked the spreadsheet for turbulence along this new route, commenting:

> "It's gonna hit some turbulence on the descent but there's not a whole lot to do about that."

In the end, he decided that:

> "Based on getting a smoother ride, I'd have to pick the slightly longer route [a western deviation]."

'Route Constraints' Subject 16

This subject looked at the fronts and weather radar for the current weather and then asked the computer to find a route with no turbulence or precipitation. The computer generated the 'eastern route' for which the subject noted:

> "The fuel burn is up about 1700 pounds. We're out of the turbulence and we're out of the thunderstorm activity. ... I'd say that looks like a good route. ... The passengers will be comfortable and it looks like a good safe flight."

In the debrief, however, when he was shown both the computer-generated suggestions for avoiding the storm (one west of the storm and the other — the one which he had selected — east of the storm) and was shown all of the relevant weather data, he completely changed his opinion:

> "I should have gone with the other one [the western deviation]. I forgot about the forecast. The fuel's not that much different and it definitely keeps you away from all the thunderstorm activity. ... Please burn this tape."

Case 3 — Route Selection

Table 1 indicates the influence that system design had on route selection. There was a overall significant effect on performance ($p<.02$), with the pilots showing a greater (but not statistically significant) tendency to be influenced by the computer to select the poor eastern deviation.

A finer-grained analysis suggests that the timing of seeing the computer's suggestion is a critical factor. All of the subjects in the 'automatic' version immediately saw the computer's suggestion. Of the 9 dispatchers and 10 pilots in the 'route constraints' version, however, only 4 dispatchers and 3 pilots immediately chose to ask for the computer's suggestion. Of these 7 subjects, 3 dispatchers and 2 pilots chose the computer's poor suggestion. Thus, 12 of 26 subjects who saw the computer's suggestion before exploring on their own were influenced to select that route, indicating some powerful biasing effect when the computer provides input before the users have time to evaluate the situation themselves.

Table 1. Final Route Choices for Case 3.

	Computer-Suggested Eastern Route
Dispatchers:	
'sketching only'	0/10
'route constraints'	3/9
'automatic'	0/8
Pilots:	
'sketching only'	1/10
'route constraints'	3/10
'automatic'	4/10

Conclusion

This study left little doubt that the introduction of computer-generated suggestions for solving a flight planning problem can have a marked impact on the cognitive processes of the user and on the plan ultimately selected. In some cases, this impact is beneficial. If the computer's model of the "world" is adequate for a particular scenario, the best route is more likely to be identified with the computer's assistance, helping the user to avoid the problems associated with searching a large solution space.

In other cases, however, the computer's suggestion can have a profound adverse impact. In situations where the computer is brittle and makes a poor suggestion (because its model of the "world" is inadequate or because it doesn't adequately consider all of the relevant factors), sizable numbers of users are likely to be induced to accept this poor suggestion.

Study 2 — Critiquing vs. Partial Automation

A second study addressed similar issues, using blood banking as the context. This case study used the field of immunohematology to illustrate certain design concepts and principles for developing an "interface" between medical technologists in blood banks and a computer system designed to help with a particular problem-solving tasks, antibody identification. In generic terms, this is an abduction task in which multiple primitive solutions or underlying causes of system behavior (antibodies) can be present at the same time. These primitive solutions can mask one another, and the diagnostic data is noisy. Furthermore, it is a high-consequence task, as errors can lead to patient deaths, and is sometimes completed under time stress.

Specifically, the goal of this antibody identification task is to determine which of a set of over 400 possible alloantibodies are present in a patient's blood, so that compatible blood can be found for a transfusion. This task is currently completed in a "paper-and-pencil world" in transfusion services laboratories. The goal of this project has been to study the potential for computerized support tools to reduce errors, as previous studies in this domain have identified errors due to slips, misconceptions, ignoring base rates, biased assimilation, perceptual distortions, and biased reviewing (Fraser, et al., 1992, 1993; Smith, et al., 1991, 1992).

Methods

32 certified medical technologists at 6 hospitals were studied using one of two versions of this expert system. (16 used each version.) The first version was designed as a

tool that would, upon request by clicking a button, automatically perform the task of ruling out antibodies based on the available data. The second version used the same knowledge base but acted as a critic, interrupting the user with a cautionary message when it disagreed with a rule-out marked by the user. After training, each subject was tested on five cases.

Results and Discussion

For four of the five test cases, the computer's knowledge base was adequate to support correct performance. On these four cases, there was no significant difference in performance between the two alternative system designs.

In one case, however, the heuristics used by the computer for rule-out led to an incorrect inference. (Many practitioners make this same error.) On that case, subjects using the automated system made 29% more errors than subjects using the critiquing system, consistent with the results of Study 1.

Study 3 — Evaluation of a Critiquing System

Guided by the results of Studies 1 and 2, the design of this critiquing system, the Antibody IDentification Assistant (or AIDA), was further refined and extended (see Fig. 1).

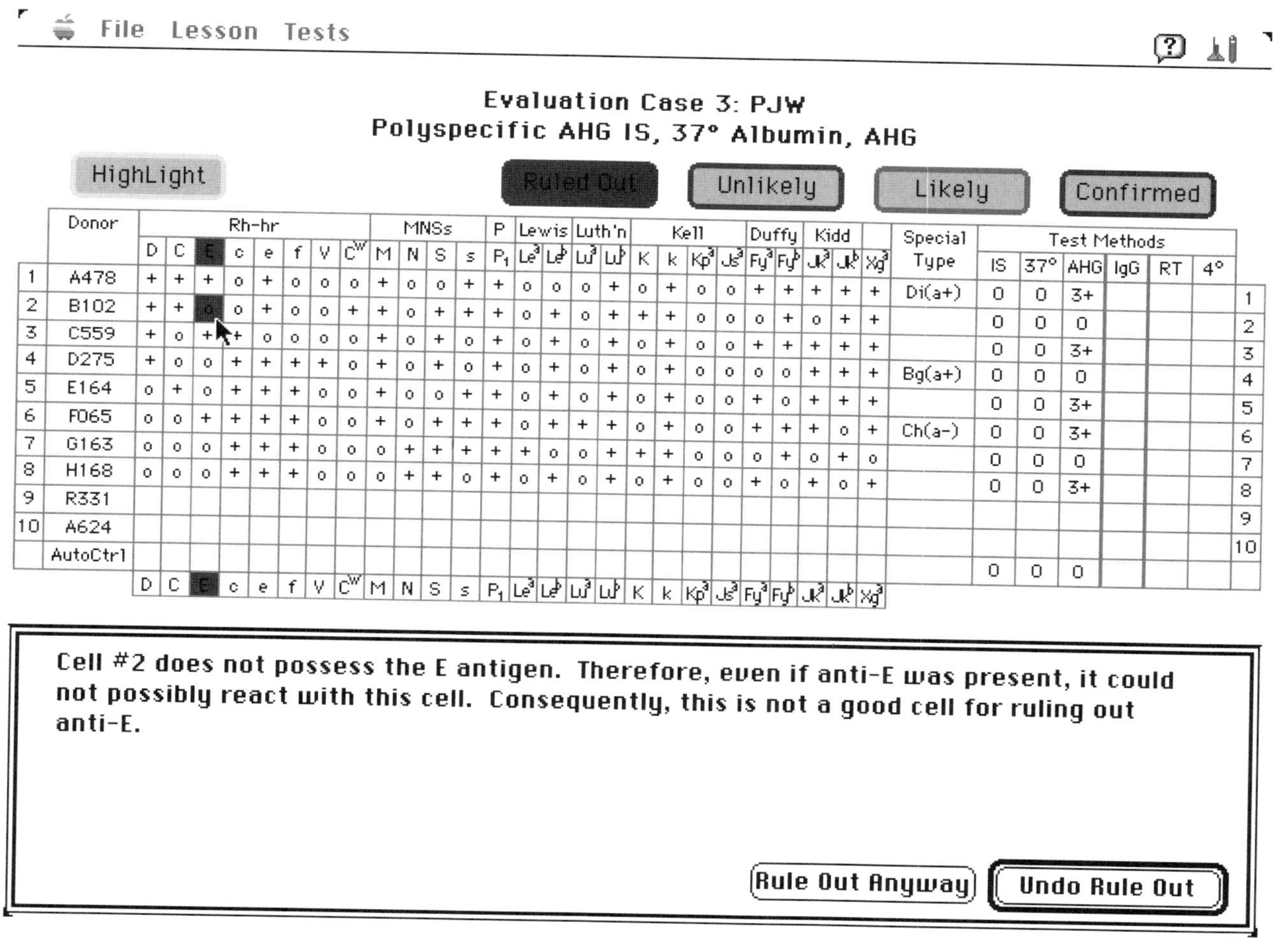

File Lesson Tests

Evaluation Case 3: PJW
Polyspecific AHG IS, 37° Albumin, AHG

HighLight Ruled Out Unlikely Likely Confirmed

	Donor	Rh-hr								MNSs				P	Lewis		Luth'n		Kell				Duffy		Kidd			Special Type	Test Methods						
		D	C	E	c	e	f	V	C^w	M	N	S	s	P_1	Le^a	Le^b	Lu^a	Lu^b	K	k	Kp^a	Js^a	Fy^a	Fy^b	Jk^a	Jk^b	Xg^a		IS	37°	AHG	IgG	RT	4°	
1	A478	+	+	+	o	+	o	o	o	+	o	o	+	+	o	o	o	+	o	+	o	o	+	+	+	+	+	Di(a+)	0	0	3+				1
2	B102	+	+	o	o	+	o	o	+	+	o	+	+	+	o	+	o	+	+	+	o	o	o	+	o	+	+		0	0	0				2
3	C559	+	o	+	+	o	o	o	o	+	o	+	o	+	o	+	o	+	o	+	o	o	+	+	+	+	+		0	0	3+				3
4	D275	+	o	o	+	+	+	+	o	+	o	+	o	+	o	+	o	+	o	+	o	o	o	o	+	+	+	Bg(a+)	0	0	0				4
5	E164	o	+	o	+	+	+	o	o	+	o	o	+	+	o	+	o	+	o	+	o	o	+	o	+	+	+		0	0	3+				5
6	F065	o	o	+	+	+	+	o	o	+	o	+	+	+	o	+	+	+	o	+	o	o	+	+	+	o	+	Ch(a-)	0	0	3+				6
7	G163	o	o	o	+	+	+	o	o	o	+	+	+	+	+	o	o	+	+	+	o	o	o	+	o	+	o		0	0	0				7
8	H168	o	o	o	+	+	+	o	o	o	+	+	o	+	o	+	o	+	o	+	o	o	+	o	+	o	+		0	0	3+				8
9	R331																																		9
10	A624																																		10
	AutoCtrl																												0	0	0				
		D	C	E	c	e	f	V	C^w	M	N	S	s	P_1	Le^a	Le^b	Lu^a	Lu^b	K	k	Kp^a	Js^a	Fy^a	Fy^b	Jk^a	Jk^b	Xg^a								

Cell #2 does not possess the E antigen. Therefore, even if anti-E was present, it could not possibly react with this cell. Consequently, this is not a good cell for ruling out anti-E.

Rule Out Anyway Undo Rule Out

Figure 1. The Antibody IDentification Assistant (AIDA).

System Description

Several points are worth highlighting:

1. The final interface to AIDA was explicitly designed to provide a one-to-one mapping between actions in the current paper-and-pencil world of the laboratory and this new computerized "world;"
2. The interface was designed to encourage users to provide the computer with highly diagnostic data about their intermediate and final conclusions about a case. This design

feature made it possible for AIDA to provide the user with rich, context-sensitive feedback about potential errors in a timely fashion;

3. To ensure that the user had an appropriate mental model of the problem-solving strategies used by the computer, a check-list was designed that enumerated the subgoals considered necessary to adequately solve a case. Users could apply additional strategies without interference from the computer and could override a critique from the computer, but the checklist (shown below) made clear the computer's expectations before completing a case. The computer also allowed the user some flexibility in deciding what order to use in completing the subgoals listed on the checklist. (i.e., the computer did not monitor for the ordering of the steps listed in the checklist except when that ordering was critical to successful problem-solving.)

   As an example of the application of this checklist, the user might be viewing the screen cells (Step 2) and get a message that she ruled out the C antibody erroneously, as it was heterozygous in the observed data and should only be ruled out homozygously.

4. Based on our studies of human experts, AIDA was designed around a broad strategy of collecting converging evidence before completing a case. This global strategy provided protection against the fallibility of the heuristic methods underlying strategies applied at different points in the case (individual steps on the checklist). To help ensure use of this strategy, AIDA monitored for both errors of commission and errors of omission;

5. The role played by AIDA was that of a critic (since our earlier studies indicated that this was the most effective way to support users).

Checklist for Alloantibody Identification

Case: ____________

Step 1. Complete ABO and Rh typing.

Step 2. Check screen cells.
  a. Mark the unlikely antibodies (usually f, V, Cw, Lua, Kpa, Jsa).
  b. Rule out antibodies (and, if appropriate, mark likely antibodies).

Homozygous: C, E, c, e, M, N, S, s, Lea, Leb, Fya, Fyb, Jka, and Jkb

Homozygous or Heterozygous: D, f, V, Cw, P1, Lua ,Lub, K, k, Kpa, Jsa, and Xga

Step 3. Check patient history if available.

Step 4. Check auto control on the Poly Panel.

Step 5. Check the Polyspecific Panel. (If necessary, use another panel to enhance reactions.)
  a. Rule out antibodies.

Antibody reactions that could be weakened in Certain test conditions:

Enzyme: M, N, S, s, Fya, Fyb, Xga

Prewarm: M, N, P1, Lea, Leb, Lua,

Eluate: M, N, P1, Lea, Leb

Room Temperature: D, C, E, c, e, f, V, Cw, s, Lub, K, k, Kpa, Jsa, Fya, Fyb, Jka, Jkb, Xga

Cold 4° C: D, C, E, c, e, f, Cw, S, s, P1, Lub, K, k, Kpa, Jsa, Fya, Fyb, Jka, Jkb, Xga

  b. Mark likely antibodies.

Step 6. If necessary, use additional cells to rule out the remaining antibodies, and to help you to confirm your answer.

Step 7. If necessary, use antigen typing to rule out the remaining antibodies.

Step 8. Use antigen typing to help confirm your answer.

Step 9. Make sure that all antibodies that have not been confirmed or marked unlikely (usually f, V, Cw, Lua, Kpa, Jsa) have been ruled out.

Step 10. Make sure the confirmed antibodies are not on any non-reacting cells.

Step 11. Make sure that at least one confirmed antibody is on every reacting cell.

Step 12. Look at your answer and ask whether it is plausible (or is it a "unicorn?").

The system contains several types of knowledge in order to support this critiquing function, including detectors for:

1. Errors of commission (due to slips or mistakes):
   a. errors in ruling out antibodies;
2. Errors of omission (due to slips or mistakes):
   a. failure to rule out an antibody for which appropriate data exists;
   b. Arriving at a final answer without first ruling out the alternative clinically significant antibodies;
   c. Failure to confirm that the patient does not have an autoimmune disorder (a different class of problems);
   d. Failure to confirm that the patient is capable of forming the antibodies in the final answer set (a check on plausibility of the final answer);
3. Errors due to masking;
4. Errors due to noisy data;
5. Answers that are implausible because the observed data is unlikely given the proposed answer set:
   a. Failure to account for all reactions;
   b. Inconsistency between the answers given and the reactions (data) typically exhibited by those antibodies;
6. Unlikely answers based on prior probabilities.

Experimental Methods

As a step in evaluating the efficacy of AIDA and its underlying design concepts, a formal empirical assessment was conducted.

Subjects

Two groups of subjects were studied. First, four practicing technologists, from three hospitals, who were identified as "highly proficient" by their supervisors, were studied. The objective of this preliminary study was to make sure AIDA did not interfere with the performances of skilled practitioners.

Second, thirty-two practicing technologists from seven hospitals were studied. Each of these technologists performed the task of antibody identification as part of their regular job, but they were identified by their supervisors as persons who "would benefit from additional experience and training." Their years of experience ranged from 1 to 35 years, with a mean of 10 years.

Experimental Design

A combined within- and between-subjects design was used. Each of the subjects completed the same set of cases, working alone. Half of the subjects used a version of AIDA with all feedback/critiquing tools turned off (the Control Group), while the other half used the full system (the Treatment Group).

Training

All thirty-two subjects were first trained on the same case regarding the use of the AIDA interface, with the critiquing turned off. Then all thirty-two subjects were given a pre-test case and asked to solve it by themselves with the critiquing turned off. (This pre-test case came from a set of two cases matched for their general characteristics. For each subject, one of these two matched cases was randomly selected for use as the pre-test case, while the other was used as the first post-test case.) Following completion of the pre-test, the Treatment Group was trained on a set of tasks designed to introduce them to AIDA's critiquing functions. The Control Group worked through the same set of tasks, but with the critiquing turned off. At the end of each training task, AIDA simply told each subject in the Control Group the correct answer vs. his or her answer.

Post-Test

Following training, four additional test cases were given to all subjects individually. The first was the case matched to the pre-test. The next three cases were the same for all subjects. For the Treatment Group, the critiquing was left on for the post-test, since the goal of this study was to evaluate performance of the technologists when assisted by AIDA. None of the subjects in this study were told the correct answers to the test cases while completing the post-test.

- The first post-test case (with its matching pre-test case) presented a problem with two alloantibodies (anti-E plus K or anti-c plus K). This was a type of case that was expected to fall within AIDA's range of competence.
- The second case was a weak anti-D. (This was the same case used in our earlier study to evaluate the effect of system role — critiquing vs. partial automation — on user performance.) This case was deliberately selected as being outside of AIDA's range of competence. (AIDA's knowledge base, if acting automatically, would have ruled out the correct answer.) A number of AIDA's critiques, however, were still expected to be helpful in solving this case.
- The third case was a masking case, with anti-E hidden under anti-Fya on the initial panel. AIDA was expected to be competent in assisting with this case.
- The fourth case was selected by an expert at a laboratory in another state who was unfamiliar with our work in developing AIDA. She was asked to contribute a real case that "many

practicing technologists would find difficult." The case she provided contained anti-E, anti-c, and anti-Jkb.

Thus, with the exception of the pair of matched pre-test and post-test cases (which were randomized for order), all thirty-two subjects experienced the same training and test cases in the same order. The only difference was that the Treatment Group was trained to regard AIDA's expectations (its problem-solving methods) and had the benefits of AIDA's checklist and critiquing functions.

Each subject was tested alone. Completing the testing session took two to three hours. At the end of the session, each subject was asked to complete a questionnaire.

Data Analysis

The resultant data were evaluated in terms of gross performance measures (subjective responses on the questionnaire and final answers to each case), and in terms of a detailed analysis of all of the behaviors which were recorded by the computer as cases were completed. This detailed analysis included studying the broad strategies adopted by technologists on different cases, the errors made, and the responses to critiquing messages (Treatment Group only). The analysis also included looking at changes in performance over time (learning) for the Treatment Group.

Results and Discussion

As discussed earlier, two populations were studied. The first was a set of "highly proficient" technologists, the second a set of thirty-two practitioners whose supervisors indicated "would benefit from additional experience and training."

Expert Practitioners

This group was tested prior to the second group of practitioners, primarily to evaluate AIDA for usability and to make sure AIDA did not create difficulties or induce new errors for skilled technologists. Two of these four technologists were tested as the Control Group and two as the Treatment Group. All four subjects found the correct answers for all five cases (pre-test and post-test).

Three points merit noting:

1. All four of these experts found the correct answers for all five cases, even though they committed a number of actions that AIDA considered wrong or suspicious;
2. For the Treatment Group it was possible to determine whether the actions flagged by AIDA on the post-test Cases 1-4 were slips, mistakes, or considered correct by the technologists. All 6 such actions were clearly slips, indicating the potential value of AIDA even for experts. (Smith, et al. (1991) found that even highly skilled practitioners make such slips on over 4% of the cases they complete.);
3. The experts using the checklist (the Treatment Group on post-test Cases 1-4) committed far fewer actions flagged by AIDA as wrong or suspicious. Furthermore, they never chose to override AIDA because they felt they knew what to do better than the system did.

The responses on the questionnaire (Treatment Group only) were very positive:

1. How would you rate this software in terms of ease-of-use?
    - "I found it easy to use, but it seemed to take me longer to do the identification."
    - "Quite easy after you get the hang of it."
2. Would you find this software useful for your job?
    - "Useful in teaching students. If put into use in the Blood Bank, all panel results would be saved on computer disc."
    - "Yes. Great for new employees and students."
3. What did you like most about this software?
    - "I like highlighting positive reactions and doing the rule-outs and the screen showing what was still left as possible antibodies."
    - "The corrections if you made an incorrect assumption."

Less Skilled Practitioners

Of the thirty-two subjects tested in the actual evaluation study, sixteen were randomly assigned to the Control Group and sixteen to the Treatment Group.

Gross Performance Measures. There was no detectable difference between the Control Group and the Treatment Group on the pre-test case ($p>0.10$). (The pre-test was identical for both groups, with the critiquing turned off.)

The Control Group showed no detectable improvement from the pre-test case to the matching first post-test case ($p>0.10$). The Treatment Group, however, showed a sizable and significant improvement ($p<0.01$) from the pre-test case (31.25% of subjects had an incorrect final answer) to the matching first post-test case (0% incorrect). Furthermore, the errors made were very significant, with subjects erroneously answering anti-Fyb alone or anti-Fya plus anti-k on the case where the correct answer was anti-E plus anti-K, and erroneously answering anti-c alone, anti-S alone, anti-K alone, anti-K plus anti-s or anti-K plus anti-Lua on the case where the correct answer was anti-c plus anti-K.

The between-subject comparisons were equally striking, as shown in the table below. All four of these differences are significant at $p<0.01$).

Table 2: Error Rates for the Post-Test Cases

Post-Test Case	1	2	3	4
Control Group (n=16)	37.5%	56.25%	37.5%	62.5%
Treatment Group (n=16)	0%	18.75%	0%	0%

These data indicate that:

1. On cases where AIDA was fully competent (Cases 1, 3, and 4), using AIDA (with the accompanying checklist) was highly effective in eliminating errors;
2. On the case where AIDA was not fully competent (Case 2), the general purpose critics contained in AIDA still helped significantly.

Conclusions

In considering the design of a decision support system, three sources of human error need to be considered:

1. Errors made by unaided practitioners (the status quo);
2. Errors made by the aided practitioners;
3. Errors made by the system designers and implementers.

Thus, a critical question in developing such a system is: Can we reduce the costs associated with the first category of errors without creating significant new costs associated with the latter two categories of errors?

Our three studies caution designers about the design of tools that automatically generate inferences before the user has evaluated the situation. Such tools can trigger a variety of cognitive biases that induce the user to accept the computer's inference even when it is incorrect. The use of the computer as a critic seems to be a powerful strategy for dealing with this problem.

In particular, our third study leaves little doubt that many practicing technologists are using incorrect knowledge and inadequate problem-solving strategies, and that even highly qualified practitioners make significant numbers of slips. Our research further shows that on cases where AIDA is competent, such errors are effectively eliminated. Finally, our data strongly suggest that the impact of a system's brittleness is substantially reduced by putting the expert system in a critiquing role.

Acknowledgments

This research has been supported by NASA Ames Research Center and the FAA under grants NCC2-615 and NCA2-701. Special thanks is given to Roger Beatty, Joe Bertapelli, Rich Milligan, Craig Parfitt and the Airline Dispatchers Federation, to Larry Earhart, Deb Galdes and Dave Williams, and to all of the dispatchers who donated their time to make this study possible. It has also been supported by the National Heart, Lung, and Blood Institute under grant HL-38776.

References

Ericsson, K. A. and H. A. Simon (1993). *Protocol Analysis: Verbal Reports as Data*. MIT Press, Cambridge, MA.

Fraser, J. M., P. J. Smith and J. W. Smith (1992). A catalog of errors. *International Journal of Man-Machine Systems*, **37**, 265-307.

Fraser, J. M., P. Strohm, J. W. Smith, D. Galdes, J. R. Svirbely, S. Rudmann, T. E. Miller, J. Blazina, M. Kennedy and P. J. Smith (1989). Errors in abductive reasoning. In *Proc. of the 1989 IEEE Intl. Conf. on Systems, Man, and Cybernetics*. pp. 1136-1141.

Hayes-Roth, B. and F. Hayes-Roth (1979). A cognitive model of planning. *Cognitive Science*, **3**, 275-310.

Hoc, J. M. (1988). *Cognitive Psychology of Planning*. Academic Press, London.

Miller, G. A., E. Galanter and K. H. Pribram (1960). *Plans and the Structure of Behavior*. Holt, New York.

Sacerdoti, E. D. (1974). Planning in a hierarchy of abstraction spaces. *Artificial Intell.*, **5**, 115-135.

Schank, R. and R. Abelson (1977). *Scripts, Plans, Goals, and Understanding*. Erlbaum, Hillsdale, NJ.

Smith, P. J., D. Galdes, J. Fraser, T. Miller, J. W. Smith, J. R. Svirbely, J. Blazina, M. Kennedy, S. Rudmann and D. L. Thomas (1991). Coping with the complexities of multiple-solution problems: a case study. *International Journal of Man-Machine Studies*, **35**, 429-453.

Smith, P. J., T. E. Miller, J. Fraser, J. W. Smith, J. R. Svirbely, S. Rudmann and P. Strohm (1992). An empirical evaluation of the performance of antibody identification tasks. *Transfusion*, **31**, 313-317.

Suchman, L. A. (1987). *Plans and Situated Actions: The Problem of Human Machine Communication*. Cambridge, New York.

Wilensky, R. (1983). *Planning and Understanding: A Computational Approach to Human Reasoning*. Addison-Wesley, Reading, MA.

SESSION SUMMARY: HUMAN-MACHINE INTERACTION

Ming Rao
Intelligence Engineering Laboratory
University of Alberta
Edmonton, Alberta, Canada T6G 2G6

Mark A. Kramer
Gensym Corporation
Cambridge, MA 02140

Introduction

Industrial development can be classified as four stages based on automation technology (Lu, 1989; Rao and Qiu, 1993). The first stage, the labor-intensive stage, mainly relies on the personal skills of human operators who use the simple non-automation tools. At the second stage, the equipment-intensive stage, automatic instrumentation plays an important role in the process operation. Now our industry is moving into the information-intensive stage, where computing facility is much more powerful and affordable. The next stage is the knowledge-intensive industry which will rely on artificial intelligence technology.

During the industrial automation development, the relationship between the human operators and the controlled process is continuously changed (Rao and Qiu, 1993). At the first stage, the operator directly controls the process by manual operation. At the equipment-intensive stage, there appears to be an intermediate facility — instrumentation, which is laid between operators and the controlled process. The information-intensive stage of industrial automation adds another intermediate facility — computer system, thus the human operators have been isolated. At the future knowledge-intensive stage, we will not increase more intermediate blocks, but keep the human operators in the control loop by providing intelligent operation support systems. The past practice has proved that keeping human operator in control loop is an advantage, rather than a disadvantage.

A session on Human-Machine Interaction was organized during the 1995 International Conference on Intelligent Systems in Process Engineering, co-chaired by Dr. Ming Rao, Intelligence Engineering Laboratory at the University of Alberta, Edmonton, Alberta, Canada and Dr. Mark A. Kramer, Gensym Corporation, Cambridge, MA 02140.

In this session, two research papers were selected for oral presentation, which included "Intelligence in Man-Machine Interaction" presented by Professor Hirokazu Nishitani, Nara Institute of Science and Technology, Japan, and "Theory and Practice of Human-Machine Interfaces" presented by Ms. Stephanie Guerlain, Ohio State University.

Professor Nishitani discussed various concepts about human-machine interaction in process control systems, covering system design, function analysis of human interface, CRT operation, human factor engineering, operator support, as well as human behavior modeling. During the process operation and control activities, the process engineers have to interpret observations on man-machine interaction. Being similar to the Conference Keynote speaker Dr. Winston's viewpoint, Professor Nishitani suggested to use visualization to better support process operator to perform their operation and decision making. In order to facilitate different user interfaces in the same process plant, an integrated man-machine interface must be developed. Finally, Professor Nishitani concluded the interdisciplinary research cooperation will be a necessity for human-machine interaction research.

Ms. Stephanie Guerlain, a Ph.D. candidate from Ohio State University, reported their discoveries in human-machine interface development with applications to air-traffic control system and blood banking in medicine. An interdisciplinary research team, including faculty members and graduate students, applies the multidisciplinary knowledge from artificial intelligence, cognitive science, human factors, aviation, control system, medicine and computer engineering, to solve the challenging problems in the real world. Why and how to keep the human being in the control loop were discussed. The role of computer system and its impact on human performance in the decision support activity were discussed. The research results based on theoretical investigation and experimental examination were reported. Three application case studies were reviewed, including a flight planning system for airline dispatchers (a prototype software system, namely Flight Planning Testbed, or FPT, was developed), and a cognitive tool to support blood bankers (the developed software was named as AIDA, which stands for Antibody IDentification Assistance). The design tools used in these three case studies can automatically generate the user interfaces before the user evaluates the situation. During the presentation, Stephanie demonstrated the developed software system.

Both presentations were well received, and generated many interested questions from audience. Human factor issue was highlighted in both presentations and audience discussion. As our manufacturing systems become more complex, the effectiveness of human-machine interaction become more important and critical. Due to the effect of human factor, the man-machine interface design is difficult since human often do not act logically; and how to prevent human errors in process operation becomes the key and the bottleneck to solve the operation safety problems in nuclear power plant and chemical industry. However, the available technology can not handle the human behavior modeling well, especially when dealing with personality, motivation, human group dynamics, as well as mental processes.

During the discussion, most people agree that man-machine interface plays a key role for computer technology applications, especially for intelligent system development. With the current computer interface facility, when we use computers, we are forced by the computers to think like computers and to reason like computers. Rather than doing this, we should develop the "natural interface" in order to have computers act like human being. Dr. Winston presented the similar viewpoint in his conference keynote speech. He stated that today, people have to adapt to computer by using display, mouse and keyboard; tomorrow, computer must enter human being's world with the natural interaction. We have realized that the success of an intelligent system development is mainly relied on whether or not it has a truly user friendly interface. The final acceptance of the intelligent system is very often associated with the user interface.

In many of today's industrial companies, the highly computerized human-machine systems have been widely installed. Therefore, it is becoming increasingly difficult for human operators to understand various signals and information from computer screens, display boards and video. The development of the intelligent multimedia interfaces can better facilitate human operator interaction with the complex real-time monitoring and control systems. The multimedia interface can communicate the human operators via natural language, graphical presentation, sound, voice, video, animation, numerical data, and so on (Rao and Qiu, 1993). Virtual reality technology may create another new environment for man-machine interaction practice. As suggested by Dr. Winston, artificial intelligence should provide a means to develop natural interaction between human being and computing machines. Finally, most audience agree that human factor engineering methodology should be emphasized in the future intelligent process system research and applications.

References

Lu, S. C.-Y. (1989). Knowledge processing for engineering automation. *Proceedings of 15th Conference on Production Research & Technology*. Berkeley, CA. pp. 455-468.

Nishitani, H. (1995). Intelligence in man-machine interaction. Presented at the *International Conference on Intelligent Systems in Process Engineering*. Snowmass Village, Colorado.

Rao, M. and H. Qiu (1993). *Process Control Engineering*. Gordon & Breach Science Publishers, London.

Smith, P., S. Guerlain, J. Smith, Jr., R. Denning, C. McCoy and C. Layton (1995). Theory and practice of human-machine interfaces. Presented at *the International Conference on Intelligent Systems in Process Engineering*. Snowmass Village, Colorado.

SUPERVISION AND DECISION SUPPORT OF PROCESS OPERATIONS

Ted Cochran
Honeywell Technology Center
3660 Technology Drive
Minneapolis, MN 55409

Duncan Rowan
DuPont Engineering
Process Control Initiative
N6545A, 1007 Market St.
Wilmington, DE 19898

Abstract

We review the state of the art in technology available for decision support, as well as the extent to which that technology has been applied in process control. We conclude that user interface hardware is less than state of the art due to the requirements of the domain, and the introduction of some user interface software technologies has been delayed as a result. On the other hand, decision support technology has been given a thorough trial, and process control applications have sometimes advanced the state of the art. Current research into the development of environments to support collaborative interaction of multiple operators with multiple tools may be the key to more cost-effective use of available technology.

Keywords

Process control, Decision support, Expert systems, On-line aiding, Supervisory control, User interface.

Introduction

This year-1995-may well be the year selected by future historians to mark the beginning of the information age. Last year, for the first time in history, more was spent by consumers on new computers than on television sets, and it is only a matter of months before computer sales exceed television sales in raw numbers, as well. The world-wide web is growing at 10-20% *per month*, and press coverage is ubiquitous. E-mail addresses and URLs are making appearances in popular publications and in television and print advertising-they are no longer confined to *Wired*. The rate of improvement to desktop computer power is accelerating: Performance is no longer doubling every three to four years, but every two to three. Networks are being installed which are capable of delivering more information, more rapidly, than can most hard disks. CD-ROMS are now standard accessories, pre-installed on the majority of new personal computers.

So, what heights have these developments enabled supervisory control of process operations to achieve?

In this paper we'll review the state of the art in supervisory control of complex processes. We'll begin by describing the roles academic and industry researchers are playing in this area. We'll discuss the state of the art for some of the key generic technologies in terms of recent research achievements, what's available commercially, and what's typical in current process control systems.

We'll then describe the current state of process-specific supervisory control and decision support technologies, offer an opinion about what's still needed, and assess the prospects for getting there in the near future.

Technology Development

Academic research, and research directed at technology development by private enterprise, tends to define the leading edge of the state of the art in most technology-driven fields. Petrochemical companies have historically been conservative about connecting the latest bells and whistles to their processes at low levels (a reliability track record is crucial, and takes time to attain), but have adopted relevant new technologies relatively quickly at the supervisory control level when cost-effective applications are evident.

Applied Research

The objectives of industrial research organizations are now more focused on core technologies essential to the business. These groups are now focused on extending process understanding and driving the development of new technologies by vendors. Special attention is being devoted to the development of process models (fundamental and empirical, steady state and dynamic), on new control technologies (e.g., non-linear model predictive control), and on control strategy improvements. Models are often developed using commercially available tools such as Aspen Technology's SPEEDUP product, or Honeywell SACDA's TRAINER. The application of mature technologies such as decision support are no longer receiving extensive attention from corporate research organizations; instead, these are used by corporate and plant engineering groups and by some groups in Information Systems organizations.

User Interfaces

The state of the art in interface hardware includes the 9000 x 3000 pixel (25 ft. x 9 ft.) composite display system assembled by the Electric Power Research Institute (EPRI) to simulate mimic boards (Fray, 1995), and by virtual scenes generated by head mounted displays (stereo color 1280 x 1024 displays throwing 60K polygons/sec). Technologies such as gesture recognition ("put that there"), unconstrained speech input and output, handwriting, and tactile feedback (gloves, manipulators with force feedback) are at varying stages of maturity, but all have been developed sufficiently to enable practical application in at least some domains. User interface "software" technologies include virtual reality, synthetic vision (e.g., realistic displays created by combining terrain models with position and attitude information and projected for pilots of windowless aircraft), augmented displays (e.g., infrared data overlaid on the users' existing view of the world), and new navigation, data visualization, intent recognition, and user modeling approaches.

Commercially available hardware includes displays with resolutions up to 2048 x 2048 pixels at 24 bits per pixel, integrated multimedia (for example, up to four simultaneous real-time 640 x 480 video windows, stereo CD-quality audio), constrained speech I/O, rudimentary handwriting recognition, and 100 mb/s networking (28 kb/s wireless). Commercially-available user interface software technology includes simple intent recognition (e.g., Microsoft Excel's assistant), low resolution virtual reality, first generation groupware (Notes), third-generation user interface building tools (Visual Basic, TAE), and second-generation graphical user interface technologies (Quickdraw 3D and VR, Scripting, Agents).

The performance of such systems is best measured from the user's perspective (we assert that CPU speed is irrelevant). Users typically demand response times of less than one second for frequent operations and will tolerate 60 second delays for e.g., file operations (save, paginate, recalc). Reliability is also important; typical software has a MTBF (crashes) on the order of hours, and a hardware MTBF (system needs repairs) on the order of months.

Typical process control applications incorporate user interface hardware that is much less capable, but much more reliable, than that described above. Displays of 640 x 480 pixels with up to 256 colors are just beginning to give way to 1280 x 1024 x 16 bit displays. Systems can, however, be configured to support many such displays simultaneously. Multimedia use is minimal (sound = horn). Software is highly customized: X-windows and other open environments are becoming available, but are being adopted only very slowly for critical operations. Graphical user interfaces lag those available elsewhere.

Reliability, however, is robust: Software MTBF is measured in months; hardware MTBF is often measured in years. The user-visible performance is also adequate: <1 sec scan rates, <1 sec updates, <4 sec screen call ups.

We conclude that while there are significant differences between the state of the art and current practice, they are not unexpected given the relatively small market, rigorous requirements, different nature of work in process control domains.

Decision Support

A working definition of decision support systems is *technology that presents data to users in ways that enable easier detection of significant patterns and easier, more rapid selection and execution of appropriate actions.* Technologies that comprise and contribute to decision support systems include knowledge-based systems, model-based systems, empirical approaches (e.g., statistical process control and neural networks), and engineering process control (Davis, 1995). Decision support can be augmented via on-line documentation, procedural advisors, scheduling assistants, and other technologies. Our overview of decision support in process control distinguishes between state of the available *technology*, which may include demonstrations of applications in restricted domains, and the typical *application* of that technology to the supervisory control domain.

Expert Systems

The promise of expert systems in the mid-80's was to insert the best operational and decision-making capability into a computer program that would then continuously give that advice on-line. In this way, the best knowledge about the process and the process operation would be available at all times, even to less experienced operating personnel. This promise has largely been left unfulfilled.

Corporate downsizing has led to the reduction or elimination of support for large expert system projects, due to lack of capital and engineering resources. Available capital and people have been focused on very basic opportunities using conventional technologies. Management focus is on quick pay back applications. The concepts of intelligent systems and expert systems are being applied in small focused applications. There have been very few "expert-in-a-box" applications. In many cases, the expertise is less well understood than initially thought and may be disbursed among various people in many groups. It can be difficult for these people to get commitment to work on a large expert system project.

The current cost of abnormal situations - at least $20B to the U.S. economy as a whole-is mostly derivative; that is, it does not directly impact the process industry, and the costs that do represent a relatively small proportion of their total operating budgets. Significant incidents occur infrequently, in which case the somewhat reduced process utility does not significantly affect the economics of the operation. Only in sold-out conditions do short-term outages have a major impact on economics [However, the continuing consolidation of the industry and increasing plant utilization are leading to such impacts more frequently.] A more serious economic consequence is degraded operation, which can produce off-specification product with attendant recycling or waste disposal cost. Even then, using cost reduction as a justification for investment is difficult: Competing opportunities leading to increased revenue are typically viewed more favorably.

Even when cost-effectiveness is proven, expert systems must also compete with other approaches for dealing with detection and correction of abnormal operation. Chronic or serious abnormal operation may be addressed by redesigning the process or process equipment to eliminate the possibility of occurrence. In some cases, this more cost effective than developing an expert system to detect or predict such an occurrence.

Most current expert system applications therefore address specific, high pay back situations, where degraded performance can be detected early, and prior to the process becoming out of specification. These small process "watchdogs" are being embedded in some distributed control systems (DCS) as well as in higher level software environments such as host computer systems or Gensym's G2 tool. The advantage of the higher level network-based platforms is the ability of the applications to communicate information via electronic mail and provide report information which is easily accessible to users.

More complex diagnostic systems have been developed by Davis (1991) of Ohio State University in collaboration with a number of petrochemical companies. In DuPont, more complex expert system applications have been targeted for process transition management (Rowan, 1992) and complex diagnostic advisory systems. One of these diagnostic systems is a control loop performance monitor which tracks several hundred loops and reports any deviation from "normal" performance statistics. The application has proven valuable in detecting several instances of loop instability caused by marginal tuning, transmitter degradation, and blocked impulse lines; in one case, using the statistics generated by the application, an engineer gained a key insight into process equipment. These results would not have been possible without this sophisticated loop monitoring. The techniques in this monitor are similar to those reported by Jofriet, et al. (1994).

Another promise of expert systems technology was that once applications were built for one location, they could be easily leveraged to other locations with the same process, therefore distributing the best knowledge throughout the business. In practice, this has been difficult. Cultural barriers exist as these sites tend to be operated independently. Also, due to control/computer systems infrastructure being different from site to site, these applications must be redesigned for a new site. This is a significant cost, but some companies have nevertheless successfully leveraged applications. Air Products & Chemicals, Inc. claims to have broadly leveraged their expert system applications to their industrial gas plants, which are similar in design (and minimally staffed), and therefore a particularly good fit for these applications.

Though large expert systems applications are infrequent, the technology is recognized as a useful component of other types of systems. For example, in new energy management technology that DuPont is evaluating, expert systems capability is needed to provide equipment diagnosis prior to optimization. Expert systems technology has also been found to be useful in other types of optimization applications (Faccenda, 1995). Finally, this capability is necessary to support robust performance in advanced model predictive control (MPC) applications.

Modeling

Model-based diagnostic and optimization techniques compare plant data with a model of a plant and identify disparities of potential interest. Some, like APACS (Benjamin, et al., 1990), attempt to identify potential causes of abnormal situations by manipulating the model until better agreement with plant data is achieved (and then announcing the manipulation that most economically explains the plant data). Others, such as Formentor

(Pennings and Saussais, 1993), use knowledge-based approaches to identify discrepancies. Strictly empirical methods use neural networks, fuzzy logic, or statistical methods to categorize plant data as normal or abnormal, and, if abnormal, to further determine what the problem is.

Process models can be used to study process behavior and therefore used as an analysis tool during development of decision-support systems. These models tend to be dynamic, mechanistic models based on first principles. Empirical or data models can also be used to detect and announce abnormal patterns in process data. These models are developed from existing process data containing the patterns which must be detected.

Model-based solutions are limited to systems for which models can be constructed (from first principles or empirical data) and maintained, and this continues to be a significant issue. Tools for model building and maintenance are becoming better, enabling models to be cost-effectively incorporated into more systems than even five years ago, but more automated approaches are required. A large model can be affected by every maintenance change made to the process, and quickly become inaccurate unless painstakingly maintained. There are some efforts to tie maintenance information systems to operational systems in ways that could enable the models to be more easily maintained, but this approach is in the earliest stages of development. An alternative is to develop methods of linking much smaller, presumably more easily maintained, models, and this approach is being studied by Honeywell, among others.

An alternate approach to decision support technology is to use model-based technologies to train operators to themselves detect abnormal situations and react appropriately. There is much interest at DuPont in improving operator skills and abilities. One plant is adding a fifth shift so that operators are in training as much as one week per month. A number of projects have been undertaken to create small dynamic simulations of critical processes (including abnormal situations). These simulations can then be connected to the plant DCS and operators can be trained in a "flight simulator" mode. This has been particularly effective for the startup of new processes and the training of inexperienced operators (Bauer and Stainback, 1995; Hawthorne, et al., 1995). Operators who are confident that they can recognize and correct abnormal situations are a great asset.

Empirical Approaches

Empirical approaches are also useful as decision support technology (Piovoso and Owens, 1991). Empirical models are linear or nonlinear correlation models derived from actual process data. These models include multiple linear regression, partially least squares, principal component regression, neural nets, and time-series models. There is a great deal of interest in industry in using existing process physical measurements to supply models which can then predict process parameters or product properties. In some cases, such as with neural nets, the "model" thus derived is the quintessential black box: It generates correct outputs for specific inputs, but it does so using techniques that do not map well onto human understanding of such processes.

Generally, a large amount of historical process data is available from which to develop these models. In practice, however, often this historical data is insufficient to make accurate models. Often, the process must be disturbed with step or pulse testing in order to have sufficient movement in all input variables in order to identify the model structure. These tests generally must be agreed upon by manufacturing management and be scheduled in advance. Therefore, these applications take more effort than might be initially expected. Nonetheless, there is a great deal of interest in generating these models because of the high perceived value.

As industry moves to on-line product release capability, these types of models will be essential in determining that the process is within specification on a continuous basis. In order to cope with the complexities of real process data, many of these models tend to be small multiple input, single output models. In order to capture process dynamics for model identification and control, time series techniques are frequently used.

Statistical/Engineering Process Control

Cooperation between Control Engineers and Statisticians over the last few years has led to better control techniques for regulating continuous process variables with infrequent laboratory sampled measurements (MacGregor, 1988). The exponentially weighted moving average (EWMA) filter has been shown to be superior to other statistical filters where regulatory action can be taken without a cost penalty. Because of the low level of complexity, these applications are of wide interest within DuPont. Also, Tennessee Eastman has reported a great deal of activity in this area (Paulonis, 1995). The EWMA-based controller can substitute for a Cusum or Shewart controller. Reductions in process variability for 30 to 60 percent have been seen.

Coordination of Multiple Approaches

All of the approaches to decision support described above are more applicable in specific, well-defined problem domains than they in general: They are well suited to specific kinds of problems, but ill-suited to determining the overall operational status of a large plant. Since no candidate exists for the general case, attention is turning to the development of coordination methods that will enable multiple disparate approaches to successfully work together to diagnosis problems and support operators' responses to those problems.

The NIST-funded Abnormal Situation Management program, undertaken by a consortium of software and petrochemical companies led by Honeywell, is a particularly ambitious effort in this area: The intent is to provide an infrastructure which will enable a wide variety of diagnostic applications, problem solving systems, planners, and decision aids to work together using a common language and a well-defined software architecture. This effort has just begun, so it will be some time before we can determine whether the approach is feasible for large plants.

Application of Future Technology

Networks

Computer and networking hardware and software are advancing rapidly and will add new dimensions to decision support system applications. One DuPont business is porting an existing batch scheduling, ingredient monitoring application to a client/server architecture. The user interface will be revised from a character cell terminal to a Windows-based package. The Windows-based user interface will allow for much easier data access and application navigation. Many of these legacy business support applications will move towards client/server and Windows interface technologies. Software tools such as VISUAL BASIC and VISUAL C++ make it feasible for corporate Information Systems programmers to develop Windows applications. For real-time applications, software environment such as Gensym's G2 allows for much more rapid software development. These tools also have the advantage of running across various hardware platforms and operating systems. Therefore, when computer hardware architecture changes, the application software will not have to be ported; this is a major cost savings.

Having real-time process data available on inexpensive PC packages allows for other inexpensive software packages to readily access the data. Applications such as spreadsheets, statistical packages, modeling and math packages are very cost effective tools.

Network technology is allowing process data and information to move over wide area networks. This is allowing for global distribution of decision support systems. Data can be pulled across the network, processed in a central computer system and the results can be returned. Current network bandwidth is sufficient for monitoring relatively slowly changing chemical processes. This allows for minimal hardware/software and application support at the local site and allows a central support staff to handle most of the work. Thus economies-of-scale are achieved via this architecture. This will be a huge advantage for global companies with sites operating in relatively remote low technology areas.

Collaborative Systems

As the capabilities of decision support techniques improve, the fundamental principles of user-system collaboration are receiving renewed attention. In some cases, the demands of industry applications are driving the development of theoretical knowledge in this area, blurring the line between academic and applied research (Cochran, 1994). It is now apparent that the success of decision aiding systems does not depend merely upon their accuracy, but on their ability to accommodate specific and sometimes unrecognized user needs.

For example, *human* experts may be ignored by an operator in critical situations due to their unfamiliarity, their interaction style, or because the operator can simply not afford the additional effort required to pay attention. Decision support systems have fallen victim to the same fate: If they are in any way misunderstood or mistrusted by users, or even if they are merely inscrutable, they tend not to be relied upon when they are really needed. Esprit's Gradient project and PRECARN's IGI project are both attempting to address this specific issue.

Even if systems are mostly correct, and mostly paid attention to, other problems emerge: If operators grow to depend upon decision aids, such systems may lead to a decrease in the alertness of operators in the short term, and to a decrease in their proficiency in the long term. This problem has received significant attention in the aerospace industry with the introduction of glass cockpits and automated flight management systems; DCS vendors are applying the knowledge thereby gained-sometimes at great cost-to industrial DCSs as well.

The Abnormal Situation Management program is a particularly ambitious effort in this area as well: It aims to provide an environment for user system collaboration, which specifies standards for the interaction of diagnosis, state estimation, scheduling, planning, and operator support systems. Such systems, once installed, will become part of a collaborative environment in which the expectations and needs of operators are explicitly supported during abnormal situations. If successful, this research project will prove the feasibility of combining the many current approaches to decision aiding so that the operations staff receives support in a comprehensive and effective way.

Conclusions

The state of the art of user interface hardware in the process industries is well behind that which is available in other markets, which is unsurprising given the extreme reliability demands placed on process-connected equipment. As a result, user interface displays and some other forms of user interface software are also lagging.

Nevertheless, the application in the process industry of decision support technology as a whole is not only at the state of the art, but is driving it, due to the same factors that drive the hardware reliability criteria-process safety, the criticality of the operator's role in the management of the process, the enormous capital investment in the process equipment and the materials being processed, and overall economic competitiveness. These factors have led to a continual search for better ways to support operations personnel, backed in some cases by significant, shared, investment in new approaches.

If these approaches pan out, the industry may finally be available to develop some comprehensive and cost-effective decision aiding systems which integrate a wide variety of the successful, but isolated, efforts to date.

References

Benjamin, M., Q. Chou, S. Mensah and J. Mylopoulos (1990). Expert systems for advanced process analysis and control of nuclear power plants. *IAEA Specialists Meeting on Analysis and Experience in Control and Instrumentation as a Decision Tool*. Arnhem.

Bauer, E. J. and J. T. Stainback (1995). History of an operator training simulator on a new polymer line project. *AICHE 1995 Spring National Meeting*. Houston.

Cochran, E. (1994). Basic versus applied research in cognitive science: a view from industry. *Ecological Psychology*, **6**, 131-136.

Davis, J. F. (1991). A Framework for implementing on-line diagnostic advisory systems in continuous process operations. *AICHE 1991 Fall National Meeting*.

Davis, J. F. (1995). Intelligent support systems in the process industries: the critical importance of structure in large-scale systems. *Abnormal Situation Management Consortium Spring Meeting*. Cleveland.

Faccenda, J. F. (1995). In C. E. Bodington (Ed.), *Planning, Scheduling, and Control Integration in the Process Industries*. McGraw-Hill, New York, pp. 201-209.

Fray, R. (1995). Compact simulators for fossil-fueled power plants. *IEEE Spectrum*, **32**, 46-51.

Hawthorne, R., T. Fu, B. Tyreus and B. Donovan (1995). Process dynamic simulation at DuPont's Corpus Christi plant. *AICHE 1995 Spring National Meeting*, Houston.

Jofriet, P., M. Harvey, C. Seppala, B Surgenor and T. Harris (1994). A G2 knowledge base for control loop performance analysis. *1994 Gensym Users Meeting*, Cambridge, MA.

MacGregor, J. F. (1988). On-line statistical process control. *Chemical Eng. Progress*, **10**, 21-31.

Paulonis, M. A. (1995). Closed-loop control of properties measured off-line. *AICHE 1995 Spring National Meeting*, Houston.

Pennings, R. and Y. Saussais (1993). Formentor: real-time safety oriented decision support using a goal tree-success tree model. *Third International Conference on Artificial Intelligence*, Avignon.

Piovoso, M. J. and A. J. Owens (1991). Sensor data analysis using artificial neural networks. In Arkun and Ray (Eds.), *Chemical Process Control (CPC-IV)*. AICHE, New York. pp. 101-118.

Rowan, D. A. (1992). Beyond FALCON: industrial applications of knowledge-based systems. *IFAC Symposium: On-line Fault Detection and Supervision in the Chemical Process Industries*, Newark, DE.

INTELLIGENT SYSTEMS IN HEAVY INDUSTRY

Adam J. Szladow
Lobbe Technologies Ltd.
Regina, Saskatchewan, Canada

Debbie Mills
REDUCT Systems Inc.
Regina, Saskatchewan, Canada

Abstract

This paper assesses the applications and benefits of intelligent systems in five heavy industry sectors: Iron and Steel, Cement, Mining and Metallurgy, Oil and Gas, and Pulp and Paper. The paper looks at implementation of intelligent systems for improvement of productivity, quality and energy efficiency. It starts with a review of the type of applications and solutions used in the industry, then it discusses the reported performance and benefits for intelligent systems implemented. The paper concludes with a summary of the results of a survey of developers and users addressing the barriers to development and implementation of intelligent systems in industry. A full report entitled: Intelligent Systems for Productivity, Quality and Energy Efficiency Improvement in Heavy Industry, is available from Lobbe Technologies Ltd. (306) 586-9400.

Keywords

Expert systems, Fuzzy sets, Neural networks, Heavy industry.

Introduction

Intelligent systems have a number of applications in heavy industry. They are used to accomplish tasks as diverse as tuning control loops, scheduling production or providing fault diagnostic advice. In general, their role is to enhance decision-making processes by turning data into decision-support rules, methods and tools.

From the application point of view, intelligent systems usually perform one of the following functions: process control, process monitoring, scheduling and planning, fault diagnosis and maintenance, and design (Fig. 1).

The following sections discuss the surveyed applications of intelligent systems in heavy industry in terms of these functions.

Applications of Intelligent Systems

Process Control

These tasks involve automation of the low level control (loop control) in a real-time system. The implemented systems are concerned with fault detection, diagnosis and alarming, and with handling control devices in the control loops. The systems can run in several modes including continuous scanning, executed only when new data are available, or activated by specific external events, e.g., change of feed properties. An integral part of intelligent systems' functions are sensor diagnostics, handling erroneous or missing data, performing temporary reasoning, make inferences and reason about the process, and implement complex control decisions. The systems often incorporate mathematical models of unit processes and operations, and empirical control rules acquired from the operators.

Examples of intelligent systems application in process control are Nippon's system for preventing breakup of continuous casters in steel processing (Hatanaka, et al., 1993), the Control International system for tuning controllers to correct for highly non-linear dynamics of semi-autogenous mills (Broussard, 1992) and the Automation Technology system for updating control models as process conditions change in brownstock operations (Beaverstock, et al., 1992).

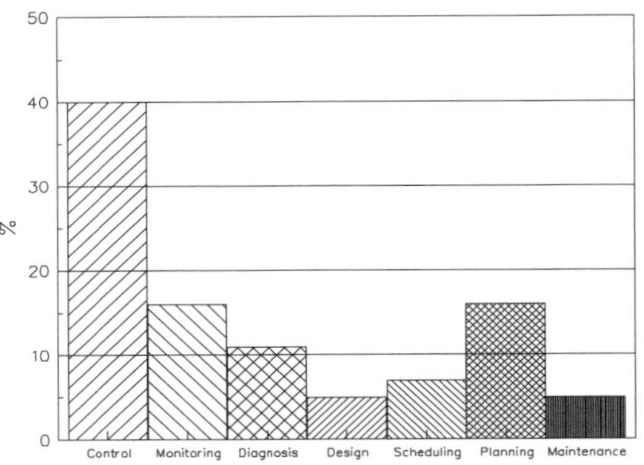

Figure 1. AI use in heavy industry by application.

Process Monitoring

Intelligent systems offer powerful solutions to use operational type of knowledge. By modeling decision-making processes and not equipment or operations, on-line decision-support systems provide consistent shift-to-shift quality, higher productivity and lower product cost. The systems monitor plant information and make recommendations only where they are needed, thus preventing cognitive overload. They may automate some tasks (usually repetitive tasks requiring less skill) to allow operators to concentrate on the major tasks. Supervisory type intelligent systems are event-driven, interactive and responsive tools for decision making. They can be built and maintained at a lower cost than real-time control systems and can access other plant information like data concerning process and equipment trouble shooting and statistical process control.

Intelligent applications for control monitoring constitute about 20 percent of the intelligent systems implemented in heavy industry and are implemented about evenly in all heavy industry sectors. Good examples of advanced, on-line supervisory control systems are POSCO's system for diagnosing abnormal blast furnace conditions (Choi, et al., 1991), the Outukump Oy system for control of flotation cells at copper mines (Karhu, 1992), the Wiggins system for control of coater/paper machines (Grigoriu, 1990), and the Kemira expert system for optimization of phosphate concentrator (Jamsa-Jounela, et al., 1989).

Scheduling and Planning

Intelligent systems offer several advantages in developing computerized scheduling systems. They can represent symbolic and non-mathematical knowledge, use expert heuristics to reduce the search space, offer easy implementation of high level reasoning, and mimic the expert's mapping out methods in scheduling. Intelligent systems can carry on distributed processing to implement specific tasks at appropriate locations; include information on limitation bottlenecks, quality variables, strategy changes; are more flexible than conventional schedules; handle uncertainties, vagueness, and conflicting constraints.

This study identified over forty intelligent scheduling and planning systems implemented in heavy industry. The systems are used in all sectors and constitute on average 20 percent of the systems implemented (Fig. 1). Examples of the diversity of intelligent scheduling systems are the Voest-Alpine system for planning operations of a continuous slab caster (Stohl, 1991), the Yukong Ltd expert system for delivering crude oil to its refineries (Lee, et al., 1991), the Tanoma Coal systems for underground mine planning (Briton, 1987), and the Oji Paper Co. scheduling expert system for paper production (Kojima, et al., 1991).

Fault Diagnosis and Maintenance (Equipment)

Artificial Intelligence systems offer a number of advantages for working with diagnostic problems. First, the systems can monitor and analyze hundreds of sensors, determine any anomalies in their functions and identify probable causes of the discrepancies between expected and actual operating conditions. The systems can contain the expertise of the best maintenance personnel, information from equipment manuals, previous records of equipment/process failures, etc. The systems can be available to the operators and maintenance personnel at all times and at the right locations. They can have embedded human expert reasoning and heuristics about diagnostic problem solving including detailed information on process and equipment performance and possible malfunctions.

Diagnostic and maintenance applications constitute about 15 percent of all intelligent applications in heavy industry. The number is quite low for all sectors with the exception of Pulp and Paper where diagnosis and maintenance constitute over 30 percent of AI applications. Examples of diagnostic expert systems in the industry are the Nippon expert system for preventive maintenance of over 500 pieces of electrical and mechanical equipment (Harita, et al., 1991), LKAB Company's use of AI for diagnosing transmission of load-haul-dump mine vehicles (Vagenas, 1991), the Wiggins Teape Appleton system for paper machine fault diagnosis (Grigoriu, 1990), and the PAPRICAN system for diagnosing pitch formation in kraft pulp plants (Kowalski, et al., 1993).

Design

Opportunities for using intelligent systems in design are the least understood and explored of all AI applications in heavy industry. There are several barriers to the introduction of computerized systems in design beyond the drafting, cataloguing, or visualization of the designs. The process of design involves a large number of combinations, and, often, subjective judgments. Design problems are similar to scheduling and planning problems, yet they also involve more than a simple configuration of objects that can satisfy defined constraints of the problem.

Progress has been made in the application of AI in design where the objective of the systems is to assist the designer in generating alternative designs for further refinement by the human expert. Examples of implemented systems are the Nippon Felt system for design of wet felts for paper machines (Kawata, et al., 1990) and the Schlumberger Slurry design system (Kelly, et al., 1991). However, in total design systems constitute less than 5 percent of the identified intelligent applications in heavy industry.

Performance and Benefits

A survey of users and developers of intelligent systems, conducted for this study, showed that the majority of intelligent applications are less than three years old. The surveys results include over 50 organizations and individuals respondents active in implementing and developing intelligent systems in industry. They applied/developed on average three intelligent systems, 60 percent of which were in heavy industry specifically.

Sixty-three percent of developers and users reported that their application resulted in quality improvement and fifty percent reported productivity improvement (Table 1). Twenty-seven percent of respondents could not determine or did not see any benefits from some of their applications. Over sixty percent of the respondents rated the success of their applications as high or very high. Less than twenty percent of respondents rated the success of their applications as low or very low.

Table 1. Percentage of Developers/Users Reporting Specific Benefits.

Benefit	Percent
Quality	63
Productivity	50
Reliability	41
Labour	32
Maintenance	23
Safety	18
Energy Use	14

The following is a summary of the benefits and performance of intelligent system applications evaluated in this study.

Productivity in Control

Improved stability of the process, quick adaptation to process changes and feed variation, and improved control of plant disturbances are the most frequent benefits reported from the implementation of intelligent systems. Other reported benefits are elimination of the effects of shift changes, reduced duration of work stoppages and increased efficiency of materials flow.

Productivity in Monitoring

The most frequently reported benefits from the implementation of intelligent systems for process diagnosis are improved, more consistent decisions by non-experts, decreased quality variation in the product, expertise being kept in the company, and reduced downtime and repair time.

Productivity in Scheduling

Increased productivity due to improved manpower scheduling, the ability to reschedule production for new, urgent orders, and improved schedule quality such as reduction of machine waiting time, are the most frequently reported benefits of intelligent scheduling.

Quality

Better training of new staff and service personnel, development of better understanding of process issues and equipment behavior, and improved diagnosis of quality problems are examples of the quality improvement gained from the intelligent systems.

Energy

Typical operational benefits reported by high energy users (Iron and Steel, Cement, and Metallurgical sectors) from implementing intelligent systems are: reduced peak and average processing temperatures, more stable and efficient furnace operations, improved consistency of heat control, reduced peak demand electricity charges, reduced discharge of nitrogen oxides and other emissions.

Conclusions

This study showed that although they are relatively new to the industry, AI technologies offer attractive solutions for development of advanced control systems, management of production workflow, and training of staff. In addition to productivity and quality improvements, intelligent systems can significantly reduce energy use in energy intensive operations, through better control and scheduling of production, and reduction of work stoppages.

There are a number of reasons for the differences in the frequency of application types (Fig. 1), e.g. accessibility of the technology, how easy it is to use or duplicate, benefits to be accrued, associated risks, complexity of the problem, etc. For example, control is the most frequent application type because it can be well defined and there are many solutions for duplication. Design, on the other hand, is more complex, benefits are more difficult to quantify, and there are few existing solutions which are easy to duplicate. In general, therefore, there is no simple correlation between the frequency of the application type (Fig. 1) and the benefits accrued (Table 1).

Acknowledgment

The support of National Resources Canada, CANMET, for this work is gratefully acknowledged.

References

Beaverstock, M., et al. (1992). Neural network helps G-P Ashdown mill improve brownstock operation. *Pulp & Paper*, **66**(9), 134-136.

Britton, S. (1987). Computer-based expert aids underground mine planning. *Coal Age*, 69-70.

Broussard, A. (1992). Model-based expert systems for mineral processing. *World Mining Equipment*, **18**(4), 34-37.

Choi, T.H., et al. (1991). An expert system to aid operation of blast furnace. *Proc. of the IFAC Workshop on Expert Systems in Mineral and Metal Processing*. Espoo, Finland. pp. 45-49.

Grigoriu, M.M. (1990). Expert systems increase competitiveness in the paper industry. *Paper Technology*, **31**(12), 18-24.

Hatanaka, K., et al. (1993). Breakout forecasting system based on multiple neural networks for continuous casting in

Jamsa-Jounela, S.L., et al. (1989). A simulation study of expert control system for a phosphate floatation process. *Proc. of the IFAC Workshop on Automation in Mining and Metal Processing*. Buenos Aires. pp. 45-52.

Steel production. *Fujitsu Scientific and Technical Journal*, **29**(3), 265-270.

Karhu, L. (1992). User's experience of Outukumpu expert systems at Outukumpu plants. *Powder Technology*, **69**, 61-64.

Kawata, H., et al. (1990). Application of expert system to wet felt design. *Proc. 1990 Engineering Conference*. Seattle, WA. pp. 367-372.

Kelly, E.B., et al. (1991). SlurryMINDER: a rational oil well completion design module. In *Innovative Applications of Artificial Intelligence*. AAAI Press. pp. 193-215.

Kojima, S., et al. (1991). A scheduling expert system for paper production. In *Operational Expert Systems Application in the Far East.* Pergamon Press. pp. 122-132.

Kowalski, A., et al. (1993). Pitch Expert: a productivity improvement tool for pulp production. *Computers in Industry*, **23**, 110-116.

Lee, J.K., et al. (1991). UNIK-PC: a crude oil delivery scheduling system. In *Operational Expert Systems in the Far East.* Pergamon Press. pp. 109-121.

Stohl, K. (1991). Development of a scheduling expert system for a steel plant. *Proc. of the IFAC Workshop on Expert Systems in Mineral and Metal Processing*. Espoo, Finland. pp. 39-44.

Vagenas, N. (1991). PROGNOS: a prototype expert system for fault diagnosis of the transmission system of load-haul-dump vehicles in Kiruna Mine, LFAB. *Proc. of the IFAC Workshop on Expert Systems in Mineral and Metal Processing.* Espoo, Finland. pp. 173-178.

PRO-SKED — A TECHNOLOGY FOR BUILDING INTELLIGENT TRANSPORTATION AND PRODUCTION SCHEDULING SYSTEMS

Llewellyn Bezanson and Robert Fusillo
Setpoint Inc.
14701 St. Mary's Lane
Houston, TX 77079-2995

Abstract

A new methodology has been developed for refinery and transportation scheduling. It is designed as an integral piece of a plant-wide information and control system. Pro-Sked is the technology developed to implement this methodology. Pro-Sked consists of an integrated environment used to assemble knowledge-based scheduling systems. It is implemented on an object-oriented, open-system platform that facilitates the management of external databases and equation solvers. At the core of a Pro-Sked application is a model of the physical plant by which the plant is simulated and scheduled. The model combines an event-driven process simulation, heuristics to guide the schedule, and a graphical user interface that allows the user to develop a schedule and iterate through scenarios. The scheduling model receives economic guidance from the plant's planning model and can call optimization submodels to calculate crude and product blends. Pro-Sked is designed to integrate tightly with a relational database that maintains plant data, oil movement and production targets, and sets of provisional schedules. At present, applications have been focused on the oil refining industry, including transportation and production scheduling. The architecture, data model integration, and modeling techniques of Pro-Sked.

Keywords

Scheduling, Hybrid knowledge-based systems, Event-driven simulation.

Introduction

The scheduler of a manufacturing plant or transportation facility faces many competing demands. His performance is judged on how well the schedule meets the economic targets set in the monthly plan. The schedule is also judged according to whether it conforms to operating objectives related to the economic targets: meeting shipping and receiving schedules, quality constraints for blended products, and purity constraints for high-value materials shipped via pipeline. The schedule must be physically feasible, satisfying the physical constraints of the facility, such as pumping and ullage limits, and operably feasible, keeping mode and tank switches to a minimum.

A decision support system for scheduling should help the scheduler keep track of these constraints over the long and short terms. Over the long term, the scheduler needs to make projections to check the suitability of feedstock receipts that have long lead times. In the shorter term, the scheduler generates a set of orders to be performed by the operating personnel: pumping line-ups, tank switches, process unit cutpoint and throughput changes, blend orders, and shipping/receiving instructions.

A schedule, like the scheduler who creates it, must be flexible. The scheduler frequently must respond to changes in the marketplace or the physical plant that affect the schedule. A design objective of Pro-Sked is to facilitate the re-scheduling process by providing an environment that allows the user to manipulate the schedule at a very high level of detail.

Artificial Intelligence Applied to Scheduling

The scheduling engine of Pro-Sked is not fundamentally rule-based, nor is it based primarily on mathematical programming technology. Pro-Sked is a hybrid technology, designed to facilitate the application of the scheduler's knowledge in an integrated environment. Pro-Sked represents and applies domain-specific knowledge in a variety of ways, including:

It provides an event-driven simulator, implemented in an object-oriented environment. Knowledge about the elements of a processing system is encapsulated in methods that are invoked to simulate the behavior of the process elements. These methods can contain knowledge applicable to a class of units or specific to an individual unit.

Through "whenever-rules," Pro-Sked can automate some low-level scheduling decisions. Whenever rules are rules that monitor a variable or class of variables. For example, whenever rules might monitor the inventories of one or more gasoline blending components and create blend orders for finished gasoline when a component's stock level gets too high.

Through its event hierarchy, Pro-Sked can represent the macro-level "handles" or "control points" through which the scheduler creates or adjusts the schedule. These macro-level events capture much domain-specific scheduling knowledge that can be applied in interactive or automatic iteration modes.

Pro-Sked consists of a graphical user interface, an object-oriented knowledge base, a relational database, and numeric equation solvers. The graphical user interface provides a high level, interactive environment for the end-user to control the flow of information and operate the system. The knowledge base is used to capture domain-specific knowledge for scheduling and operating a particular facility. The numeric equation solvers are used to calculate optimal blend recipes based upon inventory projections of materials. Pro-Sked is implemented with a mixture of substrate software. Its graphical user interface, model base, and simulation engine are implemented in the G2 object-oriented development environment. G2 provides user-interface building components, rule-based and procedural programming capabilities, and bridges to external software. Pro-Sked's standard relational database management software is Oracle, and other database management systems can and have been integrated with Pro-Sked. External optimization submodels have been implemented in the GAMS modeling language. GAMS provides continuous and mixed-integer linear and nonlinear programming capabilities, and can call a variety of solvers.

Process Simulation

At the heart of Pro-Sked is a simulation of a manufacturing plant or pipeline system. This simulation model must be different from one found in a planning (LP) model, because it explicitly represents time. It is also different from a continuous-dynamic (differential equation-based) model that might be used for design purposes, because it explicitly represents discrete decisions that are made in operating the physical system. The discrete-event architecture offers some attractive characteristics.

- The control structure is transparent, making it easy to add functionality to the modeling system.
- Each scheduling decision is explicitly represented by a data object that is available for archiving, revision, and re-use. This leads to the following benefits:
  - The scheduler produces a list of movements, operating instructions, and blend orders to pass on to operating personnel, a database, or the plant automation system.
  - Backtracking is possible. If the scheduler reaches a "dead end," the user can selectively revoke scheduling decisions, thus backtracking to a prior decision point.
  - Schedule playback and partial re-use of schedules are enabled.
- Time can be divided into arbitrarily fine slices. This is useful for scheduling oil movements in the near term, where detailed tank allocation decisions are necessary.
- Event-driven simulations are more efficient than time-driven simulations when applied to problems with discrete decision variables (Bryant, 1993).

Pro-Sked's representation of events include the following classes of event:

- *material-movement*. Describes the transfer of material from one location to another, e.g., into/out of tanks, into/out of pipelines, or between process units.
- *operating-instruction*. Describes an operating instruction for a process unit. Subclasses include *start-unit-op*, to start up a process unit, *end-unit-op*, to shut down a process unit, *change-mode*, to change the operating mode of a unit, and *change-throughput*, to change the flow rate into the unit.
- *blend-order*. Describes an instruction to make a blend, which could be an intermediate process material or a finished product. There is a parallel class of blend order for crude oil.

- *receipt events and shipment events.* Describes a proposed parcel of material to be received or shipped.
- *simulation-event.* This class exists purely for the sake of the simulation. Events of subclass *accounting-update-event* are scheduled at intervals specified by the user (e.g., once every four simulated hours). These events drive the simulation by computing the yields of process units, the volumes, properties, and compositions of tank contents, and the volumes and positions of pipeline batches.
- *production-train-event.* This represents a macro-level instruction. It is used, for example, to change the operation of a group of process units in order to implement a campaign to produce a material or group of materials. Production train events facilitate the practice of material-dominated scheduling (Taylor and Bolander, 1991).

Pro-Sked's simulation features a variable time flow mechanism (Ravindran, et al., 1987). With this next-event time flow mechanism, time is incremented between consecutive events. This was chosen over a uniform time flow mechanism because of the irregular timing of events. In the near term, when a detailed schedule is needed, the events can be timed to arbitrarily fine fractions of an hour. In the provisional period, when detailed sets of tasks are less useful, then larger time periods can be skipped between events, saving computer time.

Pro-Sked's Simulation Event Manager processes events in order of event time. The Simulation Event Manager routine selects the next event and removes it from the event queue. The Simulation Event Manager reads the event time of the next event and posts that time value on the master schedule clock. The Simulation Event Manager calls the subprocedure defined as the simulation method of this class of event. Side effects of processing this event may include updating the system or model parameters and establishing logical relations to describe the changes that the event produces. Events, when processed, also spawn new events, which are added to the event queue.

Modeling Process Operations: Control Points and Iteration

Pro-Sked applies scheduling calculations that are guided by the process structure. A process structure consists of process units, stages, and production trains (Taylor and Bolander, 1991). A process unit is a basic processing step. Process units can be combined into stages with stages separated by inventory. A combination of stages constitutes a production train, in which a set of materials are simultaneously produced. Within oil refining, several materials are produced simultaneously, and multiple combinations of process stages and production trains exist. Production trains are fixed within given periods, but may change during the course of a schedule. Hence the same set of process units may be used in several different production trains.

Pro-Sked applies a mixture of forward and reverse flow scheduling concepts. Forward scheduling is characterized by being supply side driven. Starting with the initial processing step, operations are scheduled forward through the process structure in order to meet raw material supply commitments. Reverse flow scheduling is characterized by being demand side driven. Starting with the last processing step, production is scheduled back through the process structure in order to meet product delivery commitments. In oil refining and transportation applications, combinations of supply and demand commitments exist. Furthermore, the commitments may change depending upon where these exist within the scheduling period. For example, an oil refinery may be able to project long term when products will be available for delivery, depending upon the processing requirements for a given crude supply schedule (forward scheduling). However, short term product delivery commitments may mandate when specific crude types are needed (backward scheduling).

To implement mixed-flow scheduling, Pro-Sked combines a hierarchical event structure with an interactive simulation environment. Fig. 1 depicts this hierarchy of events. The top-level events in the hierarchy are used as drivers from which several scheduling decisions, each represented by an event, can be spawned. At the top of the event hierarchy are receipt and shipment events, representing the scheduled arrival of feedstocks and the scheduled offtake of products. As these events are processed by the simulation manager, production train events are spawned in order to maintain material pool constraints. As production train events are processed, operating instructions are spawned to change process stage configurations and unit operating modes. Material movements are spawned to change feed sources and rundown destinations, and to manage individual tank usage. The hierarchical approach provides the scheduler with the ability to manipulate several operating decisions by adjusting a few, high-level, control points.

Fig. 2 depicts the iterative process of producing a schedule. The user interacts with the Scheduling Event Manager to view and edit scheduled events. The Scheduling Event Manager places selected events into the Nominated Event Queue. Upon initiating a simulation run, the Scheduling Event Manager transfers events from the Nominated Event Queue to the Master Event Queue. The Simulation Event Manager processes each event in the queue, as described above. The Simulation Event Manager places successfully processed events in the Processed Event Queue. Events that are not successfully processed may be rejected, or recycled to the Master Event Queue for processing at a later time. At any time during the simulation, the user can view or edit events in the Master and Processed Event Queues. Upon a reset of the simulation, the Scheduling Event Manager allows the user to select events from the Processed Event Queue for recycle to the Nominated Event Queue.

The scheduling loop of Fig. 2 allows the user to schedule and reschedule through iterative repair (Zweben, et al., 1993). The first pass through a schedule is not likely to satisfy all of the scheduler's objectives. The scheduler may wish to improve the operable feasibility of the schedule, or the schedule may violate inventory constraints. Iterative repair involves altering the schedules of high-level events (shipments, receipts, or production train events) in order to improve the schedule. Algorithms can be applied to automate the iterative repair of schedule.

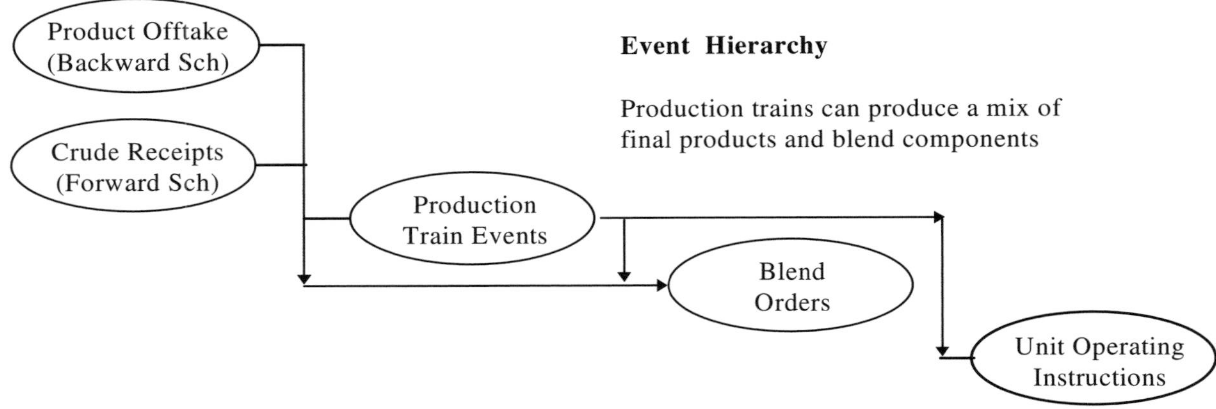

Figure 1. Event hierarchy in Pro-Sked.

Graphical User Interface

Through our conversations with refinery and transportation schedulers, we have learned that "black box" schedule solutions are not easily trusted. These schedulers want the ability to override any scheduling decision made by the system. They also want to be able to create different scenarios, testing the effect on the schedule of altering nominated receipts and shipments. For this reason, Setpoint considers the graphical user interface to be an essential piece of Pro-Sked. Pro-Sked's GUI includes the following elements:

- Motif-like controls to manage data flow and control the simulation.
- Pull-down menus for navigation and control of the system.
- Interactive Gantt charts to review and edit schedule of events.
- Interactive dialogs for editing production orders, material property specifications, maintenance activities, etc.
- Message boards used to contain status and warning messages.
- User configurable charts of inventory and property profiles.

Relational Database

Fig. 3 depicts the data integration between a typical Pro-Sked knowledge base and its relational database. The relational database helps to maintain data integrity and facilitates case management and data exchange with external systems. Fig. 3 shows these categories of data:

- Planning data, consisting of the target objectives to be achieved by the schedule. This can consist of quantities of raw materials, finished products, unit capacity utilization.
- Baseline data, representing the state of the facility at the schedule start time. This consists of tank inventories, active movements, unit operating conditions, current balances against planning targets.
- Definitive events, which make up the currently approved schedule. The blend orders, operating instructions, and material movements that comprise the definitive events can be transmitted to operating personnel for execution in the plant.
- Provisional events, which make up provisional schedules that extend beyond the approved schedule period. These events consist of receipts, shipments, production train events and blend orders. Fig. 3 shows multiple provisional schedules, each of which represents a separate case.

Producing the Schedule

The user begins to produce the schedule by retrieving the current sets of baseline data and definitive events. The user also retrieves business plan data, containing volumes of crude oil to receive and process, and volumes of finished product to make.

The scheduler uses Pro-Sked to perform material-dominated scheduling. Pro-Sked schedules material production over the entire period to meet the business plan. Once the material production is scheduled, Pro-Sked performs more detailed equipment scheduling, in terms of tank movements, operating conditions, unit feed rates, and blend orders.

Through the graphical user interface, the scheduler interacts with the Scheduling Event Manager, adjusting events in the schedules, controlling iterations of the schedule simulator, and creating new provisional schedule cases. The Scheduling Event Manager controls data flow to and from the relational database. Iterations continue until the scheduler is satisfied that the criteria of physical feasibility, operational feasibility, and satisfying the business plan are met.

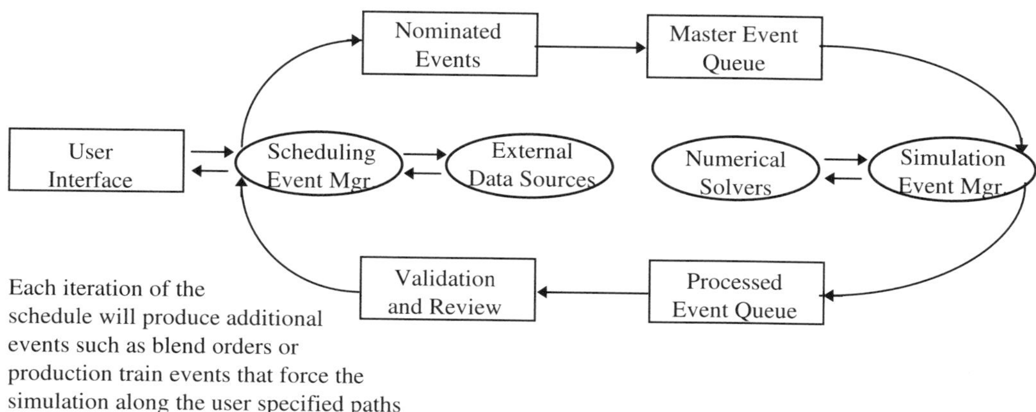

Figure 2. The iterative process of schedule production.

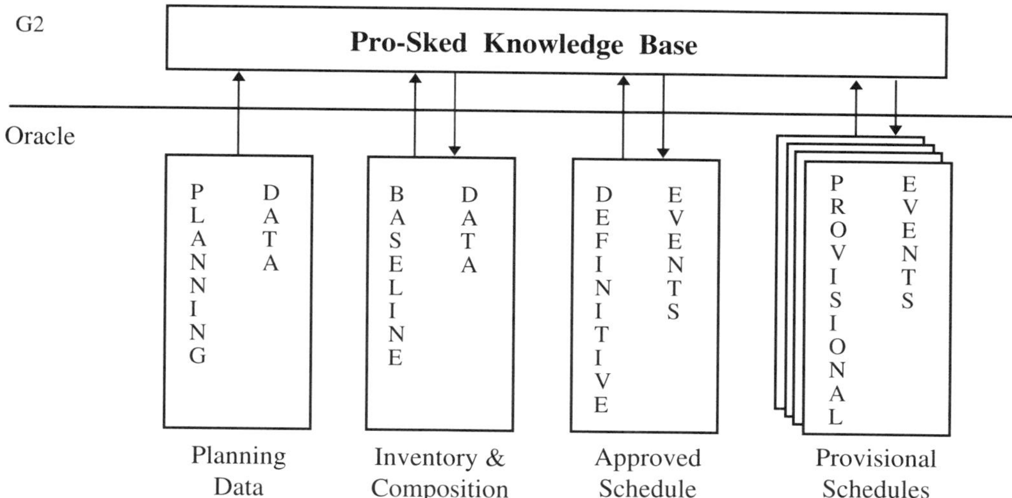

Figure 3. Data integration in Pro-Sked.

Conclusion

Pro-Sked is a hybrid of several technologies: AI, simulation, mathematical programming, and relational databases. It is designed to facilitate the application of a scheduler's knowledge in an integrated environment. The present paper has focused on the aspects of Pro-Sked that help the scheduler to focus his knowledge on producing the schedule: the event hierarchy, the iteration process, integration with an external database, and the user-interactivity. To date, Setpoint has applied Pro-Sked to petroleum refineries and pipelines. We believe, however, that this technology will also prove appropriate for product pipelines and chemical plants.

References

Ravindran, A., D.T. Phillips and J.J. Solberg (1987). *Operations Research Principles and Practice*. New York: John Wiley and Sons. p.382.

Taylor, S.G. and S.F. Bolander (1991). Process flow scheduling principles. *Production and Inventory Management Journal,* **32**, 67-71.

Bryant, G.F. (1993) Developments in supply chain management control systems design. In D.W.T. Rippin, et al. (Eds.), *Proc. Second Intl. Conf. on Foundations of Computer-Aided Process Operations.* CACHE.

PLANT MONITOR: AN ON-LINE ADVISORY SYSTEM FOR MONITORING POLYETHYLENE PLANTS

T.S. Ramesh
Mobil Research and Development Corp.
Princeton, NJ 08543

B.V. Kral
Mobil Chemical Co.
Beaumont, TX 77704

Abstract

The Plant Monitor is an on-line advisory system for monitoring and diagnosing polyethylene plants. The system monitors about 1400 controllers and sensors in the reactor and compounding zones of the plant on a 2 minute frequency, and notifies control room operators of abnormal conditions. The system can detect deviation from technological limits, process problems, and instrumentation faults. The Plant Monitor has been in operation since January, 1994 and is used by control room operators on all shift schedules. This paper describes the scope and functionality of the system, the design principles on which it is based, implementation details, and system performance. The paper also describes the way the project was managed, and the lessons learned on how to introduce advisory system technology into an operating plant.

Keywords

Advisory systems, Expert systems, On-line, Process monitoring, Diagnosis.

Introduction

In recent years, with the drive for greater profitability, the process industries have been placing greater emphasis on improving process performance, operating closer to constraints, and reducing downtime (Benson, 1992; Pekny, et al., 1991). Several technologies have helped to make these goals a reality: process modeling, statistical quality control, optimization techniques, and advisory systems (Calandranis, et al., 1990; Davis, et al., 1991; Rehbein, et al., 1992; Rowan, 1989; Sanders, 1991). The technology focus of this paper is on on-line advisory systems. These are essentially systems which have the capability to automatically collect, analyze and interpret process data, providing the results in the form of on-line advice to production staff, in a timely manner (Ramesh, 1993). This paper describes an on-line monitoring and advisory system, called the Plant Monitor, developed for the Mobil Chemical polyethylene plant located in Beaumont, Texas. The paper discusses the technical details and conceptual development of the system as well as the practical project implementation aspects.

Project Goals

The Plant Monitor project was initiated with the object of building an advisory system that could:

1. Provide early warning of faults before they develop into process upsets
2. Verify adherence of the process to technological and regulatory limits
3. Monitor a large number of variables at a resolution and time window beyond normal human capacity
4. Operate continuously and robustly, and display advisory messages to control room operators

5. Function as a server, providing results to other users on the local area network
6. Be maintained and supported almost entirely by plant personnel
7. Be migrated to other areas of the plant easily.

Project History

The Plant Monitor project was initiated in March, 1993. This was the first attempt to apply advisory system technology in the plant. The economic justification of the project was based on the objective of avoiding at least one reactor shutdown and two lots of off-grade product per year. Implementation started in May, 1993. The system was installed in the control room in January, 1994. Enhancements to the system were completed in May, 1995. All development and implementation work was done on-site. The production team was involved from the start, providing on-going feedback on scope, functionality and directions.

System Description

Scope and Functionality

The Plant Monitor covers the entire low pressure polyethylene unit, including feed metering, reactors, purge bins, product discharge, comonomer recovery, and compounding sections. The total number of sensors and controllers monitored is about 1200. The system monitors the process at a 2 minute frequency to detect the following types of events:

1. basic faults, e.g., bad input, unusually high values, abnormal rate of change, values out of statistical process control, controller output model failure
2. control system faults, e.g., controller failure, actuator failure, controller saturation
3. process problems, e.g., product transfer problem, comonomer recovery problem, reaction rate problems

The system partitions advisory messages based on their importance. Advisory messages are time-stamped and logged to file if needed. Audible alarming is provided as an option. Analytical functions are available for tracing the system's messages in detail. This is useful for the production engineer, in understanding the origin and rationale of the system's advice. The Plant Monitor also performs self-diagnostics and detects failures of its own data communications interfaces. The system includes maintenance and testing utilities to support modifications and updates.

Design Principles

The Plant Monitor is based on the following key design principles:

1. The foundation of a reliable on-line monitoring system lies in the ability to robustly identify low-level process events, called *basic faults*. The identification of basic faults is based on the causal and temporal behavior of individual sensors and controllers.
2. In order to robustly detect basic faults, data must be sampled at a resolution and time window appropriate to the sensor or controller being monitored.
3. Combinations of basic faults provide distinctive patterns that characterize the abnormal behavior of the process.
4. The identification of events should satisfy two conditions: elimination of false negatives, and minimization of false positives.
5. The structure of the system should support selective suppression of conclusions in a straightforward way. Conclusions should be suppressed: a) for sensors which have a temporary problem, b) for entire process areas when there is a unit shutdown, and c) when higher priority conclusions are found.
6. In order to be a credible tool for production staff, the system should also be able to: scan a large number of variables, beyond normal human capacity; look ahead into the future (prediction capability); filter conclusions intelligently (e.g., predicted events are suppressed if the actual event is detected); and display certain conclusions only if the corresponding events exist for certain periods of time.

Architecture

The structure of the Plant Monitor system is modular and layered. Raw data from the plant is transformed into useful advisory messages through a sequence of processing steps. Each major processing step is handled by a dedicated module. Figure 1 shows the overall architecture.

The data acquisition module retrieves current snapshot values for all the monitored variables every cycle, into an Excel spreadsheet. The module is implemented using a Mobil-developed product called Datanet, in conjunction with Excel macros. A 30-value history is maintained in the spreadsheet for each variable, but the interval between each data point is dependent on the time constant of the variable. The interval varies from 2 to 16 minutes, for a total time window of between 1 hour to 8 hours. The time constant for each variable was determined by reviewing 6 months of operating data.

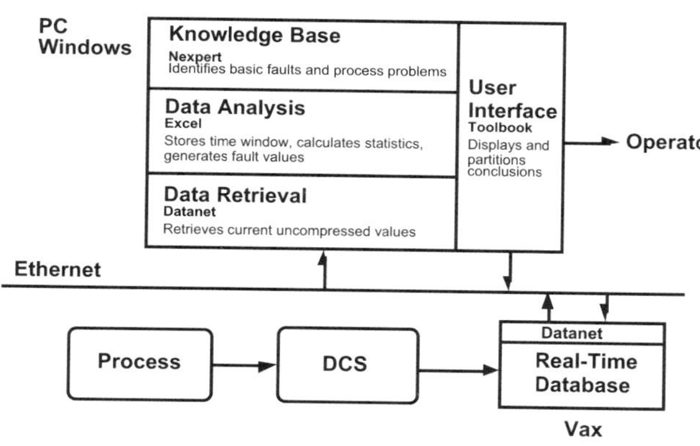

Figure 1. Overview.

The data analysis module, implemented in the Excel spreadsheet, performs validation and interpretation of the data. Validation consists of screening the raw data for retrieval errors, violation of physical limits, absence of noise, and violation of consistency checks. Consistency checks include physical and analytic redundancy checks. Data interpretation involves conversion of the validated data into descriptive labels called *fault values*. This is done by comparing the validated data against static limits, e.g., LOW_RANGE, HIGH_RANGE, MAX_CHANGE; dynamic limits, such as UCL and LCL based on 3 standard deviations; and applying process models, consisting of linear and non-linear equations that relate controller output to the value of the controlled variable. Examples of fault values are INPUT_OUT_OF_RANGE, OUTPUT_OUT_OF_RANGE, DEV_X_BAR_SPC, BLIP, and OUTPUT_MODEL_FAILURE (where BLIP refers to a sudden peak, and DEV_X_BAR_SPC refers to deviation outside the X bar SPC limits). The fault value labels can take values such as LOW, HIGH, TRUE or FALSE.

All process variables which have at least one fault value set to TRUE, HIGH or LOW are sent to the fault detection module for further evaluation. The fault detection module combines fault values into two types of conclusions: generic and process-specific. Generic conclusions are the result of applying a process-independent set of pattern-matching rules across all sensors and controllers based on their fault values. The sensors and controllers are organized into classes and sub-classes. The first layer of these rules simply generates a set of variables meeting each fault value condition. These are the basic faults referred to earlier. Examples include "Dead signals," "High rate of change," "Unusually high controller output," and "Model calculation failed." The first layer of rules also screens out basic faults that are not relevant to the context, e.g., disabling all basic faults for a reactor that is currently down. Combinations of basic faults result in *generic problems*, which capture some significant general behavior of controllers and sensors. Examples include "Predicted value out of range," "High deviation because output action failed," and "High deviation and high output." The rules associated with generic conclusions are applied to entire classes of sensors and controllers, and are therefore very few in number.

Process-specific conclusions address specific problems that have a distinguishable pattern of basic fault combinations. The system was designed to address an initial list of such problems, but many more were added based on production department feedback. Rules which detect process-specific problems use a combination of basic faults, generic problems and individual fault values which describe the process condition being detected. In many cases, the fault values used belong to pre-existing classes, so they do not have to be explicitly enumerated. Examples of process-specific conclusions are "Comonomer recovery unit not fully loaded," "Suspect plugged pressure tap," "Reactor rate-limited by rotary feeder," and "No product transfer from product purge vessel." The process-specific rules are designed conservatively, so that false negatives are avoided, and the number of false positives minimized. Figure 2 shows the structure of the rules in the knowledge base. The rule structure, together with the class-subclass structure of sensors, controllers, process areas and conclusions, supports the selective suppression and filtering of conclusions in an easy way. The total number of rules in the fault detection module is about 300; the number of classes is 50; the number of objects, excluding conclusions, is about 1400; and the number of different conclusions is about 150.

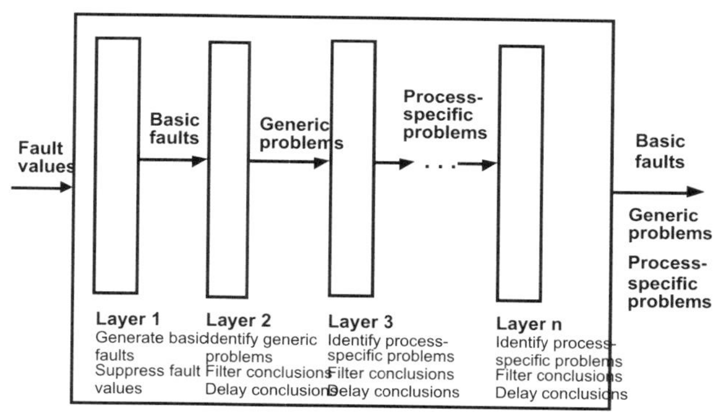

Figure 2. Knowledge base structure.

The modular approach used in the Plant Monitor is both powerful and robust. The knowledge base is structured so that increasingly detailed conclusions can be drawn from the basic faults identified. This is different from conventional rule-based expert systems where the focus is on specific problems collected from experts, leading to what is known as the *brittleness problem* (Ramesh, 1993). When brittle systems encounter situations outside a predetermined set, they are unable to generate any meaningful conclusions. By contrast, if the Plant Monitor encounters situations which do not match its predefined set of process-specific conclusions, it can still identify basic faults, controller problems and low-level process problems, which can serve as useful pointers to production staff in resolving the situation. In addition, the Plant Monitor's modular structure supports incremental updates of the knowledge, facilitating the incorporation of additional process troubleshooting expertise over time.

The user interface displays the conclusions generated by the fault detection module. Messages are partitioned into separate windows by processing zone and importance. The partitioning is defined at the knowledge base level by means of classes representing priority and process area. The two zones are Reactors and Compounding. Within each, the window categories are Problem Warnings, Faults, Low Level Faults, Data Errors and Info. The first four windows display conclusions of progressively lower importance to the operator. The Info window displays messages of general value, e.g., "Reactor kill," and "Compounder down."

Hardware and Software

The LP Monitor system is implemented on a Pentium PC running Microsoft Windows 3.1. The software tools used include Datanet (a client-server DDE-based data acquisition module developed by Mobil), Excel 4.0 (a spreadsheet from Microsoft), Nexpert 2.0 (a knowledge-based tool from Neuron Data), and Toolbook 1.5 (a user interface development tool from Asymetrix). The total cost of the hardware and software (approximately $10,000) is insignificant (less than 10%) compared to the invested development effort.

Conclusions

Validation and Testing

The system was tested in three main areas:

1. Correct generation of fault values
2. Validation of individual rules
3. Validation of final conclusions, including clarity of messages, appropriate message suppression, message filtering, and time delays.

System testing and validation became an integral part of the development effort as soon as the first functional version of the data analysis module was ready. The testing and validation effort was greatly facilitated by the modular design of the system, since individual modules could be tested independently of each other, in addition to integrated tests. A number of system utilities were also developed during the implementation to speed up the testing effort. These included test functions for simulating system behavior using hypothetical data, as well as analytic functions for detailed tracing of system conclusions. A considerable amount of testing was also done on-line during the first two months after system installation. Some errors were found and fixed several months after installation. However, due to the modularity of the system and the layered knowledge base structure, these errors appeared only in unusual combinations of events, and did not hinder the normal performance of the system.

Performance

The Plant Monitor has been operating continuously on a 24-hour basis since January, 1994, with the exception of brief outages due to computer downtime, bug fixes, and scheduled system updates. The overall uptime of the system has been better than 97%. Despite the large number of possible advisory messages, very few typically appear on the screen. This is reflective of the normal, stable operation of the process. The frequency of advisory messages usually increases for a period following a system update. This is possibly due to the fact that operators test the credibility of the system after each update. After the reliability of the new messages is established, the operators tend to run the process tighter, resulting in the messages gradually disappearing. This often results in more economical or safer operation of the process. Message frequency and operator responses were recorded in detail during the first year of system operation, and a preliminary post-audit of the application was done in March, 1995. The post-audit revealed that the actual benefits of the system far exceeded the estimated basis used as justification for the project. Based on the post-audit, the payback period of the Plant Monitor system is 3 months.

Lessons Learned

Successfully introducing advisory system technology into an operating plant for the first time involves more than good design and implementation. Here are some useful lessons we learned from the Plant Monitor project that can help guide other similar efforts.

1. Successfully developing a monitoring system requires an implementation team that has expertise in advisory technology, insight into the process, and an understanding of the plant organization. This diverse mix of skills is vitally important.
2. If the project is a first-time application of the technology, risk tolerance is likely to be low. This factor should be used to guide project decisions regarding scope, functionality and platform. The decision to use a low-cost hardware/software platform in the Plant Monitor was based on risk. There are two benefits resulting from this approach. Firstly, the performance/cost ratio on the project increases. However, this may be a temporary benefit if the system has to be re-implemented

later on more powerful hardware. Secondly, and more significantly, the software tools chosen, although low-cost, are high-level. This enabled the plant engineers to quickly grasp the structure of the system.

3. The operating plant has to invest technical resources in the implementation. This is the only way to ensure that there is plant ownership and sustained maintenance of the system over time. In the Plant Monitor project, this was accomplished through a series of implementation blocks during which the plant technical engineer was 100% dedicated to the project. In the periods between the blocks, the plant engineer typically spent a much smaller portion of his time (perhaps 15%). There were a total of about 5 blocks in the 9 months prior to system installation, and about 4 more in the year following installation. Each block was about 2 to 3 weeks long.
4. The active involvement of the production department is important right from the start. This includes clearly identifying the needs and the performance criteria on which the success of the system is judged.
5. Acceptance of the system by control room operators is heavily dependent on the amount of effort expended in the weeks following installation. During this sensitive phase, the Plant Monitor required constant nurturing, in the form of testing, validation and fine-tuning, taking up almost 40% of the plant technical engineer's time in the first month, and 20% in the second month.
6. It takes time for an operator to appreciate the value of a monitoring system. Operators' responses to system messages improved after they had tested its credibility. Those operators who had provided input into system development accepted the system more readily.
7. It is important to emphasize that the system's messages are not alarms. Alarms require immediate action, while the advisory system's messages require evaluation. In this sense, the system should be viewed as an operator assistant, not a control system adjunct.
8. Operations personnel have to be reassured that the monitoring system is intended to monitor the process, not the operators. Operator acceptance is facilitated if it is made clear that there is no attempt to measure shift performance via the system.

References

Benson, R.S. (1992). Computer-aided process engineering — an industrial perspective. *Computers Chem. Engng.*, **16**, S1-S6.

Calandranis, J., G. Stephanopoulous and S. Nunokawa (1990). DiAd-Kit/Boiler: On-line performance monitoring and diagnosis. *Chem. Eng. Prog.*, **86**, 60-68.

Davis, J.F., et al. (1991). On-line diagnosis of process and manufacturing operations: the integration of knowledge-based, neural net and conventional numeric approaches. *Proc. Int. Conf. on Probabilistic Safety and Management (PSAM)*, Vol. 1. Beverly Hills, CA. pp. 589-594.

Pekny, J., et al. (1991). Prospects for computer-aided process operations in the process industries. In L. Puigjaner and A. Espuna (Eds.), *Computer-Oriented Process Engineering*. Elsevier, Amsterdam.

Ramesh, T.S. (1993). Expert systems. In J.A. Kroschwitz (Ed.), *Kirk-Othmer Encyclopedia of Chemical Technology*, Vol. 9, 4th ed. John Wiley, New York.

Rehbein, D., et al. (1992). Expert systems in process control. *ISA Transactions*, **31**, 2.

Rowan, D.A. (1989). On-line expert systems in the process industries. *AI Expert*, **August**, 30-38.

Sanders, F.F. (1991). Application of an on-line expert system for a continuous chemical process. *Proceedings of the ISA/91*. October.

OPERATOR SUPPORT FOR DIAGNOSIS IN A FERTILIZER PLANT

Are Mjaavatten and Kjetil Fjalestad
Norsk Hydro Research Centre
N-3901 Porsgrunn, Norway

Bjarne A. Foss
Department of Engineering Cybernetics
NTH, N-7034 Trondheim, Norway

Abstract

We describe a modular architecture for process information systems. Each process unit is represented by a separate module that is connected to its neighbors in the process flow sheet through connections that represent the process streams. The resulting system is easy to configure and to maintain. The scheme forms the basis for a successful operator support system that has been in operation in a fertilizer plant since November 1992.

Keywords

Diagnosis, Modularity, Fertilizer plant.

Introduction

Reducing the emission of pollutants is a major challenge to the process industries. This challenge was the outset for the system reported in this paper. Taking an overall view on the problem every plant consists of three main parts: the process plant itself, an information and control system, and the plant operation organization. All three parts will influence the emission of pollutants, hence reduction of emission may be invoked through all three channels. A typical measure on the process plant itself is to introduce material recycling and new cleaning technology. The information system can decrease emissions by improved control performance and by providing the plant operation organization with precise information on emission status and on the cause of violations. Finally, emissions can be reduced by educating and motivating plant personnel.

The focus at the Glomfjord plant in Glomfjord, Norway was on improvement through the development of an information system. The plant is a moderately complex chemical plant consisting of a large number of process units such as mixing tanks, distillation columns and heat exchangers. To be able to think constructively about a process plant of this size one must break down the information into manageable chunks. This generally means that some kind of hierarchical decomposition is needed.

The central theme of this paper is the development of a method to enable the construction of modular process information systems. We propose to represent each process unit by a single module, comprising both estimation and diagnostic information and routines. The models and diagnostic logic to be used should be determined by the type of process unit. A library of models for the required units should be available to the system designer. Using these building blocks it will be easy to build information systems for complex processes. The goal is to make a system that can be configured by connecting unit modules in the same way that they are connected in the process flow sheet. The method will be presented in the Method section. Further, the implementation is discussed in the Implementation section.

As mentioned earlier the focus at the Glomfjord plant was on improvement through the development of an information system. To gain full advantage of the improvement of one part of the overall plant it is usually necessary to harmonize other parts of the overall system. In our case this means plant personnel. We will, hence, in the Discussion section focus on worker participation in system development and implementation.

Investments are usually initiated to achieve some primary goals, in our case the reduction of emission. An investment, especially in information systems, will trigger a chain-reaction not necessarily anticipated prior to the project. This will also be a theme in the Discussion section.

Important references to this paper are Mjaavatten (1994) and Saelid, et al. (1992).

Method

A chemical process plant may be represented by a directed graph with the process units as nodes and the process streams as edges. The directions of the edges reflect the normal flow direction in the stream. We define the term *process topology* to mean the connection structure of this graph. The process topology is independent of the geometry of the plant, and neighboring nodes may well be situated on different floors or in different buildings. In this section we describe how a diagnostic search can be structured by the process topology. The aim of the diagnostic system is to detect and help explain all disturbances that negatively affect the environment or product quality, and propose actions to remedy the situation. We will assume that well-defined quality parameters exist, which allow plant management to set clear-cut quality limits that the process should be kept within. These limits will generally encompass product quality, energy consumption and pollutant discharge. We introduce the term *episode* to denote an event where one or more of these limits are violated. The role of the detection and diagnosis system is to detect all such episodes and to find the process unit where the disturbance started (the initiating unit). This search may be followed by a more detailed local diagnosis to find one or more possible *primary cause* of the disturbance. We also want to detect disturbances before they lead to episodes. These disturbances result in a *warning*. The role of warnings is to alert the operator and thereby avoid or minimize the episode.

The diagnosis system consists of four parts: episode detection, a diagnostic procedure that tracks disturbances to the initiating process unit(s), a set of decentralized estimators for estimating process variables, and an internal diagnosis scheme for each unit.

Detection

Detection of serious process disturbances is performed by *detection monitors*. These take the form of procedures that monitor the value of some measured or estimated variable, such as total release rate of a given pollutant. There are two types of detection monitors: episode and warning monitors. The episode detection monitors trigger when an episode is registered while the warning detection monitors trigger when a warning is registered.

Topological Search

The triggering of an episode or a warning must be followed by a search for the cause of the disturbance. This search uses another kind of monitor, called a *search monitor*. This monitor may test for disturbances in (the estimates of) all components of the process state vector, or in combinations of these components. Furthermore, the search monitor tests a time window, and not only the present state. Search monitors are passive until interrogated during a diagnostic search. Before describing the search monitor in more detail, we need to explain the diagnostic search procedure. The detection of an episode or a warning triggers a diagnostic search for possible initiating units. When a detection monitor is triggered, it starts a procedure that recursively searches for disturbances in upstream units and streams. The tests for disturbances are performed by the search monitors. In process units with more than one inlet stream the search branches out to search all inlet streams. The search is pursued only along streams where the search monitors report disturbances. When the search reaches a unit where no inlet streams report disturbances, this branch of the search stops. It is decided that a primary cause is possible to be found in this unit, since it cannot "blame" the disturbance on any upstream unit. The path from the triggering detection monitor to the initiating unit is called a diagnostic path. Diagnostic paths are highlighted on the operator's computer screen by changing the color of the stream and the process unit symbols. There are some possible problems associated with the search procedure related to controlled flows, the choice of time window for the search monitor, and feedback loops in the process. Details on this can be found in Mjaavatten (1994).

Estimator

Our approach depends critically on the availability of the current *process states* (flow, composition, temperature and pressure) in every stream, since the search procedure is based on comparing the process state to nominal values. Since it is impossible to measure the composition of every stream it is necessary to introduce model-based estimators to compute the process states. To comply with the modular structure of the system we want these estimators to be local in the sense that the estimator for a unit only depends on information, i.e. estimated process states, from neighboring units together with local measurements. If we further restrict this by limiting the information to that coming from neighboring upstream units as well as local measurements we obtain a convenient one-way structure for computing the estimates of all units. A limitation is that downstream measurements cannot be used to improve estimates in upstream units. This does introduce problems as discussed in Mjaavatten (1994). It should be emphasized that all "local" estimator schemes are sub-optimal compared to a structure where every unit has access to all measurements.

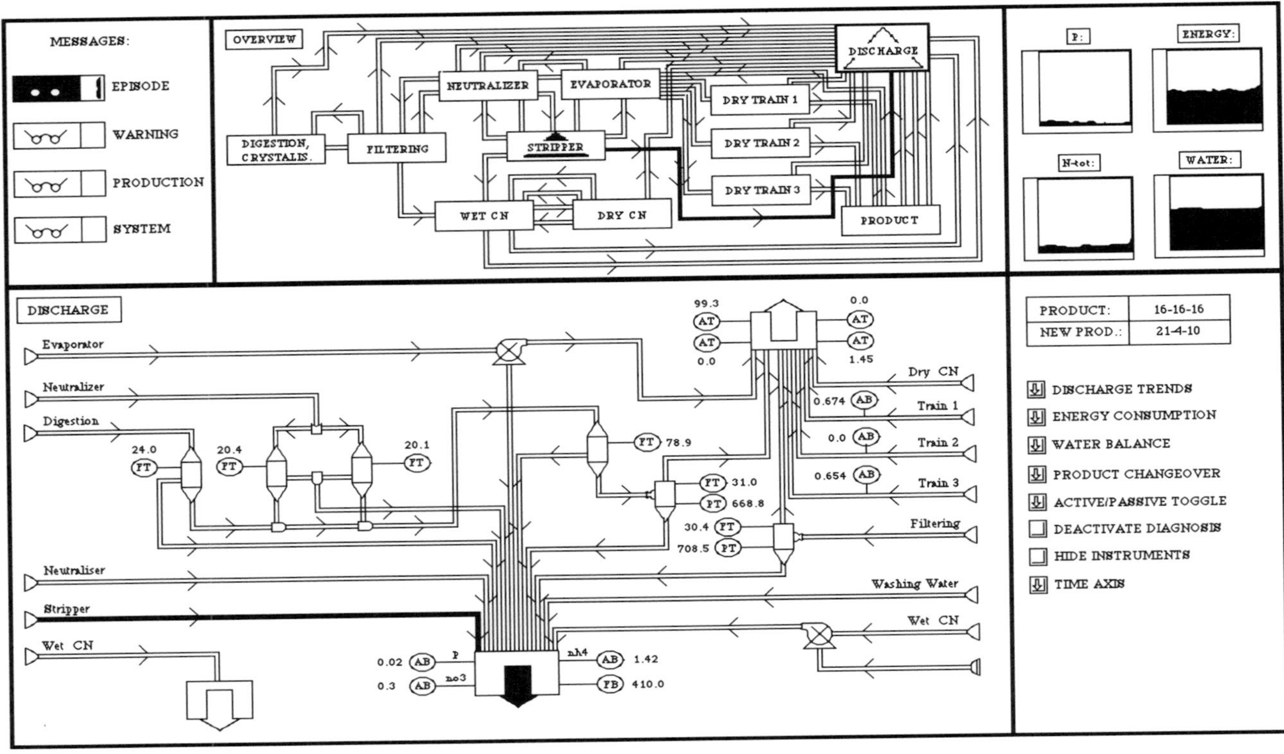

Figure 1. The OSS user interface. Process units, sections and streams involved in a pollution episode are highlighted.

Internal Diagnosis

For every initiating unit, an internal diagnosis is started to find the primary cause for the erroneous situation. The level of sophistication of this diagnosis may vary, in the simplest case the primary cause is immediately reported as "Problems with unit A," leaving the details to the operators.

Implementation

This section describes the Operator Support System (OSS) installed at Norsk Hydro's fertilizer plant at Glomfjord, Norway. The OSS is closely integrated with a Reporting and Documentation System (RDS). The compound system is called the Integrated Pollution Control system (IPC). The IPC system was described by Fjalestad, et al. (1994). The description in the present paper is limited to the implementation in the real-time expert system shell G2 (Gensym, 1992). We use the modular approach that is described in the Introduction. Each process unit is represented in the OSS as an object with a number of attributes. The attributes include estimation and diagnostic procedures and an icon for graphical representation of the unit on the computer screen. See for instance Booch (1991) for an introduction to object-oriented design.

In the OSS, both detection monitors and search monitors are objects. Each detection monitor is restricted to monitoring one value only. The detection monitors are polled every minute, to check for disturbances.

The OSS system has a graphical user interface as shown in Fig. 1. The central upper window is called the overview window. It gives a general view of all process sections. A process section may be activated by pointing and clicking in this window. A detailed view of the corresponding process section is then displayed in the area window at the lower left. The upper right hand window shows graphs of total discharge of pollutants together with energy and water consumption. The menu window at the lower right displays a set of buttons that trigger various functions. Messages from the system to the operators are displayed in the message window at the upper left.

The icons used for process equipment are modeled on the icons used in the DCS and in the P & I diagrams, so that operators can easily recognize the process sections and equipment. Most operations are done by pointing and clicking. The keyboards is used only for entering comments. In Fig. 1 an ammonia discharge episode in the sewer has been detected by the on-line analyzer. This is indicated by a bold (or actually red) arrow in the sewer symbol in the area window. The stream coming from the stripper section is also highlighted in red, and the triangle shown in the stripper section in the overview window indicates that the disturbance comes from this section. If the operator wants to show the stripper section in the area window he can either click on the triangle at the left end of the highlighted stream, or he can click on the stripper section in the overview window.

Discussion

The Operator Support System has been characterized as a success by the users and the company. Reasons for this will be reviewed in the sequel. In addition, the system's impact on plant performance as well as the company will be discussed.

System Performance

The system performs well in the sense that it gives the operators valuable advice on emissions and the cause of these. One reason for this is that the plant lends itself to the unit-process decomposition, the search procedure, and the estimator scheme which utilize local and upstream information only. This is because the plant has a predominantly one-way cause-effect flow. In plants with stronger two-way interaction between units this approach may not be viable.

The method utilizes plant models on two levels, a structural model of the process topology and mechanistic models of each unit. Hence, both structural plant knowledge and knowledge of the characteristics of each unit are captured by the models. This knowledge is merged with process measurements to form a system that efficiently utilizes all relevant knowledge. The lesson to learn from this is that it is important to develop diagnosis methods that are able to capture as much as possible of relevant knowledge of a plant.

Worker's Participation

An interactive operator support system will make little difference to plant operation without a commitment from the work force to actively utilize the system in plant operation. To obtain this, it is imperative that there is a close cooperation with the users during development commissioning, and modifications, and that a support system is based on the underlying concept that the user always should be in command.

Even though a computer based control system had been in operation for some time before the operator support system was introduced, this was a completely new tool. The system is quite complex, but this complexity is hidden from the users.

In the Glomfjord project plant personnel were informed continuously during the development phase and their views on the user interface and on which features to include were taken into account. The resulting system thus met many of the real needs of the users, as opposed to the needs perceived by the system designers. After installation, one person from the development team was stationed in Glomfjord for six months. During this period estimators and diagnosis were tuned, bugs were fixed and some new functions were added, based on discussions with the users. Items that turned out not to be useful were removed in order to avoid unnecessary confusion. This close follow-up was crucial to the success of the system.

The philosophy behind the diagnosis and advice part of the Glomfjord system is that the user always has the final say. This implied that the system was viewed as an advisory system and thereby used to the extent the individual operator felt it improved his/her decision making. Experience in Glomfjord has shown that this strategy was successful in the sense that all operators within short time actively used the system in their day-to-day work.

To elaborate further, there are some characteristics of the Scandinavian work environment that seem to favor the introduction of new and advanced concepts (Emery and Thorsrud, 1976; Qvale, 1992).

There is strong egalitarian tradition within the society. This simplifies communication between operators, system developers, and production management. Hence, it promotes the use of all relevant resources in the different phases of a project, i.e. development, commissioning and modification. Furthermore, the egalitarian educational tradition contributes to the existence of a well-educated work force. This, again, is important to take advantage of advanced concepts among workers.

The Norwegian Work Environment Act (§12) states that all workers have a right to participate in specifying his/her own work environment. This proclaims a general attitude which again is helpful in "Glomfjord-like" projects since all parties involved are obliged to contribute to the project (Gustavsen and Hunnius, 1981).

In traditional manufacturing industries, like Glomfjord, virtually all workers are members of the trade union. If used constructively this is an instrument to convey new ideas to the work force, and utilize advice from the work force. The reason is that it forms a well-established two-way communication channel between workers and management.

Impact

The impact of the project has been quite diverse.

There is a consensus among all the involved parties that a significant reduction of emission episodes can be related directly back to the diagnosis system (Mjaavatten, 1994).

The decentralized estimator scheme provides plant personnel with estimates of non-measurable variables, in particular quality variables. An example of this is the nitrogen to phosphorus ratio in selected parts of the process. These estimates are used to monitor product quality. This can be viewed as an add-on effect, caused by the choice of methodology.

The infrastructure represented by the Operator Support System will be used as a basis for implementing new functionality into the system such as planning and documentation. Since the operators have been through a period with major changes in their information system environment and this has had a positive impact on plant operation, further changes are in general welcomed. This is a major asset for the company since change will be the normal state of operation in the future.

The Glomfjord application has become an important reference for the Norsk Hydro company because of the significant reduction of pollutants. Having, in addition, received a prize from the Norwegian minister of environmental affairs for the project's environmental impact helps defining Norsk Hydro a.s. as a "Green Industrial Company" in the eyes of the public, the market, and the politicians. This, again, is an important asset in tomorrow's business environment.

The first effect can be seen as a direct consequence of the Operator Support System. The other effects can be viewed as add-on effects that were difficult to foresee. These follow-on effects, however, are significant and may, in the long run, be the most important.

Conclusions

A diagnostic tool for use in an operator support system for chemical process plants has been designed. This system uses a highly modular representation of the process, and it has been successfully implemented in an industrial operator support system.

References

Booch, G. (1991). *Object-Oriented Design with Applications*. The Benjamin/Cummings Publ. Co., Redwood City, USA.

Emery, F. E. and E. Thorsrud (1976). *Democracy at Work*. Nijhoff, Leiden, Norway.

Fjalestad, K., J. Gravklev, A. Mjaavatten, and S. Saelid (1994). A total quality management system for reduction of industrial discharge. *Comput. Chem. Eng.*, **18**, 369-373.

Gensym (1991). *G2 Reference Manual. Version 3.0*. Cambridge, MA, USA.

Gustavsen, B. and G. Hunnius (1981). *New Patterns of Work Reform: The Case of Norway*. University Press, Oslo.

Mjaavatten, A. (1994). *Topology Based Diagnosis for Chemical Process Plants*. Dr. Ing. Theses, NTH, Trondheim.

Qvale, T. U. (1992). Scandinavia: direct participation. In G. Szell (Ed.), *Concise Encyclopedia of Participation and Co-Management*. Walter de Gruyter, Berlin, New York.

Saelid, S., A. Mjaavatten, and K. Fjalestad (1992). An object oriented operator support system based on process models and an expert system shell. *Comput. Chem. Eng.*, **16S**, 97-108.

REFINERY IMPLEMENTATION OF AN OPERATOR ADVISOR SYSTEM FOR PROCEDURE MANAGEMENT

David W. Beach
BP Oil
Cleveland, OH 44114

Michael J. Knight
BP Oil
London, UK TW16 7LN

Abstract

This paper describes the implementation of an on-line knowledge-based system for improved management of planned transitions such as startups and shutdowns. The objective is to give the operator a better tool than paper-based checklists to improve the reliability of transitional operations. The system includes live process data monitoring, automatic record keeping, and a graphical user interface. It continuously monitors process data to verify normal completion or to advise the operator of unexpected conditions. The user interface provides a visual overview to the complete procedure and point-and-click access to specific details. The experience of installing the system for startup of a grass roots refinery unit is described. The attributes of the knowledge representation that make this approach attractive are discussed.

Keywords

Procedure management, Knowledge representation, Checklists, Knowledge-based systems.

Introduction

The Operator Advisor System that this paper describes has been developed and implemented for use by refinery operators and their supervisors to help manage transitional operations such as planned startups and shutdowns. These are normal yet non-routine operations and are often a demanding and stressful job for operators. Because these transitions are infrequent and involve non-steady state operations, a typical operator will not usually have a great deal of experience in executing the procedure. Potential problems with procedure execution include the lack of familiarity, widely distributed knowledge and experience, use of systems that are designed for steady state operation, and revisions to the process that result in procedure changes since the last time it was executed. The impact of improper procedure management can result in costly delays, damaged equipment, unexpected downtime, and even major disasters.

Paper checklists have traditionally been the primary tool used to manage procedures. One of the disadvantages of paper checklists is that each operator might be using his own copy of the procedure, maybe even a different revision of it. Some attempts to computerize the checklists have been made and these address some, but not all of the disadvantages of the checklist format.

The potential for problems with procedures is complicated by the increased communication load between the operating parties and the systems that are largely designed for steady state operation. The operator's attention is often focused on a task at hand, not on the rest of the process that he would be closely monitoring during steady state operation. The control system's alarm functions do not usually have the ability to dynamically adjust for the changing conditions expected during transition operations and are not as useful. In fact, the expected condition of many variables is often reversed from steady state and sometimes oscillates back and forth. Keeping up with unexpected conditions using the control system's alarm capabilities has not been successful. Increased verbal communication (via radio) is also a factor in complicating the execution of a procedure.

Many of the problem attributes that are seen with execution of a procedure directly map into the promise of knowledge based systems: scarce and distributed expertise, vigilant monitoring, centralized and maintainable knowledge, and "natural" representation of the knowledge. It was with this apparent match between characteristics of appropriate KBS applications and the problem attributes that BP chose to begin applying knowledge-based systems to the procedure management problem. Discussions began in 1990 with development of a framework in 1991 and 1992. The refinery application described in this paper was begun in 1992 and finished in 1993.

Framework Development

A rapid prototype approach with quick review by sponsors and potential users has been followed throughout the development. The user ownership is increased and the quality of the feedback is increased when there is "something to see" rather than just a concept to discuss.

The development has considered three procedures of increasing complexity. The first prototype was developed for the drum swing of a coker pilot plant. It was sufficient enough in scope to acquire sponsorship and write general specifications. The second prototype was for the drum swing of a commercial coker and focused on incorporating manual field actions. Lastly, there was a demonstration using the initial startup of a grassroots diesel hydrotreator. Each of these implementations added new requirements and the process provided a manageable evolutionary path toward a functional system.

The first prototype identified that we needed some help with the knowledge representation. A plant-scale system would quickly become a nightmare without a maintainable structure. We identified 5 levels of system functionality that addressed the general requirements. System development focused on these levels with each level becoming a foundation for the next. The 5 levels are described in Table 1. Each of these levels adds a level of complexity to the previous level. There is a significant difference between the complexity below level 3 and the complexity of levels 4 and 5. Implementation of a demonstration through level 3 using only knowledge-based technology was feasible and would provide a beneficial system with practical usefulness. Investigation of levels 4 and 5 is ongoing at Ohio State University (Davis, 1994; Marchio, et al., 1993; Marchio and Davis, 1995). It appears that the framework selected for implementation provides a sufficient platform for the diagnostic reasoning and corrective advice generation.

Table 1. Levels of Functionality.

Level 1	*Organization and computerized representation of operating procedures*
Level 2	*On-line procedure interpretation - detecting expected completion of a specification and responding appropriately*
Level 3	*On-line procedure interpretation - detecting completion failures, undesirable variable trajectories, and unexpected conditions and responding with appropriate advice.*
Level 4	*Diagnostic reasoning for root cause identification for the purpose of generating context-sensitive corrective action advice.*
Level 5	*Generation of corrective action advice based on the results of the diagnosis*

Procedure Representation

An object-oriented hierarchical organization was selected as the architecture to accomplish the Level 1 requirement for representation of an operating procedure. The levels in the hierarchy represent relative level of detail in accomplishing the operational goal that is at the top level.

The attribute template for each of the objects in the hierarchy contains fields for holding the appropriate details. Having a template greatly aids the knowledge acquisition and system building processes. After organizing instances of the objects into a hierarchy, the representation is completed by filling in the static attributes. The primary objects in the hierarchy are described in Table 2.

A hierarchical organization was chosen because it provides a somewhat natural collection of detailed executable steps into higher functional goals. It is easily updated and evolved and can be audited prior to use. We found that the object-oriented framework could serve as both the developer and operator interface. It is highly desirable not to have to build a separate operator interface. The hierarchical display reinforces the higher level functional goals. Although checklists are often written in outline form the higher level goals can be lost by the user because their focus is on the individual task at hand. An unexpected benefit of this representation was that the procedure author often revised his written procedure after reviewing the hierarchy. A higher quality procedure resulted.

Table 2. Hierarchy Object Definitions.

Plan	*A complete operating procedure for accomplishing a high level process objective*
Task	*A functional goal of a portion of the plan that collects together related sub-tasks and steps in a logical manner (also includes subtasks).*
Step	*Lowest level in the hierarchy representing an individual executable action or collection of highly related executable actions.*

There is no theoretical restriction on the number of layers in the hierarchy although we found that 4 layers (sometimes just 3) were practical and described the complexity of the hydrotreator startup. There is no reason to expect other procedures to match that complexity. An example hierarchy for the air-freeing operation is shown in the example user interface, Fig. 1.

Application construction makes use of the detailed checklists to group together related activities into a hierarchy. It was beneficial for us to construct the hierarchy in parallel with the checklist preparation. Construction of the hierarchy for an application is not an exact science but here are some rules we follow when decomposing a procedure into a hierarchy:

1. Group related activities so that steps are organized into operational goals.
2. Separate steps when they have unique data monitoring requirements.
3. Group similar steps that are to be done on separate pieces of equipment into separate tasks.
4. Combine related steps for operational convenience when they have no unique data monitoring requirements and are done within the same area at the same time.
5. Combine related steps into an artificial subtask to maintain the clarity of the display if they have no unique data monitoring requirements.

The structure of the hierarchy allows management of the large number of situations that can occur in a complex operation.

Data Monitoring for Completion Verification

We planned to specify the dynamic data monitoring for the purpose of verifying expected completion (Level 2) as an attribute of each procedure object. We found the expression of the required data monitoring takes just a few forms:

1. Has variable X reached Y desired?
2. Has variable X been steady at Y for Z period?
3. Is discrete variable A = desired state?
4. Is the rate of change acceptable?

As simple as the completion expressions and requirements appear, we found ourselves almost always writing specific rules for each verification. One of the difficulties was in transferring the different operands (>, =, <, between, increasing at, etc.) from the attribute table into generic rules. The biggest difficulty with specific monitoring rules was synchronizing the rule-firing with operator sign-offs. The process conditions will not necessarily still be true when the operator gets back to the system after completing several field actions.

We found an independent object-oriented monitoring network more practical. While it is desirable to have a monitoring design that is highly related to the hierarchy for ease of development and maintenance, the processing frequency could not be dependent on the operator to keep the monitoring in synch with the activities in the field. A sequential monitoring network was designed containing monitoring objects for any type of monitoring required. The networks can be distributed throughout the hierarchy. The monitoring sequence matches the expected order of actions. This design for monitoring preserved the programming connection between the steps yet allowed the operator some freedom in marking them complete. Very few general rules were needed.

Data Monitoring for Unexpected Conditions

As steps in the procedure are executed, the expected sensor values transition from initial ambient conditions to steady state ranges. Monitoring for expected conditions is the specification described as Level 3 of the 5 levels in the system functionality. Detecting unexpected conditions is a primary benefit of the advisor system. It is easily possible for the operator's attention to be focused on completing the task at hand when an unrelated variable deviates from an expected condition. An example is that before filling a tower with liquid it would be the absence of a low level alarm that should be brought to the operator's attention.

Within the advisor system, we added expected condition attributes to the process sensor definitions. We found high and low limit checking to be effective although richer monitoring functions could be included. The attributes are updated as conditions which are expected to be maintained are achieved. Although the diagnostics and corrective advice described in Levels 4 and 5 are desirable attributes of a system it is often sufficient just to focus the operator's attention on a developing situation that he might not have noticed. Our implementation focused only on detecting gross deviations from conditions that we expected to be maintained.

User Interface

A structure as described above is necessary for practical development and maintenance of the procedure knowledge, but an acceptable user interface is absolutely necessary for a system to be a success. The hierarchy provides the basic frame for our operator interface. With the addition of dynamic attributes that update as the procedure progresses and targets for giving the user the flexibility to selectively view more or less detail, the display is pretty intuitive to use. Figure 1 is an example of the user interface. All potential users were trained in using the advisor as part of their DCS console training and accepted it with little trouble. Both color and text are used to indicate the state of the procedure.

The user can move through the display to mark steps complete or to view the details. Using it is comparable to turning the pages of the checklist version but with a clearer view of the state of the complete procedure.

A step detail display contains an explanation of the action to take, a question confirming the process response expected, a space for a warning message if the process data does not verify the action, and an entry for the operator's initials. For us there is always a requirement for the operator to indicate that a step has been completed even if there is also a data verification requirement. We also included a pointer into the appropriate section of the electronic documentation management system. It could be printed for use by the outside operator and helped the acceptance by the operators. The step detail display covers the hierarchy display when it is selected.

Most of the display space is reserved for showing the hierarchy. The bottom section of the display is a scrollable area for messages and a small area of targets that bring auxiliary functions to the screen. The messages that are put in this area are typically confirmation of actions and for status of the system. The intent is for there to be just a few messages at a time. The auxiliary functions include a chronological log of all procedure activity and a log of all operator-entered comments. The operator log is intended to centralize the hand-scribbled notes in the margins of each operator's copy of the checklist. It's a good way to capture their experience for use in improving the procedure.

An auxiliary function that is handled a bit differently is the warning messages that are generated from monitoring for expected conditions. This is handled more like a traditional alarm system. A new warning message is annunciated by a system beep and color indication. These messages require acknowledgment and clearing. A reason for the warning and how it was corrected is put in the electronic log when it is cleared. This is another way that the system captures experience for use in improving the procedure.

Another feature supported by the user interface is multiple user access. The network architecture is set up to have more than one user connected to the knowledge base at the same time. A username security login is used to customize each session with the proper authorization.

An aspect of the user interface that we have not solved yet is the integration of manual field actions. It would be desirable for the outside operators to directly link with the system but we have not found acceptable technology for this. We have done some tests with portable RF terminals but have found the devices bulky and the text displays limited.

It took a surprisingly high percentage of the behind-the-scenes code and development time to build the user interface. It is not a new finding but bears repeating that the importance of the user interface or the time it will take to make it right should not be overlooked.

Software Validation

Validation of the operator advisor was particularly important since we were expecting to provide operating instructions to the operator. There were two components to the validation, the basic procedure that the advisor was mimicking and the data monitoring.

The structure of the hierarchy had to make sense as did the specific text used to describe the actions. A version of the advisor was set up for commissioning team to review. This was an off-line version of the system (no data monitoring) with a few extra comment entry buttons in the user interface. The reviewers' comments were captured in a file that could be read remotely and incorporated into the system. Interestingly, most comments concerned the actual procedure and not the advisor representation. These were new comments on the procedure from people that had already thoroughly reviewed the checklist version of the procedure. This reinforced the benefit that the graphical view of the procedure was delivering. Nearer the end of construction, the checklist author and the advisor developer worked together to finalize both in parallel.

Validating the data monitoring required another approach. Being a startup advisor for a brand new plant there was no ability to use live data from the plant to test the data monitoring rules. The data monitoring had to be right the first time that real process data was generated. We used an I/O simulator on a development control system at the company's R&D center to manually walk through a startup to validate the data monitoring. The plant's control system configuration was copied into the development system and the validation of the advisor monitoring then took only a few days. Since most of the data monitoring was for threshold limits, the I/O simulator was acceptable. Some simple dynamic simulation was programmed into the control system for special cases.

Conclusions

This application of the Operator Advisor System provided us with a proof-of-concept demonstration, a beneficial application, and specifications for future enhancements. While we found it lacking in some areas, it provides the operator with a useful tool. The benefits enhance safety by reducing the chance of error. The primary benefits that we observed are:

1. Graphical view of the procedure
2. Context-sensitive Data Monitoring
3. Comment capture
4. Automatic record keeping

The primary shortcomings were in two areas:

1. Integration with checklist documentation
2. User Interface (outside operators)

Beneficial applications can be delivered without resolving these limitations however they will have to be addressed for the technology to become in common use. There are some promising developments. Integration with a checklist version of the procedure may be addressed by on-line documentation tools. Perhaps mobile computing will have an impact on integrating the work that goes on outside the control rooms with the plant's computing networks. And there is promising research at Ohio State for adding diagnostics and corrective action planning into the basic hierarchical framework (Marchio, et al., 1993; Marchio and Davis, 1995).

This application had about 600 elements in the hierarchy that represented about 150 pages of traditional checklist procedure. About 35% of the elements had process data verification. There were 200 process data sensors involved with verifying completion or monitoring expected conditions. The use of a knowledge-based system is seen as having potential for reducing upsets and accidents during transitional operations. The hardware and software of the system (Gensym's G2 on a Vax) were deemed manageable by the refinery support personnel. It was a great example of a diverse team involving management, engineering, and operations working together to accomplish a difficult task.

References

Davis, J.F. (1994). On-line knowledge-based systems in process operations: the critical importance of structure for integration. *Proc. IFAC Symposium on Advanced Control of Chemical Processes, ADCHEM '94*. Kyoto, Japan.

Marchio, J., S. Thapliyal and J.F. Davis (1993). A knowledge-based system approach to during cycle diagnosis in batch operations. *Proc. 8th IEEE International Symposium on Intelligent Control*, Chicago.

Marchio, J.L. and J.F. Davis (1995). Diagnosis and procedure management in discrete process operations. *AIChE 1995 National Meeting*. Miami Beach. preprint.

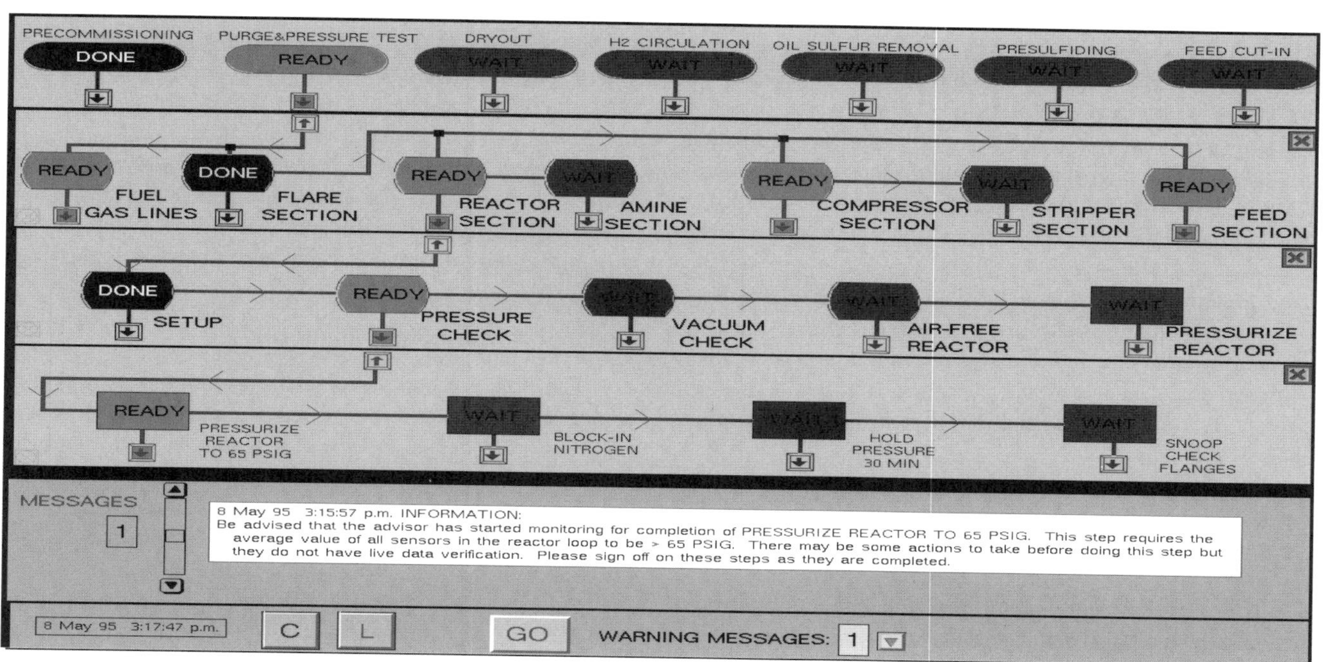

Figure 1. Advisor user interface example.

MEASURING, MANAGING AND MAXIMIZING PERFORMANCE IN PETROLEUM REFINERIES

O.A. Bascur and J.P. Kennedy
OSI Software, Inc.
The Woodlands, TX 77381 and San Leandro, CA 94577

Abstract

The implementation of continuous quality improvement is the confluence of Total Quality Management, People Empowerment, Performance Indicators and Information Engineering. The supporting information technologies allow a refiner to narrow the gap between management business objectives and the process control level. Dynamic performance monitoring benefits come from production cost savings, improved communications and enhanced decision making. A refinery workgroup information flow model is presented to help automate continuous improvement of processes, performance and the organization.

Keywords

Integrated refinery management, Performance monitoring, Continuous quality improvement, Three tier client-server architecture, Real-time historian, Refinery integration and coordination, Knowledge coordinator, Software agents, Object-oriented programming.

Introduction

In this period of challenge and change to our industry, we are going to see increased demands on the types, quantity and priority of information. The pressures of governmental regulations and self-imposed safety and environmental initiatives will require that a real time information product be made available to all our personnel. Everyone from Engineering, Maintenance, Operations, through to top management and outside authorized customers, suppliers and governmental agencies must be served.

Information technologies are emerging faster than many organizations can take advantage of them. The top priority is to align information systems with corporate goals.

Standards and technology for bringing information to the desktop has greatly evolved in the recent years with the introduction of high-speed networks that can transfer data between the various computer in a plant.

This paper will highlight practical information technologies that are being applied in the development of enhanced dynamic performance monitoring systems and integrated workflow systems. The impact on the plant organization is discussed and a high level plant work flow model is presented. This information flow adds another dimension to enhance the productivity of the plant. The concept of the knowledge coordinator is introduced. Here one's focus on the processes, not the functional aspects of the work.

The key ingredients for such a integrated system to work are the Total Quality management philosophies, Empowering People, Business Refinery Performance Strategies, and Information Engineering Technologies.

Dynamic Performance Monitoring

The objective of the plant information network is to bridge the gap between process control and business information systems, and provide adequate information to maximize day-to-day economic performance at the plant through fast, on-line and off-line optimization. A refinery data warehouse integrates the information necessary to act on opportunities in economics and planning, safety, maintenance, engineering, laboratory, environmental, advanced supervisory controls, scheduling and crude supplies.

It is amazing to discover that each business day many companies throw away a valuable corporate asset, their high fidelity history of production.

A plant information system automates the collection and archiving of this information and makes data from the all operating areas available to the whole company. Included is a real time data historian (RTDH) which scans and historizes information from different sources, e.g., Process Unit, Costs, Laboratory Assays, Tank Gauging, Manually Entered Readings, and provides viewing, reporting, analysis and communication tools for operators, engineers and managers. Plant personnel monitor performance indices to identify problems and their root causes. Process managers identify initiatives to correct problems and analyze costs and profits.

The key elements of real time performance monitoring control and monitoring system are:

1. Agreement on the critical few
2. Agreement on the methodology of measuring performance
3. Corrective Action Planning
4. Clarification of roles and Responsibilities
5. Regular progress reports and review meetings
6. Recognition of progress and achievement

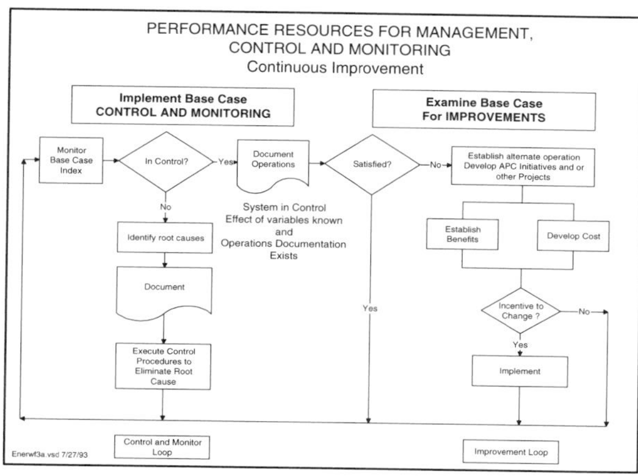

Figure 1. Continuous improvement loops.

Once these key performance indices are defined such product yields, quality indicators, costs indicators, energy indicators, safety indicators, environmental indicators, their calculations methods need to be defined and they must be monitored continuously. As such, each area take what it needs to monitor and learn how to control these indices. Each area also needs to define the cause and effects related to their major activities.

Figure 1 shows the workflow process of monitoring and reviewing these indices. This figure is an extension of the one proposed by Badavas (1993) and is based on two separate loops. One is the control and monitoring loop and the other is the improvement loop. The improvement loop is handled by the support teams (operations, engineering, maintenance and management). This metaphor simplifies process management. Operating teams are in charge of maintaining the process at a certain degree of variability while the personnel in charge of the improvement loop must ask the question if it is cost effective to improve. There are other workflows, e.g. production, quality, cost, maintenance, safety, environmental improvement, that could use similar continuous improvement loops.

The process for continuous improvement is well known in the field of quality and is sometimes called TQM (total quality management). The key to the process is accurate, timely metrics (not necessarily data). Metrics are ambiguous measurements as to what is important from the viewpoint of the customer (internal of external to the company). This system addresses specifically the issue of timeless and distribution of the information to those that need it in a form they can handle, and secondarily the generation of the metric. Use of spreadsheet add-ins not only makes the computation of the metrics viable to nearly all users (results and methods), it allows these results to be exchanged between members of a workgroup for rapid action and decision making. The spreadsheet can be either dynamic after passing (if TCP/IP network is present) or static, if only mail handling it is available on the network. Spreadsheets can also get data from many sources via ODBC or SQL and integrate them into a single presentation. Once completed, the spreadsheet can be put on a server with the links to the data and available to all.

Nearly everyone knows how to use PC based applications and the training requirements are much less. The emphasis is the integration of information at the industrial desktop.

In the next paragraph presents examples using innovative tools for continuous quality improvement.

Process Continuous Improvement

One of the main connectivity features between a real time data historian and the desktop is a tool called PI PC Datalink. The PI PC Datalink allows easy and fast interface to the familiar MS Windows spreadsheets such as Excel and Lotus. Learning is easy, requiring minimum training time. It allows reporting and analysis without having to re-enter manually data from logsheets or other data acquisition systems.

Dynamic performance spreadsheets are easily constructed (using Excel as an Example) with the ProcessBook workbook. Figure 2 shows an integrated process management of a complete refinery business area and a particular recalculated TBP curve for the Vacuum tower. The integration of process data with laboratory results organized by unit and blocks allows the user to configure these specialized spreadsheets for performance monitoring.

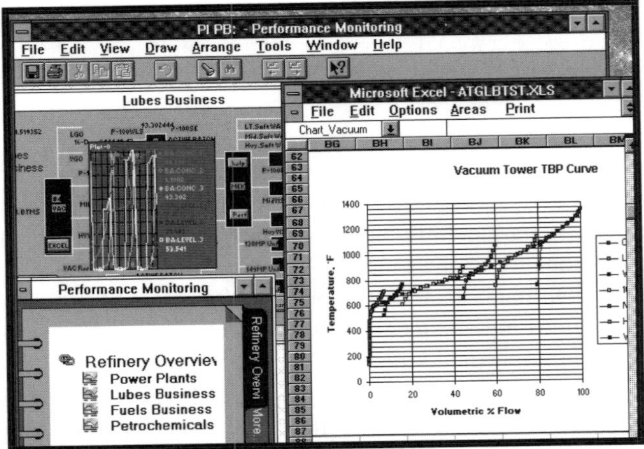

Figure 2. Integrated process management workbench.

The elegance of using a spreadsheet is that everyone in the workgroup has access to a single copy of accurate process data and can make their own analysis. Individuals may also have their own Excel spreadsheets, but they are always working with current information and from a unique data source for enhanced decision making and for continuous improvement.

The Knowledge Coordinator and Repository Manager

Setting of key business objectives and goals requires cooperation across functions and organizations. We need stronger, more timely links for integrated knowledge building by application development teams within different organizations (Bascur, et al., 1992). Figure 3 shows a three tier client server software architecture. Several servers are interconnected to an application layer and a presentation layer. The application layer uses the data from the servers layer to populate spreadsheet tools, business models, expert system shells, preventive maintenance systems, etc. The presentation layer provides a programmable graphical interface via objects. This layer provides the right information to everyone of the functions shown (suppliers, operations, maintenance, engineering, management, customers, etc.). Details of such computer architecture are described in Bascur and Kennedy (1995a).

Business information must be timely, accurate and accessible by the decision makers if the enterprise is to be successful. For this concept to work it is necessary to identify the process and its structure. The major benefits come from identifying the interrelationships and commonalties to extract the essence of the involved processes. The final benefit is an overall simplified process, communication and understanding.

Figure 3 shows key servers such as an SQL server to store relation data used for maintaining the production model of the plant, maintenance records, the Lotus Notes server for maintaining the documents and group workflow messaging information, a common file server maintaining files as used currently, plus fax, modem, Internet access.

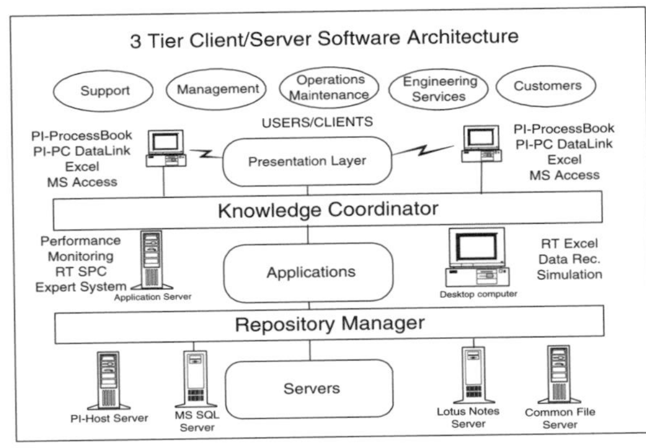

Figure 3. The knowledge coordinator and repository manager.

At the presentation layer, the displays, trends, scripts, mathematical models and others objects can call associated information found in any of the data servers and from the electronic document databases such as Lotus Notes, or ODBC complying databases (Maintenance, Material Resources, Product Specifications). As such, trends can display historical values in PI with targets or goals defined by the improvement group in an SQL Server database.

The knowledge coordinator is a metaphor that allows to integrate the way that people interact together and their access to the information required to perform their activities. The knowledge coordinator is a metaphor that allows to access not only the information but the training methods and procedures available to interpret the information presented to the users. Some of our customers are using extensively Windows Help to store operating procedures using a bookshelf metaphor (Fig. 4).

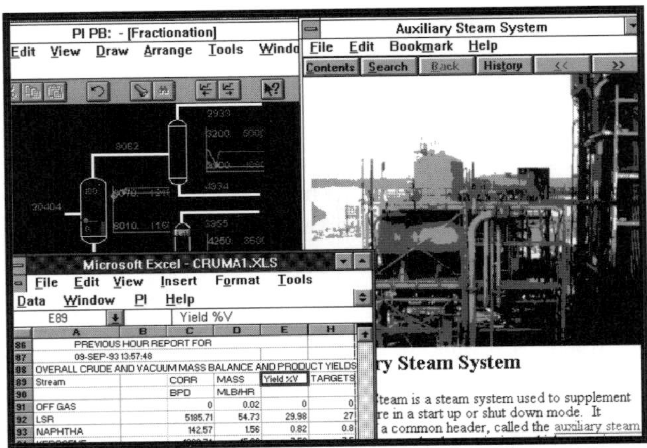

Figure 4. Live spreadsheet and graphics.

The PI ProcessBook can be used very effectively to integrate this type of information. For instance, for a energy index, we can develop a cause and effect diagram which has the relevant areas connected to real time or historical trends and variables organized to add value to the raw data and thus simplifying the process analysis as shown in Fig. 5.

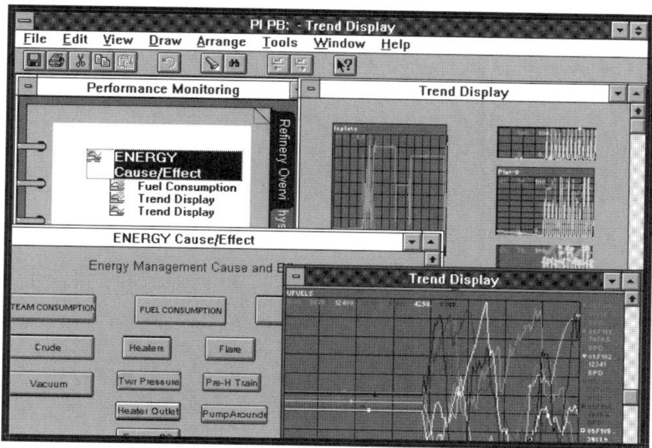

Figure 5. Energy cause and effect diagrams and dynamic trends.

The data repository manager or data warehouse maintains the integrity of the information. It ensures safe, controlled sharing of the information from several sources. Some of the functions of a repository manager are: security, data access and validation, auditing, work flow, information version and status management, automated change management, and replication.

The basic system is composed of two parts. The real time data historian (RTDH) is composed of the interfaces to the process control and automation systems, instruments, analyzers and laboratory systems. The Client programs are divided into a presentation layer and an application layer. The client programs which display and manipulate these data in the form that each performer want to see. Each user can tailor their interface according to their job functions and responsibilities. This presentation layer is a software program where the user can do the integration at the industrial desktop. Since the system is build according to the Microsoft OLE standards, it can access to over 250 OCX or OLE controls that do everything from faxing, image management, attaching a calculator to hypertext context sensitive instructions. In Fig. 3, these are all common implementation of commercial software (only some which comes from OSI Software) but the key is that they can communicate via the Microsoft OLE Technology. The application layer consists of the Application Programming interfaces and where the special application servers can use the RTDH and other sources of data to process the data into information using the business rules or methods.

Some of the key technical reasons to move to three-tier client server technologies include reducing the process load on the database server to improve performance and application integrity, facilitating applications and database management by isolating these functions, facilitating the creation and maintenance of business rules by isolating them from the data, and facilitating the creation of distributed applications and databases.

Until recently engineering, operations/maintenance, management and others could access the information only in one direction (vertical). The key is to enable an organization to manage processes with multifunction access (horizontal). The PI ProcessBook facilitates the integration of information from different sources as well as from different applications.

The PI ProcessBook is built on Microsoft Foundation Class Libraries, and is compatible with Microsoft programs or other compliant applications, including word processing, spreadsheets, electronic mail, mathematical and statistical analysis or modeling programs. OLE2, the capability to embed live applications into other programs, is supported. This allows full integration of the desktop with different similarly compatible packages (Excel, Word, Access, Notes, Visio, etc.) across several vendors.

Spreadsheets can be linked to special button objects that calculate results based on the current or predefined history. These button objects can call trends for cause and effect root cause identification or call special Troubleshooting procedures. Access to other information provided by other programs is also possible. Spreadsheets such as Excel or Visual Basic are used as programmatic interfaces.

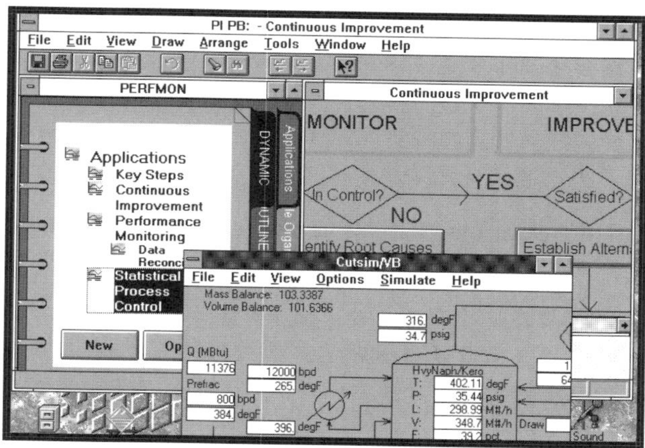

Figure 6. Continuous improvement using an on-line fractionation model.

Figure 6 shows a simplified fractionation model written in Visual Basic for fractionation optimization. As such, applications can display real and historical process data with targets or goals defined by the planning group for a relational database in an SQL Server.

Figure 7 shows the long term view of accessing the information. Here the knowledge coordinator routes specific information from the same sources to an engineer, an operator and a maintenance engineer. In near future the tools are going to be objects which will allow to be interconnected to generate highly powerful applications (Bascur, 1993).

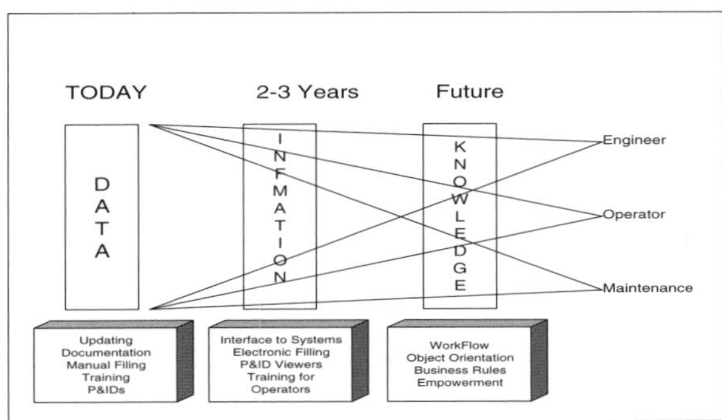

Figure 7. Long term view for access of information.

Improvements in throughput, product yield and efficiency can be realized with relatively small or no investments. Plant designers are often trying to foresee what will happen in the practice. Market changes require flexible production facilities which can cope with a variety of features and realize different product specifications. Modern data networks will bring cost, pricing and industry average indices to the plant floor.

There is a wide range of potential operating zones which opens the way to continuous optimization of process operation. Decisions on change are always present but not clearly recognized for what they are, improvements are often neglected. Examples above have shown individual actions. The next paragraphs will look at workgroup activities.

Example of An Integrated Plant Operations Workflow

Figure 8 shows the coordination and integration metaphor for a refinery. The operation/maintenance box shows real time performance management, just in time training, and access to process safety documentation. The planning group is connected to the information to examine the current performance and to plan the production using tools such as LP, Data reconciliation, supplier and customer data bases

The engineering group is empowered by having access to the real time data to identify process control improvements as well as define benefits for capital projects to enhance production. Management can visualize the work in progress from the workgroups and has access to the information for informed decision making.

The relationship between the coordination and productivity can be visualized in Fig. 8. Graphical data flows for the Operator and Maintenance, Supply and Distribution, Engineering, and Management are shown. Behind each of these interfaces are the links and filters to information for the key 20% of the data for each person according to his/her role and responsibilities. It is important to note that a person can have several roles (Covey, 1989). The described tools facilitate this process by providing with easy access to the production data to do the important which is to continuously improve.

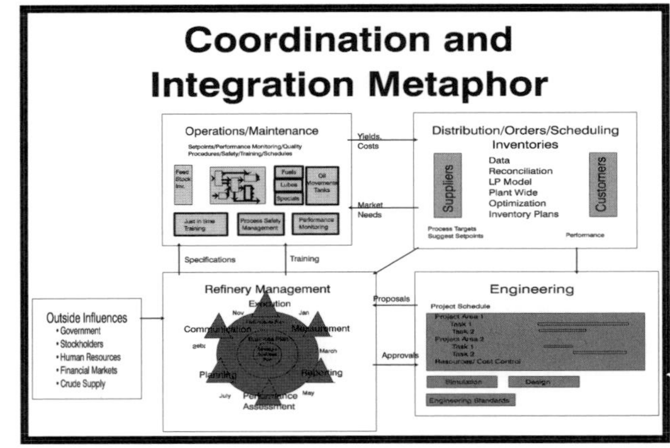

Figure 8. Refinery coordination and integration.

As these systems become widespread, the entire organization will become empowered as shown in Fig. 7. In such an organization connectivity will allow the information to be focused by the knowledge coordinator. The system should contain context sensitive, just-in-time training and assist in identifying commonalties among systems and the integrated work flow in a chemical plant or a refinery.

This simple integrated work approach produces unexpected economic benefits. There is a significant reduction in maintenance costs and improved production. Both operations and maintenance work together. The maintenance personnel because they are integrated with process management can really improve the operation, and all, for a fraction of the manpower of a manual workflow.

Other enhancements in refinery integration of blocked operations management is described by Bascur and Kennedy (1995a). The coordination between actual refinery processing and refinery planning and scheduling is being revisited as an opportunity that we called continuous recipe improvement. A production model of the refinery acts as the relationship between the control variables and the business variables organized by product or by customer.

Conclusions

These technologies are available today to reengineer plant operation and increase the performance of current productions systems. The key is linking people workflow, business processes, strategies and the best enabling technologies. The major ingredients for a successful implementation are:

1. An infrastructure that follows the suggested computer architecture, (Technology)
2. The integrated Process Management Workbench using the Knowledge Coordinator such as the PI-PC ProcessBook metaphor to unify the access to information, (Business Processes)
3. The implementation of total quality management guidelines and people empowerment (Strategies and People).

Large benefits can be obtained using current process control and information management systems, thus the incentives to implement these systems are very attractive.

The successful application should be judged on how it provides added to the overall information system, such as new ways of combining and visualizing existing data or information. Process and business information must be timely, accurate and accessible by the decision makers. This information has to have a value added to the operator, engineer, manager, otherwise the final contribution is not fully attained.

Object-oriented analysis and design engineering are emerging information tools that will enable the implementation of workgroup type of plant information systems. This paper discussed the value of integrating process data, people and work flow procedures related to the business strategies defined by the plant business needs.

Special Note

There are 1200 copies of the RTDH server in used with more than 850 copies installed and paid in operating facilities. The client products, however, are much greater because they are on a per user basis; as of last count more than 40,000 copies had been shipped.

References

Badavas, P.C. (1993). *Real-Time Statistical Process Control.* Prentice Hall, New Jersey.

Bascur, O. A. (1993). Bridging the gap between plant management and process control. In B. J. Steiner, et al. (Eds.), *Emerging Computer Techniques for the Minerals Industry.* SME, Littleton, CO. pp. 73-81.

Bascur, O. A., and Kennedy, J. P. (1995a). Maximizing performance in petroleum refineries. In M. Kanko (Ed.), *Japanese Petroleum Industry.* COSMO, Tokyo, Japan,. January.

Bascur, O. A., and Kennedy, J. P. (1995b). Performance monitoring in metallurgical complexes. *Proceedings of the XIX IMPC.* SME, San Francisco, CA. October.

Bascur, O. A., Vogus, C. B. and Bosler, W. H. (1992). Long term knowledge integration with OSHA PSM. *Proceedings of NPRA Computer Conference.* Washington D.C. November 16-18.

Covey, S. R. (1989). *The 7 Habits of Highly Effective People.* Fireside Book, Simon & Schuster Inc., N.Y.

SESSION SUMMARY: PARADIGMS OF INTELLIGENT SYSTEMS IN INDUSTRIAL PROCESS OPERATIONS

Eleni P. Patsidou
Shell Oil Products Company
P.O. Box 1380
Houston, TX 77251-1380

Sunwon Park
Department of Chemical Engineering
KAIST
373-1 Kusong-dong, Yusong-gu, Taejon 305-701 Korea

Introduction

As the title suggests, this session offers examples of successful industrial applications of intelligent systems.

Previous sessions discussed the state of the art in the research and development of artificial intelligence techniques. During these sessions many questions arose concerning the practicality of some proposals.

The practical problems encountered in the implementation of an intelligent system are many and they appear even when starting to discuss a potential application. Is this application going to be profitable? How is the profit going to be predicted and measured? Then — if there is no process historian available — how is the application going to obtain live data? Who is going to provide the knowledge, maintain and support the application? How is the operator training going to be implemented, etc. Some of these issues may seem trivial but sometimes are the limiting steps in the development of a new application.

The papers included in this session touch some of the above issues. They cover a wide area of intelligent applications in process engineering. A brief introduction to each paper follows:

- "Supervision and Decision Support of Process Operations" by Ted Cochran (Honeywell Corporation) and Duncan Rowan (DuPont Company). This paper presents the state of the art technology available for decision support with emphasis on technologies and problems related to process and supervisory control.
- "Intelligent Systems in Heavy Industry" by Adam J. Szladow (Lobbe Technologies, Ltd.) and Debbie Mills (REDUCT Systems, Inc.). This is a review of applications and benefits of intelligent systems in heavy industry, classified by their function.
- "Pro-Sked — A Technology for Building Intelligent Transportation and Production Scheduling Systems" by Llewellyn Bezanson and Robert Fusillo (Setpoint, Inc.). This describes an integrated environment for refinery and transportation scheduling, implemented on an object-oriented, open system platform that facilitates the management of external databases and equation solvers.
- "PlantMonitor: An On-line Advisory System for Monitoring Polyethylene Plants" by T. S. Ramesh and B. V. Kral (Mobil). This system detects basic sensor faults, control system faults, and process problems.
- "Operator Support for Diagnosis in a Fertilizer Plant" by A. Mjaavatten, B. A. Foss, and K. Fjalestad (University of Trondheim). The objective of this application is to reduce the emission of pollutants through a construction of modular process information.
- "Refinery Implementation of an Operator Advisor System for Procedure Management" by David W. Beach and Michael J. Knight

(BP Oil). This paper describes an on-line knowledge-based system for improved management of planned transitions, e.g., start-ups, shutdowns. The system includes live data monitoring, automatic record keeping, and a graphical user interface.

- "Measuring, Managing, and Maximizing Performance in Petroleum Refineries" by Osvaldo A. Bascur and J. P. Kennedy (Oil Systems, Inc.). The objective of the plant information network is to bridge the gap between process control and business information systems, and provide adequate information to maximize day-to-day economic performance at the plant through fast on-line and off-line optimization.

Some common characteristics in the applications presented in this session are summarized below:

- Most applications involve integration of different decision support technologies along with other technologies that have been used in modeling and control, e.g., SPC, linear programming, dynamic simulation, etc.

- Systematic grouping of the information and a good structure of the process model are very important factors in understanding the problem and potentially re-using the system in other applications.

- Software development in the areas of expert system shells as well as process historians has facilitated the development of advisory applications.

The session closed with discussion on the details of developing and using these applications.

BATCHKIT — KNOWLEDGE INTEGRATION FOR PROCESS ENGINEERING

M. Hofmeister
Lab. F. Tech. Chem., ETH Zürich
CH-8092 Zürich

Abstract

BatchKit is a knowledge integration environment for process engineering. It provides a broad range of problem representation and solving capabilities, from trivial calculations and simple inferences to complex optimization problems which are addressed using special-purpose chemical engineering solvers such as SPEEDUP process simulator and GAMS modeling language and optimization solvers. Representation is based on the integration of object-oriented, logic and constraint programming. The integration of logic programming (LP) supports logic based analysis of problem properties such as problem-specific consistency checks and backtracking search formulations. Constraints introduce equations and inequalities and constitute the first step in (re-)integrating symbolic algebra in a computer-aided, problem-oriented programming language for engineering. BatchKit currently provides prototype conceptual models of plants, processes, production plans and their various specializations. Knowledge maintenance is facilitated by providing multiple external representations for every type of object — graphs, tables or, for more detailed inspection, the assert and query language of logic programming. Human interaction widely relies upon graphic interfaces developed and implemented using the graphic building tools of the KEE system by IntelliCorp. Complete history is maintained to support chronological backtracking or checkpoint rollback and knowledge state backup.

Keywords

Knowledge integration, Process engineering, Constraint logic programming.

Introduction

The driving force behind the development of BatchKit was the recognition that the pre-requisite of the computer-aided engineering (CAE) of complex production/processing systems leading to integrated manufacturing (CIM) is the integration of all pertinent knowledge (all relevant models) about processing systems in a common framework of adequate flexibility.

This applies particularly to chemical production systems which are designed as unique artifacts and evolve over a significant time period. The growing scale and complexity accentuate the need for model-based diagnostic and control in an exponentially growing number of operating situations. The design and use of large systems overlap in time and the complete history may become relevant e.g. to the product quality control.

In this paper, we first briefly discuss the related research oriented towards developing CAE environments. Further, we present the representation layers constituting the BatchKit system and the techniques used for knowledge maintenance.

Related Work

The need for "intelligent" programs or even for a knowledge-based approach to solving complex, many-faceted problems efficiently led to the recognition that before the knowledge about such problems can be organized in a formal system, the language for the formulation of at least a relevant range of problems has to be unified to allow analysis and/or automated reasoning about them. In the BatchKit project, this recognition was stimulated by an early effort to generalise and classify the problems of batch processing systems in (Rippin, 1983).

Integrated systems for chemical engineering have long been a goal, e.g. in the following projects: Batches (Joglekar and Reklaitis, 1982) — a batch simulation language and system, ASCEND — an object-oriented modeling language for simulation and analysis, KBDS (Banares-Alcantara, et al., 1994), DesignKit — an object-oriented environment based upon the MODEL.LA language (Stephanopoulos, et al., 1987) etc.

The knowledge- or data-based approach with means similar to those of BatchKit was taken in Banares-Alcantara, et al. (1994). The AI toolkit Knowledge Craft was used to manage constraints which were, in contrast to our approach, used only in checking mode.

Representations

The open environment of the BatchKit system is organized in the following layers:

LISP

Some of the unique properties of LISP have proven to be essential to the development of the BatchKit environment:

- Integration of interpreted and compiled language.
- Integration of LISP evaluation with emacs editor allows extremely fast development (fast prototyping).
- The parsimonious syntax facilitates extensions in a uniform manner.
- General concept of textual representation and input/output of LISP objects (numbers, strings, arrays, structures etc.).
- Expressive high-level abstractions e.g. for mapping operations, iteration, exception handling, input/output and file system.

KEE

The KEE (Knowledge Engineering Environment) system by IntelliCorp provides the object-oriented programming and the graphic interface building tools — menus, ActiveImages and KEEPictures subsystems.

ActiveImages, e.g., is an intelligent, interactive GUI toolkit which proposes a selection of possible representations, default selection and the directionality of the graphic/internal representations' interconnection according to the class of the variable selected.

Logic Programming

LinX system is the low-level representation layer of the BatchKit system. It integrates LISP and KEE's objects and slot links with logic programming (for full report see Hofmeister (1992)).

The expressive power of the LISP/LP/OOP combination will be illustrated by a simple example: Given a predicate `eqitem-succ-r`, implement a function for graph generation which will derive all successors (= node children) of a given equipment item.

The resulting code is:

```
(defun Children
 (EqItem ;; The parent equipment item
  ) ;; The children of EqItem
 (let ((Children NIL)
       (Child
        ;; Prove the first
        (ask `(eqitem-succ-r ,EqItem ?X)
              ?X)));; Result form
   (loop
     (unless Child
         (return Children));; Terminate
     ;; Add to the list
     (push Child Children)
     (setq Child (ask));; Get next
     ) ;; POOL
  ) ;; TEL
 )
```

The function defined above can be supplied as an argument to the general graph generation utility. The resulting graph for an example plant is shown in Fig. 1.

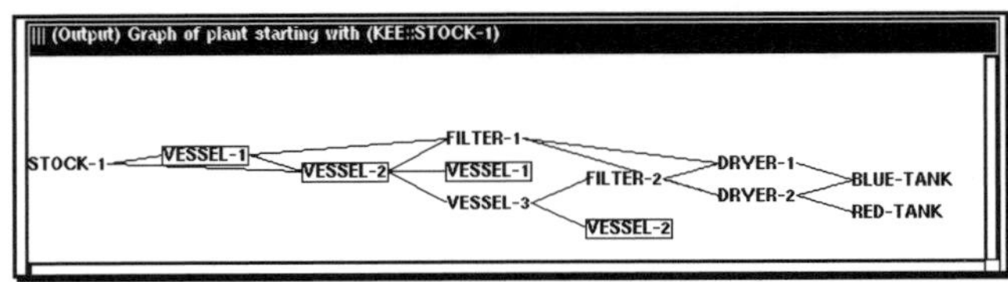

Figure 1. Graph of plant equipment items.

Constraints

The constraint concept can be generalised to subsume simple value assignments, propositions, first-order logic and expressions in any algebra over arbitrary domains.

Quantified Constraints

The constraints normally express, in some given algebra, theories of the domain and are therefore valid for all objects which fall into the scope of their validity. In the LinX system the compactness of problem formulation is enhanced by supporting quantified constraints — a concept similar to the indexed constraints used in the GAMS language but significantly more general since it allows to select the sets of objects involved in the constraints formulation by an arbitrary predicate.

E.g., to post the constraints — equations and inequalities — of the equilibrium state of a process denominated process-1 which consists of a set of tasks and to solve the resulting set of equations, the following form can be evaluated:

```
(and (forall ?task
             (process-task-r process-1 ?task)
             (task-constraints ?task))
     (speedup-solve)
     ) ;; DNA
```

The first argument in the term starting with the system predicate `forall` is the variable which will be bound by the true instances of the second form - the quantification form, the third term will become a conjunctive subgoal after instantiation of all variables (in this case, only `?task`). In our case, the predicate `task-constraints` posts the equation and inequality constraints of the task model corresponding to its argument (a process task). After the execution of the above form in the prover, the free variables of the problem will obtain the equilibrium values.

Special-Purpose Solvers

Logic programming is suited for the solution of problems over the Herbrand universe. The integration of constraints introduces the capability to solve also problems including equations and inequalities in integer and real variables characteristic for engineering problems.

For the solution of large problems characteristic of chemical engineering — such as plant design or operation - special-purpose solvers are therefore integrated within the LinX framework:

- GAMS — Modeling language and optimization including mixed-integer nonlinear programming.
- SPEEDUP — Equation-oriented mass and energy balance calculation and dynamic simulation.

For detailed report see Hazdra and Hofmeister (1994).

Representation of Process Engineering Concepts

Plants: An example plant graph representing the abstract topology of the plant is shown in Fig. 1.

Processes: Processes can be represented by graphs of varying abstraction, analogous to those of a plant graph in Fig. 1.

Production Plans: The most important information relating to the production plan — an implementation of a process on a given plant — is the occupancy of equipment over time, which is commonly represented in a Gantt-chart as implemented in the GanttKit component of the BatchKit system (see Fig. 2, for full report see Halasz, et al. (1992)).

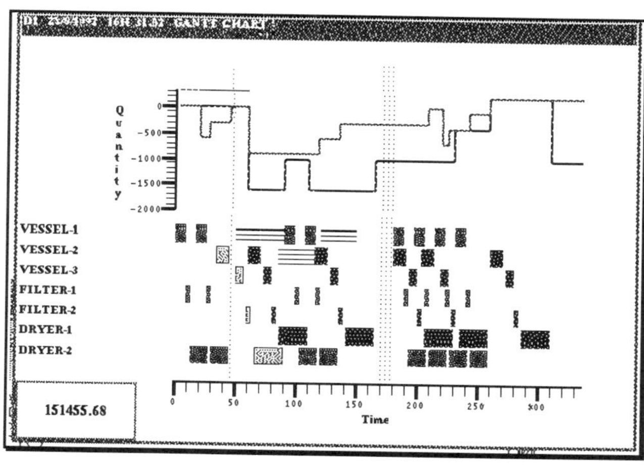

Figure 2. Production plan in GanttKit panel.

Knowledge Maintenance

- The BatchKit system offers the following knowledge browsing/editing tools:
- Tell-And-Ask language via the emacs text editor or the Tell-And-Ask panel of the LinX system (see Fig. 3).
- The LinX system's object editor/browser allows tabular representation of objects with their attributes/relations and a recursive expansion of nested objects (see Fig. 4).

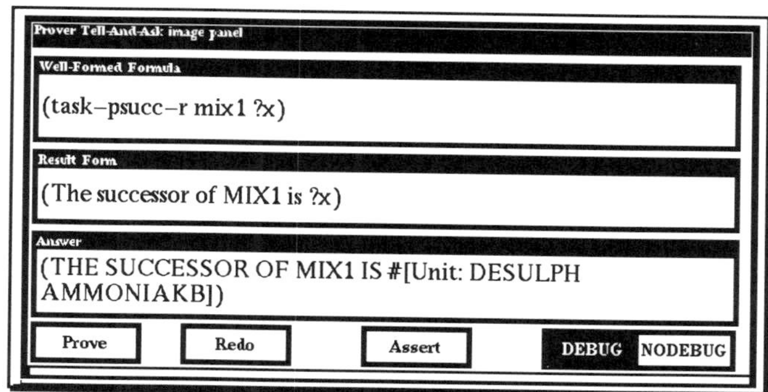

Figure 3. LinX system Tell-And-Ask panel.

Status, Open Problems and Further Work

While a range of problem formulations have been studied, important problems have not been addressed:

- Conceptual integration of optimization and constraint satisfaction
- Problem versions
- Distributed and cooperative problem solving

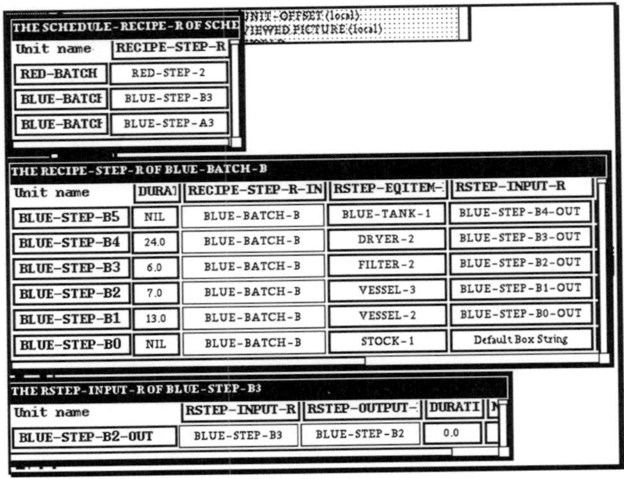

Figure 4. Graph representation.

Conclusions

The BatchKit system offers the possibility, in a particular field of engineering, of a paradigm shift:

- from individual equations and inequalities to constraints quantified by arbitrary predicates over sets of objects.
- from tabular representation to graphs.
- from reading of computer output listings to browsing of knowledge bases.
- from the simplicity of the if-then-else procedural thinking to thinking in terms of theories and constraints.

The effort to develop an open environment in the spirit of the "Dynabook for chemical engineering" has been undertaken. The present system is based on tradeoffs between ease of use and generality, whereas efficiency was generally considered secondary. A significant level of integration was achieved at low cost of development due to the use of LISP and other tools and to the open character of the system, some of which will be very expensive when also the safety and security measures have to be included.

Current state of information technology does not yet provide general solutions at acceptable cost for some of the problems encountered (concurrent engineering, general transaction processing, languages effectively and uniformly supporting all problem types).

References

Banares-Alcantara, R., H. Lababidi, G. Ballinger and J. King (1994). KBDS: a support system for chemical engineering conceptual design. *Proc. ESCAPE-4*, Rugby, UK, pp. 419-426.

Halasz, L., M. Hofmeister and D. W. T. Rippin (1992). GanttKit — an interactive scheduling tool. *Proc. NATO ASI on Batch Processing Systems*, Antalya.

Hazdra, T. and M. Hofmeister (1994). Integration of GAMS system solvers within the BatchKit environment. Technical Report 294, SEG, Lab. F. Tech. Chem., ETH Zurich.

M. Hofmeister (1992). LinX — integration of object and logic programming. Technical Report 241, SEG, Lab. F. Tech. Chem., ETH Zurich.

Joglekar, G.S. and G. V. Reklaitis (1992). A simulator for batch and semi-continuous processes. *AICHE Annual Meeting*, Los Angeles.

Piela, P. C., T. G. Epperly, K. M. Westerberg and A. W. Westerberg (1991). Ascend: An object-oriented computer environment for modeling and analysis: The modeling language. *Comp. Chem. Engng.*, **15**, 53-72.

Rippin, D. W. T. (1983). Design and operation of multiproduct and multipurpose batch chemical plants — an analysis of problem structure. *Comp. Chem. Engng.*, **7**, 463-481.

Stephanopoulos, G., J. Johnston, T. Kritikos, R. Lakshman, M. Mavrovouniotis and C. Siletti (1987). DESIGN-KIT: An object-oriented environment for process engineering. *Comp. Chem. Engng.*, **11**, 655-674.

Struthers, A. (1990). *A Knowledge Based Approach to Process Engineering Design*. PhD thesis, Univ. of Edinburgh.

Waters, A. and J. W. Ponton (1992). Managing constraints in design: Using and AI toolkit as a DBMS. *Comp. Chem. Engng.*, **16**, 987-1006.

HEURISTIC-NUMERIC PROCESS SYNTHESIS WITH PROSYN

Gerhard Schembecker and Karl Hans Simmrock
Lehrstuhl für Technische Chemie A (Prozeßtechnik)
Fachbereich Chemietechnik
Universität Dortmund, D-44221 Dortmund

Abstract

The paper summarizes the results of our research after approximately 220 years of manpower in the field of computer-aided process synthesis. The strategy presented copies the industrial procedure of inventing chemical processes. An overall manager system PROSYN-Manager applies a heuristic branch-and-bound method to administer and evaluate process synthesis alternatives. Specialized modules and service systems provide specific knowledge for solving detailed design problems. An interface to commercial flowsheet simulators (e.g. to ASPEN PLUS™) allows easy analysis of the derived results.

Keywords

Computer-aided process synthesis, Heuristic-numeric system, Flowsheet development, Branch-and-bound.

Introduction

The objective of process synthesis is to develop a conceptual flowsheet starting with basis information about the chemical reaction path, see Fig. 1.

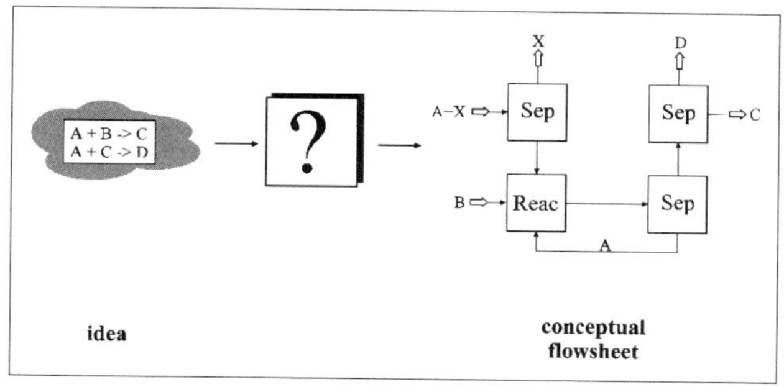

Figure 1. Aim of process synthesis.

Developing new chemical processes is characterized by three important points:

Conflict during process synthesis: Before decisions are made on an upper level no decision should be allowed on a lower level. However, no optimal decision is possible on an upper level without knowing the results of the lower level (Funder and Simmrock, 1991).

Different types of information: Process synthesis requires three different types of information. Data like physical properties, costing information etc. are stored in databases. Algorithms and numeric routines help to design specific unit operations like reactors, distillation columns etc. They are available e.g. within commercial flowsheet simulators. Design knowledge and experience can be formulated as heuristic rules.

Process synthesis is teamwork: Due to the character of process synthesis it is a common industrial practice to develop chemical processes in a team. The cooperation of several process engineers is organized by a manager who is responsible for the whole project. It is the task of the engineers to solve subproblems in detail. Additional information like physical properties are provided by service groups.

It is the intention of this paper to present a strategy which copies the industrial procedure of inventing chemical processes in a team. Due to the character of the synthesis problem the concept of cooperating heuristic-numeric systems seems to be best suited to gain promising process flowsheets. Databases, numeric routines as well as heuristics are combined within one design tool called PROSYN (Process Synthesis).

The PROSYN Environment

In the PROSYN environment three different types of systems can be distinguished.

Manager Systems

These systems are responsible for the problem solving strategy and the coordination of the corresponding specialists. They must be able to identify subproblems, find an appropriate problem definition and invoke the qualified specialists. The overall manager system for the generation of complete processes is PROSYN-Manager (Process Synthesis Manager). There are two areas of outstanding interest for the development of total processes: The selection of an appropriate reactor network and the design of unit operations, see Fig. 2. In case of a reaction engineering problem PROSYN-Manager activates the module for reaction. If it is necessary to treat a problem concerning a unit operation PROSYN-Manager delegates the task to the responsible system. In case of distillation another manager system Rectification-Manager is called. This system coordinates the use of several modules dealing with distillation.

Specialists

Specialists have to work out synthesis subproblems in detail. To find a satisfactory solution they use deep knowledge of their sphere of responsibility available. For example, the systems mentioned below act as specialists in the PROSYN environment.

READPERT (Reactor Evaluation and Design Expert System) is a heuristic-numeric system that selects and designs reactors for a given reaction path (Fried, B., 1990; Westhaus, 1995).

REKPERT (Rectification Knowledge Based Expert System) generates distillation sequences using simple and complex columns (Funder, 1995).

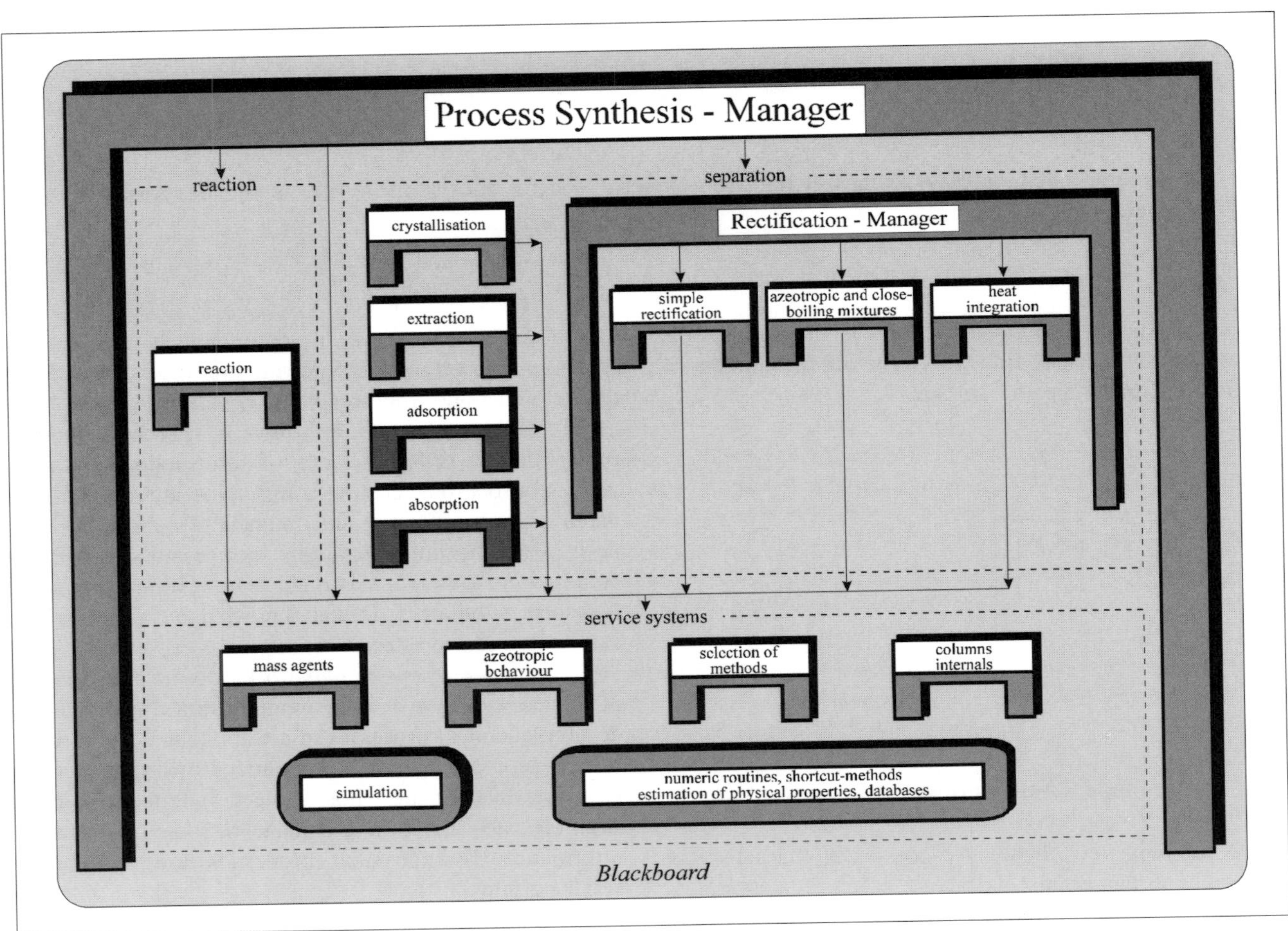

Figure 2. Structure of PROSYN.

Energy integrated distillation sequences are developed by HEATPERT (Heat Integration Expert System). HEATPERT requires an initial distillation sequence as a starting point. For example, the initial sequence may be created by REKPERT (Schüttenhelm, 1994).

TEAGPERT (Trennung engsiedender und azeotroper Gemische Expertensystem = Separation of Azeotropic and Close Boiling Mixtures) synthesizes distillation processes for the separation of azeotropic or close-boiling mixtures (Welker, 1991; Simmrock, et al., 1993). The system analyses the distillation behavior of a mixture, determines distillation areas and proposes separation units (e.g. pressure swing distillation, hetero-azeotropic distillation with mass agents ...).

Problems dealing with crystallization are treated by KRISPERT (Crystallization Expert System). KRISPERT uses basic physical property data to determine the method of producing crystals. In addition an appropriate crystallizer is selected and designed.

Service Systems

Service systems supply basic information for process synthesis. For example, service systems determine thermodynamic data, select mass agents and simulate unit operations.

SOLPERT (Solvent Selection Expert System) selects a limited number of suitable mass separating agents for extractive and azeotropic distillation (Fried, A., 1990; Bieker, 1995). Starting from a very rough classification according to Harrison and Berg SOLPERT determines specific groups which should be integrated into the solvent. Based on the information of these incremental pieces of the molecule the program select suitable mass agents.

AZEOPERT (Azeotrope Predicting Expert System) finds out the azeotropic behavior of a mixture. It uses a large database with experimental data as well as a set of numerical routines to estimate azeotropes. In addition, heuristics can be used for mixtures which cannot be calculated or for checking the results of the estimation (Schembecker and Simmrock, 1994).

SEMPERT (Selection of Methods Expert System) provides physical property data using databases and estimation routines. For unknown data it selects the best way of estimating the value.

Furthermore, algorithmic programs as well as a direct interface to the commercial flowsheet simulator ASPEN PLUS™ are available to fulfill design and cost calculations.

An important feature for the process synthesis environment is the transfer of information. It is necessary to distinguish between direct and indirect communication. The developed communication tool arranges the direct communication between individual systems. With the help of the communication tool the calling system transfers the problem description and the problem data to the qualified system. The qualified system has to work out a detailed problem solution. After having found a solution the results are returned to the calling system. The indirect communication medium of the environment is a blackboard realized as a central database and accessible via a local area network. This indirect communication guarantees the consistency of all data which are used within a certain process synthesis problem.

Heuristic Branch-and-Bound Technique

The branch-and-bound technique is a tree search method combined with a bounding method. Our approach uses the sum of the costs of all process steps within the flowsheet. At each level one calculates the objective function for all alternative nodes. The node showing the lowest value of the tree is the best solution attainable at the moment. The other nodes need not to be investigated further (bounding). They are reactivated if they are found to be more promising than the current alternative. Continuing with the best momentary solution possible alternatives are investigated for the following synthesis step (branching). The procedure stops if a complete problem solution exists which owns the lowest function value of all nodes in the tree.

In contrast to the original branch-and-bound technique the heuristic branch-and-bound method additionally uses heuristics to reduce the number of alternatives (Wolff, 1994). Instead of all possible alternatives given by the original branch-and-bound method only heuristic favorable alternatives are examined. This reduction is absolutely necessary because of the large number of potential solutions for the synthesis of total chemical flowsheets.

The branch-and-bound technique described above combined with the application of heuristics proved to be an excellent approximation of the practical problem solving strategy. Hence, complete processes are developed gradually under the control of the heuristic branch-and-bound method, see Fig. 3. Each node symbolizes a process step, each branch a partial flowsheet.

Conclusions

PROSYN had been applied successfully to various industrial synthesis problems. It is worth to summarize the most important reasons:

1. The application of a real-life solution strategy like the heuristic branch-and-bound method.
2. The integration of heuristics, algorithmic routines and databases in one flexible design tool.
3. The engineer's opportunity to influence the process development at each decision level.

4. The combination of synthesis and analysis steps.
5. The existence of interfaces to commercial flowsheet simulators like ASPEN PLUS™.

References

Bieker, Th. (1995). Ein Beitrag zur Auswahl von Hilfsstoffen und Hilfsstoffgemischen für die Extraktiv- und Azeotroprektifikation. Ph.D. thesis. Universität Dortmund, Verlag Shaker, Aachen.

Fried, A. (1990). Erstellung eines wissensbasierten Beratungssystems zur Auswahl von Zusatzstoffen für die Hilfsstoffrektifikation in der Prozeßsynthese. Ph.D. thesis. Universität Dortmund .

Fried, B. (1990). Regelbasierte Auswahl grundlegender Reaktortypen mittels wissensbasierter Programmierung. Ph.D. thesis. Universität Dortmund.

Funder, R. (1995). Entwicklung kooperierender verteilter Expertensysteme am Beispiel der Rektifikation. Ph.D. thesis. Universität Dortmund.

Funder, R. and K. H. Simmrock (1991). Prozeßsynthese mit hilfe kooperativer, verteilter expertensysteme. In Heinz (Ed.), *Dortmunder Expertensystemtage 1991.* TÜV-Rheinland Verlag, Köln.

Kim, Y. J. (1991). AZEOPERT - an expert system for the prediction of azeotrope formation. Ph.D. thesis. Universität Dortmund.

Schembecker, G. and K. H. Simmrock (1994). AZEOPERT, ein heuristisch-numerisches System zur Vorhersage azeotropen Verhaltens. *Chem. Ing. Tech.* **66**, 1273.

Schütenhelm, W. (1993). Ein Beitrag zur Synthese energieintegrierter Rektifikationsschaltungen. Ph.D. thesis. Universität Dortmund.

Simmrock, K. H., A. Fried and R. Welker (1993). Expert system for the design of separation sequences for close-boiling and azeotropic mixtures. *Int. Chem. Eng.*, **33**, 577-590.

Welker, R. (1991). Erstellung eines wissensbasierten Beratungssystems zur Auswahl destillativer Sonderverfahren und Konfigurierung zugehöriger Trennsequenzen für azeotrope und engsiedende Gemische. Ph.D. thesis. Universität Dortmund.

Westhaus, U. (1995). Beitrag zur Auswahl chemischer Reaktoren mittels heuristisch-numerischer Verfahren. Ph.D. thesis. Universität Dortmund.

Wolff, A. (1994). Heuristisch-numerisches Managersystem zur Synthese chemischer Verfahren. Ph.D. thesis. Universität Dortmund. Verlag Shaker, Aachen.

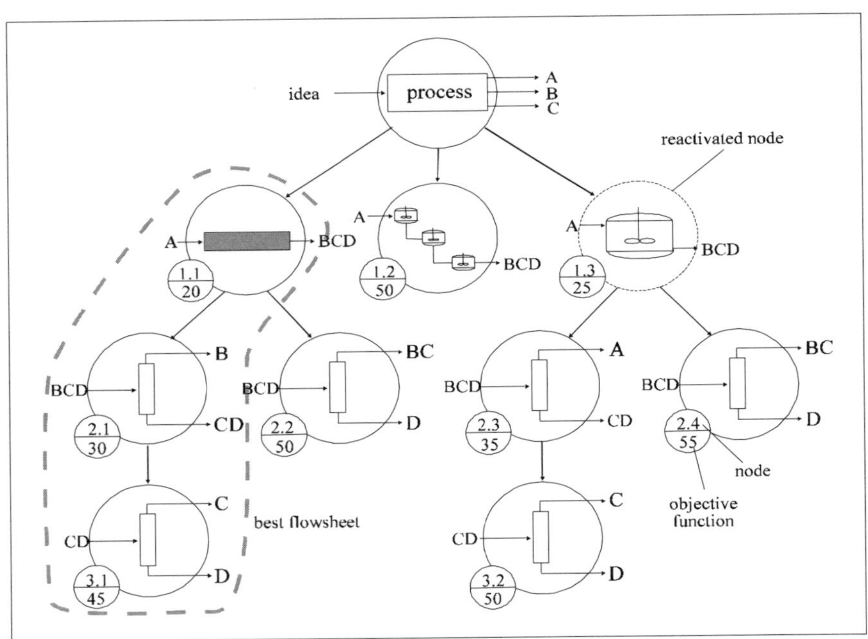

Figure 3. Heuristic branch-and-bound technique.

A NEURO-FUZZY TOOL FOR THE GENERATION AND IMPROVEMENT OF HEURISTIC RULES FOR PROCESS SYNTHESIS, DESIGN AND CONTROL

Klaus Hartmann and Rolf Hartmann
Brandenburg Technical University Cottbus
Faculty of Environmental Sciences and Process Engineering
Department of Process Systems Engineering
Karl-Marx-Str.17, D-03044, Cottbus

Vladimir Gilyarov
Sankt Petersburg Institute of Technology
Department of Process Automation and Control
Moskowski, 26, SU-198013, St. Petersburg, Russia

Abstract

Heuristic methods for process synthesis and process control are applied, when decisions have to be made based on incomplete and uncertain information. The presented Neuro-Fuzzy-Tool LINGVOGENERATOR can be used to support the "learning" process of generation and optimization of heuristic rules directed by new data about a given object. The main advantage which characterizes this tool is the flexibility in handling different data types in an automated "learning" process. This flexibility makes it possible to include a maximum of a priori information about a given object, which can be input/output data, linguistic variables, a rule-base, etc. The main elements and structure of the tool are presented and its applicability is demonstrated on two examples in the chemical engineering field.

Keywords

Neuro-fuzzy systems, Process synthesis, Process control, Heuristics, Rule generation, Data processing.

Introduction

For a great number of chemical engineering problems such as process synthesis, process design and control, decisions have to be made on the basis of incomplete and uncertain information about data, variables, models and goals. One of the suitable approaches for the problem solving under these circumstances is the application of heuristics. This approach can only be successful if comprehensive and well defined heuristic or production rules are available. However, the availability and reliability of heuristic rules is one of the bottlenecks of this approach.

The generation, selection and adaptation of heuristic rules can take advantage of the experience of design engineers, existing knowledge on the chemical/physical laws and processes related to the design problem as well as modeling and simulation tools. Heuristic rules formulated in the "language of the design engineer," represent a good basis for the dialogue engineer-computer in the design process.

However, heuristic rules are often not exact and sometimes contradict each other. Therefore a combination of fuzzy logic and heuristic rules should be a suitable basis for the implementation of rule-based process synthesizers on the computer.

Since heuristic methods provide no guarantee of optimality and the depth of knowledge on which the rules are based differs from one to another, the set of rules for a process synthesizer has to be refined and extended continuously in parallel to the design process. A possibility for the automation of this "learning process" is to use the methodology of fuzzy logic in combination with neural networks. A tool which realizes this approach is the developed Neuro-Fuzzy-Tool LINGVOGENERATOR.

Basic Idea and Description of the Program

The basic idea of the tool is the transformation of a fuzzy system into an equivalent neural network with special structure and neurons, which is illustrated in Fig. 1.

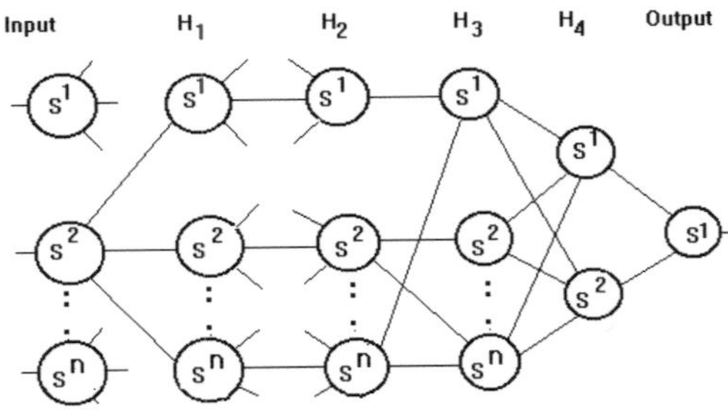

Figure 1. Neural network model of the fuzzy system.

The transfer functions of the neurons for the different layers are:

$$S^i_{H_1} = \begin{cases} 0, & x_j < a_1 \\ sfunc(a_1, b_1, x_j), & a_1 \le x_j \le b_1 \\ 1, & b_1 < x_j < b_2 \\ sfunc(b_2, a_2, x_j), & b_2 \le x_j \le a_2 \\ 0, & x_j < a_2 \end{cases} \tag{1}$$

where $S^i_{H_1}$ are the membership functions (MBF) and:

$$sfunc(a,b,x) = \begin{cases} \frac{(x-a)^2}{2(c-a)^2}, & x \le c \\ 1 - \frac{(b-x)^2}{2(c-a)^2}, & x > c \end{cases}, \tag{2}$$

with $c = \frac{a+b}{2}$

$$S^i_{H_2} = \prod_j^L \rho_j^{(1-\gamma)} \left(1 - \prod_j^L (1-\rho_j)\right)^{\gamma}, \tag{3}$$

where $0 \le \gamma \le 1, \rho_j = x_j + (1 + x_j)(1 - m_{ji})$

$$S^i_{H_3} = \prod_j^L \rho_j^{(1-\gamma)} \left(1 - \prod_j^L (1-\rho_j)\right)^{\gamma}, \tag{4}$$

where $\gamma \approx 1, \rho_j = x_j m_{ji}, m_{ji} = \text{weights of rules}$

$$S^i_{H_4} = \sum_j^L m_{ji} X_j, \text{ where } i = 1,2,$$

m_{j1} – momentums of MBF, m_{j2} – areas of MBF (5)

for output variable

$$S_{\text{Output}} = \frac{x_1}{x_2}, \tag{6}$$

(MAXDOT inference / defuzzification)

Figure 2 shows the comparison of the transfer functions of a fuzzy system with two different defuzzification methods and the presented neural network. The data for the pure mathematical example object with two input

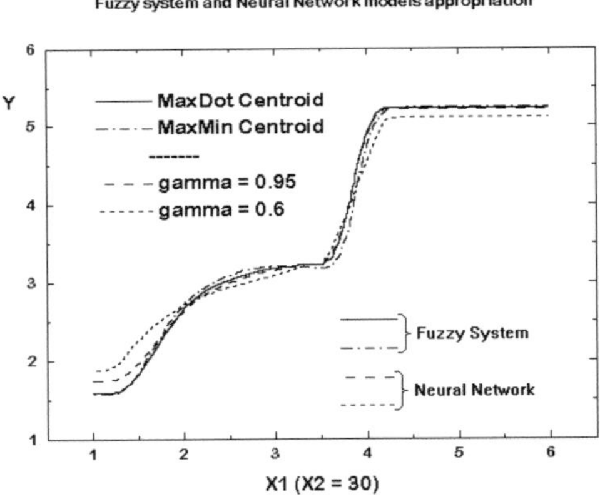

Figure 2. Transfer functions of a fuzzy system and the corresponding neural network.

variables (X1, X2) and one output variable (Y) were generated using the commercial Fuzzy-Tool TilShell. The figure shows that the best approximation is obtained for γ = 0.95 in Eqn. (4) and the Max-Dot/Centroid Defuzzification-method.

After the network has been trained on given data, the network parameters are transformed again into the corresponding parameters of the fuzzy system.

Different sources are possible from which the data for the rules and membership functions can be obtained. According to their sources the data used in this tool can be classified in:

- data from measurements on real existing and operating objects (experiments),
- data generated with physical or mathematical models (by simulation),
- data based on human knowledge and experience (knowledge acquisition),
- information in the form of approximate fuzzy linguistic models.

The main advantage which characterizes this tool is the flexibility in handling different data types for training the network. It is intended to use, as a maximum, all the available a priori information on the object to be simulated (see Table 1). There are two extreme situations possible:

- No information available on the "fuzziness" and linguistic rules that determine the relation between input and output data. Thus only "pure" input and output data available.
- Nearly complete information on the linguistic input and output variables (terms, membership functions, etc.) and on the rule-base given.

In the first case it is intended to derive a fuzzy object description including a rule-base and fuzzy input and output linguistic variables. In the second case only an optimization or improvement of a given model is needed. However, in this case a search for additional rules can be performed. Between these extremal situations intermediate information levels can occur, e.g. only part of the rule-base or parameters of some of the linguistic variables are known. All these different information levels can be handled with the developed tool. The information is transformed into fixed parameters or initial values for training the network.

The user-interface of the program package is realized as a graphical interface. The following functions are incorporated:

- primary data processing (smoothening, filtering, normalization, generation of membership functions as S-functions, square functions and singletons),
- input of the structure and initial values for parameters of a fuzzy system and/or the corresponding neural network,
- training of the neural network with different algorithms (back propagation, genetic algorithm, etc.),
- calculation of output data with a generated model and a set of test input data.

The tool runs under DOS-Windows 3.1.

The tool has been successfully tested on some examples in the chemical engineering field. As one example the fuzzy model for controlling a melting furnace for the production of corundum from aluminum oxide was improved. Lingvogenerator was applied for the optimization of the membership functions (MBF) of the four input variables (thickness of inner coating of the furnace, vapor production from cooling water flowing down the outside of the furnace (Yes/No), colour of the furnace bottom, type of raw material used) of the controller. The 12 control rules and seven terms of the output variable (levels of supplied electrical power) were held constant.

In another example Lingvogenerator was used to construct a rule for stream splitting in the design of heat exchanger networks. For the case of two hot streams and one cold stream, the cost optimal solution with and without splitting of the cold stream was evaluated for different start and target temperatures of the three streams. As input data for Lingvogenerator the values of four input variables (common temperature interval of the two hot streams divided by the temperature intervals of the hot streams, driving forces for the heat exchangers in the case of splitting) were used. The output variable "stream splitting" was assigned 1 for the case where stream splitting was cheaper and 0 otherwise. In both cases satisfactory results were obtained.

Conclusions

The actual version of Lingvogenerator is a prototype which can be applied for problems with small dimensions. For the extension of the program capacity a new SUN-UNIX-Version is in development. For the new version it is planned to extend the possibilities for primary data processing. In addition the authors are working on the improvement and the inclusion of new learning algorithms for the neural network. In the future the program will be tested on new typical examples to evaluate the possible field of application for the tool. Based on the examples, proposals for concrete solution strategies for different classes of problems will be developed.

References

Hauptmann, W. and K. Heesche (1994). Ein prototyp für ein integriertes fuzzy-neuro-system. In B. Reusch (Ed.), *Fuzzy Logic. Theorie und Praxis*. Springer Verlag, Heidelberg, New York. pp. 390-399.

Henning, H., V. Tryba and E. Muehlenfeld (1994). Automatische generierung von fuzzy-systemen mit genetischen algorithmen. In B. Reusch (Ed.), *Fuzzy Logic. Theorie und Praxis*. Springer Verlag, Heidelberg, New York. pp. 167-174.

Ishibuchi, H., K. Kwon and H. Tanaka (1993). Learning of fuzzy neural network from fuzzy inputs and fuzzy targets. *Fifth IFSA World Congress*. Seoul. pp. 147-150.

Kosko, B. (1992). *Neural Networks and Fuzzy Systems: A Dynamical Systems Approach to Machine Intelligence*. Prentice Hall, Englewood Cliffs, New Jersey.

Shann, J. J. and H. C. Fu (1993). A fuzzy neural network for knowledge learning. *Fifth IFSA World Congress*. Seoul. pp. 151-154.

Table 1. Starting Points for the Construction or Refinement of a Linguistic Model.

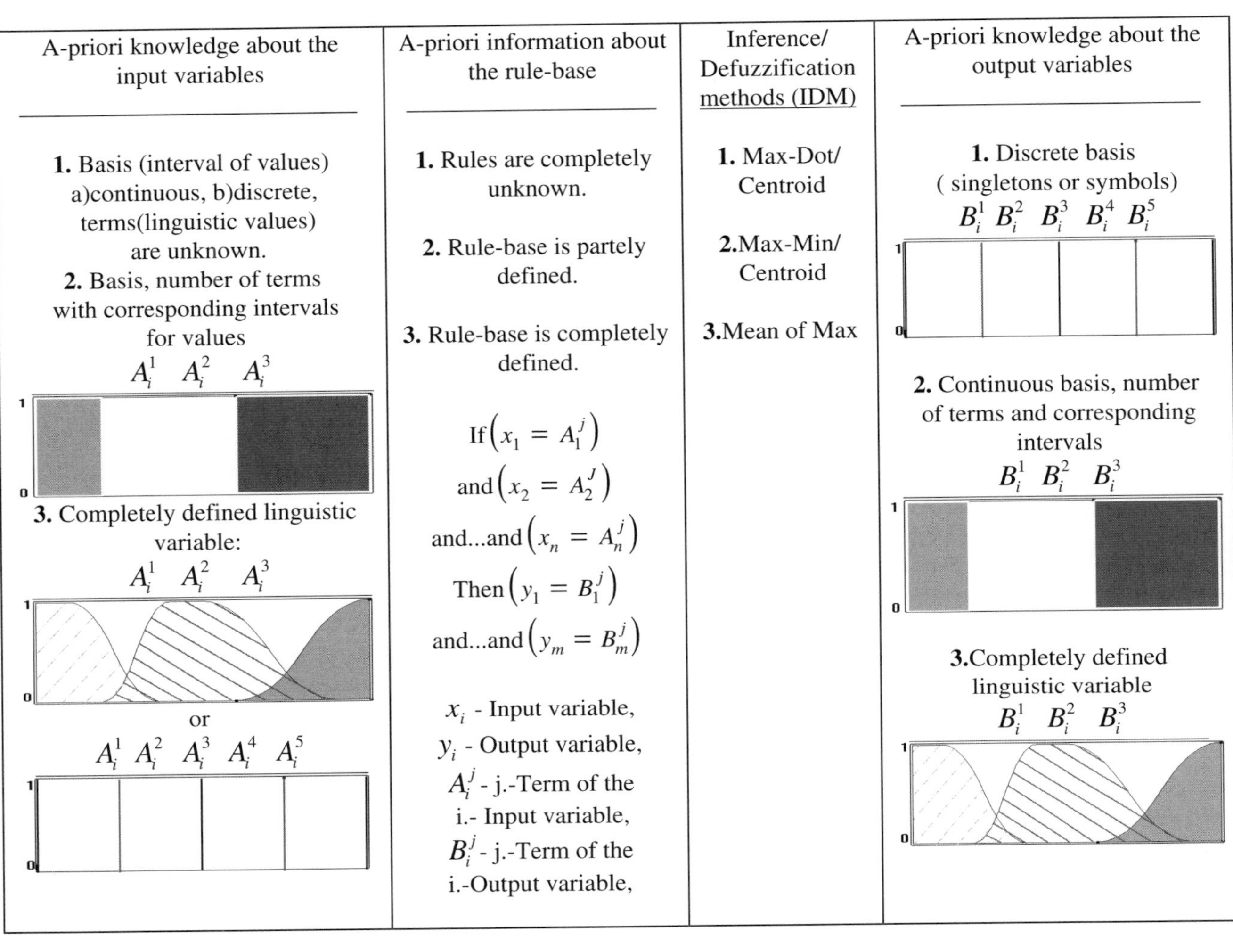

A-priori knowledge about the input variables	A-priori information about the rule-base	Inference/ Defuzzification methods (IDM)	A-priori knowledge about the output variables
1. Basis (interval of values) a)continuous, b)discrete, terms(linguistic values) are unknown.	**1.** Rules are completely unknown.	**1.** Max-Dot/ Centroid	**1.** Discrete basis (singletons or symbols) B_i^1 B_i^2 B_i^3 B_i^4 B_i^5
2. Basis, number of terms with corresponding intervals for values A_i^1 A_i^2 A_i^3	**2.** Rule-base is partely defined.	**2.**Max-Min/ Centroid	**2.** Continuous basis, number of terms and corresponding intervals B_i^1 B_i^2 B_i^3
3. Completely defined linguistic variable: A_i^1 A_i^2 A_i^3 or A_i^1 A_i^2 A_i^3 A_i^4 A_i^5	**3.** Rule-base is completely defined. $\text{If}\left(x_1 = A_1^j\right)$ $\text{and}\left(x_2 = A_2^J\right)$ $\text{and...and}\left(x_n = A_n^j\right)$ $\text{Then}\left(y_1 = B_1^j\right)$ $\text{and...and}\left(y_m = B_m^j\right)$ x_i - Input variable, y_i - Output variable, A_i^j - j.-Term of the i.- Input variable, B_i^j - j.-Term of the i.-Output variable,	**3.**Mean of Max	**3.**Completely defined linguistic variable B_i^1 B_i^2 B_i^3

PHENOMENON DRIVEN PROCESS DESIGN METHODOLOGY: FOCUS ON AUTOMOBILE EXHAUST GAS CONVERTER KINETICS

Veikko J. Pohjola and Juha L. Ahola
University of Oulu
SF-90570 Oulu, Finland

Abstract

Phenomenon driven process design methodology is a systematic approach to innovative and creative design in which design decisions are primarily made on conditions of the inherent controllability of the phenomenon to be taken under control. This work reports on application of the Phenomenon driven process design methodology in a design project, including the activities of producing experimental data in transient conditions in a laboratory scale apparatus, and building a kinetic model of the composite chemical reaction taking place in an automobile exhaust gas converter. As the control of phenomenon takes place through the control of the thermodynamic state of the material in which the phenomenon occurs, the existence of a relation between the rate and the thermodynamic state is a necessary prerequisite for a phenomenon to be controllable.

Keywords

Design methodology, Procedural model, Object-oriented, Exhaust gas conversion, Transient, Kinetics.

Introduction

Phenomenon driven process design methodology (Pohjola, 1994; Tanskanen, Pohjola and Lien 1994) is a systematic approach to innovative and creative design in which design decisions are primarily made on conditions of the inherent controllability of the phenomenon to be taken under control. Thus, in design, attention is first paid on characterizing the phenomenon, its structure and state, and on assessing the inherent controllability on that basis.

This work reports on application of the Phenomenon driven process design methodology in a design project, including the activities of producing experimental data in transient conditions in a laboratory scale apparatus, and building a kinetic model of the composite chemical reaction taking place in an automobile exhaust gas converter. According to the methodology, the kinetics are represented as a relation model and inserted into an object-oriented framework permitting a natural way of associating the knowledge of kinetics and the relevant contextual information with the converter design and the design activity itself, all in a unified format. The systematization of the modelling activity and the utilization of the knowledge of the kinetics in conceptual design of the converter are illustrated by a procedural model, to be implemented in an object oriented environment, in the form of object flow diagram.

Phenomenon Driven Process Design Methodology

Exhaust gas converter is a chemical reactor. Reactor is a chemical process where chemical reaction is in the main role. All what applies for chemical process design within the Phenomenon driven process design methodology, applies for exhaust gas converter design. According to the Phenomenon driven process design methodology, design is always viewed as a project. This is how every decision — the choice made, the assumption accepted — at any level of the task hierarchy of a design project, is put into perspective by weighting it against the goals set and the resources allocated. Especially the required credibility of the information and knowledge, in the form of data and relation models, about the real world, becomes an explicit issue. All data and every relation model should serve for achieving some goal, which specifies the expectation on the credibility, at the same time when the efforts for satisfying the expectations should agree with the constraints posed on time and other resources.

General Characteristics

The Phenomenon driven process design methodology is a conceptual framework, which is based on a carefully worked-out concept analysis, giving the methodology the following distinctive general characteristics:

(1) Human activity (design) and the target of activity (artifact) are both represented within a single framework to form a whole named project.

(2) Any object within project (activity or target), or project itself, can be characterized by a unified set of attributes: Purpose, Structure, State, and Performance, which is a sufficient set of attributes.

(3) Structure of any object has two dimensions: Topology and Unit structure. These two dimensions are sufficient to describe or prescribe the structure of an object to any degree of detail or abstraction. The structural representation of an artifact results in an object class hierarchy and is the declarative metamodel of the artifact. The structural representation of an activity results in a task hierarchy and is the procedural metamodel of the activity.

(4) Project object, including the two metamodels, forms a unified framework into which the instance level project management specific, target specific, and activity specific knowledge can be inserted.

(5) The framework can be implemented as a computer program integrating a dialog part, functioning as an interactive procedural coordinator for both the management and the designers during the project, with an object-oriented database, in which the design result can be dynamically built and documented in association with the relevant contextual knowledge.

Specific Characteristics

The distinctive specific characteristics of the Phenomenon driven process design methodology are embedded in the Unit structure attribute of process object and stem from the following explication: "(Chemical) Process is Control of (physico-chemical) Phenomena for Purpose." Thus phenomenon is given the central role in process design. It is characterized as a spontaneous event, independent of human being, but subject to becoming tamed down by the designer for making it advance in a controlled manner, and thus subject to becoming called process. The means to control become thus chosen largely on conditions of the nature of the phenomenon. The specific characteristics are:

(1) Unit structure of any (chemical) process is composed of the parts: Boundary, Interior, Exterior, and Interaction (as illustrated in Fig. 1)

(2) Each unit-structural part of process is an artifact (characterized by attributes: Purpose, Structure, State, and Performance).

(3) Phenomenon to be taken under control, and the material to be converted by the action of phenomenon, reside in process interior. Phenomenon cannot be controlled directly, but only through controlling the thermodynamic state of interior material. This is possible only if there is knowledge in possession of the designer, in the form of a relation model, about the relationship between phenomenon state and material state. This is how kinetic modelling becomes associated with process design. Control is effected by a proper design of boundary and interaction. Process design is deciding on the values of attributes of these artifacts.

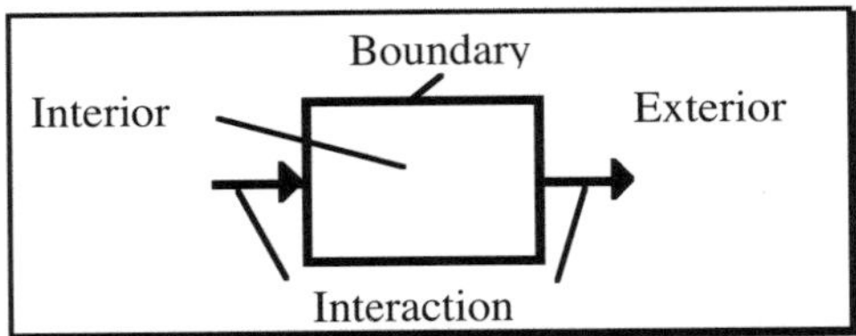

Figure 1. Unit structure of process.

(4) Process interior is disaggregated into subinteriors only if, within a single interior, such material state distributions cannot be achieved and maintained which are necessary for phenomenon control. The subinteriors are formed by applying more boundaries. This is the mechanism how process topology is generated. Typically, the reason for difficulties in phenomenon control is that the phenomenon is an aggregate of subphenomena with conflicting control requirements.

(5) While process structure is detailed by repeated disaggregation into lower level topologies, process state at a higher level of abstraction is an aggregate of the states of these topological units or subprocesses. The state of any subprocess is an aggregate of the states of its unit-structural parts (Boundary, Interior, Exterior, and Interaction). The state of any unit-structural part (like interior) is an aggregate of the states of its topological units (like subinteriors), the states of which are aggregates of the states of the corresponding unit-structural parts (like phenomenon) etc.

(6) Process performance is the follow-up of the decisions made on process structure. It is evaluated by explicit criteria like controllability, profitability and safety, and is used for steering the decision making forming the basis of "Performance-driven design strategy" (Pohjola and Alha, 1994).

Converter Design Project

In the automobile exhaust gas converter design project the goal was "a better knowledge about transient effects at the catalyst surface on the converter performance." Thus, the transient data gathered and the kinetic models built were expected to have a credibility, sufficient that, in the case that a satisfactory improvement of the converter performance would be predicted, either a decision on starting a continuation project with additional resources, or a decision to build a converter prototype, can be made.

Key Phenomena

The phenomenon to be taken under control is a heterogeneously catalyzed composite chemical reaction disaggregating into a set of simultaneous reactions: CO oxidation, reaction between CO and NO, water-gas shift reaction, reaction between NO and H2, hydrocarbons oxidation, steam reforming, and reaction between hydrocarbons and NO, and further into mechanistic steps of each of these. This is the characterization of the (topological) structure of the phenomenon. It implies the presence in the gas phase of the components appearing in the stoichiometries, as well as the presence of the active sites (occupied and unoccupied) on the solid catalyst. The rate of the composite reaction is an aggregate of the rates of the individual reactions, the rates of which again are aggregates of the rates of the mechanistic steps. The inherent controllability of the phenomenon is an assessment of the ease or difficulty by which the mechanistic steps, and thus the composite reaction, can be made advance at the desired rate.

As the control of phenomenon takes place through the control of the thermodynamic state of the material in which the phenomenon occurs (exhaust gas and catalyst), the existence of a relation (the kinetic equation) between the rate and the thermodynamic state is a necessary prerequisite for a phenomenon to be controllable. As there are now many more than a single chemical reaction occurring within the same boundaries, the requirements on thermodynamic state can certainly be expected to mutually conflict. Then it is only by distributing the thermodynamic state spatially and/or temporally that the requirements of each rate may be satisfied.

Kinetic Model

Declarative knowledge related to an object instance refers to the values assigned for the instance attributes. A value can be viewed as a solution of the corresponding relation model and can exist either in the solved (explicit) or the unsolved (implicit) form. The State attribute of Phenomenon object decomposes into subattributes Rate and Extent. The Rate attribute of a Phenomenon instance has the solution of the rate relation model as its value. Sometimes this relation model is available in an explicit form, readily solved for the rate variable.

The unified format of the object hierarchy implies that the values of attributes can be objects themselves. Thus reaction kinetics, represented as the solution of a relation model, is an object. The solution is usually a mathematical expression. Any mathematical expression can be described by the structure:

Expression = Operator[Operand],

where Operand, and even Operator can be Expression. The operation that Operator carries out on Operand can be either maintaining or manipulating. Using the *Mathematica* notation (Wolfram, 1991), the meta-level prescription of the explicit rate of a composite chemical reaction is then :

Solve[{RateRel,AuxRel},RateVar],

where each operand is a list like

RateVar = List[RateVar1,RateVar2,...]

and each individual rate relation has the form like:

RateRel1 = Equal[RateVar1,Apply[f,StateVar]]

where StateVar is a list of thermodynamic state variables.

The anonymous function f is the operator to be fixed on the basis of kinetic theories and experimental data. The auxiliary relations (AuxRel) include all the assumptions made and taken to match the credibility requirements of the goal.

Modeling Activity

The topology of kinetic modelling activity includes the following subactivities: (1) Observation (Data acquisition), (2) Explanation (Relation model generation), and (3) Prediction (Relation model solving), which each can disaggregate further. Data acquisition implies the following subactivities: Experimental apparatus design and construction, Experiment planning, and Data production. In order to get the Purpose attributes of each of these subactivities fixed, there should be initial knowledge at the kinetic modelling subproject level about what is it that the observation should be directed into. In this project it was initially known that: (a) the converter will always operate in transient conditions, (b) according to some evidence, a controlled transient operation may be beneficial (Cho, 1988), and (c) measurement of catalyst performance in transient conditions may give new insight into the mechanisms at the catalyst surface (Kobayashi, 1982).

The association of objects at the project level to objects at the kinetic model level is via their belonging to the same object hierarchy. Part of the hierarchy is depicted in Fig. 2.

The task hierarchy follows from the target object hierarchy. The generic task is: “Get value of attribute.” A task induces activities for generating knowledge, formalizing it into a relation model, and solving the relation for the attribute value. The task hierarchy is a procedural metamodel of process design and is depicted in Fig. 3 in its most abstract form: at the project level.

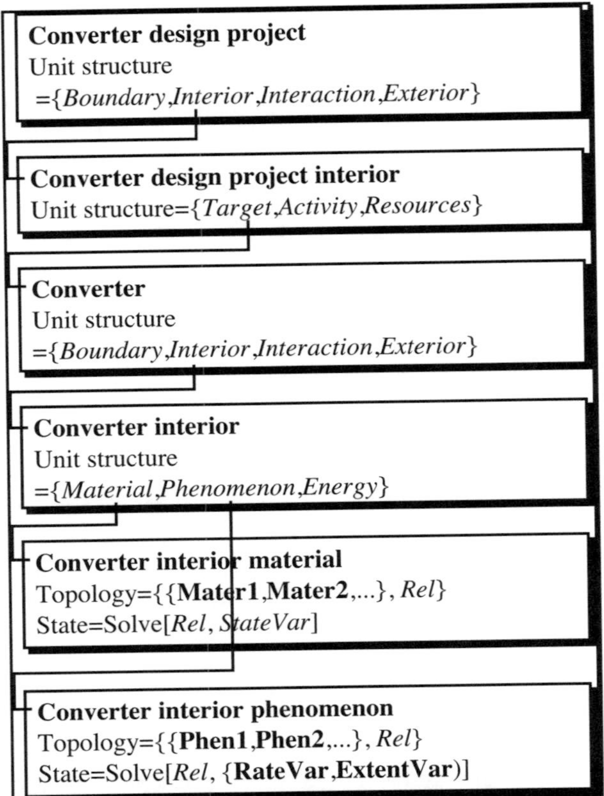

Figure 2. A fragment of object hierarchy representing declarative metamodel of process design project.

In its simplest form a relation model is an explicit assignment of a value for an attribute. The quality of the domain knowledge behind a relation model should be in agreement with the requirements posed for the credibility of the value. Kinetic data to be generated in the converter design project were to serve for solving (1) the topology of the composite chemical reaction and (2) the anonymous function f in each individual kinetic equation. The data were inspected together with domain knowledge on kinetics, that is, theories and assumptions on the prevailing relations.

Reaction topology is actually the representation of the mechanism of a composite reaction, the set of stoichiometric equations for each subreaction and for each mechanistic step. Hence, such data were generated which give clues about the mechanisms. The expectation was that data measured in transient conditions would be particularly beneficial. The relation model to be built had then the following format: IF the shape of step (or pulse) response is ..., THEN the probability of that mechanistic hypothesis is ... The common domain knowledge. normally associated, is composed of the collision theory of kinetics and the Arrhenius law, which yield:

f = gT[#1] gcA[#2] gcB[#3]&

where

gT[x_] = Times[k0Exp[-E/(Rx)]
gcA[x_] = Power[x,a]
gcB[x_] = Power[x,b]

The experimental data were also to serve for obtaining the numerical values of the parameters appearing in f.

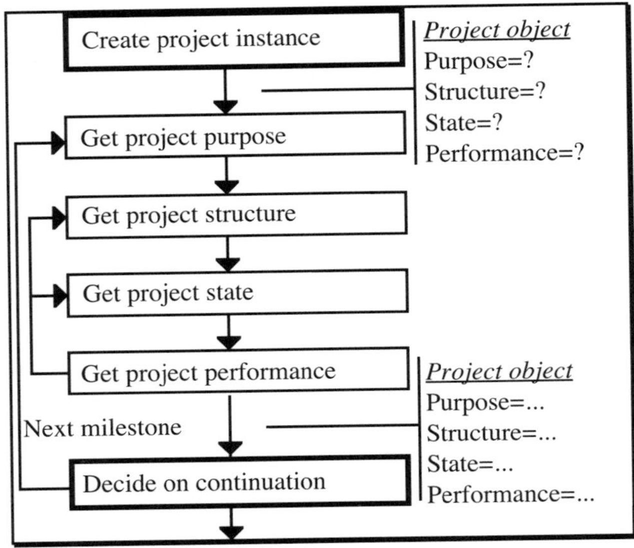

Figure 3. Procedural metamodel of running a project represented as an object flow diagram.

Conclusions

The phenomenon driven methodology gives a sound basis for building AI-tools for systematizing the utilization and application of know-how of the domain and thus for enhancing process design including all its subactivities.

References

Cho, B. K. (1988). Performance of Pt/Al_2O_3 catalysts in automobile engine exhaust with oscillatory air/fuel ratio. *Ind. Eng. Chem. Res.*, **27**, 30-36.

Kobayashi, M. (1982). Characterization of transient response curves in heterogeneous catalysis — I. classification of the curves. *Chem. Eng. Sci.*, **37**, 393-401.

Pohjola, V. J. (1994) Systematization of process design. CARD-2 project report. Oulu, Finland.

Pohjola, V. J. and M. K. Alha (1994) Performance driven strategy of process design. *Proceedings of the PSE'94*. Kyongju, Korea.

Tanskanen, J., V. J. Pohjola and K. M. Lien (1994) Phenomenon driven process design methodology: focus on reactive distillation. *ESCAPE 4*. Institution of Chemical Engineers, Rugby, UK.

Wolfram, S. (1991). *Mathematica*. Addison-Wesley, Redwood City, California.

SERO: A KNOWLEDGE-BASED SYSTEM FOR HAZOP STUDIES

A. Vecchietti and H. Leone
Instituto de Desarrollo y Diseño, Conicet, Avellaneda 3657
Facultad Regional Santa Fe, U.T.N, Lavaise 610
Santa Fe, 3000, Argentina

Abstract

This paper describes the development of a knowledge based system prototype (SERO) to aid process engineers in hazard and operability studies. Hazard and Operability Study (HAZOP) is a methodology to identify hazards and potential operation problems in process plants. In HAZOP studies, the process P&IDs are examined and possible deviations from the designer's intentions are discovered and considered in terms of their causes and consequences. Since HAZOP studies require many different kinds of knowledge sources it is necessary a flexible tool for knowledge representation. Therefore, the use of an object-oriented and rule-based environment is the natural option to capture the different types of knowledge that have to be represented in the system. The described prototype has been developed on KAPPA-PC™ (an object oriented environment, running on a DOS-WINDOWS platform, IntelliCorp Corp.).

Keywords

HAZOP studies, Object oriented programming, Qualitative modeling.

Introduction

Over the last years, there has been an increase in the scale and complexity of process plants.

The design of these plants requires a revision for possible hazards and operability problems. This is a complex and time consuming task. One of the most effective techniques in achieving it is the HAZOP study. For many years this technique was applied by a team of experts.

A HAZOP study is mainly a qualitative and symbolic task (Waters, 1989). For this reason, the development of a knowledge based system using qualitative models for process operations is the natural approach to provide computer support.

HAZOP Study

In a continuous plant, the HAZOP study is performed in a systematic way. Each pipeline is considered in turn to reduce the chance that something is missing. A pipeline, for our purpose, is a device that links or joins two main pieces of equipment, e.g., we must begin with the line that join the feed tank to the reactor, or the feed tank to the first heat exchanger (Lees, 1980; Roach, 1981). Then, a series of guide words (none, more, less, more than) is applied to process variables like flow, temperature, pressure and concentration to select a deviation. Then, questions like:

- Could there be more flow?
- If so, how could it happen?
- What are the consequences of more flow?

are asked.

The target is to identify causes and consequences in response to the deviation keywords considered (Fig. 1), and to suggest the necessary changes in the design or in the operative procedures, to prevent it.

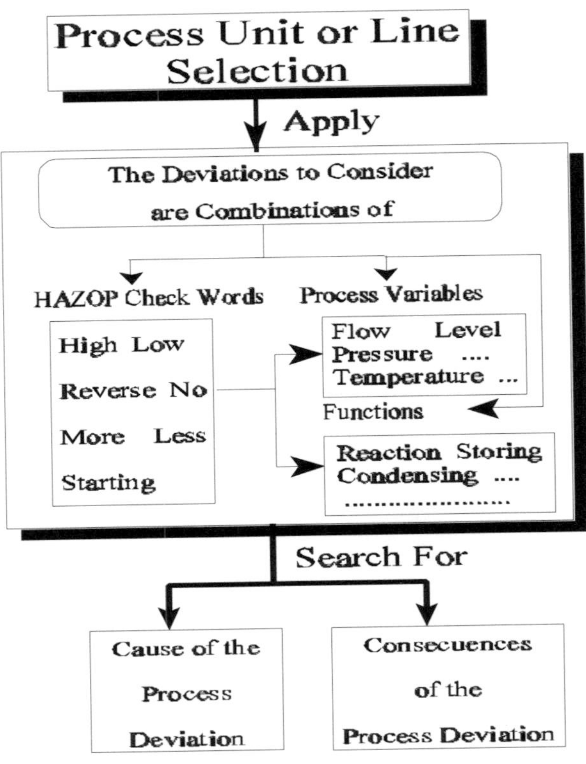

Figure 1. HAZOP study strategy.

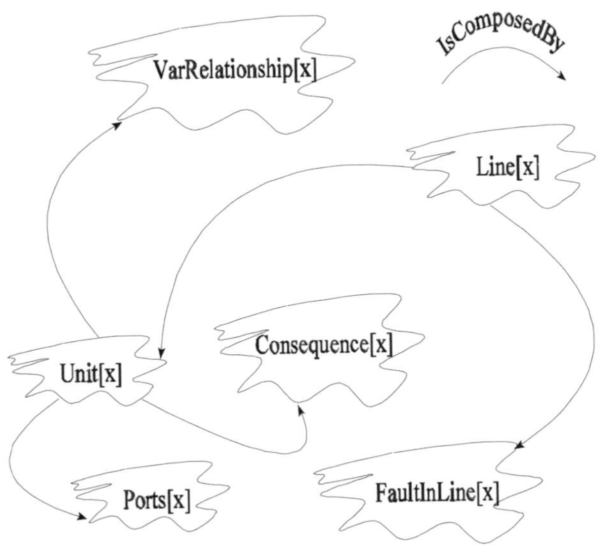

Figure 2. Class diagram.

Knowledge Base

SERO uses qualitative models of each piece of equipment that capture the cause and effect nature of the interactions that occur between the variables that describe the system (i.e., temperatures, valve positions, pressures, flow-rates) and the events that occur within the system (i.e., valve failure, operator error, weather changes, etc.).

The knowledge base of SERO was designed in the context of the object oriented programming paradigm (Booch, 1991) because it provides the natural framework for the representation of the knowledge needed in the HAZOP Study.

Each processing unit is represented by a composite object (subclass of unit) that contains an explicit qualitative model of it (Fig. 2). This object is composed of instances of *VarRelationship*, *Consequence* and *Port*. The *Line* is another HAZOP concept represented by a composite object (Fig. 2) (an instance of *Line* is ComposedBy instances of *Unit* and *FaultInLine*). The relationship between the different variables is represented by a gain value (0: independent variables, +1: variables that tend to change in the same direction, -1: variables that tend to change in opposite directions).

Each one of this relationships is an instance of the class *VarRelationship* (Fig. 4).

The relationships represent the behavior of the unit in the absence of a fault. The model also has the representation of the operation of the unit in a faulted mode.

The unit class model has the sintaxis proposed by Henning, et al. (1990)

```
<input-port> ::= <input-port-name> |
    <input-port-name> <input-port>
<output-port> ::= <output-port-name> |
    <output-port-name> <output-port>
<mod>::= <unit><var><unit><var><gain> |
    <unit><var><unit><var><gain><mod>
<unit> ::= <piece-of-equipment> |
    <input-port> | <output-port>
<var> ::= T | F| P | CompVec? | Comp? | ...
<gain> ::= -1 | +1
<fault-mode> ::= <fault-mode-name> <var> <value>
<consequence> ::= <consequence> <var> <value>
<value> ::= VeryHigh | High | Normal | ...
...
```

SERO uses the class model expressed with this language in the creation of the composite object that represents a given processing unit.

The prototype uses qualitative models with a bottom-up approach to consequence detection (from fault mode pattern to consequence). Consequently, the qualitative simulation has more information about the system state and it is possible to reduce the number of ambiguities.

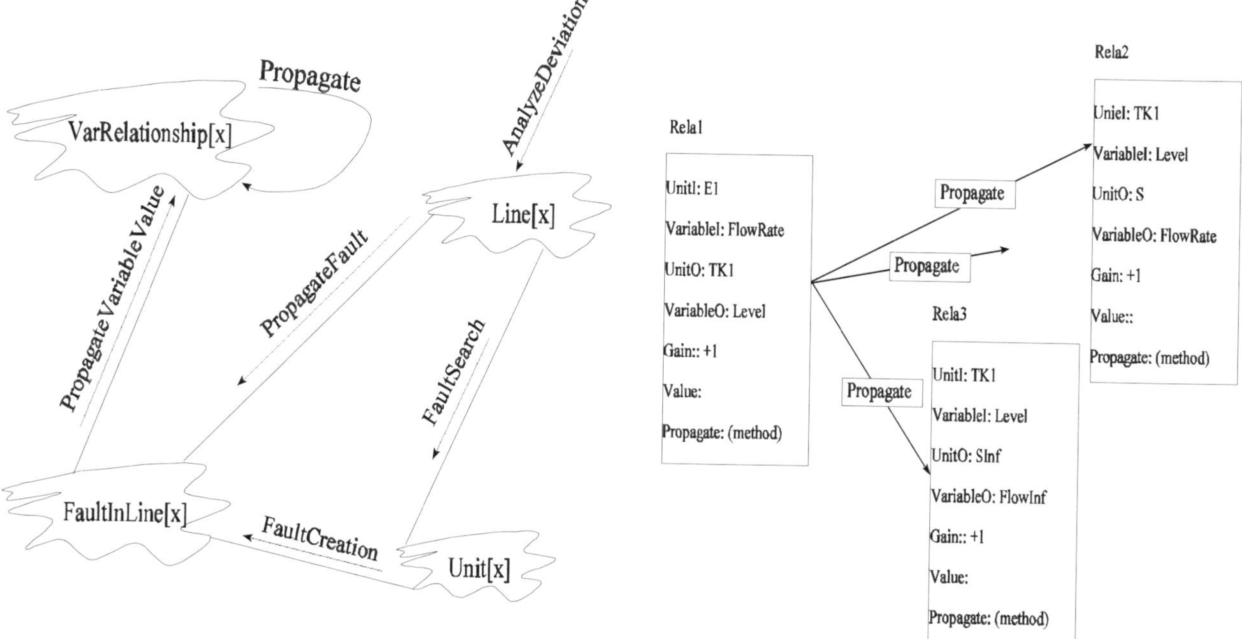

Figure 3. Message passing diagram.

Figure 4. Propagation in the model.

Inference Mechanism

SERO proposes a HAZOP Check word to a line or unit variable, and then propagates backward the variable through the qualitative model of the unit or line, looking for faults that cause the mentioned deviation (message *FaultSearch* in Fig. 3).

When a fault is detected, the system creates an instance of the class FaultInLine (message *FaultCreation* Fig. 3). After all the faults have been found, the system propagates the effects of each one (messages *PropagateFault* and *PropagateVariable-Value*) through the model represented by the objects of the class *VarRelationship*. During this process, **SERO** detects the consequence of the fault (message Propagate represented in Fig. 4).

The consequences are related to the variable values obtained from the fault propagation. In the model of each unit there is a consequence object associated with each variable value (Fig. 2).

SERO System

SERO is an environment with three main components:

- Qualitative Model Editor (**QME**)
- Process P&I configuration (**PPI**)
- HAZOP Study Session (**HSS**)

A **QME** session window is an editor where the process engineer can define a model .

QME allows the user to define:

- the topology of the unit model
- variables of the unit and input/output ports
- relations between variables (Fig. 5a)
- fault mode models (Fig. 5b)
- consequences

QME allows process engineers to modify an existing model class or tailor a process unit instance (e.g., he/she can add process specific consequences, relations or any other model component). Figure 5 shows the model editor session for gain relation and fault mode specifications.

A **PPI** window session is a graphical interface where the process engineer can introduce the P&I diagram of the line.

A **HSS** session window is the interface where the process engineer can perform the HAZOP study. He /She can select the line or process unit. Then, it is possible to start the study or choose any one of the pair HAZOP Check Word/Variable. The engineer analyzes the HAZOP study results in this session window.

Conclusions

At present, the library of qualitative models is being expanded. We hope to extend this in the future by means of a high-level language that allows the user to write a qualitative model for his problem in a more explicit way. The next step is to incorporate the HAZOP analysis for batch processes, taking into account the concept of operating procedures and other topics related to discontinuous processes.

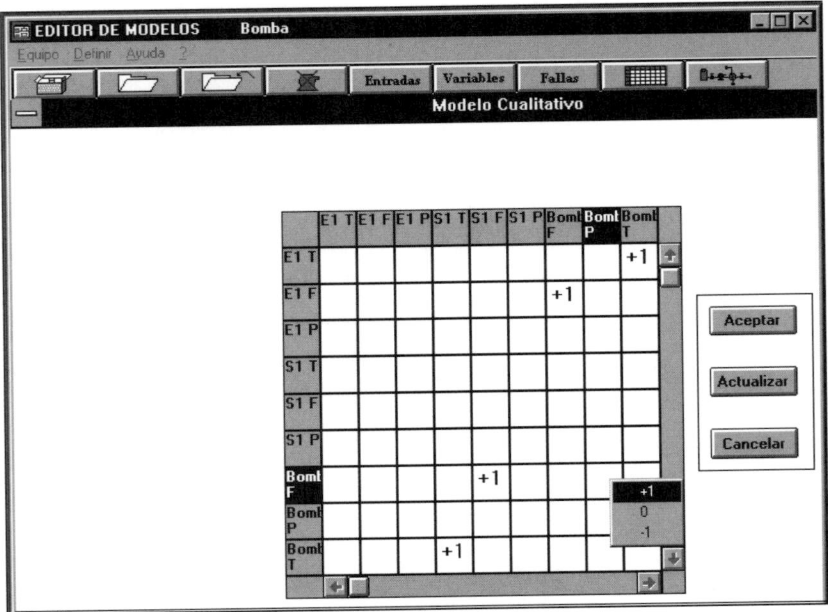

Figure 5a. Qualitative Model Editor.

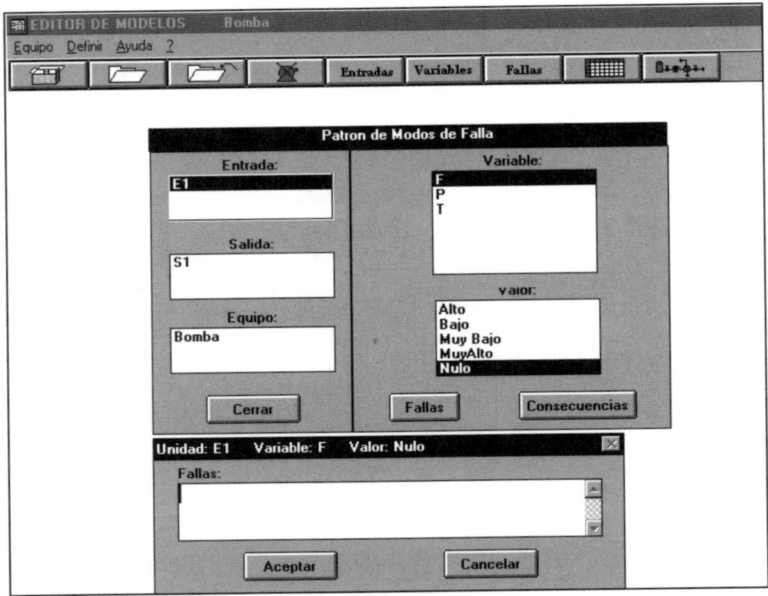

Figure 5b. Qualitative Model Editor.
(Entrada - Input, Salida-Output, Equipo-Unit, Valor- Qual.Value, Falla - FaultMode)

References

Booch, G. (1991). *Object Oriented Design with Applications*. The Benjamin/Cummings Publishing Company.

Henning, G. P., H. P. Leone and Geo. Stephanopoulos (1990). MODEL.LA. A MODELing LAnguage for process engineering. Part I: The formal framework. *Computers and Chemical Engineering*, **14**, 8.

Lees, F. (1980). *Loss Prevention in the Process Industries*. Vol. I and II. Butherworths, London.

Roach, J. R. and F. Lees. (1981). Some features of and activities in hazard and operability (HAZOP) studies. *The Chem. Eng.*, October.

Waters, A. and J. Ponton. (1989). Qualitative simulation and fault propagation in process plants. *Chem. Eng. Res. Dev.*, **67**, July.

INTEGRATED DESIGN IN POLYMER COMPOSITES: DISTRIBUTED HYPERMEDIA INFORMATION RESOURCES AND INTELLIGENT DECISION SUPPORT SYSTEMS

James K. McDowell, Timothy J. Lenz, Jon Sticklen and Martin Hawley
Intelligent Systems Laboratory and Composite Materials & Structures Center
Michigan State University
East Lansing, MI 48824-1326

Abstract

Successful design requires the inclusion and integration of downstream issues in order to achieve feasible and affordable designs. Intelligent decision support systems (IDSS's) have a role in such an integrated, concurrent view of design. The use of such systems in the context of polymer composites design facilitates the continued, consistent application of engineering design knowledge. These systems can also provide, early in the design process, criticism and advice concerning downstream issues which can impact affordability. This work discusses the use of two IDSS's in the rapid prototyping of a polymer composite missile seeker casing. The IDSS's used were: CTechSel, an IDSS for the selection of polymer composite fabrication methods, and COMADE, an IDSS for the design of polymer composite materials. Working versions of both CTechSel and COMADE are available on the World-Wide Web (WWW). Parallel to the design and production of the seeker casing was an effort to capitalize on the multimedia documentation capabilities of the WWW. The documentation generated both illustrates how these IDSS's were used in the design and production of the seeker casing and describes in detail the decisions and actions of the designers and the production staff.

Keywords

Design, Polymer composites, Intelligent decision support, Hypermedia, World Wide Web.

Introduction

Design is critical to the success of polymer composites and can be viewed from the perspectives of materials, part and processing. The interactions of these perspectives provide serious bottlenecks to affordability, requiring knowledge intensive solutions. This is particularly important as polymer composites penetrate the durable goods markets and as traditional high performance applications like aerospace focus on affordability. Success requires that designers include and integrate downstream aspects, providing designs that are feasible and affordable. Intelligent decision support systems (IDSS's) have a role in this integrated, concurrent view of design. The use of such systems in the context of polymer composites design facilitates the continued, consistent application of engineering design knowledge, thus enabling the transfer of expertise and freeing design engineers for more creative tasks. These systems can also provide criticism and advice concerning downstream issues that impact affordability, early in the design process.

Knowledge-based systems (KBS) applications for the design and manufacture of polymer composite parts/assemblies offers a fruitful problem solving domain for IDSS's and provides challenges for the continued development of KBS technology. This work discusses the use of two IDSS's in the rapid prototyping of a polymer composite missile seeker casing. These tools were used in the context of an integrated vision of composites design, which spanned initial conceptual ideas through to the physical production of actual seeker casings. One of the casings produced was then sent on to the teams that were producing the seeker optics and control electronics.

The IDSS's used were: CTechSel (McDowell, et al., 1993), an IDSS for the selection of polymer composite fabrication methods and COMADE (Lenz, et al., 1994), an IDSS for the design of polymer composite materials. Both of these systems are founded on general notions of task specific architectures and specifically on the generic task (GT) theory of KBS (Chandrasekaran, 1983). GT theory has driven the development of the tools used to construct CTechSel and COMADE and has impacted the overall problem solving vision for polymer composites design. The use of these IDSS's in an actual design and production exercise provided considerable insight and validation not possible in isolated testing. It also demonstrated many of the ideas on how these systems might be used as aides in design problem solving.

Parallel to the design and production of the seeker casing was an effort utilize the multimedia documentation capabilities of the Internet's World-Wide Web (WWW). As a result, a multimedia, hyper-indexed archive of the design and production process is available to anyone on the Internet. This documentation illustrates how the IDSS's were used in design and production of the seeker casing as well as the decisions and actions performed by designers and production staff. Additionally, working versions of CTechSel and COMADE are available on the WWW. Though users cannot directly view or change the knowledge in these versions, they are completely executable and will respond appropriately to user input. There is also documentation of the knowledge structures used for the problem solving. The use of the WWW in this manner provides an early example of how design/manufacturing information and design problem solving services can be made widely available on the Internet.

The MADEFAST Experiment

MADEFAST was an ambitious experiment in collaborative engineering over the Internet, undertaken by the Department of Defense's Advanced Research Projects Agency (ARPA) Manufacturing and Automated Design Engineering (MADE) contractors. The objective of the project was to deliver a working infrared seeker prototype for a tactical missile within a 6 month time-frame while establishing a collaborative infrastructure for the MADE community of sufficient depth to support this development. Once established, this collaborative group was to serve as the foundation for future engineering collaborations over the Internet.

The portion of the project assigned to Michigan State University's Advanced Computing Thrust (ACT) was the design and fabrication a composite casing, or housing, for the seeker electronics. This served as a high profile showcase for the agile manufacturing and rapid prototyping afforded through the use of the intelligent decision support systems developed by the MSU team.

A total of 13 different companies and academic institutions were officially involved in the MADEFAST project, and three of these (Stanford's Center for Design Research, University of Utah's Alpha-1 Project, and MSU's ACT as lead) collaborated extensively for the seeker casing portion of the project. AutoAir Composites, while outside of the official MADEFAST community, significantly contributed to the seeker casing effort as well.

IDSS and Composites Design

The ACT employs a three-pronged approach to the design of polymer composites, dealing with the design of the material, the design of the manufacturing process and the design of the part in an integrated manner. Intelligent decision support systems for polymer composites have been developed for both technology selection (CTechSel) and material design (COMADE). Where applicable, these systems were used in the design of the seeker casing. However, since these systems addressed only a portion of the overall vision of composites design, outside interaction with composites design experts was utilized to complete the design. Portions of the project which did not employ the IDSS included mold design and fabrication, determination of the specific part architecture and design of the processing parameters.

The goal of CTechSel is to provide a screening aid to design engineers who may not know the critical issues for every fabrication process available. Through a combination of qualitative economic factors and issues of the part geometry, CTechSel is able to assess the feasibility of using various composite fabrication processes. With a set of suitable fabrication processes, in hand the design engineer can apply more detailed screening and continue the evaluation design effort. The goal of COMADE is to provide the design engineer with a family of material systems that will satisfy the performance requirements and the environmental conditions that the composite part will face. Such an aid will help ensure that the most appropriate material systems are applied to a particular situation and will further the effective use of polymer composite materials.

Though both CTechSel and COMADE had been tested (granted under laboratory conditions), their use in this design exercise provided additional insight to their domain coverage, their task structure and their use and utility. It should be noted that in the scenarios described both CTechSel and COMADE produced reasonable answers.

On the fabrication technology selection side, two issues arose. One had to do with the interpretation of qualitative inputs by the users. Though notions of high, medium and low are normalized across the technologies in CTechSel's knowledge-base, such notions may not match the experience and understanding of average user of the system. Our current remedy to this is to provide documentation so that the user clearly understands the commitments that CTechSel is making and the meaning behind the qualitative ranges. The longer term solution would be to adapt CTechSel's reasoning to the profile of the user. In this way CTechSel could understand that "high process throughput" from a user in the automotive industry is different from that of a user from the aerospace industry.

The other issue for CTechSel is iterative examination of the user inputs. The goal of using CTechSel is to choose a suitable fabrication technology. There are several combinations of inputs under which no technology can satisfy the inputs. CTechSel rates them all as "against." The user is still faced with the prospect of producing a composite part and still needs advice on what fabrication technology to use. CTechSel's utility would increase if it provided suggestions on how to change the input profile so that one or more technologies would be found suitable.

In the material design arena, COMADE did well in generating possible alternatives for use in the seeker casing study. It is interesting to note however that the final material system used, the hybrid carbon/epoxy unidirectional prepreg (Hercules AS4/3501-6) and chopped glass/vinyl ester (styrene) SMC (Quantum QC-8800), was not directly generated by COMADE. In its current form COMADE cannot generate hybrid material systems. To its credit COMADE did generate these systems individually under the less severe operating conditions and the four thermoplastic systems generated under the severe specifications would have done the job. The generation of hybrid material designs produces additional concerns and requires an intelligent partitioning of the input specifications, knowledge about composite material systems compatibility and in the general case appears to be a potentially computationally explosive process.

The WWW

The World Wide Web (also know as the Web) is the fastest growing portion of the Internet. Estimates are that the Web is growing on the order of 10% per month.

From education to commerce to entertainment, new services appear on the Web every day. The capability to perform secure transmission of data is beginning to make electronic commerce a reality. Distributed hypermedia in the form of text, graphics, sound and movies allow for the presentation of sophisticated services and documentation, all of which can be navigated in a point and click manner with low cost (and in some cases free) software and an Internet connection.

Documentation of Design and Manufacturing

As mentioned previously, design in polymer composites in quite complex and involves the interaction of materials, process and part issues. Ultimately the design is realized in the actual manufacture of an artifact, also a complex enterprise. A traditional linear description of either design and/or manufacturing does not capture the complexity and interactions in the domain. Using the WWW's hyperlinked multimedia features, the non-linear aspects of design and manufacturing in polymer composites can begin to be conveyed. This is further facilitated by the use of image maps as a basis for navigation to various information sources. An example image map is shown in Fig. 1.

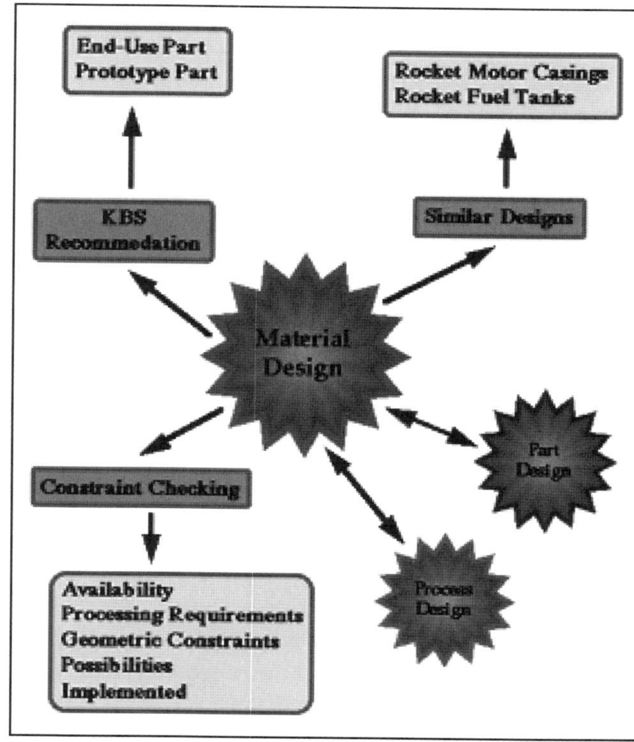

Figure 1. Image map for the material design phase.

Just by pointing and clicking on such a diagram a user is guided to various information sources. The image map in Fig. 1 shows the issues surrounding the material design, including similar designs from the literature, recommendations from the KBS and pragmatic constraints. There are also links (note the other starbursts) to the image maps that organize the part design and process design perspectives. In those image maps the person browsing is guided to descriptions of the part geometry, fiber architecture issues, diagrams of mold alternatives, graphs of the processing conditions and photographs of the part manufacturing steps (material layup, bagging) and equipment like the autoclave. Using these features of the WWW a comprehensive view of the polymer composite design and manufacturing activities can be presented.

Decision Support on the WWW

Included in the WWW documentation of the design and manufacture of the seeker casing are WWW versions of the IDSS used in the project. Using the forms features of the WWW, a user browsing can reproduce the input/output behavior of the problem solvers from the seeker casing scenario or explore scenarios of their own. The user only has to fill-in a form that includes type-in boxes or pull down menus and then submit a request by clicking on the submit button. The information on the form is transmitted to a computer at MSU where the problem solver is running. The input is processed by the IDSS and an output page is formatted and sent back to the user for display. Figure 2 shows the WWW input interface for COMADE.

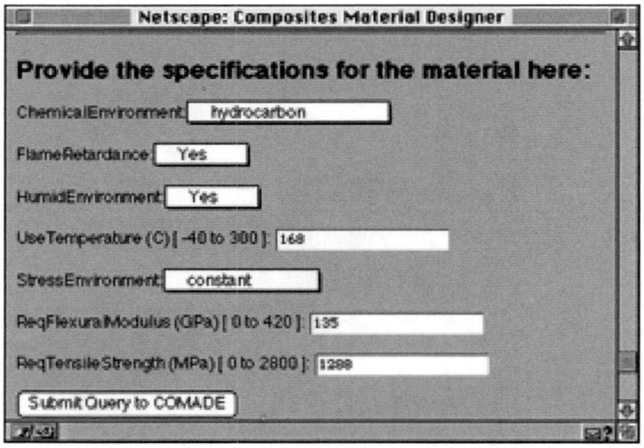

Figure 2. WWW interface for COMADE.

Providing access to IDSS on the WWW considerably expands their range of use. It makes delivery and maintenance of such systems nearly seamless. These WWW versions of COMADE and CTechSel as well as the WWW documentation of the seeker casing effort were shown at the Society for the Advancement of Materials and Process Engineering, SAMPE'95 Exposition, and were quite well received Beyond demonstration systems, it is easy to envision similar systems with more industry specific capabilities being made available via the WWW on a fee basis.

Conclusions

The WWW could significantly impact engineering and manufacturing. The WWW and its hyperlinked multimedia features provides the capability to document complex design and fabrication enterprises. This is demonstrated in the documentation of the polymer composite seeker casing. The WWW also provides access to computational services like intelligent decision support (e.g. COMADE and CTechSel). Other computational services could be made available in a similar manner.

A complete documentation of Michigan State's role in the MADEFAST is available on the World Wide Web at http://isl.cps.msu.edu/madefast.

Acknowledgments

This research is supported in part by ARPA's MADE Program under grant #8673, by the NSF Center for Low-Cost, High-Speed Polymer Composites Processing and by the State of Michigan's Research Excellence Fund. Apple Computer has also supported this research through generous equipment donations. The authors would also like to thank Pradeep Khosla, ARPA MADE program manager, for proposing the MADEFAST effort, John Scanlon of AutoAir Composites for assisting in the design and manufacture of the seeker casing and providing some of the materials used, Sam Drake and Carolyn Valiquette of the University of Utah for fabricating the mold used to produce the casing, Mike Muczynski, Brian Rook and Mike Rich of the MSU Composite Materials & Structures Center for their assistance with the fabrication of the part and finally Mark Cutkosky and Charles Petrie of Stanford's Center for Design Research for their fruitful discussions.

Although the individuals noted above have facilitated the efforts described in this paper, they should neither be considered responsible for the content of this paper nor should their support of the statements made herein be inferred. All opinions expressed in this paper are entirely those of the authors.

References

Chandrasekaran, B. (1983). Towards a taxonomy of problem-solving types. *AI Magazine*, **4**, 9-17.

Lenz, T., J. K. McDowell, B. Moy, J. Sticklen and M. C. Hawley (1994). Intelligent decision support for polymer composite material design in an integrated design environment. *Proceedings of the American Society of Composites 9th Technical Conference*, pp. 685-691.

McDowell, J., A. Kamel, J. Sticklen and M. Hawley (1993). Integrating material/part/process design for polymer composites. A knowledge-based problem solving approach. *Proceedings of American Society of Composites 8th Technical Conference*, pp. 54-63.

APPLICATION OF COMPUTER-AIDED TECHNIQUES TO PROCESS SCHEMATIC CREATION

Jim R. Baird and Stephen T. Jenkins
Eastman Chemical Company
Kingsport, TN 37662

Stephen R. Strong
Integrated Systems Engineering
Hudson, OH 44224

Abstract

A unified framework for capturing, maintaining and utilizing chemical process configuration knowledge is proposed. A specific English-like, structured language has been developed to capture process configuration knowledge. Functional analysis of standard process configurations served to identify the configuration knowledge. The working system utilizes a commercial Knowledge-Based Engineering (KBE) application development tool linked with a relational database, and a computer-aided drafting (CAD) tool. Within the KBE application, three object models were created to reasonably segregate information and functionality. The first model, referred to as the Expert Model, contains the knowledge constructs resulting from the functional analysis. The second model, referred to as the Engineering Model, contains the object representation of the configured, connected process components (equipment, piping, etc.). The third model, referred to as the Presentation Model, contains objects necessary to create various reports from the Engineering Model.

Keywords

Process schematics, Computer-aided engineering, Functional analysis, Configuration, Knowledge base.

Introduction

In business, there is the old maxim, "Good. Quick. Cheap. Choose two," and most projects are seen in that light. As information technology evolves, however, chemical companies are turning to computers and intricate networking to establish a new maxim — "Good. Quick. Cheap. Have all three!" In order to achieve this goal of reduced project cycle time with increased project effectiveness, many companies are looking into the establishment of computer systems to assist designers in the creation of various project deliverables. This has been particularly true of the creation of process schematics such as block flow diagrams (BFDs); process flow diagrams (PFDs); and piping and instrumentation diagrams (PIDs). Under the auspices of "automation," many resources have been spent creating process schematics more efficiently in a computer-aided drafting (CAD) environment. A serious shortcoming with such drawings, as noted by Colton and Pun (1994), is that they do not represent the knowledge and rationale considered during design decision-making. This problem of losing the design intent still exists even with presently available CAD systems which link graphic elements to information stored in relational databases.

Eastman Chemical Company is attempting through this project to go a step beyond the thought of simple data connections to schematic drawings and move forward to capturing and applying process configuration knowledge to the creation of design models. In this environment, process schematic views (BFD, PFD, and PID) of the design models are considered reports against the design model, just as would be a bill of material.

Application of knowledge engineering systems to the process design arena have been limited to date. This is due in part to a lack of general tools which meet industrial requirements for wide-scale utilization (Ichikawa, et al., 1986). The task is also encumbered by the wide variety of formats used by the numerous chemical process design knowledge sources. Additionally, much in-house configuration knowledge is simply not documented at all.

This project attempts to identify methods and build tools to assist in the identification, capture, and use of process configuration knowledge.

Knowledge Capture

Expressing human expertise in the form of configuration knowledge is perhaps the greatest challenge to implementing KBE systems as process design assistants. KBE systems typically do not provide the language constructs necessary to capture process configuration requirements except in a language too esoteric for enthusiastic use by domain experts (often the language is simply a common programming language). This frustrates the domain experts as they are unable to update the KBE system. It also places the company at risk of losing reasonable access to their own knowledge if the particular programming language falls into disfavor or the KBE system company goes out of business. A primary goal of the project was to place the chemical process designers in a position of control and responsibility in capturing and maintaining configuration knowledge.

Functional analysis of previously standardized process configurations served as the primary tool for identifying configuration knowledge. This analysis method provided a structured approach to formalizing user requirements in terms of functionality, focusing attention on the main purpose of the design (Kuttig, 1993), and forcing consideration of functionality rather than limiting thought to existing solutions. The resultant three-tiered organization of knowledge follows that of Kota and Lee (1993) [Fig. 1] where: the top tier contains solution independent functions (e.g. flow restricting); the middle tier contains feasible solutions which are generic physical devices (globe valve); and the bottom tier contains specific physical device information (3-inch diameter, carbon steel, etc., globe valve). The primary thrust of the project to date involves the understanding and capture of configuration knowledge at the first two tiers. Linkage to the bottom-tier knowledge will be made in the future when ties to equipment design packages, and piping and instrument component catalog information are implemented.

The functional requirements of the knowledge language are identification of:

1. Process function-solution pairs;
2. Configuration rules necessary to arrive at the best solution for a functional need;
3. Process information sources to feed the configuration rules; and
4. Relations to be established between functional solutions (e.g., connected-to, upstream-of, etc.) when they are brought into existence.

An information model to support the knowledge capture was developed and implemented in Oracle DBMS. Initial efforts using this tool indicate acceptance by the domain experts as the means for capturing and maintaining their configuration knowledge.

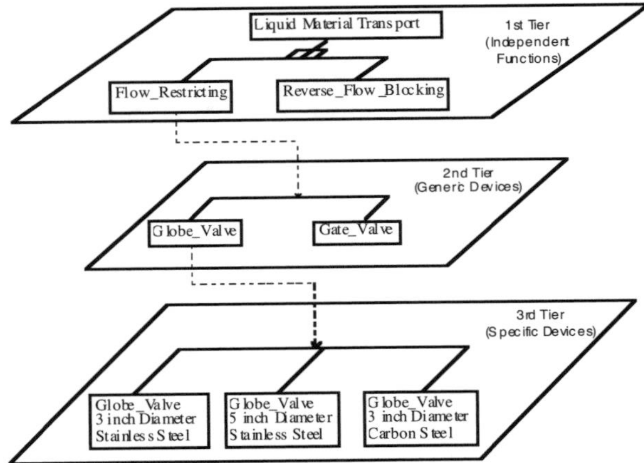

Figure 1. Three-tier knowledge breakdown.

System Architecture

An architecture, designed by Integrated Systems Engineering, and implemented in Design Power, Inc.'s D++ environment involves the segregation of information and functionality into three primary object models referred to as: Expert Model; Engineering Model; and Presentation Model. These are shown in Fig. 2. The D++ environment provides the mechanisms to establish user defined object-to-object relationships and the capability to trace through the relationships, conducting whatever reasoning is appropriate. The segregation into three models eliminates the mixing of rules used to: choose between functional solutions; logically connect resultant solutions; and create user acceptable graphic representations of the configured process.

Expert Model

The Expert Model contains information about functions which are candidates for configuration; the feasible solutions for the functions, and how to choose between the solutions. This is accomplished by:

1. Creating a copy of the configuration knowledge in various inter-related objects; and
2. Creating special objects to reason over the configuration knowledge, gather process information, and determine best solutions.

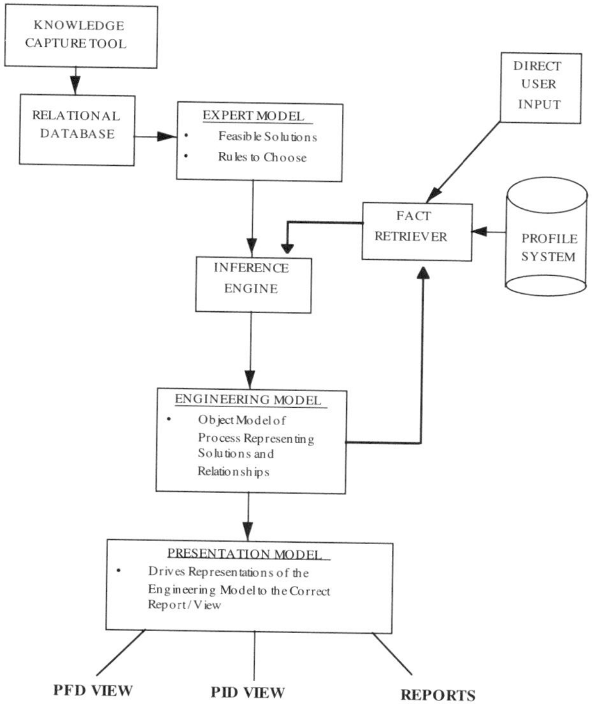

Figure 2. General system architecture.

In the future, it is likely the configuration knowledge will need to remain in the relational database for reasons of accessibility and the shear volume of data. For proof of concept purposes, having a copy of the configuration knowledge in an object model minimized system complexity and software development.

The system is extensible by either adding possible solutions to existing functions, or by the addition of new function-solution pairs. The critical point is that the system functionality is extensible via addition of knowledge by the process configuration expert using a language with which he/she is comfortable. The computer system can be used to evaluate a larger and larger number of possible solutions (Bligh and Chakrabarti, 1994), assisting the designer in migration to the optimum solution.

Engineering Model

The Engineering Model objects represent the actual, interconnected process components which are solutions based on specific process requirements. This model is built by the Expert Model as process information is supplied to the configuration rules by either the database or user interaction. Objects in the Engineering Model can be taught to adhere to basic engineering rules. Therefore, objects may be taught to size themselves given adequate process stream profile information, or to choose their material of construction, etc.

Using the D++ tool to define object-to-object relationships, it is also possible to maintain a "trail" to the function-solution pair from which the Engineering Model solution object came into existence. While this feature has not been exploited to its fullest, it does show promise for capturing the design intent as the system is enhanced.

Presentation Model

The Presentation Model contains objects whose responsibilities are to determine what Engineering Model objects require presentation for a particular view of interest, then present them in a way acceptable to the user. For example, the process functional requirement of moving a fluid from one vessel to another may exist. Out of the possible solutions (single centrifugal pump, dual centrifugal pumps, gravity flow, etc.), a single centrifugal pump is chosen and placed in the Engineering Model along with all its associated objects for piping and instrumentation. Requesting a PFD view of the pumping function from the Presentation Model may result in the user seeing a pump symbol, with in and out pipelines containing no piping components. If asked to present the PID view, the Presentation Model may include all the piping components as well as the pump controls. Since changes made to the Engineering Model would impact all Presentation Model views of the changed object, change control issues are minimized.

The Presentation Model not only serves as the mechanism to "present" the Engineering model to the outside world, but also serves as the user's "gateway" into the Engineering model. The system is designed such that design changes from the user's perspective are effected from the CAD environment. In reality, change requests from the CAD environment chain through the Presentation Model, to the Engineering model, and back to the Presentation Model and CAD environment.

The D++ product has interactive linkages with Intergraph Corporation's CAD product. This capability is utilized by the Presentation Model to create the graphic report in the user's desired format. While no tests were conducted on Presentation Model views other than standard CAD reports, it should be possible to utilize the inherent D++ relational database link and commercial database report writers to create most other reports of interest.

Conclusions

Using functional analysis of existing process configurations, it was possible to identify an appropriate knowledge language for capturing process configuration knowledge. An information model was designed and a corresponding relational database was created to contain the configuration knowledge. The interface to the relational database provides an adequate tool for use by domain experts to enter and maintain the configuration knowledge.

An architecture involving three object models was designed to reasonably segregate knowledge and functionality of the working system. An Expert Model contained a copy of the configuration knowledge and possessed the capability to reason over rules and information creating a resultant Engineering Model. The Engineering Model contained inter-related objects representing the process solution based on specific process requirements. When reports of the Engineering Model information are required, a Presentation Model is created. This object model is responsible to create a user acceptable report for the particular view of interest.

The methods and systems described have been tested in the capture and use of configuration knowledge for a liquid storage and pumping function with associated piping, and instrumentation and controls. Primary future directions will involve:

1. Linking the knowledge capture tool with a CAD system;
2. Linking the Expert and Engineering Models more closely with process information contained in a relational database;
3. Extending the function-solution pair configuration knowledge base to include other commonly used process functions; and
4. Increasing the Presentation Model capabilities to generate additional views of the Engineering Model.

References

Bligh, T. P. and A. Chakrabarti (1994). An approach to functional synthesis of solutions in mechanical conceptual design. *Research in Engineering Design*, **6**(3), 127-141.

Colton, S. J. and R. C. Pun (1994). Information frameworks for conceptual engineering design. *Engineering with Computers*, **10,** 22-23.

Ichikawa, A., J. Itoh, S. Kobayashi, K. Niida and T. Umeda (1986). Some expert system experiments in process engineering. *Chem. Eng. Res. Des.*, **64,** 372-379.

Kota, S. and C.-L. Lee (1993). General framework for configuration design: part 1-methodology. *Journal of Engineering Design*, **4**(4), 277-289.

Kuttig, D. (1993). Potential and limits of functional modelling in the CAD process. *Research in Engineering Design*, **5**(1), 40-48.

PRINCE: A KNOWLEDGE-BASED SYSTEM WHICH BINDS PROCESS SPECIFICATION TO EXPERIMENTAL DATA

François Wahl[1] and Bertrand Braunschweig
Institut Français du Pétrole
Computer Science and Applied Mathematics Department
BP 311, 92506 Rueil Malmaison, France

Abstract

PRINCE is a program designed to help in answering the calls for tenders received by Institut Français du Pétrole. For a given process, the PRINCE software helps in determining the operating conditions and the resulting product qualities. PRINCE federates and structures, within the same software environment, all the numerical and symbolic knowledge useful for preparing proposals. PRINCE has been developed as a problem-solving environment capable of determining and executing all the tasks to be performed to reach the goal formulated by the user. This task-based representation is the main characteristic of the software. We also emphasize the user-interface aspect devoted to the management of charts and correlations.

Keywords

Task, Process design, Conception, Curves, Experimental data.

Introduction

PRINCE is a program designed to help in answering the calls for tenders received by Institut Français du Pétrole (IFP). Among its multiple activities, IFP develops refining units by experimenting with scale-model pilot units, enabling it to faithfully reproduce actual behavior. Experiments produce an enormous amount of data. A set is first analyzed, and then serves to predict the behaviour of the units and thus to sell licenses for the processes developed. For a given process, the PRINCE software helps in determining the operating conditions and the resulting product qualities.

PRINCE federates and structures, within the same software environment, all the numerical and symbolic knowledge useful for preparing proposals. PRINCE has been developed as a problem-solving environment capable of determining and executing all the tasks to be performed to reach the goal formulated by the user. This task-based representation is the main characteristic of the software. We also emphasize the user-interface aspect devoted to the management of charts and correlations.

The expert system PRINCE is designed to help two groups of engineers: 1) The IFP Industrial Development Division expects PRINCE to assist them within their commercial activity in producing proposals in reply to calls for tenders from refiners throughout the world and for feasibility studies; 2) The engineers on the pilot-plant site run the system PRINCE for assessing tests performed at the IFP refining center. They analyze and validate the knowledge that will be transferred to the specialists at the Industrial Development Division. This organization is operational for the hydrocracking process. It will be extended to hydrotreatments in 1995.

In the first section, we present the management and utilization of multidimensional correlations and charts. Section 2 shortly describes the problem solving, task-based architecture. In section 3, we show how PRINCE is used with a simplified example.

Flexible Use of Multidimensional Correlations

The PRINCE program exploits empirical curves and correlations, which form an essential part of the knowledge, and which are built from experimental results.

[1] *Francois.Wahl@ifp.fr.*

In the absence of models, these results are necessary for explaining the behavior of processes. The validation of PRINCE depends on the validation of these curves. Actually, the construction of knowledge and validation are concomitant. PRINCE provides an interactive environment which is a tool for manual curve-plotting. Instead of being limited, like standard commercial software packages, to plotting curves from their points of definition, the environment also makes it possible to directly modify a curve, for example by shifting the points of definition, and to plot a curve from a formula or even from a program.

A Man-Machine Interface Problem

Users of PRINCE do not want to deal with the internal representation of correlations, charts and with smoothing algorithms. What they need is a flexible tool that combines the simplicity of pen and paper with the productivity gained from using a computer. More precisely, the graphical facilities of the software must allow them to draw a curve (as a matter of fact, to have it automatically drawn) close to experimental points, as if they were doing it on paper. This flexibility is provided in PRINCE by combining two techniques: 1) a user-friendly interface providing all the tools for displaying, selecting and changing curves and points; 2) a smoothing algorithm capable of following numerical data (passing through a set of points) and qualitative constraints (on the shape of a curve).

Interactive Curve Editing and Optimization

The charts are visualized in different coordinate systems. Interpolation and extrapolation tools are provided. Editing tools can improve the final presentation. Two functionalities are particularly useful for analyzing charts: 1) tracking, enabling the coordinates for a chosen point to be seen by clicking on a curve; 2) calculating by points: if a value is given to the abscissa, the computer will compute the value of the ordinate on the curve. For monotonic curves, the reverse operation (x as a function of y) is possible. A point may be modified, added to, or deleted from a curve.

Another feature is PRINCE's optimization module, which is capable of adjusting the parameters of pseudo-kinetic formulas for curves defined by points.

Process Design Know-How

In this section, we first give the principal reasons for choosing a problem solving approach based on tasks. Then, we outline the structure of the tasks as classes in the object representation.

Opportunistic Design of Processes

When experts are asked how they proceed in designing a refining reactor, they describe a relatively sequential process made of about thirty successive tasks, i.e. the elementary work which ends up by being organized in the form of a task and subtask tree with the following characteristics:

- Tasks are diverse: formulas, programs, rules, etc.
- At runtime, tasks are used directly and inversely if possible. For example, *the more the feedstock is difficult, the more necessary it is to raise the reaction temperature. Inversely, the more the reaction temperature increases, the more the reaction is efficient.*
- Experts skip from one subproblem to another in a opportunistic way, depending on the information available and the difficulty of the subproblems they are dealing with. Sometimes they use tasks in an isolated way, without any apparent connection with the rest of the procedure.
- When they try to determine a set of results, they perform tasks in a chained way. Experts start from the goals to be reached and cover the causalities behind the hierarchical chains. The link is made by the data. In this specific context, engineers use procedures of goals made up of elementary tasks as in a problem-solving environment (Rousseau, 1988).

The entire problem, of course, is to identify the individual tasks in the reasoning process at the right level of detail: all the intermediate stages must be masked. The resulting computing steps may, upon request, be chained to each other, particularly when one wants to follow the process flowsheet, using PRINCE as a tool box.

Other Approaches

The choice of a task-based architecture quite naturally leads to choosing knowledge-based systems development tools using rules, formulas, charts, programs, etc. The organization of tasks in PRINCE is based on a long tradition in AI (Brown and Chandrasekaran, 1989; Willamowski, 1994).

A Problem Solving Architecture for Process Design

The two main object classes used in the problem solving process are *relations* and *flows*; these correspond to *inferences* and *goals* in other frameworks such as KADS (Wielinga, 1991). Input and output data are indicated so as to allow a flexible use of each elementary reasoning step, which can be executed by itself or in conjunction with others. Particularly, 1) one can apply each reasoning step independently, if the input data are provided; 2) if several methods can be used for reaching a goal, the user or the system may decide which one is best in a given context; 3) the chaining of relations leading to the satisfaction of a goal is context-dependent.

For a *flow*, the following attributes are needed within the reasoning process:

- Goal: any output flow corresponding to an elementary task in the reasoning approach is identified as the *goal* of a flow. Other subgoals exist but they remain hidden.
- Relations: the list of relations, automatically generated, whose output_flow is Flow.
- Value: the value of the flow, given by a method.

For a *relation*, the following attributes are used:

- Input_Flows, Output_Flows, I/O_Flows: lists of flows that are input, output or both for Relation. If the relation is known as *inversible*, the I/O_Flows can be dynamically designed as input or output of a specific reasoning step.
- Pre_Condition: a predicate determining whether the current context allows the relation to be executed.
- Execute: execution method attached to the relation.

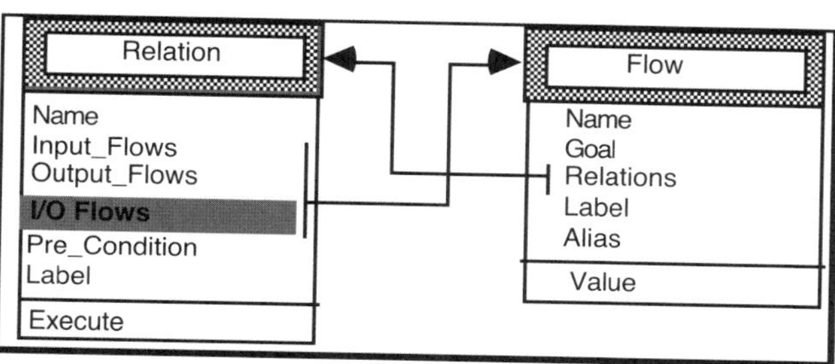

Figure 1. Flows and relations.

Flows and Relations are the upper classes of the hierarchy of problem solving objects. Both can be specialized. For example, Relations can be specialized into five types, for which the *execute* method is redefined:

- *sequence* activates a series of other relations.
- *rule* activates a first-order forward chaining inference engine with a rule packet.
- *program* starts a procedural program.
- *formula* activates a numeric formula.
- *chart* activates the multidimensional correlations system.

The major benefits from using this architecture are *modularity* and *transparence*. *Modularity* is attained with the typing of relations and flows. Pieces of knowledge are introduced with instances of the class "relation." Defining a new relation subclass simply consists in providing its specific *execution* method. For example, the solving of a linear system would be implemented as a new subclass of relations. *Transparence* is attained with the explicit definition of input and output of any relation. Not only can any user of the system understand how and why any elementary step takes place in the reasoning process, but moreover maintenance is made easy by a clear and granular representation of these elementary steps.

A Working Example for a Hydrocracking Process

Problem Definition

The problem is to specify a hydrotreatment process aimed at desulfurizing a petroleum feedstock. PRINCE's function is to help experts in the preliminary engineering design of a unit. Once the feedstocks and the objectives of the treatment are known, the work of PRINCE consists in defining, for each catalyst, the operating conditions and the quality of the products obtained.

Feed Characterization

The expert's first task is to examine the feedstock to be processed in order to assess the feasibility of the request. After having fed the descriptive elements of the feedstock available to the system, the expert user will have to:

1. *Complete the data* and find the information lacking. Some functions used for this are well known and can be found in the petroleum literature. Others make use of correlation or methods developed by IFP itself (Wahl, 1994). Especially, PRINCE is able to give every missing point of a distillation curve, by finding a similar existing distillation curve and by adapting it to the current feed, in a case-based reasoning manner.
2. *Display the feedstock* and situate among known feedstocks that have been analyzed in previous studies. These diagrams enable users to get a good qualitative evaluation of the feedstock.
3 *Qualify the feedstock,* which can be done in two ways: first, the problem is to check the coherence of the data. Call-for-tender feedstocks may have various anomalies, due in particular to lack of measurement accuracy. To check the coherence of feedstock-description data, production rules such as the following are used: *If the end boiling point is high and the viscosity is not high or is very high, then conclude "incoherence of measurement between viscosity and distillation."*
Second, the feedstock must be qualified in relation to the process. Some feedstocks may be at the limits of the domain which has been systematically explored for a given process

and may thus require a look by the expert. Likewise, a simple examination of the feedstock serves to estimate the difficulty of the processing to be applied. Here also, a production rule base is used.

Operating Conditions and Throughput

Once the feedstock has been analyzed, PRINCE can determine the operating conditions and yields of the process. The user provides a list of goals to be reached. If all the successive goals are given in the proper order, the processing that the system will follow will be represented exactly by this list. But if the goals are not given in the proper order, or if subgoals have to be fulfilled, the task-based reasoning engine will be responsible for generating the intermediate goals to be reached. The systems tries to solve the problems raised successively, skips the goals that it cannot reach and proposes results for each test. The expert judges the best test.

For example, to determine the hydrogen consumption, the sum of the consumptions required by each reaction must be determined — denitrogenization reaction to remove nitrogen, desulfurization to remove sulfur, etc. To obtain the amount of hydrogen consumed by the denitrogenization reaction CN, the level of the hydrodenitrogenization, HDN, must be known, which measures, in relation to the feedstock, how much nitrogen has to be removed. This parameter is fixed by the expert and given to the system; then CN is deduced.

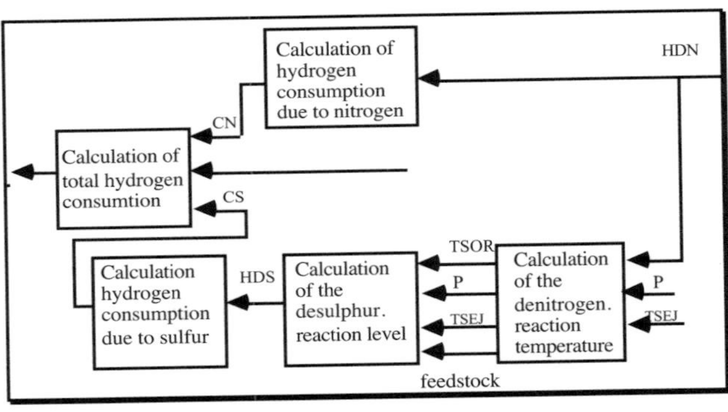

Figure 2. Backwards generation of tasks.

The engine now turns its attention towards another reaction, desulfurization, which is characterized by its HDS intensity. The engine finds that, in order to determine the HDS rate, it must first calculate the temperature TSOR of the reactor (TSOR stands for Temperature Start of Run). Since the feedstock parameters are part of the input data, and since the residence time TSEJ is known at this point, along with the pressure P and the HDN rate, the temperature TSOR of the reaction can be now evaluated. With the TSOR temperature, we now deduce the intensity of the desulfurization reaction HDS, which in turn gives the hydrogen consumption due to the sulfur CS. And so forth. Note that the reasoning processes undertaken are expressed in terms very close to the ways specialists think, i.e. declaratively. The goals do not mention the method used.

The result obtained are used in the ulterior phases of the session, which are not described in detail here.

Conclusions

The specifying of a refining process for a specific installation makes use of various techniques and tools, ranging from heuristic rules applied by experts to the kinetic model of the reactor via the scaling up of experimental data. PRINCE reasons within this multiple-faceted universe by opportunistically making use of each method when the procedure so requires. Within this framework, it makes a crucial and flexible use of multidimensional charts representing the knowledge acquired on pilot and industrial plants operated by IFP and its clients. Although PRINCE has been designed to meet a request concerning a specific process, its concepts and implementations are generic. Gradually, the task of making industrial proposals with limited use of computers will be transformed. For this, the use of knowledge-based systems, but also and above all the adopting of a man-machine cooperation model leaving the expert in command, is proving to be fundamental.

References

Brown, D. C. and R. Chandrasekaran (1989). In *Design Problem Solving: Knowledge Structure and Control Strategies*. Pitman, London.

Rousseau, B. (1988). *Vers un environnement de résolution de problèmes en biométrie. Apport des techniques de l'Intelligence Artificielle et de l'interaction graphique* (in French). Ph.D. thesis, University Claude Bernard Lyon I, Lyon, France.

Wahl, F. (1994). *Un environnement d'aide aux ingénieurs basé sur une architecture en tâches et sur un module de visualisation de courbes. Application à la conception de procédés de raffinage* (in French). Ph.D. thesis, École Nationale des Ponts et Chaussées, Paris, France.

Wielinga, B. J., Schreiber, A. Th. and Breuker, J. A. (1991). KADS: A modelling approach to knowledge engineering. University of Amsterdam Report # KADSII/T1.1/PP/ UvA/008, Amsterdam.

Willamowski, J. (1994). *Modélisation de tâches pour la résolution de problèmes en coopération système-utilisateur* (in French). Ph.D. thesis, University Joseph Fourier Grenoble 1, Grenoble, France.

A COMPUTER-AIDED SYSTEMATIC APPROACH TO CHEMICAL PLANT LAYOUT

Martina Erdwiens and Henner Schmidt-Traub
Department of Chemical Engineering
University of Dortmund
44221 Dortmund, Germany

Abstract

During conceptual plant layout many decisions are made that have great influence on the following detail planning. Problems arise due to the amount of different factors influencing the layout, deficits of input data, and the lack of standards and regulations for this design phase. Therefore, in today's practice plant layout is carried out manually by experienced senior engineers. In this paper, heuristics and algorithms are combined to lead to new methods and tools for successful conceptual layout. First, the required plant area and volume and the design of the basic (steel) structure are proposed. Space requirements are then estimated by modeling each piece of equipment using statistical and heuristic data. The evaluation of a knowledge base yields constraints that describe the placement requirements of and the dependencies between items. By representing all items as box models and their relations as 'interactions' between these models, a two-dimensional force-directed algorithm can accomplish single floor placement. For layout evaluation, a sequential three-dimensional routing algorithm is used to carry out piping studies. An example of this procedure is presented and the results are discussed.

Keywords

Chemical plant design, Conceptual plant layout, Layout algorithm, Piping study, Routing algorithm, CAD/CAE, Heuristics, Knowledge-based system.

Introduction

Plant layout involves the spatial arrangement of process equipment and their connection by pipes, ducts, and vehicle transportation. To ensure operability and adequate safety of the plant, an economical layout satisfying a multitude of factors — ranging from process needs, maintenance and operational requirements to construction and safety considerations — has to be developed. The preliminary placement of equipment results in approximate positions for vessels and machinery. This conceptual layout constitutes part of the input data for and constraints on detailed design in civil engineering, piping design and construction, electrical engineering, etc. Well-done conceptual layout can therefore effect substantial savings in the following detail planning while errors are costly to rectify later.

Problems arise due to the amount of different and sometimes competing factors and the sequential work flow of layout and detail design. Due to the early design stage, insufficient and inconsistent input data makes estimation necessary. Another problem is the lack of standards or regulations to guide the 'plant' engineer; therefore experience and subjective reasoning are conventionally used in layout design. Consequently, no CAE-tools are available for this early, creative stage of plant layout.

The aim of the research project described in this paper is to offer methodical support for the 'plant' engineer by developing methods and tools for conceptual layout. Systematic conceptual plant layout should be able to answer the following questions:

- How much space is required?
- Where is each piece of equipment to be positioned?
- How can alternative layouts be evaluated?

Process, operation and maintenance requirements as well as the demands of detail planning, e.g. piping design, must be regarded when establishing a procedure for answering these questions.

Previous Approaches

Detailed information about plant layout has been accumulated by Kern (1977), Mecklenburgh (1985) and others and is documented in company standards. These collections, usually a mixture of general rules and exact-to-the-inch distance information, contain neither methods nor the associated data needed for conceptual design.

Previous attempts to develop methods for plant layout have usually been concerned with the computation and optimization of equipment positions within a given area. Though usually no explicit distinction is made between conceptual and detailed design, most methods require detailed and exact information regarding equipment dimensions and floor sizes for input information.

Mathematical tools as different as position swapping (Gunn and Al-Asadi, 1987), cluster growth methods (Shahookar and Mazumder, 1991), graph partitioning (Jayakumar and Reklaitis, 1994) and more have been applied to reach an 'optimum' layout, usually using estimated piping lengths or other weighted distances as a means of evaluation. Since most methods are derived from floor planning or the design of manufacturing plants, the generation and integration of process, operational and site constraints have often been neglected. To avoid this deficiency of solely numerical solutions, heuristic approaches have been introduced by Malingriaux (1980), McBrien, et al., (1989), Suzuki, et al., (1991) and others. However, so far no successful method or tool has been developed to cover the complexity of conceptual layout.

Description of the APACHE System

The layout system APACHE aims at combining heuristic and numerical methods as well as conventional CAD-applications to create an interactive conceptual layout tool. Figure 1 gives an overview of the system.
To answer the questions mentioned above, the layout procedure can be divided into three steps:

1. basic concept (steel structure),
2. conceptual layout: definition of space requirements and placement constraints, positioning of equipment, and
3. evaluation.

The preparation of input data includes the evaluation of equipment lists, the process flow diagram (PFD) and site information. Incomplete and inconsistent data makes estimation necessary while information drawn from the PFD must be transferred into evaluable data. Since all data is subject to multiple change during the planning process, layout design progresses through numerous iterations before the final conceptual model is "frozen" and made available to detail engineering.

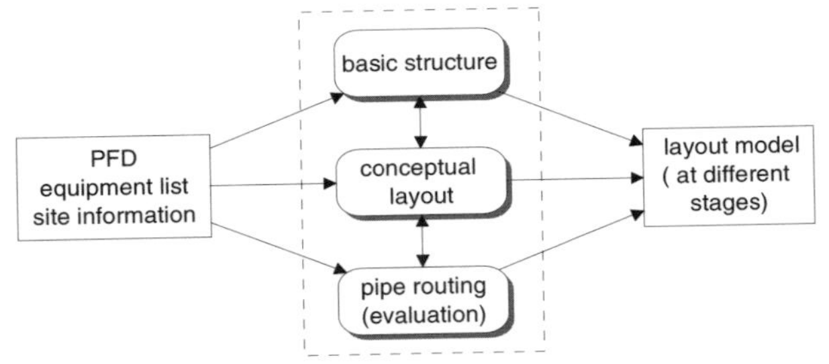

Figure 1. Systematic conceptual plant layout.

I. Basic Structure

The first step, the development of the basic structure, comprises the estimation of the required plant area and volume. For the design of the steel structure or grid mostly statistical methods are used (Fig. 2). Based on equipment types and sizes, the area and number of floors as well as the grid length are fixed. If a plant is to be installed within an existing complex, available roads, pipe racks and maintenance areas must be documented.

The optional partitioning of the PFD into sections like reaction and separation parts and their approximate positioning within the grid allow for an early consideration of safety zones and an easy adaptation to site specifications.

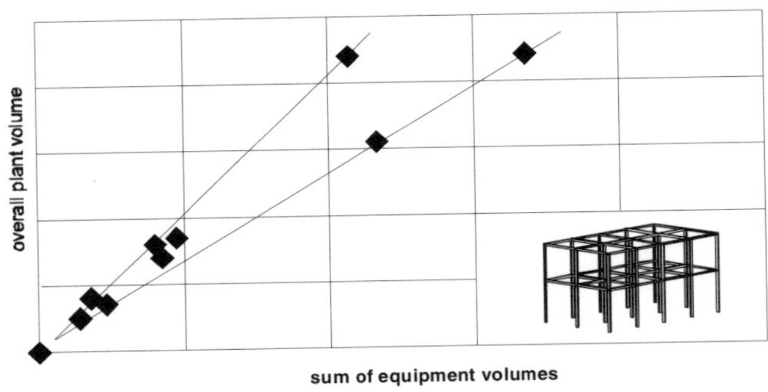

Figure 2. Statistical estimation of plant volume.

II. Conceptual Layout

For the actual placement phase, an approach in three steps is adopted: First, local space requirements are estimated for each item. Then placement constraints have to be defined to take into consideration process needs, operation and maintenance, safety regulations, etc. Finally, the collected data is evaluated mathematically. This procedure is detailed in the following paragraphs.

Space Requirements

Space requirements for each piece of equipment are estimated heuristically on the basis of the PFD and approximate equipment dimensions. The resulting box model for each item includes accommodation for: the equipment itself, equipment-assigned piping, instrumentation, safety facilities, and access space for operation and maintenance. Since each model contains 'hard' (equipment, steel, etc.) and 'soft' (access ways, etc.) volumes the models may partly overlap when placed.

Central access ways and maintenance areas must be fixed independently during basic plant structuring.

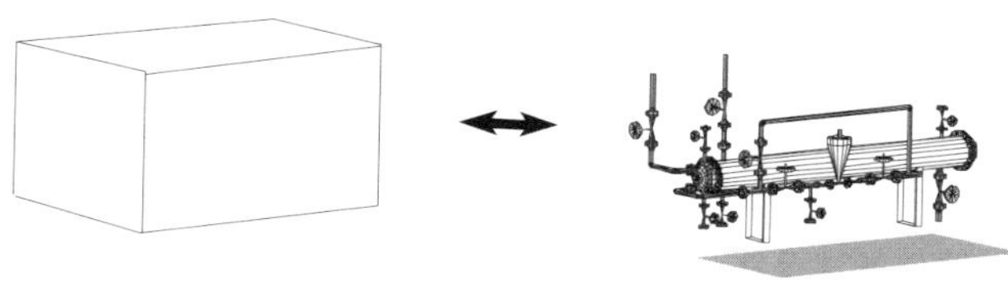

Figure 3. Representation of space requirements.

Placement Requirements and Constraints

The factors that have to be considered when positioning items within the steel grid are described as placement constraints. The 'plant' engineer determines the requirements of each piece of equipment regarding the above mentioned criteria, e.g. the association of a vessel with a column system as a process group or the necessity to dismantle a heat exchanger frequently due to extensive fouling. Default values for typical (standard) items are presented to determine these requirements heuristically. For 'non-standard' pieces of equipment detailed check lists are available.

In the following step, placement constraints are determined heuristically, e.g. "Heat-exchanger H-110 has to be placed close to column C-100 because of process flow." The constraints refer to absolute positions ("H-110 should be next to an access way for maintenance") or describe relations between pieces of equipment ("H-110 has to be above B-120 because of gravity flow").

Placement Algorithm

So far for each piece of equipment a box model with fixed dimensions and a number of constraints 'attached' to it have been defined. A placement algorithm solves the problem of positioning all box models according to their constraints within the given basic plant concept. The two-dimensional force-directed algorithm can be described as using 'springs' that are attached to the center coordinates of each item, pulling the objects together or keeping them apart. The algorithm presented by Shahookar and Mazumder (1991) and others has been modified to take into account the importance or 'weight' of different constraints.

Two different methods have been tested in this project to represent item geometry: While Sha (1987) used an algebraic description based on a geometrical representation of each object, we have developed a different function defining the relations between items as repulsive forces.

Detailed Placement

Prior to evaluating different plant layouts, 'dissolution' of the box models is necessary since so far only the positions of the box models have been fixed. The positions of the 'hard' and 'soft' volumes, i.e. equipment, instrumentation, access ways, etc., are determined heuristically within the range of each box model, allowing for the use of common access and maintenance areas between items.

III. Evaluation / Routing

To compare alternative layouts, quantitative parameters as floor area exploitation and constraint satisfaction can be used.

As a means of qualitative evaluation, an overall piping draft is generated by a three-dimensional grid routing algorithm (Hasenauer, 1994). Here routing is carried out sequentially for all relevant pipes with estimated nozzle positions and pipe diameters. A grid covering the routing volume (i.e. part or all of the plant) is generated in which equipment, steel, access way, etc. are marked and preferred piping areas, e.g. pipe racks, can be designated. By advancing stepwise from the starting nozzle, each grid cube is assigned a penalty value for each 'step' from one cube to the neighboring one. The penalty value for going straight is less than for bends, and advancing close to supporting steel structures is preferred to routing in 'free' space. Figure 4 shows the result of such a 'grid wave'.

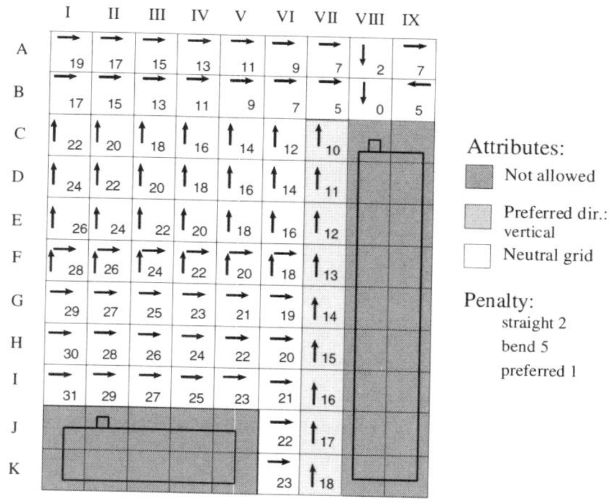

Figure 4. Grid routing algorithm.

The resulting pipe can be found by tracing the cubes backwards from finishing nozzle to the starting point following the arrows. The resulting draft of pipe positions serves to compare alternative layouts.

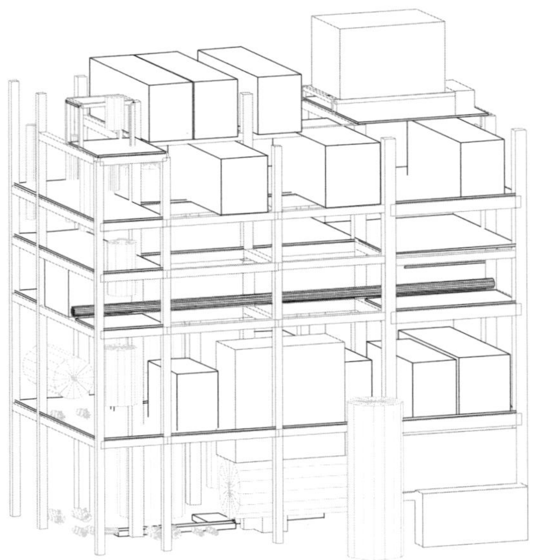

Figure 5. Box models placed in existing structure (boxes: placed models;cylinders: existing vessels).

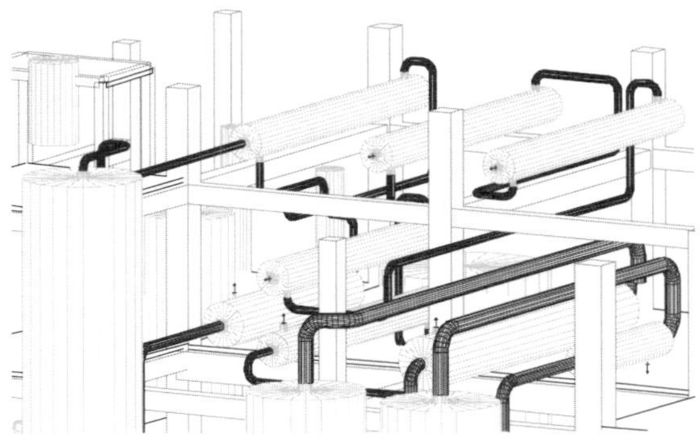

Figure 6. View on a piping study.

Example

The procedure described in this paper was successfully tested by comparing an APACHE layout to the final layout of the same plant designed manually at a chemical operating company.

The pilot study comprised a double train fluid phase process with reaction and product recovery units. Process data was documented in a PFD containing mass and heat balances as well as basic equipment dimensions. The aim of the study was to check if the process could be placed within an existing steel structure. Information about the chosen site was documented in existing plot plans.

At first, box models of all items were calculated with the help of a database. The 'plant' engineer made assumptions about the space requirement of the 'non-standard' reaction vessels. General maintenance and process requirements were extracted from the knowledge base. As much additional information as possible, especially about reactor operation, was acquired from the process engineer. With the help of a second knowledge base, placement constraints were generated automatically on the basis of the accumulated data. The equipment not yet assigned by constraints to a certain floor was assigned manually. During the floor-wise placement existing steel and vessels were taken into consideration by the algorithm. The resulting layout (Fig. 5) shows how tightly the box models fit into the existing steel structure.

All pipe connections listed in the PFD were ordered by diameter and routed sequentially to obtain data about pipe lengths for two generated layouts. The 'plant' engineer who evaluated the results found significant similarities to his manually designed plot plan and considered our results to be successful. The documentation of requirements and additional (piping) information acquired by the system proved to be a definite advantage.

Conclusions

In this paper methods for conceptual plant layout have been described. The rules and data integrated in the APACHE tools apply to continuous single product plants. They take into regard 'typical' features of European chemical plants. The comparison of layouts generated and these methods with manually designed plot plans proved satisfying and underlined the advantages of the procedure.

Acknowledgment

This AIF-project No. 8956 has been funded by the German Ministry of Economy (Bundesamt für Wirtschaft).

References

Gunn, D. J. and H. D. Al-Asadi (1987). Computer-aided layout of chemical plant: a computational method and case study. *Computer-aided Design*, **19**(3), 131-140.

Hasenauer, L. (1994). Entwicklung einer methodik zur rechnergestuetzten erstellung von rohrleitungsstudien. *Fortschrittsberichte VDI,* No.151.

Jayakumar, S. and G. V. Reklaitis (1994). Chemical plant layout via graph partitioning. *Computers Chem. Engng.*, **18**(5), 441-458.

Malingriaux, R. (1980). Zu wissenschaftlichen Grundlagen der rg. Aufstellungsplanung. *Dissertation*, WPU Rostock.

McBrien, A., et al. (1989). Artificial intelligence methods in process plant layout. *ACM*, 364-373.

Shahookar, K. and P. Mazumder (1991). VLSI cell placement techniques. *ACM Computing Surveys*, **23**(2), 143-220.

Suzuki, A., T. Fuchino, M. Muraki and T. Hayakawa (1991). An evolutionary method of arranging the plot plan for process plant layout. *Journal of Chemical Engineering of Japan*, **24**(2), 226-231.

IMPACT OF AN OPEN ARCHITECTURE ENVIRONMENT ON THE DESIGN OF SOFTWARE COMPONENTS FOR PROCESS MODELING

Peter D. Edwards
DuPont Company, Experimental Station
Wilmington, DE 19808

Gary Merkel
Scientific Computing Solutions, Inc.
Newark, DE 19702

Abstract

Dramatically new expectations and visions are emerging for software tools used in chemical process modeling and simulation. These visions result from large and pressing unmet needs of operating companies, combined with a growing belief that emerging software technology opens up huge opportunities for better ways of doing things. DuPont anticipates a rapid migration of process modeling software to an open architecture. DuPont's internal software for process modeling is now being developed as a highly modularized library of OLE software components. These components exploit object-oriented pragmatics, including abstraction, encapsulation, inheritance, and polymorphism. In this paper, we discuss the software architecture being used for DuPont's development of tools supporting process synthesis and operations optimization. We actively solicit contributions from others towards a common goal of more effective modeling software.

Keywords

Object-oriented software construction, Abstraction, Encapsulation, Inheritance, Polymorphism, Component software, Object linking and embedding (OLE).

Introduction

Process modeling and simulation tools will soon embrace major evolutionary trends in software technology, providing engineers with computerized workbenches that give ready access to the best software tools for each job. Seamless tool integration and a consistent user interface will enable engineers to focus on the problem to be solved, and not the application being used. Achieving this vision will require cooperation among industry leaders, universities, and commercial software vendors. European initiatives are already well underway, and participants are vigorously working together with the expectation that resulting tools will greatly increase global competitiveness.

The Industrial Perspective

Achieving best-in-the-industry processes and creating value in the marketplace requires enabling technology in modeling and simulation. Industry leaders procure this technology from four sources:

1. Internal Resources
2. Commercial Software Vendors
3. Universities
4. Industrial Consortia

Internal Resources

Many of the best algorithms and problem-solvers are developed internally by engineers to solve real problems. Software tools produced by such efforts are easily identifiable: the code is slick, it solves the right problem, but it operates in an isolated context and can only be used and maintained by its author. For internal commercialization a sponsor must be found.

The Information Systems group may reject ownership of the orphan because of resource limitations or because it was written in an unsupported language. In deciding whether or not to authorize resources to continue development, engineering management experiences an identity crisis and begins asking questions such as "Are we in the business of engineering or in the business of software?"

If internal resources cannot be committed to internal commercialization, two choices are left: add the slick code to the growing heap of internal legacy code or entrust commercialization to a software company. Some of the best proprietary algorithms are turned over to commercial vendors, and thus competitors, in return for a friendly user interface and ongoing support.

Commercial Software Vendors

Strong company affiliations with commercial software vendors hold a great deal of promise, especially for volume purchasers with clout. Many companies, as mentioned, are working closely with modeling and simulation vendors, and have entrusted to them the commercialization of proprietary software.

The problem with relying on vendors for commercialization is that problems of local interest are often best solved by highly specialized tools, whereas the vendor's compelling interest is to sell general solutions to a broad-based market. The company asks for a snake and the vendor delivers an elephant (some three years later) claiming that an elephant has a snake-like trunk.

Another problem with commercial products is that vendors have historically produced proprietary software architectures and interfaces. Seeking market domination, they try to capture markets by producing huge programs that out-feature the competitors. The resultant software, evocatively referred to as "monolithic fatware," precludes any vision for a heterogeneous workbench.

Universities

Strong industry-academia ties also hold great promise, as university research pioneers tremendous technological capability. The problem with this partnership has been getting research's leading thinkers working on problems of interest to industry.

Another difficulty lies in adapting and integrating university research into problem-solving tools for industry use. The wide gulf between the technical advances and useful tools for improving products and processes is exacerbated by faculty and student choice of software tools and methodologies. The tools are generally selected from the available catalog of freeware, and very few Ph.D. points are awarded for structured programming and a usable interface. The monumental task of overhauling university software into a usable industry tool is usually tackled only by commercial vendors who can sustain costly multi-year product development cycles.

Industrial Consortia

Operating companies do not, in general, have the internal resources (money and skills) to develop commercial grade process modeling and simulation software tools. Vendor-supplied commercial products have very long cycle times and often require large learning curves. Engineers may have to learn several products to determine their respective strengths and weaknesses as applied to the local problem-mix. University research provides step changes in problem-solving ability, but demonstrates the technology with software that is unfit for industry use.

To address pressing unmet needs, many companies have banded together with other companies, even competitors, to work collaboratively on solving common problems. These consortia will play an ever-increasing role in providing the thrust and economic incentive for overcoming barriers to improving modeling and simulation tools. These companies cannot wait for change, they must drive it.

Emerging Software Technologies

Two exciting developments in software technology are the advent of

1. Object-Oriented Software Construction, and
2. Component Software.

In reality, both of these technologies have been around for a long time; however, recent advances in operating system software are now enabling us to realize the promise of open architecture software systems assembled from functional components that expose usable objects with well-defined interfaces and behaviors.

"Object-oriented" is rapidly becoming the most overused adjective in all of computing. Its promotion as an end-user benefit often serves as a signal that an application is probably lacking in other important areas. As discussed in the next section, object-oriented software construction is the first punch in a one-two combination. In addition to the well-documented benefits to developers, the primary value of object-oriented software construction is how well it lends itself to the development of component software.

"Component software" is the formal term applied to the result of assembling software systems from heterogeneous functional components. Heterogeneous in this case means the components ("computer programs") are likely to be developed by different programmers, using different languages, on different computer platforms.

The combination of these two technologies *presents the possibility* of software architectures that will carry modeling and simulation into the millennium. One ingredient is still missing: the cooperative effort among industry leaders, universities, and commercial software vendors needed to make use of the technology.

Object-Oriented Software Construction

DuPont is now applying object-oriented software construction methodology to the development of modeling and simulation software tools. The general theme is a departure from the normal procedural emphasis, that is, start with a task, and develop a sequence of subroutines that accomplishes the task. This has been replaced with a paradigm that focuses on modeling objects in the chemical process engineering domain, such as streams and unit operations.

The practical result of modeling a conceptual entity is a *C++ class* that is the complete programmatic representation of the entity. The includes both *property* (variables for storing data) and *method* (functions for manipulating objects of the class) definitions. Each class property and method is defined to be public or private. Access to private parts of a class definition is restricted. An object's *exposed interface* comprises the public properties and methods in the its class definition.

The practical result of modeling all relevant conceptual entities in the chemical process engineering domain is a *C++ Class Library*. Once the class library is completed, the procedural nature of programs to accomplish specific tasks is simplified to object creation (instantiation or allocation), manipulation (changing property values and invoking methods), and destruction (deallocation).

Four important aspects of object-oriented software construction are:

1. Abstraction
2. Encapsulation
3. Inheritance
4. Polymorphism

Abstraction

Abstraction involves identifying the key objects (conceptual entities) in the chemical process engineering domain. Class definitions are then developed for each different object type. The class definition employs properties and methods to define the behavior of the object.

Abstraction forces class designers to carefully indicate the essential characteristics of objects, which involves consideration of how they relate to other objects in the domain. The key to realizing an efficient open architecture is the attainment of an agreement among industry leaders, universities, and commercial software vendors on a standardized set of objects (entity abstractions) and interfaces (exposed properties and methods).

Encapsulation

Once abstractions are formalized through class definitions, an encapsulated implementation of the class is developed. The term encapsulation simply formalizes the notion that implementation details that do not contribute to understanding the object are hidden from view.

The important characteristics and behavior of each object are set up as class properties and methods. The exposed interface is all that should matter to object users. The details of the implementation are left entirely to the class developer.

Inheritance

Inheritance provides a means for hierarchical development. The similarities of closely related objects can be used to establish the properties and methods of a *base class* (e.g., "column"). The dissimilar aspects of objects can be used to refine (by addition or replacement of) the properties and methods of a base class to establish *derived classes* (e.g., "distillation column" or "absorption column"). Objects in classes derived from the base class share a certain collection of common properties and methods.

Inheritance facilitates code sharing and improves class quality by supporting hierarchical development. By deriving classes from well-established class libraries, such as the Microsoft Foundation Class Library (MFC), programmers inherit a tremendous amount of tested and debugged functionality.

Polymorphism

Polymorphism is a programming language feature that facilitates the use of standard property and method names across inherited classes, promoting uniform naming. Invoking the proper properties and methods for each object is handled automatically in the software based on the object's type.

Component Software

Software components are applications that make objects and their exposed interfaces available to other applications. Software development will soon encompass two principal activities: writing value-added software components in areas of core competency, and software integration — assembling plug-in components to solve specific problems or to create a workbench for accessing collections of useful components.

The upshot of component software is evident in the following scenario. A process model is developed using company A's simulation program. Using the graphical editor, a button click on a palette inserts the fully functional distillation model from company B into the flowsheet, and another click inserts the reactor model from company C. These two unit operations are present in a flowsheet connected by company D's stream models.

Company A's simulation program is referred to as the *container (or client) application* because it manages a *compound document* that contains embedded objects provided by the *server applications* supplied by companies B, C, and D. Selecting an embedded object (by clicking on it) activates the server application that created it and allows end-users to modify object properties. Each software component (container or server) can be developed in a different programming language and may even be running on a different machine.

The development of extensive system support libraries is gradually making this level of software interoperability possible. Microsoft Object Linking and Embedding (OLE) technology is just one example of an emerging standard for component software. OLE comprises a Component Object Model (COM) and an Application Program Interface (API). COM provides the foundation for how components interact, and the API implements hundreds of function calls for adding OLE features (component software capability) to programs.

The arguments surrounding the choice among OLE (Microsoft), OpenDoc (CI Labs), and CORBA (OMG) are interesting, but moot. The backers of these standards are already working on compliance with the other standards through bridge software referred to as object middleware. OLE is a safe choice for the personal computing environment because of the widespread use of Microsoft Windows.

Object-oriented concepts and emerging component software standards will hasten the rapid migration to open architectures for many companies and software vendors; however, even though the physical mechanisms for software integration are quickly evolving, matters of object definition and standardization are likely to become the practical bottleneck.

The Missing Ingredient

The missing ingredient is the broad-based, coordinated cooperation among industry leaders, commercial software vendors, and university researchers to overcome the practical barriers to full exploitation of open architectures in process modeling and simulation software. DuPont is promoting its vision for component-based modeling and simulation software among willing collaborators, and is actively engaged in initiatives to define the abstractions (objects and interfaces) of interest to the chemical process engineering environment.

These abstractions will become the basis for reusable class definitions. Software developers can then implement the functionality of the class definition any way they choose, and can extend the functionality through derived classes.

The approach to developing component software needs to be documented and made available to all developers, including internal developers, university students, and commercial vendors.

Conclusions

University research must be brought into practical application faster, in order for industry to achieve its goals. Commercial vendors must be willing to participate in the opening up of software architectures. Those who attempt to draw users deeper and deeper into the folds of their monolithic fatware will find themselves writing tomorrow's legacy code.

Broad-based participation will be in the best interests of industry. We actively solicit ideas from anyone who believes they can contribute.

References

Booch, G. (1994). *Object-Oriented Analysis and Design with Applications.* 2nd ed. Benjamin/Cummings Publishing Company, California.

Brockschmidt, K. (1994). *Inside OLE 2.* Microsoft Press, Washington.

Fraga, E.S., et al. (1994). The implementation of a portable object-oriented distributed process engineering environment. University of Edinburgh, Department of Chemical Engineering, Technical Report 1994-17.

Institute for Systems Research (1994). Systems challenges for the next decade. ISR Technical Report No. TR 95-38.

Microsoft Corp. (1994). *Microsoft OLE Control Developer's Kit. Visual C++ Version 2.0 User's Guide and Reference,* Volume Six. Microsoft Press, Washington.

Toohey, J. (1994). *Using OLE 2.x in Application Development.* Que Corporation, Indiana.

INTELLIGENT SYSTEM APPLICATIONS: A TECHNOLOGY LICENSOR'S PERSPECTIVE

Sik K. Shum and Douglas R. Myers
UOP
Des Plaines, IL 60017-5017

Abstract

UOP is a leading licensor of process technologies in the refining and petrochemical industries. Process technology licensing is a knowledge-intensive business: revenues are generated by delivering technical know-how and associated products and services to customers. As a result, the cumulative proprietary process and product knowledge is the most valuable asset and the primary driving force behind the financial performance of the company. Intelligent systems provide an important means of capturing this knowledge and deploying it in ways that best address customer requirements. UOP has two main categories of intelligent systems applications: support systems used internally within UOP and external systems used by UOP's customers. This paper focuses on external systems and highlight two applications that have been installed in petroleum refineries.

Keywords

Expert system, Neural network, Process control, Malfunction diagnosis.

Introduction

Founded in 1914 as Universal Oil Products, UOP is currently the world's largest process licensing organization. UOP offers more than 60 processes in the petroleum refining and petrochemical industries. More than 4200 units are licensed worldwide, including Merox* process for mercaptan oxidation, Platforming* process for catalytic reforming, Unicracking* process for hydrocracking, and FCC for catalytic cracking. In addition to process licensing, the UOP* product line includes catalysts, molecular sieves, chemicals, food additives, process equipment, process plants, control systems, on-line analyzers, advanced process control and optimization systems, on-line expert systems, training simulators, engineering and start-up services.

Internal Systems

Intelligent system applications for internal use mainly aim to improve quality and consistency and to reduce time requirements for internal work processes. These systems also make possible the performance of complex tasks without requiring substantial training to acquire the level of expertise normally required. Other important benefits are:

- Ability to effectively evaluate multiple alternatives so that the best one is chosen
- Capture and documentation of corporate decision-making and problem-solving knowledge, thereby providing a platform for continuous improvement of the knowledge.

Opportunities in UOP's business activities for internal intelligent system applications range from process technology licensing, marketing and sales, engineering design and specification, and manufacturing to technical services. However, this paper focuses on external systems and shows applications that have been installed in refineries.

External Systems

External systems refer to intelligent system applications developed by UOP for customer use. The addition of intelligent system capabilities to UOP's existing software products and the creation of new intelligent software applications are value-added features that aim to improve the customer's process operations. The following paragraphs will discuss the incorporation of intelligent system capabilities to enhance two UOP products. Both applications have been installed in petroleum refineries.

Product: Process Control Equipment

UOP has added knowledge-based capabilities to improve the operation of its line of turnkey, specialized control systems. In these days of reduced operations staff facing an increasing amount of information, knowledge-based systems can be used to help interpret and reduce the overload of information by providing focused operational advice. Examples of UOP's knowledge-based products include real-time diagnostic and maintenance systems and intelligent human-machine interfaces with self-contained diagnostics. These products help customers with process operation and maintenance by making UOP's diagnostic expertise and advice on maintenance and repair available to operators 24 hours a day. Because the high reliability of the control system prevents end users from acquiring much diagnostic experience, significant benefits can be derived from the expert advice provided by the system.

Application: The UOP CCR LHCS Diagnostic Advisor*

UOP's Lock Hopper Control System (LHCS) is a specialized safety-based control system that maintains the safe circulation of Platforming catalyst in UOP's continuous regeneration CCR Platforming* process. Because of the relative reliability of the CCR system and the LHCS, problems occur too infrequently for CCR operators or maintenance personnel to acquire much troubleshooting knowledge of the system. As a result, an on-line expert system can provide valuable troubleshooting and operational assistance for months and years after the process start-up.

UOP has developed and commercialized an on-line expert system, the CCR LHCS Diagnostic Advisor*, which is currently installed and running at various refineries. Myers (1995) provides a detailed description of the system. Knowledge for the Diagnostic Advisor is based on UOP's start-up and operational experience in more than 125 CCR units at petroleum refineries worldwide. The system is a turnkey, on-line knowledge-based diagnostic and maintenance system for UOP's LHCS. All necessary software, computer, and communication hardware is included in the complete system. The Diagnostic Advisor is a PC-based application. The user interface consists of software based on Windows* 3.1 and displayed on a SVGA color monitor. The software structure is shown in Fig. 1. The Main Diagnostic Advisor Software performs the diagnosis and the Diagnostic Advisor Statistical Software performs data abstraction by tracking certain events and making statistical interpretation of key analog data. The diagnostic reasoning is provided by a rule and object-based Nexpert Object program, which is called by the Main Software. A communication driver is used to allow the Main Software and Statistical Software to monitor and request data from the LHCS.

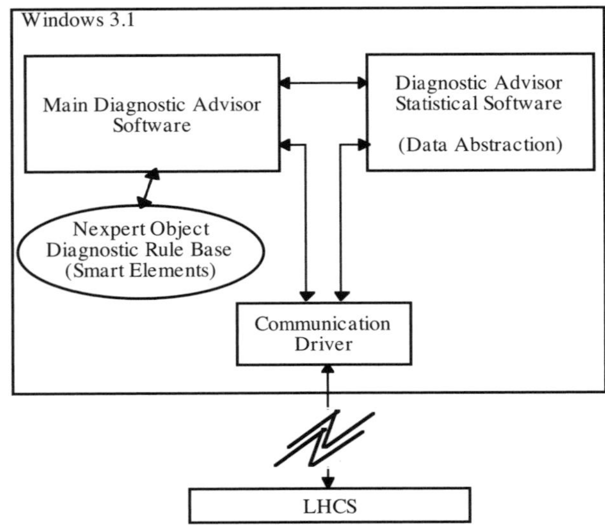

Figure 1. Diagnostic Advisor software structure.

The Diagnostic Advisor continually monitors and logs alarms and events from the LHCS to identify problems associated with process equipment, instrumentation, electrical equipment, operational errors, and control systems. Designed for use by engineers, supervisors, operators, or maintenance staff, the Diagnostic Advisor is easy to use and provides information in layers, from a description of a problem to a detailed repair or maintenance procedure for correcting the problem. The Diagnostic Advisor is an interactive system so the user can initiate a complete diagnosis of the LHCS. Using the system, the user can troubleshoot alarms or abnormal events detected by the system. During the diagnosis, the Diagnostic Advisor retrieves all data from the LHCS, searches through the rule base to identify any malfunctions, and then displays diagnostic conclusions as "problem" windows. A "problem" window shows a description of each independent problem the system has found along with the specific alarms or abnormal events caused by that problem. Detailed repair and maintenance instructions are presented in "hypertext help" windows when the user clicks on a possible cause of the problem with the mouse.

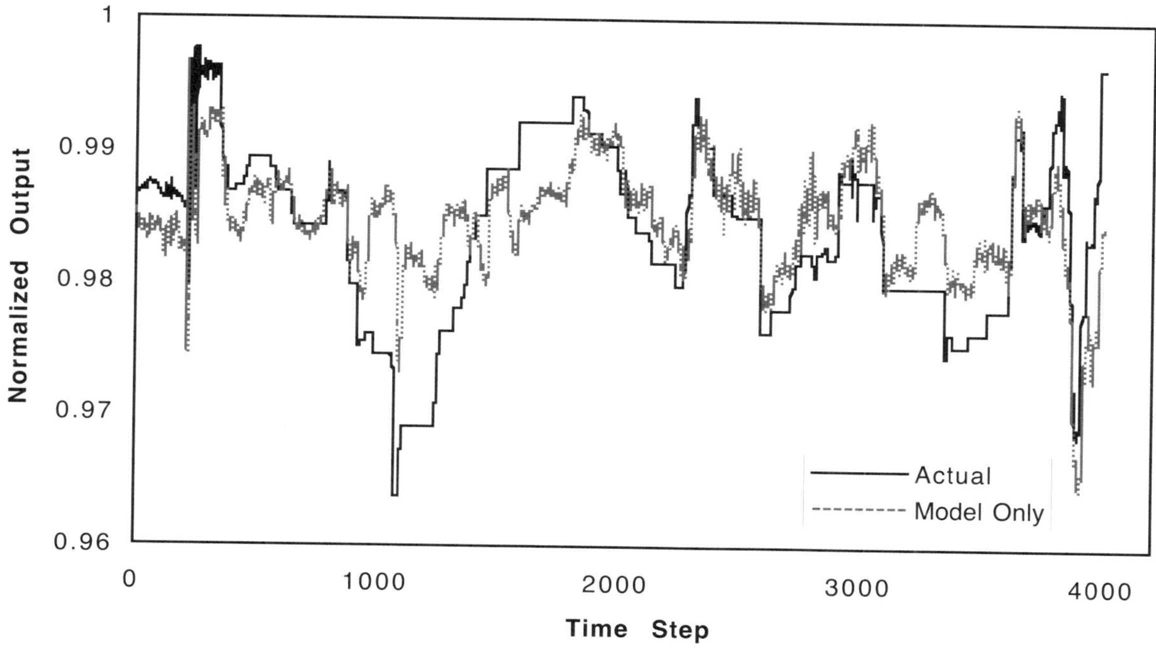

Figure 3. Model-only results.
Normalized RMS Error = 5.16 E-3.

Product: Advanced Process Control and Optimization Systems

UOP's advanced control and optimization products make use of proprietary process models to optimize and control single and multiple process units at or near to maximum profitability. These models often require analytical inputs, such as feed quality, that are generally determined by off-line laboratory analyses. Because these analyses are performed infrequently, the model may be using inputs that no longer truly represent the condition in the plant. As a result, model accuracy and, in turn, performance of the control and optimization systems can be adversely affected. This problem can be addressed in some cases by the installation of on-line analyzers. However, for certain measurements, on-line analyzer technology has not been perfected, or its application cannot be justified for various reasons. In any event, intelligent system techniques provide a viable approach to the problem.

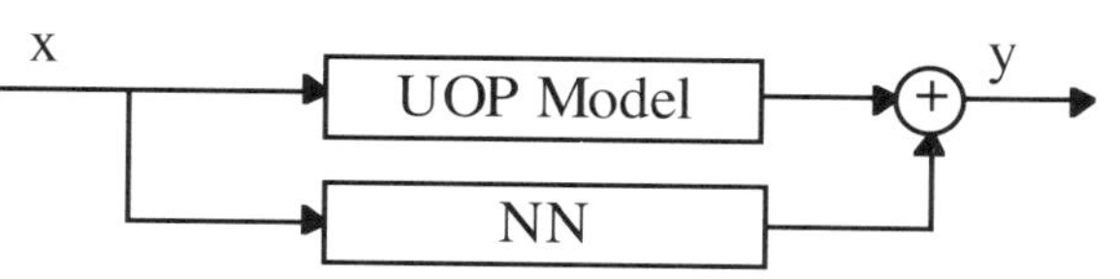

Figure 2. Neural network approach.

To enhance the capabilities of these systems, intelligent system techniques such as neural networks have been incorporated. With the historical operating data available to UOP, neural networks can be implemented to compensate for unmeasured disturbances and unmodeled complexities of the process in the model. Because an existing process model is generally used as a starting point, the function of the neural network is to adapt the model to the conditions in the plant.

Application: Neural Network Enhancement to Model-Based Octane Control

Octane control in the Platforming unit is one of the advanced control systems offered by UOP. Besl, Cusworth, and Livingston (1993) discuss Platforming advanced controls. The reformate octane is controlled by adjusting the reactor weighted average inlet temperature (WAIT). A Platforming unit generally has three or four reactors. The reactor WAIT is the sum of the reactor inlet temperatures weighted by the fraction of total catalyst in each reactor. The control strategy makes use of a proprietary model to determine the WAIT required to achieve a specified octane target. The model requires feed quality as one of the inputs on process conditions. However, the laboratory update on feed naphthenes and aromatics is often done weekly or even less frequently. In some refineries, feedstock to the Platforming unit can change significantly every few days, but the measurement of feed quality is not updated accordingly. In that event, model prediction accuracy and controller performance deteriorate.

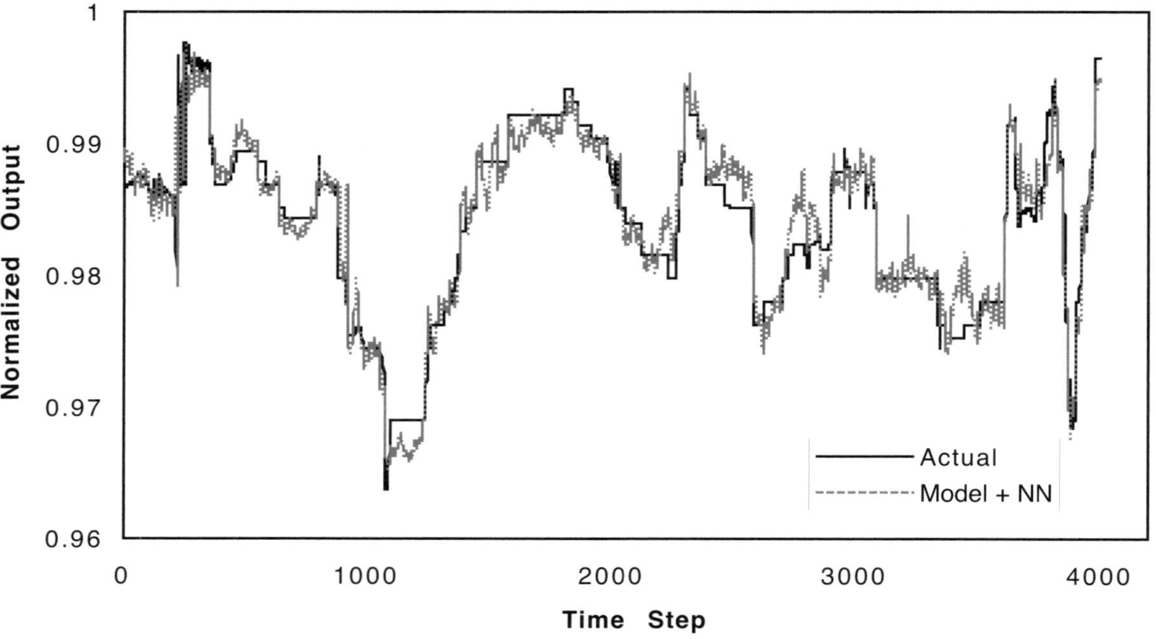

Figure 4. Results from model enhanced with neural network. Normalized RMS Error = 1.66 E-3.

UOP has developed a neural network model enhancement to compensate for the mismatch between the proprietary model prediction and the actual process conditions. A conceptual view of this approach is shown in Fig. 2. The proprietary process model determines the base estimates. The residuals obtained by subtracting the model estimates from the actual process conditions are used as outputs for the neural network training. As a result, the trained network will produce a minor correction to be added to the model output to determine the overall output. Fig. 3 shows the actual plant condition and model prediction without neural network enhancement. Fig. 4 shows the same test data set with neural network model enhancement. The combined model and neural network approach exhibits a clear improvement over the model-only results. Although the model-only results were not as good as the combined results, the model did provide a good foundation for the neural network to improve on. This application has been installed for commercial Platforming units, and plans for additional installations are being made.

Summary

This paper focuses on two external intelligent system applications that have been successfully installed in petroleum refineries. These applications make use of knowledge-based system and neural network techniques to enhance the capabilities of UOP products offered to customers. These applications are just the "tip of the iceberg" in terms of opportunities for applications. Other important internal and external applications are currently at various stages of development at UOP.

References

Besl, H., T. Cusworth and E. A. Livingston (1992). Advanced controls improve profitability of UOP CCR Platforming. *Fuel Reformulation*, Nov/Dec, 50-56.

Myers, D. R. (1995). The UOP CCR LHCS Diagnostic Advisor: an expert system for on-line diagnostics and maintenance of a safety-based control system. *AIChE Spring National Meeting*. Houston, Texas.

---

* CCR, CCR Platforming, Diagnostic Advisor, Merox, Platforming, Unicracking, and UOP are trademarks and/or service marks of UOP. Windows is a trademark of Microsoft Corporation.

A UNIFIED VIEW OF ARTIFICIAL NEURAL NETWORKS AND MULTIVARIATE STATISTICAL METHODS

Utomo Utojo and Bhavik R. Bakshi
Department of Chemical Engineering
Ohio State University
Columbus, OH 43210

Abstract

A unified framework for neural and statistical empirical modeling methods is presented based on model representation as an expansion on a set of basis functions. Various types of artificial neural networks (ANN), and linear and nonlinear multivariate statistical methods (MSM) may be obtained from the framework by selecting the appropriate transformation of the input space, type of activation function, and optimization criteria. A general, hierarchical training methodology is presented for techniques that involve projection of the inputs on a hyperplane before application of the activation function. These projection-based methods include linear and nonlinear principal component regression and partial least squares regression, backpropagation networks with a single hidden layer, and projection pursuit regression. The similarities and differences between various projection-based methods are brought out based on the changes necessary to specialize the general methodology to the selected ANN or MSM. A simple example illustrates the properties of projection-based methods within the unified framework, and guidelines are provided for selection of the appropriate method for a given problem. It is argued that among projection-based methods, PPR is most likely to provide the smallest error of approximation. The model learned by PPR is also more compact and physically interpretable than that learned by BPN.

Keywords

Empirical modeling, Artificial neural networks, Multivariate statistical methods, Projection pursuit regression, Backpropagation networks Partial least squares, Principal component regression.

Introduction

Several process operation and control tasks require development of empirical models from measured data. Among the most popular and successful empirical modeling methods are Artificial Neural Networks (ANN) such as, Back Propagation Networks (BPN) and Radial Basis Function Networks (RBFN); and Linear Statistical Methods (LSM) such as, Principal Component Regression (PCR) and Partial Least Squares Regression (PLS). The properties of ANN and LSM are complementary in nature. ANN possess the ability to approximate arbitrarily nonlinear input-output relationships, and continuously adapt the model to new data without requiring storage of previous data, but require trial and error for network construction and provide little physical insight into the data. On the other hand, LSM provide physically interpretable models, but are unable to capture nonlinear behavior, or adapt continuously to new data without extensive storage of previous data. Over the last several years, significant insight and unified frameworks have been developed for ANN such as, regularization networks by Poggio and Girosi (1989), and for LSM such as, continuum regression by Stone and Brooks (1990). Since ANN and LSM may both be used for empirical modeling, the fact that they are related is not surprising, but relatively little attention has been directed towards unifying these techniques.

In parallel with the development of ANN, Nonlinear Statistical Methods (NLSM) such as, Projection Pursuit Regression (PPR) (Friedman and Stuetzle, 1981), Multivariate Adaptive Regression Splines (MARS)

Table 1. Comparison Matrix for Empirical Modeling Techniques.

Method	Input Transformation	Activation Function	Objective Functions
PCR	Projection	Fixed shape, linear	Max. variance of projected inputs. Min. output mean squares error.
PLS	Projection	Fixed shape, linear	Max. correlation between inputs and outputs, and variance of projected inputs.
BPN	Projection	Fixed shape, sigmoid	Min. output mean squares error.
PPR	Projection	Adaptive shape, ridge	Min. output mean squares error.
RBFN	Kernel	Fixed shape, radial	Select basis functions through clustering. Min. output mean squares error.
Wave-Net	Kernel	Fixed shape, wavelet	Select basis functions from dyadic grid. Min. output mean squares error.
MARS	Kernel	Adaptive shape, spline	Partition input space recursively, and min. output mean squares error.

(Friedman, 1991), and nonlinear versions of PCR and PLS (Frank, 1990; Qin and McAvoy, 1992; Holcomb and Morari, 1992) have been developed. These methods are found to lie on the interface between MSM (includes LSM and NLSM) and ANN (Barron and Barron, 1988; Cheng and Titterington, 1994), and provide the necessary insight leading to a unified view of ANN and MSM as developed in this paper.

Our view of ANN and MSM is based on representing the model as a weighted sum of basis functions. Different empirical modeling methods may be obtained by changing the following model characteristics,

- nature of the input transformation,
- type of activation functions,
- optimization criteria.

We focus on methods that combine the inputs by projection on a hyperplane as a linear weighted sum before operation by a basis function. These "projection-based" methods include PCR, PLS, PPR, and BPN with a single hidden layer. We confirm existing insight and provide new insight about the similarities and differences between projection-based methods by deriving a general training methodology that may be specialized to each method. Finally, we present a simple example to illustrate and compare the properties of each method, and provide some guidelines for selecting the appropriate empirical modeling technique for a given task. In this paper, only a single output variable is considered. Additional details and examples may be found in Utojo and Bakshi (1995).

A Common Framework for ANN and MSM

The model determined via ANN or MSM may be represented as a weighted sum of basis functions,

$$\hat{y}(x_1, ..., x_p) = \sum_{k=1}^{m} \beta_k f_k(\alpha_k; x_1, ..., x_p) \qquad (1)$$

where, $\hat{y}$ is the predicted output for inputs x_j, β_k are the weights, f_k are the activation or basis functions, α_k are the basis function parameters. The specific nature of the model depends on decisions about the following, as summarized in Table 1.

Input Transformation. In order to fight the curse of dimensionality, empirical modeling techniques transform the inputs to take advantage of correlation and/or sparseness in the input space. Projection-based, or non-local methods try to find correlation among inputs by projecting them on a hyperplane before applying the activation function. Kernel-based, or local methods try to exploit sparseness by using a localized activation function.

Activation Functions. The activation functions may be of a fixed shape such as linear, sigmoid or radial, or may adapt their shape to the data in the projected hyperplane or kernel.

Optimization Criteria. The adjustable parameters in Eqn. (1), may be determined by optimizing an objective function consisting only of outputs, or both inputs and outputs. All methods have the common aim to minimize the output prediction error. Some methods also try to maximize the input covariance captured by each basis function and impose orthonormality on α_k.

The training procedure for each method may be hierarchical and determine the model node-by-node, or may be global and estimate all parameters simultaneously.

Projection-Based Methods

As can be seen from Table 1, projection-based methods include some of the most popular LSM and NLSM, and BPN. In general, the model learned by projection-based methods may be represented as,

$$\hat{y}(x_1, ..., x_p) = \sum_{k=1}^{m} \beta_k \; f_k\left(\sum_{j=1}^{p} \alpha_{kj} x_j\right) \qquad (2)$$

Table 2. General Methodology and its Specialization to PPR and Linear PLS.

#	Description	PPR	Linear PLS
1a	initialize	$R_1 = y; X_k = X; \hat{y} = 0$	$R_1 = y; X_k = X; \hat{y} = 0$
1b		Initial guess for α, β, f	
2	begin loop	For k = 1 to m	For k = 1 to m
3a	optimize projection	$\alpha_k^T = \min\{E[R_k - \beta_k f_k(\alpha_k^T X)]^2\}$	$\alpha_k = E(R_k X_k)$
3b	direction	$\|\alpha_k^T\| = 1$	$\|\alpha_k^T\| = 1$
4a		$Z_k = \alpha_k^T X_k$	$Z_k = \alpha_k^T X_k$
4b	activation function	$f_k(\alpha_k^T x_i) = \frac{R_{ki}}{\beta_k}$. Smooth f_k.	
5	regression coefficient	$\beta_k = \left[\frac{E(R_k f_k(Z_k))}{E(f_k(Z_k)^2)}\right]$	$\beta_k = \left[\frac{E(R_k Z_k)}{E(Z_k^2)}\right]$
6	model update	$\hat{y}_{k+1} = \hat{y}_k + \beta_k f_k(Z_k)$	$\hat{y}_{k+1} = \hat{y}_k + \beta_k Z_k$
7a	residual	$R_{k+1} = R_k - \beta_k f_k(Z_k)$	$R_{k+1} = R_k - \beta_k Z_k$
7b	update(s)		$X_{k+1} = X_k - [(E(Z_k X_k)/E(Z_k^2)]Z_k$
8	backfitting, termination	if $E[(y - \hat{y}_{k+1})^2] \leq \varepsilon$, exit	if $E(X_{k+1}^T X_{k+1}) = 0$, exit
9	end loop	end	end

For LSM such as PCR and PLS, the function, f_k is linear, for BPN, f_k is sigmoid, while for PPR f_k can take any shape depending on the projected data.

A General, Hierarchical Training Methodology

The model learned by all projection-based methods may be determined by a general methodology for hierarchical, or node-by-node training. The general methodology unifies all projection-based methods and consists of the steps shown in Column 2 of Table 2. This methodology may be specialized to give different projection-based methods by changing steps to include the specific characteristics listed in Table 1. The general methodology may be specialized to PPR as shown in Column 3 of Table 2. The projection directions are determined by minimizing the MSE, and the basis functions by smoothing the projected data and selecting the degree of smoothness via crossvalidation. The projection directions for PPR may be orthogonalized at each stage, and backfitting may be avoided, and the resulting network would be easier to modify, but larger. Specialization to linear PLS is shown in Column 4 of Table 2. Similarly, this general methodology may be easily specialized to linear PCR and several suggested versions of nonlinear PCR and PLS. For example, quadratic PLS (Wold, et al., 1989), supersmoother PLS (SS-PLS) (Frank, 1990), and neural network PLS (Qin and McAvoy, 1992; Holcomb and Morari, 1992), may each be obtained by computing the activation function by a quadratic model, supersmoother, and BPN respectively.

Specialization of the generalized methodology to BPN may be easily achieved by restricting the PPR algorithm to using only sigmoid activation functions. This does not lead to the traditional training methodology for BPN which involves selection of a network structure, and then estimation of the model parameters by training the entire network simultaneously to minimize the prediction error. This global training method requires trial and error for selection of the appropriate number of nodes. Adaptation of the hierarchical PPR training technique to BPNs would transform some of the benefits of PPR to BPN. Thus, nodes could be added to the BPN until the testing error reached a desired level, providing a convenient stopping criterion for BPN construction. Indeed, such a method has been suggested by Fahlman and Lebiere (1990) as the cascade correlation network.

Example

The features of various projection-based empirical modeling methods are illustrated by this simple example. The objective is to learn the relationship between two inputs and one output represented by a purely additive model (Hwang, et al., 1994),

$$y = 1.3356\left[1.5(1-x_1)+e^{2x_1-1}\sin(3\pi(x_1-0.6)^2) + e^{3(x_2-0.5)}\sin(4\pi(x_2-0.9)^2)\right] \quad (3)$$

The lack of interaction between inputs in each additive component of Eqn. (2) indicates that the optimal projection directions should be parallel to the input axes. Various projection-based methods were trained using 200 uniformly distributed data points drawn randomly form a large standardized data. The testing set for crossvalidation consisted of 400 data points drawn from the same distribution. The final testing mean-squares error (MSE) and the projection directions (input weights) for each method are shown in Table 3. For this example, PPR provides the smallest MSE with only two nodes. As expected, the projection directions are parallel to the input axes. The MSE of SS-PLS is larger than that of PPR due to satisfaction of additional optimization criteria. A BPN with three hidden nodes results in a larger MSE than PPR and SS-PLS. Notice that BPN has also managed to find the expected projection directions parallel to the input axes.

Table 3. Final MSE and Projection Directions for Some Projection-Based Methods.

Method	MSE	Node # 1	Node # 2	Node # 3
PPR	0.019	1, 0.002	-0.0051, 1	N/A
BPN	0.068	-1, 0.0024	-0.0018, -1	-0.99, 0.09
SS-PLS	0.066	-0.0056, 1	-1, -0.0056	N/A
SS-PCR	0.399	0.24, -0.97	-0.97,-0.24	N/A

Conclusion

We have presented a unified view of various neural and statistical empirical modeling techniques. All ANN and MSM may be represented as a weighted sum of basis functions. Their approximation quality and parameter values depend on decisions about the nature of input transformation, type of basis functions, and optimization criteria. The emphasis in this paper has been on methods that project the inputs on a hyperplane as a linear weighted sum before application of the activation function. These projection-based methods include linear and nonlinear PCR and PLS; BPN with one hidden layer; and PPR. A general methodology for hierarchical, node-by-node training of *all* projection-based models is proposed.

The unified view provides several heuristics about the characteristics of neural and statistical empirical models. Among projection-based methods, PPR is likely to provide the best approximation and most compact model due to its adaptive basis functions and emphasis on minimizing the output prediction error only. Nonlinear PLS methods are unlikely to perform as well as PPR because of the additional objectives of maximizing the input variance captured by each node, and orthonormal projection directions. BPNs are also unlikely to beat the performance, model compactness and physical interpretability of PPR. Furthermore, PPR may train faster than BPN (Hwang, et al., 1994). The formulas for the relative importance of each input and each node given by Friedman (1985) for PPR may be applied to all projection-based methods, but are likely to be most meaningful for methods that result in the most compact model and small prediction error. Despite the fact that PPR was developed before or in conjunction with BPN, and is in many ways superior to BPNs, it has received surprisingly little attention from the non-statistics community. Empirical modeling at the interface of ANN and MSM is becoming an active research topic within both, the statistics and neural network community. We expect that the unified view presented in this paper will contribute to new methods for empirical modeling that can make the best of both, neural and statistical worlds.

Acknowledgment

Support from an Ohio State University Seed Grant is gratefully acknowledged.

References

Barron, A. R. and R. L. Barron (1988). Statistical learning networks: a unifying view. In E. J. Wegman, D. T. Gantz and J. J. Miller (Eds.), *Computing Science and Statistics*. pp.192-203.

Cheng, B. and D. M. Titterington (1994). Neural networks: A review from a statistical perspective. *Stat. Sci.*, **4**(1), 2-54.

Fahlman, S. E. and C. Lebiere (1990). The cascaded-correlation learning architecture. In *Adv. Neur. Info. Proc. Sys. 2*. Morgan Kaufmann. pp. 524-532.

Frank, I. E. (1990). A nonlinear PLS model. *Chemom. Intel. Lab. Sys.*, **8**, 109-119.

Frank, I. E. and J. H. Friedman (1993). A statistical view of some chemometrics regression tools. *Technomet.*, **35**(2), 109-148.

Friedman, J. H. and W. Stuetzle (1981). Projection pursuit regression. *J. Amer. Stat. Assoc.*, **76**(376), 817-823.

Friedman, J. H. (1985). Classification and multiple regression through projection pursuit. Tech. Rep. 12, Dept. of Stat., Stanford University.

Friedman, J. H. (1991). Multivariate adaptive regression splines. *Annals Stat.*, **19**(1), 1-141.

Holcomb, T. R. and M. Morari (1992). PLS/Neural networks. *Comp. Chem. Eng.*, **16**(4), 393-411.

Hwang, J. N., S. R. Lay, M. Maechler, R. D. Martin and J. Schimert (1994). Regression modeling in back-propagation and projection pursuit learning. *IEEE Trans. Neur. Net.*, **5**(3), 342-353.

Poggio, T. and F. Girosi (1989). A theory of networks for approximation and learning. MIT AI Lab. Memo 1140.

Qin, S. J. and T. J. McAvoy (1992). Nonlinear PLS modeling using neural networks. *Comp. Chem. Eng.*, **16**, 379-391.

Stone, M. and R. J. Brooks (1990). Continuum regression. *J. Royal Stat. Soc., Ser. B.*, **52**, 237-269.

Utojo, U. and B. R. Bakshi (1995). A unified view of ANN and MSM. Tech. Rep., Dept. Chem. Eng., Ohio State Univ.

Wold, S., N. Kettaneh-Wold and B. Skagerberg (1989). Nonlinear PLS modeling. *Chemom. Intel. Lab. Sys.*, **7**, 53-65.

APPLICATIONS OF NEURAL NETWORKS FOR MULTIVARIATE PROCESS MONITORING

John Plummer[1]
Daly City, California 94014

Abstract

Standard, univariate control procedures, x-bar-, r- and s-charts being typical examples, are demonstrably inadequate for monitoring such complex processes as those encountered in semiconductor wafer fabrication. Rather, identification of process shifts, commonly referred to as 'out-of-control' conditions, requires that a number of variables, usually a very large number, be considered simultaneously. It is often necessary to look at vectors of test results that may number in the hundreds of entries. Statistical approaches to analysis of covariance matrices of this size quickly run out of computational steam. Neural networks provide an approach which emulates statistical covariance analysis and which can be applied effectively to test results of this magnitude. This paper discusses an approach to process disturbance identification and classification which incorporates a number of neural network paradigms. Cottrell-Zipser autoassociative memory, based on backpropagation, provides a filter to separate disturbance terms from normal processing signals. Kohonen learning vector quantization (LVQ) data reduction further filters the data, making clusters of performance in a historical database easier to identify. Assigning clustered exemplars as weights in a probabilistic neural network (PNN) provides a tool which can readily identify and classify out-of-control conditions in a multivariate world. This approach is applied to examples taken from the semiconductor industry.

Keywords

Statistical process control, Process condition monitoring, Principal components analysis, Analysis of variance, Pattern classification.

Introduction

One basic tenet of process condition monitoring, often called statistical process control (SPC), is that if a problem goes away, and you don't know why, it will almost certainly come back. This indicates that systems do not go out of control by sliding off the end of some Gaussian distribution. Rather, they switch from state to state with some Markovian-like probability.

Univariate control procedures, x-bar, r charts and the like, are frequently insensitive to changes of this sort. Not until rather gross process disturbances occur can variations from the norm be identified. It is then recommended practice to halt processing until the problem can be identified and fixed. This is a very expensive approach in terms of lost capacity, and lost credibility from forepersons and operators who are, frequently with justification, skeptical that a single, apparently random, outlier is sufficient cause for such extreme reaction.

Multivariate process monitoring, using neural networks, can be much more sensitive to process changes, and may help identify incipient problems before they become so major as to warrant shut-down. It is often advisable to separate a disturbance term from the underlying signal. This makes the occasion of a process change easier to detect. Furthermore, by clustering historical data, processing states which have occurred most frequently can be identified. Pattern classification techniques can then be used to signal their recurrence.

This approach involves the following steps:

[1] *johplummer@aol.com.*

1. Develop a multivariate filter by training a Cottrell-Zipser autoassociative memory using ‘good’ or ‘normal’ data.
2. Apply the trained filter to a larger historical database. Calculate disturbance vectors by comparing the input and output for each data entry.
3. Group the disturbance terms by category. This step is optional, but usually simplifies interpretation of the results.
4. Reduce the resulting disturbance dataset to a relatively small set of exemplars using a Kohonen LVQ network. This acts as a secondary filter, making clusters in the data more apparent.
5. Identify clusters in the historical database using some standard statistical clustering technique.
6. Assign the clustered LVQ exemplars as weights in a PNN. The PNN can then serve as a multivariate control chart, both identifying and classifying out-of-control conditions.

The Failure of Univariate Monitors

Univariate process condition monitors, traditionally misnomered statistical process controls (SPC), are insensitive to major changes in process conditions in such complex processes as semiconductor wafer fabrication. To illustrate this fact a SPICE simulation of a semiconductor device known as an operational amplifier was used to generate a test data set. SPICE is a well-known simulation language used to evaluate the performance of complex electronic circuitry by stepping that circuitry through a performance cycle in very small time increments. Important characteristics of the device being simulated can then be studied at each time increment.

A simulation was used for several reasons. First, process conditions could be controlled exactly, a feat difficult if not impossible to accomplish in the real world. Second, data collected contained no noise, measurement error or other distortions of that sort. Finally, if an event of interest occurred at some region in the response space that region could be studied in any desired level of detail.

As illustrated in Table 1, six input and six output parameters were measured. Input parameters reflect the condition of the wafer fabrication process. Output variables indicate the performance characteristics of the resulting device. These parameters proved to be highly correlated, with an average multiple r-square (linear) between the output and input variables of 0.824. npn Beta was linearly correlated with each of the output variables with a significance of 99%. The correlation between npn Beta and the bias currents was 0.81. One would think, then, that a disturbance in the process which affected npn beta could be identified by monitoring the bias currents. This proved not to be the case.

Table 1. Input and Output Variables Measured in the SPICE Simulation of an Operational Amplifier.

Input Variables	Output Variables
npn Beta	Offset Voltage
Lateral npn Beta	Offset Current
Vertical pnp Beta	Supply Current
Resistor Tolerance	Bias Current (+)
Resistor Mismatch	Bias Current (-)
Zener	DC Gain

Two data sets were generated, one with a normal distribution of npn beta, one with a distinct tri-modal distribution of this same parameter over the same range. The first data set then represents a process which is in control, the second a process badly out of control. Table 2. summarizes these two data sets.

The distributions of each of the output variables for these two data sets were statistically indistinguishable. Even the bias currents, which correlated strongly with npn Beta ($r > 0.8$), produced distributions which, when tested using the chi-square statistic, were not distinguishable even at a 90% confidence level.

Table 2. *Statistical Comparison of In-control and Out-of-control Processes.*

	Group 1. In Control		Group 2. Out of Control	
Electrical test variables:	mean	std dev	mean	std dev
npn Beta	218.1	67.2	213.1	82.2
Lateral pnp Beta	81.6	21.3	81.6	21.8
Vertical pnp Beta	149.8	41.8	150.7	42.9
Resistor Tolerance (%)	0.3	5.6	0.4	6.1
Resistor Mismatch (%)	0.0	0.1	0.0	0.1
Zener (Volts)	6.6	0.2	6.6	0.2
Wafer probe variables:				
Offset Voltage (μV)	0.6	13.5	2.1	12.3
Offset Current (nA)	0.0	0.1	0.0	0.1
Supply Current (mA)	1.2	0.1	1.2	0.1
Bias Current (+) (nA)	0.9	0.3	0.9	0.4
Bias Current (-) (nA)	0.9	0.3	0.9	0.4
DC Gain (dB)	135.7	2.6	134.8	3.4

Applying Multivariate Pattern Classification

Our first objective is to demonstrate that neural network based multivariate pattern classification techniques can successfully identify clusters in the data base of output parameters which are invisible to univariate statistical techniques. The following steps were taken to accomplish this goal:

1. The data in each dataset were reduced using Kohonen LVQ networks. Ten exemplars were selected, arbitrarily, for training. This

accomplished two objectives. First, the number of data vectors was reduced by 90%, making data handling easier. Second, a result not anticipated, clusters are more apparent in the reduced dataset than in the original dataset. LVQ reduction appears, then, to provide a sort of filtering, making the layout of the forest more apparent when there are fewer trees.

2. Kohonen exemplars were then clustered using simple Euclidean distances. The first indication that this approach would be successful: exemplars derived from the in-control dataset grouped into a single cluster, exemplars from the out-of-control dataset into several. This result was repeatable.
3. Using a technique described by Burrascano (1991), LVQ vectors were assigned as weights in a PNN. This approach provides a good approximation of Bayesian pattern classification as the number of exemplars in each cluster is approximately proportionate to the *a priori* probability of membership in the corresponding category.
4. The original data test vectors were classified using this PNN. The result, illustrated in Table 3: more than 90% of the original data vectors were correctly classified. This result compares quite favorably with statistical clustering of the original data, which produced only about a 70% accuracy after assuming that the number of categories and the approximate *a priori* probabilities of category membership were known, a requirement not applicable to the neural network case.

Table 3. Comparison of Actual Clusters and PNN Category Assignments.

Input	Kohonen Cluster				
Cluster	1	2	3	4	Total
1	25	2	3	0	30
2	0	50	0	0	50
3	0	5	6	9	20
Total	25	57	9	9	100

Reviewing Actual Wafer Fab Test Data

Often tools that work in the world of simulation fail to perform when faced with real-world data. The introduction of measurement error, and other sources of noise, can make patterns more difficult to discern. Also it may be necessary to increase the dimensionality of the problem space; techniques which work with six variables may fail when confronted with sixty.

To test these results wafer electrical test data for another linear semiconductor device were selected. This data is collected from test chips, especially designed devices which are placed at selected locations on a wafer for the purpose of gathering data which can be used to monitor the process. Typically these test chips comprise a large number of devices, each to test some special process characteristic.

We selected data from the npn transistors, eight of which were included in the test chip design. There were 15 test types, breakdown voltages, forward voltages, betas, and the like. Of the possible 120 variables (8 devices times 15 tests), 80 were actually measured. These 80 variables then constituted our data vector.

Wafers were sorted by yield into 5 yield categories, high to low, with approximately 20% of all wafers in each yield category. Again univariate techniques were unable to distinguish low-yield from high-yield wafers. 95% of the test results for low-yield wafers fell within the 3-sigma range of the high yield wafers. 66% of the low-yield wafers had all 80 test results within the 3-sigma range of the high-yield wafers.

Analysis of this data was undertaken using the following procedure:

1. A Cottrell-Zipser autoassociative memory was trained using data from high-yield wafers only. This provides a filter which, when applied to data from low-yield wafers can be used to separate process disturbance from the underlying normal process patterns.
2. Data from low-yield wafers was processed using the trained filter. Subtracting the input vector from the output vector provides a disturbance vector with 80 entries.
3. This disturbance vector was squared, providing a vector of variances. This enables evaluation of the disturbance pattern using standard statistical techniques.
4. Marginal variances were calculated for each row (device) and each column (test type) in the text matrix by adding the variances in the corresponding cells. This reduced the dataset from 80 variables to 23. Marginal standard errors were then calculated simply by taking the square root of the marginal variances. Finally, a ratio of the resulting standard error vectors was calculated by dividing standard errors of low-yield wafers by mean standard errors for high-yield wafers. Table 4 summarizes the results of this analysis.

It has for some time been recognized that large numbers in this disturbance vector indicate possible sources of trouble. Eryurek and Upadhyaya (1990), at the University of Tennessee, applied this technique to identification of faulty sensors in a nuclear power plant. Kramer (1991) demonstrated that properly constructed such a filter can provide nonlinear principal components analysis for this purpose. However, less attention has been paid determining an objective measure of "large" in this context. Obviously this is an important question when applying this approach to condition monitoring.

Table 4. Disturbance Index for Marginal Values Entries Represent Ratio of Disturbance by Yield Category to High-Yield Wafers.

	Yield Group			
	Hi-Avg	Avg	Lo-Avg	Low
Tests:				
Forward voltage	1.23	1.65	.148	3.07
Breakdown volt.	1.14	2.79	1.63	1.54
Beta	1.12	1.49	1.52	1.49
Vsat	1.34	1.85	2.52	1.41
Devices:				
10X per slot	1.07	4.09	1.70	1.31
Min. geometry	1.33	3.46	1.58	3.80
Large area	1.30	1.67	2.25	1.43

The process of converting the disturbance vectors to variances provides a handy statistical tool for determining objectively when a disturbance is statistically significant. Furthermore, in as much as the data has at this point been summarized to a significant degree, the assumption of normalcy in the distribution of the standard error terms is likely to be valid even when the underlying parametric distributions are not normal. This phenomena deserves further study.

Finally, summarizing the results into marginal terms provides a sort of engineered principal components analysis. Whether the marginal variables are devices and test types or temperatures, pressures and flow rates makes no difference to the arithmetic. The approach is then equally applicable in a wafer fab or in a chemical processing plant. The results make more sense to the engineers because disturbances are considered both within group (*e.g.*, all of the temperatures) and between group (*e.g.*, the flow rates given the temperatures and pressures). The dimensionality of this process is essentially unlimited as well.

In the present instance it is clear that the type of problem encountered is important in determining the extent of yield loss. Problems which affect breakdown voltage on the 10X per slot and minimum geometry devices are most likely to produce a relatively mild yield loss. Losses due to problems which affect forward voltage on the minimum geometry device are likely to be more dramatic. The process engineer is provided both clues as to where to look for a solution and the degree of urgency of a problem.

Identifying Process Shifts

Finally, using this same wafer fab data, an attempt was made to determine if the process conditions had remained constant during the data collection period. The process described above, reducing the data with Kohonen LVQ exemplars, clustering those exemplars, and assigning them as weights in a PNN was followed. This time, however, marginal disturbance data (23 variables) was clustered rather than the original dataset (80 variables). It was discovered that sometime between workweek 8 and 10 the characteristics of the low-yield wafers changed. Review of process conditions identified a change in the second-metal procedure which had gone unnoticed by the product engineers.

Different categories of low-yield wafers were successfully identified in a database for which low-yield and high-yield wafers are indistinguishable when test results are considered one test at a time.

References

Eryurek, E. and B. R. Upadhyaya (1990). Sensor validation for power plants using adaptive backpropagation neural networks. *IEEE Nuclear Science Symposium*, San Francisco, CA.

Kramer, M. A. (1991). Nonlinear principal component analysis using autoassociative neural networks. *AIChE Journal*, **37**, 233-243.

Burrascano, P. (1991). Learning vector quantization for the probabilistic neural network. *IEEE Transactions on Neural Networks*, **2**, 458-461

ON-LINE DETERMINATION OF THE DEGREE OF CURE OF EPOXY/GRAPHITE COMPOSITES WITH A NEURAL NETWORK

H.B. Su, L.T. Fan and J.R. Schlup
Department of Chemical Engineering
Kansas State University
Durland Hall, Manhattan, KS 66506

Abstract

An approach is presented for on-line determination of the degree of cure (DOC) of epoxy/graphite composites by a recurrent neural network (RNN). The RNN predicts the DOC of the composites by monitoring the Damköhler number with the so-called dual heat-flux sensor. Computer simulation has demonstrated that neural networks are indeed viable for predicting the DOC of composites.

Keywords

Epoxy/graphite composites, Degree of cure, Recurrent neural network, On-line monitoring, Heat flux sensors, Damköhler number.

Introduction

Epoxy/graphite composites possess highly desirable properties; thus, they serve widely as structural materials. Cure of composites, however, is often time consuming, energy expensive, and materials costly. The properties of the final products must be reproducible; nevertheless, the currently available sensors appear to be ineffective for on-line control based on the evaluation of the properties of composites being cured.

Several types of sensors for on-line monitoring of the cure's progress have been developed. For instance, near infrared spectroscopy has been employed to detect the disappearance of epoxy groups with time (see, e.g. Young, et al., 1989). A dielectric sensor can determine the extent of cure by measuring the response of molecular dipoles in terms of the dissipation factor as a function of time to an oscillating electric field (see, e.g. Ciriscioli and Springer, 1989). A dual heat-flux sensor monitors the cure reaction by measuring the heat flux and subsequently evaluating the Damköhler number (Perry, et al., 1992).

The degree of cure (DOC), α, can be defined as

$$\alpha = \frac{H(t) - H(0)}{H(\infty) - H(0)} \tag{1}$$

where $H(t)$ is the enthalpy of the system at time t. In the present work, the DOC of composites is predicted on-line by an artificial neural network on the basis of the Damköhler number measured with a dual heat-flux sensor (Perry, et al., 1992).

Neural Network Models

Artificial neural networks have been demonstrated to be capable of playing a variety of roles in process engineering (see, e.g. Bhat and McAvoy, 1990; Ungar, 1990; White and Sofge, 1990; Joseph and Hanratty, 1993). Among various architectures proposed for artificial neural networks, recurrent neural networks (RNN's) have proven to be effective in learning temporal patterns; thus, RNN's are adopted in this study. A RNN consists of neurons and synapses similar to those of a feedforward network, except that the information flows not only in a forward direction but also in a backward direction along the synapses.

An adaptive random search technique (Fan, et al., 1975; Chen and Fan, 1976) is adopted in the present work to mitigate the problem of severe convergence when a RNN is trained by the gradient-type training. This search technique tends to lead to the global minimum.

On-Line Monitoring with Dual Heat-Flux Sensors

Advanced composites are fabricated from continuous fiber prepregs by laminating multiple piles into the desired shape and then by curing the assembled part. Vacuum-bag molding in an autoclave is the most common technique for curing composites. A typical cure cycle is illustrated in Fig. 1. Cure cycles in the manufacture of composites evolve through trial-and-error experimentation (see, e.g. Servais, et al., 1986). That the determination of a cure cycle is tedious and costly is attributable to the difficulties involved in precisely measuring the physical properties of composites, e.g. the viscosity and DOC, as functions of process conditions.

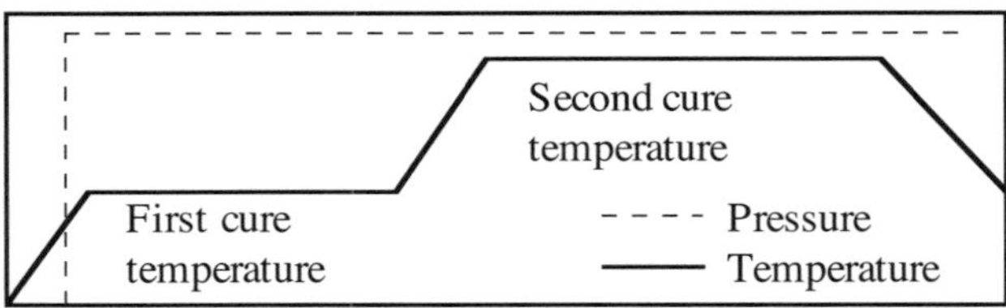

Figure 1. Typical cure cycle in an autoclave.

A dual heat-flux sensor comprising thermocouples for monitoring cure has been developed by Perry, et al. (1992). It is capable of measuring the Damköhler number defined as

$$Da = \frac{rate\ of\ heat\ generation}{rate\ of\ heat\ conduction} \tag{2}$$

The dual heat-flux sensor implemented by a pile of three thermocouples is illustrated in Fig. 2. For the case of one-dimensional heat transfer, the Damköhler number, computed on-line by the sensor, can be expressed as (Perry, et al., 1992)

$$Da = \frac{L^2}{kT}\{\rho C_p \frac{\Delta T}{\Delta t} + (\frac{-k}{\Delta x^2})[(T_i - T_0) - 2T]\} \tag{3}$$

where k is the heat conductivity; ρ, the density; C_p, the specific heat; ΔT, the temperature change during Δt.

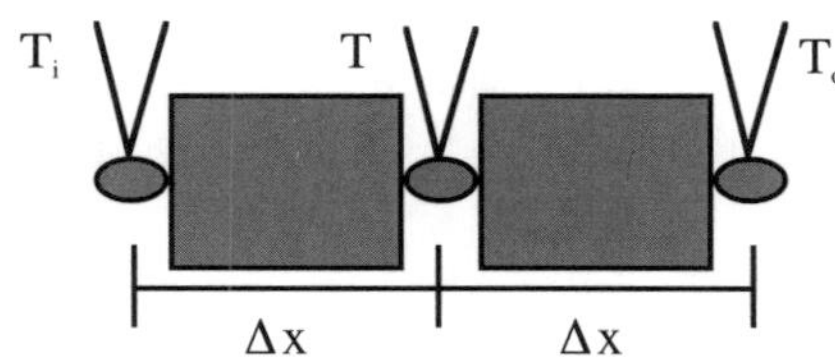

Figure 2. Dual heat-flux sensor comprising a pile of three thermocouples.

The cure reaction may involve three stages, i.e. onset, acceleration, and termination. The onset of reaction can be recognized when the Damköhler number becomes positive; the acceleration, when the Damköhler number exceeds unity; and the termination, when the Damköhler number returns to zero. The acceleration stage should be circumvented because it causes thermal run-away. A simulation program for autoclave processing of polymer matrix composites has been developed by adapting the works of Loos and Springer (1983) for the present work. Figure 3a displays the terminal profile of Damköhler number obtained by simulating the cure cycle illustrated in Fig. 3B.

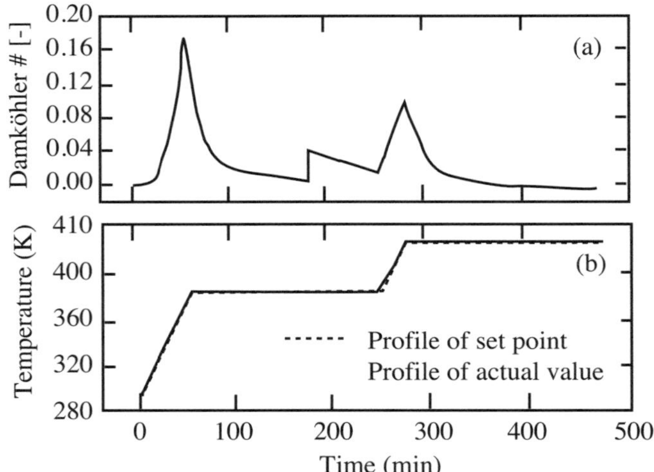

Figure 3. Computer simulation of the cure cycle: (a) the Damköhler number determined by the dual heat-flux sensor; (b) the temporal profiles of the average temperature in the sample and the set points of the cure cycle.

RNN for Predicting the DOC

Measurements of the Damköhler number reveal qualitatively the characteristic features of the three stages in the cure reaction, which should be quantitatively interpreted. Note that the Damköhler number decreases rapidly towards zero even in the second phase (see Figs. 3a and 3b), thereby falsely indicating that the cure reaction is being terminated. Moreover, the Damköhler number approaches asymptotically to zero far ahead of the completion of the cure reaction in the fourth phase, at which the DOC is equal to unity. It is extremely difficult, therefore, to determine the exact instant for terminating the cure process by simply observing the Damköhler number. This difficulty may be aggravated by the presence of noise, which is inevitable when the Damköhler number is monitored *in situ*. The interpretation of the DOC in terms of the Damköhler number, however, can resolve the difficulty.

During the cure, the DOC, α, can be expressed as

$$\alpha = f(T, P, \overline{X}, t) \tag{4}$$

where T is the temperature; P, the pressure; $\overline{X}$, the properties of the sample, such as concentration, heat capacity, heat of reaction, and viscosity; and t, the processing time. In principle, this function can be derived through first-principle modeling provided that the properties of the sample and the processing conditions can be precisely determined. Nevertheless, the majority of the properties of the cure sample are either impossible to measure, or inordinately expensive to do so. Alternatively, therefore, the function can be derived through semi-empirical or empirical modeling.

A RNN has bee trained by the data generated from a constant heating rate experiment, similar to the dynamic differential scanning calorimetry (DSC) which is the most prevailing method for studying the kinetics of cure reaction. Specifically, the average temperature of the sample (average of those at the center and top) is increased at the rate of 2°C/min from 295 K until it reaches 540 K. Meanwhile, the dual heat-flux sensor measures the Damköhler number through Eqn. (2). The training patterns, consisting of 490 data points, are linearly scaled in [0.15, 0.85]. The results of training through the adaptive random search technique are summarized in Table 1 and displayed in Fig. 4.

Table 1. Summary of the Results for Training and Testing the Recurrent Network for Predicting the Degree of Cure.

	Training cycle	Testing cycle 1	Testing cycle 2
Initial resin content (Volume %)	42	42	50
Heat of cure reaction (cal/g)	-150.2	-150.2	-215.5
Thickness (cm)	1.2	2.4	3.2
First cure temperature (K)	N/A	380.0	370.0
Period of first cure temperature (min)	N/A	200.0	120.0
Second cure temperature (K)	N/A	420.0	430.0
Average sum-squared error	0.00001	0.0000005	0.000003

The RNN is constructed with temperature, time, and the Damköhler number as input variables to predict the DOC which is represented by a first-order dynamic model (see, e. g. Loos and Springer, 1983). Specifically, this RNN comprises four nodes in the input layer. The first is for the processing time; the second, for average temperature of the sample; the third, for the Damköhler number; and the fourth, for the predicted DOC at the preceding instant. The five nodes in the hidden layer are determined by trial and error based on the average mean-squared error. There is only one node in the output layer for the DOC. The training and testing data are generated by the simulation program.

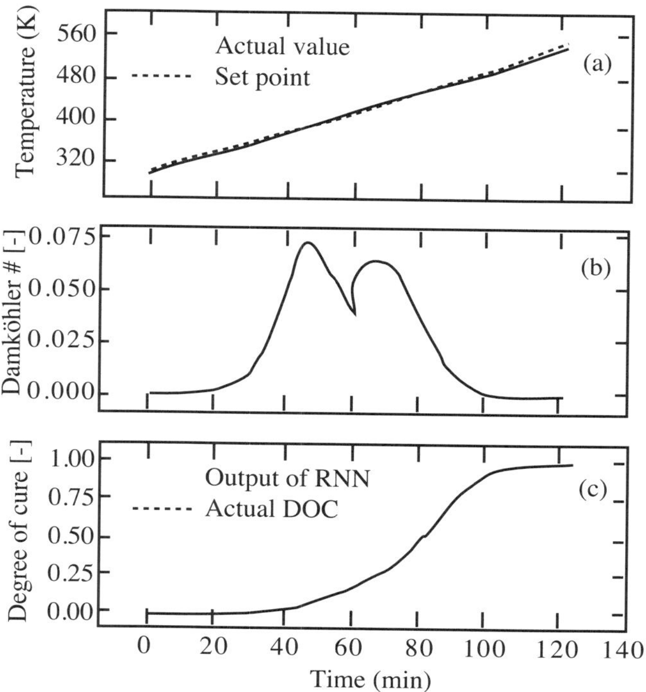

Figure 4. Cure cycle for training the RNN to predict the DOC: (a) the temporal profiles of the average temperature in the sample and the set points; (b) the Damköhler number determined by the dual heat-flux sensor; (c) the output of the RNN and the actual DOC, which overlap.

Evaluation of the RNN

The trained RNN for predicting the DOC has been tested by simulating various cure cycles with different composite samples. Two of the cycles are displayed in Figs. 5 and 6 and summarized in Table 1. In the first, the predicted DOC's are virtually identical to the actual DOC's obtained with the cure cycle completely different from that for training and for a thicker sample. The average mean-squared error is 0.0000005 for the first cure cycle of testing. In the second, the thickness of the sample, initial resin content, and the heat of the cure reaction are rendered far different from those for either training or the first cycle of testing. The average mean-squared error is 0.000003 for the second cure cycle of testing.

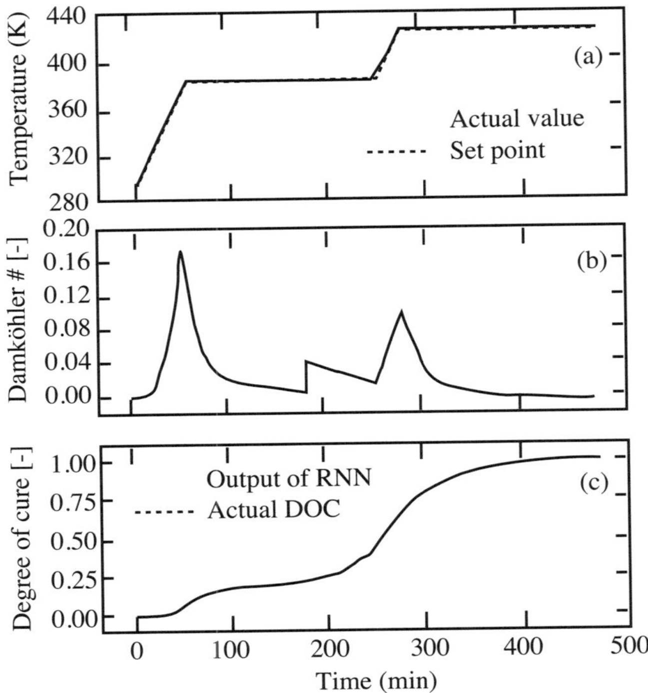

Figure 5. First cure cycle for testing the RNN to predict the DOC: (a) the temporal profiles of the average temperature in the sample and the set points; (b) the Damköhler number determined by the dual heat-flux sensor; (c) the DOC predicted by the RNN and the actual DOC, which overlap.

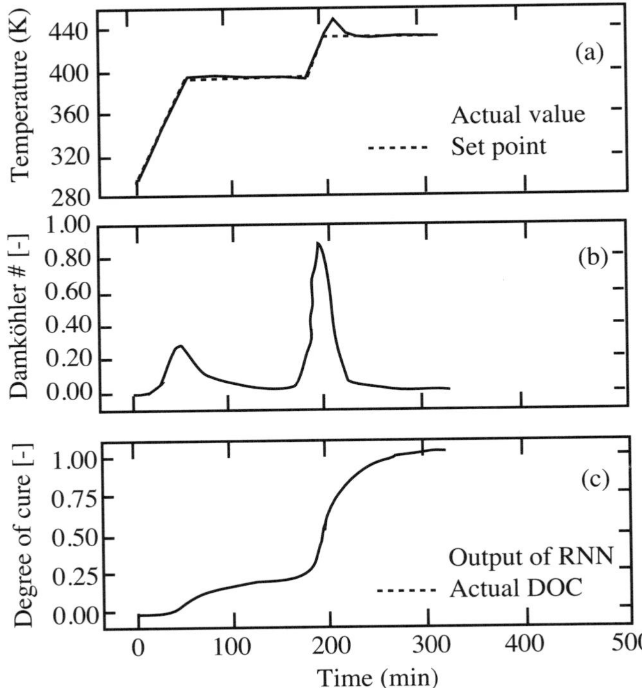

Figure 6. Second cure cycle for testing the RNN to predict the DOC: (a) the temporal profiles of the average temperature in the sample and the set points; (b) the Damköhler number determined by the dual heat-flux sensor; (c) the DOC predicted by the RNN and the actual DOC, which overlap.

Conclusions

A recurrent neural network for predicting the degree of cure has been constructed. The results of testing indicate that the recurrent neural network can satisfactorily predict the degree of cure in an autoclave with the Damköhler number measured by the dual heat-flux sensor as the input for various cure cycles with samples of diverse characteristics, such as thickness, initial resin content, and heat of cure reaction.

Acknowledgment

This work was supported by the Advanced Manufacturing Institute under Grant Number 93116.

References

Bhat, N. and T. McAvoy (1989). Use of neural nets for dynamic modeling and control of chemical process systems. *Proc. of American Control Conf.*, Pittsburgh. pp. 1342-1348.

Chen, H.T. and L.T. Fan (1976). Multiple minima in a fluidized reactor-heater system. *AIChE J.*, **22**, 680-685.

Ciriscioli, P.R. and G.S. Springer (1989). Dielectric cure monitoring — a critical review. *SAMPE J.*, **25**, 35-42.

Fan, L.T., H.T. Chen and D. Aldis (1975). An adaptive random search procedure for large scale industrial and process systems synthesis. *Proc. Symp. on Computers in the Design and Erection of Chemical Plants*. Karlovy Vary, Czechoslovakia. pp. 279-292.

Joseph, B. and F.W. Hanratty (1993). Predictive control of quality in a batch manufacturing process using artificial neural network models. *Ind. Eng. Chem. Res.*, **32**, 1951-1961.

Loos, A. C. and G. S. Springer (1983). Curing of epoxy matrix composites. *J. Composite Materials*, **17**, 135-169.

Perry, M., L.J. Lee and C.W. Lee (1992). On-line cure monitoring of epoxy/graphite composites using a scaling analysis and a dual heat flux sensor. *J. Composite Materials*, **26**, 274-292.

Servais, R.A., C.W. Lee and C.E. Browning (1986). Intelligent processing of composite materials. *31st International SAMPE Symp. and Exposition*, Las Vegas, NV, April 7-10.

Ungar, L.H. (1990). Adaptive networks for fault diagnosis and process control. *Comput. Chem. Eng.*, **14**, 561-572.

White, D.A. and D.A. Sofge (1990). Neural network based control for composite manufacturing. In *Intelligent Processing of Materials*. ASME Publication, NY..

Young, P.R., M.A. Druy, W.A. Stevenson and D.A.C. Compton (1989). In-situ composite cure monitoring using infrared transmitting optical fibers. *SAMPE J.*, **25**, 11-16.

MONITORING WITH TRACKERS BASED ON SEMI-QUANTITATIVE MODELS

Benjamin Kuipers
University of Texas at Austin
Austin, TX 78712

Abstract

MIMIC monitors continuous dynamic systems by tracking multiple hypotheses in parallel, in order to reduce the "missing model" problem that leads to system accidents. A tracker embodies a particular hypothesis as a semi-quantitative model, obtaining the expressive power and guaranteed coverage of qualitative models. The tracker assimilates an observation stream, incrementally becoming more precise or deriving a contradiction. We discuss the architecture of individual trackers, and provide a simple example.

Keywords

Monitoring, Qualitative models, Qualitative simulation, Guaranteed coverage, Trackers, System accidents.

Introduction

Human operators of complex process plants cope with complexity by using a hypothesized model of the plant to organize input signals and alarms, and to motivate actions and investigations. When the operator's initial model is incorrect, it may continue to provide a plausible explanation for observations until proper corrective action is no longer possible. Indeed, the incorrect model can recommend actions that make the situation much worse, as in the Three Mile Island nuclear disaster. These phenomena, called "system accidents," occur across domains where system complexity exceeds operator ability to consider multiple hypotheses (Perrow, 1984).

The MIMIC approach to monitoring continuous systems attempts to address this problem with multiple hypotheses considered in parallel (Dvorak and Kuipers, 1991). If several hypotheses with qualitatively different consequences are explicitly considered, observations can be collected or experiments run to distinguish them explicitly from each other and avoid the "missing model problem" that leads to system accidents.

We define a *tracker* as an object which embodies the hypothesis that the available observation stream is consistent with a particular model and behavior. The tracker assimilates observations, updates its explanations of the past and predictions for the future, and maintains its own status (consistent, superseded or refuted). MIMIC manages a set of trackers.

When knowledge is incomplete — as it must be if we hope to track fault hypotheses — it is difficult to implement the MIMIC approach to monitoring using numerical models. Fortunately, *qualitative* models make it possible to express incomplete qualitative knowledge of systems, and *semi-quantitative* models allow us to express partial quantitative knowledge as well.

In previous research (Kuipers, 1994), we have successfully developed a technology for (a) building qualitative and semi-quantitative models from libraries of model-fragments, (b) simulating these models to predict future behaviors with the guarantee that all possible behaviors are covered, (c) assimilating observations into behaviors, shrinking uncertainty so that incorrect models are eventually refuted and correct models make stronger predictions. This paper recasts the original design for MIMIC (Dvorak and Kuipers, 1991) around the concept of the tracker, in order to separate the properties of individual trackers from the problem of managing the tracking set as a whole.

Qualitative Reasoning

I assume familiarity with the terminology of qualitative reasoning, including *qualitative differential equation* (QDE), *qualitative value* (qval), *qualitative state* (QState), *qualitative behavior* (QBeh), and so on, as defined in (Kuipers, 1994).

The QSIM Guaranteed Coverage Theorem (Kuipers, 1986; Kuipers, 1994) says that qualitative simulation of a QDE predicts qualitative behaviors describing every real-valued solution to every ODE consistent with the given QDE. Semi-quantitative reasoning methods can be used to exploit *a priori* quantitative knowledge, and to interpret quantitative observations by unifying them with a qualitative behavior. We use three complementary methods.

Q2: The QDE can be annotated with bounding information — real intervals for landmark values and real-valued function envelopes for monotonic function constraints — which can be propagated to derive tighter bounds, or to detect a contradiction and filter out the behavior (Kuipers and Berleant, 1988).

Q3: The coarse grain-size of the qualitative behavior can be adaptively refined by inserting additional qualitative states, providing convergence to a real-valued function as uncertainty goes to zero (Berleant and Kuipers, 1992).

NSIM: Dynamic Envelopes. The bounding information can also be used to derive an *extremal system*: a numerical ODE whose trajectories bound all solutions to the original QDE (Kay and Kuipers, 1993).

Tracker Architecture

Figure 1 describes the architecture of an individual tracker. We review its elements.

QSIM takes a qualitative differential equation QDE) model and an initial qualitative state (QState (t_0)) and predicts a set of qualitative behaviors guaranteed to describe all real behaviors.

Each qualitative behavior defines a set of semi-quantitative constraints, which can be combined with *a priori* semi-quantitative information to derive a semi-quantitative behavior prediction. This initializes the tracker with its hypothesis.

The plant is monitored by sensors, which produce an observation stream. The observation stream is *not* assumed to be homogeneous, synchronous, or supplied in real time. Although Fig. 1 suggests that each tracker may regard the observation stream as autonomous and exogenous, observations may be produced in response to requests from the monitoring system as a whole.

For example, if we are monitoring a patient in intensive care, some observations such as heart rate and blood pressure are on-line and real-time; others such as lab reports arrive intermittently, much delayed, and possibly

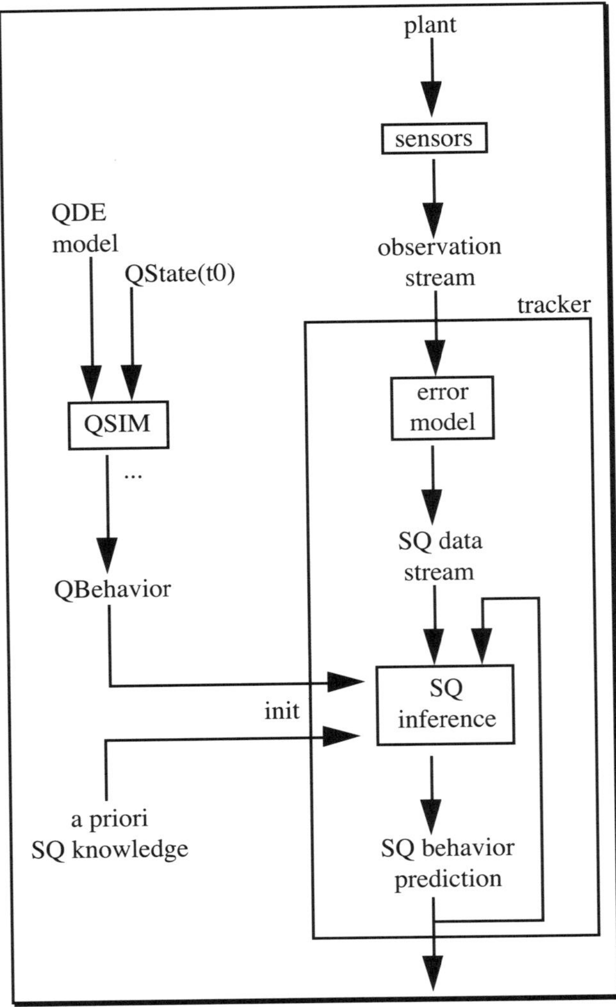

Figure 1. A tracker is initialized with a qualitative behavior and a priori semi-quantitative information, consumes an observation stream, and produces a semi-quantitative behavior prediction.

out of order; other invasive observations are undertaken only after careful cost-benefit analysis.

Many sensors produce real numbers as outputs, yet they are known to be neither perfectly accurate nor perfectly precise. Therefore, the raw observations are transformed by an error model to a stream of semi-quantitative data to reflect the uncertainty they represent.

Observations are assimilated into a behavior prediction by the same inference methods used during initialization. As each observation is assimilated, the prediction becomes more precise, or the hypothesis is refuted. When an individual tracker hypothesis becomes more precise, differential diagnosis between competing trackers becomes more effective. When an individual tracker is refuted, the tracking set as a whole (a disjunction) becomes more precise.

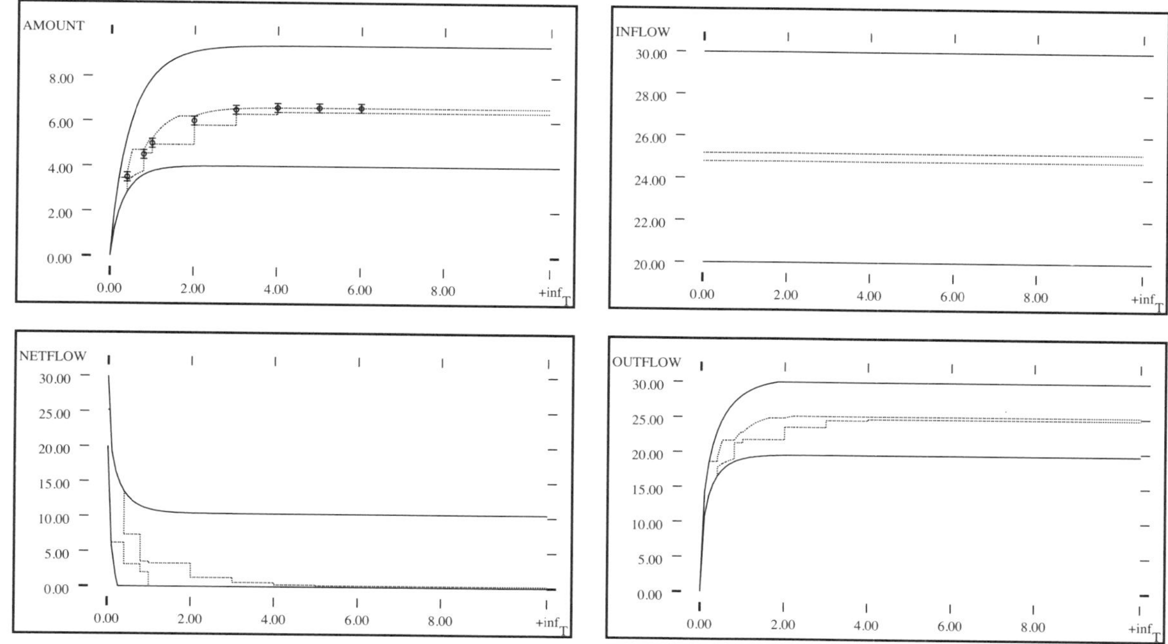

Figure 2. Tracker results after eight observations of Amount = x(t).

The tracker itself — indicated by the inner box in Fig. 1 — is a persistent object which is initialized with a hypothesis (QDE, initial qualitative state, initial semi-quantitative knowledge), consumes a stream of observations, and provides a continually updated semi-quantitative behavior prediction.

Clearly, after only a few observations, the predictions are much more precise, and the rate of inflow has been identified much more accurately. Because of this precision, future observations that deviate from these narrow bounds will be able to refute the current model quickly and suggest a fault model such as an operator action or an external event.

Example: Water Tank

We demonstrate trackers assimilating a data stream from a model of a water tank. The water tank has constant inflow, open drain, and is initially empty. There is substantial uncertainty about the constant rate of inflow.

The qualitative water tank model is x' = q - f(x), where $f \in M_0^+$ is the amount of water in the tank, q is the inflow rate, f(x) is the outflow rate, and x' is the net rate of change of amount.

The semi-quantitative bounds on the model consist of numerical *static envelopes* around the partially known function $f \in M^+$ representing ±1% uncertainty bounds, and a numerical interval [98,100] representing ±1% uncertainty bounds on the value of the landmark *full* representing the value of x at which the tank overflows.

As the observation stream is assimilated, the error model expands each observed data point by ±0.2 to obtain a numerical interval. Error bounds are not currently added to time values.

Amount Increases to Equilibrium

The rate of inflow is initially very uncertain ($q \in [20,30]$) but is low enough that equilibrium without overflow is assured. Figure 2 shows two sets of dynamic envelope predictions for each of the four model variables, given a sequence of eight observations.

The outer envelopes represent the prediction from prior information only, without the observation stream. The inner, more jagged, envelopes represent the prediction after eight data points have been assimilated, one by one.

Research Issues

The tracker architecture suggests three major areas for concern and future research, particularly to ensure the tractability of the method: tractable qualitative simulation, semi-quantitative inference, and tracking set management.

Tractable Qualitative Simulation

The failure mode of qualitative simulation is intractable branching among alternate possible behaviors. One class of branch arises when the simulator is not able to reason deeply enough to detect genuine inconsistencies. Research in qualitative simulation continually finds deeper forms of analysis to remove such branches, such as energy analysis, testing for self-intersection of phase-plane trajectories, or composition of the phase portrait.

A second class of intractable branching arises when there are very many *real* qualitative distinctions among possible behaviors, but those distinctions are not of interest to the modeler. This problem is solved by finding higher levels of abstraction, at which the offending distinction is not explicit, so many behaviors are collapsed into a single description. However, abstraction methods must balance the need for tractability against the need for specificity in describing the phenomenon of interest to the modeler.

In some cases, intractable branching can be prevented by taking advantage of constraints from the observation stream. Figure 2 treats qualitative simulation as taking place off-line, with a tracker created for each predicted behavior. However, it is also possible to create trackers for a small set of initial behavior segments, then extend the behavior segments only for trackers that represent worthwhile hypotheses, and that remain consistent with the observation stream.

Semi-Quantitative Inference

Semi-quantitative inference uses the qualitative behavior description as a framework for assimilating semi-quantitative information. The key guarantee we wish to provide is that numerical uncertainty in the prediction decreases, in the limit to zero, as uncertainty in the given information decreases.

Most of our current work has been based on a representation of uncertainty as numerically bounded intervals and function envelopes. Information from different sources is combined very simply by intersecting intervals. Predictions are envelopes that bound the set of trajectories corresponding to a given qualitative behavior. More realistic error models require representing uncertainty in terms of probability distributions such as Gaussians, combining information using Kalman filters. We are developing methods to use both kinds of information.

Tracking Set Management

The tracking set represents the set of hypotheses currently being tracked by the monitoring system. The initial set of hypotheses includes the nominal model of the system, plus additional models that should be continually considered. Models are removed from the tracking set when they are inconsistent with the observations stream, and new models are added by a model-based diagnosis method, driven by discrepancies between predictions and observations.

Tracking a hypothesis across qualitative time-points and region transitions poses problems for naive approaches, but we plan to use regression to make a best estimate of the time of transition from one qualitative state to the next. Operator actions, accompanied by a detectable signal, raise no additional issues. However, external events, anticipated in the model set but not predictable by the current model, can be a source of intractability since their occurrence cannot be anticipated in general. Model-based diagnosis with expected value cutoffs will be necessary to solve this problem.

Conclusions

Trackers based on semi-quantitative models have certain potential advantages over numerical models (Patton, Frank, and Clark 1989). First, trackers focus on individual qualitative behaviors, whereas a numerical model does not automatically classify its predictions into qualitative categories. If a numerical model attempts to provide error bounds with a guarantee of coverage, the bounds must enclose the entire set of qualitative behaviors predicted for a model, rather than just one. Second, the qualitative behavior makes explicit symbolic terms describing the qualitatively important properties of the behavior, such as landmark values and regions of monotonic change. These additional sources of knowledge can be exploited when matching observations against a hypothesis.

Acknowledgments

This work has taken place in the Qualitative Reasoning Group at the Artificial Intelligence Laboratory, The University of Texas at Austin. Research of the Qualitative Reasoning Group is supported in part by NSF grants IRI-8904454 and IRI-9216584, and by NASA contract NCC 2-760.

References

Berleant, D. and B. Kuipers (1992). Combined qualitative and numerical simulation with Q3. In B. Faltings and P. Struss (Eds.), *Recent Advances in Qualitative Physics*. MIT Press, Cambridge, MA.

Dvorak, D. and B. Kuipers (1991). Process monitoring and diagnosis: a model-based approach. *IEEE Expert*, **6**(3), 67-74.

Kay, H. and B. Kuipers (1993). Numerical behavior envelopes for qualitative models. *Proc. 11th National Conf. on Artificial Intelligence*. AAAI/MIT Press, Cambridge, MA. pp. 606-613.

Kuipers, B. (1986). Qualitative simulation. *Artificial Intelligence*, **29**, 289-338.

Kuipers, B. (1994). *Qualitative Reasoning: Modeling and Simulation with Incomplete Knowledge*. MIT Press, Cambridge, MA.

Kuipers, B. and D. Berleant (1988). Using incomplete quantitative knowledge in qualitative reasoning. *Proc. 7th National Conf. on Artificial Intelligence (AAAI-88)*.

Patton, R., P. Frank and R. Clark (1989). *Fault Diagnosis in Dynamic Systems: Theory and Applications*. Prentice Hall, New York.

Perrow, C. (1984). Normal Accidents: Living With High-Risk Technologies. Basic Books, New York.

A MACHINE LEARNING APPROACH TO DESIGN AND FAULT DIAGNOSIS

Aydin K. Sunol, Burak Ozyurt, Praveen K. Mogili and Larry Hall
University of South Florida
Departments of Chemical Engineering and Computer Science
Tampa, FL 33620

Abstract

Automation of the knowledge acquisition process in building knowledge-based systems for process design and fault diagnosis is addressed through Machine Learning techniques. A hybrid Machine Learning algorithm developed at the University of South Florida is presented as a knowledge acquisition tool for developing knowledge-based systems. The learning algorithm addresses the knowledge acquisition problem by developing and maintaining the knowledge base through instance based inductive learning from the examples.

The learning algorithm named as Symbolic-Connectionist net (SC-net) overcomes the problems associated with neural and symbolic learning systems by integrating the symbolic information into a neural network representation. The learning system allows for knowledge extraction and background knowledge encoding in the form of rules. Fuzzy logic has been made use of in dealing with uncertainty in the learning domain. The description language for the learning system consists of continuous and discrete variables along with relational and fuzzy comparators.

The applicability of the learning system for process design is illustrated through a complex column sequencing example. The performance of the learning system is discussed in terms of the knowledge extracted from example cases and its classification accuracy on the test cases.

A large fault diagnosis example, Syschem plant, is utilized to illustrate the promise of this approach in identification problems. In order to deal with uncertainty and noise, fuzzy variables were used for input domain representation. Since natural language like rules are required, linguistic fuzzy variable attributes had to be chosen. The ranges of the fuzzy variable attributes had to be determined, which was accomplished by using K-means clustering in neural network form. The approach compares favorably with neural nets and other inductive learning algorithms it was compared to.

Keywords

Complex distillation columns, Machine learning, Artificial intelligence, Fault diagnosis, Symbolic-connectionist network.

Introduction

The knowledge extraction and background knowledge encoding problems associated with connectionist learning algorithms and inability of the symbolic learning algorithms to handle continuous variables effectively, and allow parallel knowledge representation have prompted the researchers in AI to seek a hybrid approach. One development in this direction is Symbolic Connectionist network (SC-net) (Romaniuk, 1991). The SC-net, designed specifically for constructing Expert Systems, is based on a hybrid of Symbolic and Connectionist architectures. The system allows extraction of knowledge in the form of rules and can handle both scalar and fuzzy variables. The following features of SC-net make it a connectionist method

1. A highly parallel and uniform representation of knowledge
2. Fault tolerance and noise resistance.

3. A built-in ability to deal with non-crisp inputs and outputs.

and the following features make it a symbolic method

1. The ability to encode rules to support knowledge refinement.
2. Allow for rule extraction as a direct means to elicit learned knowledge and support the implementation of Expert System standards such as consultation and explanation facilities.
3. Provides means to represent symbolic constructs such as variables, comparators and quantifiers. This leads to a more powerful language for description of knowledge.

The network topology in SC-net is based on the training examples unlike the user specified ones in connectionist approaches, promising an optimal representation. Fuzzy logic (Zadeh, 1988) is used to deal with uncertainty.

The SC-net network structure consist of simple cells modeling the fuzzy operators min, max and sum. When SC-net is presented with a training set consisting of input and corresponding output information (i.e. in a supervised manner) then it invokes the Recruitment of Cells Algorithm (RCA), which generates a network to map the training set into a network representation. The network generated consists of Input Cells, Information Collector (IC) Cells, Negative Collector (NC) Cells, Positive Collector (PC) Cells, Unknown (UK) Cells and Output Cells. Input Cells are responsible for input information representation. The positive and negative collector cells collect information for and against the presence of conclusion, respectively. These collector cells are connected to every output and IC cells, which combine the input information into an intermediate form. The UK cell always propagates a fixed activation of 0.5 and, therefore acts on the positive and negative collector cells as a threshold, letting them propagate an activation representing the type of the evidence present (positive, negative or unknown).

Every training instance is presented to the network once, unlike the artificial neural networks, where for convergence many passes through the data is necessary. After the pass of the training instance, the actual and expected activation for every output are compared. Possible conditions resulting from this comparison are

1. The example was correctly identified. No modifications are necessary.
2. The example is similar to at least one previously recognized instance The bias of the cells are adjusted to incorporate the new instance.
3. The example could not be identified by the network resulting in recruitment of new cell(s).

The problem of prespecification of the network topology in other connectionist approaches like neural networks, does not arise for the SC-net, where the network topology is based on the training examples.

RCA's main disadvantage is that the network growth under worst case conditions is linear, resulting in large networks. This problem can be addressed by applying a Global Attribute Covering (GAC) algorithm. Given the RCA generated network as its input GAC generates (except for contradictions and inconsistencies in the examples) an equivalent network minimized by both the number of cells and links and as a side effect allows the extraction of highly general and simple rules for explanation purposes favored by human experts. The GAC algorithm generates sub-minimal cover of the input domain, which is not always the correct one. In the process of minimizing the network, the learning algorithm may identify the wrong input features as important especially when there is an insufficient number of examples. Preselection of Attributes (PSA) algorithm addresses this problem by introducing redundancy into the network.

Knowledge can be extracted from the trained network in the form of rules by making use of the symbolic information collected by the IC cells. Also knowledge encoding in the form of rules are possible into the network which can affect the network growth, training time and the number of rules generated.

Design of Complex Column Sequences

The design of complex distillation column sequences has been of interest to chemical engineers. We used the method proposed by Tedder and Rudd (1978) to test the capability of SC-net. The optimal/best column sequencing is based on separation and cost efficiency as given by Tedder and Rudd (1978).

For training purposes, the SC-net is presented with a set of examples covering as many different permutations of the selected attributes as possible. A subset of the training set used for this example is given in Table 1. Only 15 training cases were used for learning. The selected design, e.g. direct, or indirect,..., are given as the class, D1-D7. The input variables in the training vector are very general and represented using fuzzy numbers. This fuzzy representation helps us to represent a region (pi-shaped) as opposed to a single point.

After the training phase was over, the SC-net was given some new cases. The new cases used are shown in Table 2. The classification accuracy is very encouraging as it is able to identify all the classes correctly. Due to the fuzzy nature of the input attributes, in some cases, more than one design was identified as optimal. This feature is particularly desirable in uncertain domains where there are large number of attributes with interdependency and the attributes are best represented in a qualitative fashion.

Table 1. The Training Set Subset for the Complex Column Sequencing Example.

ESI	*MP*	*OH*	*BP*	*OHBP*	*Class*
ge_1.6	*ge_50*	*any*	*b_5_20*	*any*	*D5*
lt_1.6	*ge_50*	*any*	*lt_5*	*any*	*D6*

Table 2. An Excerpt from Comparison of SC-net Performance with Simulation.

ESI	*MP*	*OH*	*BP*	*OHBP*	*Expected*	*SC-net*
1.3	*65*	*17*	*18*	*1*	*D5*	*D5*
1.3	*80*	*18*	*2*	*16*	*D6*	*D6*

A sample of the knowledge extracted from the SC-net is shown in Table 3. In terms of simplicity and accuracy, these rules are very similar to the ones proposed by Tedder and Rudd. More detail could be found in Mogili, et al. [1994].

Table 3. Sample Rules Generated by SC-net.

If and (ESI [lt_1.6], MP[lt_50], BP[lt_5]) then D1 (1.00)

If and (MP[lt_50], OH [lt_5]) then D2 (1.00)

Fault Diagnosis Case Study-Syschem Plant

In order to test the capability and usefulness of the SC-net a chemical plant simulator, Syschem Plant (Doig, 1983) is used. The Syschem Plant simulates a hydrocarbon chlorination plant including a liquid-phase CSTR, vaporizer, adiabatic plug flow reactor (PFR), condenser, absorption column, drier, blowers, pumps, control valves and piping. There are 418 possible sensor measurements for 11 properties of 38 streams. From 21 faults which can be generated by the System Plant 19 were selected for this study.

Input Variable Preparation

In order to deal with uncertainty and noise, fuzzy variables were used for input domain representation. Since natural language like rules are required, linguistic fuzzy variable attributes had to be chosen. Linear, trapezoidal shaped fuzzy variables are used here. The ranges of the fuzzy variable attributes were determined by using K-means clustering (Moody and Darken, 1989) where the number of clusters should be set a priori. Therefore, the number of clusters are usually overestimated and several runs are made to find the number of clusters, i.e. attributes and their ranges. This procedure was repeated for each input variable with the input data from Syschem Plant along with the fuzzy variable files are created for SC-net training. Different representations of the process information, mainly sensor measurements including measurement noise, were tested. The measurements can be used directly for SC-net training, after fuzzy clustering. Alternatively, the measurements can be used for steady state mass and energy balances around each unit and the training could be done with this new set. The later approach can be useful when coupled with generalized training of the network in mass and energy balances. Here, for process measurement variables, attributes like LOW, NORMAL, HIGH, VERY_HIGH and for the mass and energy balance fuzzy variables attributes like OPEN (low), CLOSED and OPEN (high) are used for generalization.

Training and Testing Considerations for SC-net

Different training and testing data sets were used for different representation of the process information, e.g. shallow knowledge representation with only process measurements, deep knowledge representation with the aid of mass and energy balances. Only a subset of the available measurements, mainly the routine measurements, were used for training and testing data preparation. Since actual plant data would include sensor noise, for each of the 19 faults and the normal case 10 instances were generated and averaged.

SC-net training was done using the prepared input variable vector sets (1 instance/class). The training involved only a feed forward pass through all the training data. In some cases, trained GAC algorithm is activated in order to get more compact, highly general rules and to reduce the network simulation time. The training, including the GAC algorithm, takes less than a few seconds on a moderately loaded Sun SPARC IPC workstation. Test runs with 200 instances requires even shorter computing time, where training time for backpropagation neural networks with linear activation functions for all the process measurements was reported to be about 24 hours (Hoskins, et al., 1991).

The rules generated by the SC-net using RCA/GAC are quite general and simple. They indicate the most important and characteristic attributes causing the fault. For example, the SC-NET generated rule seen in Table 4

Table 4. A Rule Generated by SC-net for the SYSCHEM PLANT

if and (fuzzy (STR_11_TOT_FLOW[normal]) = 1.000, fuzzy (STR_9_TEMPERATURE[low]) = 1.000) then Insulation_damage_in_STR_7 (1.000)

recognizes the fault "Insulation damage in Stream 7" using only shallow sensor measurements of stream 9. Due to heat loss in stream 7 temperature at the stream 9 (combining stream 7 and 8) upstream to the plug-flow reactor, decreases. Mass balance based rule shows an indirect linkage of the fault with an positively open chlorine balance around the PFR, which is due to the decreased reaction rate, resulting from the loss of heat in the reactor inlet stream. Some of the rules generated share the same attributes, like the rules generated for the faults, from process measurements only, "PFR hot spot formation" and "Further reaction in the condenser shell" . They both have the attribute "Stream 11 hydrocarbon dichloride flow high," in common, which can result both from high conversion in the PFR due to hot spot formation or from further reaction in the condenser. However, the rule for the condenser fault additionally indicates high cooling water flowrate, since due to the exothermic reaction more cooling water will be needed for the condenser, which differentiates the faults from each other. The mass and energy balance based rules for these faults support also these evidence.

Since the faults range from equipment leakages to parameter deterioration, mass and energy balances, separately, are not capable of identifying all of the faults. Therefore, combination of mass and energy balances with shallow knowledge information (i.e. process measurements) will complete the picture.

For the extended variable case, performance of the SC-net is nearly perfect, which indicates that SC-net is robust under sensor noise. More detail could be found in Ozyurt, et al. [1996].

Conclusions

The rules generated by the SC-net is compact, readable and logical and they can be further refined, extended or modified by domain experts if additional information is available, or they can be used for educational purposes for operators, since they include extracted and generalized fault cause information. They can be used to construct the knowledge base of a domain specific expert system. SC-net as itself is capable to be used as an diagnostic system on line for real time process monitoring, since the response time for novel instances is very short for the SC-net.

Because of its unique properties like short training time and no network topology prespecification unlike neural networks, SC-net seems to be useful in the field. Beside these properties, symbolic knowledge processing and knowledge extraction in form of rules capabilities makes the introduced instance based learning system superior to many forms of learning approaches in structured selection problems (selection among a set of complex column schemes) and fault diagnosis.

References

Doig, I. D. (1983). Fault detection and correction in a malfunctioning plant. *CACHE Corp*., Austin.

Hoskins, J. C., K. M. Kaliyur and D. M. Himmelblau (1991). Fault diagnosis in complex chemical plants using artificial neural networks. *AIChE J.*, **37**, 137-141.

Mogili, P., S. Romaniuk, A. K. Sunol and L. Hall (1992). Machine learning in design and analysis of chemical processes. *AIChE Annual Conference*. Miami.

Moody, J. and C. J. Darken (1989). Fast learning in networks of locally-tuned processing units. *Neur. Comput.,* **1**, 281-294.

Ozyurt, B., A. K. Sunol, M. C. Camurdan, P. Mogili and L. Hall. (1996). Chemical plant fault diagnosis through a hybrid symbolic-connectionist machine learning approach. *Comput. Chem. Eng.*, D. W. T. Rippin. Memorial Issue, accepted.

Romaniuk, S. (1991). Ph.D. Dissertation, Univ. of South Florida.

Tedder, D. W. and J. D. Rudd (1978) Parametric studies in industrial distillation. *AIChE J.*, **24**, 303-334.

Zadeh, L. A. (1988). Fuzzy logic. *IEEE Computer Mag.,* April, 83-92.

DATA COMPRESSION FOR PROCESS MONITORING

Aleksandar Kudic and Nina Thornhill
Department of Electronic and Electrical Engineering
University College London
Torrington Place, London WC1E 7JE

Abstract

The paper presents an approach to data compression based on the discrete wavelet transform. The method is suitable for batch-wise compression, for example for the long term archiving of an ensemble of spot data collected throughout a shift. Time and frequency domain measures have been used to show that the compressed data are reconstructed well both in terms of trend and in terms of the spectral content. The method adjusts the compression factor in order to achieve a consistent reconstruction performance.

Keywords

Process data compression, Discrete wavelet transform, Spectrum, Performance measure, Process trend.

Introduction

The on-line data collected from the instruments on a large petrochemical process plant can fill the available storage in just a few months. Some form of data compression and archiving is therefore required. Apart from cost savings on the magnetic media and system management the motivations for data compression include rapid access to historical data and for a remote installation, the more rapid transmission of information through a satellite link.

Once the data have been compressed they lose information and the reconstructed trends are deficient in various ways compared with the originals. The purpose of this work is to study a compression algorithm based on the discrete wavelet transform (DWT) which reduces the time domain error of reconstructed data and which also has the ability to reconstruct the spectral content of the original data.

The DWT algorithm has been compared with the box-car/backward slope (BCBS) compression method (Hale and Sellars, 1982). The DWT method permits a higher level of compression to be achieved for the same reconstruction performance.

As well as the wanted signal a process measurement includes noise and other spectral features characterised by high frequencies. In this work we have assumed that the whole signal is to be reconstructed as accurately as possible. There are cases such as the identification of system dynamics where it is useful to retain the high frequency component.

Review of Data Compression Methods

There is a large literature on compression methods for images, speech and text compression (for example, Watson, 1993). Compression techniques for ECG (electrocardiogram) signals are also at an advanced stage (Philips, 1993). The motives in the ECG case are similar to those for process data compression, particularly in regard to the bandwidth limitations of transmission of the ECG signals by telephone. Process data applications have begun to develop, particularly since 1990.

Compression techniques may be divided into three functional groups, whose key features are outlined below.

Direct Methods

A direct method handles spot samples of the signal to provide compression. The BCBS method (Hale and Sellars, 1982) is a direct method in which a spot value is retained only if the subsequent point fails a box-car and backward slope test. The box car test determines whether the deviation of a new value from the previously archived value exceeds a threshold, while the backward slope test determines a deviation from a previously established trend.

Trends are reconstructed by piecewise linear approximation between the saved spot values.

The swinging door method of Bristol (1990) has provided an alternative way of handling spot data. Mah, et al. (1995) have evaluated the BCBS, the swinging door algorithm and their own new on-line compression algorithm called piecewise linear on-line trending. The new method adapts to process variability using a goodness-of-fit with a statistical confidence limit.

Direct methods are suitable for on-line application but tend to focus on the time-domain characteristics of the data. The frequency domain reconstruction of the piecewise linear segments is poor because the ramps representing the straight line segments contribute to low frequency errors.

Transform Methods

In a transform method the original data samples are subject to a linear transformation and the compression is performed in the new domain. The wavelet transform fall into this category. Such methods require a batch of data and are less suited to on-line implementation since the transform is a weighted summation of data collected over a period of time. Bakshi and Stephanopoulos (1995), however, have applied time-varying wavelet packets to provide an on-line implementation suited to feature extraction and noise removal from non-stationary signals.

Parameter Extraction Methods

Parameter extraction methods use pre-processing to extract features which can reconstruct the signal. For example, principal component analysis (PCA) can reduce colinear multivariate data to a few dimensions that contain most of the information. Storage of the PCA scores rather than of the multivariate data set provides data compression. The method is suited to on-line use if a suitable PCA model has been generated from historical data, provided the model continues to represent the process.

Methods

The Wavelet Transform

The wavelet transform can be used to determine the signal power in a time trend as a function of time and frequency (Daubechies, 1992). The advantage of the wavelet transform over other transform methods is that it matches the time localisation to the frequency of interest, using fine time resolution for higher frequencies and coarse time localisation for low frequencies.

In the wavelet analysis the signal $f(x)$ is defined on the range $0 \le x < 1$. Thus, for example, if a process data signal had 1024 samples the change of variable $x \rightarrow t/1024$ would be required. The new variable x becomes 1 when t is 1024 sample periods. A wavelet decomposition has the following form:

$$f(x) = a_0\phi(x) + a_1W(x) + a_2W(2x) + a_3W(2x-1) + \ldots + a_{2^j,k}W(2^jx-k) + \ldots \quad (1)$$

$j = 0$ to $(n - 1)$, $k = 0$ to $(2^j - 1)$

The function $\phi(x)$ is known as the scaling function. It reflects the overall trend of the data. In the discrete wavelet transform used in this paper $\phi(x)$ is unity in the range $0 \le x < 1$ and the coefficient a_0 therefore reflects a steady offset in the data trend.

The wavelets $W(2^jx - k)$ are localised functions. That is to say, they are zero except over a restricted range. The $j = 0$ wavelet $W(x)$ is the wavelet with the coarsest resolution, and the wavelets with $j = n - 1$ have the highest resolution. The value of n is related to the number of data points, m, in $f(x)$ by the relationship $m = 2^n$. The number of points in the data set must be a power of 2.

As well as the property of scaling to different resolutions the wavelet functions can also translate along the x-axis. While $W(2^{n-1}x)$ has its origin at $x = 0$, the translated wavelet $W(2^{n-1}x - k)$ has its origin at $(2^{n-1}x - k) = 0$, or when $x = k/2^{n-1}$. It is possible to fit 2^{n-1} of the highest resolution wavelets into the interval $0 \le x < 1$.

Daubechies' scaling and wavelet functions were used in this work. These wavelets form an orthogonal, normalised set of basis functions and are defined from the dilation and translation of a series of square pulses. The scaling functions have the property of self-similarity at different scales. They are defined recursively as a superposition of the scaling function at the next resolution using sets of Daubechies coefficients (Daubechies, 1988).

The discrete wavelet transform is an algorithm for computing the a-coefficients. In practice it is not necessary to derive the scaling and wavelet functions. The Mallat pyramid algorithm (Mallat, 1989) provides the required a-coefficients without evaluating $\phi(x)$ and $W(x)$. All that is required is the set of Daubechies coefficients from which the mother wavelet is to be constructed.

The Process Application

Figure 1 shows portions of two process data records (courtesy of BP Oil, Grangemouth Refinery Ltd). The data represent deviations of flow in a process stream from the mean value. It is convenient for the compression performance measures to exclude a steady non-zero offset from the trend. Since the a_0 coefficient represents the steady mean value the offset is easily re-instated when the data are reconstructed.

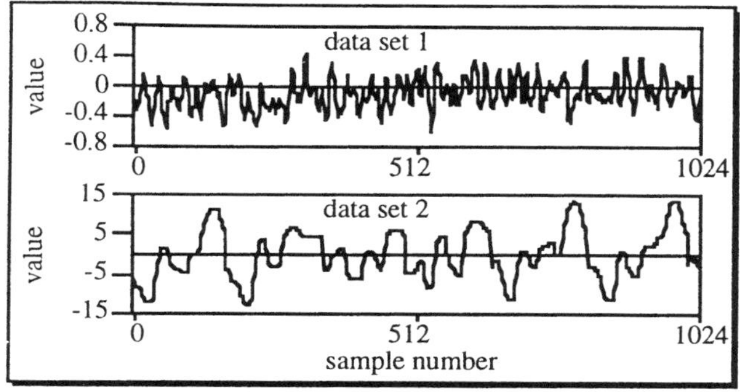

Figure 1. Process data records to be compressed.

Data set 1 shows a persistent oscillation characterised by about seventeen or eighteen samples per cycle. Figure 3a shows the spectrum of this signal up to the Nyquist sampling frequency. The spectral peak is to be retained during compression and reconstruction as well as the characteristics of the time-domain trend.

Data set 2 has a different behaviour characteristic, with a tendency to stay at a value for a time and then to move rapidly to a new level. The purpose of including data set 2 is to illustrate that the wavelet compression method can be used adaptively to achieve the highest possible compression for a given reconstruction error.

Measures to Assess Performance

The similarity between the original signal and the reconstructed one is measured in this paper using two indexes. The index for a time domain comparison was the normalised RMS difference (NRD).

$$e_{NRD} = 100\frac{1}{(f_{\max} - f_{\min})}\sqrt{\frac{1}{m}\sum_{i=1}^{m}(f(i) - f'(i))^2} \qquad (2)$$

$f(i)$: original data samples
$f'(i)$: reconstructed data samples
$f_{\max}$: maximum amplitude of original signal
$f_{\min}$: minimum amplitude of original signal

The index used for comparing the algorithms in the frequency domain was the RMS error between the frequency spectra:

$$E_{RMS} = \sqrt{\frac{1}{m}\sum_{i=1}^{m}(F(i) - F'(i))^2} \qquad (3)$$

$F(i)$: spectrum of original data
$F'(i)$: spectrum of reconstructed data

A key performance measure is the compression factor (CF), the ratio between points stored after compression and the points of original data.

Results

Visual Comparisons

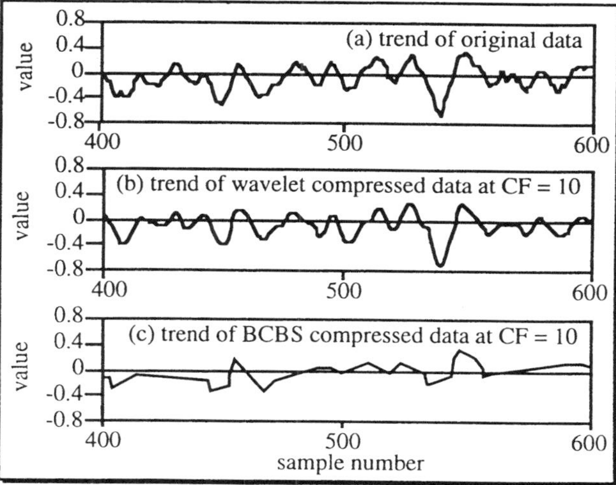

Figure 2. Original and reconstructed trends.

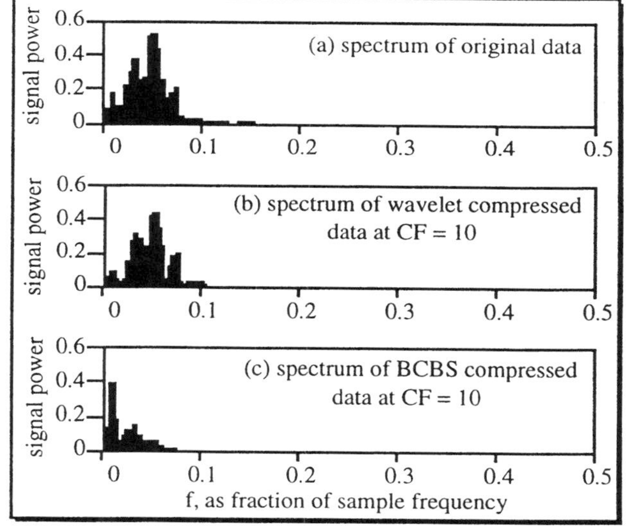

Figure 3. Original and reconstructed spectra.

Figure 2 shows a sub-section of data set 1 and the trends reconstructed after wavelet compression at CF = 10 and after compression using BCBS procedure. The mother wavelet is Daubechies Wavelet 10. The BCBS method retains, on average, every tenth spot data point. The persistent oscillations tend to be characterised by about seventeen samples per cycle. After compression with BCBS they are defined by less than two samples per cycle. The piecewise linear reconstruction therefore produces spurious triangular peaks.

The visual record shows the wavelet reconstruction has better time domain accuracy. Its frequency domain accuracy is also enhanced. The frequency component in the original signal reflecting the persistent oscillation also appears in the spectrum of the wavelet reconstruction (Fig. 3). That spectral feature is missing from the BCBS reconstruction.

Comparison of the Performance Measures

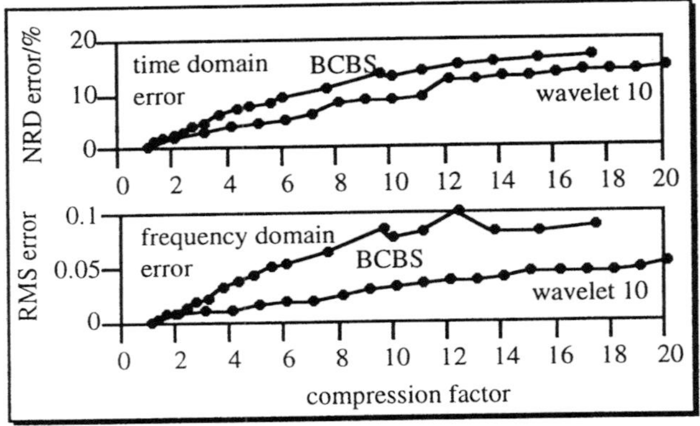

Figure 4. Time and frequency domain errors for data set 1.

Figure 4(a) shows a comparison of the time domain error e_{NRD} for different compression factors for the wavelet and BCBS compression methods. The figure shows that the BCBS compression method gives higher time domain errors. It also gives higher frequency domain errors, as indicated in Fig. 4(b). Similar behaviour (not shown) was observed for the comparison between BCBS and the wavelet compression for data set 2.

Adaptive Compression

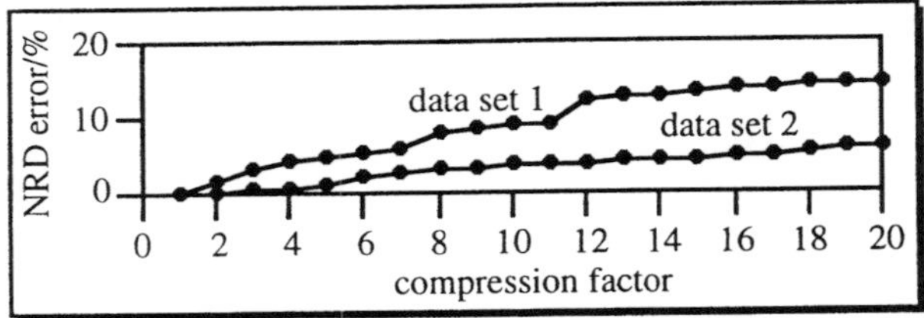

Figure 5. Time domain errors for two data sets.

Figure 5 shows a comparison of the time domain NRD measure for data set 1 and data set 2. The objective of the adaptive method is to choose a compression factor which gives a consistent time domain NRD error for each data set. For instance if an error of 5% was selected, data set 1 would achieve a CF of 6 while data set 2 would achieve a CF of 14. In that way, the compression factor will take advantage of any characteristic of the signal that aids its compressibility.

Conclusions

The results lead to the conclusion that the wavelet transform provides more compression for the same time domain error than the direct BCBS method. The adaptive approach, moreover, adjusts the compression factor for each data record to give the same dimensionless NRD value. Such an approach is not possible with the BCBS method because there is no opportunity to post-process a data ensemble in an on-line direct method.

This paper has also shown that wavelet compression retains the spectral features of the data, while spectral information about a persistent oscillation was destroyed by the BCBS method. As control engineers make increasing use of archived data for retrospective analysis of dynamic performance it is important to retain all significant spectral features, including those at high frequency.

Instances also arise where high spectral fidelity is *not* required. The reconciliation of a mass balance or the comparison of averages at different throughputs are served well by direct methods such as BCBS.

Compression using wavelet transform is acausal. It required a batch of data before the compression algorithm can be activated, or if implemented on-line requires a moving data window. Implementation and maintenance of the wavelet transform method is likely to need an expert who has the time to focus on the algorithm. Compared with transform methods on-line direct methods are easy to implement and the conceptual basis easy to understand. Such features have merit, but as the paper shows they are at the expense of fidelity in the reconstructed trends.

References

Bristol, E. H. (1990). Swinging door trending: adaptive trend recording? *ISA National Conf. Proc.*, pp. 749-753.

Bakshi, B. and G. Stephanopoulos (1995). Compression of chemical process data through functional approximation and feature extraction. *AIChE J.*, in press.

Daubechies, I. (1988). Orthonormal bases of compactly supported wavelets. *Comm. Pure and Appl. Math*, **41**, 909-996.

Daubechies, I. (1992). Ten lectures of wavelets. *CBMS regional Conference Series in Applied Mathematics*, Philadelphia, PA.

Hale, J. C. and H. L. Sellars (1981). Historic data recording for process computers. *Chem. Eng. Prog.*, Nov., 38-43.

Mah, R. S. H., A. C. Tamhane, S. H. Tung and A. N. Patel (1995). Process trending with piecewise linear smoothing. *Comput. Chem. Eng.*, **19**, 129-137.

Mallat, S. G. (1989). A theory for multiresolution signal decomposition: the wavelet representation. *IEEE Trans. Pattern Anal. Mach. Intell.*, **11**, 764-693.

Philips, W. (1993). ECG data compression with time warped polynomials. *IEE Trans. Bio. Eng.*, **40**, 1095-1101.

Watson, A. B. (Ed.) (1993). *Digital Images and Human Vision*. MIT Press.

UPDATE ON THE EUROPEAN STATE OF THE ART OF INTELLIGENT FIELD DEVICES

D. Galara
EDF Direction des Etudes et Recherches
6 quai watier 78400
Chatou, France

F. Russo
ENEL 1 Via Volta 20093
Cologno Monzese MI, Italy

G. Morel and B. Iung
CRAN-EACN Université de Nancy I BP 239 54506
Vandoeuvre, France

Abstract

To face up to the complexity and reliability constraints of the distributed process control as well as minimising the costs, the proprietary solutions or the new configurations, the users need interoperable and interchangeable intelligent field devices. The challenge is to develop these types of standardised equipment satisfying both users and vendors. This paper aims at describing a first systemic approach defined in three consecutive ESPRIT projects in order to boost a standardisation of the Intelligent Actuation and Measurement devices, based on the mutual understanding between users and vendors through a common reference model.

Keywords

Intelligent actuators and sensors, Intelligent actuation and measurement, Functional modeling, Technological modeling, Genetic modeling, Systemic approach, Common reference model, Functional companion standard, Communication companion standard.

Introduction

In the distributed process control, the emergence of new intelligent actuators and sensors linked by fieldbusses, constitutes an important step to make more efficient the operation of the plant in relation to the integrated Control, Maintenance and technical Management points of view (Capetta, et al., 1993).

The required up-grading to go from the conventional field devices to the innovative ones, cannot result from several independent proprietary solutions but from a cooperation between users and vendors. This cooperation has to be based on the need definition and to lead to products guaranteeing their interoperability and interchangability inside an application.

An European initiative started by the ESPRIT II — D.I.A.S. (Distributed Intelligent Actuators and Sensors) project n° 2172 and continued by the ESPRIT III — P.R.I.A.M. (Prenormative Requirements for Intelligent Actuation and Measurement) project n° 6188 aims at proposing a new rupture way of re-engineering facilitating the mutual understanding between users and vendors through a common reference model (Morel, et al., 1993).

The DIAS project (Galara, et al., 1993) has developed the Intelligent Actuator and Transmitter concept which advocated the intelligence distribution down to field devices within a C.M.M. System (integrated Control, Maintenance and technical Management System). In industrial sites, it has achieved its demonstrative objectives of specifying some intelligent functions, embedded into remote actuators and transmitters prototypes which were linked by an emerging standardised fieldbus.

The DIAS basic rupture way is to promote the intelligence distribution in opposition with the current hierarchical process control in order to implement an Information driven instead of a Data driven.

To ensure the interoperability and interchangability of the new intelligent field devices, the PRIAM project aims at developing the IAM (Intelligent Actuation and Measurement) concept as an integrated sub-system of the entire process system. The homogeneous PRIAM way of working (Morel, et al., 1994) is based on the prototyping of the CMM reference approach by both users and vendors through a common reference model for Intelligent Actuation and Measurement. The main methodological objective is to qualify and to quantify the IAM interchangability and interoperability gap and to boost a standardisation : the Functional Companion Standard representing the IAM system interfaces.

The second rupture way emerges from PRIAM in terms of promoting an IAM application driven independently from the implementation as opposed to an IAM technology driven dependently from the implementation.

In additional pre normative view, the on-going European Intelligent Actuation and Measurement User Group (ESPRIT project n° 8244) extends this way of thinking and working to an independent and open forum in charge of providing a more complete and reviewed normative proposal for IAM devices. The EIAMUG organisation (Russo, 1995) where contribute end users, equipment suppliers, engineering companies, scientific and standardisation committees, is based on several working groups. The first, called Group1&2, is aiming at refining the PRIAM concepts and IAM model. The second, Group3&4, is dedicated to the expression of the user needs. The last working Group (5&6) has to validate the Functional Companion Standard view.

To build the IAM Common Reference Model of understanding within an organisational CMM framework (Galara, et al., 1994), the way of working developed through these projects and more precisely in PRIAM has to refer not to a conventional cartesian approach oriented technology but to a systemic approach oriented finality (Lemoigne, 1993).

IAM System

The objective to build a multi-paternity IAM Reference Framework satisfying end-users, vendors and academic, made clear that this reference would not be restricted to the devices technological description.

Therefore, considering the IAM system (Actuation and Measurement parts composed of Actuators and Sensors devices) as an object, the IAM system description has to be analysed from three complementary points of view given rise to three models :

- functional modeling : What is the IAM system making ?
- technological modeling : What is the IAM system made of ?
- genetic modeling : How has the IAM system been made and what it will become all its life cycle long ?

According to a Systemic point of view, the construction of the multi-paternity Common Reference Model of understanding for IAM system has to result from the consistent integration of these three local models during the whole of the iterative way.

A first analysis of the PRIAM return of experience underlined the difficulty of implementing this type of approach and models to obtain a common basis for the whole of the IAM actors. Indeed these actors do not speak the same language (syntax, semantic, knowledge more or less advanced, ...) and do not have the same interest : the user has to use the IAM system, on the contrary the vendors have to supply the IAM system.

Therefore the building of the local models cannot be individual and linear but has to result from a strong collaboration and cooperation between the IAM actors through a concurrent engineering. That allows to validate one model in relation to the two others and to ensure the model perennity (procedure made thanks to the EIAMUG project).

IAM System Functional Modeling

Independently of the IAM system technology, the functional model is a definition of the IAM expected services. These services represent the needs, the operation of the system, and have to be expressed as occurrences of the seven generic types of services: Documentation, Configuration, Parameterisation, Test, State, Status and Mode of operation. This external description linked to the finality of the system in relation to its environment, is materialised on the one hand by agents (e.g. control operator) and on the other hand by the interactions between agents and the IAM system in terms of Requests and Reports. The agents accesses to the IAM system through the IAM channel. So, the requests are sent from an agent to the IAM channel (e.g. To open the valve, To recover the alarms) and the reports are sent from the IAM channel to an agent (e.g. The valve is open, The actuator is available, Torque Curve). An agent can be an operator or a device and interact directly with the system really implemented : the "plant user sub functional model," or with its tools to create the system : the "genetic sub functional model."

The "plant user sub functional model" restricted in the PRIAM project to the plant agent's needs for control and maintenance, has normally to integrate both the end-users (operators and automation devices) and the vendors within a CMM architecture.

The genetic sub functional model takes into account the vendor's needs linked to the life cycle of each device integrated into IAM system (specification, development, commissioning).

IAM System Technological Modeling

As opposed to the functional view, the technological and genetic modeling is technology dependent because it is internal description of the IAM system devices.

Therefore, the technological model can be identified by:

- a process part : the electromechanical part,
- an information processing part :
  - the software application functions distributed into the IAM system devices and satisfying the user's needs,
  - the basic hardware and software supporting IAM system application functions,
  - the communication system inter-connecting the IAM system devices.

The technological models are concretely realised in PRIAM through a laboratory mock-up (intelligent devices integration) and in DIAS through the implementation on industrial sites of intelligent devices prototypes. Inside an IAM application, the current description of the technological model is based on the cooperation between vendors dedicated to the definition of a communication system to interconnect their IAM devices by a selected fieldbus, and on a standalone action of vendor to develop his own equipment. This vendor development of each device (field equipment, maintenance station,...) is made from the functions identified in the genetic modeling.

To operate, in accordance with the IAM application finality, the devices through their interconnection have to exchange data produced or consumed by the distributed software functions of each equipment. Therefore to achieve the interoperability and interchangability IAM objective, the data have to be firstly identified independently of the communication support : the Functional Companion Standard, and then adapted to the fieldbus : the Communication Companion Standard.

The EIAMUG project proposes a first validation of the Functional Companion Standard extending the interface mechanism to a generalisation/particularisation principle to create new IAM functional taxonomies linked to the types of process, of users, of vendors, …

From the device operating in PRIAM, one major difficulty of the technological model is that a functional description of the functions implemented into the IAM device is missing. This means that the end-users need an equipment definition independent of the technology because it is difficult to understand how the equipment is operating from the technological description made by the suppliers. So the technological model has to evolve to two sub-models :

- the "technological sub model" which is a description of the technology, made by suppliers, embedded into IAM devices
- the "technological dual sub model" which is the description of the behaviour and of the performances of the IAM devices using a neutral language. This neutral language allows both end-users and engineering agents to understand the devices operation and to compare devices by selecting those suiting best to application needs.

IAM System Genetic Modeling

To face up to the constraints of a large variety of industrial environments, the genetic model generally restricted to the activities performed to create the IAM system processing, has to be composed of a "genetic creation sub-model" and a "genetic implementation sub-model." The genetic creation sub-model provides a development methodology to deliver IAM devices then available on the schleps. The genetic implementation sub-model supplies an implementation methodology to perform, to configure, to parameterise, to integrate and to commission the devices satisfying plant application and user's needs.

Once the process part of the field devices has been selected, the genetic implementation sub-model is based on four steps:

- the description of the Functional Requirement Diagram using a language (as IEC 1131-3) independent of the language of implementation of a supplier. This diagram defines what the "users" want (Galara, et al., 1994),
- the designing of the architecture of the IAM system distributing the Functional Requirement Diagram into different subsets taking profit of the devices available on the schleps.
- the selecting of the IAM devices among the available devices (basic hardware and software support) in order to translate the Functional Requirement Diagram subsets into the programming language of each device.
- the integration of each device into IAM system and the validation of the IAM system operation.

The partition between the first two phases dedicated to the study of the processing part and the two others dedicated to the implementation of the processing part, clearly identified the limits of scope user-vendor ensuring the reusability of interchangeable and interoperable equipment (standardised and industrial).

Conclusions

From the PRIAM return of experience, the IAM common reference model (Fig. 1) should be built on the integration of :

- the Functional model,
- the Technological model,
- the Genetic model,

knowing that each model has to be consistent with the two others during the entire system approach.

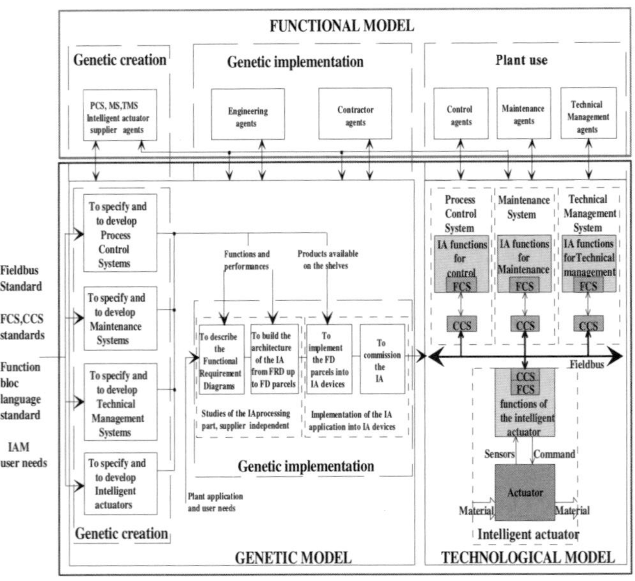

Figure 1. Shape of a IA common reference.

From a dissemination point of view of this new way of thinking, all these Systemic paradigms have to be considered as basic guidelines for an IAM Education and Training (Morel, et al., 1995): a new Education and Training Challenge. So, it is necessary to promote, in the EIAMUG project, an innovative and interactive IAM Education Engineering based on a common IAM Systemic Academia through telematics technologies to bring together a wide range of academic and industrial skills.

Acknowledgments

This paper has been prepared by the authors involved in these three ESPRIT projects on behalf of the PRIAM consortium

Nomenclature

CCS	Communication Companion Standard
CMMS	Control Maintenance and technical Management System
DIAS	Distributed Intelligent Actuators and Sensors
EIAMUG	European Intelligent Actuation and Measurement User Group
ESPRIT	European Strategic Program for Research and development in Information Technology
FCS	Functional Companion Standard
FRD	Functional Requirement Diagram
IAM	Intelligent Actuation and Measurement
IAMS	Intelligent Actuation and Measurement System
IAT	Intelligent Actuators and Transmitters
IFD	Intelligent Field Devices
PRIAM	Pre normative Requirements for Intelligent Actuation and Measurement

References

Capetta, L., D. Galara, J. Graefer, G. P. Lovischek, L. Mondeil and M. Sanguinetti (1993). From current actuators and transmitters towards IAM — PRIAM approach. *Proceedings of Automation 1993-BIAS.* Milan, November 23-25. pp. 1171-1186.

Galara, D. , J. M. Favennec, B. Iung and G. Morel (1993). Distributed intelligent actuators and sensors. *Proceedings of the 7th COMPEURO'93-IEEE conference.* Paris, May 24-27. pp. 79-85.

Galara, D., B. Iung, G. Morel and F. Russo (1994). Intelligent actuation and measurement system-based modeling : the PRIAM way of working. *Proceedings of 2nd IFAC Workshop.* Lund, Sweden, August 10-12.

Lemoigne, J. L. (1993). La Modélisation des systèmes complexes. *Afcet Systemes, Dunod Editor*, (in French).

Morel, G., P. Lhoste, B. Iung, J. F. Pétin, F. Corbier and O. Douchin (1993). Discrete event automation engineering: outline for the PRIAM project. *Proceedings of Automation 1993 — BIAS.* Milan, November 23-25. pp. 1105-1117.

Morel, G., B. Iung, D. Galara and F. Russo (1994). Prototyping a sub-concept of computer integrated manufacturing engineering : the integrated control, maintenance and technical management system. *Proceedings of ISRAM'94.* Wailea-Maui, USA, August 18.

Morel, G., F. Mayer, J. F. Pétin, and P. Lhoste (1995). System-oriented education engineering. *Proceedings of the IEE/IFAC/INTERNATIONAL Conference on CAD/CAD, Robotics and Factories of the Future.* Pereira, Colombie, August 28-30.

Russo, F. (1995). IAM as users need it. *EIAMUG Issue 1.* March.

A ROBUST RULE-BASED SYSTEM FOR ON-LINE DIAGNOSIS OF NITRIFICATION PROBLEMS IN ACTIVATED SLUDGE TREATMENT

Michael W. Barnett and Bruce Gall
Hydromantis, Inc.
Hamilton, Ontario, Canada, L8S 1G5

Abstract

Conventional rule-based systems developed for diagnosis of process problems in activated sludge treatment have not fulfilled expectations. Rules are typically unstable over time and across applications. These problems can be solved by taking a model-based approach to development of rule-based systems. Models provide a structure for knowledge. They can be used to define a complete, consistent set of rules and to establish causality rigorously. This paper describes a model/rule-based system for diagnosing nitrification problems in the activated sludge process. The diagnosis makes use of the signed digraph technique to identify causal paths. In this system a valid diagnosis is guaranteed as long as the model is a good representation of the real system and its calibration is maintained.

Keywords

Decision support, Modeling, Simulation, Wastewater treatment, Activated sludge, Nitrification, Rule-based system.

Introduction

This paper describes work on development of a decision support system (DSS) for diagnosis of nitrification problems in activated sludge wastewater treatment. The DSS is one component of an integrated system, referred to as the integrated computer control system or IC^2S, as shown in Fig. 1. The IC^2S is being developed as part of a US$1M project funded by Hydromantis and the Ministry of Environment and Energy of Ontario, Canada. The objective of this project is to develop key modules and to demonstrate the system at full-scale. Most modules of the IC^2S system have been completed or are nearing completion with full-scale testing scheduled for late 1995.

Background

The key bottleneck in development of a DSS for wastewater treatment process operation is knowledge acquisition. Developers have found it difficult to structure and write the rules needed for a truly robust DSS. Rules developed based on heuristics or experiential knowledge are difficult to verify, often incomplete, sometimes spurious, and can even conflict. There is no general approach for design of a robust rule base for wastewater treatment process diagnosis other than the ill-structured interviewing and knowledge organization strategies used in knowledge engineering.

Knowledge needed for detection, diagnosis and control of problems in the activated sludge process can be categorized as either mechanistic or heuristic. Mechanistic knowledge has been formalized to a high degree. In wastewater engineering, mathematical models describe many processes with a high degree of fidelity when compared to real plant data. With specialized optimization tools, it is possible to maintain this calibration and thus use the model as a repository for knowledge about the current state of the process.

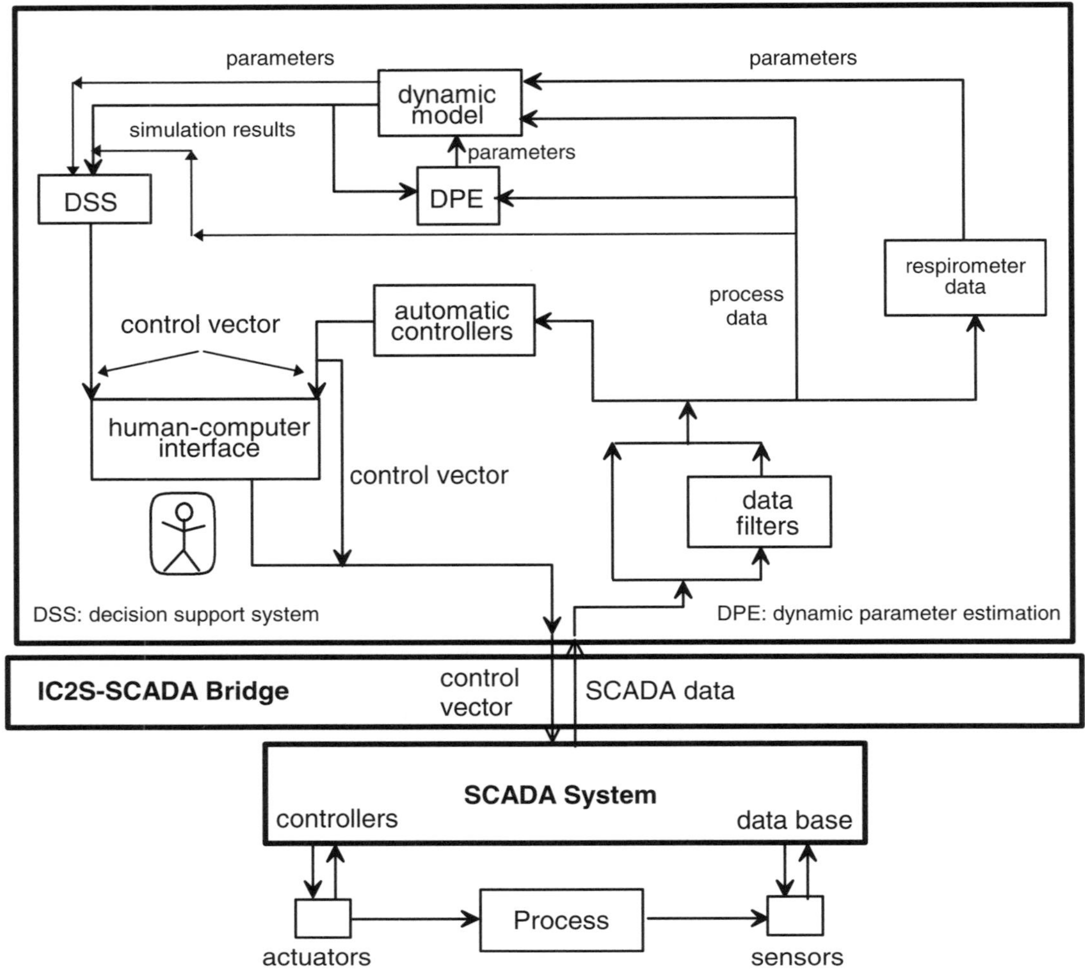

Figure 1. Information flow in the integrated computer control system (IC^2S).

Objectives

This investigation concentrates on the manner in which mechanistic knowledge can be used to prepare systems for detection and diagnosis of process problems in activated sludge treatment. This project applies the dynamic modeling and simulation tools available with the General Purpose Simulator, or GPS-X, (Hydromantis, Inc.) and the intelligent decision support tools available in G2 (Gensym Corporation) together with a special bridge developed to link these two applications.

A dynamic model prepared with the GPS-X based on the International Association on Water Quality (IAWQ) Activated Sludge Model 1 provided the basis for this analysis. The expected outcomes of this investigation are:

1. an architecture for integrating a model-based DSS with supervisory control and data acquisition (SCADA) systems in a wastewater treatment plant,
2. the logic for data flow in the integrated system,
3. a technique for employing dynamic models in decision support,
4. a formal basis for rule development, and,
5. a knowledge-based system design that allows for incorporation of heuristic knowledge as well as mechanistic knowledge.

The key result to be demonstrated is that as long as the model is a valid representation of the real plant and the model remains well-calibrated, a correct detection and diagnosis is guaranteed. Once a proper diagnosis is made, the selection of appropriate control actions is simplified.

This paper describes the DSS using a specific example — the detection and diagnosis of nitrification problems in activated sludge treatment. Extensions to other types of process problems are straightforward.

Description

As shown in Fig. 1, the DSS takes as input process data as well as model simulation results. The DSS is model-based and relies extensively on results generated by the simulation model. It is assumed that the simulation model is valid and is properly calibrated at all times.

The DSS is invoked automatically and contains a simple program which compares important simulated and real water quality parameters. The information flow in this program is shown in Fig. 2. The DSS described in this section concerns only that portion of Fig. 2 which leads to either "normal operation" or "identifiable upset." The remaining paths (lower left quadrant of Fig. 2) signal the need for model tuning and re-calibration.

If the error (ε) between actual data and simulation results is small, then the model is tracking the plant. Under these circumstances, the model can be used with confidence to infer the real process condition. If the data fall outside the normal range then the plant may be in an upset state. If it is an upset state, the model can be used to identify the source of the upset. Moreover, since all possible upset behaviors for the model can be determined, a correct detection and diagnosis is guaranteed.

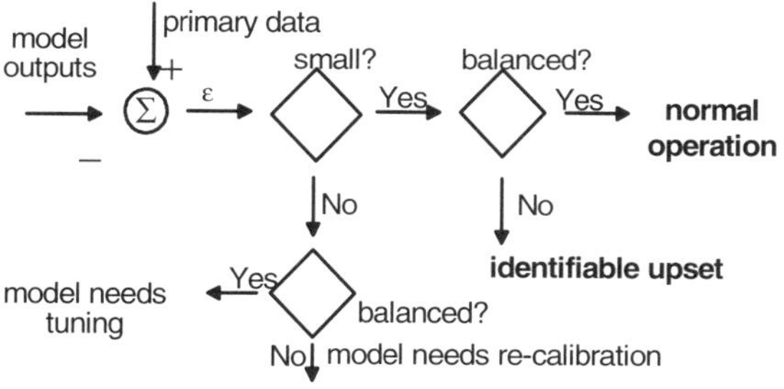

Figure 2. Interpreting the error between simulation values and real data.

DSS Modules

The DSS has the following three functions:

- Detection - determine the diagnostic category.
- Diagnosis - determine the cause of a diagnosis.
- Control Decision - determine (when necessary) appropriate control actions.

A *diagnostic category* is a qualitative description of the current plant condition. These categories can be defined quantitatively in terms of the model's state variables. An example of a diagnostic category is "nitrifying," which indicates that some degree of ammonia removal is occurring. Non-normal diagnostic categories are referred to as *process faults*.

Detection Module

The purpose of the detection module is to determine the diagnostic category. This determination is based on an analysis of simulation results and model parameter data. Rule bases have been developed to examine options for detection of diagnostic categories based on simulation results. The most robust of these makes use of a mechanistic understanding of nitrification processes in activated sludge treatment.

Mechanistic Rule Development

In this approach, the detection module focuses on the definition of nitrification process *potential*. An analysis of the diagnostic categories shows that in activated sludge treatment, many of the categories are expressed in terms of the removal of a certain wastewater component. For example, the removal of organic components measured in terms of biochemical or chemical oxygen demand (BOD or COD), the removal of particulate material or the removal of nitrogen.

Nitrification is defined as "the conversion (removal) of ammonia nitrogen in the influent wastewater to nitrites and nitrates by autotrophic nitrifying bacteria." The nitrification concept must be represented in such a way as to capture these ideas and permit the definition of degrees of nitrification. Nitrification cannot be properly defined in terms of the plant effluent concentration, since low concentrations can be associated with low influent concentrations. Similarly, high effluent concentrations may also be present when nitrifying bacteria are growing at the maximum rate and influent ammonia concentrations are high. In order to properly define nitrification it is necessary to identify the nitrification process *potential*.

The process rate is defined in the IAWQ model as follows:

$$rN = \mu A\left(\frac{snh}{knh + snh}\right)\left(\frac{snh}{kna + snh}\right) \times\left(\frac{so}{koa + so}\right)xba \tag{1}$$

Note that this equation could be expanded to include the effects of other influences such as toxicity. The nitrification potential (NP) is then defined as:

$$NP = \frac{rN}{\mu A(xba)} \tag{2}$$

The denominator on the right-hand side is the maximum rate at which ammonia can be removed, therefore, the value of NP ranges from 0 to 1 depending on the degree to which the rate is limited by the ammonia concentration (S_{nh}) and dissolved oxygen concentration (S_o). Notice that since the model is the source of information, values for all variables in these equations are available. Use of the NP permits development of a simple and complete set of rules for detecting all nitrification diagnostic categories.

Diagnosis Module

The diagnostic category determined by the detection module is passed to the diagnosis module. The goal of the diagnosis module is to validate the detection and to determine the cause of the fault condition. A fault condition may have one of two causes:

1. a sensor fault, or
2. a process fault.

The DSS includes logic to test the primary sensors for drift, fouling and other faults. Process faults are diagnosed by checking appropriate causal variables. Inspection of the model defining equations enables the determination of all estimated parameters and input variables, i.e., the causal variables, that can affect a given measured variable. Causal chains can be constructed through a structured search of causal variables using signed digraphs. An example is shown in Fig. 3. In this figure, an increase (+) in ammonia concentration (S_{nh}) can be traced back to a decrease (-) in autotrophic bacteria (X_{ba}) which is caused by an increase in influent flow (*Flow*).

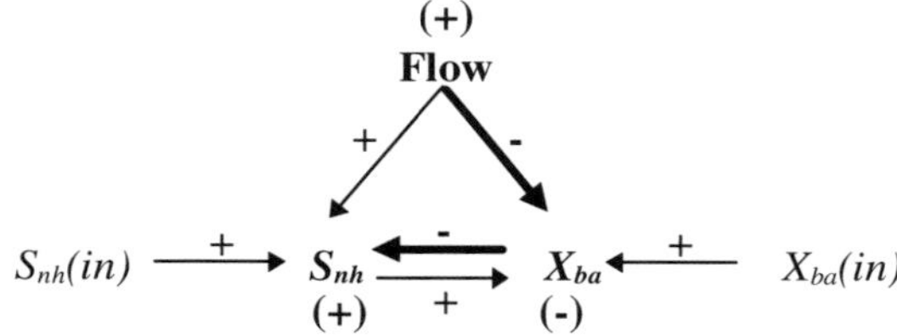

Figure 3. Simplified signed digraph for ammonia removal.

Conclusions

A decision support system has been prepared that uses a dynamic, mechanistic model of a wastewater treatment process to derive robust rules for diagnosis of nitrification problems in activated sludge treatment. The system assumes the model is valid and it is possible to maintain calibration of the model in real time. Given this assumption one can use the structure of the mathematical model to derive a complete and consistent set of rules for detection and diagnosis of nitrification problems.

Acknowledgments

Partial funding for this work was provided by the Environmental Technologies Program of the Ministry of Environment and Energy, Ontario, Canada.

Nomenclature

r_N = nitrification rate, mg/L-hr
μ_A = autotrophic maximum specific growth rate, 1/d
s_{nh} = ammonia concentration, mg/L
$s_{nh}(in)$ = influent ammonia concentration, mg/L
k_{nh} = ammonia switching parameter, mg/L
k_{na} = half saturation parameter, mg/L
k_{oa} = oxygen switching parameter, mg/L
x_{ba} = autotrophic bacteria concentration, mg/L
$x_{ba}(in)$ = influent autotrophic bacteria concentration, mg/L
NP = nitrification potential, dimensionless

References

Barnett. M. W. (1991). Qualitative modeling of a biological reactor system: steady-state simulation based on confluences. *Env. Sanit. Eng. Res. Japan*, **5**, 43-58.

Henze, M., C. P. L. Grady, W. Gujer, G. vR. Marais and T. Matsuo (1987). A general model for single sludge wastewater treatment systems. *Water Research*, **21**, 505-515.

A REAL-TIME KNOWLEDGE-BASED SIMULATION FOR PROCESS CONTROL AND MONITORING OF A NUCLEAR POWER PLANT

Kyoungho Cha
Human Factors Research, Advanced Research Group
Korea Atomic Energy Research Institute
P.O. Box 105, Yusong, Taejon, 305-600, Korea

Abstract

A real-time knowledge-based simulation, for which studies man-machine interaction and develop dynamic process control strategy in the main control room of a pressurized water reactor nuclear plant, is described. The real-time knowledge-based simulation is approached by integrating real-time with knowledge-based agents. agents. Real-time is designed to simulate a generic pressurized nuclear power plant while knowledge-based agents are designed for process control, monitoring and user interface. Hayes-Roth's real-time performance architecture in intelligent agents is considered in the design of the real-time knowledge-based simulation and a direct manipulation process interface is prototyped as the user interface of the real-time knowledge-based simulation. A scenario-based experiment is tested for validating the real-time knowledge-based simulation prototype with a simulated environment.

Keywords

Real-time expert system, Process control and monitoring, Deep knowledge.

Introduction

Real-time expert system technology has been applied to control and monitoring of complex processes in engineering plant (Ingrand, et al., 1992; Moore, et al. -) and direct Manipulation Process Interface (DMPI) design (Beltracchi, 1990) has also been approached to the hard real-time domain including a nuclear power plant. The design of the real-time knowledge-based simulation is approached by integrating real-time with knowledge-based agents. Hayes-Roth's real-time performance architecture in intelligent agents (Hayes-Roth, et al., 1991; Hayes-Roth, 1993) is also used as the real-time knowledge-based simulation architecture. Multiple tasks including process control, monitoring, procedural operator support (e.g., operating procedure), real-time diagnosis for control automatics, are designed as knowledge-based agents. Real-time consists of mathematical process models of plant systems including steam generator, feedwater system, chemical volume and control system, electrical system, etc. Safety-critical functions and DMPI are prototyped for improving operational goals. The real-time knowledge-based simulation will be used to establish the experimental research test facility for process control, monitoring, and Man-Machine Interface (MMI) in the Main Control Room (MCR) of a nuclear plant.

Functional Requirements

NPP processes are continually monitored on CRT and Large-Scale Project Panel (LSDP), and controlled by supervisory controllers and human operators. Supervisory controls are being simulated by automating control tasks and a set of initial conditions including startup, shutdown, power increase and decrease, and RCS (reactor coolant system) pressure control, etc. Major functions of the real-time knowledge-based simulation are classified as process control and monitoring, fault diagnosis with setpoints and domain knowledge, execution of simulated emergency operating procedures, analysis of control automatics with control and instrumentation diagrams, real-time data interface with external environments, and real-time execution of a simulated plant model. The real-time knowledge-based simulation consists of three major subsystems of cognition system, perception system, and action system. Each subsystem communicates through Communication Interface (CI) in real-time. CI manages overall data communications between subsystems of the real-time knowledge-based simulation. CI gets interaction data from external simulation and human operators, and transfers the perceived data into knowledge-bases of cognition system or directly to action system. Agenda manager identifies executable reasoning operations by using recent cognitive/perceptual events, and records on agenda. Scheduler determines which operation of executable reasonings is executed and when it is executed. Finally, executor performs the chosen operations

Design and Implementation

Hayes-Roth's second generation real-time performance in intelligent agents and real-time expert system shell G2 are used in the design and implementation of the real-time knowledge-based simulation. Figure 1 illustrates the real-time knowledge-based simulation architecture. Control and diagnostic knowledge, multi-reasoning skills and control automatics analysis are used for these tasks. Control logics are implemented as a graphical object-oriented representation for control automatics. Another knowledge is emergency operating procedure which can execute the predetermined procedure for simulated accidents. The knowledge is also represented as graphical objects, and a set of rules and procedures. Overall knowledge-based agents are implemented with G2, and real-time for simulating a generic PWR NPP are implemented using FORTRAN 77 and C.

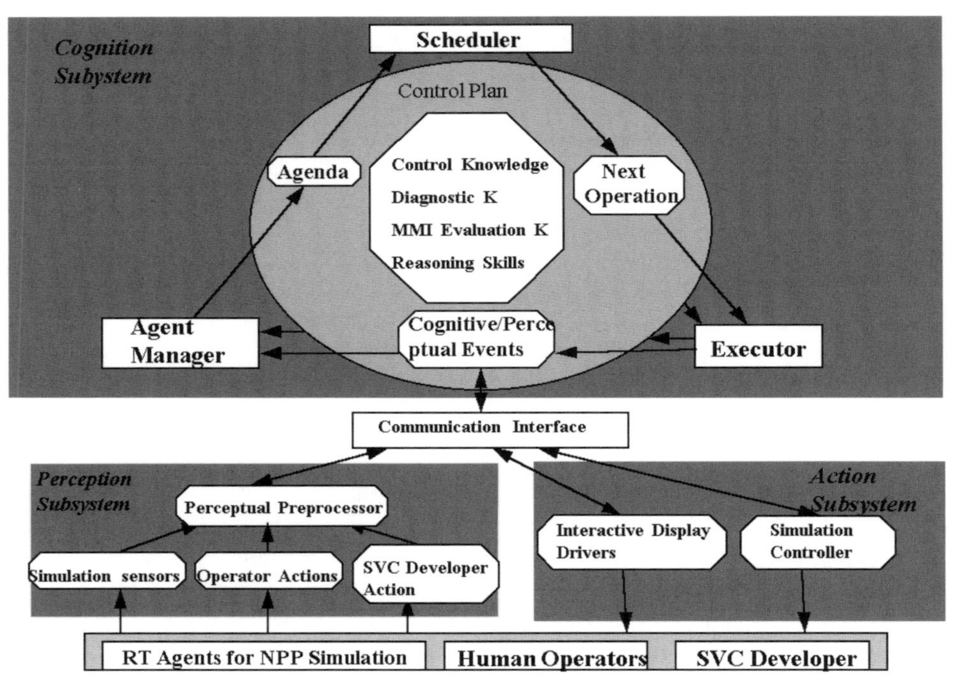

Figure 1. Real-time knowledge-based simulation architecture.

Cognition System

The cognition system architecture is shown in Fig. 2. The real-time knowledge-based simulation has the internal knowledge of safety-critical functions, control logics, dynamic processes, emergency operating procedures for simulated accidents, and fault diagnosis. Control logics are referenced for control automatics analysis. Current control automatics analyzer performs its function as on-line control logic diagrams and monitors control automatics status. Control logic diagrams (Fig. 3) are structured with lower-level components of NPP, i.e., AND, OR logic gates, valves, etc. The control automatics analyzer will be extended with a textual message for their contextual request such as "why a valve/pump will not open/start?" (Winsnes, 1991). Event-based and temporal reasonings are applied both to determine trends using real-time simulation data of a generic PWR NPP, and to diagnose safety-critical states. Emergency operating procedure prototypes are implemented as object-oriented and with a set of rules and procedures of G2.

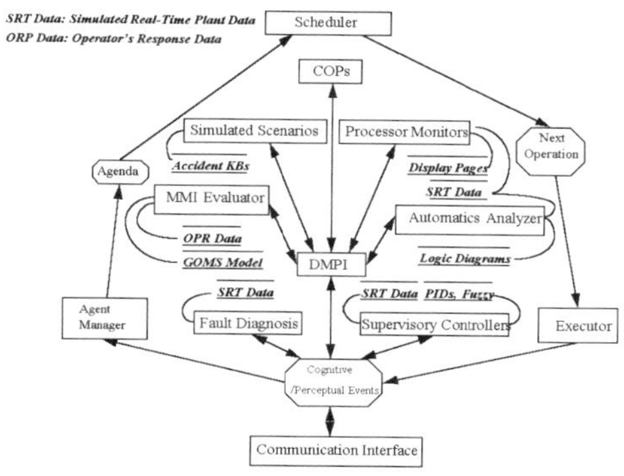

Figure 2. Cognition system architecture.

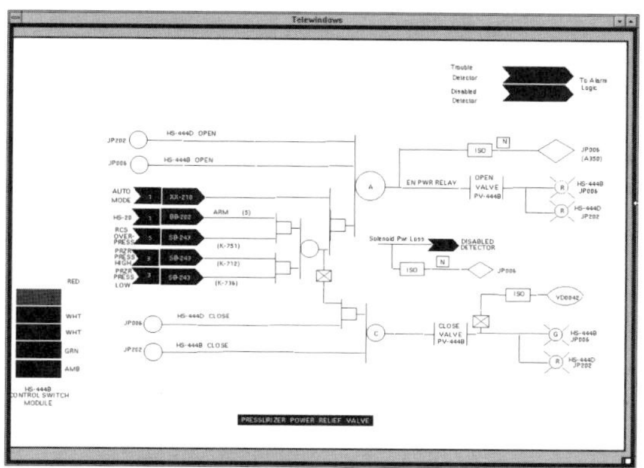

Figure 3. Control logic diagram.

Perception System

About 1000 plant parameters are perceived at the rate of 0.2 second while operator's response data are immediately perceived by conceptual preprocessor. Operator's response data are maintained with time-tag, display page number and object- or parameter-name while plant parameter data are maintained with time-tag and parameter-name for selected one.

Action System

A G2 prototype for an Integrated Process Status Overview (IPSO) of a NPP is designed with a projection display. DMPIs are designed for human operators, and control functions are integrated into displays. Human operators can diagnose and control an activated function with DMPI which several kinds of CRT displays are interfaced with DMPI.

Performance Evaluation

Experimental scenarios with accidents and incidents are supplied to evaluate the real-time knowledge-based simulation. Representative accident scenarios of small break LOCA and steam generator tube rupture, and initial conditions (e.g., start-up, 10% power, full power, etc.) are used to evaluate the real-time knowledge-based simulation. An operator support for emergency operating is prototyped for Youngkwang 1 & 2 units of a pressurized water reactor and is utilized to test and evaluate complex process control in the main control room environment. Validation of process control is performed with control logic diagrams, experimental scenarios, and DMPI. A safety-critical event is simulated and then the operator's interaction for the control task is evaluated with GOMS knowledge.

Conclusions

The real-time knowledge-based simulation, which is developed by applying the Hayes-Roth real-time performance architecture in intelligent agents, enables new control concepts and operator support systems to be easily prototyped and evaluated with the dynamic simulation environment. Our experience thus far shows that knowledge-based simulation is very useful for prototyping and evaluating various tasks of process control and monitoring, MMI in a highly automated MCR and for testing a new control concept.

Acknowledgments

I would like to thank Dr. Bong-Sik Sim for porting CNS code on an HP9000/755 Unix workstation, and POSCON for maintenance support of G2.

References

Beltracchi, L. (1990). A direct manipulation system-process interface. *Proc. Human Factors'90*. pp. 286-292.

Folleso, K. and F. S. Volden (1993). Lessons learned on test and evaluation methods from test and evaluation activities performed at the OECD Halden Reactor Project. *OECD Halden Reactor Project*, Report No. HWR-336.

Gensym Corp. (1993). *G2 Reference Manual*, V3.0 and 4.0.

Hayes-Roth, B., et al. (1991). Frameworks for developing intelligent systems: the ABE systems engineering environment. *IEEE Expert*, June, 30-40.

Hayes-Roth, B. (1993). Architectural foundations for real-time performance in intelligent agents. In J. M. David, et al. (Eds.), *Second Generation Expert Systems*. Springer-Verlag, Colorado, pp. 643-672.

Ingrand, F. F., et al. (1992). An architecture for real-time reasoning and system control. *IEEE Expert*, December, 34-44.

KAERI (1989). *Advanced Compact Nuclear Simulator Textbook*. KAERI.

Mark, W. S., et al. (1991). Knowledge-based system: an overview. *IEEE Expert*, June, 12-17.

Moore, R., et al. (-). Process control using a real-time expert system. Source unknown.

Musliner, D. J., et al. (1995). The challenges of real-time AI. *IEEE Computer*, January, 58-66.

Penalva, J. M., et al. (1993). A supervision support system for industrial processes. *IEEE Expert*, October, 57-65.

XPERT QUALITY EVALUATION: A HYBRID SYSTEM FOR QUALITY PREDICTION ON PRODUCTION PROCESSES

B. Henze
Mannesmann Datenverarbeitung GmbH
D-40885 Ratingen, Germany

U. Falkenreck and P. Monheim
Mannesmann Demag Hüttentechnik
47053 Duisburg, Germany

Abstract

No human expert really works like an expert system. The brain of a real life expert is a hybrid tool using elements of expert system reasoning, fuzzy logic, cluster analysis, data acquisition, strategies to draw parallels to past experiences and a number of other methods. Most of the time, the expert is not aware of this — his brain has a user-friendly interface. Realizing this, the Xpert Quality Evaluation System has been developed. By being a hybrid tool, it avoids frequent problems of pure AI paradigms such as Expert Systems and Artificial Neural Networks. Online diagnosis/prediction and offline data analysis are supported in a two-step approach: first a set of raw data representing a case are compressed into causes which are defined by the user's concept of influences on quality, then the actual quality of the case is predicted by it's own position and positions of learnt cases in a multidimensional space spanned by the causes. Thus, both the user's knowledge, expressed in various ways, and experience from the production process cooperate in a flexible way without forcing the user to fully rely on one or the other. The system is designed to grow with the user's human knowledge and production case collection; it also circumvents customers' frequent fear of giving their production secrets to external knowledge engineers, since the user himself customizes the system's prediction methods via a graphical user interface, even without knowing much about Artificial Intelligence at all. Xpert Quality Evaluation was originally developed for steel industry processes, but, having no process-specific features itself, is usable in a much wider area.

Keywords

Quality prediction, Hybrid tool, Knowledge vs. experience, Online, Offline, User-friendly.

Introduction

In this context, knowledge means a clear concept of relationships between physical process conditions and resulting product quality, in other words, understanding why this quality is produced by that process under these circumstances. Knowledge can be transferred between human beings. Different from this, experience is a matter of bulk data storage. Many historical cases of product qualities resulting from corresponding production conditions build experience which can be measured in years, but cannot be transferred from one person to another. Any human expert works with an ever-changing combination of knowledge and experience: new experience is collected, sometimes experience is converted into knowledge. This can be displayed in a two-dimensional graph (Fig. 1). Well-known human intelligence paradigms are shown in black, some AI paradigms are shown in white.

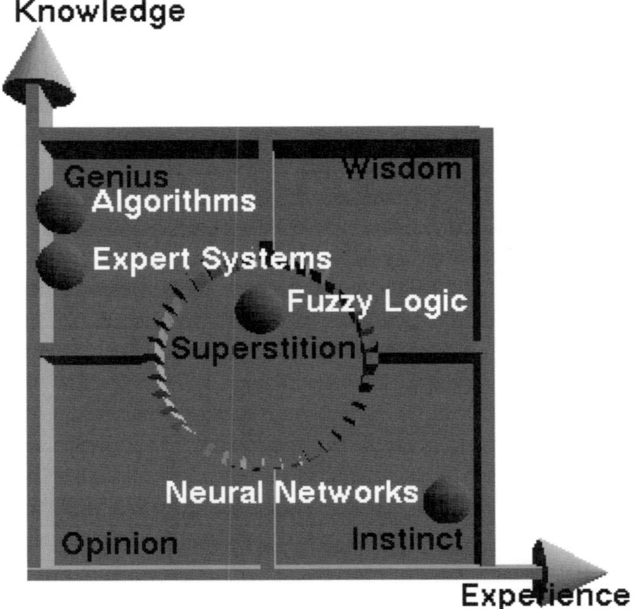

Figure 1. Intelligence paradigms.

The general problem of a diagnosis tool built on one paradigm alone is it's fixed position in this graph, while the user is probably somewhere else: algorithms and expert systems do nothing for him until he puts perfect process knowledge into them (upper left); a neural network will not be reliable until it's training set contains cases for all possible situations (lower right); both goals will probably never be reached. Most users gather in a central area intentionally called superstition (here, superstition means to follow a set of rules without fully understanding their background; this is independent of the truth they may contain. Such rules are often derived from case collections, e.g. via statistical means, cluster analysis algorithms or personal experience. One possible way to express superstition is fuzzy logic, and it's success is an indication how widespread superstition really is). But even if a user happens to be near his tool in the beginning, he will move away from it with growing experience and knowledge; or he is near it with one diagnosis problem, but far away with another. This leaves him with two choices: to use different tools in different development stages and problems and end up with a motley crew of systems and licenses, or use a hybrid tool.

A Hybrid Approach

In the Xpert Quality Evaluation System (XQE), quality is a vector of defects, with an integer number assigned to each one telling how strongly this defect has occurred; this number is the so-called severity (the higher a severity, the worse the quality with respect to the corresponding defect; severity=0 means the defect has not occurred). The relationship between quality and it's defect vector representation remains outside XQE, i.e. it is up to the user to know what a certain severity really means.

Every defect has causes. Here, causes are defined as variables that influence a defect's severity. These causes are calculated from raw measured data via so-called transformations. These transformations contain the user's knowledge. A set of measured data that belong to one piece of product is called a case. Old product pieces whose real quality is known are called old cases and represent the process experience; data sets coming in for quality prediction are called actual cases. Both types of cases go through transformation and into prediction or analysis (Fig. 2).

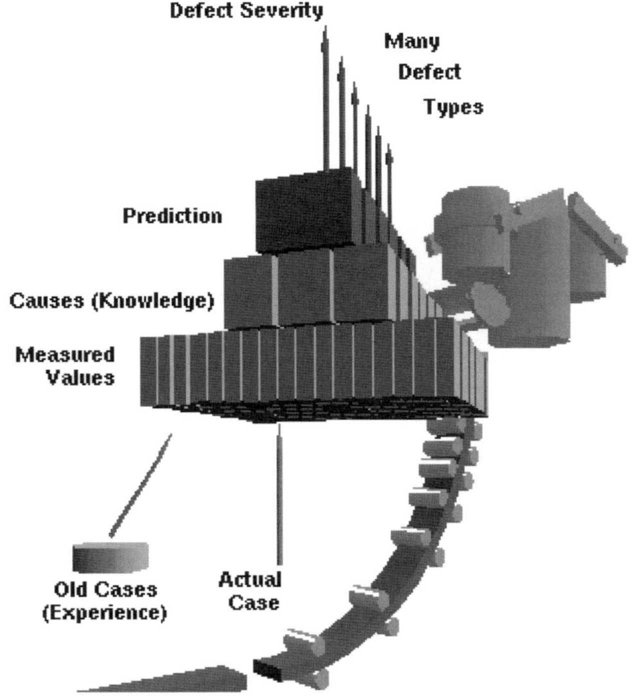

Figure 2. Transformation and prediction.

For prediction, the causes of a defect span a multidimensional space where a case is represented by a point. Prediction is reduced to giving that point a severity number; this is done via two methods called Basic Prediction and Experience Prediction.

Basic Prediction evaluates the severity from the case's position (shown as a little ball) versus a system of thresholds put on each axis, i.e. for each cause pairs of upper/lower limits for every severity are placed within each other, thus forming a system of rectangular boxes. The innermost box gets severity 0, surrounded by a severity 1 box and so on (Fig. 3).

Experience Prediction evaluates the severity from the verified severities of old cases near it's position (shown as little cubes). To prevent unwanted effects from empty or unreliable areas, the user defines hypercuboids around areas he allows for this prediction method. The actual case is then given a severity by the majority of old cases within the same hypercuboid (Fig. 4).

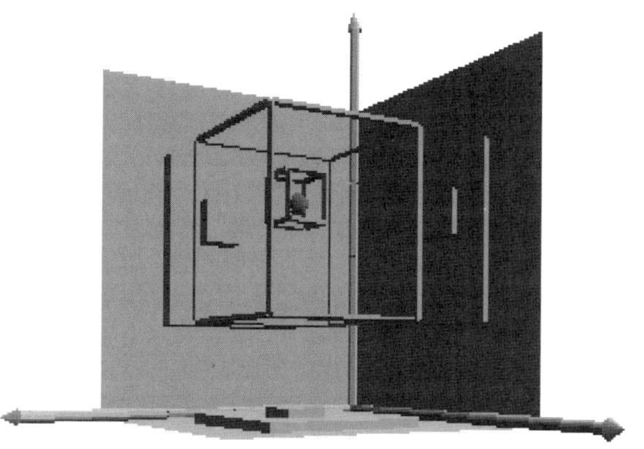

Figure 3. Basic prediction.

Figure 4. Experience prediction.

For every actual case, Experience Prediction is tried first; if that gives no result, the case is passed to Basic Prediction.

The flexibility of this approach becomes clear from two extreme scenarios:

- With no knowledge but full experience, all raw measured data that may have some influence on the current defect are used directly as causes, i.e. the cause space has many dimensions; practically all cases are predicted by experience.
- With full knowledge but no experience, the user knows the very formula to calculate defect severity and makes it a transformation; the cause space is one-dimensional, Basic Prediction gives excellent results.

Forms of Knowledge and Superstition

Xpert Quality Evaluation offers a variety of what a transformation can be. In general, anything having many inputs and one output will do:

A mathematical formula can be expressed directly in an easy language.

A Fuzzy Logic controller can be created via a graphical interface to maintain input/output variables, draw membership functions and control rule results and completeness. The controller can be done either full-size or with certain restrictions supporting linear interpolation.

A built-in linear/logarithmic regression algorithm can generate a formula from experience.

A built-in decision tree generator can generate a transformation from experience. Geometrically, this is equivalent to building a structure of rectangular rooms into the cause space, the decision cuts being walls, floors and ceilings (Fig. 5).

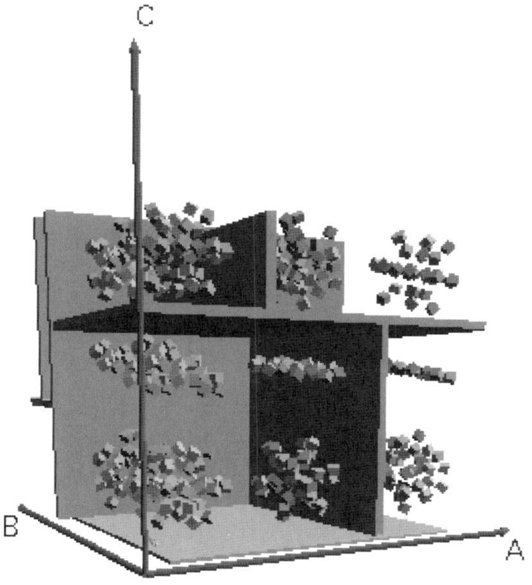

Figure 5. Decision tree generation.

The last two transformation types represent a way to let a machine do what a human expert is doing every day: build up superstition in the sense defined before. But there are more examples: The decision tree generator is also used to create a hypercuboid geometry for Experience Prediction, and the thresholds for Basic Prediction may be adapted to experience by a built-in adaptation algorithm.

Prediction Errors

Experience is hardly ever accurate. So, the verified severities of old cases in a cause space will show a certain scatter, even if the causes are very good ones (bad or forgotten causes usually generate a total scatter over whole subspaces). Therefore, prediction errors are unavoidable, and a certain allowance of them must be specified for all algorithms and methods that are using experience. Prediction errors mean that prediction is either too careless (the predicted quality is better than the real one) or too anxious (the predicted quality is worse than the real one). Because of the scatter, it is possible to put limits on one error type only; since the too careless prediction errors are generally regarded to be the more dangerous ones, upper limits for them can be specified by the user for threshold adaptation, decision tree generation and experience prediction.

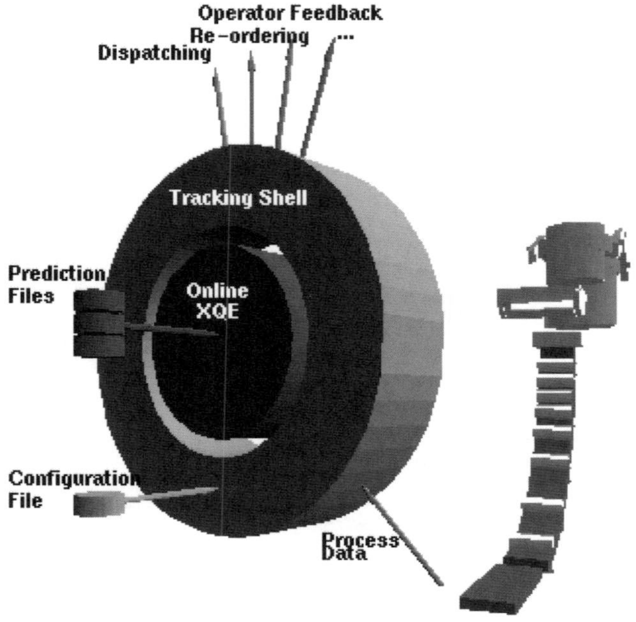

Figure 6. Online environment.

Online Environment

Online XQE runs as a black box, receiving actual case datasets and sending prediction results, thereby storing the cases for later verification/analysis. Online XQE uses three types of prediction files: experience, transformations and a model file containing the defect/cause structure, threshold and hypercuboid geometries etc. These prediction files are maintained within the Offline Environment. The case data, supposed to represent the product's voyage through the production process, are collected and sent to Online XQE by the Tracking Shell, an object-oriented forward chaining module specialized on tracking problems that reads it's process-specific configuration from a file at startup (Fig. 6). The Tracking Shell may also use prediction results for product dispatching, re-ordering or operator feedback.

Offline Environment

Offline Xpert Quality Evaluation offers two ways to maintain the prediction files (Fig. 7):

- Case data files and model files are accessed via a special software layer and a quasi-database language. In this language the user may write little access programs via special editors. Also online prediction is such a program, automatically generated by Online XQE. This language also includes a macro preprocessor (e.g. all transformations are stored as macro definitions).
- A set of convenient model windows present graphical views on model structures, two-dimensional subspace navigation through cause spaces for both prediction methods and standard prediction error rate statistics. This access is controlled by the model manager, an embedded expert system containing the knowledge about this tool (but no process-specific knowledge).

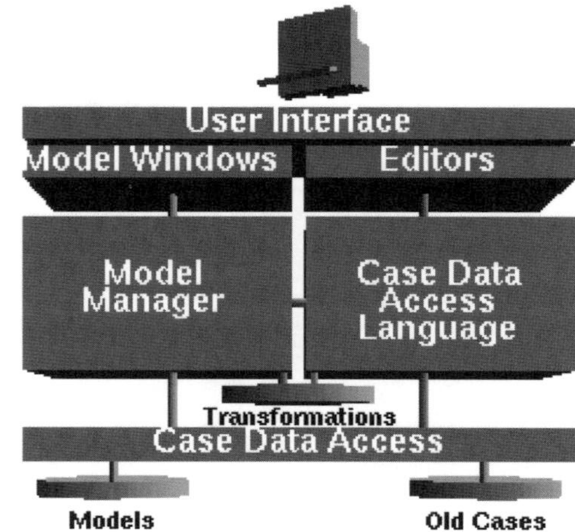

Figure 7. Offline environment.

Platforms

XQE runs on Workstations with no less than 32 MB memory. Operating Systems are major UNIX platforms (OSF/1, SOLARIS, HP-UX, etc.) and OpenVMS.

Conclusions

The Xpert Quality Evaluation system is a workstation-based hybrid tool for online quality prediction and offline data analysis and tuning. It uses a flexible combination of user knowledge and process experience for prediction and prediction reliability improvement. It has a graphical user interface for all analysis and tuning tasks, including cause transformation maintenance, cause space navigation and statistics.

SUPERVISORY AND DECISION-SUPPORT SYSTEM FOR INTELLIGENT MONITORING AND CONTROL OF A PIPELINE NETWORK

Weerapong Kritpiphat and Paitoon Tontiwachwuthikul[1]
Process Systems Laboratory, Faculty of Engineering/Energy Research Unit
University of Regina
Regina Saskatchewan, Canada S4S 0A2

Aijn An, Christine W. Chan and Nick Cercone
Department of Computer Science, Faculty of Science
University of Regina
Regina, Saskatchewan, Canada S4S 0A2

Abstract

This paper presents the development of an expert system for supervisory and decision-support of a municipal water distribution system. Manual knowledge acquisition resulted in construction of a conceptual model of the problem domain which formed the basis for prototypes implemented in two expert system shells, G2 (Gensym Corporation™) and Comdale/X (Comdale Technologies, Inc.™). Knowledge discovery and a fuzzy set mechanism were used to model aspects of the problem not describable by conceptual modeling. The information generated can potentially be incorporated into the rule base.

Keywords

Supervisory and decision-support, Expert systems, Monitoring and control, Water network processes, Object-oriented approach.

Introduction

A supervisory and decision-support system for monitoring and control of a municipal water network process is useful when the most experienced operators of the municipal water distribution control station will soon resign or retire, and their knowledge and expertise have not been documented and transferred to aspiring operators. Some advantages to developing such a system include:

- Operating costs will be minimized in the long run.
- Proven decisions, operating procedures and expert operator knowledge will be documented.
- The knowledge required for process operations is consolidated and is applied consistently to provide a more reliable monitoring and control system.
- Response time and actions in emergency cases will be improved.
- There is an opportunity to re-deploy staff into more productive jobs.

This paper describes our efforts at developing an intelligent supervisory and decision support system for monitoring and controlling a pipeline network.

Problem Domain

The domain addressed is a municipal water network process of a typical moderate-sized prairie city in North America. The major sources of water include a lake and a number of underground wells (Karius, 1993). The lake water is treated at a water treatment plant to an acceptable standard before being delivered to the system, while the

[1] Author to whom all correspondence should be addressed.

well water is directly drawn from underground wells located in and adjacent to the city. Water is pumped from its sources to fill in reservoirs located at different elevations in the city and is pumped from the reservoirs to the distribution pipeline, or to another reservoir when it is necessary to adjust water levels (Brindle, et al., 1993). Distribution pressures and rates of water transfer throughout this process can be controlled by means of pumps and valves housed in remote hubs and pumping stations.

Monitoring and control functions are required at all times to maintain smooth operations of the process since conditions are ever-changing. Parts of these functions are automated by means of Programmable Logic Controllers (PLCs) at remote sites and the Supervisory Control And Data Acquisition (SCADA) system at a main control center. Although they have some "intelligence" built-in like sequencing actions for turning on/off the process equipment, and automatic starting of diesel generators and engine-driven pumps in case of electrical failure, a human operator is currently required for the decisions that are based on economic, environmental and sociological factors. The operator employs heuristics to maintain the distribution pressure, to predict the long-range and short-range water demand and to keep the water level of reservoirs within a reasonable range. The expert operator can also make decisions to maintain consistent water quality in terms of residual chlorine level and hardness, to prevent water from standing and becoming stale in reservoirs, to minimize the risk of icing in reservoirs and pipeline for winter operations, to avoid low suction pressure at each pumping station and to lower operation costs of the entire process. The operator usually uses information obtained from the SCADA system (e.g., equipment status and data collected from field instruments) and outside sources (e.g., weather forecast and special events from TV, radio or newspaper) for decision-making and sends his/her control commands to SCADA. The SCADA system then transmits the commands through modems and leased telephone lines to the PLCs, which in turn, issue their own commands to the equipment they control.

However, different operators do not control the water distribution operations in the same way. There are generally many possible alternative paths or sequences of actions which would result in similar control functions, but they may not all be equally effective from an engineering and business point of view. For example, one alternative may achieve the same acceptable performance but with lower power costs than the others. Inconsistent operations would reduce the overall process efficiencies and even degrade the performance and life-time of equipment. Documenting operators' heuristics and developing a consistent and optimal operating style are therefore beneficial for the water distribution process.

A supervisory and decision-support system is suitable for this process since the water distribution operations involve heuristics and human expertise; it will enable electronic distribution of expert-level knowledge and problem-solving capabilities. A supervisory and decision-support system renders the assistance of expert consultants readily accessible at times when, and in locations where, human experts are not available. Where human experts are available, such a system can also handle the more routine problems, freeing the experts to be re-deployed with more challenging and productive cases.

We have developed prototypes of the supervisory and decision-support system for monitoring and control of a municipal water network process. The prototypes advise users who are less experienced operators on how to properly control the SCADA system. An overview of the supervisory and decision-support system is shown in Fig. 1.

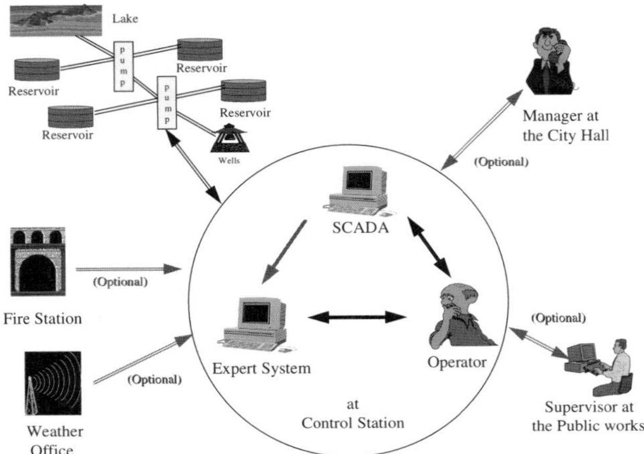

Figure 1. Overview of the supervisory and decision-support system.

System Development

Knowledge acquisition is a crucial phase for expert system development; this process can be divided into the three subphases of (1) knowledge elicitation (2) knowledge analysis and (3) knowledge representation (Bell and Hardiman, 1989). We conducted manual knowledge acquisition by interviewing a total of 7 city employees including the expert operators, supervisor, engineer and manager. The information obtained was analyzed using the Inferential Modeling Technique (Chan, 1992), and the analysed knowledge was decomposed into knowledge items and represented with an object-oriented modeling approach. The product of the knowledge acquisition stage is a conceptual model of the supervisory and decision-support system which embodies a logical and maintainable structure of the system. It consists of four different models: interface model, object model, functional model and control model. Each of these models has its own purpose and describes one specific aspect of the system. The interface model represents all the objects that exist within the system which are responsible for communicating with the entities that need to exchange information with the system. The object model represents the static data structure of the real world system. Object modeling consists of identifying the objects of interests in the system, grouping the objects into classes, specifying attributes for the classes and organizing a class hierarchy from their common properties and inter-relationships.

A class hierarchy for our supervisory and decision-support system is shown in Fig. 2. This figure illustrates a segment of the object model, showing a super-class, its associated classes, sub-classes and objects. For example, pump *FP2 (Engine)* is an object under the "Pumps" class and the "Diesel Engine" sub-class; and the "Pumps" class belongs to the "Process Equipment" superclass.

The functional model shows how values are computed without regard for sequencing, decision, or object structure. The functional model shows which values depend on which other values and the functions that relate them. Data flow diagrams are used for showing functional dependencies and functions are expressed by mathematical equations or formulas. Values refer to the attribute values of objects in the object model. In this water distribution system, there are two kinds of values. The first kind represents quantities such as ground levels, reservoir areas, reservoir depths, maximum levels, performance (H-Q) curves for pumps, etc. The second kind represents the time variant data, e.g., flow rates, reservoir current levels, discharge pressures, weather conditions, etc. The time variant data can be further subdivided into the data gathered from external sources, such as field instruments and the weather office, and the data that are calculated according to the relationships with other data, for example, expected demand. Usually the functional model is used to describe the second kind of time variant data. However, the expert system presented here is an off-line system in which the time variant data theoretically derivable from external sources in fact obtain their values through simulation. Therefore mathematical simulation formulas are included in the functional model along with the non-simulation formulas which calculate the second kind of time variant data, A sample of the simulation formulas is

$$L_{new}=L+dt*(inflow-outflow)/area \quad (1)$$

The control model describes the sequences of operations that occur in response to external stimuli. We used decision trees to describe the control model of our supervisory and decision-support system, an example of which is shown in Fig. 3. The main operating decisions include setting up cut-in and cut-out levels in a reservoir, choosing a small, medium or large pump to start pumping water into a reservoir when the level of the reservoir reaches its cut-in level, and stopping pumping when the reservoir reaches its cut-out level. These decisions depend on many factors such as demand, reservoir storage capacity, required emergency storage, input capacity and also on the economics of supply such as the cost of pumping. Different operators may make different decisions in the same situation. In building the control model, different operating styles need to recognized and rules belonging to the same operating style are grouped together to form a rule set. For a detailed discussion of the four models, see (Chan, et al., 1994). Together the four models serve as the basis for the implementation stage in the development of the supervisory and decision-support system.

Based on the conceptual model, we have developed prototypes of the supervisory and decision-support system for monitoring and control of the water pipeline network operations in the expert system shells, Comdale/X under the Windows (PC based) environment and G2 under the UNIX (work station based) environment. The prototype in Comdale/X was designed with "information-driven" menus on the text workspaces so that it is easier for the knowledge engineers to observe and inspect the output responses. More user-friendly and "graphic-driven" menus will be added in the near future after the knowledge base of the prototype has been extended and refined. The prototype in G2 was based on a small scale version of the pipeline system. Classes in the object model and interface model were created under the G2 pre-defined class hierarchy (Gensym Corporation, 1992). Attributes are specified through creation of a definition item for each class that associates with a table where the common properties and characteristics can be described in standard formats. Using the G2 icon editor, an icon or a graphic image of an object class can also be drawn within its definition item. Thus by creating and cloning instances for different object classes and placing them on a workspace and connecting them according to their physical and logical relationships by a defined connection class, system schematics can be built. The functional model was implemented with generic and specific formulas in G2. The decision trees in the control model were translated into structured English-like rules with the help of the G2 editor for rule definitions.

Analysis of the knowledge obtained from the domain experts indicates that prediction of consumer demand is a prerequisite for optimal control of the pipeline network for the water distribution system. However water demand prediction is currently poorly understood even by the experts, who estimate daily or hourly water demand based on their experience. The often inaccurate estimations result in inefficient operations of the water distribution system. The lack of knowledge on water demand prediction also translates into a gap in the knowledge base of the supervisory and decision-support system. In other words, manual knowledge acquisition by itself is inadequate for handling all the situations that arise in our application. Three alternative methods for water demand prediction are considered: (1) a fuzzy set method that is used to assign a certainty factor to the logical value of objects, e.g. high vs. low, hot vs. cold and good vs. bad, on the basis of data obtained from experts; (2) automated learning from observed data, that is, to design an algorithm which can learn and refine decision rules from a set of training samples; and (3) generating rules based on engineering principles on mass and energy balances, fluid mechanics and dynamic process control.

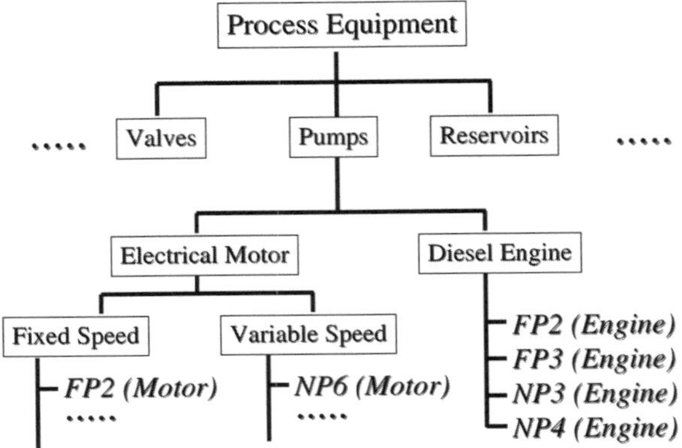

Figure 2. Class hierarchy.

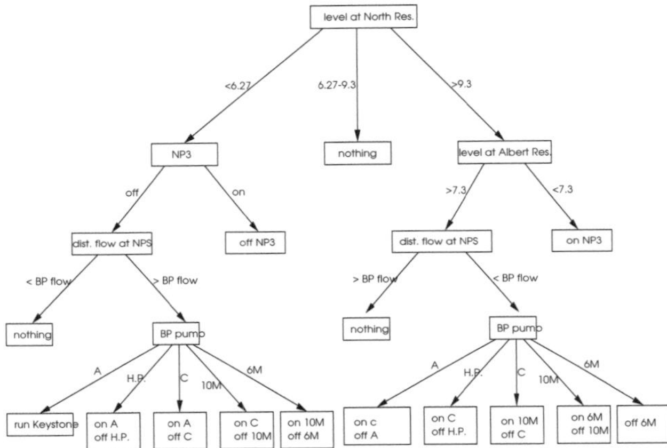

Figure 3. Decision tree for maintaining water level in one of the reservoirs.

The first alternative has been implemented in Comdale/X (Comdale Technologies, 1993). Primarily we used the fuzzy set mechanism and two stages of implementation are involved. First, we used the fuzzy sets' approach for assigning how sure we are that the input date is in the summer by plotting a temperature vs. month graph which is normalized into the range of 0-100. We called this the "season" certainty factor. We then apply the "season" certainty factor on the input date to select whether the hourly water demand patterns of summer or winter should be used in the next stage. The fuzzy data here is based on previous records of average temperature from January to December. The scale of zero is assigned wherever the average temperature is below -10°C (winter). On the other hand, the full scale of 100 is assigned whenever the average temperature is above 18°C (summer).

Secondly, we used the fuzzy set mechanism in Comdale/X to express the correspondence between time within a 24 hour period and water demand flow rate which are represented as the value and rank respectively in Fig. 4. The experts indicated that the water demand flow rate varies from hour to hour and its pattern is different for each day of the week. The experts classify the patterns into four types of water flow rate versus hours for (1) Monday, (2) Tuesday to Friday, (3) Saturday and (4) Sunday, for each pumping station and each season. We then scaled the water flow rate by dividing the water flow rate with a "scaling" constant so that the data can be fitted into the fuzzy set table in the range of 0-100.

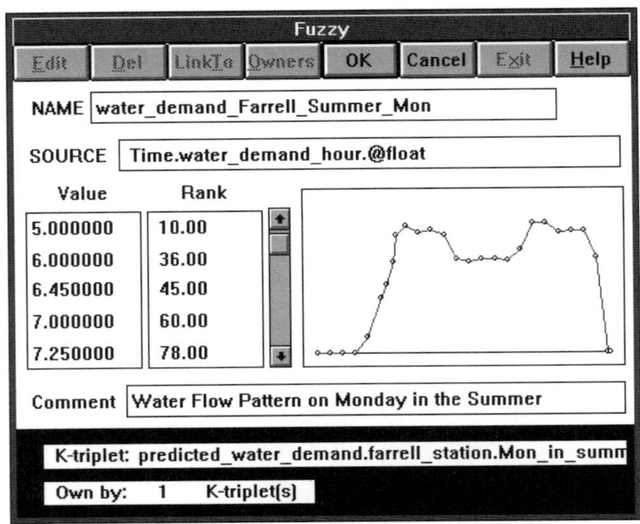

Figure 4. A sample screen showing the fuzzy set representation of scaled water flow rate versus 24-hour in Comdale/X.

After coding the water demand patterns and the "Summer vs. Month" fuzzy set into Comdale/X, water demand can be estimated using the following steps:

1. Convert the input date and month from the user or clock to a floating number of month, e.g., April 15 becomes 4.5
2. Assign a "season" certainty factor to the input date using the "Summer vs. Month" fuzzy set
3. Select the season pattern of Summer or Winter based on the certainty factor (The Summer pattern is chosen when the certainty factor is equal or greater than 50.)
4. Find the day of week from the input using the function of WEEKDAY available in Comdale/X
5. Select the water demand pattern for the input weekday
6. Get the input time from the user or some external utility program

7. Interpolate the scaled water flow rate from the selected water demand pattern pre-defined in the fuzzy set for each pumping station
8. Multiply the scaled water flow rate by the "scaling" constant to obtain the estimate of water demand flow rate at that time
9. Assign the certainty factor in Step 2 to be the certainty factor of the water demand flow rate in Step 8
10. Use the water demand flow rate and its certainty factor to fire rules in the rule-base until it reaches the goal of maintaining the distribution pressure in the water network process.

The second alternative for water demand prediction is automated learning from training samples. We proposed a rough-set based knowledge discovery method to automatically generate decision rules from a set of observed data. The proposed rough set method is based on an extension of the standard rough set model. The salient feature of this method is that it makes use of the statistical information inherent in the data to handle incomplete and ambiguous training samples. The database of observed data include the City's daily water consumption and weather conditions which include temperature, humidity, precipitation and wind speed. These data were obtained from the City of Regina and Environment Canada. We conducted the knowledge discovery experiment by using 306 training samples containing information on 14 environmental and sociological factors and their corresponding daily volume of distribution flow from March 1994 to December 1994. Nearly 150 rules were obtained based on these data. A sample of these rules is:

> *If day of the week is Sunday, Monday or Tuesday and minimum humidity is greater than 64% and average speed of wind is less than or equal to 10.84 km/h and mean temperature is less than or equal to -3.36°C, then the water demand is between 53 ML and 60 ML with an uncertainty factor being 1.*

In order to evaluate the rules derived by our method, we conducted a *Leave-Ten-Out* experiment by using 90% of the data for training and the remaining 10% of the data for testing. The error rate depends on the selection of training samples. We conducted the experiment ten times. The best error rate of prediction is 6.67% and the average error rate of prediction is 10.27%. We have worked on applying the knowledge discovery algorithm to a larger set of training data. We have also conducted validation of the generated rules by human experts before attempting to incorporate them into the knowledge base. For a detailed discussion of the knowledge discovery method, see (An, et al., 1995).

Conclusions

We have presented our efforts at developing an expert system for monitoring and control of the water distribution system at the City of Regina. Manual acquisition of human expertise has resulted in the construction of a conceptual model of the problem domain which is the core model of the system. However, this core model is far from complete and aspects of the problem not describable by conceptual modeling need to be addressed. We present here two alternative approaches for predicting water demand by using the fuzzy set mechanism provided in Comdale/X and knowledge discovery from data on weather conditions and water usage. Both approaches have been implemented and yielded results that can potentially be incorporated in the rule base of our prototypes.

The future work in this project include validation of the conceptual model developed using human experts and simulation, further application of neural network and knowledge discovery techniques for water demand prediction, and implementation of the full systems when gaps in the conceptual model have been closed using the various Artificial Intelligence techniques.

Acknowledgments

The generous support of Telecommunications Research Laboratories (TRLabs) and the cooperation of personnel at the City of Regina's Municipal Water Engineering Department are deeply appreciated.

References

An, A., N. Shan, C. Chan, N. Cercone and W. Ziarko (1995). Discovering rules from data for water demand prediction. *Proc. of IJCAI'95 Workshop on Machine Learning in Engineering*. Montreal, Canada.

Bell, J. and R. J. Hardiman (1989). The third role — the naturalistic knowledge engineer. In D. Diaper (Ed.), *Knowledge Engineering: Principles, Techniques, and Applications*. Chichester: Ellis Horwood Ltd.

Brindle, D., W. Tunison and L. R. Martin (1993). Water distribution with TIRS. *Proc. of SHARE-80/Winter Conference*. San Francisco USA.

Chan, C. W., N. Cercone, A. An and P. Tontiwachwuthikul (1994). Object-oriented modeling and simulation of an expert system for monitoring and control of a water distribution system. *Proc. of the 1994 International Conference on Object-Oriented Information Systems*. London UK. pp. 130-133.

Chan, C. W. (1992). Inferential model and inferential modeling technique: a systematic technique for knowledge analysis. *Proc. of the 7th Banff Knowledge Acquisition for Knowledge Base System Workshop*. Banff Canada. pp. 6.1-6.15.

Comdale Technologies (1993). *COMDALE/X User's Manual and Reference Guide*, Revision 5.12.930121. Comdale Technologies, Toronto, Canada.

Gensym Corporation (1992). *G2 Reference Manual Version 3.0.* Gensym Corporation, New York, USA.

Karius, R. H. (1993). Long term water utility study. Report No. 6: Operational considerations and Report No. 7 Major facilities. prepared for the City of Regina. Associated Engineering (Sask.) Ltd., Regina, Canada.

CASE-BASED REASONING INTELLIGENT SYSTEM FOR PROCESS OPERATION SUPPORT

Qijun Xia and Ming Rao
Dept. of Chemical Engineering, University of Alberta
Edmonton, AB, Canada T6G 2G6

Christer Henriksson and Hassan Farzadeh
Slave Lake Pulp Corporation, P.O. Box 1790
Slave Lake, AB, Canada T0G 2A0

Abstract

Process operations in pulp and paper production are very complicated. An individual mill operator may find it difficult to contribute a quick solution when facing undesirable situations. To approach this problem, an intelligent system for operation support in a pulp mill is developed. The intelligent system is integrated with a distributed computer system and a management information system that are available in most of modern pulp and paper mills. The system evaluates the performances of the pulp production process, detects undesirable situations and provides operation suggestions to operators. Case-based reasoning (CBR) method is applied in the intelligent system, which makes knowledge acquisition easier and makes the system consistent with natural reasoning operators do in problem solving. The standard activities of case-based reasoning are implemented in meta level knowledge bases with an object-oriented tool. The process knowledge is implemented in domain-specific knowledge bases. The techniques proposed in developing the system have shown great advantages.

Keywords

Intelligent system, Case-based reasoning, Fault diagnosis, Operation support, Pulp and paper industry.

Introduction

Most of the nowadays pulp and paper companies have successfully installed distributed computer systems (DCS) and information management system (IMS). However, process operations still rely on individual operators' experience. Operators need to deal with a large amount of raw data and supplemental information in order to contribute an effective solution. Operation support is very important to improve process operations. Artificial intelligence technology can play an important role in achieving this purpose. To approach the problems, an intelligent operation support system is being developed (Xia, et al., 1993). The intelligent system is integrated with DCS and MOPS, a mill wide management information system developed by MoDo Chemetics. It receives raw data from DCS and value added data from MOPS. The system monitors these input data to evaluate operating conditions and to decide the corrective actions for improving the production. These results are then sent to MOPS for graphical display and to DCS for necessary automatic corrections.

Fault diagnosis and handling is one of the most important parts of the intelligent operation support system. This paper will focus on this part. Fault diagnosis and handling has a weak domain theory. It mainly relies on the experience of individual operators. Problem solving in fault diagnosis and handling requires creativity and common sense knowledge. Case-based reasoning (CBR) systems have a number of advantages to solve such kind of problems (Kolodner, 1991). Many software systems have been developed using CBR, such as CHEF for preparing dishes (Hammond, 1989), KRITIC for design (Goel and Chandrasekaran, 1989), PLEXUS for plan (Alterman, 1988), PROTOS for diagnosis (Bareiss, 1989). However, very few have been developed for process industries, especially pulp industry. In this paper, the application of

CBR for fault diagnosis and decision making in a pulp mill will be presented.

Overview of the Pulp Process

We focus our discussion on the bleach plant in Slave Lake Pulp Corporation. The bleach plant can be divided up into four components: first stage washing, interstage bleaching and washing, third stage washing and bleaching, and fourth stage washing. The most important quality variable affected by the bleach plant is pulp brightness. However, the operating conditions in the bleach plant affect the other pulp quality variables as well, such as freeness, bulk, breaking length, tear index, burst index, opacity. The operation conditions that affect pulp brightness are caustic charge, peroxide charge, silicate charge and DTPA, pulp consistencies and temperatures in the bleach towers, wood species and chip quality.

For such a complex process, developing a fault diagnosis and decision making system is extremely difficult. We need to find out the natural reasoning operators and engineers do to solve a problem. The purpose of artificial intelligence research is to search a way to model the human recognition processes with computer programs.

During our research, we found that both novice and experienced operators and engineers tend to use previous cases for generating the hypothesis for operation problems. They have experienced a large number of different events in the operation and derived their knowledge from these events (cases) to solve future problems. The basic knowledge unit in their memory is case rather than rule. When an abnormal situation occurs, they rely heavily on memory of the past cases in solving the problem rather than creating a solution from scratch. In the knowledge acquisition, they usually present us with the previous incident records.

These facts reveal that human experts are good at using analogs or past cases in solving new problems. Pulp production implies a large number of potential abnormal situations (cases). A novice operator may lack enough experience of these cases. He needs to use the experience from others. An experienced operator may have experienced most of these cases, however, it is difficult for him to remember the cases that happen very seldom.

CBR is an analogical reasoning method that simulates human expert's problem solving. It reasons from the past experience for problem solving. A CBR system can store all the past cases from various mill experts for all mill personnel to share. It has capability of self learning from problem solving. When a new problem is satisfactorily solved, the CBR system can store the solution for future use.

These arguments suggest that CBR is consistent with the natural problem solving mill operators and engineers do. A CBR system is an extension of human experts' memory. Using CBR will make operators feel more comfortable to use intelligent systems in operation.

System Organization

The reasoning process of operators in process operations suggests the following structure for the intelligent system (Fig. 1).

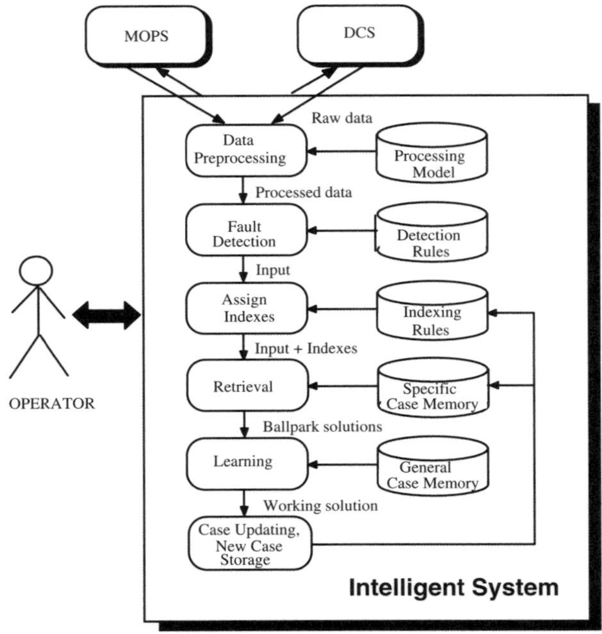

Figure 1. Structure of the CBR system.

The intelligent system first receives data from MOPS through a system interface. These data are passed to the data preprocessing module to be transformed into {normal, abnormal} form. The processed data are then evaluated by the fault detection module to determine whether there is any process failure. If the operation is in normal situation, the system will return to routine monitoring. If there is anything abnormal detected, the system will call the indexing rules to define the index for the new event; retrieve the relevant cases from the specific case memory; calculate the similarity to select the best past case; and adapt the solution of the past case for the new situation. If there is no relevant case in the specific case memory, the system will call the general case memory and interact with operators to create a solution. The new solution will be automatically stored in the specific case memory and the indexing rule will be changed to incorporate the new solution. The details of main modules are described as follows:

Data preprocessing: This module transforms the absolute values (measurement) of process variables into the corresponding state values: {0} — normal, {+1} — high, {-1} — low. For some finely controlled variables, we define two additional states: {+10} — extreme-high, and {-10} — extreme-low.

The transformation from absolute values to state values is done by two methods. In the first method, the standard limits are defined for each process variable. The state values are determined by simply comparing the measurement of the process variables with their standard limits. Considering that the standard limit method may not be stable for noisy process data, a time series bias detection method based upon the cumulative sum (CUSUM) is also applied. If these two methods give conflicting answers, the system will assume the process variable be in normal condition.

Fault detection: The system detects the presence of abnormal situations by simply checking the states of a number of important process variables. If all these variables are in {0} — normal, the production is normal; otherwise, something abnormal has happened. Further diagnosis and decision making are necessary.

Specific case memory: A case in the specific case memory is the detail information about a specific abnormal situation or failure. These cases are identified from the mill incident records. The system stores these cases to describe the situations of and the solutions to a number of known abnormal process events. The information captured in these cases includes:

- mechanism to calculate the similarity;
- critical rate of the event;
- last occurrence time of the event;
- frequency of the occurrence;
- root cause;
- consequences to the production;
- suggested correction actions for the event;
- predicted outcome after corrective actions.

General case memory: The cases in the specific case memory can not cover all the potential events in a real world pulp plant. To approach this problem, a general case memory is developed as the supplemental knowledge. The general case memory describes the general knowledge about the process variable interactions and control loops. For each controlled variable, a so-called control unit is defined which consists of the following elements:

1. controlled variables (measurement);
2. controller output;
3. control mode;
4. setpoint;
5. control valve;
6. sensor;
7. material supply and pump;
8. disturbances.

The information for the first four items is provided by a control data point (CTL) in MOPS. The material supply represents the control medium for the control loop, such as steam for temperature control. The disturbances represent the independent affecting variables in the control loop and the upstream process influences. If the controlled variable is off its standard limits, the problem definitely comes from the malfunction of the elements (3)-(8). A simple example of a general case is:

IF State of controlled variable $\neq 0$
AND Control Mode="Auto"
AND $|\text{Measurement - Setpoint}| < \varepsilon$
THEN Wrong setpoint setting

We define a group of important process variables as the monitoring variables. A control unit is defined for each monitoring variable. For those monitoring variables that are not the directly controlled variables of any control loops, a controlled variable is assigned to each of them as follows

$$C \xrightarrow{opt(+,-)} M \tag{1}$$

where M is the monitoring variable and C is the controlled variable. The above relation implies that M is positively (or negatively) affected by C. In this way, the system can use the knowledge about the variable C to diagnose the problem presenting in M.

The general case memory can not provide complete information about a presenting event. However, it can give important clues for operators to make quick decision about what has happened and what to do.

Assign indexes: When a new event happens, the system identifies the features from the inputs and assigns them to the event. The features provide appropriate indexes into the case memory to retrieve the best matched past case. The features of an event are the state values (that is, {-1, 1, 0}) of important quality variables and operating variables.

Retrieval: The retrieval of past cases from case memory is based on the features of the new event and the indexes of the past cases. The system first tries the specific case memory, retrieve the relevant cases and calculate the similarity of the relevant cases to the presenting event using the mechanism included in each case. The solution of the best past case is assumed to be the solution to the current situation. If no case in the specific case memory is relevant to the presenting event, the general case memory will be utilized.

Learning: Learning includes two aspects: the modification of the existing past cases and the creation of new cases in the specific case memory. The modification of the past cases includes the following activities: (1) Updating the frequency and the most recent occurrence time of the case. Whenever the occurrence of a case has been proven, the time and frequency of the occurrence of the case will be updated in the specific case memory. (2) Modifying the index rules. If a past case is wrongly accessed, the index rules for this case are considered improper and should be changed. (3) Changing the content of cases. If the solution of a past case is proven not to be exactly right, on the request of operators, the system will correct the solution to the case.

The learning for creating new cases is triggered when there is no relevant case for the current event in the specified case memory. A new case that gives the solution to the current situation should be added to the specific case memory so that in the future when the event happens again, the system can easily retrieve the solution. The creation of a new case is fulfilled by using the general case memory with the help of operators. The system first finds a controlled variable that is related to the event according to the state values of the monitoring variables, finds all the information about the controlled variable, i.e., the eight elements of the control unit as defined previously, and then retrieves a case from the general case memory. Provided with the information from the general case memory, the operator will be able to find a more detail explanation of and solution to the current event. After the solution is tested and finalized, it will be stored as a new case in the specific case memory.

System Implementation

We apply an object-oriented intelligent system building tool, Meta-COOP, which is developed in the Intelligence Engineering Laboratory at the University of Alberta, to implement the CBR intelligent system. The standard activities of CBR are implemented with the object oriented tool in several meta level modules: a meta level knowledge model for data preprocessing, a meta level knowledge model for fault detection, a meta level knowledge model for case retrieval and learning. The general knowledge about control units is installed in a meta level general case memory. The two domain specific knowledge bases, i.e., the specific case memory and the knowledge base about the configuration of the control loops and the process variable interactions, make each process-specific system unique. These two domain specific knowledge bases are installed in ASCII files. The mill engineers can easily implement new applications by modifying the ASCII files.

An operator interface is provided for operators to participate in the decision making in fault detection and diagnosis. The possible causes of the situation are listed in the display and deleted one by one as they are ruled out. The interface presents the historical trend curves, which can be expanded, zoomed and shifted, of the monitoring variables and the system's judgment. An operators can agree or disagree the system's judgment. He can also choose "quick decision" to allow the system to make its decision without his approval.

For each past case in the specific case memory, an index (features) is defined. The system compares the current event features with the indexes of the past cases. The cases whose features are matched will be accessed and the similarity are calculated using the mechanism included in each case. The priority of the retrieved cases are ordered according to the values of the similarity, critical rate, last occurring time and the frequency. The operator interface provides the detail information about all the retrieved cases (Fig. 2).

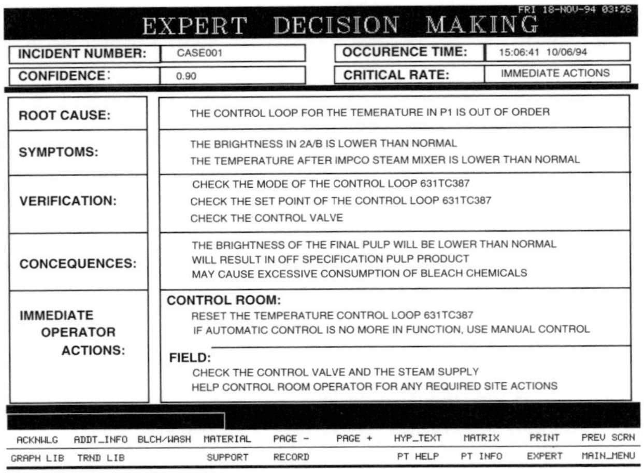

Figure 2. Operator interface for solution.

The intelligent system has a mechanism to record the time and confidence (similarity) of the last 20 occurrences of all cases. An operator interface enable operator to scroll through these records when necessary.

Conclusions

The case-based reasoning system presented in this paper has many advantages. Its reasoning mechanism is consistent with the problem solving operators do in process operation. Its self learning capability makes the system keep improving. Another important advantage is that mill engineers can easily implement new applications without knowing the detail mechanism of the intelligent system.

References

Alterman, R. (1988). Adaptive planning. *Cognitive Science*, **12**, 393-422.

Bareiss, E. R. (1989). *Exemplar-Based Knowledge Acquisition: A Unified Approach to Concept Representation, Classification, and Learning.* Academic, Boston.

Goel, A. and B. Chandrasekaran (1989). Use of device models in adaptation of design cases. *Proc. DARPA Workshop on Case-Based Reasoning*, Volume 2. pp. 100-109.

Hammond, K. (1989). *Case-Based Planning: Case-Based planning: Viewing Planning as a Memory Task.* Academic, New York.

Kolodner, J. L. (1991). Improving human decision making through case-based decision aiding. *AI Magazine*, Summer, 52-68.

Xia, Q., H. Farzadeh, M. Rao, C. Henriksson, K. Danielson and J. Olofsson (1993). Intelligent operation support system for Slave Lake Pulp Corporation. *Proc. IEEE Conf. on Control Application.* Vancouver. pp. 591-596.

MODEL LEARNING IN MPC: AN INTELLIGENT CONTROL FORMULATION

Alexandros Koulouris and George Stephanopoulos
Laboratory for Intelligent Systems in Process Engineering
Massachusetts Institute of Technology
Cambridge, MA 02139

Abstract

Model Predictive Control reinforced with on-line model learning is presented as an example of intelligent control. In the presence of generic unstructured process uncertainty, the performance and stability needs of the closed-loop can be better served when model adaptation is performed locally in a multi-scale hierarchy. This is satisfied by the Wave-Net identification algorithm that constructs an expanding family of process models with increasing accuracy but variable stability characteristics with respect to the closed-loop. By treating the model as a tuning parameter taking values in the learned family of models, stability of the system can be safeguarded without the need to compromise performance.

Keywords

Learning, Model predictive control, Wave-Net.

Introduction

During the past few years, the Artificial Intelligence (AI) community has become increasingly aware of the importance of learning as an essential element of intelligence. The resurgence of learning is bound to bring some new fertile ideas in the area of control where intelligence has so far been recognized to only non-algorithmic representations of the control law. The modern viewpoint on AI forces us to recognize intelligence even in the more traditional mathematical controllers as long as they possess the capability to learn.

Learning can be defined as the ability of the controller to autonomously improve its performance. Although various learning formulations are possible, *model* learning is of particular importance considering the central role that the process model plays in control, especially in the framework of Model Predictive Control (MPC). The use of nonparametric neural network (NN) maps has been proposed in literature (Narendra and Parthasarathy, 1990; Hernandez and Arkun, 1992) as a way to represent nonlinear plant dynamics. Although NNs hold the promise of on-line model learning, very few studies have been reported where their on-line adaptability is exploited.

In this paper, MPC equipped with model learning is proposed as an example of learning (and therefore, intelligent) control formulation. The incorporation of on-line model learning is motivated by the presence of large process uncertainties stemming from the complexity of the chemical plants. With the plants represented by empirical NN maps, model uncertainty is necessarily unstructured and non-parametric. This is contrary to the established adaptive control practice where uncertainty is usually associated with a fixed number of unknown parameters. In adaptive control, model adaptation is performed continuously in time and globally in the model space. In this paper, we will advocate the idea that model learning in an unstructured uncertainty environment has to be discrete in time, local and exhibit scale multiplicity. Model learning is therefore better served by the L^{∞}-based Wave-Net (Koulouris, Bakshi and Stephanopoulos, 1995) which supports the multiresolution character of learning.

In the presence of unstructured uncertainty it is impossible to develop adaptive laws which can a priori guarantee the stability of the closed-loop, unless one imposes severe restrictions on the expected structure of the model derivative. In this paper, our interest will lie in revealing the role of the model and its uncertainty in the stability and performance of MPC, and formulating model learning accordingly.

The Wave-Net Learning Algorithm

Let the following difference equation model represent a nonlinear SISO process:

$$y(t+1) = f(y(t), \cdots, y(t+1-n_y), u(t), \cdots, u(t+1-n_u)) \quad (1)$$

A Wave-Net (Koulouris, Bakshi and Stephanopoulos, 1995) model of this process is recursively given as a hierarchical ladder of approximations, $\hat{\mathbf{f}}_{\mathbf{j}}$, according to:

$$\hat{f}_j(\mathbf{x}) = \hat{f}_{j-1}(\mathbf{x}) + \sum_{\mathbf{k}} c_{j\mathbf{k}} \psi_{j\mathbf{k}}(\mathbf{x}) \quad (2)$$

where, $\mathbf{x}$, is the vector of past input-output values in Eqn. (1), and, $\psi_{j\mathbf{k}}(\mathbf{x})$, is a wavelet function at scale j and position $\mathbf{k}$. The use of the localized wavelet functions results in a multiresolution decomposition of the input space of the model, as shown in Fig. 1 for a first-order process. The square regions in Fig. 1 correspond approximately to the support of the corresponding wavelets covering the input space in different scales and translations.

For every data point available during on-line process operation, model learning in the wavelet multiresolution hierarchy proceeds by the following steps:

Step 1: Update the Wave-Net coefficients by solving the minimax optimization problem:

$$\min_{\mathbf{c}} \max_{i=1,\dots,n} \left| y_i - \sum_j \sum_{\mathbf{k}} c_{j\mathbf{k}} \psi_{j\mathbf{k}}(\mathbf{x}_i) \right| \quad (3)$$

Step 2: Calculate the worst-case (l^∞) model error on the data.

Step 3: If the error is greater than a predefined error threshold, ε, introduce locally into the Wave-Net the coarsest scale wavelet whose support covers the mispredicted data point. Repeat all steps until error threshold is satisfied.

Compared with other NNs, the distinguishing characteristics of the Wave-Net learning mechanism are:

1. The pointwise presentation of the empirical data to simulate real-time implementations.
2. The use of an L^∞-based error threshold as a criterion for model adaptation.
3. The localized structural adaptability of the Wave-Net model.

In the multiresolution structure of the model, worst-case local error bounds can be estimated as:

$$b_{j\mathbf{k}} = \max_{i=1,\dots,n} (abs(y_i - \hat{f}(\mathbf{x}_i))) \quad (4)$$

where the error is calculated over all data covered by the corresponding wavelet at scale, j, and position, $\mathbf{k}$.

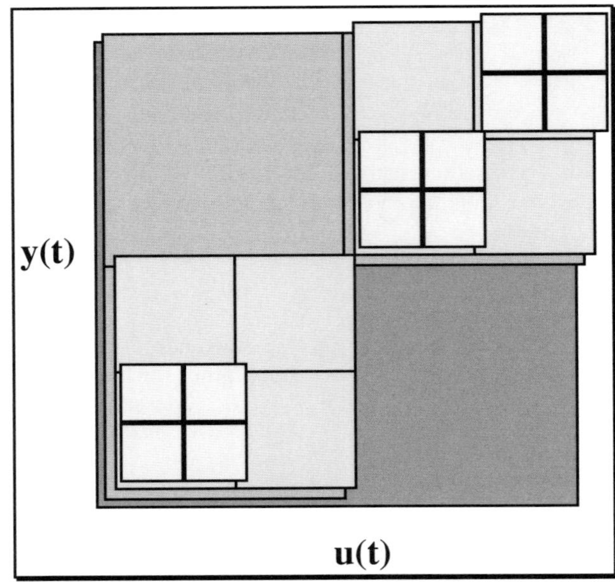

Figure 1. Input space multiresolution decomposition for a first-order model.

Nonlinear MPC with Wave-Net Models

MPC with the use of NN models can be formulated as a natural extension of the linear MPC. The controller is given as the solution to the following optimization problem:

$$\min_{\mathbf{u}} \frac{1}{2} \sum_{i=1}^{P} \{\gamma_i (y_{sp}(t+i) - \hat{y}(t+i) - e(t+i))^2 + \lambda_i \Delta u(t+i)^2\} \quad (5)$$

where the output predictions, $\hat{\mathbf{y}}$, are given by the Wave-Net model. Using on-line process data, model adaptation can be formulated following the Wave-Net learning algorithm. In the resulting hierarchical family of models, each member exhibits different stability and performance characteristics with respect to the closed-loop. In this framework, the model can be considered as a *tuning parameter* taking values within the hierarchy of updated models. In the following sections we will attempt to reveal the properties of the model(s) that result in stable and good set-point tracking performance.

Model Learning: Performance Considerations

Improved and more demanding control performance is possible with the help of an increasingly accurate model. Learning can undertake the task of improving the model accuracy, especially when the process operation shifts to previously unexplored model spaces. Such a case is when the set-point exhibits time-variations in order, for example, to accommodate variable product specifications. When such shifts occur, the existing model might be incapable of capturing the plant dynamics around the new steady-state.

When only parametric uncertainty is considered, model adaptation is global. This lack of localization is responsible for the "memoryless" performance of adaptive controllers that have to relearn the same parameters every time the operation shifts to a new regime. The localized structural adaptation of the Wave-Net learning mechanism can accommodate the need for preserving the existing model knowledge, as shown in the following example.

Example 1

Let the plant be given by the following discrete-time equation:

$$y(t+1) = f(y(t)) + u(t) = -0.8\frac{y^3(t)}{1+y^2(t)} + u(t) \qquad (6)$$

A Wave-Net model of this plant is constructed from historical data. The model which is of the form:

$$\hat{y}(t+1) = \hat{f}(y(t)) + u(t) \qquad (7)$$

is shown by the solid line in Fig. 2 in comparison to the

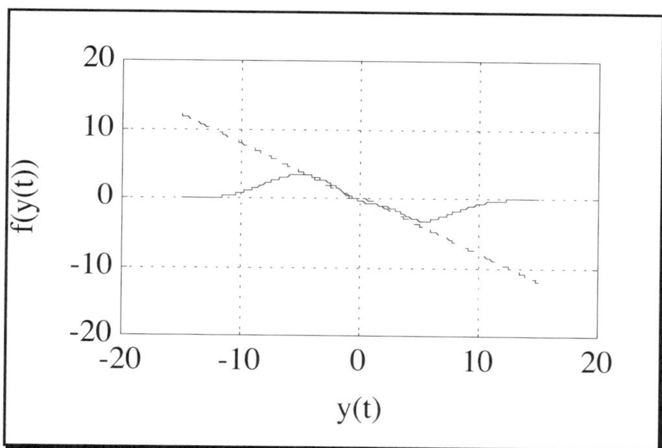

Figure 2. The plant vs. the model of example 1.

true plant nonlinearity (dashed line). Outside the training region (-5,5), the model settles to zero and, consequently, the modeling error can take arbitrarily large values. If we implement a deadbeat MPC controller of the form:

$$u(t) = r - (y(t) - \hat{y}(t)) - \hat{f}(y(t)) \qquad (8)$$

and we set the set-point, r, at 10, the response of the closed-loop is unstable due to the large modeling error around the new set-point. If, however, we allow on-line adaptation of the process model following the Wave-Net learning algorithm, the closed-loop is stabilized as a result of the model's learning of the process dynamics in the new operation range. The model improvement can be seen by comparing the resulting, after learning, model (Fig. 3) with the initial model in Fig. 2. In Fig. 3, we can recognize the benefits of localized adaptation that is able to increase the reliability of the model without sacrificing its accuracy in the original operating range.

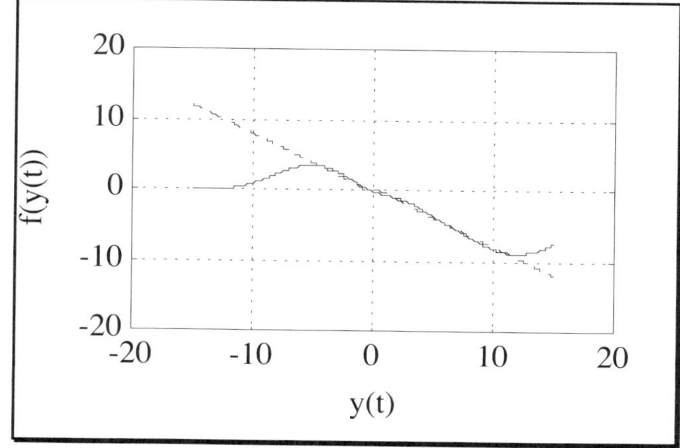

Figure 3. The updated model of example 1.

Model Learning: Stability Considerations

It was recognized by Skelton (1989) for linear systems, and easily verified for nonlinear systems as well, that closed-loop instability can occur even when the modeling error is arbitrarily small. On the one hand, this observation points to the fact that the ability of NNs to present arbitrarily well the behavior of chemical processes is not a guarantee for stable closed-loop performance. On the other hand, it suggests that other properties of the model, such as its derivative, are more straightforward to relate to stability. Figure 4 (Koulouris, 1995) presents the closed-loop stability condition for a first-order system with a deadbeat MPC controller. The set-point values, r, for which the model derivative (solid line) is within the bounds (dashed lines) around the real plant derivative, correspond to stable closed-loop operation. It can be concluded that stability is a "local" property, and also that the model derivative is the quantity of interest. In order to avoid deterioration of the local stability characteristics of the closed-loop, on-line model adaptation should be local, as it is the case with the Wave-Net learning mechanism. The Wave-Net model allows explicit associations between its structure and the model derivative to be made. This is because fine wavelets recruited during structural adaptation amplify the derivative content of the model since the derivative of the wavelets increases exponentially with their scale. Worst-case estimates of the derivative uncertainty of the model can also be obtained from the local modeling bounds under the assumption that the real plant function does not contain local features of finer scale in the corresponding region of the model space.

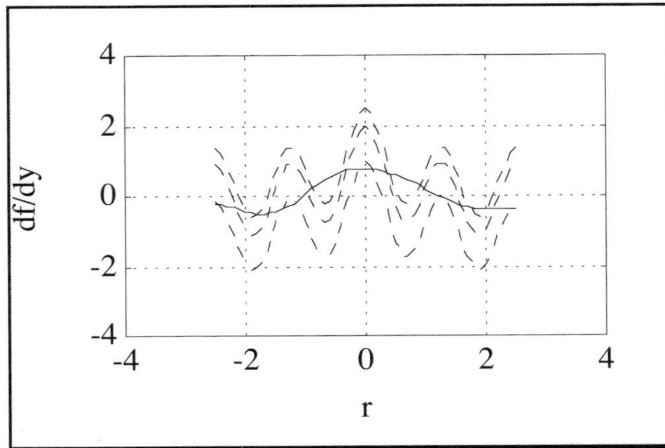

Figure 4. The stability condition for a simple closed-loop system.

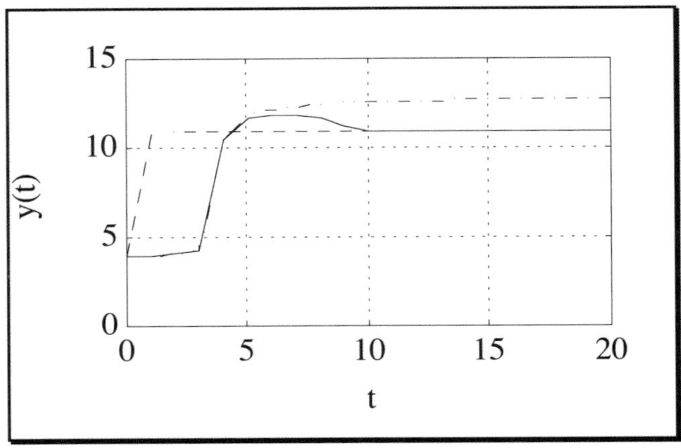

Figure 5. The closed-loop response of example 2.

For a feature of scale, j, and error bound, b_{jk}, local margins of the plant derivative values can be estimated as:

$$\left[\hat{f}(\mathbf{x}) - \frac{b_{j\mathbf{k}}}{2^{-j-1}}, \hat{f}(\mathbf{x}) + \frac{b_{j\mathbf{k}}}{2^{-j-1}}\right] \tag{9}$$

Example 2

The following pH control example is adopted from Li and Biegler (1990). The open-loop process can be represented by a first-order difference equation model:

$$pH(t) = f(pH(t\text{-}1), u(t\text{-}1)) \tag{10}$$

where the pH of the solution and the flow rate of the base, u(t), are the only on-line measured quantities. A Wave-Net model is constructed and an MPC controller with prediction horizon P=2 is developed with the task of bringing the system from pH = 4 to pH = 11. A hard velocity constraint of $\delta u_{max}=0.25$ is implemented to allow smooth transition through the fast changing dynamic region around pH = 7. With these settings, the controller based on an off-line constructed Wave-Net model fails to stabilize the system at pH = 11 (dash-dot line in Fig. 5). While "traveling," however, along the path towards the new steady-state, the system acquires useful information captured in a higher resolution model. With the learned model, stabilization of the closed-loop at the desired equilibrium point can be achieved (solid line in Fig. 5). In view of the previous discussion on stability, it can be concluded that the model through localized multiscale adaptation has learned the "appropriate" derivative values along the path.

Conclusions

By locally adapting the model in the presence of unstructured, nonparametric uncertainty, the closed-loop can enjoy the benefits of both its newly acquired and the old knowledge. Closed-loop stability, which seems to be weakly related to model accuracy, can be served by a multiresolution family of Wave-Net models where it is straightforward to relate the scale of the model with its derivative content. With the model as a local tuning parameter, the MPC tuning coefficients can be exploited to boost the loop performance rather than protect stability.

The challenge emerging from this work is to develop a deeper understanding of the role of the form of the model in the stability of a closed-loop system. A controller equipped with such knowledge will be able to learn appropriate model forms and make autonomous, intelligent decisions on the model used in real-time control.

References

Hernandez, E. and Y. Arkun (1992). A study of the control relevant properties of backpropagation neural net models of nonlinear dynamical systems. *Computers Chem. Engng.*, **16**, 227.

Koulouris, A. (1995). *Multiresolution Learning in Nonlinear Dynamic Process Modeling and Control.* PhD Thesis. Massachusetts Institute of Technology.

Koulouris, A., B. R. Bakshi and G. Stephanopoulos (1995). Empirical learning through neural networks: the Wave-Net solution. In G. Stephanopoulos and C. Han (Eds.), *Advances in Chemical Engineering*, **22**. Academic Press.

Li, W. C. and L. T. Biegler (1990). Newton-type controllers for constrained nonlinear processes with uncertainty. *Ind. Eng. Chem. Res.*, **29**, 1647.

Narendra, K. S. and K. Parthasarathy (1990). Identification and control of dynamical systems using neural networks. *IEEE Trans. on Neural Networks*, **1**, 1.

Skelton, R. E. (1989). Model error concepts in control design. *Int. J. Control*, **49**, 1725.

FUZZY NEURAL NETWORK APPROACH FOR NONLINEAR PROCESS CONTROL

Atsushi Aoyama[1], Francis J. Doyle III and Venkat Venkatasubramanian[2]
Laboratory for Intelligent Process Systems
School of Chemical Engineering
Purdue University
West Lafayette, IN 47907

Abstract

This paper proposes an internal model control (IMC) strategy using a fuzzy neural network for nonlinear process control. This strategy is useful for systems where first principle-based descriptions are difficult to obtain, but partial knowledge about the process and input-output data are available. The fuzzy neural network is trained using steady-state as well as transient data by back-propagation. The inverse model of the process is obtained by a simple interval halving method. The performance of the proposed approach for the modeling and control of a pH neutralization process is presented.

Keywords

Neural networks, Fuzzy logic, Process control, Internal model control, Nonlinear control.

Introduction

In recent years, several neural network-based controllers have been proposed for nonlinear processes. Typically, the neural networks are trained to model processes and subsequently used in model-based control schemes (Narendra and Parthasarathy, 1990; Narendra and Mukhopadhyay, 1992; Lee and Park, 1992; Saint-Donat, et al., 1991; Bhat and McAvoy, 1990; Ydstie, 1990; Psichogios and Ungar, 1992; Nahas, et al., 1992). While a neural network model-based controller can be constructed by making a neural network learn the relationship between inputs and outputs, it is rather crude and unsatisfactory as it does not use a priori knowledge of the process which might be available. This paper proposes a fuzzy neural network approach where a fuzzy logic model of the knowledge of the process is integrated with neural networks in the internal model control (IMC) framework. The use of this scheme makes it possible to obtain the inverse model by a simple interval halving method, and the global existence and uniqueness of a solution is guaranteed. This way, one has directly incorporated the known aspects of the process behavior, leaving the neural network to capture the unknown aspects empirically. This strategy is useful for nonlinear systems where first principle-based descriptions are difficult to obtain, but partial knowledge about the process and input-output data are available.

Fuzzy Neural Network

Fuzzy inference systems with adaptive capability have received considerable attention in recent years. A fuzzy neural network (Kandel, 1992; Yager and Filev, 1993; Horikawa, et al., 1992; Sugeno and Yasukawa, 1993; Lin and Lee, 1994; Li-Xin, 1993; Jang, 1992; Berenji, 1992) is one of the more promising approaches in this category. Figure 1 shows the basic configuration of a fuzzy process model which is composed of three major components: a fuzzifier, a fuzzy inference engine with rule bases, and a defuzzifier. The fuzzifier performs the function of fuzzification that converts input data from the observed input space into proper linguistic values of fuzzy sets through predefined input functions. The fuzzy inference matches the output of the fuzzifier with the fuzzy logic

---

[1] Present address: Centre for Process Systems Engineering, Imperial College of Science, Technology and Medicine, London, SW7 2BY, U.K.

[2] Author to whom all correspondence should be addressed.

rules and performs the approximate reasoning. Finally, the defuzzifier performs the function of defuzzification to yield a crisp output through predefined output membership functions.

A major problem in designing a fuzzy logic model is the determination of the proper membership function and fuzzy logic rules. The fuzzy neural network system used in this work is a relatively simple scheme in which the consequence parts of the fuzzy rules are constants and the input space is partitioned using a fuzzy grid as shown in Fig. 2. In this figure, shaded strips express the condition parts of rules and the cross areas of strips express the rules.

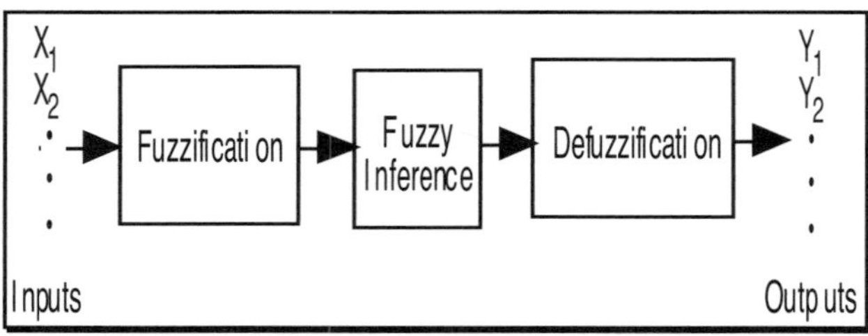

Figure 1. General structure of fuzzy logic system.

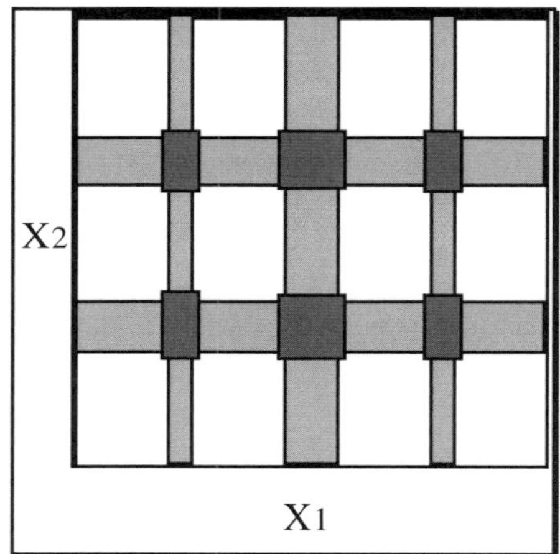

Figure 2. Fuzzy grid for input space.

Figure 3 shows the proposed fuzzy neural network structure. The system has four layers. In the following description, systems with two inputs and single output are considered, but the results are easily generalized for systems with more than two inputs and one output. The functions of the nodes in each layer is described as follows:

Layer 1: The nodes of this layer transmit input values directly to the next layer. The nodes takes the input from each corresponding dimension. The number of nodes is equal to the number of input dimensions.

$$U_i = X_i \tag{1}$$

Layer 2: This layer performs fuzzification. The nodes take the outputs of the corresponding nodes in layer 1 as the input. Each node in this layer represents the condition part of rules and the node output is equal to the degree of match for each condition. The node function of the j-th node is

$$U_j = \exp\left(-\frac{(\mu_j - U_i)^2}{\sigma_i^2}\right) \tag{2}$$

where μ_j and σ_j are, respectively, the mean and the standard deviation of the Gaussian-like function of the j-th node.

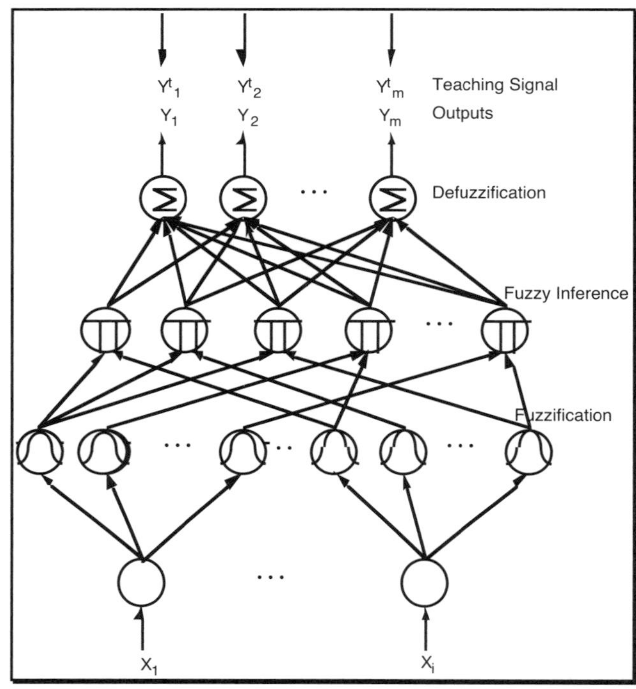

Figure 3. Fuzzy neural network structure.

Layer 3: This layer performs a fuzzy inference which is carried out by multiplication.

$$U_{kj} = U_k \times U_j \tag{3}$$

where × represents arithmetic multiplication. Each node in this layer corresponds to a rule and the node output is equal to the firing strength of each rule. The number of nodes in this layer is equal to the number of rules.

Layer 4: The nodes in this layer transmit the decision signal out of the network. These nodes act as the defuzzifier. The number of nodes in this layer is equal to the output dimension. The following function is used to simulate the center of gravity defuzzification method:

$$Y = \frac{\sum_{k=1}^{n} \sum_{j=1}^{m} \mu_{kj} U_{kj}}{\sum_{k=1}^{n} \sum_{j=1}^{m} U_{kj}} \tag{4}$$

where n and m are the number of input nodes of each dimension, respectively and μ_{kj} are the centers of the membership functions. The basic learning rule to minimize the error function is the back-propagation type gradient descent method as described in Horikawa, et al., (1992).

Process Model Using Fuzzy Neural Network

In the present work, only the class of single-input single-output, minimum-phase, nonlinear processes are considered. Also, the one-to-one relationship between the input and output in a considered operating region is assumed. Such processes can be represented by the following discrete time expression,

$$\begin{aligned} \Delta x_k &= F_1(u_k, x_k) \\ y_k &= F_2(x_k) \end{aligned} \tag{5}$$

where x_k is an n-dimensional state vector, y_k is a scalar control output, u_k is a scalar control input. F_1 and F_2 are nonlinear functions. The proposed model structure is expressed as follows and considered to be a nonlinear extension of the control-affine model proposed by the authors (Aoyama, et al., 1995)

$$\Delta y_k = f(u_k - g(\mathbf{y}, \mathbf{u}, \mathbf{x}), \mathbf{y}, \mathbf{u}, \mathbf{x}) \tag{6}$$

In this scheme, the function f is implemented by a fuzzy neural network. If no constraints are imposed on the fuzzy neural network, then the resultant model is no different than a traditional black-box model.

A novel point of the proposed approach is the imposition of constraints on the parameters of the fuzzy neural network. A major advantage comes from the fact that the function g represents the steady state gain. In order to design the function g as an accurate approximation of the steady state gain, the following rules are necessary. The following example is for the case where the gain is positive. When the gain is negative, the unequal relationships in the consequence part of the rules are inverted.

Rule 1:

IF $\{u_k - g(\mathbf{y}, \mathbf{u}, \mathbf{x}) = 0\}$ THEN $\{\Delta y_k = 0\}$

Rule 2:

IF $\{(u_k - g(\mathbf{y}, \mathbf{u}, \mathbf{x}))_1 \geq (u_k - g(\mathbf{y}, \mathbf{u}, \mathbf{x}))_2\}$
THEN $\{(\Delta y_k)_1 \geq (\Delta y_k)_2\}$

Rule 1 comes from the definition of the steady state and Rule 2 represents the monotonic relationship between the controlled output and the manipulated input. The corresponding constraints imposed on the parameters of the fuzzy neural network are:

Constraint 1:

$\mu_{ki} = 0$ for all k where $\mu_i = 0$

Constraint 2:

$\mu_{ki} \geq \mu_{ki}$ for all k where $\mu_i \geq \mu_i$

The function g approximates the steady state gain because of the above constraints. In order to train fuzzy neural networks, the errors are back-propagated to both f and g, g is also trained by using steady state data simultaneously. In every training iteration, if updated parameter values violate the constraints, those values are corrected to keep the constraints. For example, if $\mu_{ki} < \mu_{kj}$ happens for $\mu_i \geq \mu_j$ then μ_{ki} is increased to be equal to μ_{kj}.

In order to carry out the control action, the value of u_k satisfying the following relationship has to be calculated.

$$y_{SD} - y_k = f(u_k - g(\mathbf{y}, \mathbf{u}, \mathbf{x}), \mathbf{y}, \mathbf{u}, \mathbf{x}) \tag{7}$$

The monotonic relationship expressed by Rule 2 between the controlled output and the manipulated input guarantees the unique solution u_k which is obtained by a simple interval halving method (Carnahan, et al., 1969).

Fuzzy Neural Network Model-Based IMC

Figure 4 shows the schematic of the internal model control scheme based on the fuzzy neural network. In this work, the process model is given by the fuzzy neural network model described in the previous section and the nonlinear controller is the inverse of the model obtained by the internal halving method. The existence of dead time is not assumed here, but the proposed scheme can be combined with a Smith predictor-type dead time compensation scheme (Nahas, et al., 1992).

Because the controller inverts the model exactly, steady state offset is eliminated. To ensure the stability, both a setpoint filter and the usual IMC filter are added to the control loop. Both filters are implemented as first order digital filters. The parameters of the filters are tuned to provide a suitable compromise between performance and robustness.

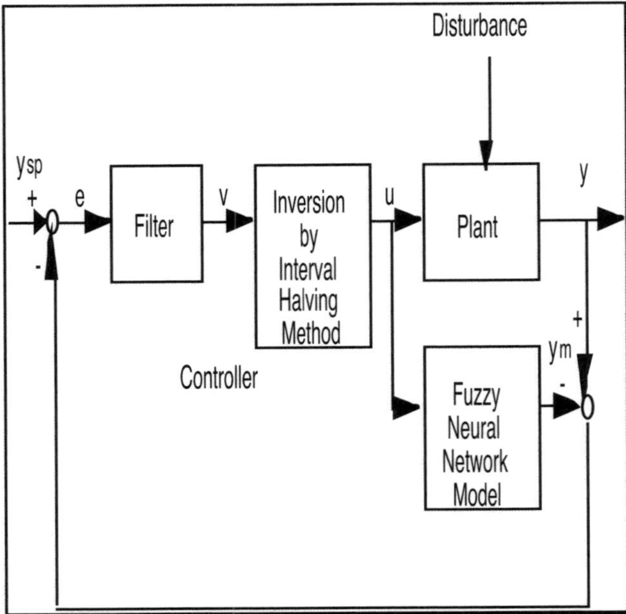

Figure 4. Fuzzy neural networks model based IMC.

This IMC strategy was tested on a simulated pH neutralization process shown in Fig. 5. This problem has been studied by Nahas, et al., (1992) earlier. The objective is to control the pH in the tank by manipulating the base flow rate. Open loop responses for ± 10% step changes in q3 are shown in Fig. 6. The process exhibits highly nonlinear dynamics where the process gain varies by more than 250% for these small input changes. The sampling time is chosen as 0.25 min. The relative degree of this system is calculated as one, and therefore the previous control output value pH is chosen as the sole input to the function g. In addition to this input, q3-g is an input to the function f. A fuzzy neural network modeling the function f has 55 rules where the q3-g input space is partitioned to 5 regions and the pH input space is partitioned into 11 regions. A fuzzy neural network modeling the steady state gain has 10 rule nodes. The identification required approximately 500 data points representing 25 steady states and 600 transient data points. Figure 7 shows the process model for various conditions captured by a fuzzy neural network. As intended, the output increment of pH is close to 0 when the input, q3-g, is 0, and the relationship between the increment of pH and q3-g is monotonic.

The performance of the fuzzy neural network model based IMC is compared to that of a PID controller. Two controllers are compared for a step change of from 7 to 4 and from 7 to 9 occurring at t=0. Set-point tracking performance of controllers are illustrated in Fig. 8. The control-affine IMC yields a fast response with a little overshoot for both changes. In the contrast, the PID controller is very sluggish when the setpoint is decreased.

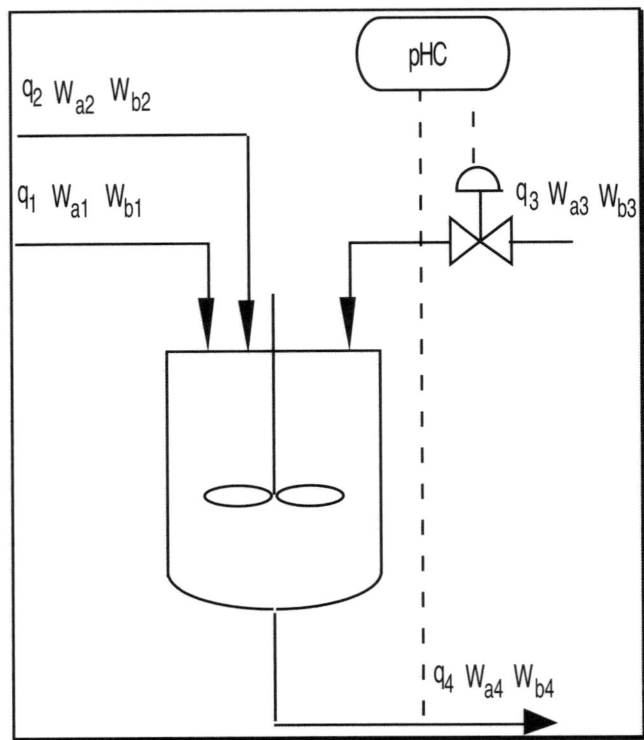

Figure 5. pH neutralization system.

Conclusions

A fuzzy neural network applicable for nonlinear process modeling and control is proposed. It has the ability to include a priori knowledge as rules. A novel modeling scheme is proposed using this fuzzy neural network. The inverse of the process model is obtained by a simple interval halving method to ensure zero steady state offset performance. An internal model control strategy using this modeling scheme was proposed for the SISO nonlinear process control and applied to the modeling and control a pH neutralization process which exhibits highly nonlinear behavior. The results show significantly better performance than that of a PID controller.

References

Aoyama, A., et al. (1995). Control-affine fuzzy neural network approach for nonlinear process control. *J. Proc. Control*, in press.

Berenji, H. (1992). Learning and tuning fuzzy logic controllers through reinforcements. *IEEE Trans. on Neural Networks*, **3**, 724-740.

Bhat, N. and T. J. McAvoy (1990). Use of neural nets for dynamic modeling and control of chemical process systems. *Computers Chem. Engng.*, **14**, 573-583.

Carnahan, B., et al. (1969). *Applied Numerical Methods*. John Wiley and Sons.

Horikawa, S., et al. (1992). On fuzzy modeling using fuzzy neural networks with the back-propagation algorithm. *IEEE Trans. on Neural Networks*, **3**, 801-806.

Jang, J.R. (1992). Self-learning controllers based on temporal back propagation. *IEEE Trans. on Neural Networks*, **3**, 714-723.

Kandel, A. (1992). *Fuzzy Expert Systems*. CRC Press, Boca Raton, FL.

Lee, M. and S. Park (1992). A new scheme combining neural feedforward control with model predictive control. *AIChE J.*, **38**, 193-200.

Lin, C-T. and C.S.G. Lee (1994). Reinforcement structure/parameter learning for neural-network-based fuzzy logic control systems. *IEEE Trans. on Fuzzy Systems*, **2**, 46-63.

Li-Xin, W. (1993). Stable adaptive fuzzy control of nonlinear systems. *IEEE Trans. on Fuzzy Systems*, **1**, 146-149.

Nahas, E.P., et al. (1992). Nonlinear internal model control strategy for neural network models. *Computers Chem. Engng.*, **16**, 1039-1057.

Narendra, K. and K. Parthasarathy (1990). Identification and control of dynamical systems using neural networks. *IEEE Transactions on Neural Networks,* **1**, 4-27.

Narendra, K. and S. Mukhopadhyay (1992). Identification and control of dynamical systems using neural networks. *IEEE Control Systems,* April, 11-18.

Psichogios, D.C. and L.H. Ungar (1992). Direct and indirect model based control using artificial neural networks. *Ind. Eng. Chem. Res.*, **30**, 2564-2573.

Saint-Donat, J., et al. (1991). Neural net based model predictive control. *Int. J. Control*, **54**, 1453-1468.

Sugeno, M. and T. Yasukawa (1993). A fuzzy-logic-based approach to qualitative modeling. *IEEE Trans. on Fuzzy Systems*, **1**, 7-32.

Yager, R.R. and D.P. Filev (1993). SLIDE: a simple adaptive defuzzification method. *IEEE Trans. on Fuzzy Systems*, **1**, 69-78.

Ydstie, B.E. (1990). Forecasting and control using adaptive connectionist networks. *Computers Chem. Engng.*, **14**, 583-599.

DESIGN OF A FUZZY LOGIC CONTROLLER WITH SHRINKING-SPAN MEMBERSHIP FUNCTIONS

Cheng-Liang Chen[1] and Chung-Tyan Hsieh
Department of Chemical Engineering
National Taiwan University
Taipei 10617, Taiwan
R.O.C.

Abstract

A new means for designating membership functions in a fuzzy logic controller is presented in this article. This method constructs a set of membership functions systematically by using only two parameters: *number of fuzzy subsets* and *shrinking factors*. The membership functions generated by this method, *Shrinking-Span Membership Functions* (SSMFs) have different spans in the universe of discourse and, therefore, are more rational and more practical from the human expert's point of view. The fuzzy logic controller equipped with such membership functions can be implemented with little difficulty. In addition, the satisfactory performance of such a controller can be acquired without laborious optimization calculations of the tuning parameters.

Keywords

Fuzzy sets, Shrinking-span membership functions, Fuzzy logic control.

Introduction

People very often make decisions in their daily lives based on qualitative information. Zadeh's *fuzzy sets* (Zadeh, 1965) theory was thus proposed to enable people to describe and formulate the linguistic mental models apparent in daily life behaviour.

Mamdani and his coworkers (Mamdani, 1974; Mamdani and Assilian, 1975) were pioneers in applying fuzzy techniques to process control. Their results, as well as those of many other researchers, have demonstrated the potential value of the fuzzy logic control system on simple process dynamics. A comprehensive review of the classical design and implementation of the fuzzy logic controller (FLC) can be found in Lee (1990). More advanced design techniques have also been reported in the literature, such as the adaptive hierarchical fuzzy controller (Raju and Zhou, 1993) and the fuzzy logic controller with multiple inputs (Wong, et al., 1993).

In this article, a novel approach for allocating membership functions is proposed. By using the new method, the designer of a FLC assigns only the *number of subsets* and the *shrinking factor* for a specific fuzzy variable and the obtains well-located membership functions. Such *Shrinking-Span Membership Functions* (SSMFs) not only make themselves more reasonable to the eyes of human operators but also provide the FLC the ability to easily adapt to different control systems by slight modification.

Basic Design Procedures of the FLC

Unlike a regular analytic controller, a fuzzy logic controller is comprised of four principal components (Lee, 1990): (1) a *fuzzification interface*, (2) a *rule base*, (3) *decision-making logic*, and (4) a *defuzzification interface*.

For a fuzzy logic controller treating multiple input signals and giving single output control action, the input linguistic variables $\tilde{x}_i$'s and the output $\tilde{y}$ are respectively collections of a various number of linguistic values (in

[1] Author to whom all correspondence should be addressed.

numerical sequences $A_{(i,\ell_i)}$'s and B_ℓ's) and each of these linguistic values is a fuzzy set,

$$\tilde{x}_i = \left\{ A_{(i,\ell_i)} \equiv \left\{ \left(x_i, A_{(i,\ell_i)}(x_i) \right) \middle| x_i \in X_i \right\} \middle| l_i \in I_m \right\} i \in N_n \equiv \{1,\ldots,n\} \tag{1}$$

and

$$\tilde{y} = \left\{ B_\ell \equiv \left\{ (y, B_\ell(y)) \middle| y \in Y \right\} \middle| \ell \in I_{m_y} \right\} \tag{2}$$

here, $I_{m_i} = \{-m_i,\cdots,-1,0,1,\cdots,m_i\}$ is the index set with $2m_i + 1$ terms for linguistic variable $\tilde{x}_i$ and I_{m_y} is the index set with $2m_y + 1$ terms for linguistic variable $\tilde{y}$; $A_{(i,\ell_i)}$'s and B_ℓ's are the linguistic values, and $A_{(i,\ell_i)}(x_i)$'s and $B_\ell(y)$'s are membership functions for $\tilde{x}_i$ and $\tilde{y}$, respectively. X_i's and Y are universes of discourse.

The control rule base can be represented by $\Re(\ell_1,\ldots,\ell_n;\ell)$

$$\begin{aligned} &\text{IF} \quad \tilde{x}_1 \text{ is } A_{(1,\ell_1)} \text{ AND} \cdots \text{AND } \tilde{x}_n \text{ is } A_{(n,\ell_n)} \\ &\text{THEN} \quad \tilde{y} \text{ is } B_{\ell = f(\ell_1,\ldots,\ell_n)} \\ &\qquad \forall \ell_i \in I_{m_i}, \ell = f(\ell_1,\ldots,\ell_n) \in I_{m_y} \end{aligned} \tag{3}$$

Note that if $f(\ell_1,\ldots,\ell_n) = \ell_1 + \cdots + \ell_n, \forall \ell_i \in I_{m_i}, i \in N_n$ then it is the simple rule mapping (Wong, et al., 1993).

Given any $x_i \in X_i$ the membership value of the consequent $B_{f(\ell_1,\ldots,\ell_n)}$ for the specific rule $\Re(\ell_1,\ldots,\ell_n;\ell)$ can be obtained by performing logical operations on the membership values of the antecedents of each rule, $A_{(i,\ell_i)}(x_i)$'s, i.e.,

$$B_{f(\ell_1,\ldots,\ell_n)}(y) = \mathcal{T}\left(A_{(i,\ell_i)}(x_i) \middle| x_i \in X_i, \ell_i \in I_{m_i}, i \in N_n \right) \tag{4}$$

where $\mathcal{T}$ is a t-norm operator.

Finally, the most widely used COA strategy is adopted here to carry out the defuzzification.

If the crisp inputs x_i's are given, the final crisp control action y for the whole rule base can be expressed as

$$y = \frac{\sum_{\forall \ell_i \in I_{m_i}} c\left(B_{\ell = f(\ell_1 \ldots, \ell_n)}(y) \right) \cdot \mathcal{T}\left(A_{(i,\ell_i)}(x_i) \right)}{\sum_{\forall \ell_i \in I_{m_i}} \mathcal{T}\left(A_{(i,\ell_i)}(x_i) \right)} \tag{5}$$

where $x_i \in X_i$, $i \in N_n$, $c(B_\ell(y))$ denotes the centroid value of the membership function $B_\ell(y)$.

The Design of an FLC with Shrinking-Span Membership Functions

To accomplish a better performance and to devise a more rational FLC, *Shrinking-Span Membership Functions* are proposed.

The most notable distinction between an ordinary FLC and an FLC with SSMFs (SSMFs-FLC for short) lays in the designation of the membership functions.

For any triangular membership function, $A_{(i,\ell_i)}(x_i)$, the principal value (or geometric center), $A^*{}_{(i,\ell_i)}$, left span, $S^L{}_{(i,\ell_i)}$, and right span, $S^R{}_{(i,\ell_i)}$, are the only elements needed to define the function. With these values, the membership functions $A_{(i,\ell_i)}(x_i)$'s can be expressed as

$$A_{(i,\ell_i)}(x_i) = \begin{cases} 1 - \dfrac{A^*{}_{(i,\ell_i)} - x_i}{S^L{}_{(i,\ell_i)}} & \text{for } x_i \in \left(A^*{}_{(i,\ell_i)} - S^L{}_{(i,\ell_i)}, A^*{}_{(i,\ell_i)} \right) \\ 1 - \dfrac{x_i - A^*{}_{(i,\ell_i)}}{S^R{}_{(i,\ell_i)}} & \text{for } x_i \in \left(A^*{}_{(i,\ell_i)}, A^*{}_{(i,\ell_i)} + S^R{}_{(i,\ell_i)} \right) \\ 0 & \text{otherwise} \end{cases} \tag{6}$$

The principal values $A^*{}_{(i,\ell_i)}$'s of the SSMFs-FLC are defined by

$$A^*{}_{(i,\ell_i)} = \frac{\ell_i}{m} s_i^{m-|\ell_i|}, \quad \ell_i \in I_m \tag{7}$$

where s_i is the *shrinking factor* for linguistic variable $\tilde{x}_i$. By applying various shrinking factors to the same linguistic variable, different membership functions can be obtained to examine which is the most suitable for a specific application process. In the meantime, the support of a membership function, namely $\left(A^*{}_{(i,\ell_i)} - S^L{}_{(i,\ell_i)}, A^*{}_{(i,\ell_i)} + S^R{}_{(i,\ell_i)} \right)$, can be elicited from Eqn. 7.

The left span, $S^L{}_{(i,\ell_i)}$, and right span, $S^R{}_{(i,\ell_i)}$, of the SSMFs-FLC can be acquired by

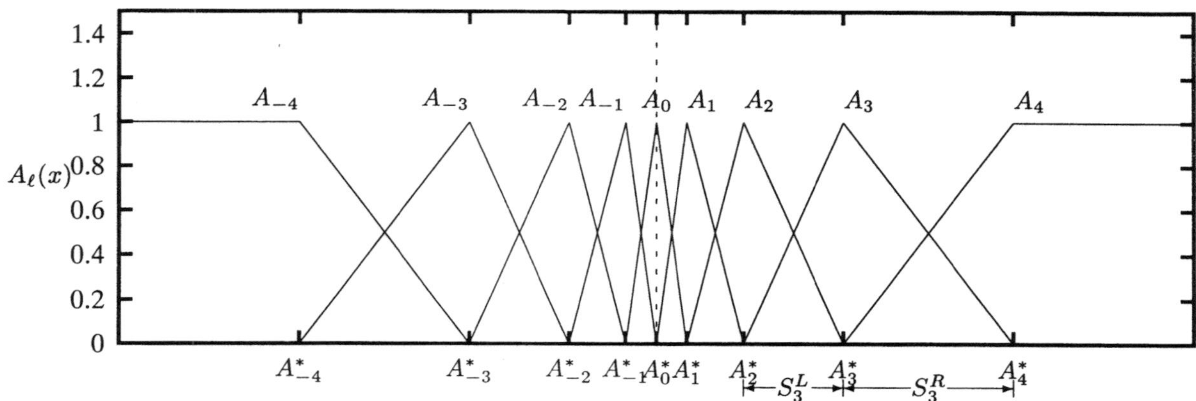

Figure 1. Typical shrinking-span triangular membership functions for m = 4, s = 0.7.

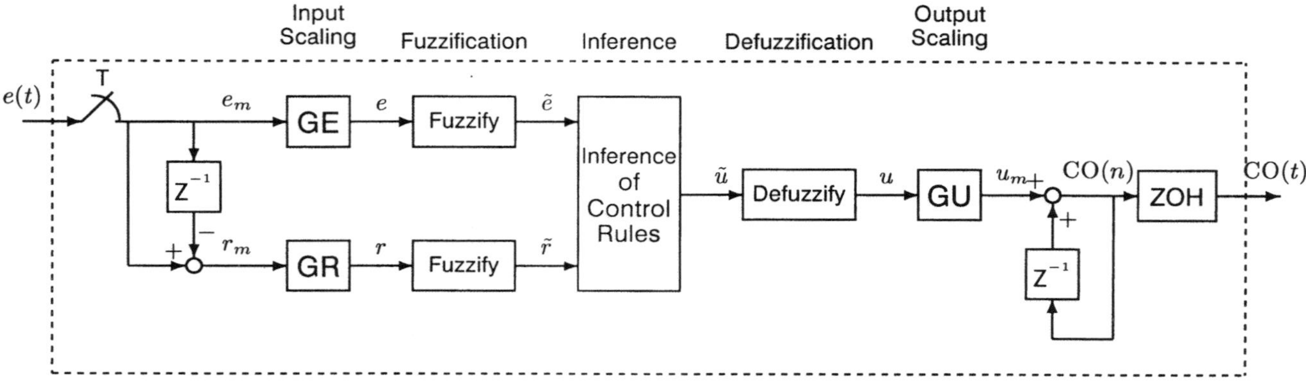

Figure 2. A typical fuzzy feedback control system.

$$S^L{}_{(i,\ell_i)} = \begin{cases} A^*{}_{(i,\ell_i)} - A^*{}_{(i,\ell_i-1)} \\ \quad \text{for } \ell_i \in I_m - \{-m\} \\ \infty \\ \quad \text{for } \ell_i = -m \end{cases} \tag{8}$$

and

$$S^R{}_{(i,\ell_i)} = \begin{cases} A^*{}_{(i,\ell_i+1)} - A^*{}_{(i,\ell_i)} \\ \quad \text{for } \ell_i \in I_m - \{m\} \\ \infty \\ \quad \text{for } \ell_i = m \end{cases} \tag{9}$$

A set of typical membership functions in the SSMFs-FLC defined by Eqns. 7 through 9 is shown in Fig. 1.

One should note that when shrinking factors s_i equals 1, the shrinking-span membership function $A_{(i,\ell_i)}$ is degraded to ordinary equal-span membership function. Therefore, the conventional equal-span membership function is a special case of the shrinking-span membership function.

The definitions of both the left and right spans as above have the following advantages. They reduce the design parameters of a fuzzy logic controller. Conventionally, one who designs a fuzzy logic controller must decide the principal values and spans of various triangular membership functions of input and output linguistic variables. This often ends with arbitrary assignment of parameters due to the lack of thorough domain knowledge about the process. With the definitions given above, however, one only needs to specify the number of linguistic values, m, and numerical value of the shrinking factor, s_i. This dramatically reduces the effort in obtaining reasonable membership functions.

Simulation Results of the SSMFs-FLC

In the previous section, the SSMFs-FLC is established as a variant of the fuzzy logic controller, and it is more reasonable in designating the membership functions and is easier to implement. In this section, the SSMFs-FLC is applied to a pendulum-car system to examine its performance.

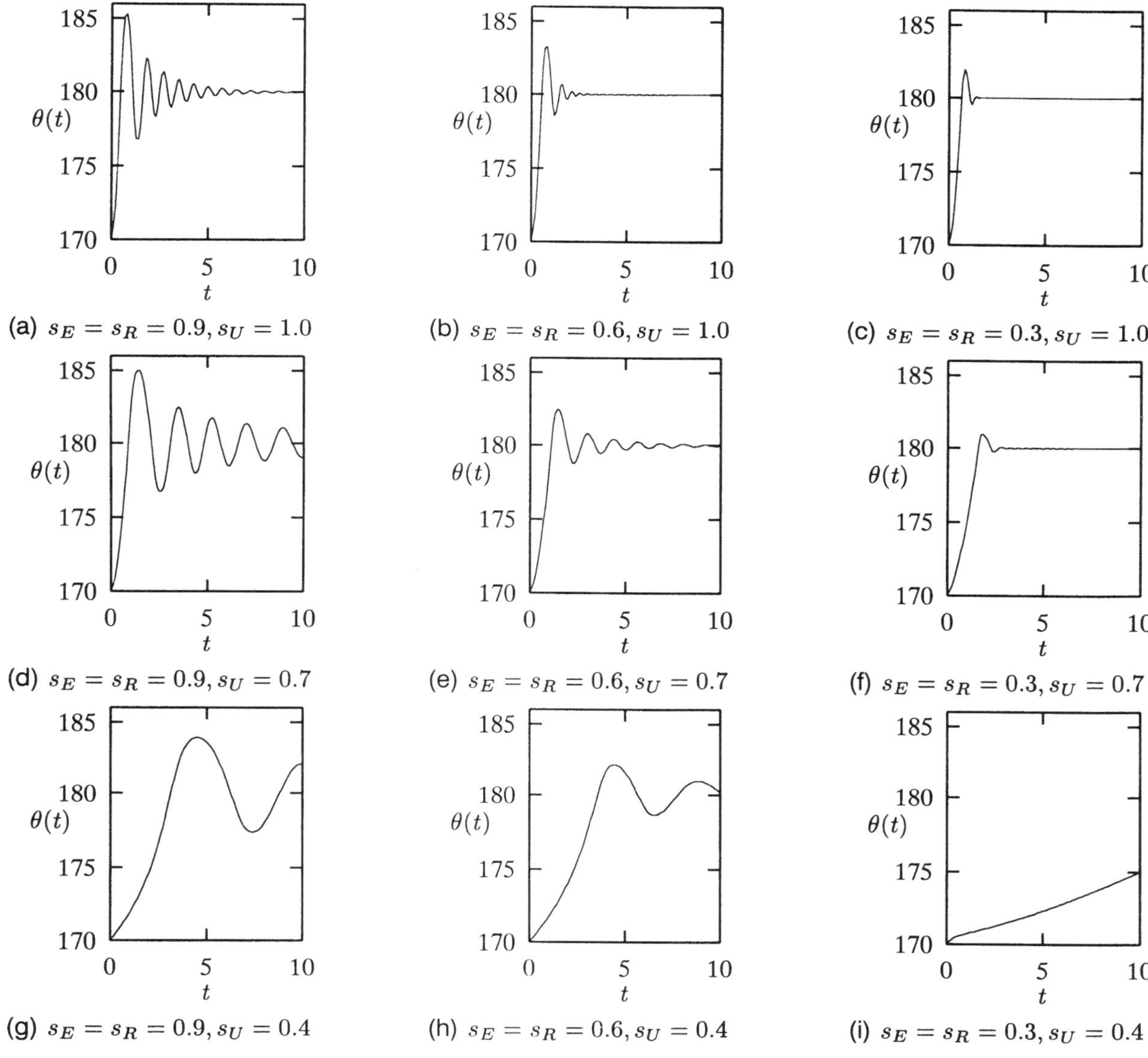

Figure 3. Response of θ(t) using different shrinking factors (θ(0) = 170⁰).

The typical 2-input/1-output fuzzy feedback control system and its detailed computational steps are depicted in Fig. 2.

Here, T is the sampling interval; $e_m(n)$, $r_m(n)$, and $u_m(n)$ are error, rate of change in error, and change in controller output, respectively, at nth sampling point; GE, GR, and GU are scalars for $e_m(n)$, $r_m(n)$, and $u_m(n)$, respectively.

Due to space limitations, the model and diagram of the pendulum-car system is omitted here. Readers should refer to Lin and Sheu (1992) and Czogala, et al. (1995) for details.

The control objective is to stabilize the pendulum in an upright (180^0) position from an initial deflect position (say 170^0) by changing the driving force u.

Due to the characteristics of the pendulum-car system, the control scheme is slightly modified so that the output linguistic variable is the *controller output*, rather that the *change in controller output*. It means that the controller output CO(t) no longer equals CO(t - 1) + $u_m(n)$ but equals $u_m(n)$ directly.

The main aim of the simulation is to demonstrate both the effects of the shrinking factors and robustness of the SSMFs-FLC for nonlinear systems.

First, the effects of the shrinking factors are demonstrated by applying different combinations of s_E, s_R, s_U to the SSMFs-FLC with constant parameters (m = 4, GE = 0.5, GF = 0.01, GU = 300), the initial position of the pendulum is 170^0, the initial control value is 0, and the sampling interval in this simulation is set to 10 ms. In order to make the differences between the responses clearer, s_E and s_R are set to 0.9, 0.6, and 0.3; s_U is changed from 1.0 to 0.7 to 0.4. Fig. 3 shows the responses of the nine combinations of shrinking factors.

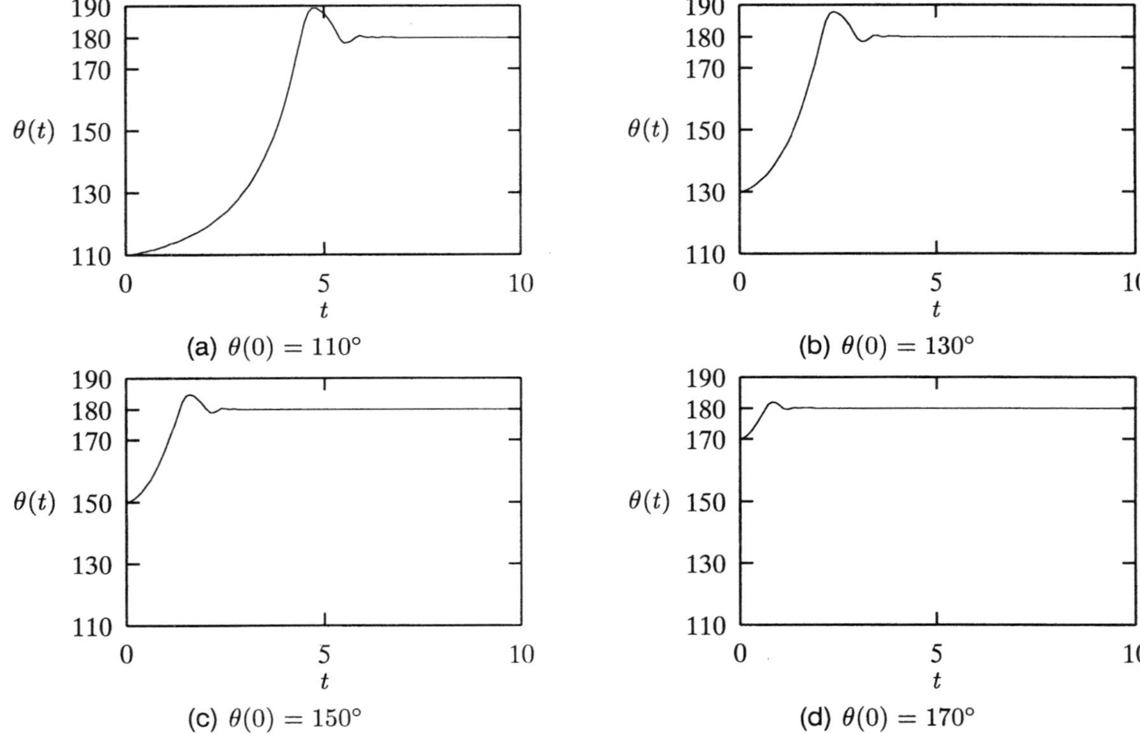

Figure 4. Response of θ(t) starting with different initial positions ($s_E = s_R = 0.3$, $s_U = 1.0$).

One can find that the responses differ markedly between these simulations. Some trends can be observed from the responses: for the pendulum-car system, (1) the less the s_E and s_R, the less the overshoot and oscillation; (2) the less the s_U, the slower the response. This is due to the fact that modifying the shrinking factors not only alters the layout of the membership functions of the linguistic variables but also changes the rules fired for the identical inputs. Certainly, the trends mentioned above may not occur for other nonlinear systems. However, the effects of shrinking factors are already elucidated by this case.

Second, the robustness of SSMFs-FLC is shown by changing the initial position of the pendulum. In this example, the shrinking factors of the SSMFs-FLC are $s_E = s_R = 0.3$ and $s_U = 1.0$. The other parameters are the same as those in Fig. 3. These values are originally used to control the pendulum with the initial position of 170^0. However, the same values are applied to control the other three pendulums with initial positions 150^0, 130^0, and 110^0, respectively, to examine the robustness of the SSMFs-FLC. The simulation results are plotted in Fig. 4. As we can see, except the length of time spent, all four pendulums reach the set point with little oscillation and then maintain their positions at 180^0 perfectly.

Conclusions

This article proposes a new scheme, SSMFs, to specify the membership functions for a fuzzy logic controller. The SSMFs provide a systematic method for designating membership functions and are more instinctive to human perception. With the SSMFs, FLC possesses the capability to adapt itself to different system output deviations. It is not necessary thus to employ heavy optimization for obtaining best performance of the FLC, and this significantly lessens the implementation efforts. Since it reduces the computation load in devising a fuzzy logic controller, SSMFs narrow the gap between a theoretical fuzzy logic controller and a practical one.

References

Czogala, E., Mrózek, A. and Pawlak, Z. (1995). The idea of a rough fuzzy controller and its application to the stabilization of a pendulum-car system. *Fuzzy Sets and Systems*. **72**, 61-73.

Lee, C.C. (1990). Fuzzy logic in control systems: Fuzzy logic controller, Part I, II. *IEEE Transactions on Systems, Man and Cybernetics*. **20**(2), 404-435.

Lin, E.C. and Sheu, Y.-R. (1992). A hybrid-control approach for pendulum-car control. *IEEE Transactions on Industry Electronics*. **39**, 208-214.

Mamdani, E.H. (1974). Applications of fuzzy algorithms for simple dynamic plant. *IEE*. **121**(12), 1585-1588.

Mamdani, E.H. and Assilian, S. (1975). An experiment in linguistic synthesis with a fuzzy logic controller. *Int. J. Man-Machine Study*. **7**, 1-13.

Raju, G. and Zhou, J. (1993). Adaptive hierarchical fuzzy controller. *IEEE Transactions on Systems, Man and Cybernetics*. **23**(4), 973-980.

Wong, C., Chou, C. and Mon, D. (1993). Studies on the output of fuzzy controller with multiple inputs. *Fuzzy Sets and Systems*. **57**, 149-158.

Zadeh, L.A. (1965). Fuzzy sets. *Information and Control*. **8**, 338-353.

EXPERIMENTAL STUDY OF A NEURAL LINEARIZING CONTROL SCHEME USING A RADIAL BASIS FUNCTION NETWORK

Suk-Joon Kim and Sunwon Park
Korea Advanced Institute of Science and Technology
Dept. of Chemical Engineering
373-1 Kusong-dong, Yusong-ku, Taejon 305-701, Korea

Abstract

Experiment on a lab-scale pH process is carried out to evaluate the control performance of the neural linearizing control scheme (NLCS) using a radial basis function (RBF) network. The NLCS was developed to overcome the difficulties of the conventional neural controllers which occur when they are applied to chemical processes. Since the NLCS is applicable for the processes which are already controlled by a linear controller and of which the past operating data are available, we first control the pH process with a PI controller. Using the operating data with the PI controller, a linear reference model is determined by analyzing the past operating data. Then, an internal model controller (IMC) using the linear reference model replaces the PI controller as a feedback controller. Thus the NLCS consists of the IMC controller and a RBF network. After the training of the neural network is fully achieved, the dynamics of the process combined with the neural network becomes linear and close to that of the linear reference model. During the training, the NLCS maintains the control performance of the closed loop system. Experimental results show that the NLCS performs better than the PI controller or the IMC controller for both the servo and the regulator problems.

Keywords

Neural control, Linearization, Radial basis function network, pH process, Internal model control, Linear reference model.

Introduction

Neural networks, inspired from the human nervous system, hold great promise for solving the current difficulties in modeling and control areas. Some researchers proved that they can be used as a universal function approximator (Honik, et al., 1989). Neural networks also find the proper values of the parameters, called as the weights, by a learning process. Originated from their highly parallel distributed architecture, several benefits are generated (Hunt, et al., 1992): easy implementation in VLSI hardware, robustness against the imperfection of the input data, input data fusion and etc.

Among the characteristics of neural networks, their learning capability attracts the system engineer's attentions. Just by sequentially applying input and output data of the process, we can construct the neural networks to produce the desired outputs. This property of neural networks spurs the control engineer to apply the neural networks in modeling and control of chemical processes. Therefore, the past plentiful engineering studies, stimulated by the above promise, have been performed to apply the neural networks in chemical process control.

However, when developing a control scheme using neural networks for chemical processes, we must consider the special characteristics of chemical plants. Before the chemical plants are constructed, the real information of them is not known accurately. Furthermore, after being constructed, they must be operated in a relatively narrow range for safety and economic reasons. Therefore, a control system must be developed under the following restrictions:

1. It must handle the regulatory problem as well as the servo one.
2. It must hold the control performance within the acceptable range even during training of neural networks.
3. It must obtain the training data from the process without the serious abnormal field tests.

The neural linearizing control scheme (NLCS) using a radial basis function (RBF) network was proposed by Kim, et al. (1994) to overcome the difficulties of the conventional neural controllers. The NLCS showed superior control performance for both the servo and the regulatory problems to that of the PI controller in the previous simulation study. This study is carried out to evaluate the control performance of the NLCS on the real process.

This paper begins with a brief review of the NLCS and shows the components and the characteristics of the experimental apparatus which is a laboratory-scale pH control unit. Finally, we discuss the result of implementation of the NLCS to a pH control unit. The experimental results show the good control performances of the NLCS for simple acid/base systems in a laboratory setting.

Neural Linearizing Control Scheme

The NLCS consists of a linear controller, a RBF network and a linear reference model as shown in Fig. 1. G_M and G_C, RBF and f(•) represent for the linear reference model (LRM), the linear controller, the RBF network and the nonlinear process respectively.

In NLCS, a RBF network, a kind of the neural networks, is trained to linearize the relation between the outputs of the linear controller and the process. Training the RBF network is to minimize the difference between the outputs of the LRM and the process with the following objective function, E:

$$E = \frac{1}{2}(\tilde{y} - y)^2 \tag{1}$$

where $\tilde{y}$ and y are the outputs of the LRM and the actual process.

The LRM is determined in order to approximate the process response showing the largest process gain among the past operating data. Therefore, there is no need of any specific test to construct the LRM. Also, based on analyzing past operating data, we can design the LRM in order that the target of the RBF network can be unique and physically realizable. If the process output is invertible, the difference between the outputs of the process and the LRM is unique.

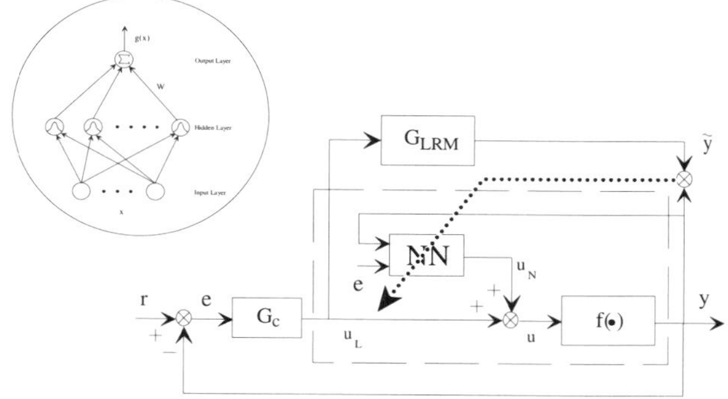

Figure 1. Schematic block diagram of NLCS.

The training algorithm is the modified Hierarchically Self-Organizing Learning (HSOL) algorithm of which the detail explanation is in Kim, et al. (1994, 1995) and Lee and Kil (1991). The training of the RBF network is performed during the finite interval after setpoint changes.

The control performance of the NLCS gradually improves from that of the existing linear controller to that of the proper nonlinear controller during the training because the output of the neural network is added to that of the existing linear controller. After the RBF network is fully trained, the dynamics of the boxed area in Fig. 1 becomes linear such as that of the LRM and also the overall control performance of the linear controller will improve. Additionally, the tuning parameters of the linear controller can be tightened to control the linearized dynamics, based on the linear control theory.

Experimental Apparatus

The lab-scale experimental apparatus consists of four sections: the process feed (acidic) section, the titration feed (basic) section, the reactor section and the control computer section. Its schematic diagram is shown in Fig. 2. Additionally the parameters and the steady-state values used in this experiment are given in Table 1.

The process feed section consists of two feed tanks, a three-way valve and a feed pump. The reactants in feed tanks have different concentrations and the three-way valve provides the way to select the feed from feed tanks. In this way it was possible to introduce disturbances in the concentration of the feed stream in a manner as close to a step change as possible. The concentration of the secondary feed stream is 0.0033N CH_3COOH. In the titration feed section, there are a feed tank and a metering pump. The feed tank contains NaOH solution. The metering pump which receives a control signal from the computer controls a base flow.

The reactor section consists of a polyethylene vessel and a pH measuring system. The inlet tubes go down to the bottom of the vessel and the pH probe is located at the effluent stream. The effluent stream tube is located at 10 cm high from the bottom and so the liquid over this

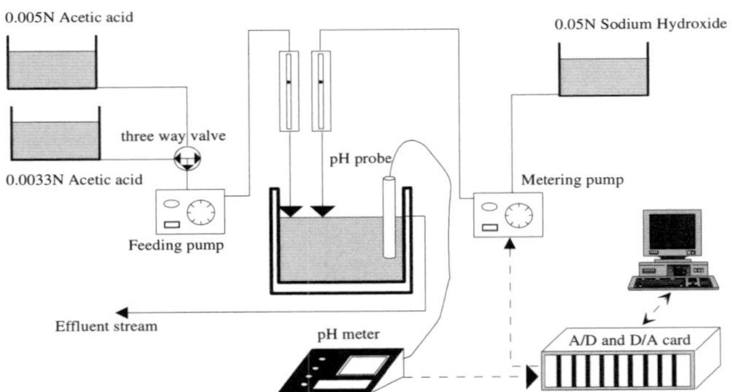

Figure 2. Schematic diagram of the experimental apparatus.

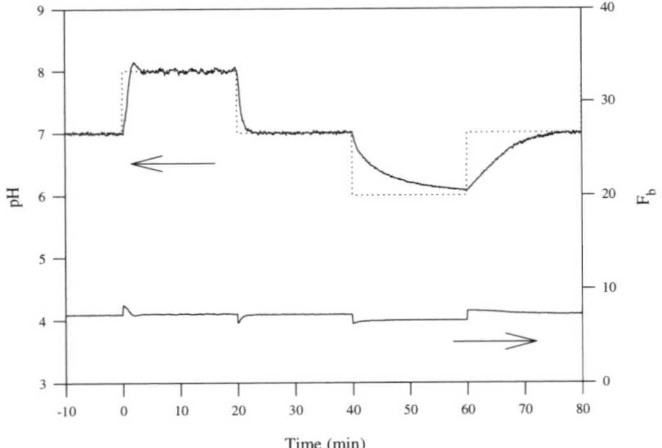

Figure 3. Closed-loop response and the corresponding control action of PI control.

Table 1. Steady-state Values.

Variable	Description	Value
C_A	concentration of acidic feed	0.005 N
C_B	concentration of basic feed	0.05 N
F_A	flow rate of acidic feed	85 ml/min
F_B	flow rate of basic feed	8.5 ml/min
V	volume of reactor	875 ml
pH	pH	7

Table 2. Parameters Used in the Experiment.

Step	Description	Value
	sampling time	Δt=5 sec
PI control	proportional gain	K_C= 1 ml/min
	integral time	T_C= 600 sec
Linear Reference Model	model gain	K_M= 13 (ml/min)$^{-1}$
	model time constant	T_M= 631 sec
	model time delay	T_D= 10 sec
IMC	filter time	T_F= 30 sec
NLCS	learning rate	η=0.3
	initial width of RBF	σ=0.1
	initial filter time	T_F= 30 sec
	final filter time	T_F= 10 sec

height overflows without pumping. A magnetic agitator is also used to ensure proper mixing. Finally, the pH measuring system consists of a pH probe and a pH meter.

In the control computer section, there are an IBM PC 386SX and a set of A/D and D/A converters.

Results

The parameters used in the experiment are summarized in Table 2. Since the NLCS can be implemented on the process which would be controlled by a linear controller, we carry out a pre-experiment to construct the control loop using a PI controller. The tuning parameters of the PI controller are roughly determined by on-line tuning because it is difficult to clearly find the optimum tuning parameters due to the severe process nonlinearity. Figure 3 shows the process responses and the corresponding control actions for various setpoint changes using the PI controller. The control performance for the setpoints between 7 and 8 is acceptable but that between 6 and 7 is very poor.

Among the historical data of PI control, we choose the transient responses of the process with respect to the setpoint change from 7 to 8 to determine the LRM, because they represent the largest gain among the past data. After regression of the LRM to fit these data, the resulting LRM is

$$\tilde{y}(k+1) = e^{-5/631} y(k) + 13\{1 - e^{-5/631}\} u(k-2) \qquad (2)$$

An internal model controller (IMC) having the minimum-phase of the LRM in Eqn. (2) as the process model is substituted for the PI controller and is implemented on the process. The reason for introducing the IMC is that the IMC can effectively handle the time delay and perfectly control the process if the perfect model is given. Then although the LRM has the time delay, the IMC can control the process better than the PI controller after the RBF network is trained. Figure 4 shows the control performance of the IMC controller with the filter time constant of 30 sec. Although the control performance of the IMC is better than that of the PI control, the intrinsic limitation of the linear controller still appears in IMC.

Then we start training the RBF network in the NLCS which consists of the IMC controller as a linear feedback controller and a RB'F network. After about 10 hours or 20 iterations, the NLCS shows the control performance for the servo problem as shown in Fig. 5. This figure shows the superior control performance in spite of the severe nonlinearity of the process. During the training the filter time constant is gradually changed from 30 sec to 10 sec.

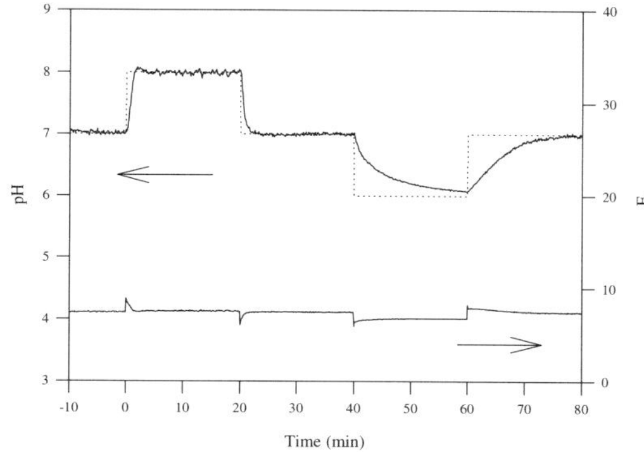

Figure 4. Closed-loop response and the corresponding control action of IMC.

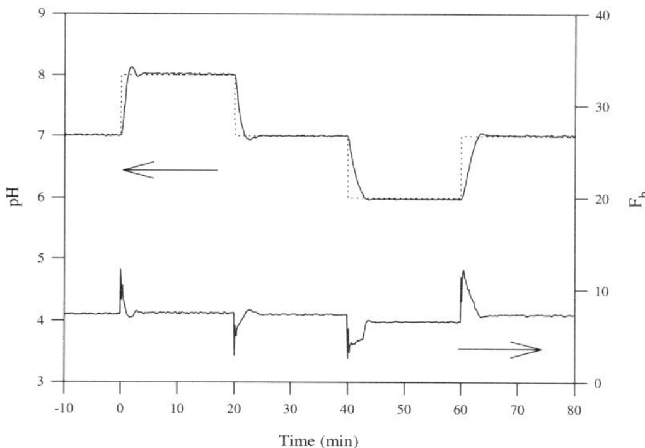

Figure 5. Closed-loop response and the corresponding control action of NLCS for the servo problem.

Finally, we test the control performance of the NLCS for a regulatory problem and compare it with those of the PI and the IMC controllers. Using the three-way valve, we make the step changes in the concentration as an unmeasured disturbance. Figure 6 shows the results of this experiment. Among three controllers the NLCS shows the best control performance.

Conclusions

The NLCS is implemented on a lab-scale pH process to evaluate its performance. Before introducing the NLCS to the process, we construct a rough PI controller by on-line tuning to evaluate the process data. Using these data, the linear reference model is determined by analyzing the past operating data and an IMC controller is designed. Then, the NLCS is trained to linearize the relation between the outputs of the IMC controller and the process while the IMC controller with the LRM mainly controls the process. After the training of the RBF network is fully achieved, the NLCS shows better control performance than the PI controller or the IMC controller for both the servo and the regulator control situations. Additionally, during the training, the NLCS maintains the closed-loop control.

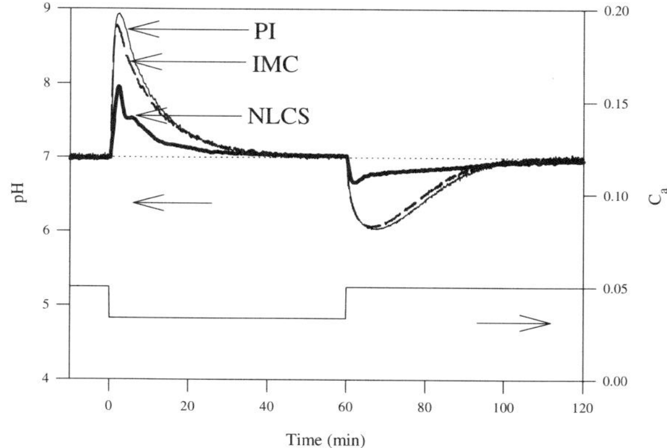

Figure 6. Closed-loop response of NLCS for the regulatory problem.

Acknowledgment

Partial financial support from the KOSEF through the Automation Research Center at POSTECH is gratefully acknowledged.

References

Bhat, N. and T. J. McAvoy (1990). Use of neural nets for dynamic modeling and control of chemical process systems. *Computers Chem. Engng.*, **14**, 573-583.

Hornik, K., M. Stinchcombe and H. White (1989). Multilayer feedforward networks are universal approximators. *Neural Networks*, **2**, 359-366.

Hunt K. J., D. Sbarbaro, R. Zbikowski and P. J. Gawthrop (1992). Neural network for control systems — A survey. *Automatica,* **28**, 1083-1112.

Kawato, M., Y. Uno, M. Isobe and R. Suzuki (1988). Hierarchical neural network model for voluntary movement with application to robotics. *IEEE Control System Magazine*, **8**, 8-17.

Kim, S.-J., M. Lee, S. Park, S. Lee and C.-H. Park (1995). A neural linearizing control scheme for nonlinear chemical processes. *Computers Chem. Engng.*, in press.

Kim, S.-J., M. Lee, S. Park, S. Lee and C.-H. Park (1994). Neural linearizing control with radial basis function network for chemical processes. *World Congress on Neural Networks,* Volume 2. San Diego. pp. 94-98.

Lee S. and R. M. Kil (1991). A Gaussian potential function network with hierarchically self-organizing learning. *Neural Networks,* **4**, 207-224.

Lee, M. and S. Park (1992). A new scheme combining neural feedforward control with model-predictive control. *AIChE Journal*, **38**, 193-200.

CONTROLLER VERIFICATION FOR POLYMERIZATION REACTORS

E. Gazi, W.D. Seider and L.H. Ungar
Department of Chemical Engineering
University of Pennsylvania
Philadelphia, PA 19104-6393

Abstract

A methodology for verifying the stability and performance of a controller, when the process model is incompletely known, is demonstrated for a polymerization reactor. Model uncertainty arises when either the parameters are known only approximately (parametric uncertainty), or when the form of the equations is not known exactly (non-parametric uncertainty). Two techniques are used to solve the semi-quantitative models to predict all of the possible behaviors. One involves qualitative analysis, implemented in the NSIM program, that uses symbolic manipulation to build bounding equations, and the other is a non-parametric Monte-Carlo technique. Temporal logic operators are used to formalize the posing of and automatic answering of qualitative questions about the response of the system.

Keywords

Controller verification, Non-parametric uncertainty, Monte-Carlo simulations, Qualitative analysis, Styrene polymerization.

Introduction

Most chemical processes are modeled by ordinary differential equations (ODEs) that exhibit substantial uncertainties in the model parameters, inputs and initial conditions. Often, the terms and the form of the equations are uncertain because the mechanisms are not well understood. As an example, the dependence of the rates of reaction on temperature and composition is often not accurately known for complex reaction mechanisms. In these cases, experimental data can help to determine bounding envelopes, using the methods described by Kay and Ungar (1993). The true functions are not likely to be coincident with either of the envelopes, but confidence levels can be established that they lie between them. Although techniques exist to deal with parametric uncertainty, the representation and simulation of nonlinear models with inexact functional relationships has not been addressed as effectively.

Several qualitatively distinct behaviors may be consistent with a single such *semi-quantitative* model. For example, an increase in the reactor temperature may result in either a desirable increase in the reaction rates or undesirable instability, depending on the values of the parameters. Two techniques are introduced to solve semi-quantitative models to predict all of the possible behaviors. One involves qualitative analysis, implemented in the NSIM program, that uses symbolic manipulation to build bounding equations, and the other is a non-parametric Monte-Carlo technique. Then, *temporal logic* operators are used to formalize the posing of and automatic answering of questions of the type: "Can the system operate adequately if a disturbance/fault occurs?" This methodology applies for the automatic verification of control actions to guarantee that processes will not move into hazardous operating regimes when faults are encountered; that is, in the design of *safety net* control actions. With such off-line verification, large overdesign factors, that are often used to compensate for model uncertainties, can be avoided.

Simulation of Uncertain Models

In this section, the two techniques are described. It is assumed that the unknown functions, $y = f\{x\}$, are bounded between envelopes, such that $f_l\{x\} \le y \le f_u\{x\}$, and are monotonic (Fig. 1). The monotonicity assumption is very important for the analysis, but is not overly restrictive, since any non-monotonic function can be constructed by combining monotonic ones. In addition, it is often known that physical quantities are related to each other monotonically, even though the relations may not be known accurately; e.g., reaction rates increase monotonically with temperature, and friction factors increase monotonically with velocity.

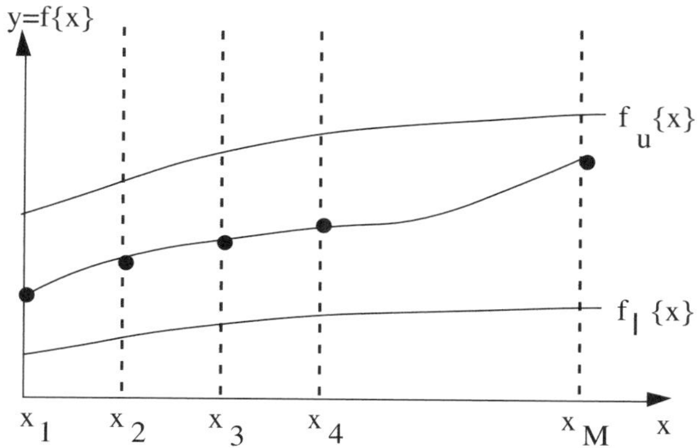

Figure 1. Bounding envelopes for f{x}.

The two techniques presented herein are based on entirely different philosophies. The first involves a "worst-case," deterministic analysis, that derives ODEs whose solution is guaranteed to bound all of the solutions of the inexact model, and is based on the NSIM algorithm. The second is a statistical analysis, that extends the Monte-Carlo technique for systems with non-parametric uncertainty. The advantages and disadvantages of each approach are discussed below.

The NSIM Algorithm

The QSIM algorithm (Kuipers, 1986) provides a framework for developing models in the form of qualitative differential equations (QDEs), which are abstractions of ordinary differential equations (ODEs), where the values of the variables are described qualitatively and the functional relationships between the variables may be known incompletely. The constraints are qualitative expressions involving the common mathematical operations, such as addition, multiplication, and differentiation as well as monotonicity. Given a QDE and a qualitative description of an initial state, QSIM derives a tree of qualitative state descriptions, where the paths from the root to the leaves of the tree represent the possible behaviors of the system.

In NSIM (Kuipers, 1989; Kay and Kuipers, 1993), an extension of QSIM, parameters may be assigned lower and upper bounds and monotonic functions may be bounded by lower and upper envelopes, thus forming Semi-Quantitative Differential Equations (SQDEs). These permit NSIM to eliminate behaviors that are inconsistent with the numerical information, while describing the remaining behaviors more quantitatively. NSIM produces bounds on the trajectories of the state variables that are guaranteed to include *all* of the behaviors consistent with the approximate model.

Given a SQDE and an initial condition, NSIM derives and numerically integrates an "extremal" system of ODEs, whose solution is guaranteed to bound all of the solutions of the SQDE. This is referred to as NSIM *simulation*. The extremal system is generated automatically, using symbolic manipulation, based on a set of rules (Kay and Kuipers, 1993). An extremal equation is a bound on the first derivative of a state variable. Minimal and a maximal equations are derived for the state variables, defining a *dynamic envelope* for each variable, and hence, an n-th order system is transformed to a 2n-th order system. As a result, the extremal ODEs are not generally members of the class of ODEs represented by the SQDE. Therefore, the dynamic envelopes do not necessarily have the same shape as the behaviors of the SQDE.

The extremal system is integrated using a standard Runge-Kutta technique. It is proven theoretically (Kay, 1991) that, when the extremal system bounds the solution of the SQDE at $t = 0$, it bounds the solution for all times.

A SQDE is an abstraction of an ODE with incompletely specified functions and parameter values. One NSIM "simulation" covers an infinite number of models. Since the NSIM bounds are guaranteed to bound all the solutions to the SQDE, NSIM can be used for automatic proof construction. However, the main disadvantage of this technique is that the bounds are conservative, as illustrated by Gazi, et al. (1994).

The Non-parametric Monte-Carlo Technique

To extend the Monte-Carlo technique to handle non-parametric uncertainty, a large number of monotonic functions are randomly generated within the envelopes that bound the incompletely known monotonic functions. Each generated function is incorporated into the ODEs that model the process, which when integrated produce trajectories of the state variables that are systematically checked for "interesting" behaviors, as described in the next section. This is accomplished using the following algorithm:

1. M intervals in x are formed.
2. At $x = x_1$, y_1 is selected randomly such that $f_l\{x_1\} \le y_1 \le f_u\{x_1\}$.
   At $x = x_k$, y_k is selected randomly such that $f_l\{x_k\} \le y_k \le f_u\{x_k\}$ and $y_k \ge y_{k-1}$, $k = 2, \ldots, M$.
3. The points (x_k, y_k), $k = 1, \ldots, M$, are connected with straight lines.
4. The system of ODEs is integrated.
5. Steps (2)-(4) are repeated.

In this analysis, it is important to ensure a satisfactory coverage of the space of possible functions. Gazi, et al. (1995) show that the monotonicity constraint clusters the generated functions close to the upper envelope, as x increases, when the y's are selected from uniform distributions. To avoid this, y_1 and y_M are selected randomly first, and then the x coordinate is subdivided recursively.

A large number of simulations (one with each generated function) is required, and the results are guaranteed to include all possible behaviors only in the limit as the number of samples approaches infinity.

Controller Verification

During process design, with substantial uncertainties present, the specific trajectories produced by numerical simulations may be unreliable. It should be helpful to the engineer, in carrying out process and controller design, to prepare qualitative descriptions of the possible closed-loop behaviors. With this, the engineer would establish the bounds on a desired *nominal region of operation* and automatically verify off-line whether, for example, the controller would return the process to a steady state, within the nominal region, in the face of anticipated disturbances and faults.

Given the nominal, semi-quantitative, model for the process and the controller, as well as the potential faults, the proposed verification scheme is as follows. The engineer asks a question of the type, "What may happen if this fault occurs?" In response, the fault model is automatically built and integrated (using either of the techniques described above). Then, *temporal logic operators* (Emerson, 1990) are used to express formally the user-supplied questions about the system behavior, and provide qualitative answers.

A similar scheme has been applied for the verification of discrete-event control systems by several authors. For example, Moon, et al. (1992) build a tree that includes all of the possible behaviors of the system (a succession of discrete events in time) and check it systematically to determine the truth of the question. Herein, the trajectories of the state variables produced by each Monte-Carlo simulation may be viewed as one behavior.

Table 1. Temporal Logic Operators.

modal	temporal	logical
necessarily	always	and
possibly	eventually	or
	until	not

The temporal logic operators used are shown in Table 1. Through the modal operators, the user can specify whether the question must be true for all Monte-Carlo simulations (necessarily) or at least one (possibly). In the case of NSIM simulations, the modal operator is set to "necessarily" by default. For each individual trajectory, a question might refer to the whole time range (always), or only over time intervals (eventually, until). The logical operators are the conventional Boolean connectives. An example question expressed using the temporal logic operators is given in the next section.

Application to Styrene Polymerization

This methodology is illustrated in Fig. 2 for the free-radical polymerization of styrene in a jacketed CSTR (Higaldo and Brosilow, 1990). The objectives are to stabilize the reactor at an unstable steady-state and control the average molecular weight and the molecular weight distribution of the polymer. The flow rate of the cooling water, F_c, is used as the primary action for the control of the reactor temperature, with the monomer flow rate, F_m, the secondary control action in response to large disturbances. The flow rate of the initiator, F_i, is manipulated to keep the weight-average molecular weight, M_w, close to its setpoint. A combination of PID controllers, using sliding-gain scheduling, regulates the reactor in the face of uncertainties and large disturbances.

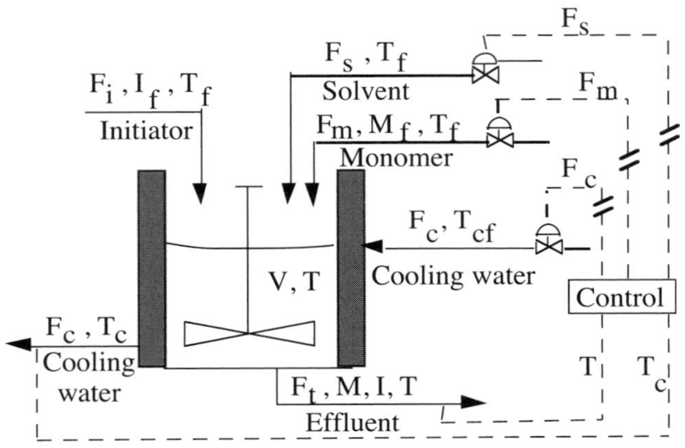

Figure 2. Schematic of a jacketed CSTR for styrene polymerization.

The models for the styrene polymerization reactor contain substantial uncertainties. As an example, the rate constant for the termination reaction is of the form, $k_t = k_{t0}\, g_t$, where k_{t0} is the value of the rate constant at very low monomer conversions, and is assumed to have an Arrhenius dependence with parametric uncertainty. g_t is a function that monotonically increases with temperature and decreases with reactor composition, reflecting the fact that k_t falls due to diffusion limitations at higher conversion. This functionality is unknown, but can be bounded by envelopes based upon experimental data (Fig. 3). Moreover, the weight-average molecular weight cannot be measured on-line. Rather, the viscosity is measured, but its relationship to M_w is uncertain.

Figures 4 and 5 show the response of the closed-loop system, as predicted by NSIM and Monte-Carlo simulations, respectively, in answer to the question: "If the reactor temperature set-point increases by 6-8K, is it

(necessarily ((always (in nominal region)) *and eventually* (steady-state))) ?"

where the "nominal region" of operation is defined such that, 340 ≤ T ≤ 360 K, and 30,000 ≤ M_w ≤ 35,000.

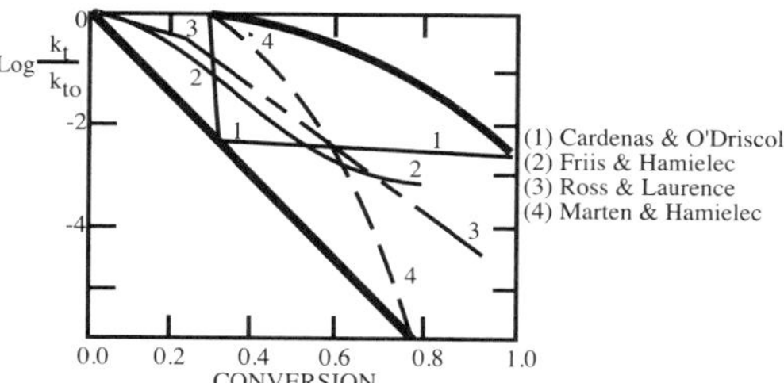

Figure 3. Bounds (bold lines) on g_t, based on experimental studies.

The NSIM bounds diverge, showing that unstable trajectories are possible, and hence, the answer "false" is returned. However, this is attributed to the conservatism of the NSIM bounds. On the contrary, the Monte-Carlo simulations predict tight bounds for all of the state variables, and hence, provide an affirmative response.

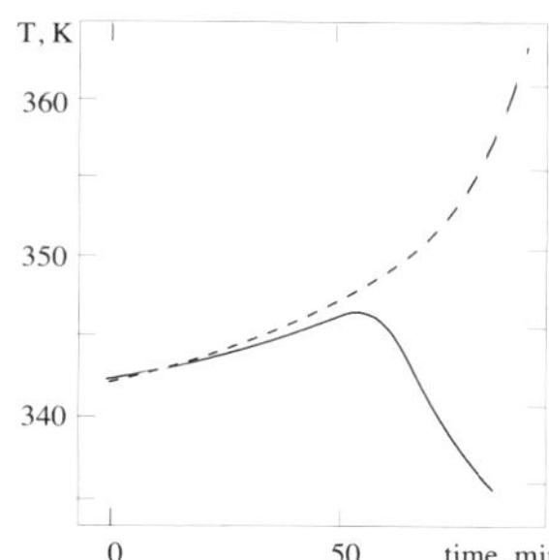

Figure 4. NSIM bounds on the reactor temperature, for a set-point increase by 6-8 K.

Conclusions

A methodology is introduced for the off-line verification of the stability and performance of controllers, when the process model has substantial, parametric and non-parametric, uncertainty. Two techniques are presented for the simulation of such semi-quantitative models. Although more expensive computationally than NSIM, with guarantees only in the limit of infinite samples, the Monte-Carlo method produces bounds that are not conservative, as illustrated for the design of a styrene polymerization reactor and its control system.

Acknowledgments

Partial funding provided by NSF Grant Nos. IRI-9216714 and DDM-9114080 is gratefully acknowledged.

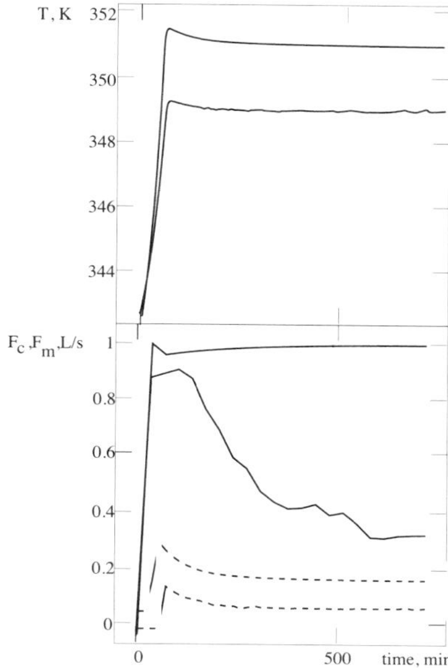

Figure 5. Monte-Carlo bounds on the reactor temperature, T, and the manipulated cooling water and monomer flow rates, F_c and F_m for a 6-8 K increase in the temperature set-point.

References

Emerson, E. A. (1990). Temporal and modal logic. In J. van Leeuwen (Ed.), *Handbook of Theoretical Computer Science*. Elsevier Science Pub. B.V./MIT Press. pp. 995-1072.

Gazi, E., W. D. Seider and L. H. Ungar (1994). Controller verification using qualitative reasoning. *Proc. of 2nd IFAC Workshop on Comp. Soft. Struc. Integ. AI/KBS Sys. in Proc. Cont.*, Lund, Sweden.

Gazi, E., W. D. Seider and L. H. Ungar (1995). A non-parametric Monte-Carlo technique for controller verification. *Automatica*, submitted.

Higaldo, P. M. and C. B. Brosilow (1990). Nonlinear model predictive control of styrene polymerization at unstable operating points. *Comput. Chem. Engng.*, **14**, 481-494.

Kay, H. (1991). Monitoring and diagnosis of multi-tank flows using qualitative reasoning. Master's Thesis. The University of Texas at Austin, Austin, Texas.

Kay, H. and B. J. Kuipers (1993). Numerical behavior envelopes for qualitative simulation. *Proc. of the National Conference on Artificial Intelligence (AAAI-93)*. AAAI/MIT Press. pp 606-613.

Kay, H. and L. H. Ungar (1993). Deriving monotonic function envelopes from observations. In *Working Papers from the Seventh International Workshop on Qualitative Reasoning about Physical Systems*. Rosario Resort, Washington. pp. 117-123.

Kuipers, B.J. (1986). Qualitative simulation. *Artificial Intelligence*, **29**, 289-338.

Kuipers, B.J. (1989). Qualitative reasoning: modeling and simulation with incomplete knowledge. *Automatica*, **25**, 571-585.

Moon, I., G. J. Powers, J. R. Burch and E. M. Clarke (1992). Automatic verification of sequential control systems using temporal logic. *AIChE J.*, **38**, 67-75.

EXPERIENCES OF DEVELOPING PROCESS MODEL LIBRARIES IN OMOLA

Bernt Nilsson[1]
Department of Automatic Control
Lund Institute of Technology
P.O. Box 118, S-221 00 Lund, Sweden

Abstract

Three different sets of process model libraries have been developed in the object-oriented modeling language, OMOLA. This paper presents the basic ideas behind the different libraries and the experiences of the development. The first is a chemical plant model library called CPM, which covers a number of unit operations in continuous chemical plants. BatchLib is the second and it is a small application to batch reactors with control systems. The third set of OMOLA libraries is the K2 thermal power plant model database. K2 has been used to model and simulate a heat recovery steam generation plant, HRSG, in an industrial case study. The decomposition of the model and class tree and the organization of model object libraries are discussed.

Keywords

Computer-aided modeling, Process model libraries, Simulation, Object-oriented methodology.

Introduction

Object-oriented modeling facilitates modeling through decomposition, reuse and high-level model description. The OMOLA language supports structure decomposition, class inheritance and equation-based model expressions. A brief presentation is found in Mattsson and Andersson (1992). Similar concepts are found in other modeling languages under development like ASCEND (Piela, et al., 1991), DYMOLA (Elmqvist, 1993), gPROMS (Barton and Pantelides, 1991), MODEL.LA. (Stephanopoulos, et al., 1990), and VEDA (Marquardt, 1993). For a survey see Marquardt (1994).

The two hierarchies of OMOLA, decomposition hierarchy and inheritance hierarchy, facilitate structuring and reuse of models. The decomposition of large and complex models into smaller submodels increases readability and understanding of the model and it is done in a number of levels. In a typical process application a number of physical objects can be described by the same model. This model is defined as the super class for the descriptions of the physical objects and they inherit the properties from their common super class.

The experiences from the development of three process oriented OMOLA model libraries are discussed in this paper. The libraries are CPM, BatchLib and K2. Chemical Plant Model Library, CPM, is presented in Nilsson (1993) and contains a number of common unit operations of continuous process operations. BatchLib, used in Nilsson (1994b), is a library for batch reactors with sequential and continuous control systems. Finally, the K2 model database is a set of model libraries for modeling of thermal power plants (Nilsson and Eborn, 1995). The K2 database has been used to simulate a Heat Recovery Steam Generation Plant, HRSG (Eborn and Nilsson, 1994).

Modeling Concepts and OMOLA

OMOLA is an object-oriented modeling language, and models developed in OMOLA can be simulated in the interactive simulation environment OMSIM. The OMSIM simulator can handle differential-algebraic equation systems, DAE, and discrete events (Andersson, 1994).

OMOLA models are primitive or composite. Primitive models are described by equations and/or events

[1] *bernt@control.lth.se*.

in a textual editor. Composite models are structures of submodels connected together by connections. This is done in a graphical editor. It is possible to mix textual and graphical descriptions. A submodel in a composite model can itself be a composite model, creating a structure hierarchy of submodels.

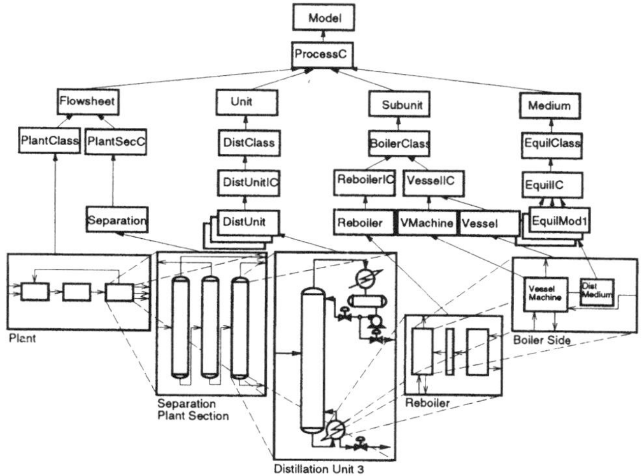

Figure 1. The CPM library organization and the structure hierarchy of a chemical plant application.

All models are classes in a single inheritance hierarchy in OMOLA. All attributes of a model class are inherited by its model subclass. Attributes are variable, parameter and terminal definitions, equations, evens, submodels, connections, etc. It is possible to overwrite inherited attributes. A submodel inside a composite model can be a subclass of a model class outside the composite model, i.e., in a model library. This makes it possible to reuse library model classes in application structures.

Chemical Plant Model Libraries

A library for chemical process applications is developed in Nilsson (1993) and it is called CPM.

Structure Decomposition

Chemical plants are often decomposed into plant sections. These plant sections are composed of a number of unit operations, e.g., distillation units, reactors, buffer tanks, etc. Unit operations are often composed of a set of processing objects, e.g., pumps, valves, tubes, vessels, sensors, etc. The plant model and plant section models are application dependent and are not a part of a general model library. Other model classes are found in the model libraries. Structure decomposition of a chemical plant examples is seen in Fig. 1 and guidelines are given in Nilsson (1994a).

Medium and Machine Decomposition

One problem in general modeling languages is the separation of the medium descriptions from the unit model descriptions. In flowsheeting packages, i.e., ASPEN PLUS this separation is almost orthogonal. This can be created in an object-oriented modeling language by the use of structure decomposition, single inheritance and polymorphic submodels. It can also be created by the use of multiple inheritance without overwriting and communication super classes but this is not implemented in the current OMOLA. A processing object is decomposed into one machine description and one medium description, i.e., *medium and machine decomposition*. The machine contains the dynamic description, i.e., balance equations, and the medium model all mall medium specific calculations. The two model classes are connected to each other by a connection in a processing object composite class. This is illustrated in Fig. 2. This decomposition makes it possible to change the medium description of a processing object without rewriting the models. In CPM there are a number of medium models for vapor-liquid equilibrium and for chemical reaction descriptions.

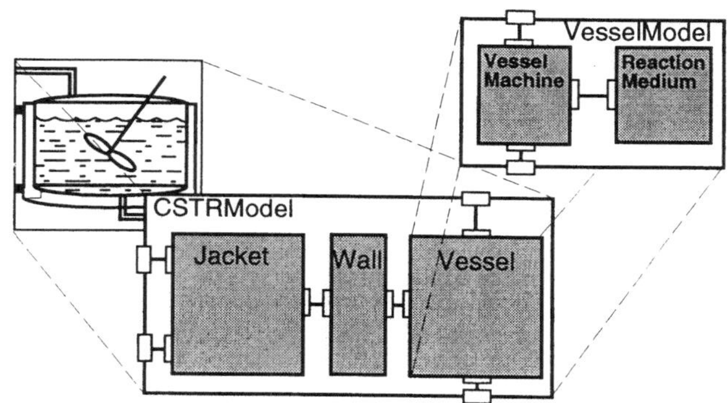

Figure 2. The medium and machine decomposition of a vessel which is part of a CSTR.

Library Structure

The CPM library is actually a set of OMOLA sublibraries. An OMOLA library contains a set of model classes and has local name space. The user has to refer to the library name if no search order is specified. The libraries contain classes for the development of unit operations, e.g., `DistillationLib`, CSTRLib, etc. This means that these libraries also contain other classes, e.g., library specific process objects, common super classes for different units, etc. Important processing objects, e.g., pumps, heat exchangers, etc., are found in specific libraries for energy supply and for flow equipment. This means that the libraries contain classes in different branches of the class tree, seen in Fig. 1.

Parameterized Classes

The medium and machine decomposition results in the possibility to change the medium description of a model class by redefinition of the medium submodel class. In unit operation classes, with an internal structure of submodels and with their own medium models, this becomes complicated. All medium models must be change d at the same time. Convenient parameterization of structures is therefore important. There is a need for *abstract class definitions* that can be done in the structure hierarchy and which can only be used for inheritance and cannot be instantiated in the simulation model. They are therefore not seen during simulation. One example is the need for medium model specification in a distillation column. The model library user only changes the medium model definition once at the top level and does not want to bother with the internal details. There are alternative ways to create this functionality (Nilsson, 1993).

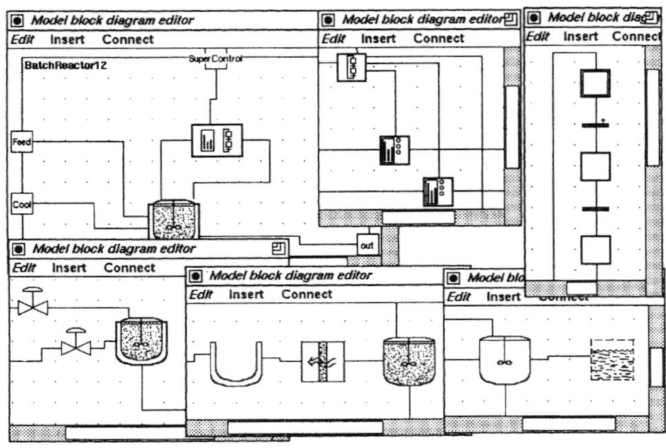

Figure 3. The decomposition of a controlled batch reactor.

Batch Model Libraries

BatchLib is the result of the application of OMOLA on chemical batch reactors. OMOLA and OMSIM can handle hybrid models in an efficient way and this is explored in this application.

Modeling and simulation of batch processes require the possibility to describe both continuous behavior and state dependent discrete events. A typical batch control system is composed of one discrete part making the decisions of changing batch phases. The other part is continuously controlling the reactor through the batch phases. The first is a finite state machine which can be described in different ways. In BatchLib it is graphically developed by the use of Grafcet, i.e., sequential function chart, SFC. The second part is composed of ordinary controllers with or without set point trajectory generators.

A controlled batch reactor and its internal structure is illustrated in Fig. 3. The batch reactor structure, shown at the top left, is composed of one reactor and one control system. The control system contains one sequential controller and three PID controllers A part of the Grafcet sequence is also seen on the right. It is easy to examine and change the batch sequence. At the bottom of Fig. 3 the reactor decomposition is seen which follows the ideas discussed in the CPM section.

The K2 Model Database

The K2 model database is a set of libraries that are used to model thermal power plants. An OMOLA model database is a set of OMOLA model libraries which allow classes to be loaded on request to the OMSIM environment. This means that OMSIM only loads model classes that are used in an application and not entire libraries. The K2 organization is similar to CPM with some important differences. The libraries are branches of the class tree (see Fig. 5). The units and processing objects are not decomposed in medium and machine classes. Instead they are described by phenomena subunits that have internal medium models.

K2 has been used in an industrial application for the modeling and simulation of a heat recovery steam generation plant, HRSG (Ordys, et al., 1994). The K2 model database and its applications were presented in Nilsson and Eborn (1995) and Eborn and Nilsson (1994).

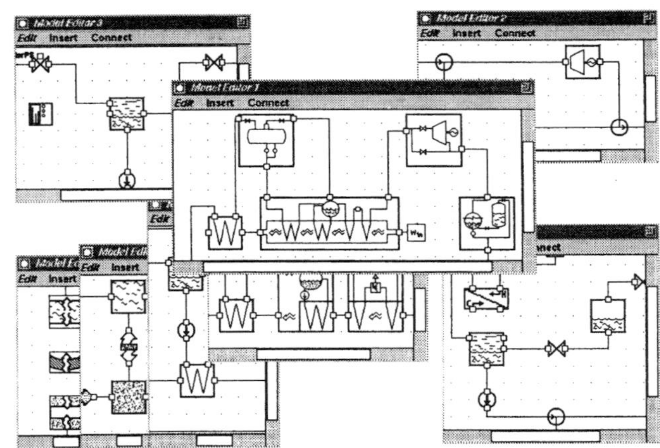

Figure 4. The HRSG plant decomposition.

Decomposition

An application is structurally decomposed as in CPM, namely into plant model, plant section submodels, subsystems or "unit operations" and finally into units with corresponding physical objects. The units are composed of subunits and these subunits describe different physical phenomena. Subunits can be divided into two major groups, *compartments* and *flow resistors*. Compartments describe the dynamics in a control volume. Flow resistors describe the medium flow or heat flow between compartments. Each subunit has an internal subclass that contains a medium model description. There are medium models for subcooled water, saturated water, water/steam mixtures, superheated steam and flue gas.

One HRSG plant configuration is illustrated in Fig. 4. The window in the middle is the plant configuration and around it are four plant sections, deaerator section, steam turbine section, pan section and the condenser section. The pan is further decomposed into subsystems, i.e., economizer, boiler and superheater. The boiler is decomposed into units, e.g., evaporator, recycled pump, drum, etc. In Fig. 4 the evaporator structure is shown in the two bottom left windows. The evaporator is decomposed in two compartments, water and flue gas, and on heat flow resistor. The heat flow resistor has an internal structure of three heat resistance descriptions.

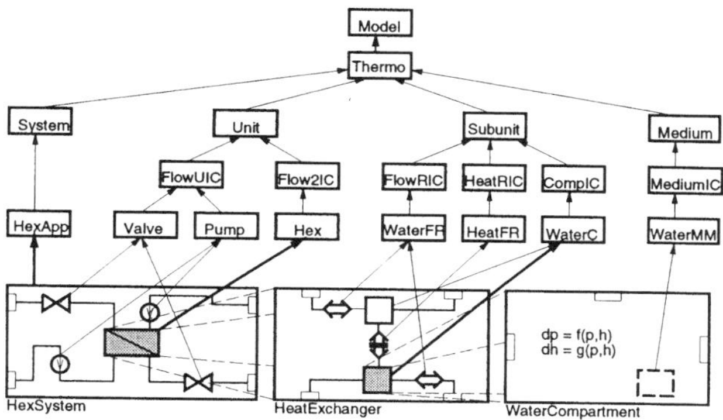

Figure 5. A part of the K2 class tree for a heat exchanger example.

Library Abstraction

The K2 model database is decomposed into three groups of model libraries, namely libraries for units, subunits and model components. Unit classes are models that describe physical objects and they are composed of subunits. Subunits describe phenomena inside units. The model component libraries contain classes for development of models, like variables, functions, terminals, etc. This library organization supports multiple users and the abstraction of model use. The basic idea is that the system developer uses the unit libraries to develop plant and plant section configurations. The unit model developer uses the subunit libraries. An advanced model developer can create new phenomena subunits using the equation based OMOLA language.

The structure of a simple heat exchanger system is shown in Fig. 5. On the left a heat exchanger application is modeled. It is composed of two pumps, two valves and one countercurrent heat exchanger. These units are all found in unit libraries of the `Unit` branch of the class tree. The heat exchanger is described by a set of subunits, namely two water compartments, two medium flow resistor one heat flow resistor, which are reused from the `Subunit` branch. A water compartment is a primitive model described by a set of equations. It has an internal class describing media specific parameters, i.e., steam table calculations.

Conclusions

Object-oriented modeling languages can be used to define process oriented model libraries that are powerful and well structured. It creates an environment with the advantages of both general and open modeling languages and from closed application oriented modeling tools. Abstraction of the model libraries is important in order to increase readability. To create a modeling layer below the unit level is also an important step towards increased reusability of open object-oriented models. The use of general modeling language, like OMOLA, has shown that there is no need for particular process model language constructs but there is a need for sophisticated model parameterization concepts.

References

Andersson, M. (1994). *Object-oriented modeling and simulation of hybrid systems*. Ph.D. thesis. Dept. of Automatic Control, Lund Institute of Technology. Lund, Sweden.

Barton, P. and Pantelides, C. (1991). The modeling and simulation of combined discrete/continuous processes. *Proc. from Process System Engineering '91*. Montebello, Canada.

Eborn, J. and Nilsson, B. (1994). Object-oriented modeling and simulation of a thermal power plant. *Technical Report TPRT757*. Dept. of Automatic Control, Lund Institute of Technology. Lund, Sweden.

Elmqvist, H. (1993). Object-oriented modeling and automatic formula manipulation in KYMOLA. *Proc. of Scandinavian Simulation Society*. Kongsberg, Norway.

Marquardt, W. (1993). An object-oriented representation of structured process models. *European Symposium on Computer-Aided Process Engineering — 1*.

Marquardt, W. (1994). Trends in computer-aided process modeling. *Process Systems Engineering, PSE '94*, Vol. 1. Kyongju, Korea. pp. 1-24.

Mattsson, S.E. and Andersson, M. (1992). The ideas behind OMOLA. *Proc. of the 1992 IEEE Symposium on Computer-Aided Control System Design, CADCS '92*. Napa, California.

Nilsson, B. (1993). *Object-oriented modeling of chemical processes*. Ph.D. thesis. Dept. of Automatic Control, Lund Institute of Technology. Lund, Sweden.

Nilsson, B. (1994a). Guidelines for process model libraries using an object-oriented approach. *European Simulation Multiconference*. pp. 349-353.

Nilsson, B. (1994b). Modeling and simulation of a batch process in OMOLA. *Proc. of the 36th SIMS Simulation Conference*. pp. 35-40.

Nilsson, B. and Eborn, J. (1995). An object-oriented model database for thermal power plants. *Proc. of the 1995 EUROSIM Conference, EUROSIM'95*. Elsevier. Vienna, Austria. pp. 747-752.

Ordys, A., Katebi, R., Johnson, M. and Grimble, M. (1994). *Modeling and simulation of power generation plants*. Springer-Verlag.

Piela, P., Epperly, T. Westerberg, K. and Westerberg, A. (1991). ASCEND: An object-oriented computer environment for modeling and analysis: The modeling language. *Comp. and Chem. Engin.* **15**(1), 53-72.

Stephanopoulos, G., Henning, G. and Leone, H. (1990). MODEL.LA.: A modeling language for process engineering — I. The formal framework. *Comp. and Chem. Engin.* **14**(8), 813-846.

AN INTELLIGENT PROCESS OPTIMIZER

Angelo Lucia and Jinxian Xu
Department of Chemical Engineering
Clarkson University
Potsdam, NY 13699-5705

Abstract

Difficulties with nonlinear programming nonconvexities and multiple quadratic programming (QP) solutions in the context of full-space successive quadratic programming (SQP) methods are studied. It is shown that some indefinite quadratic programming solutions are not descent directions for the SQP method and can lead to additional work and/or failure in the nonlinear programming calculations. A simple heuristic methodology based on pruning, mirroring and back-tracking is used to find descent directions and improve reliability. A nonlinearly constrained chemical process example and geometric illustration are used to show that the proposed methodology is capable of finding desired solutions.

Keywords

Nonconvexities, Indefinite quadratic programs, Multiple solutions, Descent directions, Heuristics.

Introduction

There are many algorithms for solving nonlinear programming problems that arise in science and engineering applications. These include projected and reduced gradient methods, augmented Lagrangian algorithms, projected Newton-like methods, range and null space decomposition SQP methods and full-space successive quadratic programming algorithms. We prefer to use SQP methods (see Powell, 1978; Schittkowski, 1981) and recent trends in engineering have shifted to full-space SQP methods with all analytical or finite difference second derivatives because of their potential to solve large problems (see, for example, Lucia and Xu, 1990, 1992; Sargent, et al., 1992; Lucia, et al., 1993; Xu, 1993; Betts, 1994; Lucia, et al., 1994). However the use of analytical or finite difference second derivatives can cause difficulties associated with nonconvexities (i.e., projected Lagrangian Hessian matrices with mixed eigenvalues or curvature) and this, in turn, can give rise to indefinite quadratic programming subproblems with multiple Kuhn-Tucker points, nondescent in the nonlinear programming calculations, line searching difficulties and algorithmic failure. To avoid the potential difficulties caused by the use of analytical or finite difference second derivatives, Sargent, et al. (1992) and Betts (1994) modify the factors of the Lagrangian Hessian to force projected positive definiteness. That way the quadratic programming subproblems are convex, have a unique solution and no ambiguities arise. However, what is not pointed out clearly in either paper is the fact that convexifying the quadratic program does not necessarily force descent in the nonlinear program. It can happen that the convex quadratic approximation is a poor approximation of the Lagrangian function, for various reasons, and this can result in nondescent directions, even though the QP solution is unique (see Lucia and Xu, 1995 for a simple example of this). This in turn can cause line searching and algorithmic failure. Our experience also shows that forcing positive definiteness can cause termination at trivial phase equilibrium solutions.

If, on the other hand, indefinite quadratic programs are permitted, then some way of handling multiplicity must be included because not all QP solutions will necessarily give descent in the nonlinear programming calculations and failures similar to those described for forcing positive definiteness will occur. The primary difficulty here is not necessarily multiplicity but rather that of measuring the ultimate usefulness of a given QP solution. Because descent is a local property, forcing descent does not always provide convergence to the desired solution, particularly if the problem at hand has multiple local and global optima at the nonlinear programming level of the calculations. Some sort of back-tracking algorithm must be incorporated in the event that a given QP solution ultimately leads to undesired results, even though it provides descent in a local sense. The particular approach that we use is described in the next section.

A Heuristic Methodology

In this section, an approach to solving nonlinear programming problems based on a full-space SQP method and heuristics to resolve ambiguities from multiple QP solutions is described. The basic features of this heuristic methodology, which include pruning, mirroring and back-tracking, are presented using a nonlinearly constrained chemical process example and a geometric illustration.

A Full-Space SQP Algorithm

The heart of our current approach for solving nonlinear programming (NLP) problems is the trust region-based SQP method described in Lucia and Xu (1990) with enhancements and refinements that have been developed over the last five years. This algorithm is a full-space SQP method that uses asymmetric trust regions (i.e., trust regions that are geometrically similar to the shape of the feasible region instead of elliptical) to bound the step taken in the NLP phase of the calculations. The algorithm does not use line searching, even though that option is available, and it does not force iterative norm-reduction. The QP subproblems are solved by a modified Bunch-Parlett (1971) factorization technique that uses constrained pivoting to ensure numerical stability (see Lucia, et al., 1993), a numerical updating scheme to update the factors as the active set changes and an active set strategy that monitors and uses projected indefiniteness to guide changes in the active set (see Xu, 1993). At all levels of the calculations, sparsity of the constraint Jacobian and sparsity and symmetry of the Lagrangian Hessian matrix are exploited to reduce storage and computation. The reader is referred to the papers of Lucia and co-workers for all algorithmic details.

Descent

We use the NLP norm defined by the rule

$$\text{NLP norm} = \text{norm}(\nabla_x L) + \text{norm}(\lambda^T c) \qquad (1)$$

to determine whether a particular QP solution is an acceptable step or not. Here L is the Lagrangian function defined by $L = f + \sum \lambda_j c_j$, f is the objective function, c is a vector of equality and active inequality constraints and λ is a vector of multipliers associated with the active constraint set (i.e., equalities plus inequalities that hold as equalities). If the Kuhn-Tucker point (or step) calculated for the quadratic program results in a decrease in the NLP norm from one iteration to the next, then we accept that QP solution and continue the SQP calculations. If not, we attempt to find a different QP solution that provides descent using the heuristics outlined in the next section.

Heuristics

The rules for searching for other solutions to a given indefinite quadratic programming problem when all QP variables are bounded (as they are when trust regions are used) are based on the observations given in Lucia and Xu (1995) and are summarized briefly below.

1. Sometimes certain subsets of a nondescent final active set are both Kuhn-Tucker points for the QP problem and descent directions for the NLP.
2. Active sets for Kuhn-Tucker points show some mirroring characteristics.
3. The empty set often corresponds to a Kuhn-Tucker point for the QP problem and a descent direction for the NLP.

These heuristics allow us to search the set of initial active sets for any indefinite quadratic programming problem in an organized manner. Note that the subsets used to reinitialize the QP problem can be generated very easily and fathomed by keeping track of the active set changes.

A Numerical Example

In our opinion, the simplest way to illustrate the usefulness of the heuristics given in the previous section is through a numerical example. Consider then the maximum entropy calculation for the liquid-liquid equilibrium of a mixture of 5 moles of ethanol, 6 moles of isobutane, 5 moles of isopentane, 6 moles of n-pentane, 67 moles of benzene and 28 moles of water at 0.1013 MPa and fixed product enthalpy (or heat duty) of 1.37957×10^4 J/h. Note that this problem is nonlinearly constrained because it involves an energy balance and has multiple NLP solutions because one fluid phase model (i.e., the UNIQUAC equation) is used to model both equilibrium phases (see Prausnitz, et al., (1980) for a description of the UNIQUAC model and the binary interaction coefficients and other physical property data used).

This particular problem has thirteen unknown variables (the six component molar flow in each of two phases and the temperature) and obviously can not be illustrated geometrically. However, Fig. 1, which gives the entropy level curves on the combined mass/energy balance surface for benzene-water, shows the generic characteristics of this class of NLP problems. Note the constraint surface is curved and that there is a ridge of trivial solutions as well as a desired nontrivial solution and its mirror image.

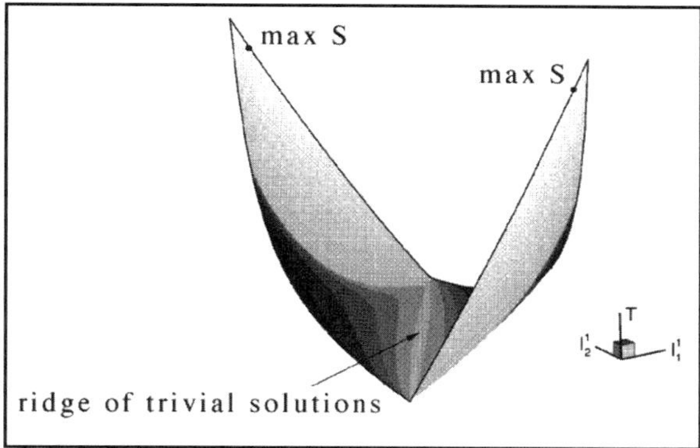

Figure 1. Level curves of entropy on combined mass/energy balance constraint surface.

Table 1. QP Solutions for a Maximum Entropy LLE Calculation for a Six Component Mixture at 0.1013 Mpa and 13795.7 J/h.*

Solution	Active Set	f_{QP}	NLP norm
1	{6,18}	-0.00313	0.9857
2	{5,19}	-0.00295	0.8404
3	{18}	-0.00204	0.6386
4	{19}	-0.00196	0.5635
5	{18,19}	-0.00225	0.9655

* the norm for the NLP at the beginning of this subproblem was 0.8106

For values of the 13 unknown variables of x = (0.00639464, 0.00712722, 0.00647935, 0.00777522, 0.110470, 0.0461719, 4.99361, 5.99287, 4.99352, 5.99222, 66.8895, 27.9538, 334.089), in the component order given above for liquid phase 1, followed by liquid phase 2, then temperature and multiplier values of λ = (-4.08286, -1.63500, -3.05906, -3.10377, -0.827286, -1.39418, -0.0003600, 0.0930928, 0.0220531), which correspond to the Lagrange multipliers for the six component mass balances, the energy balance and the two Kuhn-Tucker multipliers for the active lower bounds on the changes in the ethanol and isobutane component flows in liquid 1, we can find five solutions to the associated indefinite QP subproblem. These are shown in Table 1. In this table, the upper bounds on the changes in the liquid flows in phase 1 are denoted by 1 through 6 (in the given component order), those for changes in liquid phase 2 by 7 through 12 and the upper bound on the change in temperature is denoted by 13. Corresponding lower bounds on the changes in the variables are given by 14 through 26.

Note that QP solution 1 (the global minimum of this QP), as well as solutions 2 and 5, result in an increase in the NLP norm while solutions 3 and 4 produce a decrease in the norm. Suppose, for the sake of illustration, we calculated the global minimum of the given QP. The heuristics given in the previous section generate the initial active sets used to find other solutions to the same indefinite QP subproblem. In particular, the subset property, mirroring and heuristic 3 give the following initial active sets associated with QP solution 1 in Table 1: {6}, {18}, {13}, {25}, {13, 25} and { }. These initial active sets are first generated with one set members, then two set members, etc. until the number of inequalities in the nondescent QP solution have been exhausted combinatorially. Using {6} as the initial active set, our indefinite QP code converges to {5, 19} or QP solution 2. The subset {18} is QP solution 3. Both {13} and the null set also converge to QP solution 3. The initial active set {25} returns to QP solution 1 and {13, 25} converges to {5, 19}. For this illustration, we followed all active set changes for all subsets and mirror image sets associated with QP solution 1. In practice, however, we stop once we have found a descent direction but we do keep track of active set changes in order to terminate calculations that will ultimately find the same undesirable QP solution. In this case, the algorithm stops with the subset {18} since it is both a Kuhn-Tucker point for the QP and a descent direction. This decision, in turn, leads to convergence of the trust region-based SQP algorithm of Lucia and co-workers in 18 iterations to an accuracy of 10^{-6} in the NLP norm. Other NLP starting points that give multiple QP solutions often lead to convergence to trivial solutions (see Fig. 1) without the proposed heuristics.

Conclusions

A heuristic methodology was developed to improve the chances of finding descent directions from indefinite quadratic programming subproblems with multiple solutions. The rules for finding a descent direction for any successive quadratic programming method are based on pruning, mirroring and back-tracking. A nonlinearly constrained entropy maximization example was used to demonstrate the usefulness of the proposed heuristics. This numerical example clearly shows that the proposed methodology is capable of improving the reliability of SQP methods.

Acknowledgment

The material on which this work was based was supported by the National Science Foundation under Grant No. CTS-9312066 and is greatly appreciated.

Nomenclature

c = vector of equality and inequality constraints
f = objective function
L = Lagrangian function
S = entropy
x = vector of unknown variables
λ = multipliers

References

Betts, J. T. (1994). The application of optimization to aerospace systems. In L. T. Biegler and M. F. Doherty (Eds.), *Foundations of Computer-Aided Process Design*. Elsevier, in press.

Bunch, J. R. and B. N. Parlett (1971). Direct method for solving symmetric indefinite systems of linear equations. *SIAM J.Num. Anal.*, 639-661.

Lucia, A. and J. Xu (1990). Chemical process optimization using newton-like methods. *Comput. Chem. Engg.*, 119-138.

Lucia, A. and J. Xu (1992). Sparse quadratic programming in chemical process optimization. *AIChE Annual Meeting*, paper no. 137a.

Lucia, A., J. Xu and G. C. D'Couto (1993). Sparse quadratic programming in chemical process optimization. *Annals Oper. Res.*, 55-83.

Lucia, A., J. Xu and K. M. Layn. (1994). Nonconvex process optimization. *Comput. Chem. Engg.*, in press.

Lucia, A. and J. Xu (1994). Methods of successive quadratic programming. *Comput. Chem. Engg., Suppl.*, S211-S215.

Lucia, A. and J. Xu (1995). Nonconvexity and descent in nonlinear programming. Paper presented at *State of the Art: Computational Methods and Applications*. Princeton Univ., Princeton, NJ, in press.

Powell, M. J. D. (1978). A fast algorithm for nonlinearly constrained optimization problems, In G. A. Watson (Ed.), *Numerical Analysis*. Springer-Verlag. pp. 1-22.

Prausnitz, J. M., T. F. Anderson, E. Grens, C. Eckert, R. Hsieh and J. P. O'Connell (1980). *Computer Calculations for Multicomponent Vapor-Liquid and Liquid-Liquid Equilibria*. Prentice-Hall.

Sargent, R. W. H., M. Ding and J. L. Morales Perez (1992). A new SQP algorithm for large-scale nonlinear programming. *AIChE Annual Meeting*. paper 137b.

Schittkowski, K. (1981). The nonlinear programming method of Wilson, Han and Powell with an augmented lagrangian type line search function. *Numerische Math.*, 83-114.

Xu, J. (1993). *Sparse Quadratic Programming in Chemical Process Optimization*. Ph.D. Thesis, Clarkson University.

AN EXPERT SYSTEM FOR THE SCHEDULING OF MULTISTAGE MULTIPRODUCT PLANTS MANUFACTURING ASSORTED PRODUCTS

Gabriela P. Henning and Jaime Cerdá
Instituto de Desarrollo Tecnológico para la Industria Química
INTEC (CONICET — UNL)
Güemes 3450 (3000) Santa Fe, Argentina

Abstract

This paper presents a knowledge-based system for the predictive and reactive scheduling of a real industrial multistage multiproduct plant manufacturing assorted products. The proposed tool can be categorized as a second-generation expert system that provides a user-friendly interactive decision support environment. Appropriate knowledge representation and effective domain modeling were vital to the design of the system. In order to incorporate and integrate the several kinds of knowledge available on the industrial plant, hybrid representations implemented by means of object-oriented programming, were adopted. The system has been implemented in the KAPPA-PC (IntelliCorp, 1993) development environment and is currently being used in the facility.

Keywords

Short-term scheduling, Multiproduct plants, Second-generation expert systems, Object-oriented programming, Task-oriented decision-making, Predictive scheduling, Reactive scheduling.

Introduction

Short term production scheduling is one of the most challenging tasks process engineers face nowadays. Effective scheduling systems can make a major contribution to the competitive power of a company to meet customer demands as well as to utilize properly its assets/resources.

Despite recent advances (Reklaitis, 1992; Rippin, 1993), most industrial scheduling problems are difficult to solve by mathematical programming techniques. They may require such a computation time as to produce results which are of little or no use at the time of availability. Besides the prohibitive computing time, these techniques still have difficulty (i) to represent some of the qualitative, though very significant, features of the scheduling problem, (ii) to address the time varying nature of the objectives and the resources' availability and (iii) to adapt the model to changing processing networks as well as to product structures and Bill of Materials (BOM) often modified due to short product life cycles and recipe changes. The nature of industrial problems makes them not only complex to solve but also difficult to specify. Scheduling systems should include menus and editors enabling problem definition in a high-level fashion. Sophisticated graphic interfaces should help the planner at analyzing a solution from different perspectives. Graphic tools like Gantt charts should enable the direct interaction between user and system in order to try different "what if" scenarios.

The requirements placed above call for a non-conventional computer-aided tool. Knowledge-based (KB) approaches represent an important step toward the required expressiveness and flexibility in handling scheduling problems in live environments. These approaches provide rich knowledge representation forms for describing multiple knowledge sources such as processing networks with a changing topology, complex product structures, as well as a diversity of resources. KB systems allow a variety of problem decomposition and solution strategies to assist in finding a satisfactory solution (Atabakhsh, 1991; Rodammer, 1988). Moreover, they facilitate the easy development of more sophisticated user-system interfaces.

This paper describes a KB system that provides an interactive decision support environment.

Problem Domain

The system was developed to solve predictive and reactive scheduling problems in a real candy production facility that manufactures over 200 different final products. The plant has several production lines with multiple processing stages. The different lines are tightly coupled by a highly complex product structure.

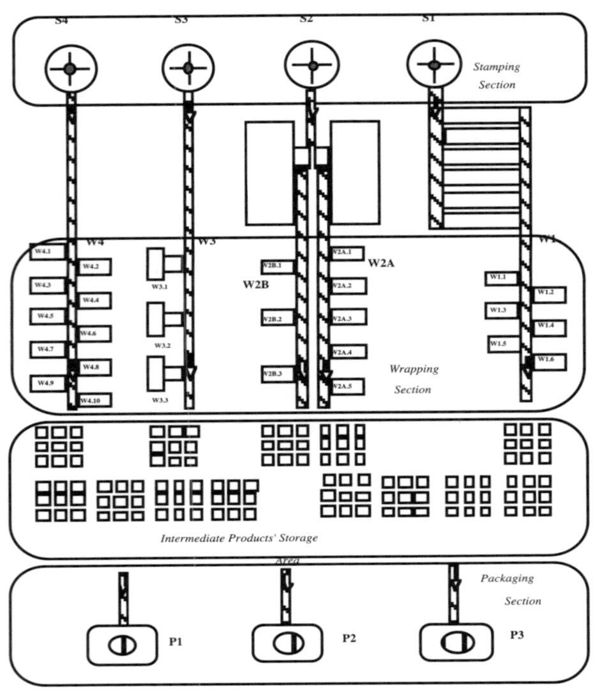

Figure 1. Simplified plant layout.

Processing Network

An schematic representation of the plant layout is illustrated in Fig. 1. As seen, the plant has three main sections. The first two, the candy stamping and wrapping sections, are highly integrated. They are followed by a storage area, from which wrapped candies are fed to the packaging section. The stamping stage features various equipment items (S1, S2, S3, S4), while the wrapping one has several lines (W1, W2A, W2B, W3, W4), each one having a set of wrapping machines operating in parallel. Figure 1 shows that certain stamping machines (S3, S4) directly feed their associated wrapping lines, while others (S1, S2) convey the unwrapped material to buffer storage trays which load the corresponding wrapping lines. These processing lines have unequal capacities and processing rates depending on the product being manufactured. Their distinctive characteristics define their suitableness to manufacture certain products. Once a stamping line is assigned to a given family of unwrapped products, all its members will be manufactured in a "one after the other" fashion. Therefore, only after all the members are produced, the manufacture of a distinct family can be started. Changeover time/cost between different families can be very important. Moreover, changeovers may be significant between family members if appropriate flavor and color sequences are not observed. Following the Intermediate Products' Storage Area, there is a Packaging Section having three unequal capacity machines, each one able to pack only a subset of final products. An extended object-oriented language (Stephanopoulos, et al., 1990) was adopted to represent the process knowledge.

Product Structure

Figure 2 presents an object-oriented hierarchical representation of the product structure showing inheritance relationships. The plant manufactures two types of intermediate products. The stamping section produces unwrapped material (over 50 different ones) and the wrapping stage manufactures around 150 distinct wrapped candies. As shown in Fig. 2, intermediates are grouped into families (around 15 families of unwrapped and 20 of wrapped candies) having not necessarily equal number of members, each one featuring a different color and flavor. There are single and multiple-flavor families of unwrapped candies. The elements of an unwrapped candy family are characterized by very similar production recipes with some differences in flavor and color. Figure 3 shows that unwrapped and wrapped intermediates are related by *"part-of"* links. These relationships, established at the level of families, define isomorphic links between instances. As seen, a family of unwrapped candies (e.g. U-F2) can participate in the composition hierarchy of more than one family of wrapped products.

Final products (over 200 different ones) consist of different types of cardboard boxes containing a variable number of small packages of distinct type (bags, boxes, etc.), each one holding within itself a certain amount of wrapped product. Thus, they can be seen as aggregate products formed by a set of wrapped candy families. Figure 3 shows that *"made-out"* relationships link families of final products with families of wrapped intermediates. The difference between this relation and the *"part-of"* one stems from the fact that *"made-out"* establishes a "1 to n" link at the level of product instances, while *"part-of"* always indicates a "1 to 1" relationship.

If the final product requires only a single-flavor family, it belongs to a category named "ONE-FLAVOR-FINAL-PRODUCT" (OFFP). For instance, OFFP-F1 is *"made-out"* of W-F1. (See Figs. 2 and 3). The class OFFP is further classified according to the involved single-flavor wrapped-product family (e.g. OFFP-F1, OFFP-F2, etc.) If instead, the final product is made out of a multiple-flavor family, it fits into a class referred as "FAMILY-TYPE-FINAL-PRODUCT"

(FTFP) (e.g. FTFP-F4 is "*made-out*" of W-F2),. This subclass is further categorized into subclasses. Assorted products involving wrapped intermediates from different families (i.e. family mixins) belong to a category called "FAMILY-MIXIN-FINAL-PRODUCT" (FMFP). As seen in Fig. 3, FMFP-F7 is made out of families W-F3, W-F5, W-F1, etc.; therefore, the proportions of the different families are to be defined. Products belonging to this last category introduce an extra degree of complexity to the problem since the production of the required intermediates, generally manufactured at different processing lines, need to be properly coordinated. When all the components required for an aggregate product become available, its packing can be started only if an alternative line is, at the time, idle. However, if some of the required intermediates arrive to the storage area much earlier than others, the in-process inventory level will increase.

Since product life cycles are generally short and recipe changes are frequent, the model of the product structure need to be continuously updated.

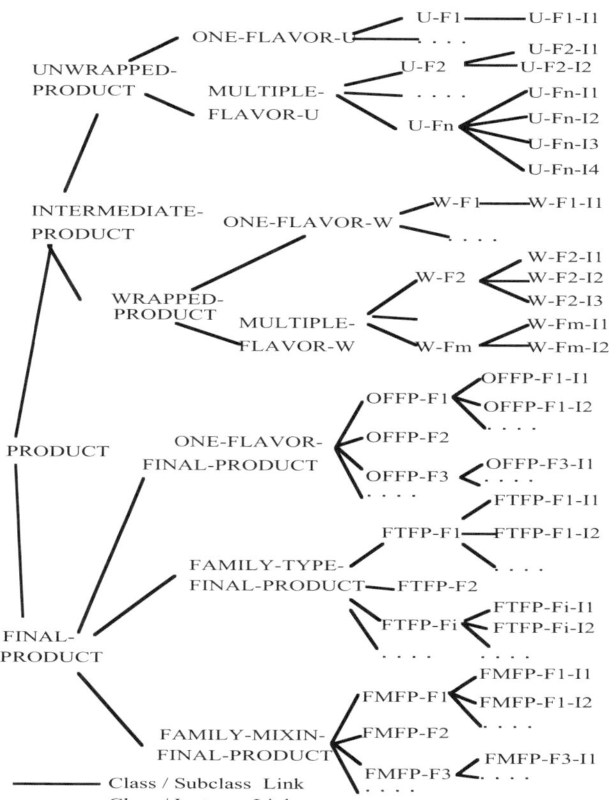

Figure 2. *Model of the product structure showing inheritance relationships.*

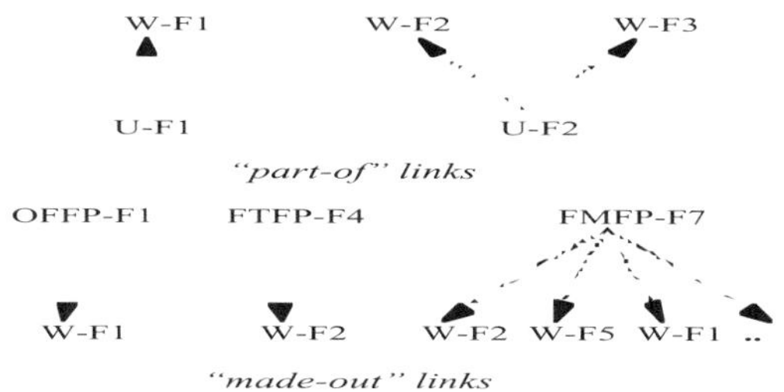

Figure 3. *Model of the product structure showing composition relationships.*

Scheduling Strategy

Given the complexity of the overall problem, it is decomposed into a sequence of more tractable subproblems. It is proposed to first find the wrapping section schedule, and then, by propagating it backwards elaborate the one of the stamping section. Once the schedules of the intermediate products' stages are obtained, the release-time for each final product becomes known and the schedule of the packaging section can be generated. The suggested methodology takes into account two important issues: (a) the wrapping section always behaves as the bottleneck stage, independently of the working load type and amount, and (b) the intermediate storage area decouples the two intermediate product manufacturing steps from the final packaging section. This strategy has been implement by means of a task-oriented solution approach (Henning and Cerdá, 1995).

Schedule of the Intermediate Products' Stages

The generation of the wrapping section production plan is tackled by means of an evolutionary approach having two main steps: (a) Development of a basic schedule that attempts to reduce work-in-process inventory and setup cost/time so as to maximize resource utilization and production throughput; (b) evolutionary improvement of the basic schedule to further increase the packaging section performance through the interactive application of corrective actions. They allow to change the timing and sequence of campaigns to be run in a given line, or to move campaigns to alternative ones.

During the generation of a basic schedule the following scheduling tasks are accomplished:

Orders' preprocessing — The starting point for the system is a list of final product requirements containing the identification, priority and amount of each one. Such products are assumed to be delivered at the end of the period. The list is "exploded" by using the product structure knowledge base to automatically generate requirements of wrapped and unwrapped intermediates, having similar priorities as the final products to which they are associated. Since distinct products may originate requisitions for the same intermediate, they are grouped into a joint demand that may be satisfied as a whole by running one or more campaigns. Nevertheless, information on the associated final products and their corresponding requests is still maintained.

Resource definition — Time is one of the most important resources. A weekly planning horizon having a reduced working pattern during the weekend is generally considered. The default pattern can be altered on a shift basis by means of an interactive graphic interface. Other resources such as manpower, equipment availability (i.e. number of wrapping machines working in each wrapping line) and operation policy for each line are also defined. In fact, some processing lines operate continuously (i.e. three shifts per day) while others work discontinuously.

Assignment of intermediates to processing lines and capacity analysis — The set of intermediates to be processed at each line is selected by taking into account their priorities, the preferred and alternative processing lines associated to each product and lines' remaining capacities. If the overall capacity is insufficient to process the whole intermediate demand, the manufacture of some low-priority products is delayed to the next period.

Campaign length and sequence — After selecting the set of intermediates and the amounts to be processed at each line, it is still necessary to establish the number, length and sequence of campaigns to be run. Depending on the scheduler's goals, different sequencing policies could be implemented by applying heuristic criteria. These criteria account for: (1) intermediate priority, (2) flavor and color compatibility between families and, also, between the members of a family, (3) intermediate type (single or multiple flavor family, and number of flavors in the last case), (4)amount of product, etc. The scheduler can define the subset of heuristics to be applied and the way they will be combined. Campaign length is another important issue. Processing large size orders in just one campaign tends to reduce scrap material and changeover time/cost. However, a long campaign may act as a bottleneck, delaying latter ones in the sequence. The system has user-modifiable criteria to establish the most appropriate campaign length for each product type and processing line.

Completion times and performance measure calculation — Having fixed the campaign sequence the start and finish time of each campaign is calculated. Total intermediate production and scrap level are two of the performance measures used by the system.

The capacity of the stamping machines always exceeds the one of the wrapping lines; therefore, the stamping section production plan is generated from the wrapping section schedule rather easily by grouping wrapped intermediate campaigns bound to the same unwrapped family, and subsequently adjusting the campaigns' timing.

Schedule of the Packaging Section

During the generation of the packaging section schedule a procedure similar to the one presented before is followed. The wrapping section schedule determines the time at which each final product is ready for packing. Different criteria are used by the system to properly assign and sequence a set of products to each line (i.e. increasing release-times, lower setup time, shortest processing times, etc.). If the required amount of a final product is relatively large, several packaging campaigns, in just one or several lines, are permitted to maintain low work-in-process inventory. Different evaluation indexes are also provided.

Schedule Evolutionary Improvement

The schedule's Gantt chart may be altered via point and click mouse actions. The scheduler can interactively change the size, split or move a campaign to another position (in the same or in alternative lines). The system does not allow the user to perform unfeasible actions. Predecessor and successors are automatically rearranged.

Conclusions

The paper presents an intelligent scheduling system for a real industrial plant, currently in operation. The system combines the power of knowledge-based technology to model, manipulate and reason with different types of knowledge, including the creative thinking of the scheduler, to generate and test different scheduling alternatives by using an interactive visual tool. In real industrial environments schedules are subject to disruptive events. Though not discussed in the paper, the system includes reactive scheduling capabilities.

Acknowledgments

The authors acknowledge financial aid from "CONICET" and "Universidad Nacional del Litoral."

References

Atabakhsh H. (1991). A survey of constraint based scheduling systems using an artificial intelligence approach. *Artificial Intelligence in Engineering*, **6**, 58-73.

Henning, G.P. and J. Cerdá (1995). An expert system for predictive and reactive scheduling of multiproduct batch plants. *Latin Am. Appl. Research* (In press).

IntelliCorp Inc. (1992). *KAPPA-PC User's Guide*, Version 2.0.

Reklaitis, G.V. (1992). Overview of scheduling and planning of batch process operations. *Proc. NATO Advanced Study Institute on Batch Process Systems Engineering*. Antalya, Turkey.

Rippin, D.W.T. (1993). Batch process systems engineering: a retrospective and prospective view. *Computers Chem. Engng.*, **17S**, S1-S13.

Rodammer, F.A. (1988). A recent survey of production scheduling. *IEEE Trans. on Syst. Man & Cyber.*, **18**, 841-851.

Stephanopoulos, G., G. P. Henning and H. P. Leone (1990). MODEL.LA. a modeling language for process engineering — I. the formal framework *Computers Chem. Engng.*, **8**, 813-846.

DECISION SUPPORT TOOLS FOR POLICY AND PLANNING

Paul Jacyk, Dave Schultz and Lois Spangenberg[1]
Los Alamos National Laboratory
P.O. Box 1663, Mail Stop E518
Los Alamos, NM 87545

Abstract

A decision support system (DSS) is being developed at the Radioactive Liquid Waste Treatment Facility, Los Alamos National Laboratory (LANL). The DSS will be used to evaluate alternatives for improving LANL's existing central radioactive waste water treatment plant and to evaluate new site-wide liquid waste treatment schemes that are required in order to handle the diverse waste streams produced at LANL. The decision support system consists of interacting modules that perform the following tasks: rigorous process simulation, configuration management, performance analysis, cost analysis, risk analysis, environmental impact assessment, transportation modeling, and local, state, and federal regulation compliance checking. Uncertainty handling techniques are used with these modules and also with a decision synthesis module which combines results from the modules listed above.

We believe the DSS being developed can be applied to almost any other industrial water treatment facility with little modification because in most situations the waste streams are less complex, fewer regulations apply, and the political environment is simpler. The techniques being developed are also generally applicable to policy and planning decision support systems in the chemical process industry.

Keywords

Decision support system, Uncertainty handling, Decision synthesis, Distributed problem solving, Intelligent user interface, Waste water treatment.

Introduction

A decision support system (DSS) is being developed at the Radioactive Liquid Waste Treatment Facility, Los Alamos National Laboratory (LANL), for use by managers engineers, regulators, and the public. This project will utilize techniques from the intelligent systems domain through a distributed framework of interacting information processing tasks. The project is driven by the need to improve LANL's existing central radioactive waste water treatment plant and also design a flexible new site-wide liquid waste treatment scheme that will effectively handle the wide variety of waste streams produced at specific sites around the laboratory. The site's radioactive waste streams can vary considerably and may require localized treatment operations or segregation to achieve the best processing for a given stream.

Radioactive liquid waste treatment at LANL is the primary focus of the DSS, but it is also necessary to consider processing, transporting, storing, and disposal of secondary waste streams when evaluating total impact or cost.

The DSS is scenario oriented. A scenario has the following characteristics: 1) concepts to be investigated, and 2) defining parameters. Specific problem solving tasks are associated with the concepts being investigated and can be considered as an additional characteristic of a scenario. A master controller module coordinates communication with the sub-tasks and synthesizes sub-task results into a set of results which address the scenario's key concepts.

[1] *jacyk@lanl.gov, schultz@esa.lanl.gov* and *lspangenberg@lanl.gov.*

The DSS consists of coordinated modules that perform the following tasks: rigorous aqueous chemistry and physical modeling and process simulations (supplemented by real data from skid mounted test units such as reverse osmosis units), configuration management, performance analysis, cost analysis, risk analysis, environmental impact assessment, transportation modeling, and local, state, and federal regulation compliance checking.

The DSS can reach valid conclusions even with minimal information, soft data, conflicting data and different human perspective on what is useful or preferred. Frequently conclusions reached with this type of information are called subjective or are not defensible; however explanations and certainties of findings generated by the DSS will help people understand how complex decisions are made and therefore increase confidence in the decisions.

Decision Support System

Decision Making

At the simplest level DSS's are used for day-to-day operations. This kind of decision making is characterized by a well defined domain and a relatively small set of decision options. Other levels of decision making in order of increasing degree of abstraction are tactical , resource allocation, and strategic. The DSS discussed here operates mainly in the last three categories where the focus is on tendencies and directions which have much associated uncertainty. Uncertainty is especially high at the strategic level where cause-effect relations are blurred by unexpected future events.

Uncertainty Handling

By its nature, decision making is an attempt to reduce uncertainty between multiple options. We are proposing that certain types of uncertainty, such as those related to partial information, unreliable information, and conflicting information from multiple sources, can be reduced or resolved before higher levels of decision making (decision synthesis) takes place. This uncertainty preprocessing approach changes the abstraction level of the uncertainties and aids the modules (mentioned in the introduction) to perform their tasks effectively. The result of this initial uncertainty handling is a set of data and uncertainties that can be effectively used by the decision synthesis module.

Our model for handling uncertainty (I-SORE) is related to Finlay's model of decision making (Finlay, 1994).

The stages of I-SORE are:

- Identify: Identify the uncertainty and classify it.
- Select: Select method(s) to resolve or reduce the uncertainty.
- Operate: Operate on the uncertainty
- Report: Report the results as internal system messages to the DSS's master controller module and/or to the user at an appropriate levels of detail.
- Evaluate: Evaluate the results of the uncertainty handling method. The evaluate stage occurs in the DSS's master controller module.

We are examining fuzzy logic, Bayesian approaches, Shafer's evidence theory, Cohen's theory of endorsements, and other qualitative approaches for uncertainty manipulation (Bhatnagar, 1986). Multiple methods can be used for the resolution and reduction of the same target uncertainty. We are using the method or methods that work best for the identified uncertainty, rather than globally using one method for all uncertainty handling (Chandrasekaran, 1986). For example, risk data is often handled with Bayesian approaches because the data frequently is in the form of probabilities, however fuzzy logic also works in this case and can be used with risk data to account for the possibility aspect of risk. The use of multiple uncertainty handling methods allows each uncertainty to be manipulated in ways which are consistent with its fundamental representation.

Decision Synthesis

The decision synthesis module passes parts of a scenario to appropriate modules then accepts results from all the modules and combines them to generate a set of conclusions. The decision synthesis module operates in two modes: "what-if" mode and "design" mode. In "what-if" mode, the user defines scenarios and goals of a study. The DSS returns results in the form of comparisons, sensitivity analysis, impact analysis, and feasibility checks. In "design" mode, the user defines a set of "design" parameters and the DSS returns a set of possible scenarios, alternatives, and recommendations.

Predefined alternatives, state descriptions, and relationships (all three referred to together as elements) are the heart of the decision synthesis module. These elements determine how variables and decisions influence other decisions. Useful models for representing and manipulating these elements are influence diagrams, decision trees, and causal stories.

For elements that are not known and can not be predefined, an enumeration approach is being used on a small portion of the decision space for generating new elements. We are also investigating abductive inference, "inference to the best explanation," for generation, criticism, and acceptance of new elements (Josephson, 1994). Abduction is appropriate for evidence combining tasks of the DSS's decision synthesis module. We believe capabilities of an abstract abduction machine, such as manipulating hypotheses, justifying conclusions, forming confident partial explanations, handling incompatibilities and many others (Josephson, 1994) can be applied to DSS decision synthesis.

We are using combinations of fuzzy logic, qualitative methods (Singh, 1991), objective functions, aspiration levels (Lewandowski, 1991), and heuristics for evaluating, screening, and judging scenarios. These methods utilize preferences identified in the scenario to focus the decision synthesis process on a few major issues out of many possible issues which are relevant.

Implementation Framework

Interacting Modules

We are currently running the DSS on four workstations. Each computer is dedicated to a group of DSS support programs as follows:

1. decision synthesis and control
2. scenario management and search
3. fuzzy logic and chemical simulation
4. database server.

Support Programs

External support programs, such as the chemical process simulator, databases, and other calculation modules, interact with the master controller through software bridges. Whenever possible, an object-passing bridge is being used because it offers a convenient way to transfer related information. A bridge was created at LANL to link the main controller module created in G2 (Gensym) to a chemical process simulator called ESP (OLI Systems, 1995). This bridge allows G2 control and integration of batch, semi-batch, and continuous process simulations; and offers a convenient way to directly pass entire process streams back and forth between G2 and ESP as objects. An object-passing bridge will also be used with object-oriented fuzzy logic tools being developed.

Library Search and User Interface

An intelligent user interface guides the operation of the DSS based on information that is available and the type of analysis required. It also helps manage a library of processed scenarios and guides the user through new scenario creation. Since the DSS is used by people with many backgrounds and interests, it is important that the user interface displays results in an understandable form which can be tailored specifically to the needs of each type of user.

A library of historic data, decision scenarios, other support data, and knowledge is being created that will provide the user and DSS modules with a flexible way to add, query, and retrieve information.

A natural language interface is being developed for human interaction with the DSS. It coexists with form-like interfaces. The natural language interface can handle a wide variety of traditional library searches, fuzzy searches, and similarity searches. The same interface can also be used to enter fuzzy logic rules. Advantages of the natural language interface are ease of operator use, syntax and consistency checking, and convenient ways to capture uncertainties (fuzzy logic hedges) associated with variables.

The user interface not part of the natural language interface also guides the user through scenario creation, DSS operation, and provides consistency and completeness checking. The user has the ability to override variable levels of checking in order to enter scenarios that may be outside of "normal" limits.

The DSS offers a flexible system for showing library search results and DSS scenario results. We offer the user a variety of tabular and graphical output styles at user selected levels of detail for the presentation of results. Multidimensional data representation and data reduction techniques are important for conveying relationships between results which are difficult to visualize in tables and simple graphs. Some examples are multidimensional scaling and biplots (Lewandowski, 1991). Clustering data is another kind of "data reduction" technique. If clusters of related data are shown on a "map," the map functions as a display for results and also as a means for the user to enter a range of scenario parameters.

Conclusions

A decision support system is being developed that will utilize techniques from the intelligent systems domain through a distributed framework of information processing tasks. Uncertainty handling and decision synthesis are key components to the success of the DSS. A layered approach to uncertainty handling and decision synthesis is being proposed. Data, knowledge, and uncertainties are manipulated with tools at levels appropriate for their abstraction level and representation.

DSS library search tools have been created for the intelligent user interface that are also used internally by DSS modules. As knowledge acquisition continues, we are able to test an increasing number of methods mentioned in this paper. We believe the techniques being developed are generally applicable to policy and planning decision support systems in the chemical process industry.

References

Bhatnagar, R. and L. Kanal (1986). Handling uncertain information: a review of numeric and non-numeric methods. In L. Kanal and J. Lemmer (Eds.),

Uncertainty in Artificial Intelligence. Elsevier Science Publishers, New York.

Chandrasekaran, B. and M. Tanner (1986). Uncertainty handling in expert systems: uniform vs. task-specific formalisms. In L. Kanal and J. Lemmer (Eds.), *Uncertainty in Artificial Intelligence.* Elsevier Science Publishers, New York.

Finlay, P. (1994). *Introducing Decision Support Systems.* NCC Blackwell Ltd., Oxford.

Lewandowski, A. (1991). In A. Lewandowski, P. Serafini, and M. G. Speranza (Eds.), *Decision Support Systems and Multiple-criteria Optimization: Methodology Implementation and Applications of Decision Support Systems.* Springer-Verlag-Wien, New York.

Josephson, J. R. and S. Josephson (1994). *Abductive Inference, Computation, Philosophy, Technology.* Cambridge University Press, Cambridge.

OLI Systems, Inc. (1995). *ESP (Environmental Simulation Program) Users Manual.* Morris Plains, NJ.

Singh, M. G. and L. Travé-Massuyès (1991). *Decision Support Systems and Qualitative Reasoning.* North-Holland, New York.

INTELLIGENT DECISION SUPPORT SYSTEM FOR SOURCE WASTE REDUCTION IN METAL FINISHING PLANTS

Y.L. Huang[1] and K.Q. Luo
Department of Chemical Engineering and Materials Science
Wayne State University
Detroit, MI 48202

Abstract

Virtually every manufacturer of precious metal products engages in metal finishing processes. These processes generate a large volume of waste streams that contain a variety of chemical, metal and non-metal contaminants. In the present work, an intelligent decision support system, namely WMEP-Advisor, is developed for source reduction in the metal finishing industry, with specific applications in electroplating plants. The system is capable of performing detailed process analysis on waste generation mechanism, evaluating waste minimization practice in process units, identifying source reduction opportunities, and providing adequate decision support.

Keywords

Waste minimization, Intelligent decision support, Expert systems, Fuzzy logic, Metal finishing.

Introduction

Waste minimization (WM) in metal finishing plants is one of major tasks in pollution prevention in the manufacturing industries. These over 4,000 metal finishing and electroplating plants in the nation utilize hundreds of chemicals to finish and plate parts with any one or a combination of over 200 different metallic coatings. This has given rise to the generation of a huge amount of waste (Melzer, et al., 1990). Waste streams contain numerous hazardous or toxic chemical, metal, and non-metal contaminants that are regulated by the EPA. To effectively prevent pollution and reduce end-of-pipe treatment cost, the quantity and toxicity of waste streams must be significantly reduced. According to the EPA's WM hierarchy, source reduction has the highest priority, since it aims at minimizing wastes in the first place. While a variety of source reduction strategies have been developed, they have not fully permeated the industry (Noyes, 1993; Duke, 1994). One of the major reasons is that they have not been well classified and compared in terms of cost, efficiency, and restriction. Moreover, the implementation of the source reduction strategies requires extensive knowledge and sufficient experience in diverse fields. The experience is usually heuristic, problem specific, and locally available. Apparently, to achieve WM goals in this industry, it is highly desirable that a computer-aided tool be developed to help plants make smart decisions on WM.

Huang, et al. (1991) developed an expert system, namely MIN-CYANIDE, to assist engineers to reduce waste solutions in electroplating plants. Although it was capable of providing decision support to the minimization of only cyanide-containing waste streams, it has evidently shown that the expert system technique is a very attractive approach to solving environmental problems at an expert level. In this paper, a much more sophisticated decision support system, namely WMEP-Advisor, is developed for source reduction in the metal finishing industry, with a specific application in electroplating plants. This system is capable of performing detailed process analysis on waste generation mechanism, evaluating WM practice in process units, identifying WM bottlenecks, and prioritizing source reduction strategies in terms of technical feasibility, WM efficiency and total cost.

[1] Author to whom all correspondence should be addressed.

Process Waste Source

A plating process involves the application of thin metal coating of one metal on another. In this process, workpieces are placed on racks or in barrels, and then are shifted through a series of process baths and rinse tanks. As illustrated in Fig. 1, a process bath is usually followed by one or two rinse tanks for removing residual process solution on the surface of workpieces, and preventing cross contamination.

Wastewater, spent solvents, spent process solutions and treatment residues are major forms of waste streams. A major portion of wastewater comes from rinsing steps. Being removed sequentially from each process tank, workpieces retain some process solution due to surface tension and geometric configurations. This results in a considerable amount of drag-out. Various solvents and process bath solutions have to be replaced after exceeding their useful lives, due to the build-up of miscellaneous contaminants in the baths. The contaminants contain high levels of metals and compounds that are always difficult to be separated. These solutions can be bled into on-site waste treatment facilities for pretreatment and recovery, or otherwise be encapsulated for off-site disposal. Treatment residues are degreaser sludge, filter sludge, and wastewater treatment sludge. The sludge contains more than 65% water in average. Its volumes can be reduced through waste segregation, water deionization, sludge dewatering, etc.

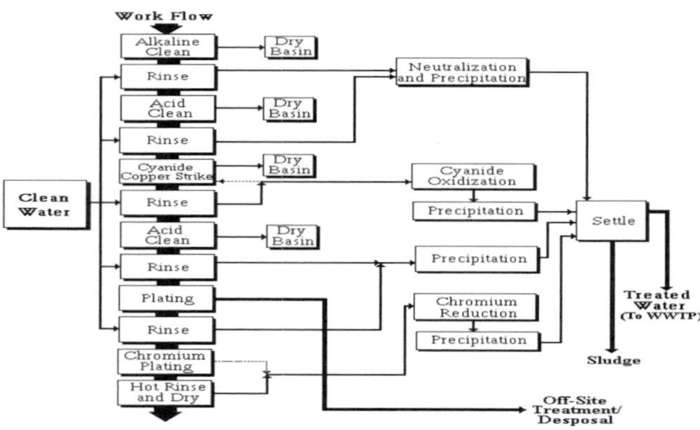

Figure 1. Typical electroplating process.

Source Reduction Strategies

The effectiveness of source reduction is usually reflected by the achievement of: (i) the minimization of drag-out, (ii) the extension of bath life, (iii) the reduction of the consumption of rinse water, (iv) the use of cyanide-free solutions, (v) the adoption of alternative plating metals, and (vi) the improvement of process operations (Freeman, 1988). These six sub-goals can be realized through changing technologies, modifying process and equipment, improving operational procedures, and substituting chemicals and materials. This requires the generation of numerous detailed strategies (Luo and Huang, 1995). For specific applications, these strategies must be carefully selected based on the trade-offs in their respective efficiencies, limitations, affect on production rate and product quality, and costs.

Intelligent Decision Support

WM in the metal finishing industry is a multi-disciplinary area involving engineering chemistry, biology, fluid mechanics, mathematics, statistics, economics, and law. The implementation of source reduction strategies requires knowledge in these fields. The knowledge can be divided into two classes: deep knowledge and shallow knowledge. The former is the law of nature; it mainly includes mass and energy balances, reaction kinetics and chemistry, and process dynamics. The latter includes various heuristic experiences in problem identification and problem solving. Usually, the available information pertaining to WM is uncertain, imprecise, incomplete, and qualitative. This hinders the use of conventional mathematical approaches. Artificial intelligence-based decision support, however, is viable in this endeavor (Huang and Fan, 1993a).

An intelligent decision support system (IDSS) can integrate different fields of knowledge in an interdisciplinary manner. It can help, but not replace, decision makers utilize the information in a smart way to solve poorly or insufficiently structured problems, as is often the case in environmental situations. The IDSS can provide a combination of several tools necessary to support the process of structuring WM problems, to gain new insights about them, to look for examples of problems that have already been solved, or otherwise to derive alternative solutions by means of AI techniques. In the end, the time and steps necessary to identify the best solutions to WM problems can be significantly shortened. Figure 2 shows the structure of the IDSS developed for source reduction. In the IDSS, expert system techniques and fuzzy logic are employed to represent and manipulate knowledge and information. Expert system techniques are used to build the frame work and user interface of the IDSS. The IDSS also contains a knowledge base and a data base. Numerous data are encoded in the data base, such as, process data, waste stream data, and chemical property data. Fuzzy logic is the logic that is much closer in spirit to human thought and language than the conventional logical system; nevertheless, in compliance with the spirit of logic, it attempts to be precise. Thus, fuzzy logic is, perhaps somewhat paradoxically, a precise system for imprecise reasoning (Zadeh, 1965). It allows vague and linguistic data (e.g. the representation of surface tension in this application) to be expressed as precise mathematical forms.

System Development

The knowledge base is the core of the IDSS. The amount and quality of acquired domain knowledge are keys to the system's functional capability (Huang and Fan, 1993b). Thus, knowledge engineers need to conduct extensive literature survey and work with experienced engineers and operators throughout the project. The expertise elucidated is a collection of problem definitions, process variables and their relations, specialized facts, algorithms, strategies, and heuristics about the process and WM re-enforcement. The acquired knowledge is represented as a large number of IF-THEN rules. Most of the rules contain imprecise information. For instance, a rule for drag-out minimization says:

IF the surface tension of a solution is very low,
AND the temperature of the solution is much lower than the optimal setting,
AND no wetter is added,
THEN excessive drag-out will be generated.

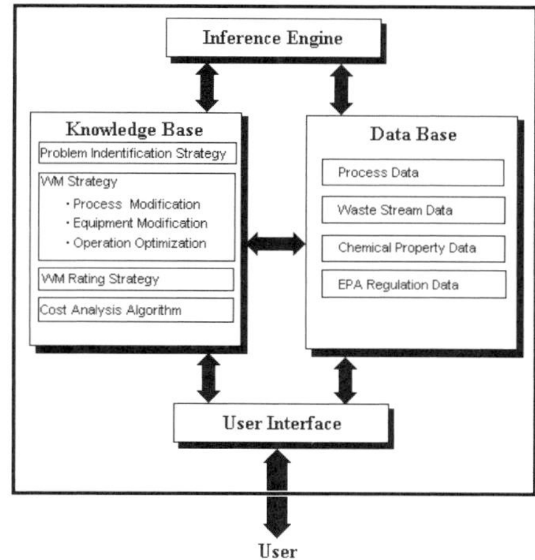

Figure 2. Structure of the intelligent decision support system for source reduction.

It is nearly impossible to define the boundary between the linguistic concepts of very low, low and high surface tension, lower and much lower than optimal temperature setting, and less excessive and excessive drag-out. Thus, fuzzy logic is employed to quantify these linguistic terms. The rules are activated by a fuzzy MIN-MAX algorithm. MIN operation selects rules providing a variety of decisions on source reduction. MAX operation ranks decisions in a prioritized order. Figures 3 and 4 demonstrates an example of rule evaluation and prioritization by fuzzy logic and MIN-MAX algorithm.

Implementation and Simulation

Based on the acquired knowledge, collected data, and adopted knowledge reasoning algorithms, an IDSS, namely WMEP-Advisor, has been built on an IBM PC compatible (486). The knowledge base is hierarchically structured and occupies 6 MB of memory. It contains numerous rules representing engineers and operators' experience, process principles, environmental regulations, and source reduction strategies. The system resides in Level5 Object, a window-based system development environment. It has been repeatedly tested by experienced engineers. Their valuable suggestions have been adopted to enhance the knowledge base and improve the system transparency. This makes the WM analysis and decision making process more practical and understandable. As an illustration, source reduction through bath-life extension is given below.

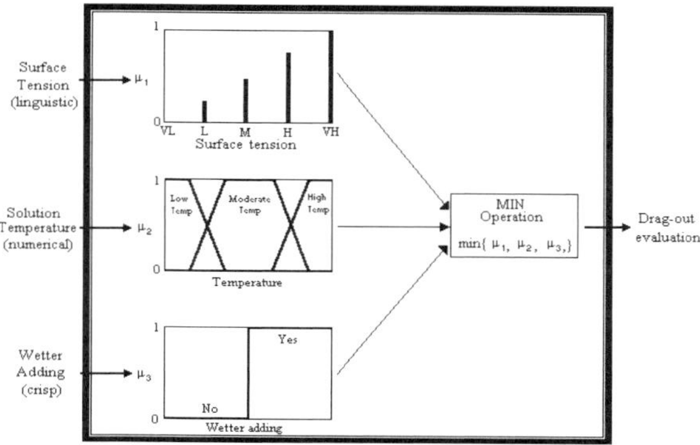

Figure 3. Rule evaluation by the MIN operator.

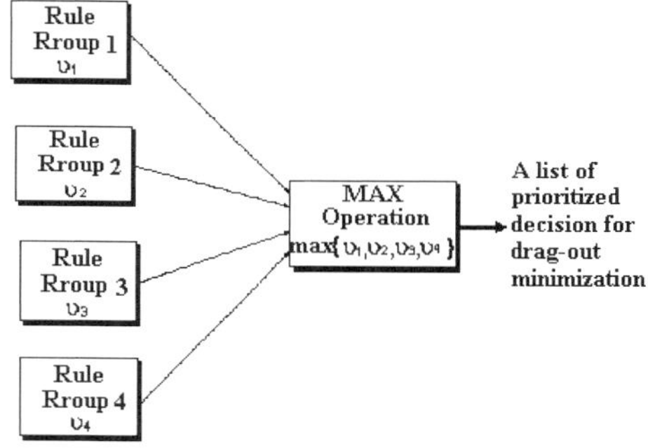

Figure 4. Rule prioritization by the MAX operator.

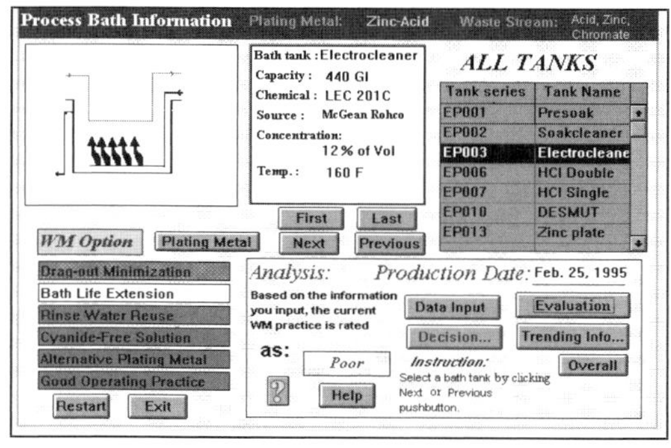

Figure 5. Main window for source reduction.

The evaluation of bath-life extension can be initiated by selecting the Bath-Life Extension option in the WM Option panel of the main window (Fig. 5). This allows the system to load all relevant process units into the ALL TANKS panel. After the user clicks push-button *Data Input*, the Data Input window appears (Fig. 6). The user can input all the data required through keyboard and mouse. As an example showing here, the system gives the rating of *"Poor"* after the user clicks push-button *Evaluate*. The user can click push-button *?* to view the supporting reasons for the rating (Fig. 7), and click push-button *Decision* to view the detailed suggestions with cost information (Fig. 8).

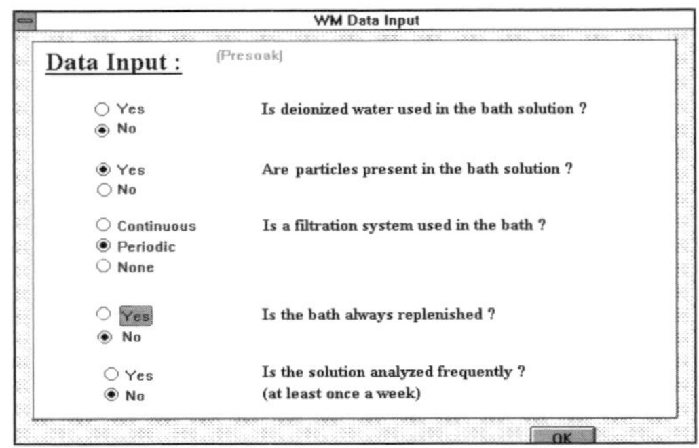

Figure 6. Data input window for drag-out minimization.

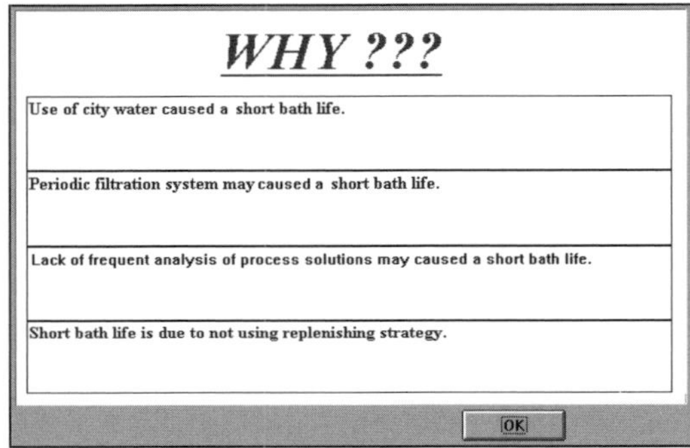

Figure 7. Decision analysis window for drag-out minimization.

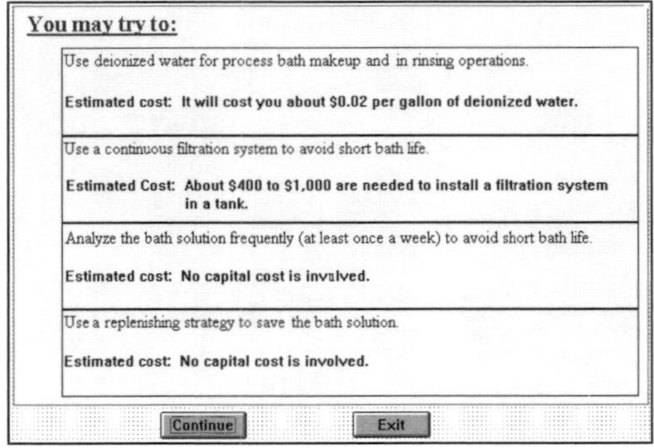

Figure 8. Decision support window for drag-out minimization.

Conclusions

A PC-based intelligent decision support system, WMEP-Advisor, has been successfully developed for the source reduction in metal finishing plants. The system contains a variety of source reduction strategies for minimizing various chemical, metal, and nonmetal waste streams. It can provide most desirable decisions to help plants reduce the quantity and toxicity of the waste streams. Currently, it is being enhanced by adding additional capabilities for recycling/reuse and source waste treatment and is being introduced to plants.

Acknowledgments

This work was supported in part by NSF (CTS-9414494), Hughes Aircraft Company, American Electroplaters and Surface Finishers Society, and Wayne State University. We greatly appreciate strong technical support from Reilly Plating Company in Detroit.

References

Duke, L. D. (1994). Hazardous waste minimization: is it taking root in U.S. industry? waste minimization in metal finishing of the San Francisco Bay Area, California. *Waste Management*, **14**, 49-58.

Freeman, H. M. (1988). Waste minimization audit report: case studies of minimization of cyanide waste from electroplating operations. EPA/600/S2-87/056, 24-62.

Huang, Y. L., G. Sundar and L. T. Fan (1991). MIN-CYANIDE: an expert system for cyanide waste minimization in electroplating plants. *Envir. Prog*., **10**, 89-95.

Huang, Y. L. and L. T. Fan (1993a). A fuzzy logic based approach to building efficient fuzzy rule-based expert system. *Comp. Chem. Eng*., **17**, 181-192.

Huang, Y. L. and L. T. Fan (1993b). Artificial intelligence technique for waste minimization in the process industry. *Int. J. Comp. in Ind*., **22**, 181-192.

Luo, K. Q. and Y. L. Huang (1995). Intelligent decision support on process modification and operational enhancement for source reduction in electroplating plants. *Int. J. of Comps. Chem. Eng*., submitted.

Melter, M., M. Callahan, and T. Jensen (1990). *Metal-Bearing Waste Streams: Minimizing, Recycling, and Treatment*. Noyes Data Corporation, Park Ridge, NJ.

Noyes, R. (1993). *Pollution Prevention Technology Handbook*, Chpt. 28. Noyes Publ., Park Ridge, NJ.

Zadeh, L. A. (1965). Fuzzy set. *Information and Control,* **8**, 338-353.

ACTIVITY MODELING IN PROCESS ENGINEERING

Kenneth A. Debelak
Dept. of Chemical Engineering
Vanderbilt University
Nashville, TN 37235

Gabor Karsai, Samir Padalkar and Janos Sztipanovitz
Dept. of Electrical Engineering
Vanderbilt University
Nashville, TN 37235

Frank DeCaria
E.I. DuPont DeNemours
Old Hickory Works, Old Hickory, TN 37138

Abstract

Activity modeling is a new functionality which seeks to integrate information which will be provided from different process monitoring and control functions and describes exactly what is done with this data. Activity modeling allows the process, monitoring, and control system to function according to the needs of different users, operators, control engineering and plant managers. The Intelligent Process Control System with its rich modeling paradigm for plant and activity modeling and automatic program synthesis capabilities offers a powerful new tool for process engineering.

Keywords

Activity modeling, Model-based systems, Fault diagnosis, Process modeling and simulation.

Introduction

Plant operation requires a diversity of problem-solving activities which must occur in an information-rich environment. Current and historical plant data, static and dynamic models of processes, characteristics and state of equipment and other information must be available and integrated at the points of decision. Efficient support for accessing relevant information and integrating it into problem-solving activities is critical for economical and reliable plant operation.

The Intelligent Process Control System (IPCS) is a domain-oriented, model-based programming environment for developing monitoring, control, simulation, diagnostic, and recovery applications. Process engineers with little or no software engineering can easily specify models of the plant. IPCS then creates executable applications from them automatically. The important point here is that the models are domain specific, i.e. built from concepts characteristic of the field. IPCS has the following capabilities:

- Collect, maintain and provide access to key plant information typically generated in design time. This information provides a plant-specific context for decision procedures.
- Model and execute high-level activities that use the available process monitoring and control (PM&C) functionalities and provide direct and relevant support for decisions.

The first capability provides support in building and managing plant models. Models are presented in the form of various diagrams and are stored in an object-oriented data base (OODBMS). The basic plant-related information, which is typically expressed in process flow sheets, P&I diagrams, and in a variety of models, is explicitly represented. The second capability provides support for building models of activities that combine various PM&C functionalities, associate the activity models with plant models and automatically synthesize executable systems in a real-time environment from the activity models. In its simplest manifestation, modeling activities might involve the gathering and interpretation of data from one of the functional modules such as the event historian, comparing the data to the standard operating conditions that are expressed in plant models (solutions of material and energy balances, outputs from a process simulator, e.g. ASPEN) and making a decision as to the possible faults in the plant. In a broad sense, the modeling of activities would describe what plant information is to be retrieved (historical data, on-line data, recipe data, etc.), what will be done with the plant information (comparison against plant models, display of the data for an operator, statistical analysis of the data, etc.), and what conclusions will be reached from an analysis of the data (determination of faults, change in control strategies or parameters, determination of production schedule, etc.).

Model Categories

IPCS offers a broad selection of *basic modeling concepts* which can be used to model various properties of a plant and related operation support activities. The basic modeling concepts are obviously not enough to model all possible plants or activities. IPCS provides a set of *model organization principles* which can be used to build more specialized, or more complex models using the basic concepts as building blocks.

There are two basic categories of IPCS models:

- *PLANT models*, which describe what is known about a plant, and
- *ACTIVITY models*, which describe the models of operation support procedures.

Users can build various models of the above categories using IPCS' graphical model builder. Once they are created, the PLANT models serve as a rich source of information about the plant, which can be easily maintained or modified. PLANT models can be built in design time as a documentation of the design process. In the case of existing plants, the models can be created retroactively as required by the ongoing need of plant operation. ACTIVITY models represent arbitrary operations, processing activities that are built from basic activity blocks supplied with IPCS. The ACTIVITY models are transformed into executable programs, called the IPCS APPLICATIONS, which execute the modeled activities (as they were specified in the models). ACTIVITY models may have multiple references to PLANT models and are in close relationship to them. During design time, ACTIVITY models can be used to specify simulation studies using the actually available PLANT models. During plant operation, ACTIVITY models may represent a variety of operation support systems, such as real-time simulators to predict the behavior of processes, plant model verification procedures, or fault recovery procedures that use the IPCS diagnostic system. These activities are also closely linked to the PLANT models, since they generally use plant information. IPCS can also model the relationship between PLANT models and ACTIVITY models, therefore directly supporting their consistency during the lifetime of the plant.

Plant Models

The basic plant modeling concepts are divided into three categories:

- Process models
- Stream models
- Equipment models.

Process models represent interactions in the plant in terms of processes and various material, energy and information streams. Processes perform some transformation on input streams and produce output streams. Streams are characterized by their components and their attributes (e.g. pressure, temperature). Equipment models describe the physical structure of the plant, which implement the processes, i.e. the hardware.

Activity Models

Activity models are divided into two basic categories, SYSTEM ACTIVITIES and USER ACTIVITIES. SYSTEM ACTIVITIES exist in one unique copy in an IPCS activity configuration, which may comprise a large variety of activities. The models of SYSTEM ACTIVITIES are part of the IPCS model library, and are used to model their connections to USER ACTIVITIES. SYSTEM ACTIVITIES are automatically generated by the IPCS program synthesis system using the information in the PLANT MODELS. USER ACTIVITIES may exist in many copies in an activity configuration, and are defined by the users.

Currently supported SYSTEM ACTIVITIES in the IPCS are:

- *Diagnostic Activity*: refers to the IPCS real-time diagnostic system, which receives and processes alarms and generates hypotheses for the underlying fault causes and future fault events in the plant.
- *System State Identifier Activity*: follows the state transitions in the processes and process equipment and generates events for other activities that are influenced by the states.

Currently supported USER ACTIVITIES in the IPCS are:

- Algorithmic Activities: are simple data processing activities which can execute a piece of code (their script) written in some high-level language (currently C).

- *Timer Activities*: can be used to generate timed delays and timed event sequences for triggering other activities.
- *Finite State Machine (FSM) Activities*: are for building simple state machines. They can be used to define FSM controllers in terms of states and transitions which can be triggered or can generate triggers for other activities.
- *Simulation Activities*: include algebraic and differential equation solvers that can compute the static and/or dynamic behavior of processes using the equation models.
- *External Interface Activities*: model communication channels to the outside world of an IPCS applications. They are typically used to configure the plant data acquisition system, retrieve data from a process historian, or start a simulation (e.g. ASPEN or SPEEDUP) on a remote node.
- *Operator Interface Activities*: model operator interactions. They describe what is shown to the operator and how the operator can interact with the activities.
- *Compound Activities*: are for building hierarchies of activities; they can contain any of the above activities and other compound activities.

The IPCS modeling methodology supports a variety of model organization principles for managing complexity. Among the various organizational principles used are: (1) multiple aspect to capture a particular kind of relationship or abstraction, (2) hierarchical composition to build deep hierarchies, and (3) connections in which the meaning of objects is determined by the objects they couple. Further reading on modeling and model organization principles in Multigraph-based systems can be found in Karsai, et al. (1991, 1992), Padalkar, et al. (1991), Abbot, et al. (1993), and Wilkes, et al. (1993).

Implementation

A graphical model editor provides a visual programming environment for building models. The use of graphics is not exclusive: whenever reasonable text is used (e.g. equations). Models are stored on an OODBMS, which supports sophisticated data structures and network access. A system integrator tool facilitates the creation of executable programs from the models, see Fig. 1. The executables contain: (1) model interpreters to create run-time objects, (2) special run-time support modules, e.g. equation solvers, diagnostic reasoning engines, (3) user-defined subroutines compiled and linked, (4) the Multigraph Kernel to implement low-level scheduling, (5) external data interfaces for plant data and other external packages, e.g. ASPEN+ and SPEEDUP, and (6) a graphical operator interface. When an executable is started, the model interpreters create and configure the computations implementing the required functionalities. The models are read from the database. After interpretation, the database is not needed, since all required information is in the running code.

Several types of applications will be presented to illustrate the use of plant and activity models. The first describes a simple display of 16 critical processing variables to an operator console. The second demonstrates the retrieval of plant data, the simulation of a pollution control device with the data, and the display of the information on an operator interface. The third describes a dynamic simulation of a train of distillation columns. Previously, Debelak, et al. (1994) and Padalkar, et al. (1995) described applications for diagnosing faulty sensors (level or flow), and predicting the actual readings from faulty sensors. The technique is based on a generic fault modeling and fault diagnostic methodology developed by Padalkar, et al. (1991) and applied in power generation plants (Karsai, et al., 1992), chemical plants (Karsai, et al., 1991), and aerospace vehicles (Carnes, et al., 1991). Debelak, et al. (1994) also presented an ASPEN+ simulation activity integrated into an IPCS application.

IPCS AS A MODEL-BASED ENVIRONMENT FOR:

- SYSTEM INTEGRATION
- BUILDING SOPHISTICATED SIMULATION, MONITORING & DIAGNOSTIC SYSTEMS
- DESIGN REPOSITORY

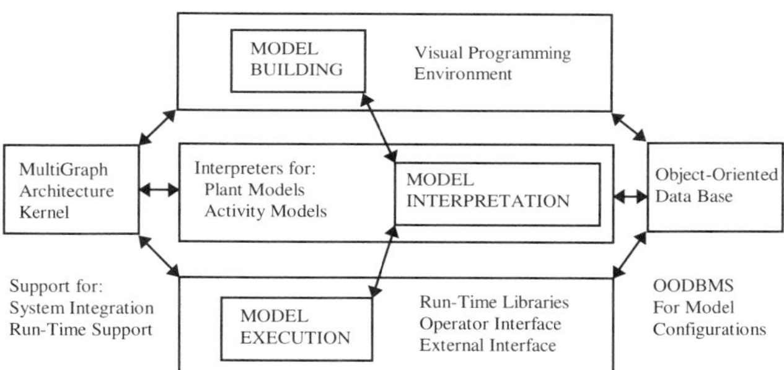

Figure 1. IPCS based on multigraph architecture.

All of the activities described below use a communication structure as shown in Fig. 2. For the first example, 16 critical process variables for the operation of a terephthalic acid (TPA) plant were identified. Plant data are retrieved from a process data base every two minutes. Some of the process variables are displayed directly. Other variables are combined in an algorithmic activity to calculate a key process variable, e.g. reactor yield. The algorithmic activity in this case is a script (written in C). The data are displayed in the control room on an x-window terminal, but can also be displayed on an x-window terminal anywhere on the network. Previously these data were not available in real-time. The operators could only determine the state of the plant in an after-the-fact mode.

Example two is a simulation activity for pollution assessment and prevention. An ASPEN+ simulation activity (ASA) provides an interface to an ASPEN+ simulation. At modeling time one creates an ASA model, which embeds and specifies an ASPEN+ simulation. The simulation is executed when all the input conditions for the ASA activity are satisfied. From the user's point of view, it is similar to an algorithmic activity: it executes a subroutine (i.e. an ASPEN+ simulation run) when the required input data are present. After execution, the results are propagated through the output ports of the activity to an operator interface. The ASA model configures an ASPEN+ simulation using an already existing ASPEN+ simulation file prepared with Model Manager, and then modifies that file based on the incoming process data. The ASPEN Summary File Toolkit is used to retrieve data for display. This application models a high pressure absorber which controls VOC emissions from a polyester intermediates plant. Process data for the simulations are retrieved from a process historian running on a VAX. A server program running on the VAX supports this data retrieval. IPCS is currently running on a HP 710 workstation. After the data are received, a request is made from another server running on a CRAY or VAX to execute the ASPEN simulation. After the simulation is completed, the data are retrieved and displayed on an operator interface. All communication is over TCP/IP connections. The engineers can use the information from the simulation to estimate the composition of the discharge stream for which there is no on-line measurement of total VOC emissions. Some of the components are at the part-per-million level. Prediction of trace contaminants is within the capability of existing steady-state simulators. The accuracy of the simulated compositions is highly dependent on the quality of the physical property data available. This limitation is faced by all process simulation. In this case, an extensive, experimentally verified physical property data base has been developed. Stack sample analyses and temperature measurements have verified the model predictions. The operators and engineers currently use this system to determine whether or not they are meeting air emission requirements. The current application has not only been used to identify and prioritize this emission stream, but the information from the simulation is used as an aid in controlling the process. Figure 2 shows a schematic of the information flow for this simulation activity. Both of these applications are beyond the development stage, and are "commercialized" (commercialized meaning on-line 24 hours a day and being used by operators to make process decisions). This is a steady-state simulation. However, a dynamic simulation could also be configured in this same manner.

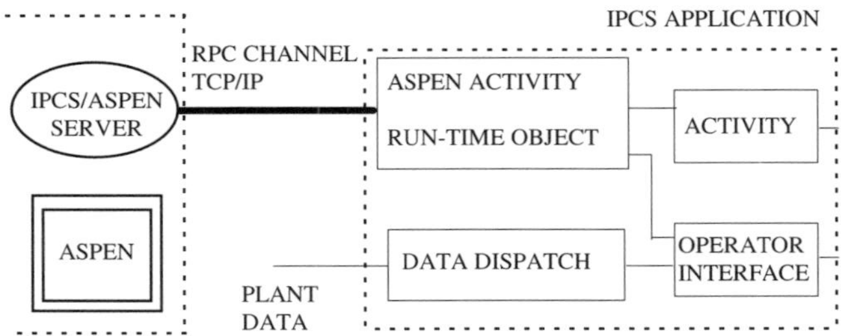

Figure 2. IPCS/ASPEN interface: run-time.

Our third example is a dynamic simulation of a four column refining train in a DMT polyester intermediates plant. This is an example of a SPEEDUP simulation activity (SSA). Similar to an ASA, at modeling time one creates a SSA model, which embeds and specifies a SPEEDUP simulation. The simulation is executed when all the input conditions for the SSA are satisfied. The results are propagated through the output ports of the activity to an operator interface. The results, however, could be propagated to other activities, e.g. one connected to the DCS for closed-loop control. The SSA model configures a SPEEDUP simulation using an existing SPEEDUP executable which has been developed in the SPEEDUP environment using the ASPEN/SPEEDUP interface. SPEEDUP's External Data Interface is used to input data to drive the simulation, and to retrieve data for display through IPCS. The same server running on the VAX retrieves process data from the historian. Another server, running on a CRAY, executes the SPEEDUP simulation. The data are retrieved and displayed on an operator interface. The SPEEDUP simulation is linked to a physical property data base through Properties Plus. The simulation of the refining train contains 4 interconnected large diameter columns with 6 components and over 100 theoretical trays. Over 7000 equations are solved for each time step. The data to drive the simulation are retrieved every two minutes. It takes approximately 30 seconds to simulate the next two minute interval. Therefore, it is possible to run the simulation synchronously with the plant in real-time. Currently, parameter estimation and optimization studies are being conducted.

References

Abbott, B. A., et al. (1993). Model-based approach for software synthesis. *IEEE Software*, May, 42-52.

Carnes, J., W. Davis, C. Biegel and G. Karsai (1991). Integrated modeling for planning, simulation and diagnosis. *IEEE Conference on AI Simulation and Planning in High Autonomy Systems*. April.

Debelak, K. A., G. Karsai, S. Padalkar, J. Sztipanovitz and F. O. DeCaria (1994). Activity modeling for process engineering. *FOCAPD'94*. Snowmass, CO.

Karsai, G., J. Sztipanovitz, S. Padalkar, C. Biegel, K. Debelak, S. Droes, N. Miyasaka, K. Okuda, F. DeCaria and M. Lopez (1991). A model-based approach for plant-wide monitoring, control, and diagnosis. *AIChE Annual Meeting*. Los Angeles.

Karsai, G., et al. (1992). Model based intelligent process control for cogenerator plants. *J. Parallel and Distributed Computing*, June, 90-103.

Padalkar, S. J., et al. (1991). Real-time fault diagnosis. *IEEE Expert*, June, 75-85.

Padalkar, S. J., G. Karsai, J. Sztipanovitz and F. O. DeCaria (1995). On-line diagnostics makes manufacturing more robust. *Chemical Engineering*, March, 80-83.

Wilkes, D. M., et al. (1993). The multigraph and structural adaptivity. *IEEE Trans. On Signal Processing*, 2695-2717.

AN EXPERT SYSTEM FOR ENHANCING COMPLIANCE WITH HAZARDOUS MATERIALS PACKAGING AND TRANSPORTATION REGULATIONS

J.J. Ferrada, R.D. Michelhaugh, R.B. Pope and R.R. Rawl
Oak Ridge National Laboratory
P.O. Box 2008, MS-6495
Oak Ridge, TN 37831-6495

Abstract

The Transportation Management Division (TMD) of the U.S. Department of Energy (DOE) has supported the development of an expert system for enhancing compliance with hazardous materials packaging and transportation regulations at the Oak Ridge National Laboratory (ORNL). This system, the Hazardous Material Expert System (HaMTES), was developed with the objective of providing a computerized expert system which (a) is easy to use, (b) will provide straightforward and consistent application of the hazardous material transportation regulations, and (c) will reduce the potential for human error in applying the regulations to both packaging and transportation of hazardous material activities.

HaMTES is based on an analysis of what an expert in hazardous materials shipping regulations could and should do when preparing a shipment of such materials. The system's proof-of-concept was demonstrated using the radioactive material regulations, and it was then expanded to include regulations for all hazardous materials. It can be used to determine for a given set of contents, the proper shipping name, hazard class, packing group, whether the material is a hazardous substance as defined by the U.S. Environmental Protection Agency (EPA), and whether it is a marine pollutant as defined by the International Maritime Organization. Specific modal requirements, packaging requirements, marking, labeling, and placarding requirements are also identified for the specified hazardous materials contents. The system has been developed based upon U.S. regulations, has been tested by experts, and is being upgraded based upon the results of these tests. Coupling of HaMTES to other transportation operations management software is underway, and methods for maintaining and upgrading the program have been developed.

Keywords

Hazardous materials, Expert system, Decision making, Multimedia, Hypertext.

Introduction

The ability of the computer to simulate and make decisions as a subject matter expert (expert system) has greatly expanded potential computer applications in complex decision-making situations. With this type of computing power available in an increasingly user-friendly format (using object-oriented programming and hypermedia capabilities), the development of a transportation packaging expert system that is based on the hazardous material regulations was possible. Under the sponsorship of DOE's TMD (EM-261), the Transportation Technologies Group at ORNL has designed and developed an expert system application of the U.S. hazardous materials transportation regulations.

The strategy to develop the expert system prototype was to, first, develop modules to capture the knowledge of different areas of the transportation regulations; and, second, append these different modules into one final package. The individual module development focused on one prototype for transporting and packaging radioactive material and another for transporting and packaging hazardous chemical materials. The two modules are integrated into a single system being implemented to support DOE-wide transportation operations.

It was found that the capability to scan the pertinent regulations in an interactive mode is invaluable. The regulations are always at hand for further consultation using a regular menu or a hypertext search mechanism. Graphic files with package information (such as dimensions, authorized contents, etc.) may also be accessed. Additional assistance may be provided by means of graphic files and/or video images.

From the analysis of the different features required for the expert system prototype, it was concluded that the developmental efforts should be directed to a Windows™ 3.1 Multimedia environment. Multimedia technology usually works as an interactive software system that gives personal computer users the ability to organize, manage, and present information in a number of formats — text, graphics, sound, and full-motion video.

The verification and validation of the hazardous materials transportation expert system (HaMTES) has been performed by comparing the operation of the HaMTES system with known shipment operations of transporting hazardous materials. Model validation was performed by trained and experienced traffic specialists at ORNL using regulation 49 CFR. HaMTES was also sent as a beta-test version to transportation experts in several DOE and contractor locations who then tested whether the computer model accurately represents the real system. In addition, experts from the U.S. Department of Transportation (DOT) and the Nuclear Regulatory Commission (NRC) reviewed and commented on the logic used to build the system.

Development

The strategy to develop the expert system was to, first, demonstrate the feasibility of developing an expert system prototype by developing modules to capture the knowledge of different areas of hazardous materials transportation based on the shipper's perspective; second, select an appropriate environment in which to deploy the expert system; third, analyze the feasibility of appending these different modules in one final full package; and fourth, develop the full-scale expert system.

Initial Prototype and Knowledge Acquisition

The DOT and EPA have stringent regulations regarding compliance of the shipment of hazardous materials before these materials are shipped. These regulations are extremely complicated. To achieve compliance, the shipper must sift through hundreds of regulations, search large tables, and perform calculations, depending on the applicable regulations. Only when all the applicable information has been determined, depending on the regulations and the tables, can the shipper properly prepare a package for shipment. Transportation specialists and packaging engineers must have the expertise to know which regulations to apply to a given shipment. It is this knowledge — that of the correct and complete path through the regulations — that this expert system based its operating logic.

The initial requirement for proper knowledge acquisition was to identify the regulations pertinent to hazardous and radioactive material transportation. From those identified, it was decided to implement in the HaMTES the 49 CFR 100-173, the transportation sections of 10 CFR, and the transportation sections of 40 CFR.

These regulations state the mechanism by which the transportation of hazardous and radioactive materials must be done. It was also agreed that understanding the mechanism of how these regulations should be applied may signify a large task and representing them computationally undoubtedly poses a formidable endeavor. Consequently, the first obstacle faced by this project was to choose an appropriate way of representing the regulations and how to navigate through them. The solution was found in the use of logic diagrams. Logic diagrams allow the representation of a knowledge base by means of a graphic tree where a decision point is found at the beginning of each branch. An IF-THEN rule mechanism applied to the decision points in the logic diagram made it possible to convert the logic tree into a computer program.

Figure 1 illustrates a portion of the logic diagram developed for HaMTES. The target of the decision paths shown in Fig. 1 is the determination of the proper shipping name for a hazardous materials shipment.

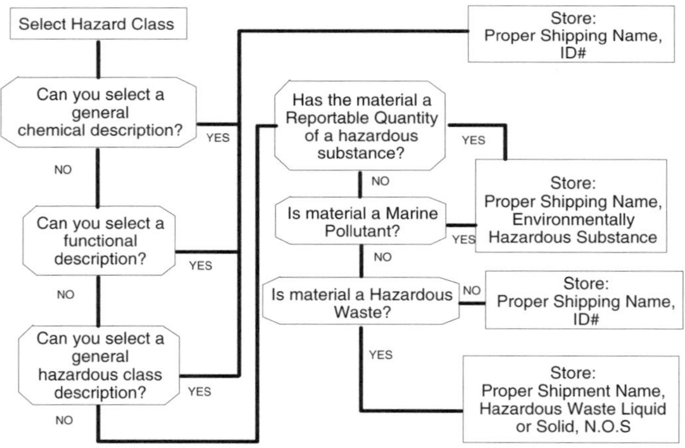

Figure 1. Example logic diagram from HaMTES.

Environment to Deploy the Program

To determine the appropriate deployment environment, some developmental stages were necessary. The first stage was to produce a proof-of-concept version of the program. This preliminary prototype was an aid in recognizing the basic requirements for development and the required features of the program.

The basic requirement of the prototype development was to run the expert system on the personal computer (PC) platform. It was clear that one requirement of the expert system was it should provide the ability to access the regulations from the commercial programs which update the regulations on regular basis. The friendliness of the user interface, along with the ability to navigate throughout the regulations and display graphics and full-motion video information on the screen, were other important requirements. The first concern of the knowledge engineer was to develop an interface between the user and all the specific program(s) that were transparent regarding the manipulations that are required to go from (a) one set of input data, to (b) calculation programs, then, to (c) decision-maker programs, etc. Thus, a straightforward system of a question-answer relationship between the computer and the user was highly desirable. The environment selected had to be able to implement a rule-based system that represents the regulations.

Other requirements, including access to several types of information sources, were suggested by the logic diagram. Behind every question asked of the user there is a set of regulatory requirement(s) which may influence the answer given by the user. Thus, in some cases, it was obvious that the user had to access the regulations before an answer could be given to the question. This feature provides the less knowledgeable user with enough information to give an appropriate response. It is common that regulations normally refer to other regulations or parts of regulations, which in turn may refer to other regulations and so on. In such complex cases, the expert system need only to access regulations and be able to browse through them. Additional explanations about regulations, interpretation of regulations, or any other aspect of the decision mechanism to determine types of packages were required to be available in a form of video images or audio. Consequently, multimedia elements, such as hypertext navigation, visual aids (whether as graphic or full-motion form), and mouse-driven interface elements, were considered to be essential parts of the expert system.

To decide which software tool would be used to create the expert system prototype, five possible tools using six criteria were rated. These criteria were (1) multimedia capabilities, (2) rule-making capabilities, (3) flexibility of the environment, (4) user interface provided by each tool, (5) data-handling capabilities of each tool, and (6) ease of use of each tool. The following tools were rated: OWL™ Industries Guide™ (a multimedia document presenter), the C/C++ programming language (a general programming language), the Prolog programming language [a logic-based disk operating system (DOS™) programming language], general expert system shells (tools used to create expert systems), and the Visual Basic™ programming language (a general Windows programming language).

Ratings on these criteria suggested that Visual Basic was the best environment in which to create the expert system. Although a Prolog-based code had already been created during the proof-of-concept stage, it was clear that putting the Prolog version together with multimedia features would be difficult. Consequently, this option was abandoned. The solution found for the prototype stage was to translate the Prolog code into the Visual Basic code, which is a Windows application.

Module Integration and Implementation of the Full Program

The radioactive and hazardous chemicals modules have been finalized and integrated. Consequently, the final version has incorporated both modules with the appropriate links so the user can prepare shipments involving radioactive, hazardous, and most importantly, mixtures of both types of materials.

Validation and Verification

During the first phase of verification and validation, experts from ORNL reviewed the program and submitted their written and oral comments. Their comments varied from changes to the screens to changes in the logic diagram. All these comments have been incorporated into the computer program. During the second phase of verification and validation, the HaMTES program and a review package were sent to experts at several DOE and contractor sites.

The DOE Transportation Management Division has different software applications under development. Joint Application Development (JAD) sessions are being held to analyze, improve, and discuss the integration of these software packages for transportation purpose. The presentation and discussion of HaMTES during a JAD session resulted in suggestions on user interface as well as changes in logic. As DOE funding allows, it is anticipated that HaMTES will be the front end for the hazardous module of the Automated Transportation Management System (ATMS) being developed under the sponsorship of DOE's Transportation Management Division.

ATMS is a modular system to automate the various functions involved in a typical transportation department. The major modules are: Rating and Routing shipments, Prepayment Auditing of Freight Bills, Electronic Data Interchange (EDI) with carriers, Shipping Document generation, Household Goods Moving, Motor Carrier Evaluation, Over-Short and Damage Claim, Historical shipment records and reporting system, Management Reporting system and Hazardous Materials Shipment preparation. The System has proven to save staff time, and save shipping costs by multimodal cost comparison and prepayment auditing of freight bills. Cost avoidance is realized in automated hazmat shipping compliance verification. DOT hazmat regulation violation penalties can be substantial.

A meeting was held with regulatory experts from the NRC and the DOT. At these meetings, the HaMTES logic diagram for transporting materials according to the current regulations was thoroughly reviewed. During the NRC review of the logic diagram, the work was concentrated in the radioactive module of HaMTES. The review with the DOT experts was essentially focused on the hazardous materials portion of HaMTES. The findings of these two meetings were incorporated into HaMTES version 1.0.

Maintenance

Regulations are affected by frequent changes. Changes to the regulations that do not affect the logic of the program are easy to implement. HaMTES can link to commercial products that update the electronic version of the regulations on a monthly basis. Furthermore, the text changes are automatically incorporated into HaMTES. When the changes to the regulations affect the logic of the program there is a need to modify the computer code. A quarterly update of the program has been considered. Normally, changes to the regulations become law after a period of 3-4 months. Consequently, a quarterly update seems to be sufficient to keep HaMTES current.

HaMTES has been licensed to the private sector with the commercial name of HMSolutions™. Both, HaMTES and HMSolutions, can link to the commercial regulation scanning program RegScan™ and are updated quarterly.

Future of HaMTES

As previously mentioned, DOT regulations determine the shipping requirements for the transportation of hazardous materials, including radioactive materials. The logic involved in determining these requirements for a shipment has been successfully implemented as a decision tree, using HM-169A and the collateral proposed changes in NRC's 10 CFR 71. Since HM-169A has not yet been published, HaMTES version 1.0 has been built on a modified logic diagram which reflects the current regulations. The version of HaMTES based on the proposed regulations has been filed for use when the regulations are finalized.

Version 1.0 of HaMTES, which is a stand-alone version, will be distributed to DOE facilities during early 1995. Concurrently, methods for making HaMTES available to users outside of DOE are being investigated. Future versions of HaMTES are expected to include more detailed data bases of package designs (Type A, Type B, UN packages, etc.), waste material shipment capabilities, and audio and video help to users. Eventually, when resources are available, HaMTES will be integrated into ATMS. Modifications to fit the ATMS systems requirements will be necessary. Mechanisms of integration have already been analyzed and tested.

Conclusions

As previously mentioned, the DOT regulations determine the correct packaging for the transportation of hazardous materials. The logic involved in determining the correct package for a shipment has been successfully implemented as a decision tree, using the Revised Radioactive Materials Transportation Regulations as proposed in HM-169A[5] and the current 49 CFR 100-173. This work was performed in anticipation of the publication of the Final Rule HM-169A this year.

The development of both modules, radioactive and hazardous chemicals, produced positive results in that it was concluded that the pertinent regulations can be translated into a logic diagram and that this logic diagram can be translated into a computer code. In addition, it was concluded that for presentation purposes, better memory utilization, and a larger portfolio of computer features, it was best to develop the hazardous materials modules completely in Windows 3.1. The language selected for developing the user interface and the rule-based system was Visual Basic.

A quality assurance plan has been developed and implemented for the validation and verification of the expert systems. The purpose of this plan is to establish specific responsibilities and methods for the validation and verification of the HaMTES in order to do quality work to support of the DOE Transportation Management Division (EM-261).

HaMTES has been selected as the front end for the HAZMAT module of ATMS. Work has been done to define and test mechanisms of linking HaMTES and ATMS. A JAD session was also held to define the functional requirements of HaMTES as related to ATMS.

Work for the FY 1995 has been scheduled around producing version 1.0 of HaMTES. Maintenance and enhancement of HaMTES has been programmed to be performed throughout the next fiscal year.

AUTHOR INDEX

SUBJECT INDEX